IB DIPLOMA

MATHEMATICS
HIGHER LEVEL

COURSE COMPANION

Jim Fensom
Josip Harcet
Lorraine Heinrichs
Palmira Mariz Seiler
Marlene Torres Skoumal

eBook

OXFORD

OXFORD
UNIVERSITY PRESS

Great Clarendon Street, Oxford OX2 6DP

Oxford University Press is a department of the University of Oxford. It furthers the University's objective of excellence in research, scholarship, and education by publishing worldwide in

Oxford New York

Auckland Cape Town Dar es Salaam Hong Kong Karachi Kuala Lumpur Madrid Melbourne Mexico City Nairobi New Delhi Shanghai Taipei Toronto

With offices in

Argentina Austria Brazil Chile Czech Republic France Greece Guatemala Hungary Italy Japan Poland Portugal Singapore South Korea Switzerland Thailand Turkey Ukraine Vietnam

British Library Cataloguing in Publication Data

Data available

ISBN: 978-0-19-912934-8
10 9 8 7 6 5 4 3 2 1

Printed by Vivar Printing Sdn Bhd, Malaysia

Acknowledgments

We are grateful to the following for permission to reprint copyright material:

Guardian News and Media Ltd for '618 is the Magic Number' by Marcus Chown, *The Guardian*, 16.1.2003, copyright © Guardian News & Media Ltd 2003.

Professor Alan Tennant from the Helmholtz Association of German Research Centres for extract from press release 'Golden ratio discovered in a quantum world', 7 January 2010.

Internet Systems Consortium, Inc (ISC) for figure: 'The growth of internet domains on the world wide web since 1994' from www.isc.org.

Although we have made every effort to trace and contact copyright holders before publication this has not been possible in all cases. If notified, the publisher will rectify any errors or omissions at the earliest opportunity.

The publishers would like to thank the following for permission to reproduce photographs:

P3: Mary Evans Picture Library/Alamy; P4: Professor Peter Goddard/Science Photo Library; P12: Photo Researchers, Inc./ Science Photo Library; P24: Science Photo Library; P46: Associated Sports Photography/Alamy; P46: Trinity Mirror/ Mirrorpix/Alamy; P46: Sheila Terry/Science Photo Library; P47: Dvmsimages/Dreamstime.Com; P47: Science Photo Library; P49: Taily_Sindariel/Dreamstime; P51: New York Public Library Picture Collection/Science Photo Library; P94: Af Archive/ Alamy; P94: Stanalin/Dreamstime.Com; P94: Elnur Amikishiyev/ Dreamstime.Com; P97: Vitaly Titov & Maria Sidelnikova/ Shutterstock; P164: F9photos/Dreamstime.Com; P165: Pictorial Press Ltd/Alamy; P165: Christoph Weihs/Dreamstime.Com; P167: Jp Laffont/Sygma/Corbis; P231: Condor 36/Shutterstock; P231: Alhovik/Dreamstime.Com; P231: Paulpaladin/Dreamstime. Com; P231: Fromoldbooks.Org/Alamy; P231: Mediacolor's/ Alamy; P233: Pseudolongino/Dreamstime.Com; P250: George Bernard/Science Photo Library; P276: Sheila Terry/Science Photo Library; P277: Science Source/Science Photo Library; P279: Martin Barraud/Alamy; P280: Will & Deni Mcintyre/Corbis; P280: Ivan Hafizov/Dreamstime.Com; P281: Brett Critchley/Dreamstime. Com; P281: Ivan Hafizov/Dreamstime.Com; P294: Sergio Azenha/ Alamy; P299: Mwaldrum/Dreamstime.Com; P301: Roman Sigaev/ Shutterstock; P303: Francesco Abrignani/Shutterstock; P312: 3dimentii/Shutterstock; P341: Dvmsimages/Dreamstime.Com; P341: Nito/Shutterstock; P343: Count/Dreamstime.Com; P357: Science Photo Library; P371: Uatp1/Dreamstime.Com; P380: North Wind Picture Archives/Alamy; P380: North Wind Picture Archives/Alamy; P381: Lonely Planet Images/Alamy; P381: Mypokcik/Dreamstime.Com; P381: Neo Edmund/Shutterstock; P383: Sculpies/Shutterstock; P383: Jim Barber/Shutterstock; P383: D_V/Shutterstock; P384: Boris Rabtsevich/Shutterstock; P386: Imagebroker/Alamy; P388: Photolibrary; P389: Robert Pernell/Shutterstock; P419: Wisconsinart/Dreamstime.Com; P422: Steve Bower/Shutterstock; P432: Interfoto/Alamy; P432: Photohedgehog/Shutterstock; P432: Rudy Balasko/Shutterstock; P432: Robinh; P435: Antoniomp/Dreamstime.Com; P493: Lauren Jade Goudie/Shutterstock; P493: Sergey Mikhaylov/Shutterstock; P495: Corbis/Photolibrary; P495: Brad Miller/Alamy; P496: Gemphotography/Shutterstock; P503: Feng Yu/Shutterstock; P509: High Voltage/Shutterstock; P513: Science Photo Library; P515: Orientaly/Shutterstock; P518: Colette3/Shutterstock; P535: Andrew Aitchison/In Pictures/Corbis; P544: Vishnevskiy Vasily/ Shutterstock; P546: Ivan Cholakov Gostock-Dot-Net/Shutterstock; P546: Nomad_Soul/Shutterstock; P552: Gl Archive/Alamy; P552: Ninell/Shutterstock; P552: Er_09/Shutterstock; P553: Reuters Pictures; P553: Nicemonkey/Shutterstock; P555: Andrei Teodorescu/Dreamstime; P556: Alistair Scott/Shutterstock; P559: Imagebroker/Imagebroker/Flpa; P587: Olgysha/Shutterstock; P594: Paul Tobeck/Shutterstock; P606: Thinkomatic/Dreamstime. Com; P627: The Art Gallery Collection/Alamy; P627: Dr. Richard Kessel & Dr. Gene Shih, Visuals Unlimited/Science Photo Library; P627: Alenavlad/Dreamstime.Com; P627: Hipgnosis/ Dreamstime.Com; P627: Anton Balazh/Shutterstock; P629: Cflorinc/Dreamstime.Com; P629: Hank Morgan/Science Photo Library; P645: North Wind Picture Archives/Alamy; P653: Science Photo Library; P658: Patrick Wang/Shutterstock; P659: Mark Garlick/Science Photo Library; P670: Scott Camazine/ Science Photo Library; P670: Nasa/Science Photo Library; P671: Beboy/Shutterstock; P671: Buslik/Shutterstock; P671: Dadek/ Shutterstock; P671: Mikkel Juul Jensen/Science Photo Library; P679: De Agostini/Getty Images; P684: Reeed/Shutterstock; P738: Georgios Kollidas/Shutterstock.

Cover image: Clive Nichols/Photo Library

Every effort has been made to contact copyright holders of material reproduced in this book. If notified, the publishers will be pleased to rectify any errors or omissions at the earliest opportunity.

Course Companion definition

The IB Diploma Programme Course Companions are resource materials designed to provide students with support through their two-year course of study. These books will help students gain an understanding of what is expected from the study of an IB Diploma Programme subject.

The Course Companions reflect the philosophy and approach of the IB Diploma Programme and present content in a way that illustrates the purpose and aims of the IB. They encourage a deep understanding of each subject by making connections to wider issues and providing opportunities for critical thinking.

The books mirror the IB philosophy of viewing the curriculum in terms of a whole-course approach; the use of a wide range of resources; international-mindedness; the IB learner profile and the IB Diploma Programme core requirements; theory of knowledge, the extended essay, and creativity, action, service (CAS).

Each book can be used in conjunction with other materials and indeed, students of the IB are required and encouraged to draw conclusions from a variety of resources. Suggestions for additional and further reading are given in each book and suggestions for how to extend research are provided.

In addition, the Course Companions provide advice and guidance on the specific course assessment requirements and also on academic honesty protocol.

IB mission statement

The International Baccalaureate aims to develop inquiring, knowledgable and caring young people who help to create a better and more peaceful world through intercultural understanding and respect.

To this end the IB works with schools, governments and international organizations to develop challenging programmes of international education and rigorous assessment.

These programmes encourage students across the world to become active, compassionate, and lifelong learners who understand that other people, with their differences, can also be right.

The IB Learner Profile

The aim of all IB programmes is to develop internationally minded people who, recognizing their common humanity and shared guardianship of the planet, help to create a better and more peaceful world. IB learners strive to be:

Inquirers They develop their natural curiosity. They acquire the skills necessary to conduct inquiry and research and show independence in learning. They actively enjoy learning and this love of learning will be sustained throughout their lives.

Knowledgable They explore concepts, ideas, and issues that have local and global significance. In so doing, they acquire in-depth knowledge and develop understanding across a broad and balanced range of disciplines.

Thinkers They exercise initiative in applying thinking skills critically and creatively to recognize and approach complex problems, and make reasoned, ethical decisions.

Communicators They understand and express ideas and information confidently and creatively in more than one language and in a variety of modes of communication. They work effectively and willingly in collaboration with others.

Principled They act with integrity and honesty, with a strong sense of fairness, justice, and respect for the dignity of the individual, groups, and communities. They take responsibility for their own actions and the consequences that accompany them.

Open-minded They understand and appreciate their own cultures and personal histories, and are open to the perspectives, values, and traditions of other individuals and communities. They are accustomed to seeking and evaluating a range of points of view, and are willing to grow from the experience.

Caring They show empathy, compassion, and respect towards the needs and feelings of others. They have a personal commitment to service, and act to make a positive difference to the lives of others and to the environment.

Risk-takers They approach unfamiliar situations and uncertainty with courage and forethought, and have the independence of spirit to explore new roles, ideas, and strategies. They are brave and articulate in defending their beliefs.

Balanced They understand the importance of intellectual, physical, and emotional balance to achieve personal well-being for themselves and others.

Reflective They give thoughtful consideration to their own learning and experience. They are able to assess and understand their strengths and limitations in order to support their learning and personal development.

A note on academic honesty

It is of vital importance to acknowledge and appropriately credit the owners of information when that information is used in your work. After all, owners of ideas (intellectual property) have property rights. To have an authentic piece of work, it must be based on your individual and original ideas with the work of others fully acknowledged. Therefore, all assignments, written or oral, completed for assessment must use your own language and expression. Where sources are used or referred to, whether in the form of direct quotation or paraphrase, such sources must be appropriately acknowledged.

How do I acknowledge the work of others?

The way that you acknowledge that you have used the ideas of other people is through the use of footnotes and bibliographies.

Footnotes (placed at the bottom of a page) or endnotes (placed at the end of a document) are to be provided when you quote or paraphrase from another document, or closely summarize the information provided in another document. You do not need to provide a footnote for information that is part of a "body of knowledge". That is, definitions do not need to be footnoted as they are part of the assumed knowledge.

Bibliographies should include a formal list of the resources that you used in your work. "Formal" means that you should use one of the several accepted forms of presentation. This usually involves separating the resources that you use into different categories (e.g. books, magazines, newspaper articles, Internet-based resources, CDs and works of art) and providing full information as to how a reader or viewer of your work can find the same information. A bibliography is compulsory in the extended essay.

What constitutes malpractice?

Malpractice is behavior that results in, or may result in, you or any student gaining an unfair advantage in one or more assessment component. Malpractice includes plagiarism and collusion.

Plagiarism is defined as the representation of the ideas or work of another person as your own. The following are some of the ways to avoid plagiarism:

- Words and ideas of another person used to support one's arguments must be acknowledged.
- Passages that are quoted verbatim must be enclosed within quotation marks and acknowledged.
- CD-ROMs, email messages, web sites on the Internet, and any other electronic media must be treated in the same way as books and journals.
- The sources of all photographs, maps, illustrations, computer programs, data, graphs, audio-visual, and similar material must be acknowledged if they are not your own work.
- Works of art, whether music, film, dance, theatre arts, or visual arts, and where the creative use of a part of a work takes place, must be acknowledged.

Collusion is defined as supporting malpractice by another student. This includes:

- allowing your work to be copied or submitted for assessment by another student
- duplicating work for different assessment components and/or diploma requirements.

Other forms of malpractice include any action that gives you an unfair advantage or affects the results of another student. Examples include, taking unauthorized material into an examination room, misconduct during an examination, and falsifying a CAS record.

About the book

The new syllabus for Mathematics Higher Level is thoroughly covered in this book. Each chapter is divided into lesson-size sections with the following features:

The Course Companion will guide you through the latest curriculum with full coverage of all topics and the new internal assessment. The emphasis is placed on the development and improved understanding of mathematical concepts and their real life application as well as proficiency in problem solving and critical thinking. The Course Companion denotes questions that would be suitable for examination practice and those where a GDC may be used. Questions are designed to increase in difficulty, strengthen analytical skills and build confidence through understanding. Internationalism, ethics and applications are clearly integrated into every section and there is a TOK application page that concludes each chapter.

It is possible for the teacher and student to work through the book in sequence but there is also the flexibility to follow a different order. Where appropriate the solutions to examples using the TI-Nspire calculator are shown. Similar solutions using the TI-84 Plus and Casio fx-9860GII are included on the accompanying interactive CD which includes a complete ebook of the text, extension material, GDC support, a glossary, sample examination papers, and worked solution presentations.

Mathematics education is a growing, ever changing entity. The contextual, technology integrated approach enables students to become adaptable, lifelong learners.

Note: US spelling has been used, with IB style for mathematical terms.

About the authors

Jim Fensom has been teaching IB mathematics courses for nearly 35 years. He is currently mathematics coordinator at Nexus International School in Singapore. He is an assistant examiner for Mathematics Higher Level.

Josip Harcet has been teaching the IB programme for 20 years. After teaching for 11 years at different international schools he returned to teach in Zagreb. He has served as a curriculum review member, deputy chief examiner for Further Mathematics, assistant and senior examiner, as well as a workshop leader.

Lorraine Heinrichs has been teaching IB mathematics for the past 12 years at Bonn International School. She has been the IB DP coordinator since 2002. During this time she has also been senior moderator for HL Internal Assessment and workshop leader for the IB. She was also a member of the curriculum review team.

Palmira Mariz Seiler has been teaching mathematics for 22 years. She joined the IB community 11 years ago and since then has worked as Internal Assessment moderator, in curriculum review working groups, and as a workshop leader and deputy chief examiner for HL mathematics.

Marlene Torres-Skoumal has been teaching IB mathematics for the past 30 years. She has enjoyed various roles with the IB over this time, including deputy chief examiner for HL, senior moderator for Internal Assessment, calculator forum moderator, and workshop leader.

Contents

What's on the CD?

The material on your CD-ROM includes the entire student book as an eBook, as well as a wealth of other resources specifically written to support your learning. On these two pages you can see what you will find and how it will help you to succeed in your Mathematics Higher course.

The whole print text is presented as a user-friendly eBook for use in class and at home.

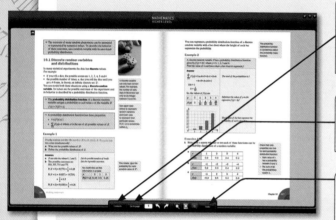

Extra content can be found in the Contents menu or attached to specific pages.

This icon appears in the book wherever there is extra content.

Navigation is straightforward either through the Contents Menu, or through the Search and Go to page tools.

A range of tools enables you to zoom in and out and to annotate pages with your own notes.

The glossary provides comprehensive coverage of the language of the subject and explains tricky terminology. It is fully editable making it a powerful revision tool.

Extension material is included for each chapter containing a variety of extra exercises and activities. Full worked solutions to this material are also provided.

Term	Definition	Notes
commutative law	A law that states that the order of the terms does not affect the result of the operation. For example $2 + 3 = 3 + 2$	
compass points	The directions on a compass. The four cardinal compass points are North (N), South (S), East (E) and West (W)	
complement	If A is a set, then the subset of the sample space U containing all elements not in A is the complement of A, denoted A'	
completing the square	Rearranging a function into the form $(ax + b)^2 + c$ where c is a constant	
complex nth roots of the unity	Complex numbers z that are solutions of the equation $z^n = r\cos\theta$	
component	The part of a vector which gives the movement of the vector parallel to one of the coordinate axes	
composite function	The resultant function when two or more functions are combined	
compound interest	Interest which is not calculated only on the original sum, but on the accumulated sum	
compound statement	A statement made up of simple statements joined together by connectives	
concave down	If $f''(x) < 0$ for all x in (a, b) then f is concave down on (a, b)	
concave up	If $f''(x) > 0$ for all x in (a, b) then f is concave up on (a, b)	
conclusion	What you believe to be true at the end of an experiment	
concurrent	Lines that all pass through a certain point	
conditional probability	The likelihood of an event after taking account of what is known about another event	
cone	A solid figure with a circular base connected to a point or vertex	
confounding factor	An additional variable that may have an effect on the data	
congruent	With the exact same form	

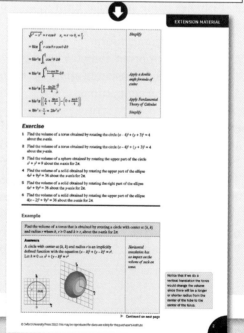

EXTENSION MATERIAL

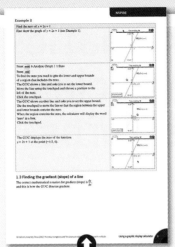

A GDC chapter gives advice on using your calculator and is provided for TI-Nspire, TI-84 Plus and Casio fx-9860GII.

Practice exam papers will help you to fully prepare for your examinations. Worked solutions can be found on the website *www.oxfordsecondary.co.uk/ibmathhl*

Alternative GDC instructions for all material in the book is given for the TI-84 Plus and Casio-9860-GII calculators, so you can be sure you will be supported no matter what calculator you use.

Powerpoint presentations cover detailed worked solutions for the practice papers in the book, showing common errors and providing hints and tips.

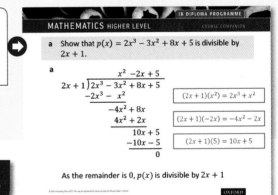

What's on the website?

Visit *www.oxfordsecondary.co.uk/ibmathhl* for free access to the full worked solutions to each and every question in the Course Companion.

www.oxfordsecondary.co.uk/ibmathhl also offers you a range of GDC activities for the TI-Nspire to help support your understanding.

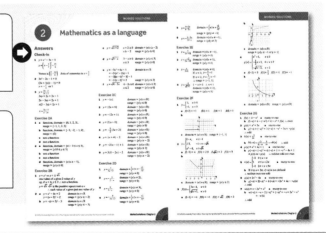

1 Mathematics as the science of patterns

Before you start

You should know how to:

1 Represent number sets.
$x \in \mathbb{Z}$ means that x is an integer
$x \in \mathbb{Z}^+$ means that x is a positive integer, an element of the set $\{1, 2, 3, \ldots\}$

2 Solve linear algebraic equations.
e.g. Solve $2(x - 4) = x + 5$
$$2x - 8 = x + 5$$
$$x = 13$$

3 Simplify surds. e.g. simplify $\dfrac{2}{\sqrt{3} + 1}$
$$\frac{2}{\sqrt{3} + 1} = \frac{2(\sqrt{3} - 1)}{(\sqrt{3} + 1)(\sqrt{3} - 1)} = \sqrt{3} - 1$$

4 Solve equations that involve fractions.
e.g. solve $\dfrac{x + 2}{x + 1} - 1 = \dfrac{2}{x}$ (multiply by $x(x + 1)$)
$$x(x + 2) - x(x + 1) = 2(x + 1)$$
$$x^2 + 2x - x^2 - x = 2x + 2 \text{ so } x = -2$$

5 Find the nth term of a sequence.
e.g. $32, 36, 40, 44, \ldots \to 4n + 28$
$2, 4, 8, 16, \ldots \to 2^n$

Skills check

1 List the numbers represented by each of these.
  a $x \le 5, x \in \mathbb{Z}^+$ **b** $-4 \le x < 2, x \in \mathbb{Z}$
  c $x \le 6, x \in \mathbb{Z}^+$

2 Solve these equations.
  a $3(x - 4) - 2(x + 7) = 0$
  b $3x - 2(2x + 5) = 2$
  c $5x + 4 - 2(x + 6) = x - (3x - 2)$

3 Simplify:
  a $2(\sqrt{3} - 2) + \sqrt{3}(1 - \sqrt{3})$
  b $\dfrac{3}{\sqrt{2}} + 5\sqrt{2}$ **c** $\dfrac{(1 + \sqrt{3})}{(1 - \sqrt{3})}$

4 Solve these equations.
  a $\dfrac{1}{(x - 2)} = -\dfrac{3}{(1 - 2x)}$
  b $\dfrac{2x}{2x^2 + 1} = \dfrac{1}{x - 1}$

5 Find the nth term:
  a $5, 11, 17, 23, 29, \ldots$
  b $25, 18, 11, 4, -3, \ldots$

From conjecture to proof

In 1637, Pierre de Fermat wrote that no three positive integers a, b and c, can be found to satisfy the equation

$a^n + b^n = c^n$ for $n > 2$

He wrote this formula and the quotation in the margin of his copy of a book written by the mathematician Diophantus.

The formula became known as Fermat's Last Theorem.

Pierre de Fermat
(1601–65)
French mathematician and physicist
"I have discovered a truly remarkable proof which this margin is too small to contain."

Andrew Wiles read about Fermat's Last Theorem when he was 10 years old and decided that he would find the proof. Little did he realize that he had undertaken a lifelong challenge. He managed to prove it in 1995, more than three centuries after Fermat claimed he had the proof.

Andrew Wiles
(1953–) British mathematician
"I don't believe Fermat had a proof."

When Fermat wrote the statement in the book he claimed to have a **proof**. No proof was ever found and, although the statement was not disproved, it remained just a **conjecture**.

> A conjecture is a rule generalizing findings made by observing patterns.

Testing different cases may verify the conjecture.

Fermat may have come up with his statement intuitively or by looking for solutions. Wiles raised the status of the conjecture to a theorem in 1995 by finding a complete proof.

In this chapter we will be looking for patterns to help us make conjectures. In order to prove a conjecture we have to show that the rule holds for all possible values and to do this we need formal proof.

Investigation – curious numbers

The diagram represents the floor of a square room, tiled with square tiles. It has a total of 9 tiles along the diagonals and 5 tiles along each side. 25 tiles are used to cover the floor completely.

Another square room has a total of 13 square tiles along the diagonals.
How many tiles are there along each side?
How many tiles are needed to completely cover the floor?

What if the total number of tiles along the diagonals is 133?
What if there is a total of 1333 tiles along the diagonals?

Can you guess what happens if there is a total of 13 333 tiles along the diagonals?
Continue to generate data to help you form a **conjecture**. Can you explain why this rule holds true?

1.1 Number patterns: sequences, series and sigma notation

A collection of numbers in a defined order, following a certain rule, is called a **sequence.** Each of the numbers in a sequence is called a **term** of the sequence.

These sets of numbers are all sequences:

i $2, 7, 12, 17, \ldots$ *start with 2 and add 5 to previous number*

ii $1, \dfrac{1}{2}, \dfrac{1}{3}, \dfrac{1}{4}$ *reciprocals of natural numbers*

iii $3, 6, 9, 12, \ldots$ *start with 3 and add 3 to previous number*

Adding all the terms in a sequence gives a **series.** If the sequence has a fixed number of terms then it is a **finite** series. If the sum of the sequence continues indefinitely the series is said to be **infinite**.

So $2 + 7 + 12 + 17 + 22$ is a finite series with five terms. The sum $1 + \dfrac{1}{2} + \dfrac{1}{3} + \dfrac{1}{4} + \ldots$ is an infinite series.

> $1 + \dfrac{1}{2} + \dfrac{1}{3} + \dfrac{1}{4} + \ldots$
> is known as the **harmonic series**.

The set of positive integers $\mathbb{Z}^+$ can be written as $\{1, 2, 3, 4, \ldots, r, \ldots\}$ where r represents the general term.

The harmonic series can be written as $1 + \dfrac{1}{2} + \dfrac{1}{3} + \dfrac{1}{4} + \ldots + \dfrac{1}{r} + \ldots$ where $\dfrac{1}{r}$ is the general term.

The general term of the finite series $3 + 6 + 9 + 12 + 15 + 18 + 21$ is $3r$ but r can only take the values from 1 to 7 because the series is finite.

The sigma (Σ) notation is a compact form to represent a series. Here is how you write the series

$1 + \dfrac{1}{2} + \dfrac{1}{3} + \dfrac{1}{4} + \ldots + \dfrac{1}{r} + \ldots + \dfrac{1}{20}$ using sigma notation:

> The largest value r can take

$$\sum_{r=1}^{20} \frac{1}{r}$$

> The smallest value r can take

Read this as 'The summation of $\dfrac{1}{r}$ from r equals 1 to 20.'

So $\displaystyle\sum_{r=1}^{4} \frac{1}{3^r} = \frac{1}{3^1} + \frac{1}{3^2} + \frac{1}{3^3} + \frac{1}{3^4} = \frac{1}{3} + \frac{1}{9} + \frac{1}{27} + \frac{1}{81}$

The series $4 + 8 + 12 + 16 + \ldots + 48$ can be written as

$4 \times 1 + 4 \times 2 + 4 \times 3 + 4 \times 4 + \ldots + 4 \times 12 = \displaystyle\sum_{r=1}^{12} 4r$

> Σ is the Greek capital letter sigma, which is used to represent 'the sum of'. **Leonhard Euler** (1706–83) was the first mathematician to use this notation.

Example 1

Write the next three terms and the general term:

a $2, 4, 8, 16, \ldots$ **b** $\dfrac{1}{2}, \dfrac{1}{6}, \dfrac{1}{12}, \dfrac{1}{20}, \ldots$ **c** $1, \dfrac{1}{2}, \dfrac{1}{4}, \dfrac{1}{8}, \ldots$

Answers

a For the sequence $2, 4, 8, 16, \ldots$
the next three terms in the sequence
are $32, 64, 128$.
The terms in the sequence can be written
$2^1, 2^2, 2^3, 2^4, \ldots, 2^r, \ldots$
General term is 2^r
where r can take the values $1, 2, 3, \ldots$

Find the general term.

b For the sequence $\dfrac{1}{2}, \dfrac{1}{6}, \dfrac{1}{12}, \dfrac{1}{20}, \ldots$ the next three

terms are $\dfrac{1}{30}, \dfrac{1}{42}, \dfrac{1}{56}$

The sequence can be written

$\dfrac{1}{1 \times 2}, \dfrac{1}{2 \times 3}, \dfrac{1}{3 \times 4}, \dfrac{1}{4 \times 5}, \ldots, \dfrac{1}{r \times (r+1)}, \ldots$

Look for the patterns in the denominators.

So the general term is $\dfrac{1}{r \times (r+1)}$

where r can take the values $1, 2, 3, \ldots$

c For the sequence $1, \dfrac{1}{2}, \dfrac{1}{4}, \dfrac{1}{8}, \ldots$ the next three

terms are $\dfrac{1}{16}, \dfrac{1}{32}, \dfrac{1}{64}$

The sequence can be written

$\dfrac{1}{1}, \dfrac{1}{2}, \dfrac{1}{4}, \dfrac{1}{8}, \dfrac{1}{16}, \dfrac{1}{32}, \dfrac{1}{64}, \ldots$

$= \dfrac{1}{2^0}, \dfrac{1}{2^1}, \dfrac{1}{2^2}, \dfrac{1}{2^3}, \dfrac{1}{2^4}, \dfrac{1}{2^5}, \dfrac{1}{2^6}, \ldots, \dfrac{1}{2^{r-1}}, \ldots$

Since the sequence has 2^0 in the first term,

The general term is $\dfrac{1}{2^{r-1}}$

the general term is $\dfrac{1}{2^{r-1}}$.

where r can take the values $1, 2, 3, \ldots$

Example 2

Write the first three terms of each series.

a $\displaystyle\sum_{r=1}^{7}(3r + 6)$ **b** $\displaystyle\sum_{r=1}^{\infty}(-1)^r \, r^2$

∞ represents infinity.

Answers

a $\displaystyle\sum_{r=1}^{7}(3r + 6) = 9 + 12 + 15 + \ldots$

*Substitute $r = 1$ for the first term,
$r = 2$ for the second term, and $r = 3$
for the third term.*

b $\displaystyle\sum_{r=1}^{\infty}(-1)^r \, r^2$

$= (-1) \times 1^2 + (-1)^2 \times 2^2 + (-1)^3$
$\times 3^2 + \ldots = -1 + 4 - 9 + \ldots$

Example 3

Write these series using sigma notation.

a $2 + 12 + 22 + 32 + \ldots + 102$

b $-8 + 4 - 2 + 1 - \dfrac{1}{2} + \ldots$

> Find the general term and the value of r that gives the last term.

Answers

a $2 + 12 + 22 + 32 + \ldots + 102$

$= \displaystyle\sum_{r=1}^{11}(10r - 8)$

The general term is $10r - 8$
For the term: 102
$10r - 8 = 102$
$10r = 110$
$r = 11$

b $-8 + 4 - 2 + 1 - \dfrac{1}{2} + \ldots$

$= (-1)^1 \times \dfrac{8}{2^0} + (-1)^2 \times \dfrac{8}{2^1}$

$+ (-1)^3 \times \dfrac{8}{2^2} + (-1)^4 \times \dfrac{8}{2^3}$

$+ (-1)^5 \times \dfrac{8}{2^4} + \ldots$

$= \displaystyle\sum_{r=1}^{\infty}(-1)^r \times \dfrac{8}{2^{r-1}}$

Alternating signs suggest that you need to multiply each term by -1.

$(-1)^n$ is positive when n is even and negative when n is odd.

The general term is $(-1)^r \times \dfrac{8}{2^{r-1}}$

Exercise 1A

1 Write the next three terms for each sequence.

a $-6, -4.5, -3, -1.5, \ldots$ **b** $\dfrac{1}{2}, \dfrac{3}{4}, \dfrac{5}{6}, \dfrac{7}{8}, \ldots$ **c** $\dfrac{1}{3}, \dfrac{1}{15}, \dfrac{1}{35}, \dfrac{1}{63}, \ldots$

2 Write the general term for each sequence.

a $2, 6, 12, 20, \ldots$ **b** $\dfrac{1}{2}, \dfrac{1}{5}, \dfrac{1}{10}, \dfrac{1}{17}, \dfrac{1}{26}, \ldots$ **c** $-1, 1, 3, 5, \ldots$

3 Given that r can take the values $1, 2, 3, \ldots$, write the first four terms of the sequence with a general term:

a $4r - 3$ **b** $\dfrac{r}{2r+1}$ **c** $\dfrac{1}{r^2}$

4 Expand these series in full.

a $\displaystyle\sum_{r=1}^{4} r(r+1)$ **b** $\displaystyle\sum_{r=1}^{5} \dfrac{r}{2r+1}$ **c** $\displaystyle\sum_{r=1}^{5}(-1)^r r^2$

5 Write these series using sigma notation.

a $-1 + 3 + 7 + 11 + \ldots$

b $-1 + 1 - 1 + 1 - 1 + 1 - 1 + 1 - 1 + 1$

c $6 - 12 + 24 - 48 + 96 - 192$

Investigation – quadratic sequences

Use your GDC to look at the numbers generated by $T = n^2 - 2n + 3$ where $n \in \mathbb{Z}^+$.

Enter the data in the GDC:

Use a table to list the first differences and the second differences of the numbers generated by this quadratic formula.

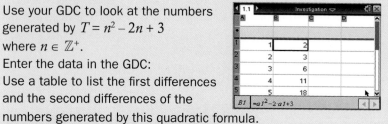

GDC help on CD:
Alternative demonstrations for the TI-84 Plus and Casio fx-9860GII GDCs are on the CD.

$n^2 - 2n + 3$	First difference	Second difference
2		
3	1	2
6	3	2
11	5	2
18	7	2
27	9	2
38	11	2
51	13	

> First differences are obtained by subtracting two consecutive terms of the original sequence.

> Second differences are obtained by subtracting consecutive terms of the first differences.

In general three consecutive integers are $p - 1$, p and $p + 1$.
Use these values to calculate $n^2 - 2n + 3$ for each one.
Use algebra to justify that the second difference is always constant.

Now see what happens if you repeat the task using:
$T = 2n^2 + 2n + 1$
$T = -n^2 + 3n - 4$
Try to generalize your results.

> A general form for any quadratic equation is $T = an^2 + bn + c$ where a, b and c are constants.

Investigation – triangular numbers

The patterns of dots represent the first five triangular numbers.

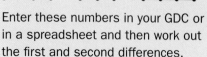

Enter these numbers in your GDC or in a spreadsheet and then work out the first and second differences.

You can continue generating consecutive numbers as shown in the table.

Use your findings from the quadratic sequences investigation to find a formula that generates the triangular numbers.

Triangle numbers	First difference	Second difference
1		
3	2	1
6	3	1
10	4	1
15	5	1
21	6	

$21 = 15 + 6$ $6 = 5 + 1$

Investigation – more number patterns

The diagrams show the first five terms of the square,
the pentagonal, the hexagonal and the heptagonal numbers. Use
lists and the method of differences to obtain formulae that generate
these numbers.

Square numbers

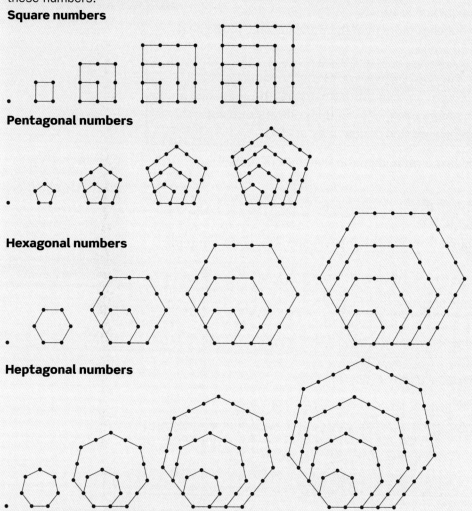

Pentagonal numbers

Hexagonal numbers

Heptagonal numbers

Use your results from the investigations to complete this table.

Term	1st	2nd	3rd	4th	5th	6th	7th	8th	nth
Triangular numbers	1	3	6	10	15	21			
Square numbers	1	4	9	16	25				
Pentagonal numbers	1	5	12						
Hexagonal numbers	1	6	15						
Heptagonal numbers									
Octagonal numbers									
Nonagonal numbers									

Use the table to make a conjecture about the nth term of any polygonal number.

1.2 Arithmetic sequences and series

The picture represents the front of a building with
arches on each level.
To find out how many arches there will be on the fourth
level you first need to count the arches on each level.
This gives the sequence 5, 10, 15.
Each floor has 5 arches more than the previous one, so the next
floor will have 20 arches.

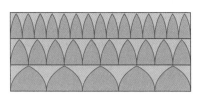

> → If the difference between two consecutive numbers in a
> sequence is constant then this is an **arithmetic sequence** or an
> **arithmetic progression**. We call this constant difference the
> **common difference** and denote it by d.

If you write the first term u_1 then the sequence of arches is

$u_1 = 5$
$u_2 = 5 + 5 = u_1 + d$
$u_3 = 10 + 5 = u_2 + d$

Each term is found from the previous one by adding the common
difference d.

$u_n = u_{n-1} + d$

Look again at the terms:

$u_1 = 5$
$u_2 = 5 + 5 = u_1 + d$
$u_3 = 10 + 5 = u_2 + d = u_1 + 2d$
$u_4 = 15 + 5 = u_3 + d = u_1 + 3d$

This leads to the general term $u_n = u_1 + (n - 1)d$

> → An arithmetic sequence with the first term u_1 and common
> difference d is $u_1, u_1 + d, u_1 + 2d, \ldots, u_1 + (n - 1)d$.
>
> The general term is $u_n = u_1 + (n - 1)d$

This is called a
recursive equation.
You can work out
any term using this
equation only if you
know or can generate
the previous term.

Example 4

Three numbers are consecutive terms in an arithmetic sequence.
Their sum is 48 and their product is 2800.
Find the three numbers.

Answer

Let the three numbers be
$u - d, u, u + d$

$u - d + u + u + d = 48$ *Write the sum of the three numbers.*

$ 3u = 48$

$ u = 16$

▶ Continued on next page

$(u - d)u(u + d) = 2800$	*Write the product.*
$u(u^2 - d^2) = 2800$	
$16^2 - d^2 = 175$	*Substitute u = 16 and divide*
$d = \pm\sqrt{256-175}$	*by 16.*
$d = \pm 9$	*The two values of d give two possible*
The three numbers are	*sequences:*
7, 16 and 25.	*7, 16, 25 or 25, 16, 7*

Example 5

Find the number of terms in these arithmetic sequences:

a $\quad$ 50, 47, 44, ..., 14

b $\quad$ $a, 3a, 5a, ..., 21a$

Answers

a $\quad u_1 = 50, d = -3$	*Using $u_n = u_1 + (n - 1)d$*
$u_n = 53 - 3n$	
$53 - 3n = 14$	*Using $u_n = 14$ and solving for n*
$3n = 39$	
$n = 13$	
b $\quad u_1 = a, d = 2a$	
$u_n = (2n - 1)a$	
$(2n - 1)a = 21a$	
$2an = 22a$	
$n = 11$	

Example 6

The second term of an arithmetic sequence is 15 and the fifth term is 21. Find the common difference and the first term of the sequence.

Answer

$u_2 = u_1 + d = 15$	*Using $u_n = u_1 + (n - 1)d$*
$u_5 = u_1 + 4d = 21$	
$3d = 6$	*Solving simultaneously*
$d = 2$	
$u_1 = 15 - 2 = 13$	

Exercise 1B

1 Find the nth term of these sequences.

    a $\quad$ 5, 11, 17, 23, ... $\qquad\qquad$ **b** $\quad$ 10, 3, −4, −11

    c $\quad$ $a, a + 2, a + 4, a + 6, ...$

2 Find the terms indicated in each of these arithmetic sequences:

    a $\quad$ 2, 11, 20, ... $\qquad$ 15th term

    b $\quad$ $-1, \dfrac{1}{4}, \dfrac{3}{2}, ...$ $\qquad$ 12th term

    c $\quad$ 3, 7, 11, ... $\qquad$ $(n + 1)$th term

3 The fourth term of an arithmetic sequence is 18 and the common difference is –5. Find the first term and the nth term.

4 The fourth term of an arithmetic sequence is 0 and the fourteenth term is 40. Find the common difference and the first term.

5 A job advertisement states that the job carries a salary of €48000 per year rising by annual increments of €500. How much would the salary be after 15 years? How many years would a person have to hold this position for a 50% salary increase on the initial salary?

The sum of an arithmetic series

When Gauss was 11 years old his teacher challenged him to find the sum of the numbers from 1 to 100. To the teacher's surprise Gauss gave the correct answer almost immediately as 5050. Here is how he did it:

$$S = 1 + 2 + 3 + 4 + \dots\ 99 + 100$$
$$S = 100 + 99 + 98 + 97 + \dots\ 2 + 1$$

$$2S = 101 + 101 + 101 + 101 + \dots 101 + 101$$

$$2S = 101 \times 100$$

$$\therefore S = \frac{10\,100}{2} = 5050$$

$\mathcal{C.F.}\,\mathcal{Gauss}$

Carl Friedrich Gauss (1777–1855) German mathematician "Mathematics is the queen of all sciences"

The numbers 1 to 100 are an arithmetic series with first term 1 and common difference 1. Gauss had found a method for calculating the sum of a finite arithmetic series.

Generalizing this method for a series containing n terms, with first term u_1 and common difference d gives

$$S_n = u_1 + u_1 + d + u_1 + 2d + \dots + u_1 + (n-2)d + u_1 + (n-1)d$$
$$S_n = u_1 + (n-1)d + u_1 + (n-2)d + u_1 + (n-3)d + \dots + u_1 + d + u_1$$

$$2S_n = 2u_1 + (n-1)d + 2u_1 + (n-1)d + 2u_1 + (n-1)d + \dots + 2u_1 + (n-1)d + 2u_1 + (n-1)d$$

$$2S_n = n[2u_1 + (n-1)d]$$

$$S_n = \frac{n}{2}[2u_1 + (n-1)d]$$

This formula can also be written as

$$S_n = \frac{n}{2}[u_1 + u_1 + (n-1)d]$$

$$= \frac{n}{2}[u_1 + u_n]$$

$u_1 + (n-1)d = u_n$

> → The sum of a finite arithmetic series is
>
> $$S_n = \frac{n}{2}[2u_1 + (n-1)d]$$
>
> $$= \frac{n}{2}(u_1 + u_n)$$
>
> where n is the number of terms in the series, u_1 is the first term, u_n is the last term and d is the common difference.

Example 7

The first term of an arithmetic series is 2 and the last term is 26. The series has 9 terms.
Find the sum of the series.

Answer

$u_1 = 2$, $u_9 = 26$, $n = 9$

$S_9 = \frac{9}{2}(2 + 26) = 126$ *Using $S_n = \frac{n}{2}(u_1 + u_n)$*

Example 8

The first term of an arithmetic series is 25 and the fourth term is 13.
The sum of the series is –119.
Find the number of terms in the series.

Answer

$u_1 = 25$

$u_1 + 3d = 13$ *$u_4 = u_1 + 3d$*
Find d.

$\quad 3d = 13 - 25$

$\quad\quad d = -4$

$S_n = \frac{n}{2}[2u_1 + (n-1)d]$

$-119 = \frac{n}{2}[50 - 4(n-1)]$ *Solve for n.*

$-238 = n(54 - 4n)$

$4n^2 - 54n - 238 = 0$ *Divide by 2.*

$2n^2 - 27n - 119 = 0$

$(2n + 7)(n - 17) = 0$ *Factorize.*

Since $n \in \mathbb{Z}^+$, $n = 17$ *$\mathbb{Z}^+$ is the set of positive integers.*

The natural numbers $\mathbb{N}$ is the set {0, 1, 2, ...} It differs from $\mathbb{Z}^+$ because it includes zero.

Example 9

Calculate $\displaystyle\sum_{r=1}^{10}(5r-7)$

Answer

$u_1 = -2$
$d = 5$
$u_{10} = 43$
$S_{10} = \dfrac{10}{2}(-2 + 43) = 205$

or

Use the formula for the sum of a finite arithmetic series

Using a GDC:

$\displaystyle\sum_{r=1}^{10}(5r-7) = 205$

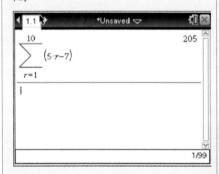

Exercise 1C

1 Evaluate these series.

   a $6 + 19 + 32 + \ldots + 110$

   b $52 + 41 + 30 + \ldots - 25$

   c $-78 - 82 - 86 - 90 - \ldots - 142$

2 Calculate:

   a $\displaystyle\sum_{r=1}^{10}(5r+7)$ **b** $\displaystyle\sum_{r=1}^{15}(5-3r)$

3 Find the sum of an arithmetic series with 16 terms, given that the first term is 60 and the 10th term is –3.

EXAM-STYLE QUESTIONS

4 The fourth term of an arithmetic series is 8.
The sum of the first five terms is 25.
Find the first five terms.

5 The sum of an arithmetic series is given by $S_n = n(2n + 3)$. Find the common difference and the first four terms of the series.

1.3 Geometric sequences and series

The diagram shows the first three steps in constructing Sierpinski's triangle, named after the Polish mathematician **Waclaw Sierpinski** who first described it in 1915.

| 1 | 2 | 3 | 4 |

If you count the white triangles in each of the figures 1 to 4 you get the sequence 1, 3, 9, 27 Figure 5 would have 81 white triangles. At each stage the number of white triangles is three times the number in the previous stage.

If the sides of the original triangle are 1 unit long, then the length of each side of the green triangle will be $\frac{1}{2}$ unit; each side of the orange triangles will be $\frac{1}{4}$ unit and each side of the blue triangles is $\frac{1}{8}$ unit.

If the area of the first triangle is 1 unit^2, then the area of the green triangle is $\frac{1}{4}$ unit^2, since four of the green triangles make up the original triangle. The area of each orange triangle will be $\frac{1}{16}$ unit^2 and of each blue triangle will be $\frac{1}{64}$ unit^2.

Claudia Zaslavsky (1917–2006) Ethnomathematician Researched expressions of mathematics in African culture, including number words and signs, reckoning of time, games, and patterns. Her book *Africa Counts* was published in 1973.

Hidden in Sierpinski's triangle are the sequences:

1, 3, 9, 27, ...   To get next term multiply previous term by 3

$1, \frac{1}{2}, \frac{1}{4}, \frac{1}{8}, ...$ To get next term multiply previous term by $\frac{1}{2}$

$1, \frac{1}{4}, \frac{1}{16}, \frac{1}{64}, ...$ To get next term multiply previous term by $\frac{1}{4}$

The ratio of consecutive terms is a constant for each sequences.

→ If the ratio of two consecutive terms in a sequence is constant then this is a **geometric sequence** or a **geometric progression**. We call this ratio the **common ratio** and denote it by r.

So for the sequence 1, 3, 9, 27, ... $u_1 = 1$, $r = 3$ and the recursive equation is:

$$u_n = u_{n-1} r$$

For this sequence:

$u_1 = 1$
$u_2 = 1 \times 3 = u_1 \times r$
$u_3 = 3 \times 3 = u_2 \times r = u_1 \times r^2$
$u_4 = 9 \times 3 = u_3 \times r = u_1 \times r^3$

This leads to the general term $u_n = u_1 \times r^{n-1}$

A recursive equation is one in which the next term is defined as a function of earlier terms.

→ In a geometric sequence with first term u_1 and common ratio r the general term is given by $u_n = u_1 \times r^{n-1}$, $r \neq -1, 0, 1$.

Why have these values of r been omitted?

Example 10

Find the common ratio of each of these sequences and write the next two terms of each sequence.

a $10, 25, 62.5, \ldots$

b $\dfrac{1}{2}, -\dfrac{1}{6}, \dfrac{1}{18} \ldots$

c $a, 2a^3, 4a^5, \ldots$

Answers

a $r = \dfrac{25}{10} = 2.5$

The next two terms are 156.25 and 390.625

b $r = -\dfrac{1}{6} \div \dfrac{1}{2} = -\dfrac{1}{3}$

The next two terms are $-\dfrac{1}{54}$ and $\dfrac{1}{162}$

c $r = \dfrac{2a^3}{a} = 2a^2$

The next two terms are $8a^7$ and $16a^9$

Example 11

Find the number of terms in each of these geometric sequences.

a $2, 4, 8, \ldots, 256$

b $5, 10, 20, \ldots, 5 \times 2^k$

Answers

a $u_1 = 2, r = 2$

$u_n = 2 \times 2^{n-1} = 2 \times 128 = 2 \times 2^7$

$n = 8$

b $u_1 = 5, r = 2$

$u_n = 5 \times 2^{n-1} = 5 \times 2^k$

$n = k + 1$

Exercise 1D

1 Write down the 6th term and the nth term of each sequence.

    a $1, 2, 4, \ldots$ **b** $9, 3, 1 \ldots$ **c** $x^3, x^2, x, \ldots$

2 Find the common ratio and the terms indicated in each of these sequences.

    a $48, 24, 12, \ldots$ 10th term

    b $\dfrac{16}{3}, -\dfrac{8}{9}, \dfrac{4}{27}, \ldots$ 5th term

3 Find the number of terms in each of these sequences.

    a $0.03, 0.06, 0.12, ..., 1.92$ **b** $81, 27, 9, ..., \dfrac{1}{81}$

4 The third term of a geometrical sequence is 2 and the fifth is 18. Find two possible values of the common ratio and the second term in each case.

5 The first term of a geometric sequence is 16 and the fifth term is 9. What is the value of the seventh term?

6 The numbers $a - 4$, $a + 2$ and $3a + 1$ are three consecutive terms of a geometric progression. Find the two possible values of the common ratio.

The sum of a geometric series

How could you find the sum of the first 10 terms of the series
$1 + 3 + 9 + ... + 3^{n-1} + ...$?

Write
$$S_{10} = 1 + 3 + 3^2 + 3^3 + ... + 3^9$$

Multiply the whole series by 3 and subtract from S:

$S_{10} =$	1	+	3	+	3^2	+	3^3	+	...	+	3^9		
$3S_{10} =$			3	+	3^2	+	3^3	+	...	+	3^9	+	3^{10}
$(1 - 3)S_{10} =$	1											$-$	3^{10}

This gives

$$S_{10} = \frac{1 - 3^{10}}{1 - 3} = \frac{1}{2}(3^{10} - 1)$$

To find the sum of the first n terms of a geometric series use the same process by multiplying by the common ratio r and subtracting:

$S_n =$	u_1	+	$u_1 r$	+	$u_1 r^2$	+	$u_1 r^3$	+	...	+	$u_1 r^{n-1}$		
$rS_n =$			$u_1 r$	+	$u_1 r^2$	+	$u_1 r^3$	+	...	+	$u_1 r^{n-1}$	+	$u_1 r^n$
$(1 - r)S_n =$	u_1											$-$	$u_1 r^n$

> → The sum of a geometric series is given by
> $$S_n = \frac{u_1(1 - r^n)}{1 - r}, r \neq 1$$
> where n is the number of terms, u_1 is the first term and r is the common ratio.

Investigation – infinite sums

In the diagram, AB represents a piece of string 11.28 cm long.

The string is cut in half and one half, CD, is placed underneath. The remaining half is again cut in half and one half, DE, is placed next to CD. The process is repeated twice more and the total length of the pieces placed side by side is noted.

A _____ B

C _____ D

C _____ D _____ E

C _____ D _____ E ___ F

C _____ D _____ E ___ F _ G

AB = 11.28 cm

CD = 5.64 cm

CD + DE = 8.46 cm

DE = 2.82 cm

CD + DE + EF = 9.87 cm

EF = 1.41 cm

CD + DE + EF + FG = 10.575 cm

FG = 0.705 cm

If this process is continued indefinitely, the total length will continue increasing BUT can never exceed 11.28 cm, the original length. Mathematically:

$$CD = u_1 = 5.64 \qquad DE = u_2 = 5.64 \times \frac{1}{2}$$

$$EF = u_3 = 5.64 \times \left(\frac{1}{2}\right)^2 \qquad FG = u_4 = 5.64 \times \left(\frac{1}{2}\right)^3$$

$$u_n = 5.64 \times \left(\frac{1}{2}\right)^{n-1}$$

This is a geometric sequence, first term 5.64 and common ratio $\frac{1}{2}$.

GDC help on CD: *Alternative demonstrations for the TI-84 Plus and Casio fx-9860GII GDCs are on the CD.*

The sum of n terms of this series will therefore be

$$S_n = \frac{u_1(1-r^n)}{1-r} = \frac{5.64\left(1-\left(\frac{1}{2}\right)^n\right)}{\left(1-\frac{1}{2}\right)}$$

Enter this on the GDC to see what happens as n gets bigger.

Note that when you sum 15 terms the result is very close to 11.28, the original length.

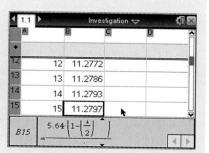

Convergent and divergent series

A geometric series is **convergent** when the sum tends to a finite value as the number of terms gets bigger. In the investigation with the string, the series converges to a sum $S = 11.28$ cm.

If a geometric series does not converge it is **divergent**.

Investigation – convergent series

In order to understand the condition for convergence now look at the formula for the sum of n terms:

$$S_n = \frac{u_1(1-r^n)}{(1-r)}$$

Use your GDC to calculate these values of r^n for $n \in \mathbb{Z}^+$, $1 \le n \le 10$:

> What happens as n increases?

a 3^n **b** 2^n **c** $\left(\dfrac{3}{2}\right)^n$

d $\left(\dfrac{1}{2}\right)^n$ **e** $\left(-\dfrac{1}{5}\right)^n$ **f** $\left(\dfrac{3}{4}\right)^n$

Try other values of r.

Use your results to justify these statements:

- $r > 1 \Rightarrow r^n$ increases as n gets larger. The larger the value of r, the faster the value of r^n increases.
- $0 < r < 1 \Rightarrow r^n$ decreases as n gets larger.
- $r < -1 \Rightarrow r^n$ has a large absolute value but its sign oscillates.
- $-1 < r < 0 \Rightarrow r^n$ has a very small absolute value but its sign oscillates.
- When the value of r is close to (but still less than) 1, the value of r^n decreases more slowly but still gets close to zero when n is large enough.

→ The sum of n terms of a geometric series is

$$S_n = \frac{u_1(1-r^n)}{(1-r)}$$

When $-1 < r < 1$, r^n approaches zero for very large values of n. The series therefore converges to a finite sum given by

$$S = \frac{u_1}{1-r}$$

> Convergence is covered in more detail in Chapter 4.

Example 12

Calculate the geometric series given by $\displaystyle\sum_{r=1}^{6}(3 \times 2^r)$

Answer

$u_1 = 6$, $r = 2$

$$\sum_{r=1}^{6}(3 \times 2^r) = \frac{6(1-2^6)}{1-2} = 378$$

or

using GDC $\displaystyle\sum_{r=1}^{6}(3 \times 2^r) = 378$

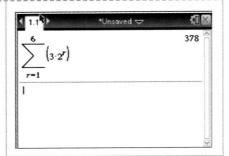

GDC help on CD:
Alternative demonstrations for the TI-84 Plus and Casio fx-9860GII GDCs are on the CD.

Example 13

Find two possible geometric sequences where the sum of the first two terms is 9 and the sum of the first four terms is 45 and write down the general term of each sequence.

Answer

$S_2 = u_1 + u_1 r = 9$

$\Rightarrow u_1 = \dfrac{9}{1 + r}$

Find an expression for u_1.

$S_4 = \dfrac{u_1(1 - r^4)}{1 - r} = 45$

Using $S_n = \dfrac{u_1(1 - r)}{1 - r}$

$\Rightarrow \dfrac{9(1 - r^4)}{(1 + r)(1 - r)} = 45$

Substitute $u_1 = \dfrac{9}{1 + r}$

$9(1 - r^4) = 45(1 - r^2)$

$(1 - r^2)(1 + r^2) = 5(1 - r^2)$

$\Rightarrow 1 + r^2 = 5$

$\Rightarrow r^2 = 4$

$r = \pm 2$

Divide both sides by 9 and factorize using the 'difference of two squares' $(r \neq 1)$.

$r = 2 \Rightarrow u_1 = \dfrac{9}{3} = 3$

$r = -2 \Rightarrow u_1 = \dfrac{9}{-1} = -9$

Substitute $r = \pm 2$ in $u_1 = \dfrac{9}{1 + r}$

The two possible geometric sequences are:

$3, 6, 12, \ldots, 3 \times 2^{n-1}, \ldots$

$-9, 18, -36, \ldots, -9 \times 2^{n-1}, \ldots$

Example 14

The sum of the first n terms of a geometric series is given by $S_n = 5^n - 1$. Find the first term and the common ratio of this series.

Answer

$S_1 = 5^1 - 1 = 4$

$u_1 = 4$

$S_2 = 5^2 - 1 = 24$

$u_1 + u_2 = 24$

$u_2 = 20$

$r = \dfrac{20}{4} = 5$

Using $u_2 = u_1 r$

Exercise 1E

1 Evaluate these sums for the number of terms stated.

a $2 + 1 + \dfrac{1}{2} + \dfrac{1}{4} \dots$ 6 terms

b $2 - 3 + 4.5 - 6.75 + \dots$ 8 terms

c $1 - \dfrac{1}{2} + \dfrac{1}{4} - \dfrac{1}{8} + \dots$ 10 terms

d $0.1 + 0.02 + 0.004 + 0.0008 + \dots$ 15 terms

2 Calculate:

a $\displaystyle\sum_{r=0}^{5} 5^{3-r}$

b $\displaystyle\sum_{r=0}^{n-1} \left(9 \times 10^{r}\right)$

3 Find the sum of a geometric series with six terms given that the third term is 2 and the seventh term is $\dfrac{1}{128}$.

4 The sum of n terms of a certain series is given by $\left(\dfrac{3}{2}\right)^{n} - 1$

a Find the first three terms of the series.

b Show that the terms of the series are in geometric progression.

EXAM-STYLE QUESTIONS

5 S_n is the sum of the first n terms of a geometric sequence with first term a and common ratio r. Let P_n represent the product of these terms. Write P_n in terms of a and r.

Show that the sequence formed by taking the reciprocals of the terms is also geometric. Write R_n, the sum of the first n terms of the reciprocals, in terms of a and r and hence show that

$$\left(\dfrac{S_n}{R_n}\right)^{n} = (P_n)^{2}$$

6 The second term of a geometric progression is 24 and the third term is $12(p - 1)$. Find the common ratio of this progression given that the series is convergent and the sum of the first three terms is 76.

7 A rope of length 2 m is divided into three pieces whose lengths are in geometric sequence. The longest piece is twice as long as the shortest piece. Find the common ratio of the sequence and the exact length of the shortest piece of rope.

8 Write the first four terms of the series $\displaystyle\sum_{r=0}^{\infty} \dfrac{(x+1)^{r}}{3^{r}}$. For what values of x does this series converge? Evaluate the sum when $x = -1.5$

9 An infinite geometric series is such that $S - S_n = ku_n$, $k \in \mathbb{Z}^{+}$. Find the common ratio and hence show that $S = (k + 1)u_1$

Problem solving using geometric progressions

Example 15

The number of Facebook users in August 2008 was 100 million and growing at a rate of 3% per week. In August 2010 the number of Facebook users was 500 million.

a If the rate of growth had remained constant at 3% what would the number of users in August 2010 have been?

b The growth rate of 3% per week actually remained steady for 6 months and then dropped to 1.1% per week. Show that this model better describes the recorded numbers.

c If the growth rate of users remained at a steady 1% per week after August 2010, how long would it take for the number of users to reach 1 billion?

Answers

a The number of users after 2 years is
$100 \times (1.03)^{104} \approx 2160$ million (3 sf)

2 years = 104 weeks

> ≈ means 'approximately equal to'.
> The numbers have been rounded to 3 sf.

b The number of users after 6 months is
$100 \times (1.03)^{26} \approx 215.7$ million
The number of users in August 2010 would be
$215.7 \times (1.011)^{78} \approx 506$ million (3 sf)
This is close to the recorded value of 500 million.

6 months = 26 weeks
6 months at 3% growth p.w.

26 + 52 = 78 weeks

18 months at 1.1% growth p.w.

c Number of users in August 2010 is 500 million
Growth model given by
$500 \times (1.01)^n$
$500 \times (1.01)^n = 1000$
Use GDC to solve for n.

1 billion = 1000 million

> In some countries
> 1 billion = 10^{12}

1.1 ▶	Investigation ▽			
A	B	C	D	E
1	66	964.23		
2	67	973.872		
3	68	983.611		
4	69	993.447		
5	70	1003.38		
B5	$=500 \cdot (1.01)^{a5}$			

It will take 70 weeks to reach 1 billion, approximately in December 2011.

> **GDC help on CD:**
> *Alternative demonstrations for the TI-84 Plus and Casio fx-9860GII GDCs are on the CD.*

Mathematics as the science of patterns

Example 16

When Sofia was born, her grandparents invested € 1000 at a compound interest rate of 4% each year.
They then added € 1000 to the account on Sofia's birthday every year.
How much money was in Sofia's account just after her 18th birthday?

Answer

At Sofia's birth € 1000	*Amount in bank + interest for one*
Just before 1st birthday	*year 1000 + 0.04 × 1000*
1000 (1.04)	*= 1000 (1.04)*
On 1st birthday	*Value after 1 year + second*
1000 (1.04) + 1000	*payment*
Just before 2nd birthday	
(1000 (1.04) + 1000)1.04	
Just after 2nd birthday	
$1000 (1.04)^2 + 1000 (1.04) + 1000$	
Just after 18th birthday the amount is $1000(1.04)^{18} + 1000(1.04)^{17} +$	*Using the pattern from years 1*
$1000(1.04)^{16} + \ldots + 1000$	*and 2*
	This is a geometric series with
$S_n = \dfrac{1000(1-(1.04)^{19})}{(1-1.04)}$	$u_1 = 1000, r = 1.04$ *and* $n = 19$
$= € 27671$ (to nearest €)	

> Including the first
> € 1000 there are
> 19 terms in total.

Exercise 1F

1 The sum to infinity of a geometric series is 4 times the second term.

  a Find the common ratio.

The first term of the series is 32.

  b What is the percentage error in the approximation $S_5 \approx S$?

> Percentage error
> $= \dfrac{V_A - V_E}{V_E} \times 100\%$
> where V_E = exact
> value
> V_A = approximated
> value

2 On his 21st birthday, Prince Abdul started receiving annual payments from a trust fund. On each succeeding birthday he received one and a half times as much as in the previous year. By the age of 25 he had received a total of $52 750.
How much did he receive on his 21st birthday?

3 Vivek can trace his family tree back five generations. Vivek's parents are the first generation back and his first set of ancestors. Vivek's four grandparents are the second generation back.

  a How many ancestors are in his family tree?

  b How many generations would he have to trace to find more than one million ancestors??

4 Yi-Ching takes out a loan of $1000 to buy a new computer.
The terms of the loan are that Yi-Ching will pay equal monthly
installments. Interest is calculated monthly and is charged
at 12% p.a. The loan is to be repaid in 2 years.

> p.a. stands for 'per
> annum' or 'yearly'.

a Calculate the amount Yi-Ching has to pay each month if
the first repayment is made one month after the money is
borrowed and after interest is calculated.

b How much, to the nearest dollar, does Yi-Ching actually pay
for the computer?

1.4 Conjectures and proofs

In most subjects knowledge is acquired from observation. This is also true in
mathematics but proof gives absolute rigor and certainty.

Mathematical proof is a logical sequence of steps which establishes the
truth of a statement beyond any doubt.

Mathematicians look for patterns and then use intuition to make a
conjecture to describe that pattern in mathematical terms. Only when the
conjecture has been proved does it become a **theorem** – a truth.

There are many types of **proofs**. Earlier in this chapter we used logical
deductive reasoning to establish formulae for sums of arithmetic and
geometric series.

Alan Turing
(1912–54) British
mathematician
"Mathematical
reasoning may be
regarded rather
schematically as
the exercise of a
combination of two
facilities, which we
may call intuition and
ingenuity."

Direct proof

Example 17

Use a direct proof to show that the sum of two odd numbers is always even.

Answer

Let a and b be two odd numbers
$a = 2p + 1$, $b = 2q + 1$
where $p, q \in \mathbb{Z}$.
$a + b = 2p + 1 + 2q + 1$
$= 2p + 2q + 2$
$= 2(p + q + 1)$
Since $p, q \in \mathbb{Z}$, $p + q + 1 \in \mathbb{Z}$
Therefore $a + b$ is even.

An odd number can be written as
$2n + 1$, $n \in \mathbb{Z}$.
An even number can be written as
$2n$, $n \in \mathbb{Z}$.

Exercise 1G

1 Show that

  a the sum of an odd and an even number is always odd

  b the product of two odd numbers is always odd.

2 Prove that: $\dfrac{1}{x-2} - \dfrac{2}{2x+5} = \dfrac{9}{2x^2 + x - 10}$

3 Use the diagram to prove that $a^2 + b^2 = c^2$

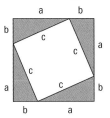

This is Pythagoras' theorem.

4 Complete the table to test if the following conjecture is true.
'The product of two consecutive integers plus the larger of the two integers is always a perfect square.'

3	4	3 × 4 + 4	16
7	8		
−6	−5		
11	12		
8	9		

Use your results to prove the conjecture.

Use algebra for the proof.

1.5 Mathematical induction

The principle of mathematical induction is a method mathematicians use to prove statements about sequences.

To illustrate the principle of proof by induction, imagine two dominoes placed at a distance less than half their length. If the first domino is tilted as shown it will fall and cause the second domino to fall with it. This is the starting point of the process.

Now imagine n dominoes arranged in a row in the same way. If the first domino falls it will cause the next dominoes to fall with it. This assumption is the second step in the process.

If we add another domino at the end of the n dominoes, this last domino also falls. This is the third and final step of the process.

The method of induction described and used in this section is called 'weak' induction. You could investigate the difference between 'strong' and 'weak' induction and their uses.

At the start of the process the first domino will cause the second domino to fall. It then follows using the second and third steps that a third domino placed behind the first two will also fall. Knowing that three dominoes will fall and again using steps two and three a fourth domino will also fall. We can go on repeating this process as many times as we want; in other words we have shown that we can have as many dominoes as we like and they will all fall if the first one is tilted towards the second domino.

The dominoes analogy can be applied to mathematics.

> → We start with a statement $P(n)$ which we want to prove is true for all values of n, $n \in \mathbb{Z}^+$.
>
> Step 1   Prove that the statement is true for a starting value $P(1)$.
>
> Step 2   Assume that $P(k)$ is true, where k is a particular value of $n \geq 1$.
>
> Step 3   Use the assumption that $P(k)$ is true to show that $P(k + 1)$ is then also true.

The starting value is not always 1.

Since it was proved true (the first domino will fall) for $n = 1$ in step 1, and steps 2 and 3 show that if one domino falls then its neighbor will also fall, then using the principle of mathematical induction it is true for all values of $n \geq 1$ (all the dominoes will fall).

Example 18

Use mathematical induction to show that the sum of the first n terms of an arithmetic sequence with first term u_1 and common difference d is given by $S_n = \dfrac{n}{2}(2u_1 + (n-1)d)$

Answer

$P_n = S_n = \dfrac{n}{2}(2u_1 + (n-1)d)$

P_1:

LHS: $S_1 = u_1$

RHS: $= \dfrac{1}{2}(2u_1 + (1-1)d) = u_1$

Notation:

LHS = left-hand side

RHS = right-hand side

LHS = RHS $\Rightarrow P_1$ is true

Assume P_k is true i.e.,

$S_k = \dfrac{k}{2}(2u_1 + (k-1)d)$

Substitute k in statement.

Show P_{k+1} is true i.e.,

$S_{k+1} = \dfrac{k+1}{2}(2u_1 + kd)$

LHS $= S_k + u_1 + kd$

$= \dfrac{k}{2}(2u_1 + (k-1)d) + u_1 + kd$

$= ku_1 + \dfrac{k}{2}(k-1)d + u_1 + kd$

$= u_1(k+1) + kd\left(\dfrac{(k-1)}{2} + 1\right)$

Use assumption to prove that

$S_{k+1} = \dfrac{k+1}{2}(2u_1 + kd)$

Substitute n = k + 1 in statement $u_n = u_1 + (n-1)d$ using assumption in step 2.

▶ Continued on next page

$$= u_1(k + 1) + kd\left(\frac{(k+1)}{2}\right)$$
$$= \frac{(k+1)}{2}(2u_1 + kd)$$
$$= \frac{(k+1)}{2}\left(2u_1 + ((k+1)-1)d\right)$$

Since the statement was shown to be true for $n = 1$, and it was also proved (steps 2 and 3) that if the statement is true for $n = k$ it is also true for $n = k + 1$, it follows by the principle of mathematical induction that the statement is true for all positive integers n.

Example 19

Use mathematical induction to prove that $n^2 > 7n + 1$ for all $n \geq 8$.

In this example the starting value is 8, not 1.

Answer

Let $P(n)$ be the statement $n^2 > 7n + 1$

Step 1

When $n = 8$, $n^2 = 64$ and $7n + 1 = 57$

Since $64 > 57$, $P(8)$ is true.

Step 2

Assume that for some $k \in \mathbb{Z}^+$

$k^2 > 7k + 1$

Step 3

Use assumption to prove that $P(k + 1)$ is true:

$(k + 1)^2 > 7(k + 1) + 1$

Proof:

$(k + 1)^2 = k^2 + 2k + 1 > 7k + 1 + 2k + 1$

$= 7(k + 1) + 2k - 5$

$= 7(k + 1) + 1 + (2k - 6)$

$2k - 6 > 0$ since $k > 8$

$\therefore 7(k + 1) + 1 + (2k - 6) > 7(k + 1) + 1$

It follows that $(k + 1)^2 > 7(k + 1) + 1$

$\therefore P(k + 1)$ is true.

Since $P(8)$ was shown to be true, and it was also proved that if $P(k)$ is true, $P(k + 1)$ is also true, it follows by the principle of mathematical induction that the statement is true for all positive integers $n \geq 8$.

If $k^2 > 7k + 1$ (step 2) then $k^2 + 2k + 1 > 7k + 1 + 2k + 1$

Example 20

Use the principle of mathematical induction to prove that

$$\sum_{r=1}^{n} r(2r+6) = \frac{2}{3}n(n+1)(n+5)$$

Answer

$$P_n = \sum_{r=1}^{n} r(2r+6) = \frac{2}{3}n(n+1)(n+5)$$

P_1:

LHS $= 1 \times 8 = 8$

RHS $= \frac{2}{3} \times 1 \times 2 \times 6 = 8$

LHS = RHS $\Rightarrow P_1$ is true.

Assume P_k is true, i.e.,

$$\sum_{r=1}^{k} r(2r+6) = \frac{2}{3}k(k+1)(k+5), \ k \in \mathbb{Z}^+$$

Add on the next number in the sequence

(k + 1)(2(k + 1) + 6)

Show P_{k+1} is true, i.e.,

$$\sum_{r=1}^{k+1} r(r+1) = \frac{2}{3}(k+1)((k+1)+1)((k+1)+5)$$

Using the assumption from step 2.

$$= \frac{2}{3}(k+1)(k+2)(k+6)$$

LHS:

$$\sum_{r=1}^{k+1} r(2r+6) = \left(\sum_{r=1}^{k} r(2r+6) \right) + u_{k+1}$$

$$= \frac{2}{3}k(k+1)(k+5) + (k+1)(2(k+1)+6)$$

Expand and simplify.

using the assumption

$$= \frac{2}{3}k(k+1)(k+5) + (k+1)(2k+8)$$

$$= (k+1)\left(\frac{2}{3}k(k+5) + (2k+8) \right)$$

This is the expression obtained when n = k + 1 is substituted in P(n).

$$= \frac{2}{3}(k+1)(k+2)(k+6)$$

= RHS:

Since $P(1)$ was shown to be true, and it was also proved that if $P(k)$ is true, $P(k+1)$ is also true, it follows by the principle of mathematical induction that the statement is true for all values of $n \geq 1$.

Exercise 1H

1 Use mathematical induction to prove that for a geometric series with first term u_1 and common ratio r, the sum of the first n terms S_n is given by

$$S_n = \frac{u_1(1-r^n)}{1-r}$$

2 Use mathematical induction to prove these statements.

a $\displaystyle\sum_{r=1}^{n} r^2 = \frac{n}{6}(n+1)(2n+1)$

b $\displaystyle\sum_{r=1}^{n} 2^{r-1} = 2^n - 1$

c $1^3 + 2^3 + 3^3 + \ldots + n^3 = \dfrac{n^2}{4}(n+1)^2$

d $\displaystyle\sum_{r=1}^{n} r(r+2) = \frac{n}{6}(n+1)(2n+7)$

Example 21

Use mathematical induction to prove that $3^{2n} + 7$ is divisible by 8 for all $n \in \mathbb{N}$.

> This means that $3^{2n} + 7$ is a multiple of 8.

Answer

$P(n)$: $3^{2n} + 7 = 8A$

Step 1

When $n = 0$, LHS $= 3^0 + 7 = 8$
so $P(0)$ is true.

> *Recall that the natural numbers include 0 so we must start with $n = 0$.*

Step 2

Assume that $P(k)$ is true for some value $k \geq 0$, $k \in \mathbb{N}$.

$3^{2k} + 7 = 8A$, $A \in \mathbb{Z}^+$

Step 3

Prove that $P(k + 1)$ is true.

$3^{2(k+1)} + 7 = 8B$, $B \in \mathbb{Z}^+$

Proof:

> *Using assumption*

$$\begin{aligned}
3^{2(k+1)} + 7 &= 9 \times 3^{2k} + 7 \\
&= 9(8A - 7) + 7 \\
&= 72A - 63 + 7 \\
&= 8(9A - 7) \\
&= 8B
\end{aligned}$$

> *$9A - 7 \in \mathbb{Z}^+$, since $A \in \mathbb{Z}^+$*
> *B is a positive integer.*

Since $P(0)$ was shown to be true, and it was proved that if $P(k)$ is true, $P(k + 1)$ is also true, it follows by the principle of mathematical induction that the statement is true for all natural numbers n.

Example 22

Prove that for all values of $n \in \mathbb{Z}^+$, $n^3 + 5n$ is a multiple of 6.

Answer

Let $P(n) = n^3 + 5n$

$P(1) = 1^3 + 5 \times 1 = 6$
Statement is true for $n = 1$.

Assume $P(k)$ is true for $k \geq 1$,
$k \in \mathbb{Z}^+$ then $k^3 + 5k = 6A$, $A \in \mathbb{Z}$.

Use assumption to show that $P(k + 1)$
$(k + 1)^3 + 5(k + 1) = 6B$, $B \in \mathbb{Z}^+$
LHS
$= k^3 + 3k^2 + 3k + 1 + 5k + 5$
$= 6A - 5k + 3k^2 + 3k + 1 + 5k + 5$ *Using $k^3 = 6A - 5k$ from step 2.*
$= 6A + 3(k^2 + k + 2)$
$= 6A + 3[k(k + 1) + 2]$
$= 6(A + C)$ *$k(k + 1) + 2 = 2C$, $C \in \mathbb{Z}$ since*
$= 6B$ *the product of any two consecutive*
$= $ RHS   *integers is even and the sum of two*
Since $P(1)$ was shown to be true,   *even numbers is also even.*
and it was also proved that
if $P(k)$ is true, $P(k + 1)$ is also
true, it follows by the principle of
mathematical induction that the
statement is true for all positive
integers n.

Exercise 1I

1 Use mathematical induction to prove that $7^n - 1$ is divisible by 6 for all $n \in \mathbb{Z}^+$

2 Prove by mathematical induction that
$$1 + 3 + 5 + 7 + \ldots + (2n - 1) = n^2 \quad \text{for } n \in \mathbb{Z}^+$$

3 Prove by mathematical induction that $9^n - 1$ is a multiple of 8 for $n \in \mathbb{Z}^+$

4 Prove by mathematical induction that $n^3 - n$ is a multiple of 6 for $n \in \mathbb{Z}^+$

5 Show using mathematical induction that $\displaystyle\sum_{r=1}^{n} \frac{1}{r(r+1)} = \frac{n}{n+1}$

6 Prove by mathematical induction that for all positive integer values of n, $2^{n+2} + 3^{2n+1}$ is exactly divisible by 7.

7 Find the first five terms of the sequence given by $u_1 = 1$,
$$u_{r+1} = \frac{2u_r - 1}{3}$$
Prove using mathematical induction that $u_n = 3\left(\frac{2}{3}\right)^n - 1$

1.6 Counting methods

Some mathematical problems about arrangements and combinations involve large numbers so you will need to develop a number of counting techniques.

Factorial notation

Look at the first four terms of this sequence.

$$u_n : \begin{cases} u_0 = 1 \\ u_n = n \times u_{n-1} \end{cases}$$

$u_0 = 1$
$u_1 = 1 \times u_0 = 1$
$u_2 = 2 \times u_1 = 2 \times 1$
$u_3 = 3 \times u_2 = 3 \times 2 \times 1$

The general term of this sequence is:

$$u_n = n \times (n-1) \times (n-2) \times \ldots \times 3 \times 2 \times 1$$

A simpler way to denote this sequence is to use factorial notation where $u_n = n!$

It follows that $u_0 = 0! = 1$

This is read as 'u_n equals n factorial'.

Working with large numbers is easier using factorial notation. Here are the first few factorial numbers.

$0! = 1$
$1! = 1 = 1 \times 0!$
$2! = 2 \times 1 = 2 \times 1!$
$3! = 3 \times 2 \times 1 = 3 \times 2!$
$4! = 4 \times 3 \times 2 \times 1 = 4 \times 3!$

Christian Kramp (1760–1826), a French mathematician, introduced factorial notation.

→ $n! = n \times (n-1) \times (n-2) \times \ldots \times 3 \times 2 \times 1 = n \times (n-1)!$

You can use this pattern to calculate expressions such as $\dfrac{8!}{6!}$

$$\frac{8!}{6!} = \frac{8 \times 7 \times \cancel{6 \times 5 \times 4 \times 3 \times 2 \times 1}}{\cancel{6 \times 5 \times 4 \times 3 \times 2 \times 1}} = 8 \times 7 = 56$$

Example 23

Evaluate $\dfrac{10! \times 5!}{7! \times 6!}$

Answer

$$\dfrac{10! \times 5!}{7! \times 6!} = \dfrac{10 \times 9 \times 8 \times 7! \times 5!}{7! \times 6 \times 5!}$$

$$= \dfrac{10 \times 9 \times 8}{6}$$

$$= 120$$

1.1 2.1 ▶	*Unsaved ▽	
$\dfrac{10!\cdot 5!}{7!\cdot 6!}$		120

1/99

GDC help on CD:
Alternative demonstrations for the TI-84 Plus and Casio fx-9860GII GDCs are on the CD.

Example 24

Simplify

a $\dfrac{(n+1)!}{(n-1)!}$ **b** $\dfrac{(n+1)! + n!}{(n-1)!}$

Answers

a $\dfrac{(n+1)!}{(n-1)!} = \dfrac{(n+1) \times n \times (n-1)!}{(n-1)!}$

$$= n(n+1)$$

b $\dfrac{(n+1)! + n!}{(n-1)!}$

$$= \dfrac{(n+1)n(n-1)! + n(n-1)!}{(n-1)!}$$

$$= \dfrac{(n+1)n + n}{1} = n(n+2)$$

Rewrite (n + 1)!
as (n + 1) × n × (n – 1)!
and n! = n × (n – 1)!

Exercise 1J

1 Copy and complete this table simplifying the expressions.

$8! - 7!$	
$10! - 9!$	
$5! - 4!$	
$95! - 94!$	
$(n+1)! - n!$	

2 Evaluate:

a $\dfrac{4!}{6!}$ **b** $\dfrac{5! \times 3!}{6!}$ **c** $\dfrac{8! \times 6!}{5!}$

3 Simplify:

a $\dfrac{n! + (n-1)!}{(n+1)!}$ **b** $\dfrac{n! - (n-1)!}{(n-2)!}$ **c** $\dfrac{(n!)^2 - 1!}{n! + 1}$

EXAM-STYLE QUESTION

4 Show that $\dfrac{(2n+2)!(n!)^2}{[(n+1)!]^2 (2n)!} = \dfrac{2(2n+1)}{(n+1)}$

Arrangements

Alma lays the breakfast table on Christmas Day for 6 people with an eggcup, a glass and a cup for each person. She cannot decide how to arrange them in a row, so she decides to arrange them differently for each person, since she thinks that there are six ways of arranging three different objects. Here is her reasoning:

Starting from the left she can choose the eggcup, the glass or the cup. Having made her first choice she is left with two objects to choose from, which then leaves her with one way to choose the third object.

Here are the different arrangements:

EGC	CEG	GEC
ECG	CGE	GCE

So there are $3 \times 2 \times 1 = 6$ ways of arranging three distinct objects in a row.

E represents eggcup
G represents glass
C represents cup

Similarly with four objects to arrange in a row there are 4 ways of choosing the first object and for each of these ways there are 3 ways of choosing the second. Having chosen the first two objects there are 2 ways of choosing the third object and one way of choosing the last object, giving a total of $4 \times 3 \times 2 \times 1 = 24$ different arrangements.

This reasoning can be extended to deduce that the number of ways in which n distinct objects can be arranged in a row is

$$n \times (n-1) \times (n-2) \times \ldots \times 3 \times 2 \times 1 = n!$$

> → The number of ways of arranging n distinct objects in a row is $n!$

The different ways in which objects can be arranged are called **permutations**.

Alma arranges some Christmas decorations in a line: two identical angels, a snowman and a bell. She can arrange them in $4! = 24$ ways. She reasons, however, that since the two angels are indistinguishable the number of different arrangements is less than this. Here is a list of all possible arrangements:

AASB	AASB	ABSA	ABSA	BSAA	BSAA
AABS	AABS	ASBA	ASBA	SABA	SABA
ABAS	ABAS	BAAS	BAAS	SAAB	SAAB
ASAB	ASAB	BASA	BASA	SBAA	SBAA

A – angel
B – bell
S – snowman

The arrangements in the left-hand column are only different from the arrangements on the right because the two angels are shown differently. Because the angels are identical there are only 12 ways of arranging the four objects.

Fan Rong K Chung Graham

(1949–) Taiwanese mathematician "... many problems from combinatorics were easily explained, you could get into them quickly but getting out was often very hard ..."

The number of ways of arranging 4 objects, two of which are the same, is $\dfrac{4!}{2!} = \dfrac{4 \times 3 \times 2 \times 1}{2 \times 1} = 12$

2! is the number of ways of arranging the two identical objects.

> → The number of permutations of n objects, k of which are identical is: $\dfrac{n!}{k!}$

No questions will be asked in your exam about permutations of identical objects.

Alma decides to arrange only two *different* ornaments. She has three ornaments to choose from (snowman, bell and angel). She can choose her first ornament in 3 ways and her second ornament in 2 ways so there are $3 \times 2 = 6$ different arrangements.

Alma changes her mind again and decides to arrange two of the objects: snowman, bell, angel and candle. Now she has four *distinct* objects to choose from. She can choose her first ornament in four ways and her second in three giving $4 \times 3 = \dfrac{4!}{(4-2)!} = 12$ different arrangements.

The candle is one of the four distinct objects.

Using the same reasoning if she had n different ornaments and wanted to arrange 4 ornaments in a line, she could do this in $n \times (n-1) \times (n-2) \times (n-3) = \dfrac{n!}{(n-4)!}$ ways.

> → The number of permutations of r objects out of n distinct objects is $P_r^n = \dfrac{n!}{(n-r)!}$

The number of ways of arranging n objects in a row is simply the numbers of permutations of n objects out of n. The formula is then

We define 0! as 1.

$$P_n^n = \dfrac{n!}{(n-n)!} = \dfrac{n!}{0!} = n!$$

Alma chooses 6 different napkins out of the 10 patterns that she has.

If she wanted to arrange 6 napkins out of 10 in a line she could do it in $\dfrac{10!}{(10-6)!}$ ways, but this includes the 6! ways of arranging the 6 napkins in a line. Now the order is not important, so she excludes the equivalent arrangements by dividing by 6!

So Alma has $\dfrac{10!}{(10-6)!6!}$ ways of choosing the six napkins.

$\binom{n}{r}$ and C_r^n are two different notations for combinations. Both are equally correct but $\binom{n}{r}$ is used throughout this book.

When the order of arrangements is not relevant they are called **combinations** and you can generalize the result.

> → The number of ways of choosing (order is not important) r objects from n is
> $$\binom{n}{r} = C_r^n = \frac{n!}{(n-r)!r!}$$

$$C_r^n = \frac{p_r^n}{p_r^r}$$

You can also think of combinations as a selection. To select two letters from A, B, C, you can have three different selections as AB is the same as BA. However, you could arrange them in six ways as AB is a different arrangement (permutation) to BA.

Having chosen her six different napkins Alma would now like to find out the number of ways of arranging them round the table. She realizes that the number of ways is no longer 6! because they are to be arranged in a circle.

> Note that we are interested in arrangements of napkins relative to each other.

These two arrangements are different because they are in a straight line.

However, arranged round a table the same arrangement is just rotated by one place.

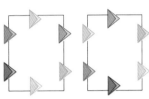

> Another way of reasoning would be to fix the first napkin. Then you can think of arranging the remaining 5 napkins in a line with each end finishing on either side of the first napkin. The number of ways of arranging 5 distinct objects in a line is 5!

In fact you could rotate an arrangement six times round the table without repeating positions. So the number of distinct ways of arranging 6 distinct objects in a circle will be $\frac{6!}{6} = 5!$

Hence the number of ways of arranging n distinct objects around a circle is $(n-1)!$

Example 25

> **a** In how many ways can the letters of the word *special* be arranged?
> **b** In how many ways can the letters be arranged taking them two at a time?

- -

Answers

a There are 7 different letters in the word *special* which can be arranged in 7! ways.

b
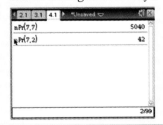

Example 26

How many four-digit numbers can be made using the digits
a 1, 2, 3 and 4
b 0, 1, 2 and 3?
Digits may be used more than once.

Answers

a There are four ways of choosing each of the four digits giving a total of
$4^4 = 256$

Since we are only concerned with distinct numbers, digits can be used more than once.

b There are only 3 ways of choosing the first digit but there are four ways of choosing each of the other digits giving a total of
$3 \times 4^3 = 192$

A four-digit number cannot start with zero.

Example 27

In a Model United Nations (MUN) club at school there are 7 girls and 8 boys. In how many ways can a delegation of 10 students be chosen from the 15 students if:
a there are no gender restrictions
b the delegation is to be made up of 5 boys and 5 girls
c at least three of each gender are included in the delegation?

Answers

a $\dbinom{15}{10} = \dfrac{15!}{10! \times 5!} = 3003$

Calculate combination, as the order of choosing is not important. Use a GDC.

b $\dbinom{7}{5} \times \dbinom{8}{5} = 1176$

Now we need to choose 5 girls out of 7 and 5 boys out of 8. Use a GDC.

c The number of ways of having a delegation with only two girls is:
$\dbinom{7}{2} \times \dbinom{8}{8} = 21$

It is impossible to have no girls or only 1 girl.
Since there are only 7 girls all possible combinations will include at least three boys.

$\dbinom{15}{10} - 21 = 3003 - 21 = 2982$

Cannot have a delegation with 0, 1 or 2 girls OR 0, 1, 2 boys. All other possible delegations are allowed.

Subtract the unwanted combinations from the total number of ways.

1.1	2.1	3.1 ▶ *Unsaved ▽	
nCr(15,10)			3003
nCr(7,5)·nCr(8,5)			1176
nCr(7,2)·nCr(8,8)			21
nCr(15,10)−21			2982
			4/99

GDC help on CD:
Alternative demonstrations for the TI-84 Plus and Casio fx-9860GII GDCs are on the CD.

Exercise 1K

1 To open your school locker, you must punch in a code consisting of three distinct letters. There are 26 letters on the lock.
   How many different locker codes are there?

2 Three maths books, four science books, two geography books and three history books are to be placed on a bookshelf.
   The books are all different.
   a In how many different ways can the books be arranged on the shelf?
   b In how many ways can the books be arranged so that books of the same subject are grouped together?

EXAM-STYLE QUESTION
3 As part of his cross-country training Mark runs a 10 km route four times a week. There are eight different routes along which he can run. He calculates that he will just manage to run a different set of routes each week leading up to his next race.
   How many weeks are there before Mark's next race?

4 A team of 4 students is to be selected for a mathematics competition. There are 8 boys and 12 girls to choose from.
   a In how many ways can a team be chosen?
   b If the team is to include at least one girl and one boy, in how many ways can a team be selected?

EXAM-STYLE QUESTION
5 a How many four-digit even numbers can be made using the digits 0, 1, 2, 3, 4, 5 and 6?
   b How many of these four-digit even numbers are divisible by 5?
   c How many of these four-digit even numbers have no repeated digits?

6 In the UK between 1932 and 1945 car registration numbers contained three letters of the alphabet followed by three digits.
   How many different number plates in this format were possible?

1.7 The binomial theorem

Repeated algebraic multiplication gives

$(1 + x)^0 = 1$
$(1 + x)^1 = 1 + x$
$(1 + x)^2 = 1 + 2x + x^2$
$(1 + x)^3 = 1 + 3x + 3x^2 + x^3$
$(1 + x)^4 = 1 + 4x + 6x^2 + 4x^3 + x^4$
$(1 + x)^5 = 1 + 5x + 10x^2 + 10x^3 + 5x^4 + x^5$

> These are called binomial expansions because the sum of two algebraic terms $(1 + x)$ is known as a binomial.

In these expressions for powers of $(1 + x)$ you can see that:

- The indices of x form an arithmetic sequence with first term 0 and common difference 1.
- The highest index of x is the same as the power to which $(1 + x)$ is raised.
- The coefficients of the first and last term are always 1.

If you write all the coefficients like this you will recognize a number of patterns.

> This triangular pattern of numbers is called Pascal's Triangle after the French mathematician **Blaise Pascal** (1623–62). It was actually recorded as early as the 11th century by the Persians and the Chinese.

```
row 0:                    1
row 1:                1       1
row 2:            1       2       1
row 3:        1       3       3       1
row 4:    1       4       6       4       1
```

There is a vertical line of symmetry. Each number in a row is obtained by adding the two numbers above it to either side. You can extend this pattern by counting or using technology.

Looking at the numbers, the two 1's in the first row can be written as combinations: $\binom{1}{0} = 1$ and $\binom{1}{1} = 1$

In fact the first and last coefficients of all the rows can be written as $\binom{n}{0}$ and $\binom{n}{n}$ where n is the row number, which is the same as the power to which $(1 + x)$ is raised.

Looking at the second row $2 = \dfrac{2!}{1!(2-1)!} = \binom{2}{1}$

so the second row is actually

$\binom{2}{0} \binom{2}{1} \binom{2}{2}$

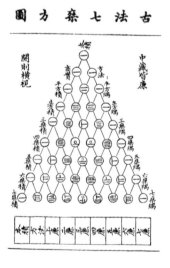

圆 方 糅 七 法 古

▲ This illustration is from a book by **Chu Shih-Chieh**. It was written in China in 1303, more than 300 years before Pascal. It shows the triangle in Chinese numerals.

As each number is obtained by adding the two numbers immediately above it, then in the third row

$$3 = \binom{2}{0} + \binom{2}{1}$$

$$= \frac{2!}{0!(2-0)!} + \frac{2!}{1!(2-1)!}$$

$$= \frac{2!}{0!2 \times (2-1)!} + \frac{2!}{1 \times 0!(2-1)!}$$

$$= \frac{2!}{0!(2-1)!}\left[\frac{1}{2} + \frac{1}{1}\right]$$

$$= \frac{3 \times 2!}{1 \times 0!2 \times (2-1)!}$$

$$= \frac{3!}{1!(3-1)!}$$

$$= \binom{3}{1}$$

To summarise

$$(1+x)^2 = \binom{2}{0} + \binom{2}{1}x + \binom{2}{2}x^2$$

$$(1+x)^3 = \binom{3}{0} + \binom{3}{1}x + \binom{3}{2}x^2 + \binom{3}{3}x^3$$

This pattern leads to a conjecture for the binomial theorem which states that

$$(1+x)^n = \binom{n}{0} + \binom{n}{1}x + \binom{n}{2}x^2 + \ldots + \binom{n}{r}x^r + \ldots + \binom{n}{n}x^n$$

A formal proof of the binomial theorem is beyond the scope of this course.

$(1 + x)^n = (1 + x) \times (1 + x) \times (1 + x) \ldots (1 + x)$

When you multiply these n factors you get a polynomial of degree n. You can work out how to obtain the expansion by considering different powers of x:

x^0: multiplying all the 1's $\Rightarrow C_0^n = \binom{n}{0}$

x^1: choose one x from the n factors $(1 + x)$ and multiply it by the 1's

in all the other factors $\Rightarrow C_1^n x = \binom{n}{1}x$

> Remember, we define 0! as 1

> Show that $\binom{2}{1} + \binom{2}{2} = \binom{3}{2}$ and that the third row of Pascal's triangle can be written as
> $$\binom{3}{0} \binom{3}{1} \binom{3}{2} \binom{3}{3}$$

> Coefficient of general term is $\binom{n}{r}$

> Number of ways of choosing 0 x's out of n factors.

x^2: choose two x's from the n factors and multiply them together and by the 1's in all the other factors $\Rightarrow C_2^n x^2 = \binom{n}{2} x^2$

x^r: choose r of the x's from the n factors, multiply them together and by the 1's in all the other factors $\Rightarrow C_r^n x^r = \binom{n}{r} x^r$

For the expansion of $(a + x)^n$ you can deduce the result like this:

$$(a+x)^n = \left(a \left(1 + \frac{x}{a} \right) \right)^n = a^n \left(1 + \frac{x}{a} \right)^n$$

$$= a^n \left(\binom{n}{0} + \binom{n}{1} \frac{x}{a} + \binom{n}{2} \frac{x^2}{a^2} + \ldots + \binom{n}{r} \frac{x^r}{a^r} + \ldots + \binom{n}{n} \frac{x^n}{a^n} \right)$$

Distributing a^n over the expansion.

$$= \binom{n}{0} a^n + \binom{n}{1} a^{n-1} x + \binom{n}{2} a^{n-2} x^2 + \ldots + \binom{n}{r} a^{n-r} x^r + \ldots + \binom{n}{n} x^n$$

Substituting $\frac{x}{a}$ for x in the expansion above.

→ The binomial theorem states that:

$$(a + x)^n = \binom{n}{0} a^n + \binom{n}{1} a^{n-1} x + \binom{n}{2} a^{n-2} x^2 + \ldots + \binom{n}{r} a^{n-r} x^r + \ldots + \binom{n}{n} x^n$$

$$= \sum_{r=0}^{n} \binom{n}{r} a^{n-r} x^r$$

Example 28

Find the values of a, b and c in these identities.

a $(1 + 2x)^8 \equiv 1 + ax + bx^2 + cx^3 + \ldots + 256x^8$

b $\left(1 + \frac{x}{2} \right)^a = 1 + bx + cx^2 + \ldots + \left(\frac{x}{2} \right)^{10}$

c $(2 - ax)^6 = b - 64x + cx^2 + \ldots$

Use the binomial theorem to write down the expansion. Then simplify and compare coefficients.

Answers

a $(1+2x)^8 = \binom{8}{0} + \binom{8}{1}(2x) + \binom{8}{2}(2x)^2 + \binom{8}{3}(2x)^3 + \ldots + \binom{8}{8}(2x)^8$

$$= 1 + \frac{8!}{1!7!} \times 2x + \frac{8!}{2!6!} \times 4x^2 + \frac{8!}{3!5!} \times 8x^3 + \ldots + 256x^8$$

$$= 1 + 16x + 112x^2 + 448x^3 + \ldots + 256x^8$$

$$\therefore a = 16, \ b = 112 \text{ and } c = 448$$

∴ means 'therefore'.

▶ Continued on next page

b $\left(1+\dfrac{x}{2}\right)^a = 1 + bx + cx^2 + \ldots + \left(\dfrac{x}{2}\right)^{10}$

$\Rightarrow a = 10$

$\left(1+\dfrac{x}{2}\right)^{10} = 1 + \dbinom{10}{1}\left(\dfrac{x}{2}\right) + \dbinom{10}{2}\left(\dfrac{x}{2}\right)^2 + \ldots + \left(\dfrac{x}{2}\right)^{10}$

$= 1 + \dfrac{10!}{1!9!}\left(\dfrac{x}{2}\right) + \dfrac{10!}{2!8!}\left(\dfrac{x}{2}\right)^2 + \ldots + \left(\dfrac{x}{2}\right)^{10}$

$= 1 + 5x + \dfrac{45}{4}x^2 + \ldots + \left(\dfrac{x}{2}\right)^{10}$

$\therefore a = 10,\ b = 5 \ \text{and} \ c = \dfrac{45}{4}$

c $(2 - ax)^6 = 2^6 + \dbinom{6}{1}2^5 \times (-ax) + \dbinom{6}{2}2^4 \times (-ax)^2 + \ldots$

$= 64 + 6 \times 32 \times (-ax) + 15 \times 16\,(-ax)^2 + \ldots$

$= 64 - 192ax + 240a^2x^2 + \ldots$

$\therefore b = 64$

$192a = 64$

$\Rightarrow a = \dfrac{1}{3}$

$c = 240a^2 = \dfrac{240}{9} = \dfrac{80}{3}$

$(a - b)^n = (a + (-\,b))^n$

Example 29

Use the binomial theorem to expand $(a + 3x)^5$.
Hence find the value of $(1.03)^5$ correct to 5 decimal places.

Answer

$(a + 3x)^5 = a^5 + 5a^4(3x) + 10a^3(3x)^2$
$\qquad\qquad + 10a^2\,(3x)^3 + 5a(3x)^4 + (3x)^5$

$\qquad\quad = a^5 + 15a^4x + 90a^3x^2 + 270a^2x^3$
$\qquad\qquad + 405ax^4 + 243x^5$

$(1.03)^5 = (1 + 3(10^{-2}))^5$

$\qquad\quad = 1 + 15 \times (10^{-2}) + 90 \times (10^{-2})^2$
$\qquad\qquad + 270 \times (10^{-2})^3 + 405 \times (10^{-2})^4$
$\qquad\qquad + 243 \times (10^{-2})^5$

Substituting $a = 1$,
$x = 0.01 = 10^{-2}$

$\qquad\quad = 1 + 0.15 + 0.009 + 0.000\,27$
$\qquad\qquad + 0.000\,004\,05 + \ldots$

$\qquad\quad \approx 1.159\,27 \ (5 \text{ dp})$

Only consider the first 4
terms to give the answer
correct to 5 decimal places.

Example 30

Find the term that is independent of x in the expansion

of $\left(3x^2 - \dfrac{1}{2x}\right)^9$

Answer

The general term, T_r of the expansion is given by:

$$T_r = \binom{9}{r}(3x^2)^{9-r}\left(-\frac{1}{2x}\right)^r$$

$$= \binom{9}{r}3^{9-r} \times x^{18-2r} \times \left(-\frac{1}{2}\right)^r \times x^{-r}$$

$x^{18-2r} \times x^{-r} = x^0$

$18 - 3r = 0$

$r = 6$

For term independent of x, index of x must be 0.

$$T_6 = \binom{9}{6} \times 3^3 \times \left(-\frac{1}{2}\right)^6 = \frac{567}{16}$$

Give answer as an exact fraction.

Exercise 1L

1 Show that

a $\dbinom{n}{r} = \dbinom{n}{n-r}$

b $\dbinom{n+1}{r} = \dbinom{n}{r} + \dbinom{n}{r-1}$

2 Write down the first four terms in the binomial expansion of:

a $(1 + 2x)^{11}$ **b** $(1 - 3x)^7$ **c** $(2 + 5x)^5$ **d** $\left(2 - \dfrac{x}{3}\right)^9$

3 Write down the required term in each of these binomial expansions.

a 4th term of $(1 - 4x)^7$ **b** 3rd term of $\left(1 - \dfrac{x}{2}\right)^{20}$

c 4th term of $(2a - b)^8$

EXAM-STYLE QUESTION

4 Find the term independent of x in the expansion of $\left(2x + \dfrac{1}{x^2}\right)^{12}$

5 Use the binomial theorem to expand $\left(2 + \dfrac{x}{5}\right)^5$. Hence find the value of $(2.01)^5$ correct to 5 decimal places.

6 a Express $\left(\sqrt{2} - \sqrt{3}\right)^4$ in the form $a + b\sqrt{6}$ where a, $b \in \mathbb{Z}$.

b Express $\left(\sqrt{2} + \dfrac{1}{\sqrt{5}}\right)^3$ in the form $a\sqrt{2} + b\sqrt{5}$ where a, $b \in \mathbb{Q}$.

c Express $\left(1 + \sqrt{7}\right)^5 - \left(1 - \sqrt{7}\right)^5$ in the form $a\sqrt{7}$, where $a \in \mathbb{Z}$.

7 Let $a = x + y$ and $b = x - y$

   a Write $a^2 - b^2$ in terms of x and y and hence show that
$$a^2 - b^2 = (a - b)(a + b)$$

   b Use the binomial theorem to write a^3 and b^3 in terms of
x and y and use your results to show that
$$a^3 - b^3 = (a - b)(a^2 + ab + b^2)$$

   c Use the binomial expansion to write a^4 and b^4 in terms of x
and y and use your results to factorize $a^4 - b^4$.

   d Use your results to make a conjecture for the factors of $a^n - b^n$.

   e Prove your conjecture using mathematical induction.

Review exercise

1 Show that there are two geometric sequences such that the second
term is 16 and the sum of the first three terms is 84.

2 Find the sum of the series.
$$1 + 3 + 4 + 6 + 7 + 9 + 10 + 12 + \ldots + 46$$

3 Three numbers a, b and c form an arithmetic sequence.
The numbers c, a and b form a geometric sequence.
If the sum of the numbers is $-\dfrac{9}{2}$, find the three numbers.

4 Write down the first six terms of the sequence given by:
$$\begin{cases} u_1 = 1 \\ u_{n+1} = 2u_n + 1, \ n \in \mathbb{Z}^+ \end{cases}$$
Use mathematical induction to prove that $u_n = 2^n - 1$.

5 Prove by mathematical induction that $3^{2n} - 8n - 1$, $n \in \mathbb{Z}^+$, is a
multiple of 64.

6 Write in factorial notation:

   a the coefficient of x^4 in the expansion of $(1 + x)^{n+1}$

   b the coefficient of x^2 in the expansion of $(1 + x)^{n-1}$

   c Find n given that the coefficient of x^4 in the expansion of
$(1 + x)^{n+1}$ is six times the coefficient of x^2 in the expansion
of $(1 + x)^{n-1}$

7 Evaluate these by choosing an appropriate value
for x in the expansion of $(1 + x)^n$

   a $\dbinom{n}{0} + \dbinom{n}{1} + \dbinom{n}{2} + \ldots + \dbinom{n}{r} + \ldots + \dbinom{n}{n}$

   b $\dbinom{n}{0} - \dbinom{n}{1} + \dbinom{n}{2} - \ldots + (-1)^n \dbinom{n}{r} + \ldots + (-1)^n \dbinom{n}{n}$

Review exercise

EXAM-STYLE QUESTION

1 The diagram shows a sequence of squares. Starting with the largest square, the midpoints are joined to form the second square of the sequence. This process can be continued infinitely.

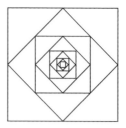

  a If the sides of the largest square have length 1, calculate the lengths of the second, third and fourth squares.

A spiral is formed by joining segments shown as red lines in the diagram.

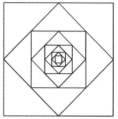

  b Use your answers to part **a** to find the length of the spiral shown.

  c What happens to the length of the spiral if we continue the process infinitely?

A different spiral is formed by shading triangles as shown in the diagram.

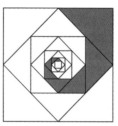

  d Find the total area of the shaded triangles.

  e What is the total area of the spiral formed if the process of forming squares and shading triangles is continued infinitely?

2 **a** In how many different ways can the letters of the word 'characteristic' be arranged?

  b How many numbers bigger than 20 000 and divisible by 5 can be formed using the digits 0, 1, 3, 5, 7 and 9?

  c Four married couples are to be in a group photo. In how many different ways can they stand in a line so that each person is next to his or her spouse?

3 In how many ways can a committee of five people be selected from six men and four women, so that there is at least one male and one female and there are more women than men on the committee?

4 Write down and simplify the term independent of x in the expansion of $\left(x^3 - \dfrac{3}{x}\right)^8$

EXAM-STYLE QUESTION

5 Given that the coefficients of x^{r-1}, x^r, x^{r+1} in the expansion of $(1 + x)^n$ are in arithmetic sequence, show that
$$n^2 + 4r^2 - 2 - n(4r + 1) = 0$$
Hence find three consecutive coefficients of the expansion of $(1 + x)^{14}$ which form an arithmetic sequence.

Extension material on CD:
Worksheet 1

CHAPTER 1 SUMMARY

Sequences and series

- An arithmetic sequence, or arithmetic progression, with the first term u_1 and common difference d is
  $$u_1, u_1 + d, u_1 + 2d, \dots, u_1 + (n-1)d$$
  and the general term is $u_n = u_1 + (n-1)d$
- The sum of n terms of a finite arithmetic series is
  $$S_n = \frac{n}{2}\left[2u_1 + (n-1)d\right] = \frac{n}{2}\left[u_1 + u_n\right]$$
  where n is the number of terms in the series, u_1 is the first term and d is the common difference
- A geometric sequence or geometric progression with first term u_1 and common ratio r is
  $$u_1, u_1 r, u_1 r^2, u_1 r^3, u_1 r^4, \dots$$
  and the general term is $u_n = u_1 \times r^{n-1}$, $r \neq -1, 0, 1$
- The sum of n terms of a geometric series is $S_n = \dfrac{u_1(1-r^n)}{(1-r)}$, $r \neq 1$
- When $-1 < r < 1$, a geometric series converges to a finite sum $S = \dfrac{u_1}{1-r}$

Proof by mathematical induction

- Let $P(n)$ be a statement for all values of n, $n \in \mathbb{Z}^+$ to be proved by induction. (Sometimes the statement may not be true for all positive integers, so the starting value may not be 1.)
  Step 1   Prove that the statement is true for a starting value, usually $P(1)$.
  Step 2   Assume that $P(k)$ is true, where $k \in \mathbb{Z}^+$.
  Step 3   Use the assumption that $P(k)$ is true to show that $P(k + 1)$ is then also true. Write a final statement quoting the principle of mathematical induction.

Counting methods

- Factorial notation $n! = n(n-1)(n-2)\dots \times 3 \times 2 \times 1$
- The number of ways of arranging n distinct objects in a row is $n!$
- The number of permutations of n objects, k of which are identical is $\dfrac{n!}{k!}$
- The number of permutations (arrangements where order matters) of r objects from n distinct objects is $P_r^n = \dfrac{n!}{(n-r)!}$
- The number of combinations (selections where order not important) of r objects out of n objects is: $C_r^n = \begin{pmatrix} n \\ r \end{pmatrix} = \dfrac{n!}{r!(n-r)!}$

The binomial theorem

$$(a+x)^n = \begin{pmatrix} n \\ 0 \end{pmatrix}a^n + \begin{pmatrix} n \\ 1 \end{pmatrix}a^{n-1}x + \begin{pmatrix} n \\ 2 \end{pmatrix}a^{n-2}x^2 + \dots + \begin{pmatrix} n \\ r \end{pmatrix}a^{n-r}x^r + \dots + \begin{pmatrix} n \\ n \end{pmatrix}x^n$$

$$= \sum_{r=0}^{n} \begin{pmatrix} n \\ r \end{pmatrix}a^{n-r}x^r$$

Searching for the truth

All about primes

David Beckham plays in the number 7 shirt for England and number 23 shirt for Real Madrid and Los Angeles Galaxy – and both 7 and 23 are prime numbers. Oxford Professor Marcus du Sautoy pointed out that the key members of Real Madrid all played in prime number shirts: Carlos No. 3, Zidane No. 5, Raul No. 7, Ronaldo No. 11.

> You can read more about Prof du Sautoy's theory here: www.plus.maths.org/content/beckham-his-prime-number.

> In 2011 Beckham named his newborn daughter Harper Seven.

If all the football players in the world decided to play with unique prime numbers, would we ever run out of primes? The Greek mathematician Euclid (c. 300BCE) provided us with a very neat proof that we wouldn't.

Euclid started by assuming that there is a finite number of primes. To simulate his method, let us start by assuming that there are only three primes: 2, 3 and 5.

Multiply these three primes together and add 1, to get 31.

31 is larger than any of our first three primes but is not divisible by 2, 3 or 5. Since any number is either prime or can be written as a product of primes, there must be a prime number bigger than 2, 3 and 5 which is a factor of 31.

In this case the prime number is 31 itself, but this is not always the case.

- Use your GDC to show that $(2 \times 3 \times 5 \times 7 \times 11 \times 13) + 1$ is not a prime number. Write this number as a product of two prime numbers.

- Use Euclid's method to show that the number of primes is infinite. Start by assuming that there are a finite number of primes, $p_1, p_2, ..., p_n$. Multiply them all together and add 1.

- Why would it not be practical for every footballer in your country to play with a unique prime number?

- What are Sophie Germain Primes? What was Sophie Germain's contribution to the proof of Fermat's last theorem?

▼ Sophie Germain, French mathematician (1776–1831)

The Riemann Hypothesis

German mathematician Bernhard Riemann (1826–66) first formulated this hypothesis in 1859. In simple words it states that there is an underlying order in the way that prime numbers are distributed.

It was included in David Hilbert's list of challenging problems for 20th-century mathematicians, and is widely believed to be true. To date no proof exists.

"If I were to awaken after having slept for a thousand years, my first question would be: Has the Riemann hypothesis been proven?"

David Hilbert

▲ David Hilbert, German mathematician (1862–1943)

Axioms vs truth

An axiom is a claim which is taken to be true without needing any proof.

- In the study of probability, the first axiom states that the probability of an event happening is a number between 0 and 1.

- Here are some axioms of number theory:

  If $a, b \in \mathbb{N}$ then:
  - $a + b \in \mathbb{N}$
  - $a \cdot b \in \mathbb{N}$
  - $a + b = b + a$

- What do we mean by axiomatic truth in mathematics?

- Mathematics is said to be an axiomatic system of knowledge. How is this different from other knowledge systems?

- Euclid wrote his thirteen volumes of *The Elements* based on five basic postulates (axioms). What were they?

- Do different axiomatic truths define different worlds? Do Euclidean axioms hold in non-Euclidean geometry?

Reasoning

Mathematicians may make insightful discoveries or intuitively recognize a mathematical truth. However this is not enough – they need formal verification or proof.

The Celsius temperature scale is based on the freezing point and boiling point of water. You may have carried out a simple experiment to show that water boils at 100 °C. But how is this proved scientifically? Does water always boil at 100 °C?

- How does reasoning in mathematics differ from reasoning in other areas of knowledge?

- Compare and contrast the meaning of induction in science to mathematical induction.

- Mathematics is a system of logical deductions through reasoning. Can the same be claimed for any other area of knowledge?

2 Mathematics as a language

CHAPTER OBJECTIVES:

2.1 Concept of function $f : x \mapsto f(x)$; domain, range image (value); odd and even functions; composite functions $(f \circ g)$; identity function; one-to-one and many-to-one functions; inverse function f^{-1} including domain restrictions; self-inverse functions

2.2 The graph of a function, its equation $y = f(x)$; investigations of key features of graphs such as maximum and minimum values, intercepts, horizontal and vertical asymptotes, symmetry, and consideration of domain and range; the graphs of $y = |f(x)|$ and $y = f(|x|)$; the graph of $y = \dfrac{1}{f(x)}$ given the graph of $y = f(x)$

2.3 Transformations of graphs; translations; stretches; reflections in the axes; the graph of the inverse function as a reflection in $y = x$

2.4 The rational function $f : x \mapsto \dfrac{ax + b}{cx + d}$ and its graph

Before you start

You should know how to

1 Change a quadratic function into the form $(x - h)^2 + k$, and identify the vertex and axis of symmetry. e.g. Rewrite $y = x^2 + 2x - 3$ as $y = (x + 1)^2 - 4$. The vertex is $(-1, -4)$ and the axis of symmetry is $x = -1$

2 Find the zeros of linear and quadratic functions. e.g. Find the zero(s) of

  a $y = 2x - 1$ **b** $y = 2x^2 + 5x - 3$

  a $2x - 1 = 0$ **b** $2x^2 + 5x - 3 = 0$

      $x = \dfrac{1}{2}$ $(2x - 1)(x + 3) = 0$

                       $x_1 = \dfrac{1}{2}$ and $x_2 = -3$

3 Change the subject of a formula. e.g.

Make x the subject of $y = 2 - \sqrt{1 - x}$

$\sqrt{1 - x} = 2 - y$

$1 - x = 4 - 4y + y^2$ squaring both sides

$x = -3 + 4y - y^2$

Skills check

1 Change $y = x^2 - 3x - 1$ into the form $(x - h)^2 + k$, and determine the vertex and axis of symmetry of the quadratic.

2 Find the zeros of

  a $y = 3x + 4$

  b $y = 3x^2 - 2x - 1$

3 Make x the subject of each formula:

  a $y = 3 + \sqrt{x - 2}$

  b $y = \dfrac{2x - 1}{3x + 2}$

Exploring the power of symbolic language

"Mathematics is the language with which God has written the universe."
Galileo Galilei (1564–1642)

You are already familiar with the expansion of algebraic expressions, for example,

$$(a + b)^2 \equiv a^2 + b^2 + 2ab$$

You would probably agree that this algebraic language is more concise and precise than saying, for example,

'If one takes a straight line and cuts it randomly into segments, then the area of the square on the line is equal to the area of the sum of the squares on the segments and twice the area of the rectangle contained by these segments.' (from Euclid's *Elements*, 300 BCE)

This example shows that mathematical language with its symbolism, notation and terminology is more effective in expressing mathematical concepts than a cultural language is.

The language of mathematics has spanned as many centuries as people have been using and studying mathematics, and it continues to grow and adapt to new discoveries. In this chapter we will explore some of the mathematical language of functions and transformations.

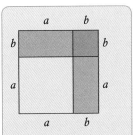

The area of this square is $(a + b)^2$. This is the same as the sum of the areas of the a by a and b by b squares, and the two a by b rectangles.

Abu Abdallah Muhammad ibn Musa al-Khwarizmi was a Persian mathematician, astronomer, geographer, and scholar in the House of Wisdom in Baghdad. He is considered to be the 'Father of Algebra', since he wrote the first known recorded work on balancing equations. In addition, 12th century Latin translations of his work on Indian numerals introduced the decimal positional number system to the Western World.

Texts from *al-Khwarizmi* seem to be the source of three common mathematical words in English:

- 'algebra', derived from *al-jabr*, one of the two operations he used to solve quadratic equations
- 'algorithm', used to refer to a sequence of routine arithmetic operations
- 'zero', which seems to derive from the Arabic *sifr* that meant empty, translated into Latin as *zephirum*.

2.1 Relations and functions

Rene Descartes, French mathematician (1596–1650)

According to legend, the invention of the coordinate system was due to a fly buzzing in René Descartes' bedroom on a sultry summer day. As it landed somewhere on the ceiling, Descartes asked himself how he could uniquely describe the position of the fly. He imagined the ceiling divided into four quadrants, with the right and upward directions being positive, and left and downward being negative. Then, no matter where the fly landed, a unique ordered pair of numbers would define its position.

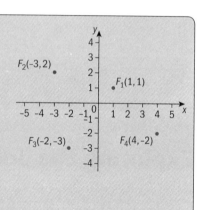

The positions of the fly are examples of a **relation**. A relation is a set of ordered pairs (x, y). A **function** is a special type of relation.

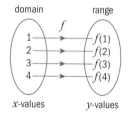

→ A function is a set of ordered pairs in which no two ordered pairs can have the same x-value. In other words, every x-value has a unique y-value.

The set of x-values is called the **domain** of a function. A function assigns, or maps, to each x-value in the domain, a unique y-value. The set of assigned values is called the **range** of the function. Since the value of y, or output of the function, depends on the value of x, or input, we call y the **dependent** variable, and x the **independent** variable. In summary:

→ A relation is a function, f, if

- f acts on all elements of the domain (x-values), and
- f is well defined, i.e. it pairs each element of the domain with one and only one element of the range (y-values).

→ In general, if y is a function of x, you can write $y = f(x)$. You can also write $f : x \mapsto f(x)$, where f is the function that maps x into $f(x)$. The independent variable, x, is called the **argument** of the function.

The term argument, used for the independent variable, stems from the 13th century, and it refers to a quantity from which another quantity can be deduced.

Traditionally, x and y are the variables used for ordered pairs. Any variables, however, can be used in defining a function. For example, the area of a circle depends upon the size of the radius. The variable for the input is r (radius), and for the output is A (area). The rule of the function is $A = \pi r^2$

You could also write the area function as $A(r) = \pi r^2$. The domain is the set of all possible radii, that is, $r \in \mathbb{R}^+$. Hence the range is the set of all possible areas, that is, $A(r) \in \mathbb{R}^+$. This function maps $\mathbb{R}^+$ into $\mathbb{R}^+$.

It is important to note here that the domain of the function $A(r) = \pi r^2$ is restricted by the **context** of the problem. Length and area are positive, hence the domain and range must be non-negative values. If a domain for a particular function is not restricted by its context, or otherwise, then the domain of a function is assumed to be the largest set of x-values for which the range will have real values. This set is called the natural, or implied, domain of the function f.

In this book, for any given function, its domain is assumed to be the natural or implied domain, unless otherwise stated.

Example 1

Determine, with reasons, which of these relations are functions. For those that are functions, write the domain and range.

a $\{(-1, 1), (-2, 4), (-3, 9), (1, 1)\}$
b $\{(4, -2), (1, 1), (4, 2), (9, 3)\}$
c $y = 2x - 1$
d $y^2 = x$
e f maps set A to set B where both A and B are the set of real numbers, and $f : x \mapsto \dfrac{1}{x - 1}$

Did you know that **Gottfried Leibniz**, one of the first to develop calculus, was the first to use the mathematical term 'function' in 1673? Almost one hundred years later, **Leonhard Euler** was the first to write a function as $y = f(x)$.

Answers

a This relation is a function as no two y-values have the same x-value. | *Both conditions are satisfied.*
Domain = $\{-1, -2, -3, 1\}$ | *Write the x-values in a set.*
Range = $\{1, 4, 9\}$ | *Write the y-values in a set.*

b This relation is not a function since both $(4, -2)$ and $(4, 2)$ have the same x-value. | *The relation is not well-defined.*

c For every x-value there is only one y-value, hence it is a function. | *Both conditions are satisfied. All non-vertical lines are graphs of functions.*
Domain = $\{x \mid x \in \mathbb{R}\}$ | *Since this is the equation of a straight line, neither x nor y have any*
Range = $\{y \mid y \in \mathbb{R}\}$ | *restrictions.*

$x \mid x \in \mathbb{R}$ means the set of x-values where x is a real number.

▶ Continued on next page

d Not a function, since $x = 1$ maps to $y = -1$ or $y = 1$	*f is not well-defined. You only need to find one counter-example.*
e f is not a function since $1 \in \mathbb{R}, 1 \mapsto \dfrac{1}{0}, \dfrac{1}{0} \notin \mathbb{R}$	*The first condition is not met because the number 1 is not mapped onto any real number.*

In Example 1, part **e** is not a function since not all elements of A are mapped onto an element of B. In this case, in order for the relation to be a function, you would have to **restrict the domain**, i.e. exclude the number 1 from the domain. Hence, if $y = \dfrac{1}{x-1}$ is a function, the domain is $A = \{x \mid x \in \mathbb{R}, x \neq 1\}$. The range of the function must exclude 0 since the numerator of the rational expression is non-zero. The range is therefore $B = \{y \mid y \in \mathbb{R}, y \neq 0\}$.

Vertical line test

The vertical line test is a practical way of determining if a relation is a function. If a vertical line intersects the graph of a relation at more than one point, then the relation is not a function.

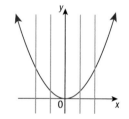

▲ This relation (red line) passes the vertical line test (blue lines).

The graph of the relation in Example 1 part **b** is

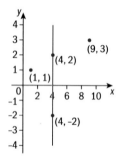

It clearly does not pass the vertical line test.

The graph of the relation in the Example 1 part **c** is

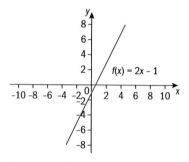

It passes the vertical line test.

Example 2

Determine, with reasons, which of these graphs show relations that are functions. For those that are functions, write the domain and range of the function.

a

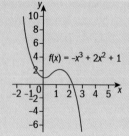

$f(x) = -x^3 + 2x^2 + 1$

b

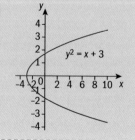

$y^2 = x + 3$

▶ Continued on next page

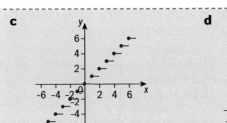

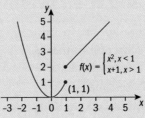

$$f(x) = \begin{cases} x^2, x < 1 \\ x+1, x > 1 \end{cases}$$

Answers

Graphs **a** and **c** pass the vertical line test, hence they are graphs of functions.

Graphs **b** and **d** do not pass the vertical line test, hence they are not graphs of functions.

a Both domain and range are the set of real numbers.
c Domain is the set of real numbers.
   Range is the set of integers.

Exercise 2A

1 Determine if these relations are functions. For those that are, state the domain and range.
   a $\{(0, 1), (1, 2), (2, 3), (3, -1)\}$
   b $\{(-3, 0), (-2, 0), (-1, 0), (0, 0)\}$
   c $\{(1, -3), (1, -2), (1, -1), (1, 0)\}$
   d $\{(\pi, \pi^2), (-\pi, \pi^3), (\pi, \pi^\pi)\}$

2 Determine which of these graphs represent a function. For those that are functions, state the domain and range.

a

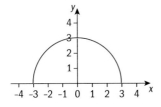

b

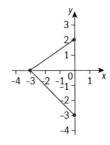

c

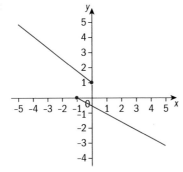

d

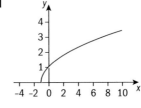

2.2 Special functions and their graphs

A quadratic function has the general form $y = ax^2 + bx + c$, where a, b, and c are real numbers, and $a \neq 0$. x and y are variables. The constants a, b, and c are called **parameters** of the function. They determine the shape of a particular quadratic function.

Here is the graph of the quadratic function
$y = x^2 + x - 5$

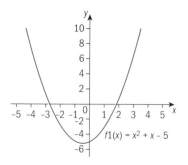

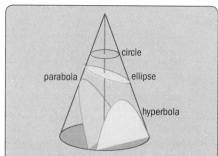

By taking slices of a cone at different angles you obtain different shapes or curves called conic sections. An instrument for drawing conic sections was first described in the year 1000 CE by the Islamic mathematician **al-Kuhi**. The shape of a quadratic function is called a parabola. You will find the cone useful as you meet different functions whose shapes are conic sections.

From the graph, the domain is the set of real numbers. The range has a minimum value.

Using the appropriate menu on your GDC you can find the minimum.

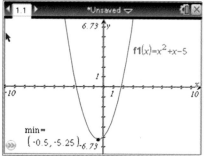

From now on the domain will be from the set of real numbers, unless otherwise specified.

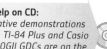

GDC help on CD:
Alternative demonstrations for the TI-84 Plus and Casio fx-9860GII GDCs are on the CD.

In $y = x^2 + x - 5$, $a = 1$ and $b = 1$. The axis of symmetry is indeed $x = -\dfrac{1}{2}$, and the vertex is $\left(-\dfrac{1}{2}, -5.25\right)$

Hence, the range $= \{y \mid y \geq -5.25\}$

Another way of finding the minimum value is to write the quadratic $y = x^2 + x - 5$ in the vertex form, by completing the square, so $y = (x + 0.5)^2 - 5.25$. The coordinates of the vertex are $(-0.5, -5.25)$. Since the leading coefficient, a, is positive, the quadratic will be concave up, and hence the vertex will be a minimum point.

→ For a quadratic in the form $y = ax^2 + bx + c$, the axis of symmetry is $x = -\dfrac{b}{2a}$, hence the vertex is $\left(\dfrac{-b}{2a}, f\left(\dfrac{-b}{2a}\right)\right)$.

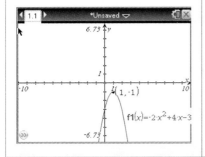

Example 3

Find the domain and range of the function $y = -2x^2 + 4x - 3$.
Confirm your answers graphically.

Answer

The domain is the set of real numbers.

Vertex = (1, −1)

Range = $\{y \mid y \le -1\}$

Use your GDC to graph the function.

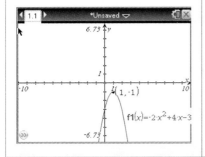

$a = -2, b = 4$

$\dfrac{-b}{2a} = \dfrac{-4}{4} = 1$

$f(1) = -2 + 4 - 3 = -1$

$Vertex = \left(\dfrac{-b}{2a}, f\left(\dfrac{-b}{2a}\right)\right)$

Since $a < 0$, the quadratic is concave down, hence the vertex is the maximum point or the absolute maximum.

GDC help on CD:
Alternative demonstrations for the TI-84 Plus and Casio fx-9860GII GDCs are on the CD.

Investigation – quadratic graphs

Consider quadratics of the form $y = (x - h)^2 + k$, where h and k are real numbers. Graph quadratics of this form for different values of h and k. What effect do these parameters have on the graph of $y = x^2$?

Radical functions

A function of the type $y = \sqrt{ax+b}$; $a, b \in \mathbb{R}$ is a square root function whose radicand (the expression in the square root) is linear. The radicand must be non-negative so the domain is restricted. The domain is $\left\{ x \mid x \geq \dfrac{-b}{a} \right\}$ since $ax + b \geq 0$

$$ax \geq -b$$

$$x \geq \frac{-b}{a}$$

The range is the set of non-negative real numbers, $\{y \mid y \geq 0\}$

$\sqrt{}$ The symbol for square root dates back to 1850 BCE from Babylonian clay tablets. The Rhind Mathematical Papyrus from 1650 BCE shows that the Egyptians obtained this symbol from the Babylonians. The first usage of the same symbol in the western world appeared in 1669 CE in an *Introduction to Algebra*, edited by **John Pell**. The word 'radical' comes from the Latin *radix*, which means root.

Example 4

Determine the domain and range of $y = \sqrt{4x-3}$ and confirm your results graphically.

Answer

$4x - 3 \geq 0 \Rightarrow x \geq \dfrac{3}{4}$

For the function to be real, the radicand must be non-negative.

Domain: $\left\{ x \mid x \in \mathbb{R}, \ x \geq \dfrac{3}{4} \right\}$

This is the restricted domain.

Range: $\{y \mid y \in \mathbb{R}, \ y \geq 0\}$

The range is the set of non-negative real numbers.

Confirmed on the GDC.

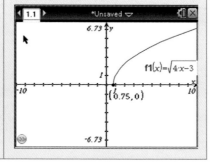

GDC help on CD:
Alternative demonstrations for the TI-84 Plus and Casio fx-9860GII GDCs are on the CD.

Exercise 2B

1 Explain why $y^2 = x$ is not a function, and $y = \sqrt{x}$ is a function.

2 Determine the domain and range of these functions, and confirm your results graphically.

 a $y = x^2 - 4x + 2$ b $y = -(x + 2)^2 - 3$

 c $y = \sqrt{x+2}$ d $y = \sqrt{3-x}$

 e $y = -3x^2 + 6x - 1$ f $y = \sqrt{4-2x}$

Absolute-value functions

For any real number a, its absolute value, denoted by vertical bars, is defined as

$$|a| = \begin{cases} a, & a \geq 0 \\ -a, & a < 0 \end{cases}$$

The absolute value of a real number is non-negative.

> Another term for absolute value is the **modulus** of a number. It comes from the French word *module*, which means unit of measure, and has been used by mathematicians since the early 1800s. **Karl Weierstrass** first used the vertical bar notation in 1841.

Geometrically, the absolute value is the distance between the point representing a number on the real number line, and the origin of the real number line. More generally, the absolute value of the difference of two real numbers is the distance between the points that represent them on the real line.

→ From the definition of absolute value and its geometrical reasoning, these useful fundamental properties follow, for a, any real number:

1. $|a| \geq 0$
2. $|-a| = |a|$
3. $|a| = 0 \Leftrightarrow a = 0$
4. $|a - b| = 0 \Leftrightarrow a = b$
5. $|ab| = |a||b|; \left|\dfrac{a}{b}\right| = \dfrac{|a|}{|b|}, b \neq 0$
6. $|a + b| \leq |a| + |b|;$
7. $|a - b| \geq |a| - |b|$
8. $|a| \leq b \Leftrightarrow -b \leq a \leq b; |a| \geq b \Leftrightarrow a \leq -b$ or $a \geq b$

The definition of an absolute-value function follows from the definition of the absolute value of a number. The absolute-value function is a piecewise defined function, meaning that it has different definitions within disjoint subsets of its domain.

$$f(x) = |x| = \begin{cases} x, & x \geq 0 \\ -x, & x < 0 \end{cases}$$

Graphically,

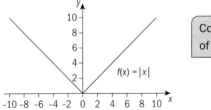

> Compare the graphs of $y = x$ and $y = |x|$.

The domain of the absolute value function is the set of real numbers, and its range is the set of non-negative real numbers.

Example 5

Determine the domain and range of $y = |2x + 3|$, and confirm graphically.

Answer

Domain $= \{x \mid x \in \mathbb{R}\}$

Range $= \{y \mid y \geq 0\}$

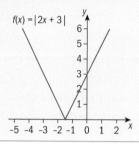

Example 6

Determine the domain and range of $y = |2 - 3x| - 1$

Answer

The domain is the set of real numbers.

Range $= \{y \mid y \geq -1\}$

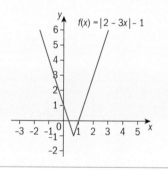

> Since the least value of $|2 - 3x|$ is 0, the least value, or absolute minimum of the function, is $y = 0 - 1 = -1$

Investigation – absolute-value functions

Consider absolute-value functions of the form $y = |x - h| + k$, $h, k \in \mathbb{R}$.

a Graph functions like this for different values of h and k.
What effect do these parameters have on the graph of $y = |x|$?

b Investigate how the parameter a in the general form $y = a|x - h| + k$ affects the shape of the graph of the function.

c Determine the coordinates of the maximum or minimum point of $y = |x - h| + k$ and the condition for it being a maximum or minimum point.

Exercise 2C

Determine the domain and range of these functions, and confirm graphically.

1 $y = -|x|$

2 $y = |2x + 1|$

3 $y = -|2x + 1|$

4 $y = 2|x - 1|$

5 $y = -\dfrac{1}{2}|3x + 2|$

6 $y = |x + 4| - 2$

7 $y = -2|x - 1| + 1$

8 $y = 3|1 - 2x| - 2$

Rational functions

A **rational function** is a function consisting of a rational algebraic expression, for example, $y = \dfrac{1}{x}$. Analyzing this function, you can see that you must restrict the domain, since $x \neq 0$. Also, a rational expression is equal to zero only if the numerator is zero. Since the numerator of this function is not zero, y will never assume the value of zero, and you must eliminate zero from the range.

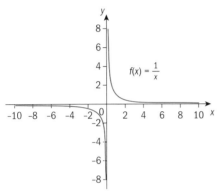

▲ This shape in a conic section is called a **hyperbola**.

Hence, the domain of the function $\dfrac{1}{x}$ is the set of real numbers excluding 0, and the range is also the set of real numbers excluding 0. You can confirm this graphically.

> Degenerate cases of conic sections depart from the generic properties of a class of shapes. For example, a point is a degenerate circle – a circle with radius 0, a circle is a degenerate ellipse, etc. A line is a degenerate parabola if it resides on a tangent plane. You can research other forms of degenerate cases of known shapes.

The function $y = \dfrac{1}{x}$ is also called the reciprocal function.

You can see clearly from the graph that the values of the function increase or decrease rapidly as you take values close to $x = 0$. For example, when $x = \dfrac{1}{100}$, $y = \dfrac{1}{\left(\dfrac{1}{100}\right)} = 100$. Similarly, when $x = -0.01$, $y = -100$. We say that as x gets close in value to 0, the values of the function, y, grow (or shrink) without bound. The lines $x = 0$ and $y = 0$ i.e. x-and y-axes respectively, are called the **asymptotes** of the graph of the function.

The term 'asymptote' will be defined carefully in Chapter 4. For now, think of an asymptote of the graph of a function as a line, such that as the distance between the graph and the line decreases, and gets closer to zero, the further the graph and the line are extended.

Example 7

Determine the domain and range of the function $y = \dfrac{1}{1-2x}$

Confirm your solution graphically.

Answer

$1 - 2x = 0 \Rightarrow x = \dfrac{1}{2}$

Since the denominator cannot be 0, to find the value that must be excluded from the domain, set the denominator equal to 0.

Domain: $\left\{ x \mid x \in \mathbb{R},\ x \neq \dfrac{1}{2} \right\}$

Range: $\{ y \mid y \in \mathbb{R},\ y \neq 0 \}$

The numerator is not equal to 0, hence $y \neq 0$.

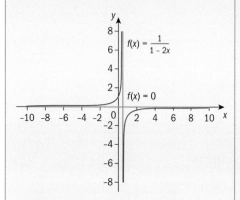

Draw the graph of $f(x) = \dfrac{1}{1-2x}$ on your GDC.

Make a sketch graph.

We can see from the graph that the lines $x = \dfrac{1}{2}$ and $y = 0$ are quite special. The graph gets close to these lines, but does not intersect them. These lines are the asymptotes of the function.

The asymprotes are $x = \dfrac{1}{2}$ and $y = 0$.

What about rational functions in which both numerator and denominator are linear expressions: $y = \dfrac{ax+b}{cx+d},\ x \neq \dfrac{-d}{c}$?

For example, the function $y = \dfrac{2x+1}{3x-3}$

The domain has to be restricted such that $x \neq 1$, as this value of x makes the denominator zero. Rearange to make x the subject in order to see any restrictions on the values of y, the range.

$y = \dfrac{2x+1}{3x-3} \Rightarrow y(3x-3) = 2x+1$

$\Rightarrow 3xy - 3y = 2x + 1$

$\Rightarrow 3xy - 2x = 1 + 3y$

This process of solving for x is the same process that you will use in finding the inverse of a function.

Factorize the left hand side

$$x(3y - 2) = 1 + 3y \Rightarrow x = \frac{1 + 3y}{3y - 2}$$

Since $3y - 2 \neq 0$, $y \neq \dfrac{2}{3}$

Hence, the domain of the function

is $\{x \mid x \neq 1\}$ and the range is

$$\left\{ y \mid y \neq \frac{2}{3} \right\}$$

Confirm graphically:

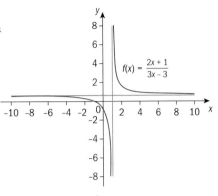

The reciprocal function is defined as $f : x \mapsto \dfrac{1}{x}$. A rational function can be obtained from the reciprocal function by applying certain **transformations**. You will look at this function after you have studied function transformations.

The asymptotes of the function are $x = 1$ and $y = \dfrac{2}{3}$.

Example 8

Determine the domain and range of the function $y = -\dfrac{1 - 3x}{2x - 1}$ and confirm your answers graphically.

Answer

$2x - 1 = 0 \Rightarrow x = \dfrac{1}{2}$

Domain $= \left\{ x \mid x \neq \dfrac{1}{2} \right\}$

$2xy - y = 1 - 3x$

$\Rightarrow 2xy + 3x = 1 + y$

$\Rightarrow x(2y + 3) = 1 + y$

$x = \dfrac{1 + y}{2y + 3}$

$2y + 3 \neq 0 \Rightarrow y \neq -\dfrac{3}{2}$

Range: $\left\{ y \mid y \neq -\dfrac{3}{2} \right\}$

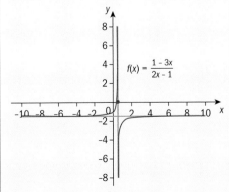

The asymptotes are $x = \dfrac{1}{2}$; $y = -\dfrac{3}{2}$

Put the denominator equal to zero.

Rearrange to make x the subject.

Find the values for which the denominator would be zero and exclude them.

Draw the graph on your GDC. Make a sketch graph. State the asymptotes.

Exercise 2D

Find the domain and range of these functions, and confirm graphically.

1 $y = \dfrac{1}{3x+2}$

2 $y = -\dfrac{1}{2-x}$

3 $y = \dfrac{3}{3-x}$

4 $y = -\dfrac{5}{6x+3}$

5 $y = \dfrac{1+2x}{1-2x}$

6 $y = -\dfrac{2-3x}{1+x}$

Now look at functions combining the special functions you have just studied, for example, $y = \dfrac{1}{\sqrt{x+1}}$

For the denominator to be non-zero, $x + 1 > 0$, or $x > -1$
Since the numerator is not zero, $y \neq 0$. Furthermore, since the denominator is always positive, the values of y will always be positive. Hence, the domain is $\{x \,|\, x > -1\}$, and the range is $\{y \,|\, y > 0\}$. You can confirm this graphically.

The asymptotes are $x = -1$ and $y = 0$.

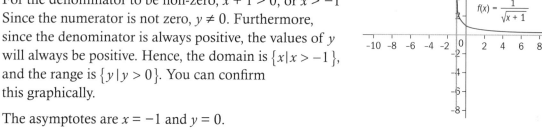

Example 9

Find the domain and range of $y = \dfrac{1}{|2x+1|}$, and confirm graphically.

Answer

$2x + 1 \neq 0, \ x \neq -\dfrac{1}{2}$

Domain $= \left\{ x \,\middle|\, x \neq -\dfrac{1}{2} \right\}$

Range $= y \,|\, y > 0$

Find value for which denominator = 0 and exclude.

Alternative notation:

domain =

$\left\{ x \,\middle|\, x \in \mathbb{R} - \left\{ -\dfrac{1}{2} \right\} \right\}$

$y \neq 0$ and always positive

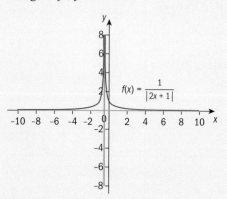

The asymptotes are $x = -\dfrac{1}{2}, \ y = 0$

Exercise 2E

Find the domain and range of these functions, and confirm graphically.

1 $y = \dfrac{1}{|x+1|}$ **2** $y = -\dfrac{2}{|x-1|}$ **3** $y = \dfrac{x}{|x|}$ **4** $y = \dfrac{-2}{\sqrt{1-x}}$

5 Let $f : x \mapsto \sqrt{\dfrac{1}{x^2} - 2}$. Find

    a the set of real values of x for which f is real.

    b the range of f

Piecewise defined functions

You have seen a piecewise defined function already – the absolute-value function. Now look at other piecewise defined functions.

Let $y = \begin{cases} 3-x, & x < 3 \\ x-3, & x \geq 3 \end{cases}$

For the first branch of the function, the domain is $]-\infty, 3[$, and for the second branch, the domain is $[3, \infty[$. The domain of the function is therefore $]-\infty, 3[\cup [3, \infty[$, or the set of real numbers. The range is the set of non-negative numbers.

You will probably have recognized that this function is equivalent to the absolute-value function $y = |x - 3|$. You can confirm this graphically.

> $]-\infty, 3[$ means the set of values from $-\infty$ up to, but not including, 3. $[3, \infty[$ means the set of values including 3 and all values greater than 3.

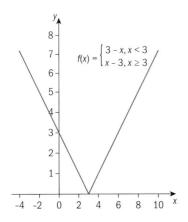

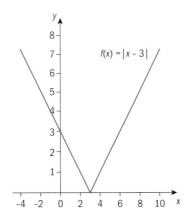

Example 10

Consider the function $f(x) = \begin{cases} -(x-2)^2, & x < 3 \\ x - 4, & x \geq 3 \end{cases}$

a Find $f(0), f(3), f(4)$.   **b** Sketch $f(x)$.

c Write down the domain and range of f.

Answers

a $f(0) = -(0-2)^2 = -4$

   $f(3) = 3 - 4 = -1$

   $f(4) = 4 - 4 = 0$

Evaluate $f(0)$ using the first branch since $0 < 3$. Evaluate $f(3)$ using the second branch, since its domain is $x \geq 3$. Evaluate $f(4)$ using the second branch since $4 > 3$.

Draw the graph on your GDC. Make a sketch.

b

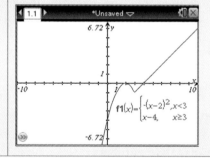

c Domain $\{x : x \in \mathbb{R}\}$

   Range $\{y : y \in \mathbb{R}\}$

GDC help on CD:
Alternative demonstrations for the TI-84 Plus and Casio fx-9860GII GDCs are on the CD.

Example 11

Consider the function $f(x) = \begin{cases} -(x^2 + 1), & x \leq 0 \\ 2, & 0 < x < 4 \\ 2 + \sqrt{x}, & x \geq 4 \end{cases}$

a Find $f(-1), f(1), f(4), f(9)$.   **b** Sketch the function.

c Write down the domain and range of the function.

Answers

a $f(-1) = -((-1)^2 + 1) = -2$

   $f(1) = 2$ $f(4) = 4$ $f(9) = 5$

b

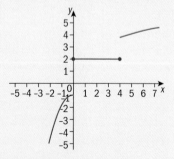

Draw the graph on your GDC. Make a sketch

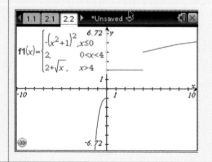

c Domain $= \{x \mid x \in \mathbb{R}\}$.

   Range $=$

   $\{y \mid y \leq -1,\ y = 2,\ y \geq 4\}$

GDC help on CD:
Alternative demonstrations for the TI-84 Plus and Casio fx-9860GII GDCs are on the CD.

Exercise 2F

1 Consider the function $y = \begin{cases} 1, x \geq 0 \\ -1, x < 0 \end{cases}$

   a Find $f(-3), f(0), f(\pi), f(4)$.

   b Sketch the function.

   c Write down the domain and range of the function.

2 Consider the function $y = \begin{cases} 1-x, x \leq -1 \\ 2.5, -1 < x < \sqrt{6} \\ x^2 - 4, x \geq \sqrt{6} \end{cases}$

   a Find $f(-3), f(0), f\left(\sqrt{6}\right), f(3)$.

   b Sketch the function.

   c Write down the domain and range of the function.

3 Consider the function $f(x) = \begin{cases} 3x - 1, x \leq 0 \\ \sqrt{x+1}, x > 0 \end{cases}$

   a Find $f(-1), f(0), f(1), f(8)$.

   b Sketch the function.

   c Write down the domain and range of the function.

EXAM-STYLE QUESTION

4 Consider the function $g(x) = \begin{cases} x^2 + 1, x \leq 0 \\ -\dfrac{1}{2}x + 1, 0 < x \leq 2 \\ -\sqrt{x-2}, x > 2 \end{cases}$

   a Find $f(-2), f(1), f(2), f(3)$.

   b Sketch the function.

   c Write down the domain and range of the function.

Classification of functions

More than one element of the domain of a function can have the same image as shown in the diagram. For example, the constant function $f(x) = 2$ maps all real numbers to the number 2. Hence, the domain is the set of all real numbers, and the range is the set containing the single element 2. This function is called a **many-to-one** function.

Graphically, you can recognize a many-to-one function by the horizontal line test. If you draw horizontal lines through the graph of the function and they intersect the graph at more than one point, then the function is many-to-one as shown in the diagram.

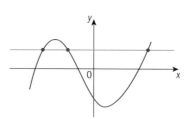

A function that does not allow an element of the range to be the image of more than one element in the domain is a **one-to-one function**. Examples of such functions are linear functions. Using the horizontal line test, if horizontal lines are drawn through the graph of a one-to-one function they intersect the graph in only one point.

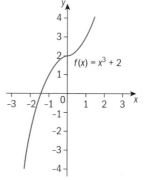

$f(x) = x^3 + 2$

Use the horizontal line test on the graphs of absolute and quadratic functions from Examples 3, 5 and 6. What do you notice?

Example 12

Which of the following functions are one-to-one and which are many-to-one?

a $\{(-0.2, 0), (1, 0), (2.4, 0), (\pi, 0)\}$

b $\left\{(-2, 0.2), (-1, 0.1), (3, -\frac{1}{3}), (\pi, -\frac{1}{\pi})\right\}$

c

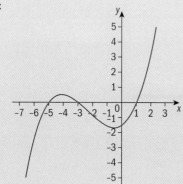

d

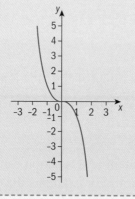

There are different meanings of the term 'curve' both in mathematics and common language. Explore the differences between algebraic and transcendental curves familes of curves (e.g. conics), and curves with special properties (e.g. cycloids).

Answers

a many-to-one

b one-to-one

c many-to-one

d one-to-one

a *All x-values map to y = 0*

b *Each x-value maps to different y-value (its reciprocal).*

c, d *Use the horizontal line test.*

Some graphs of many-to-one functions are symmetrical about the y-axis, for example, $f(x) = x^2$ and $g(x) = |x|$.

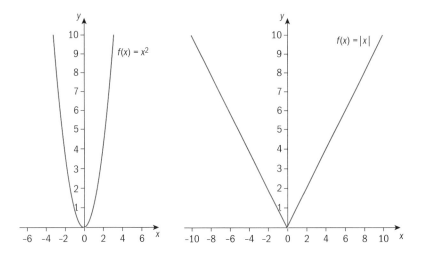

If you substitute either $\pm a$ for x, where a is a real number, you get the same result for y: $f(a) = a^2$ and $f(-a) = a^2$; $g(a) = a$ and $g(-a) = a$ These functions are **even functions**.

> → A function is **even**, if for all x in the domain, $-x$ is in the domain, and $f(x) = f(-x)$ for all values of x.

Even functions are symmetrical about the y-axis.

Other functions have a different kind of symmetry, for example, $f(x) = x(x - 2)(x + 2)$. The graph of this function has rotational symmetry about the origin.

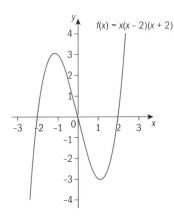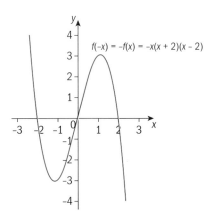

For any real value a

$$f(-a) = -a(-a - 2)(-a + 2)$$
$$= a(a + 2)(-1)(a - 2)$$
$$= -a(a + 2)(a - 2)$$
$$= -f(a).$$

so, for the function $f(x) = x(x - 2)(x + 2)$, $f(-x) = -f(x)$ for all values of x.

The graph of $f(-x)$ also has rotational symmetry about the origin.

> → A function is **odd** if for all x in the domain, $-x$ is in the
> domain and $f(-x) = -f(x)$ for all values of x.

The function $f(x)$ is many-to-one as you can see if you apply the horizontal line test.

The function $y = x^3$ has rotational symmetry about the origin and so it is an odd function but this function is also one-to-one.

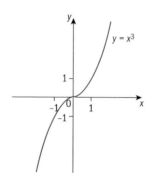

Example 13

Determine algebraically if the following functions are even, odd, or neither. Confirm your answers graphically.

a $f(x) = -3x^2 + 5$

b $g(x) = \begin{cases} x - \pi, & -\pi \leq x < 0 \\ -x + \pi, & 0 \leq x < \pi \end{cases}$

c $h(x) = \begin{cases} (x + \pi)^2, & -\pi \leq x < 0 \\ -(x - \pi)^2, & 0 \leq x < \pi \end{cases}$

Answers

a $f(-x) = -3(-x)^2 + 5$

   $= -3x^2 + 5$

   $= f(x)$

hence $f(x)$ is even.

Evaluate $f(-x)$
$f(-x) = f(x)$ means f is even.
Use your GDC to draw the graph.

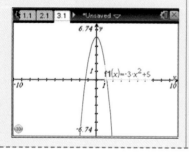

▶ Continued on next page

b $g(-x) = -x - \pi$
$\neq g(x)$ and $-x - \pi$
$\neq -g(x)$

hence $g(x)$ is neither
even nor odd.

Evaluate g(x)

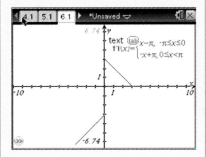

c $h(-x) = (-x + \pi)^2$
$= x^2 - 2\pi x + \pi^2$
$= (x - \pi)^2$
$= -h(x),$

hence $h(x)$ is odd.

Evaluate h(−x)
h(−x) = −h(x) means f is odd.

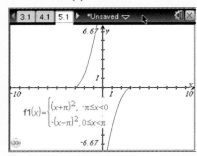

GDC help on CD:
*Alternative demonstrations
for the TI-84 Plus and Casio
fx-9860GII GDCs are on the
CD.*

Exercise 2G

Determine algebraically if these functions are even, odd or neither.
Confirm your answers graphically and state if the funtions are
many-to-one or one-to-one.

1 $f(x) = 4 - x^2$

2 $g(x) = x^3 + 3x$

3 $h(x) = -\dfrac{3}{2x}$

4 $p(x) = x^3 + 4x + 1$

5 $r(x) = \begin{cases} -1, 0 \leq x < \pi \\ 1, \pi \leq x < 2\pi \\ -1, 2\pi \leq x < 3\pi \end{cases}$

6 $q(x) = 2x^3 - 4x$

7 $w(x) = x - 2x^3 + x^5$

8 $t(x) = 4x^4 - x$

9 Find a function that is both even and odd.

2.3 Operations with functions

As with numbers, you can add, subtract, multiply and divide functions. These properties hold for operations with functions.

> → Let f and g be two real-valued functions in x. Then
>
> **1** $(f + g)(x) = f(x) + g(x)$ **2** $(f - g)(x) = f(x) - g(x)$
>
> **3** $a(f(x)) = af(x), a \in \mathbb{R}$ **4** $(fg)(x) = f(x)g(x)$
>
> **5** $\left(\dfrac{f}{g}\right)x = \dfrac{f(x)}{g(x)}, g(x) \neq 0$

Let D_1 be the domain of f and D_2 the domain of g.
Then the domain of the sum, difference, and product of f and g is $D_1 \cap D_2$.
The domain of the quotient of f and g is
$D_1 \cap D_2 - \{x \mid g(x) = 0\}$

> Exclude from the intersection of the domains the element(s) that make the denominator zero.

For example, let $f(x) = 2 - x$,

and $g(x) = \dfrac{1}{x}$. The domain of

$f(x)$ is the real numbers,

and the domain of $g(x)$ is the

set of non-zero real numbers.
$(f + g)(x) = f(x) + g(x) = 2 - x + \dfrac{1}{x}$
and the domain is the set of
non-zero real numbers.
The domain of $(f + g)(x)$ is $\{x \mid x \neq 0\}$

The graph confirms this.

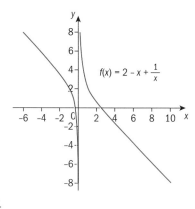

$f(x) = 2 - x + \dfrac{1}{x}$

The domain of the difference of f and g will also be $\{x \mid x \neq 0\}$.

The product of f and g, is
$(fg)(x) = f(x)\,g(x)$

$= (2 - x)\dfrac{1}{x} = \dfrac{2 - x}{x}$

From the graph, the domain
of $(fg)(x)$ is $\{x \mid x \neq 0\}$

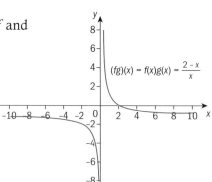

$(fg)(x) = f(x)g(x) = \dfrac{2 - x}{x}$

The quotient of f and g is $\left(\dfrac{f}{g}\right)(x) = \dfrac{f(x)}{g(x)} = \dfrac{2-x}{\dfrac{1}{x}} = x(2-x) = 2x - x^2$

$$\dfrac{2-x}{\dfrac{1}{x}} = (2-x) \div \dfrac{1}{x}$$
$$= x(2-x)$$

The domain of $\dfrac{f}{g}$ is the set of non-zero real numbers, since $\dfrac{1}{x} \neq 0$, for all real values of x.

Graphically

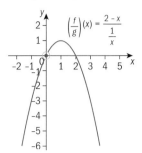

The blue circle at $x = 0$ shows this function is not defined at $x = 0$.

If you draw this graph on a GDC and use the 'trace' function, you will see that the GDC gives no value at $x = 0$.

At $x = 0$ there is a 'hole' in the graph of the function.

Example 14

Functions f, g, and h are defined as
$f = \{(-2, 0), (1, 1), (2, 0), (3, 1)\}$,
$g = \{(-2, -2), (-1, -1), (2, 2), (4, 4))$, and
$h = \{(-3, 0), (0, 0), (1, 0), (2, 4)\}$.
Find the domain of
a $f + g$,　　**b** $g - h$　　**c** fh　　**d** $\dfrac{f}{h}$

Answers

a　$\{-2, 2)$

Write the values of f, g, and h.
Domain of $f = D_f = \{-2, 1, 2, 3\}$
Domain of $g = D_g = \{-2, -1, 2, 4\}$
Domain of $h = D_h = \{-3, 0, 1, 2\}$

b　$\{2\}$　$D_f \cap D_g = \{-2, 2\}$
c　$\{1, 2\}$　$D_g \cap D_h = \{2\}$
d　$\{2\}$　$D_f \cap D_h = \{1, 2\}$

$D_f \cap D_h - \{x \mid h(x) = 0\}$
$= \{1, 2\} - \{1\}$ ($h(1) = 0$)
$= \{2\}$

Investigation – odd and even functions

By testing different examples of even and odd functions, determine whether the following are even, odd, or neither.

a　The sum (or difference) of two even functions.
b　The sum (or difference) of two odd functions.
c　The sum (or difference) of an odd and an even function.
d　The product of two even functions.
e　The product of two odd functions.
f　The product of an odd and an even function.

Composition of functions

If f and g are functions, then the notation for the composition of functions f and g is $(f \circ g)$ and $(f \circ g)(x) = f(g(x))$.

For $(f \circ g)$ to be defined, the range of g must be a subset of the domain of f.

For $(g \circ f)$ to be defined, the range of f must be a subset of the domain of g.

If $f(x) = 2x - 1$ and $g(x) = x^2$, then letting $g(x)$ be the argument of f, $(f \circ g)(x) = f(x^2) = 2x^2 - 1$ The range of g is the set of non-negative reals and the domain of f is the set of real numbers, hence $(f \circ g)(x)$ is defined. The domain of $(f \circ g)$ is the domain of g, the set of real numbers.

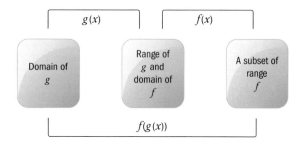

Similarly, $(g \circ f)(x) = g(2x - 1) = (2x - 1)^2$.

You can obtain specific values of composite functions by first finding the composite function, or by evaluating the inner function first at the desired x-value, and then evaluating the outer function. For example, here are the two methods for finding $f(g(0))$ for $f(x) = 2x - 1$ and $g(x) = x^2$

> You have seen composite functions before. Example 9 can be written as
> $(f \circ g)(x) = \dfrac{1}{|2x+1|}$ where
> $f(x) = \dfrac{1}{x}$;
> $g(x) = |2x + 1|$

Method 1
Find the composite function first.

$f(g(x)) = f(x^2) = 2x^2 - 1$

$f(g(0)) = 2(0^2) - 1$

$\qquad = -1$

g is the outer function. $g(f(x))$ f is the inner function.

Method 2
Find $g(0)$ first.

$g(0) = 0$

and $f(0) = -1$

Example 15

Consider $f(x) = 3x^2$ and $g(x) = \sqrt{x-2}$
a Determine the domain and range of f and g.
b Find: **i** $g(f(1))$ **ii** $f(g(3))$
c Find: **i** $(f \circ g)(x)$ **ii** $(g \circ f)(x)$, and determine the domain and range of the composite functions.

Answers

a f: Domain = real numbers
      Range = $\{y \mid y \geq 0\}$

    g: Domain = $\{x \mid x \geq 2\}$
       Range = $\{y \mid y \geq 0\}$

b **i** $f(1) = 3$, $g(3) = 1$
   ii $g(3) = 1$, $f(1) = 3$

c **i** $f(g(x)) = 3\left(\sqrt{x-2}\right)^2 = 3x - 6$

    Domain = $\{x \mid x \geq 2\}$
    Range = $\{y \mid y \geq 0\}$

  ii $g(f(x)) = \sqrt{(3x^2) - 2}$
             $= \sqrt{(9x^2 - 2)}$

    Domain
      $= \left\{x \mid x \geq \dfrac{\sqrt{2}}{3}, \ x \leq -\dfrac{\sqrt{2}}{3}\right\}$

    Range = $\{y \mid y \geq 0\}$

On the GDC you can display the composite function by entering it as seen here.

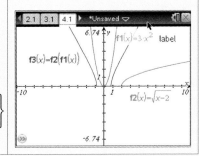

GDC help on CD:
Alternative demonstrations for the TI-84 Plus and Casio fx-9860GII GDCs are on the CD.

Investigation – composite functions

a By investigating different functions, determine if the composition of functions is commutative, that is, does $(f \circ g) = (g \circ f)$ for any functions f and g?

b By investigating different functions, determine if the composition of functions is associative, that is, does $(f \circ (g \circ h)) = ((f \circ g) \circ h)$, for any functions f, g, and h?

c By investigating different functions, determine if the composition of
  i an even function with an even function
  ii an odd with an odd
  iii an even with an odd (or an odd with an even)
  is an even or an odd function, or neither.

d Determine the kind of function that results when any function is composed with an even function.

  Attempt to justify your conclusions.

> Odd and even functions were defined on page 67–68.

Composition of functions is not limited to two functions.

For example, $y = \dfrac{1}{\sqrt{x^2 - 1}}$ is $(f \circ (g \circ h))$ when $h(x) = x^2 - 1$,

$g(x) = \sqrt{x}$ and $f(x) = \dfrac{1}{x}$. Can you create a function which is the

composition of four functions?

Exercise 2H

1 If $f(x) = 2x$ and $g(x) = \sqrt{x}$, find the domain of:

   a $2g(x) - f(x)$

   b $f(x)\,g(x)$

   c $\left(\dfrac{g}{f}\right)(x)$

2 If $f(x) = |x + 1|$ and $g(x) = \sqrt{x^2 - 4}$, find the domain of $\left(\dfrac{f}{g}\right)(x)$

3 Let $f(x) = x^2 + 2x - 1$ and $g(x) = 1 - 2x - 3x^2$
   Find

   a $f(g(0))$ **b** $gf(-1)$ **c** $f(f(0))$ **d** $g(g(x))$

4 Let $f(x) = 1 - 2x$, $g(x) = x^2 - 1$ and $h(x) = \sqrt{2x + 4}$

   a Find **i** $f(g(x))$ **ii** $g(h(x))$ **iii** $f(h(x))$ **iv** $h(g(x))$

   b Find the domain and range of the functions in **a**.

   c Confirm your results to **a** and **b** graphically.

   d Find $f(g(1))$, and hence $h(f(g(1)))$.

5 Find expressions for $f(x)$ and $g(x)$ such that $f(g(x)) = x^2 - 2$

6 Find expressions for $g(x)$ and $h(x)$ such that $(h \circ g)(x) = \sqrt{2x - 3}$

Identity function

A function $f(x)$ that, when composed with $g(x)$, leaves $g(x)$ unchanged is called the **identity function**.

For example, consider $g(x) = 2x - 1$ and find $f(x)$ such that $(f \circ g)(x) = (g \circ f)(x) = g(x)$.

To meet this condition $f(x)$ must equal its argument x, i.e. $f(x) = x$ for any $x \in D_f$.

If $f(x) = x$ and $g(x) = 2x - 1$, then
$f(g(x)) = 2x - 1 = g(x)$, and $g(f(x)) = 2x - 1 = g(x)$

Inverse of a function

A function, h that, when composed with g, results in the identity function f is called an **inverse function**. So
$(g \circ h)(x) = (h \circ g)(x) = f(x) = x$

> You have seen this process on page 72.

The final output is the argument x, so $h(x)$ is a function that maps y-values into x-values, or elements of the range into elements of the domain.

Since the function g maps x into $2x - 1$, reversing the process will map y into x. This reverse process is called finding the inverse of a function (if the inverse exists).

> How do you know if the inverse of a function exists?

To find the inverse function h, solve the equation $y = 2x - 1$ for x

$$x = \frac{y+1}{2}$$

Now swap the x and y:

The inverse function is $y = \frac{x+1}{2}$

Let $h(x) = \frac{x+1}{2}$, and see if when $h(x)$ is composed with $g(x)$ you do indeed get the identity function.

$$g(h(x)) = 2\left(\frac{x+1}{2}\right) - 1 = x + 1 - 1 = x$$

and

$$h(g(x)) = \frac{(2x-1)+1}{2} = \frac{2x}{2} = x$$

Hence, g and h are inverses of each other.

The notation for the inverse of a function g is g^{-1}.

The inverse of $g(x) = 2x - 1$ is $g^{-1}(x) = \frac{x+1}{2}$

> → Two functions g and h are inverses of each other if their
> composition results in the identity function, $f(x) = x$, i.e.,
> $(g \circ h)(x) = (h \circ g)(x) = x$
>
> The functions g and h are also said to be invertible functions.

Graphical properties of inverse functions

> → The graphs of a function and its inverse are reflections of each
> other in the line $y = x$

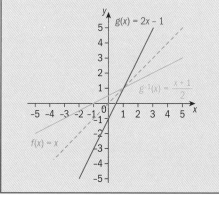

The index $^{-1}$ when applied to functions means the inverse function; when applied to numbers it means the reciprocal of a number.
Why is the same symbol used for these different mathematical objects?
What do they have in common to justify the use of the same symbol?
Are there other mathematical symbols that are used differently in a particular context?

Example 16

If $f(x) = \dfrac{x}{2} + 5$, find $f^{-1}(x)$, if it exists.

Justify your result algebraically, and confirm graphically.

Answer

$y = \dfrac{x}{2} + 5 \Rightarrow x = 2(y - 5)$

$\qquad$ *Solve for x, and then swap the x and y.*

or

$x = \dfrac{y}{2} + 5 \Rightarrow y = 2(x - 5)$

$\qquad$ *Swap x and y first, then solve for y.*

$f^{-1}(x) = 2(x - 5)$

$\qquad$ *Use inverse notation.*

$f(f^{-1}(x)) = \dfrac{2(x - 5)}{2} + 5$

$\qquad = (x - 5) + 5 = x$

$\qquad$ *Justify your answer by showing that $f(f^{-1}(x)) = x$, and $f^{-1}(f(x)) = x$*

$f^{-1}(f(x)) =$

$2\left(\left(\dfrac{x}{2} + 5\right) - 5\right) = 2\left(\dfrac{x}{2}\right) = x$

Since $f(f^{-1}(x)) = x = f^{-1}(f(x))$, f and f^{-1} are inverses of each other.

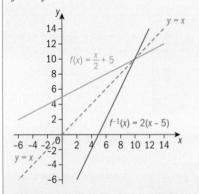

$\qquad$ *The graph of f and f^{-1} are symmetrical about the line $y = x$.*

The graphs of f and f^{-1} are symmetrical about the line $y = x$.

Since the inverse of a function is a mapping from the range to the domain of the function, the range of the function will be the domain of its inverse, and the domain of the function the range of its inverse. Example 16 is a linear function, and its inverse is a linear function. Hence the domain and range of both functions are the real numbers.

You can look at the domain and range of a function to determine when a function is **invertible**, that is, has an inverse.

Consider $f(x) = x^2$. Then $f^{-1}(x) = \pm \sqrt{x}$. The inverse is not a function, as each x-value maps to two values, e. g. $f^{-1}(4) = \pm 2$

The part of $f(x)$ which has an inverse is therefore the set of non-negative real numbers.

From the graphs of the function $y = x^2$ and the relation $y = \pm\sqrt{x}$, you can see that the relation and the function are inverses of one another, but the relation is not itself a function.

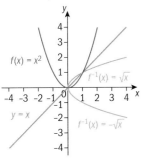

GDC help on CD:
Alternative demonstrations for the TI-84 Plus and Casio fx-9860GII GDCs are on the CD.

The graphs of the function and the relation are reflections of each other in the line $y = x$.

For $f(x) = x^2$ to have an inverse its domain must also be restricted to the set of non-negative real numbers.

When you restrict the domain of a function, you do so to ensure that its inverse is also a function.

You can also restrict the domain of f to the set of non-positive real numbers. In this case $f^{-1}(x) = -\sqrt{x}$, and the graph of f and f^{-1} would be

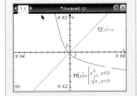

Unless you are told otherwise, the convention is to restrict the inverse to the set of non-negative real numbers.

Example 17

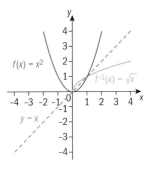

> Find the inverse of $f(x) = x^2 - 2x + 2$, if it exists, and its domain and range. Confirm your result graphically.

Answer

$x = y^2 - 2y + 3 \Rightarrow y^2 - 2y = x - 3$ *Solve for x then swap x and y.*

$\Rightarrow y^2 - 2y + 1 = x - 3 + 1 = x - 2$ *Complete the square to isolate y.*

$\Rightarrow (y - 1)^2 = x - 2$

$\Rightarrow y - 1 = \sqrt{x - 2} \Rightarrow y = 1 + \sqrt{x - 2}$ *Solve for y.*

Domain: $x \geq 2$ *Determine the domain for the*
Range: $y \geq 1$ *inverse to exist.*

    Use inverse notation.

$f^{-1}(x) = 1 + \sqrt{x - 2}$

    Graph $f(x)$, $f^{-1}(x)$ and $y = x$ on your GDC.

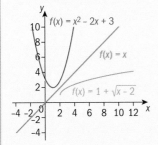

    In order for f to have an inverse, restrict its domain to $x \geq 1$, the range of the inverse.

→ Only functions that are one-to-one have inverse functions.

Exercise 2I

Find the inverse of each function, if it exists.
Justify your answer algebraically, and confirm graphically.

1 $y = 3x - 1$

2 $y = \dfrac{x-1}{3}$

3 $y = x^2 - 2,\ x \geq 0$

4 $y = x^2 + 1,\ x \leq 0.$

5 $y = x^2 + 4x - 1,\ x \geq -2$

EXAM-STYLE QUESTIONS

6 Determine algebraically if $y = 1 - 2x$ is its own inverse.

7 Given that $f(x) = 3x$ and $g(x) = 2x + 1$, show that
$(f^{-1} \circ g^{-1})(x) = (g \circ f)^{-1}(x)$

8 Given $f(x) = \dfrac{2x+1}{x-1},\ x \neq 1$, find the inverse function of f, clearly stating its domain.

Investigation – self-inverse functions

You have met the function $y = \dfrac{1}{x}$ before.
It has special properties:

- it is odd

- its axes are asymptotes of its graph

- it has no x- or y-intercepts.

It is also its own inverse. To prove this algebraically,

swap y and x, to get $x = \dfrac{1}{y}$

Then solve for y:

$y = \dfrac{1}{x}$

or $f^{-1}(x) = \dfrac{1}{x}$

$f(f^{-1}(x)) = \dfrac{1}{\left(\dfrac{1}{x}\right)} = x = f^{-1}(f(x))$

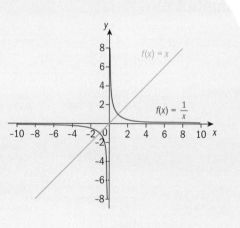

Use your GDC to find functions whose graphs are reflections of each other in $y = x$

Find other such functions by graphing functions that are their own reflection in the line $y = x$.
Is there a rule, or rules, for a function to be its own inverse?

2.4 Transformations of graphs of functions

Axis symmetry

In the previous sections you looked at some of the symmetries of graph of functions.

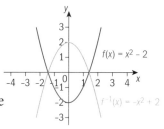

- The graph of an even function is symmetrical about the y-axis, hence $f(-x) = f(x)$.
- The graph of an odd function has rotational symmetry about the origin, hence $f(-x) = -f(x)$.
  Look at the graphs of $y = f(x)$ and $y = -f(x)$.

If $f(x) = x^2 - 2$, then $-f(x) = -x^2 + 2$.

- The graphs are reflections of each other in the x-axis.

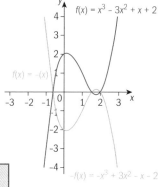

The graph of the cubic function $f(x) = -x^3 - 3x^2 + x + 2$ and its negative produce the same result – they are reflections of each other in the x-axis,
e.g. if $f(x) = x^3 - 3x^2 + x + 2$ then

$$-f(x) = -x^3 + 3x^2 - x - 2$$

> → In general, therefore, the graphs of $y = f(x)$ and $y = -f(x)$ are reflections of each other in the x-axis.

Now look at the graph of

$$f(x) = x^3 - 3x^2 + x + 2$$

and the graph of
$$f(-x) = (-x)^3 - 3(-x)^2 + (-x) + 2$$
$$= -x^3 - 3x^2 - x + 2.$$

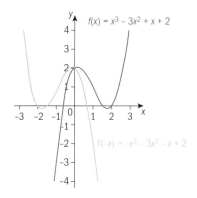

> → In general, the graphs of $f(x)$ and $f(-x)$ are reflections of each other in the y-axis.

Graphing $y = |f(x)|$ from $y = f(x)$

Take the linear function $y = 2x - 1$ and form the function which is the absolute value of this linear function, $y = |2x - 1|$. Since $y = 2x - 1$ is linear, the domain is the set of real numbers, and the range of the absolute-value function will be the non-negative real numbers.

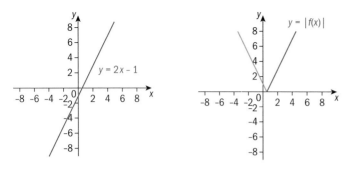

From the graph you can see that the graph is unchanged for any positive y-values, and negative y-values are reflected in the x-axis.

Example 18

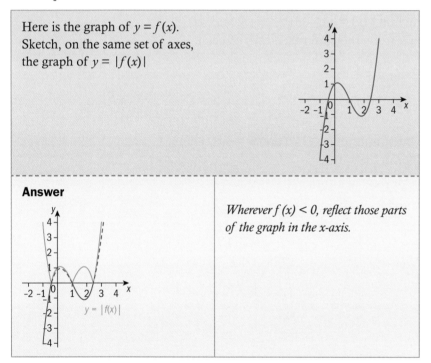

Here is the graph of $y = f(x)$.
Sketch, on the same set of axes,
the graph of $y = |f(x)|$

Answer

Wherever $f(x) < 0$, reflect those parts of the graph in the x-axis.

Graphing $y = f(|x|)$ from $y = f(x)$

Look again at the function $f(x) = 2x - 1$. The domain and range is the set of real numbers. What is the domain and range of $y = f(|x|)$? The domain will still be the set of real numbers.

The graph is unchanged where $x \geq 0$. For $x < 0$ the graph for $x \geq 0$ is reflected in the y-axis.

The range is $\{y \mid y \geq -1\}$

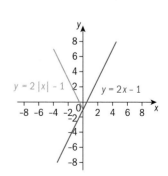

Example 19

Here is the graph of $y = f(x)$.
On the same set of axes,
sketch the graph of $y = f|x|$.

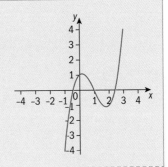

Answer

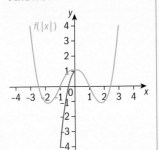

The graph is unchanged where $x \geq 0$.
For $x < 0$ the graph for $x \geq 0$ is reflected in the y-axis.

Exercise 2J

For each function $y = f(x)$, draw **a** $y = |f(x)|$; **b** $y = f(|x|)$.
Draw each part on the same set of axes as $y = f(x)$.

EXAM-STYLE QUESTIONS

1 $y = 1 - 4x$ **2** $y = x^2 - 2x$ **3** $y = -3x^2 + x - 1$

4 $y = x^3 + 3x - 2$ **5** $y = \sqrt{3x+1} - 4$

Graphing $\dfrac{1}{f(x)}$ from $f(x)$

Now look at the graph of the function $y = 2x^2 - 1$,
whose domain is the set of real numbers and whose
range is $y \geq -1$. What is the relationship between this
graph and its reciprocal, $y = \dfrac{1}{2x^2 - 1}$?

The domain is $x \neq \pm \dfrac{1}{\sqrt{2}}$

The range is $y \leq -1$ or $y \geq 0$.

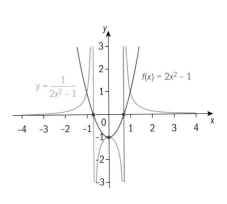

The graph of $y = 2x^2 - 1$ crosses the x-axis at $(\pm 0.707, 0)$
and the y-axis at $(0, -1)$.

Example 20

Copy the graph of $f(x)$ and sketch $\dfrac{1}{f(x)}$ on the same set of axes. Label any intercepts, asymptotes and minimum and maximum points of both graphs.

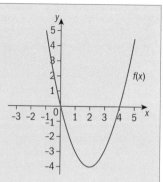

Answer

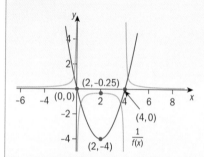

Asymptotes

$x = 0, x = 4, y = 0$

→ From the graphs

- Where they exist, the zeros of $f(x)$ are the vertical asymptotes of $\dfrac{1}{f(x)}$ and the zeros of $\dfrac{1}{f(x)}$ are the vertical asymptotes of $f(x)$.

- If $y = b$ is the y-intercept of $f(x)$, then $\dfrac{1}{b}$ is the y-intercept of $\dfrac{1}{f(x)}$

- The minimum value of $f(x)$ occurs at the same value of x as the maximum of $\dfrac{1}{f(x)}$, and the minimum of $\dfrac{1}{f(x)}$ occurs at the same value of x as the maximum of $f(x)$.

- When $f(x) > 0$, $\dfrac{1}{f(x)} > 0$; when $f(x) < 0$, $\dfrac{1}{f(x)} < 0$

- When $f(x)$ approaches 0, $\dfrac{1}{f(x)}$ will approach $\pm\infty$; when $f(x)$ approaches $\pm\infty$, $\dfrac{1}{f(x)}$ approaches 0

Exercise 2K

For $y = f(x)$, sketch $y = \dfrac{1}{f(x)}$ on the same set of axes, including asymptotes, intercepts, and any maximum and minimum points.

1 $y = x + 4$ **2** $y = 1 - 3x$

3 $y = x^2 - 4$ **4** $y = -x(x - 2)$

5 $y = \dfrac{1}{2x - 3}$

EXAM-STYLE QUESTION

6 Copy the graph of $y = f(x)$ and sketch the graph of $\dfrac{1}{f(x)}$ on the same set of axes.

a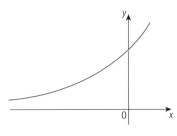

b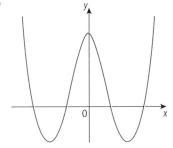

Translations

In the Investigation of absolute-value functions on page 58 you noticed the effect that the parameters h and k had on the absolute-value function $y = |x - h| + k$ when compared to $y = |x|$:

k produces a vertical shift and h produces a horizontal shift on the graph of $y = |x|$.

> Translations are sometimes referred to as rigid transformations as the shape of the graph of the function is unchanged.

Horizontal translation

- If $h > 0$, then the graph of $y = f(x - h)$ is the graph of $y = f(x)$ translated h units in the positive x-direction, or h units to the right

- If $h < 0$, then the graph of $y = f(x - h)$ is the graph of $y = f(x)$ translated h units in the negative x-direction, or h units to the left.

This is a horizontal translation by a **column vector** $\begin{pmatrix} h \\ 0 \end{pmatrix}$.

> In a chapter 11, you will explore column vectors more. In the meantime you need to know that the top number represents a horizontal translation and the bottom number represents a vertical translation.

Vertical translation

- If $k > 0$, then the graph of $f(x) + k$ is the graph of $f(x)$ translated k units in the positive y-direction, or k units upwards.

- If $k < 0$, then the graph of $f(x) - k$ is the graph of $f(x)$ translated k units in the negative y-direction, or k units downwards.

This as a vertical translation with column vector $\begin{pmatrix} 0 \\ k \end{pmatrix}$

You can combine the horizontal and vertical translations using the column vector $\begin{pmatrix} h \\ k \end{pmatrix}$

Example 21

Describe the transformations necessary to obtain the graph of $y = (x - 3)^2 + 1$ from the graph of $y = x^2$, and state the coordinates of the image of the vertex under this translation.
Sketch both graphs on the same set of axes.

Answer

The graph of $y = x^2$ is translated 3 units to the right and 1 unit upward, represented by the column vector $\begin{pmatrix} 3 \\ 1 \end{pmatrix}$

The image of the vertex is (3, 1)

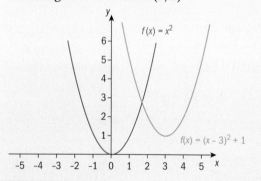

If unsure as to the order in which to perform the transformations, begin with the argument and work your way outward.

Dilations

A **dilation** of a graph occurs when x (horizontal dilation) or y (vertical dilation) is multiplied by a constant a.
A dilation can be a stretch or compression of the graph, depending on the value of a.

Vertical dilation

Here is the graph of $y = ax^2$ for different values of a, $a > 0$.

The graph is stretched or compressed by a factor of a, parallel to the y-axis.

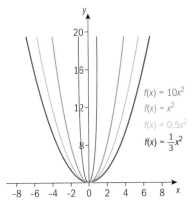

→ For $y = af(x)$, $a > 0$, the graph is vertically stretched or compressed by a factor of a.

- If $a > 1$ the graph is stretched vertically by a factor of a, i.e., it moves further from the x-axis.

- If $0 < a < 1$ the graph is compressed vertically by a factor of a, i.e., it moves close to the x-axis.

Horizontal dilation

Here is the graph of $y = (ax)^2$ for different values of a, $a > 0$.

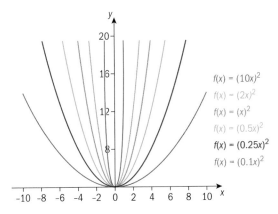

$f(x) = (10x)^2$
$f(x) = (2x)^2$
$f(x) = (x)^2$
$f(x) = (0.5x)^2$
$f(x) = (0.25x)^2$
$f(x) = (0.1x)^2$

The graph is stretched or compressed by a factor of $\dfrac{1}{a}$, parallel to the x-axis.

→ For $y = f(ax)$, $a > 0$, the graph is horizontally stretched or compressed.

- If $a > 1$ the graph is compressed by a factor of $\dfrac{1}{a}$, i.e., the graph moves closer to the y-axis.

- If $0 < a < 1$ the graph is stretched by a factor of $\dfrac{1}{a}$, i.e., the graph moves further from the y-axis.

Dilations are sometimes referred to as non-rigid transformations as they change the shape of the graph of the function.

Example 22

Consider the graph of $y = \begin{cases} -2x-8, & -4 \le x < -3 \\ 2x+4, & -3 \le x < -1 \\ -2x, & -1 \le x < 0 \\ x, & 0 \le x < 4 \end{cases}$

a Sketch the graph on your GDC, and find the coordinates of any maximum, minimum, and zeros of the function.

b Find the coordinates of the points in **a** under these transformations, and confirm your answers using your GDC.

  i $y = 2f(x)$ **ii** $y = f(2x)$ **iii** $y = -\dfrac{1}{2}f(x)$

  iv $y = f(-0.5x)$

Answers

a   Graph of $y = f(x)$

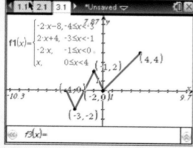

zeros: $(-4, 0)$, $(-2, 0)$, $(0, 0)$
minimum: $(-3, -2)$
maximum: $(-1, 2)$, $(4, 4)$

b   **i**   Graph of $y = 2f(x)$

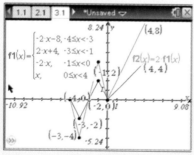

zeros: $(-4, 0)$, $(-2, 0)$, $(0, 0)$
minimum: $(-3, -4)$
maximum: $(-1, 4)$, $(4, 8)$

$y = 2f(x)$ stretches $f(x)$ vertically by a factor of 2, hence all y-coordinates are multiplied by 2, and x-coordinates are unchanged.

> Mathematics has a rich and evolving culture of language. How do the symbols relate to the concepts and analytical skills they represent, and how has evolution in mathematical concepts and skills affected changes in its language?

> GDC tip! Enter $f2(x) = 2f1(x)$

▶ Continued on next page

ii Graph of $y = f(2x)$

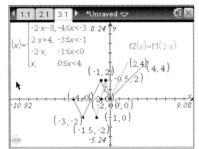

zeros: $(-2, 0), (-1, 0), (0, 0)$
minimum: $(-1.5, -2)$
maximum: $(-0.5, 2), (2, 4)$

$y = f(2x)$ compresses the graph horizontally by a factor of $\frac{1}{2}$, hence each x-coordinate is $\frac{1}{2}$ the x-coordinate of $y = f(x)$, and the y-coordinates are unchanged.

GDC tip! Enter
$f2(x) = f1(2x)$

iii Graph of $y = -\frac{1}{2} f(x)$

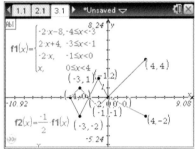

zeros: $(-4, 0), (-2, 0), (0, 0)$
minimum: $(-1, -1), (4, -2)$
maximum: $(-3, 1)$

$y = -\frac{1}{2} f(x)$ is a reflection in the x-axis and compression of $y = f(x)$ parallel to the y-axis by a factor of $\frac{1}{2}$.

GDC tip! Enter
$f2(x) = -\frac{1}{2} f1(x)$

iv Graph of $y = f(-0.5x)$

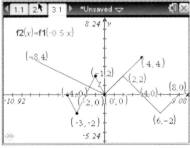

zeros: $(0, 0), (4, 0), (8, 0)$
minimum: $(6, -2)$
maximum: $(-8, 4), (2, 2)$

$y = f(-0.5x)$ is a reflection in the y-axis of $y = f(x)$ and a stretch parallel to the x-axis by a factor of 2. Hence each x-coordinate of $f(x)$ is multiplied by a factor of -2, and the y-coordinate remains unchanged.

GDC tip! Enter
$f2(x) = f1(-0.5x)$

GDC help on CD:
Alternative demonstrations for the TI-84 Plus and Casio fx-9860GII GDCs are on the CD.

You can get a rational function by performing certain transformations on the reciprocal function $y = \frac{1}{x}$. For example, if you first perform a horizontal stretch of factor $\frac{1}{3}$ followed by a vertical stretch of factor 2, then a translation of $\begin{pmatrix} 2 \\ -1 \end{pmatrix}$, what will be

Remember: if, when performing multiple translations you are not sure of their order, start with the argument and work your way outward.

the resulting function? A horizontal stretch factor of $\frac{1}{3}$ becomes $\frac{1}{3x}$,
then a vertical stretch of 2 becomes $2\left(\frac{1}{3x}\right)$, and the translation

results in

$$y = \frac{2}{3(x-2)} - 1$$

Hence, $y = \dfrac{2}{3(x-2)} - 1 = \dfrac{8-3x}{3(x-2)}$

$$\frac{2}{3(x-2)} - 1 = \frac{2-3(x-2)}{3(x-2)}$$
$$= \frac{2-3x+6}{3(x-2)}$$
$$= \frac{8-3x}{3(x-2)}$$

As seen above, the rational function you first saw on page 60
of the form $y = \dfrac{ax+b}{cx+d}$, $x \neq -\dfrac{d}{c}$ can be written in the form

$y = \dfrac{A}{B(x-h)} + k$, $x \neq h$ where A is the vertical stretch factor,
B is the reciprocal of the horizontal stretch factor, and $\begin{pmatrix} h \\ k \end{pmatrix}$ is

the translation.

Exercise 2L

1 Describe each transformation from $f(x)$ to $g(x)$ in terms of x.

a

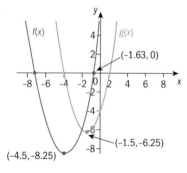

b

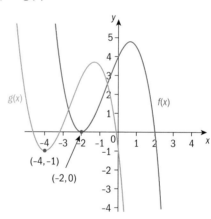

c

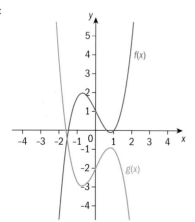

d

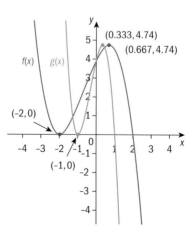

e

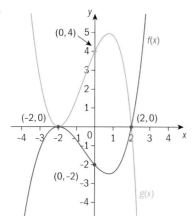

f

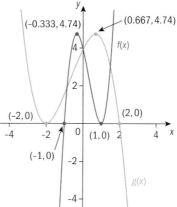

2 Using the graph of $h(x)$, (red graph), and its transformation (blue graph), find an expression for the transformation in terms of $h(x)$.

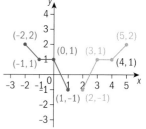

3 Given the graph of $g(x)$, make separate sketches of the following transformations of $g(x)$. In each sketch include $g(x)$. Label the coordinates of the image points under the transformation.

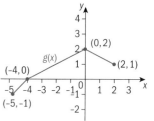

a $g(x + 1) - 2$ **b** $-2g(x)$ **c** $g(2x)$

d $\dfrac{1}{2}g(x)$ **e** $g\left(\dfrac{1}{2}x\right) + 1$

4 On a separate diagram for each pair, sketch the graphs of:

a $f(x) = x^2$, $y = -\dfrac{1}{2}f\left(\dfrac{x}{3} - 1\right) + 2$

b $f(x) = |x|$, $y = 2|2x + 3| - 2$

c $f(x) = x(x - 1)(x + 1)$, $y = f(|x|) + 25$

5 Find the rational function when $y = \dfrac{1}{x}$ is transformed by a vertical stretch of 2, then stretched horizontally by a factor of $\dfrac{1}{3}$, followed by a translation of $\begin{pmatrix} -2 \\ 3 \end{pmatrix}$. Find the domain and range of the new function.

6 Consider $f(x) = \dfrac{4x + 5}{2x + 1}$

a Find the asymptotes of $y = f(x)$

b Find the intercepts

c Sketch the graph

d Describe the transformation that changes $y = \dfrac{1}{x}$ to $y = f(x)$

Extension material on CD:
Worksheet 2

Review exercise

1 Determine which of the following is a graph of a function. For those that are functions, determine the domain and range.

a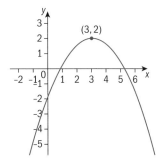

b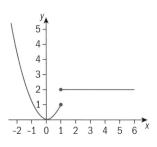

c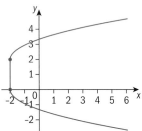

d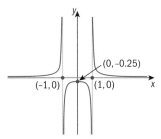

2 If $f(2) = 3$, $g(4) = 2$, and $h(3) = 4$, find $f(g(h(3)))$. Assume the inverses of f, g, and h exist, and find $h^{-1}(g^{-1}(f^{-1}(3)))$.

3 If $f(x) = \dfrac{1}{x-1}$, find $(f \circ f)(x)$, and $(f \circ f)^{-1}(x)$.

EXAM-STYLE QUESTIONS

4 Given the graph of $f(x)$, find the images of the given points under the transformation $y = -2f(x + 1) - 3$

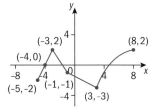

5 The graph of $y = f(x)$ is shown below.

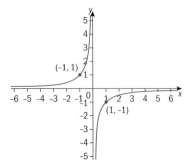

Using this graph, sketch the graphs of the following, indicating the image of the given points in each sketch.

a $y = f(x + 1)$ b $y = f(x - 1) - 2$
c $y = 1 - f(x)$ d $y = 1 + f(-x)$
e $y = 2f(2x)$

6 The graphs show the functions $y = f(x)$ (red line) and a
transformed graph $y = g(x)$ (green line). Express g in terms of f.

a

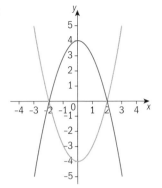

b

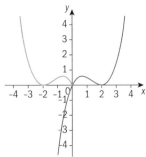

c

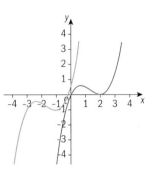

d

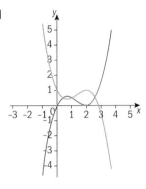

e

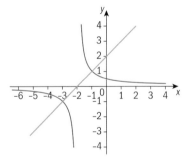

f

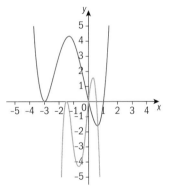

g

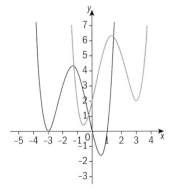

7 The domain of an odd function f is the set of real numbers.
Prove that the graph passes through the origin.

Review exercise

1 $f(x) = 3x - 1$, $g(x) = x^2$, $h(x) = \dfrac{1}{(x+2)}$

Find these composite functions.

a $f(g(x))$ **b** $h(g(x))$ **c** $g^{-1}(f(x))$ **d** $f(h^{-1}(x))$

2 Find two functions, f and g, such that $g(f(x)) = \left(\dfrac{x-1}{x+3}\right)^2$

3 If $f(x) = |x|$ and $g(x) = 4x^2 + 2x - 5$, sketch the graphs of $g(f(x))$ and $f(g(x))$.

4 The function $C = \dfrac{5}{9}(F - 32)$ describes degrees in Centigrade as a function of degrees in Fahrenheit. Describe the transformations necessary on this function to arrive at a function of F in terms of C, and find the values for which the two functions are the same.

EXAM-STYLE QUESTION

5 Find the rational function $y = f(x)$ which is the result of transforming the reciprocal function $f: x \mapsto \dfrac{1}{x}$ by a vertical stretch factor of 3, followed by a horizontal stretch factor of $\dfrac{1}{2}$, followed by a translation of $\begin{pmatrix} -1 \\ -2 \end{pmatrix}$.

CHAPTER 2 SUMMARY

- A function is a set of ordered pairs in which no two ordered pairs can have the same x-value. In other words, every x-value has a unique y-value.
- A relation is a function, f, if
  - f acts on all elements of the domain (x-values), and
  - f is well defined, i.e. it pairs each element of the domain with one and only one element of the range (y-values).
- In general, if y is a function of x, you can write $y = f(x)$. You can also write $f: x \mapsto f(x)$, where f is the function that maps x into $f(x)$. The independent variable, x, is called the **argument** of the function.
- In the form $y = ax^2 + bx + c$, the axis of symmetry is $x = \dfrac{-b}{2a}$, hence the vertex is $\left(\dfrac{-b}{2a}, f\left(\dfrac{-b}{2a}\right)\right)$.
- A function is even if, for all x in the domain, $-x$ is in the domain and $f(x) = f(-x)$ for all values of x.

Continued on next page

- A function is odd if, for all x in the domain, $-x$ is in the domain and $f(-x) = -f(x)$ for all values of x.
- Let f and g be two real-valued functions. Then
  - $(f + g)(x) = f(x) + g(x)$
  - $(f - g)(x) = f(x) - g(x)$
  - $a(f(x)) = af(x)$, $a \in \mathbb{R}$
  - $(fg)(x) = f(x)g(x)$
  - $\left(\dfrac{f}{g}\right) x = \dfrac{f(x)}{g(x)}, g(x) \neq 0$
- Two functions g and h are inverses of each other if their composition results in the identity function, $f(x) = x$, i.e., $(g \circ h)(x) = (h \circ g)(x) = x$. The functions g and h are also said to be invertible functions.
- The graph of a function and its inverse are reflections in the line $y = x$
- Only functions that are one-to-one have inverse functions.
- In general, the graphs of $f(x)$ and $-f(x)$ are symmetrical about the x-axis.
- In general, the graphs of $f(x)$ and $f(-x)$ are symmetrical about the y-axis.
- Where they exist, the zeros of $f(x)$ are the vertical asymptotes of $\dfrac{1}{f(x)}$, and the zeros of $\dfrac{1}{f(x)}$ are the vertical asymptotes of $f(x)$.
- If $y = b$ is the y-intercept of $f(x)$, then $\dfrac{1}{b}$ is the y-intercept of $\dfrac{1}{f(x)}$
- The minimum of $f(x)$ occurs at the same x-value as the maximum of $\dfrac{1}{f(x)}$, and the minimum of $\dfrac{1}{f(x)}$ occurs at the same x-value as the maximum of $f(x)$.
- When $f(x) > 0$, $\dfrac{1}{f(x)} > 0$; when $f(x) < 0$, $\dfrac{1}{f(x)} < 0$
- When $f(x)$ approaches 0, $\dfrac{1}{f(x)}$ will approach $\pm \infty$; when $f(x)$ approaches $\pm \infty$, $\dfrac{1}{f(x)}$ approaches 0.
- For $y = af(x)$, $a > 0$, the graph is vertically stretched or compressed by a factor of a.
  - If $a > 1$ the graph is stretched vertically by a factor of a, i.e., it moves further from the x-axis.
  - If $0 < a < 1$ the graph is compressed vertically by a factor of a, i.e., it moves closer to the x-axis.
- For $y = f(ax)$, $a > 0$, the graph is horizontally stretched or compressed.
  - If $a > 1$ the graph is compressed by a factor of $\dfrac{1}{a}$, i.e., the graph moves closer to the y-axis.
  - If $0 < a < 1$ the graph is stretched by a factor of $\dfrac{1}{a}$, i.e., the graph moves further from the y-axis.

The language of mathematics

A universal language?

In the science fiction movie *Contact*, the heroine played by Jodie Foster explains that aliens would use sequences of prime numbers as their initial attempt at communication because "Mathematics is the only truly universal language".

The language of mathematics is certainly one of the oldest, and has been the international language of scientists for many centuries.

It is unique in that it describes not only the real world, but also abstract structures, and the mathematics itself.

Language:
a systematic means of communicating ideas using conventionalized signs or marks having understood meanings.

- Does the language of mathematics have its own grammar, vocabulary and structure?

- Is it truly a 'language', or simply the manipulation of symbols following a set of rules?

"The miracle of the appropriateness of the language of mathematics for the formulation of the laws of physics is a wonderful gift which we neither understand nor deserve."

E.P. Wigner, Hungarian–American physicist and mathematician (1902–95)

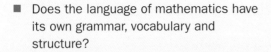

Evolving language

New words are added to the dictionary every year. Some are genuine new words – such as 'app' and 'blog'. Some are old words with a new meaning – such as 'tweet' or 'scroll'.

Languages evolve over time.

- How have new discoveries in mathematics, or new applications of mathematics, changed the language of mathematics?

Using language to convey knowledge

In *The Chicken From Minsk*, by Y. Chernyak and R. Rose, the owner of a pet shop, a retired mathematician, tells a customer that a parrot is extremely intelligent and repeats every word it hears. The customer buys the parrot, but returns a few days later wanting his money back. He claims not to have heard a single word from the 'stupid' parrot.

The pet shop owner will not return the money because he never lies (a true claim).

- What if the bird is deaf?
- Did the shop owner say when the bird would repeat what it had heard?
- Is the customer telling the truth?

Perhaps the bird is much more intelligent than the owner, and decides not to repeat such boring conversation.

- In the possible explanations you have considered:
  - ▶ Which would a physicist describe as 'no initial condition'?
  - ▶ Which would a mathematician describe as a translation along the *t* (time) axis?
- How was the cultural language in the story imprecise? Is mathematics a precise language?
- Describe some other situations where the lack of precision in cultural languages is an obstacle to gaining knowledge.
- How is the language of mathematics similar to cultural languages? And how is it different?

> *"To those who do not know Mathematics it is difficult to get across a real feeling as to the beauty, the deepest beauty of nature. … If you want to learn about nature, to appreciate nature, it is necessary to understand the language that she speaks in."*
>
> Richard Feynman, American physicist (1918–88).

You saw this quotation at the start of this chapter:

"Mathematics is the language with which God has written the universe." Galileo Galilei

- Is the universe mathematical?
- Has the language of mathematics been invented or discovered?

3

The long journey of mathematics

CHAPTER OBJECTIVES:

1.5 Complex numbers: the number $i = \sqrt{-1}$; the terms real part, imaginary part, conjugate, modulus and argument; Cartesian form $z = a + ib$; sums, products and quotients of complex numbers

1.6 The complex plane

1.7 Powers of complex numbers; nth roots of a complex number

1.8 Conjugate roots of polynomial equations with real coefficients

1.9 Solutions of systems of linear equations (a maximum of three equations in three unknowns), including cases where there is a unique solution, an infinity of solutions or no solution

2.5 Polynomial functions and their graphs; the factor and remainder theorems; the fundamental theorem of algebra

2.6 Solving quadratic equations using the quadratic formula; use of the discriminant $\Delta = b^2 - 4ac$ to determine the nature of the roots; solving polynomial equations both graphically and algebraically; sum and product of the roots of polynomial equations

2.7 Solutions of $g(x) \geq f(x)$: graphical or algebraic methods, for simple polynomials up to degree 3; use of technology for these and other functions

Before you start

You should know how to:

1 Solve quadratic equations by factorization.
e.g. $x^2 - 3x - 4 = 0$
$\Rightarrow (x-4)(x+1) = 0$
$\Rightarrow x = 4$ or $x = -1$

2 Find a linear combination of two polynomials.
e.g. $f(x) = x^2 - 3x + 1$ and
$g(x) = x^3 + 7x - 3$
$5f(x) + 2g(x) = 5(x^2 - 3x + 1) + 2(x^3 + 7x - 3)$
$= 2x^3 + 5x^2 - x - 1$

Skills check

1 Solve these quadratic equations:
a $x^2 + 2x - 3 = 0$ **b** $x^2 - 11x + 10 = 0$
c $2x^2 + x - 3 = 0$

2 Given the polynomials $f(x) = x^2 - 3x + 1$, $g(x) = 2x^3 - x^2 + 3x - 4$ and $h(x) = 3x^4 - 2x^2 - 5$, find:
a $f(x) + g(x)$
b $2h(x) - 4g(x) + 5f(x)$
c $\frac{1}{2}h(x) - \frac{2}{5}g(x)$

Important problems that challenged great minds

The Italian mathematician Leonardo of Pisa, best known as Fibonacci, made his most important contribution to mathematics by spreading the use of the Hindu-Arabic numeral system throughout Europe. In the next centuries, first in Italy, and then in other parts of Europe, bursts of mathematical creativity led to incredible developments and discoveries in mathematics and science in general.

Over the centuries generations of mathematicians have helped the scientific community to achieve great insight into nature, moving us forward in our understanding of the world and allowing the remarkable development of science and technology. Throughout this history, scientific progress has always been related to revolutions in mathematical thought.

In this chapter we are going to take a close look at the evolution of the most fundamental mathematical concept – the concept of number. Using modern methods we are going to discover and explore the properties of a new set of numbers. These are the set of complex numbers.

3.1 Introduction to complex numbers

Solving quadratic equations using the quadratic formula

Zero is in many ways a mysterious number. Medieval mathematicians could not decide whether or not it really was a number! Nowadays, however, zero has high status in mathematics due to its algebraic properties. One is the zero factor property, that can be used to solve some polynomial equations.

→ Zero factor property: $a \times b = 0 \Rightarrow a = 0$ or $b = 0$

A quadratic equation has the form, $ax^2 + bx + c = 0$, where $a, b, c \in \mathbb{R}$, and $a \neq 0$. When one of the coefficients is zero there is a special case that you can solve without using the general quadratic formula.

Special cases:

i $b = 0, c \neq 0 \Rightarrow ax^2 + c = 0$

$$\Rightarrow x^2 = -\frac{c}{a}$$

$$\Rightarrow x = \pm\sqrt{-\frac{c}{a}} \Rightarrow x = -\sqrt{-\frac{c}{a}} \quad \text{or} \quad x = \sqrt{-\frac{c}{a}}$$

> The solutions are real and opposite if $-\frac{c}{a} > 0$.
>
> When $-\frac{c}{a} < 0$ the solutions are not real.

If a function vanishes for a particular value of its argument, $f(x) = 0$, then x is called a zero or root of $f(x)$.

ii $b \neq 0, c = 0 \Rightarrow ax^2 + bx = 0$ *Factorize and apply the zero product*
$$\Rightarrow x(ax + b) = 0 \qquad\qquad property.$$

$$\Rightarrow x = 0 \quad \text{or} \quad x = -\frac{b}{a}$$

> The solutions are always real and distinct and one is always zero.

iii $b = 0, c = 0 \Rightarrow ax^2 = 0$
$$\Rightarrow x^2 = 0$$
$$\Rightarrow x = 0 \quad \text{or} \quad x = 0$$

> This is the only case where there is only one (double) real solution – which is zero.

The method for finding a **general formula for the solutions of a quadratic equation** is called 'completing the square'. This method can be used directly as in case **i**, or again by factorization.

Method I: Completing the square

$ax^2 + bx + c = 0$ *Divide the equation by a.*
$$\Rightarrow x^2 + \frac{b}{a}x + \frac{c}{a} = 0$$

$$\Rightarrow x^2 + 2 \cdot x \cdot \frac{b}{2a} + \left(\frac{b}{2a}\right)^2 = \left(\frac{b}{2a}\right)^2 - \frac{c}{a}$$ *Add $\left(\dfrac{b}{2a}\right)^2$ to both sides in*

order to apply the formula

$$(A \pm B)^2 = A^2 \pm 2AB + B^2$$

See Chapter 14, section 2.2

$$\Rightarrow \left(x + \frac{b}{2a}\right)^2 = \frac{b^2}{4a^2} - \frac{c}{a}$$ *Factorize and simplify.*

$$\Rightarrow x + \frac{b}{2a} = \pm\sqrt{\frac{b^2 - 4ac}{4a^2}}$$

Take the square root of both sides and simplify.

$$\Rightarrow x = -\frac{b}{2a} \pm \frac{\sqrt{b^2 - 4ac}}{2a}$$

$$\Rightarrow x = \frac{-b \pm \sqrt{b^2 - 4ac}}{2a}$$

$$\Rightarrow x = \frac{-b - \sqrt{b^2 - 4ac}}{2a} \quad \text{or} \quad x = \frac{-b + \sqrt{b^2 - 4ac}}{2a}$$

Method II: Completing the square and factorization

$$ax^2 + bx + c = 0$$
$$\Rightarrow a^2x^2 + abx + ac = 0$$

Multiply the equation by a.

Add and subtract

$$\Rightarrow \overbrace{(ax)^2 + 2 \cdot ax \cdot \frac{b}{2} + \left(\frac{b}{2}\right)^2}^{\text{perfect square}} - \left(\frac{b}{2}\right)^2 + ac = 0$$

$\left(\dfrac{b}{2}\right)^2$ *in order to apply the formula*

$$\Rightarrow \left(ax + \frac{b}{2}\right)^2 - \frac{b^2 - 4ac}{4} = 0$$

(A ± B)²
= A² ± 2AB + B²

$$\Rightarrow \left(ax + \frac{b}{2}\right)^2 - \left(\frac{\sqrt{b^2 - 4ac}}{2}\right)^2 = 0$$

Apply A² − B²
= (A − B) (A + B).

$$\Rightarrow \left(ax + \frac{b}{2} - \frac{\sqrt{b^2 - 4ac}}{2}\right)\left(ax + \frac{b}{2} + \frac{\sqrt{b^2 - 4ac}}{2}\right) = 0$$

Apply the zero product property.

either $ax + \dfrac{b}{2} - \dfrac{\sqrt{b^2 - 4ac}}{2} = 0$

$$\Rightarrow ax = -\frac{b}{2} + \frac{\sqrt{b^2 - 4ac}}{2} \quad \Rightarrow x = \frac{-b + \sqrt{b^2 - 4ac}}{2a}$$

Solve for x and simplify.

or $ax + \dfrac{b}{2} + \dfrac{\sqrt{b^2 - 4ac}}{2} = 0$

$$\Rightarrow ax = -\frac{b}{2} - \frac{\sqrt{b^2 - 4ac}}{2} \quad \Rightarrow x = \frac{-b - \sqrt{b^2 - 4ac}}{2a}$$

Solve for x and simplify.

→ You can use the **quadratic formula** $x = \dfrac{-b \pm \sqrt{b^2 - 4ac}}{2a}$ to find the **solutions** or **roots** of a quadratic equation.

The Babylonians (2000–1600 BCE) knew how to solve a quadratic equation by using a quadratic formula in a slightly different form from the one we use today. They were essentially using the standard formula in two different types of quadratic equation $x^2 + bx = c$ and $x^2 - bx = c$, where b and c were positive but not necessarily integers.

Why did the Babylonians need to consider two different types of quadratic equations? You may wish to explore their methods for solving these equations and their contributions to the progress of mathematics.

Example 1

Use the quadratic formula to solve these equations. Check your answers with a GDC.

a $3x^2 + 11x + 6 = 0$

b $5x^2 - 9x - 3 = 0$

c $3px^2 + (p - 6)x - 2 = 0$

Answers

a $\underset{a}{3x^2} + \underset{b}{11x} + \underset{c}{6} = 0$ $\qquad \Rightarrow x = \dfrac{-11 \pm \sqrt{11^2 - 4 \cdot 3 \cdot 6}}{2 \cdot 3}$

$$= \dfrac{-11 \pm \sqrt{121 - 72}}{6}$$

$$= \dfrac{-11 \pm \sqrt{49}}{6}$$

$$= \dfrac{-11 \pm 7}{6}$$

$$\Rightarrow x = \dfrac{-11 - 7}{6} = -3 \quad \text{or} \quad x = \dfrac{-11 + 7}{6} = -\dfrac{2}{3}$$

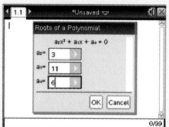 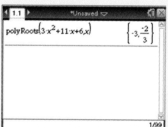

b $\underset{a}{5x^2} - \underset{b}{9x} - \underset{c}{3} = 0$ $\qquad \Rightarrow x = \dfrac{-(-9) \pm \sqrt{(-9)^2 - 4 \cdot 5 \cdot (-3)}}{2 \cdot 5}$

$$= \dfrac{9 \pm \sqrt{81 + 60}}{10}$$

$$= \dfrac{9 \pm \sqrt{141}}{10}$$

$$\Rightarrow x = \dfrac{9 - \sqrt{141}}{10} \quad \text{or} \quad x = \dfrac{9 + \sqrt{141}}{10}$$

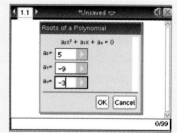 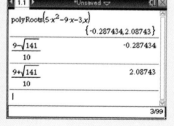

GDC help on CD:
Alternative demonstrations for the TI-84 Plus and Casio fx-9860GII GDCs are on the CD.

▶ Continued on next page

c $\underbrace{3px^2}_{a} + \underbrace{(p-6)x}_{b} \underbrace{-2}_{c} = 0$

$$\Rightarrow x = \frac{-(p-6) \pm \sqrt{(p-6)^2 - 4 \cdot 3p \cdot (-2)}}{2 \cdot 3p}$$

$$= \frac{-(p-6) \pm \sqrt{p^2 - 12p + 36 + 24p}}{6p}$$

$$= \frac{-(p-6) \pm \sqrt{p^2 + 12p + 36}}{6p}$$

$$= \frac{-(p-6) \pm \sqrt{(p+6)^2}}{6p} = \frac{-(p-6) \pm (p+6)}{6p}$$

$$\Rightarrow x = \frac{-p+6-p-6}{6p} = -\frac{1}{3} \quad \text{or} \quad x = \frac{-p+6+p+6}{6p} = \frac{2}{p}$$

> This problem cannot be solved by a GDC because it requires a Computer Algebra System.

Exercise 3A

1 Solve these quadratic equations, giving your answers in exact form.

    a $2x^2 - 3x = 0$ **b** $3x^2 - 75 = 0$

    c $5x^2 - 4x = 0$ **d** $7 + 28x^2 = 0$

    e $242x^2 + 2x = 0$ **f** $\sqrt{2}x^2 - \sqrt{8} = 0$

    g $\pi x^2 - 11x = 0$ **h** $ex^2 - \sqrt{3} = 0$

2 Use the quadratic formula to solve these equations. Check your answers with a GDC.

    a $2x^2 + 5x + 2 = 0$ **b** $3x^2 - 10x + 3 = 0$

    c $5x^2 + 3x - 2 = 0$ **d** $21x^2 + 5x - 6 = 0$

    e $9x^2 - 6x + 35 = 0$ **f** $122x = 143x^2 + 24$

3 Solve these equations and write the solutions in exact form. Check your answers with a GDC.

    a $x^2 + 4x + 2 = 0$ **b** $5x^2 - 6x - 1 = 0$

    c $3x^2 - x - 3 = 0$ **d** $2x^2 + 11x + 13 = 0$

    e $11x^2 = 23x - 7$ **f** $29x = 5x^2 - 41$

4 Solve for x:

    a $x^2 + px - 2p^2 = 0$ **b** $kx^2 + (k+2)x + 2 = 0$

    c $2ax^2 + 6 = ax + 12x$ **d** $x^2 - 2a^2 = b^2 - ax - 3ab$

Discriminant of a quadratic equation

A quadratic equation can have:

- two real roots
- one repeated real root
- no real roots.

Example 2

Solve these equations.

a $3x^2 + 5x - 2 = 0$

b $4x^2 + 12x + 9 = 0$

c $5x^2 + x + 4 = 0$

Answers

a $3x^2 + 5x - 2 = 0 \Rightarrow x = \dfrac{-5 \pm \sqrt{(5)^2 - 4 \cdot 3 \cdot (-2)}}{2 \cdot 3} = \dfrac{-5 \pm \sqrt{25 + 24}}{6} = \dfrac{-5 \pm \sqrt{49}}{6}$

$\Rightarrow x = \dfrac{-5 - 7}{6} = -2 \ \text{ or } \ x = \dfrac{-5 + 7}{6} = \dfrac{1}{3}$

b $4x^2 + 12x + 9 = 0 \Rightarrow x = \dfrac{12 \pm \sqrt{(-12)^2 - 4 \cdot 4 \cdot 9}}{2 \cdot 4}$

$= \dfrac{-12 \pm \sqrt{144 - 144}}{8} = \dfrac{-12 \pm \sqrt{0}}{8} = -\dfrac{3}{2}$

c $5x^2 + x + 4 = 0 \Rightarrow x = \dfrac{-1 \pm \sqrt{1^2 - 4 \cdot 5 \cdot 4}}{2 \cdot 5} = \dfrac{-1 \pm \sqrt{1 - 80}}{10}$

$\Rightarrow x = \dfrac{-1 \pm \sqrt{-79}}{10} \notin \mathbb{R}$

Investigation – the general quadratic function

A general quadratic function can be written $y = ax^2 + bx + c$, with $a, b, c \in \mathbb{R}$, $a \neq 0$. By using completing the square find the location of the minimum ($a > 0$) or maximum ($a < 0$) point on this curve. Hence, or otherwise, find the conditions on the coefficients a, b, c which determine how many solutions there are to the equation $ax^2 + bx + c = 0$.

The nature of the **roots** in Example 2 depends on the expression under the square root, that is, $b^2 - 4ac$. The expression $\Delta = b^2 - 4ac$ is called the **discriminant** because it acts to *discriminate* between the three different types of solutions.

> The symbol used for the discriminant $b^2 - 4ac$ is the Greek letter Δ (delta).

i $\Delta = b^2 - 4ac > 0$

If the discriminant is positive, you can add $\sqrt{b^2 - 4ac}$ to $-b$ and subtract $\sqrt{b^2 - 4ac}$ from $-b$. In this case, you obtain two different numbers so there are **two distinct real roots**.

ii $\Delta = b^2 - 4ac = 0$

If the discriminant is equal to zero, adding zero to $-b$ and subtracting zero from $-b$ gives the same solution so there is **one repeated real root**.

iii $\Delta = b^2 - 4ac < 0$

If the discriminant is less than zero, the expression under the square root is negative, and therefore the square root is not a real number. There are **no real roots**.

> Why do we use Greek letters to represent so many quantities in mathematics? You may wish to explore the ancient Greeks' contributions to number, geometry or algebra.

Example 3

Without solving the equations, determine the nature of their roots.

a $x^2 - x + 1 = 0$

b $3x^2 + 30x - 75 = 0$

c $5x^2 + 4x - 1 = 0$

Answers

a $x^2 - x + 1 = 0 \Rightarrow$ $\Delta = (-1)^2 - 4 \cdot 1 \cdot 1 = 1 - 4 = -3 < 0$ No real roots.	*Find the discriminant* $\Delta < 0$
b $3x^2 + 30x + 75 = 0 \Rightarrow$ $\Delta = 30^2 - 4 \cdot 3 \cdot 75 = 900 - 900 = 0$ One real root.	*Find the discriminant* $\Delta = 0$
c $5x^2 + 4x - 1 = 0 \Rightarrow$ $\Delta = 4^2 - 4 \cdot 5 \cdot (-1) = 16 + 20 = 36 > 0$ Two real roots.	*Find the discriminant* $\Delta > 0$

Example 4

Find the value(s) of the real parameter m so that:

a $x^2 - 6x + m = 0$ has two real roots

b $x^2 - mx + m - 1 = 0$ has one repeated real root

c $mx^2 + (2m - 1)x + 1 = 0$ has no real roots

Answers

a $x^2 - 6x + m = 0 \Rightarrow$ $\Delta = (-6)^2 - 4 \cdot 1 \cdot m$ $\Delta = 36 - 4m$ $36 - 4m > 0$ $36 > 4m \Rightarrow m < 9$	*Find the discriminant* *Simplify Δ and set $\Delta > 0$* *Solve the inequality for m*

▶ Continued on next page

$x^2 - mx + m - 1 = 0 \Rightarrow$

$\Delta = (-m)^2 - 4 \cdot 1 \cdot (m - 1)$ *Find the discriminant*

$\Delta = m^2 - 4m + 4$

$m^2 - 4m + 4 = 0$ *Set $\Delta = 0$*

$(m - 2)^2 = 0 \Rightarrow m = 2$ *Solve the equation for m*

c $mx^2 - (2m - 1)x + m = 0 \Rightarrow$

$\Delta = (2m - 1)^2 - 4 \cdot m \cdot m$ *Find the discriminant*

$\Delta = 4m^2 - 4m + 1 - 4m^2 \Rightarrow$ *Simplify Δ and set $\Delta < 0$*

$1 - 4m < 0$

$1 < 4m \Rightarrow m > \dfrac{1}{4}$ *Solve the inequality for m*

Exercise 3B

1 Without solving the equations, determine the nature of the roots.

  a $x^2 - 2x - 3 = 0$ **b** $x^2 + 10x + 25 = 0$

  c $4x^2 - 3x + 2 = 0$ **d** $5x^2 - 11x + 6 = 0$

  e $\dfrac{3}{5}x^2 - \dfrac{4}{7}x + \dfrac{2}{3} = 0$ **f** $2x^2 + 2\sqrt{26}x + 13 = 0$

2 Find the value(s) of the real parameter k so that:

  a $x^2 - 2x - k = 0$ has one real root

  b $kx^2 + 3x - 2 = 0$ has two real roots

  c $3x^2 + 5x + 2k - 1 = 0$ has no real roots

  d $x^2 - (3k + 2)x + k^2 = 0$ has one real root

  e $kx^2 + 2kx + k - 2 = 0$ has two real roots

  f $2kx^2 + (4k + 3)x + k - 3 = 0$ has no real roots

Sum and product of roots of a quadratic equation

Investigation – Viète's theorem

A general quadratic equation $ax^2 + bx + c = 0$, with $a, b, c \in \mathbb{R}$, $a \neq 0$ has two solutions, x_1 and x_2. By using the quadratic formula find expressions for the sum, $x_1 + x_2$, and product, $x_1 \cdot x_2$, of the two roots in terms of the coefficients a, b, c.

François Viète
(1540–1603)
discovered a relationship between the parameters a, b and c of a quadratic equation and the solutions x_1 and x_2.

The expressions you found in the investigation are known as Viète's theorem.

→ For a quadratic equation $ax^2 + bx + c = 0$, $a, b, c \in \mathbb{R}$, $a \neq 0$ and solutions x_1 and x_2, then the sum of the roots,

$x_1 + x_2 = -\dfrac{b}{a}$ and the product of the roots, $x_1 \cdot x_2 = \dfrac{c}{a}$

Example 5

The roots of a quadratic equation $3x^2 - 5x + 2 = 0$ are x_1 and x_2.
Without solving the equation, find:

a $\dfrac{1}{x_1} + \dfrac{1}{x_2}$ **b** $x_1^2 + x_2^2$ **c** $\dfrac{2}{x_1^3} + \dfrac{2}{x_2^3}$

Answers

a $\dfrac{1}{x_1} + \dfrac{1}{x_2} = \dfrac{x_2 + x_1}{x_1 \cdot x_2}$

$= \dfrac{\frac{5}{3}}{\frac{2}{3}} = \dfrac{5}{2}$

Apply the theorem:

$x_1 + x_2 = \dfrac{5}{3}$ *and* $x_1 \cdot x_2 = \dfrac{2}{3}$

b $x_1^2 + x_2^2 = (x_1 + x_2)^2 - 2x_1 x_2$

$= \left(\dfrac{5}{3}\right)^2 - 2 \cdot \dfrac{2}{3}$

$= \dfrac{25}{9} - \dfrac{4}{3} = \dfrac{13}{9}$

Use the binomial formula
$(A + B)^2 \equiv A^2 + 2AB + B^2$

c $\dfrac{2}{x_1^3} + \dfrac{2}{x_2^3}$

$= 2 \cdot \dfrac{x_2^3 + x_1^3}{x_1^3 x_2^3}$

$= 2 \cdot \dfrac{(x_1 + x_2)^3 - 3x_1 x_2 (x_1 + x_2)}{(x_1 x_2)^3}$

$= 2 \cdot \dfrac{\left(\dfrac{5}{3}\right)^3 - 3 \cdot \dfrac{2}{3} \cdot \dfrac{5}{3}}{\left(\dfrac{2}{3}\right)^3}$

$= 2 \cdot \dfrac{\dfrac{125}{27} - \dfrac{10}{3}}{\dfrac{8}{27}} = 2 \cdot \dfrac{\dfrac{35}{27}}{\dfrac{8}{27}} = \dfrac{35}{4}$

Use the binomial formula
$(A + B)^3 \equiv A^3 + 3A^2B + 3AB^2 + B^3$

> The binomial formula is discussed in Section 1.7

Exercise 3C

1 Given a quadratic equation whose roots are x_1 and x_2, find the indicated expression without solving the equation.

a $x^2 - 3x + 2 = 0$, $\dfrac{2}{x_1} + \dfrac{2}{x_2}$

b $3x^2 - 5x + 1 = 0$, $3x_1^2 + 3x_2^2$

c $5x^2 + x + 3 = 0$, $\dfrac{1}{x_1^2} + \dfrac{1}{x_2^2}$

d $x^2 - 2x + 4 = 0$, $(x_1 - x_2)^2$

e $2x^2 - 4x + 3 = 0$, $x_1^3 + x_2^3$

f $x^2 + 3x + 1 = 0$, $\dfrac{1}{x_1^4} + \dfrac{1}{x_2^4}$

g $4x^2 - 7x + 1 = 0$, $x_1^3 x_2^2 + x_1^2 x_2^3$

h $7x^2 + 4x - 5 = 0$, $(x_1 - x_2)^4$

Algebraic vs. geometric introduction to complex numbers

Algebraic approach

Historically, complex numbers were first encountered when solving cubic equations. However, in modern mathematics, these numbers appear naturally as solutions of quadratic equations as we shall see in this section.

Since the square of a real number is always a non-negative number, a quadratic equation of the form $x^2 = c$, $c \in \mathbb{R}^-$ has no real solution. If you say that the simplest such equation $x^2 = -1$ has solutions you can develop a whole new algebra starting from $x = \pm\sqrt{-1}$.

The first person to mention the square root of a negative number was **Heron of Alexandria** (c.10–c.60 CE) when discussing the volume of frustum of a pyramid whose side lengths were impossible.

In medieval Italy, mathematical tournaments were very popular and solving cubic equations distinguished the winners. These mathematicians discovered the formula for solutions of cubic equations and basically introduced complex numbers.

Scipione dal Ferro (1465–1526) solved a cubic equation with no quadratic term which helped **Niccolò Fontana Tartaglia** (1499–1557) to discover the formula. He shared his knowledge with **Gerolamo Cardano** (1501–1576) who published it in his algebra book *Ars Magna*. Cardano introduced complex numbers of the form $a + \sqrt{-b}$, $a \in \mathbb{R}$, $b \in \mathbb{R}^+$. Mathematicians realised that the two parts could not be combined and the second part was called an imaginary or even impossible part.

René Descartes (1596–1650) was the first person to establish the term imaginary part and **John Wallis** (1616–1703) made huge progress in giving a geometric interpretation to $\sqrt{-1}$.

Leonhard Euler (1707–1783) was the first mathematician to use the symbol $i = \sqrt{-1}$ and he called it an **'imaginary unit'**.

Today complex numbers are used in many real world applications.

You can write all the solutions of the equation $x^2 = c$, $c \in \mathbb{R}^-$ as $x = \pm i\sqrt{-c} = \pm id$, $d \in \mathbb{R}^+$. Numbers like $\pm id$ are purely imaginary numbers.

Complex numbers have the form $z = a + ib$, $a, b \in \mathbb{R}$, where a is called a **real part**, $\mathrm{Re}(z) = a$, and b is called an **imaginary part**, $\mathrm{Im}(z) = b$, of the complex number z.

Dose the terminology 'complex' and 'imaginary' make these numbers seem unnatural? Are they simply the inventions of mathematical minds?

$$c = -1 \times -c$$
$$\sqrt{c} = \sqrt{-1 \times -c}$$
$$= \pm i\sqrt{-c}$$

c is negative so $-c$ is positive and has a real square root.
$$\pm\sqrt{-c} = \pm\sqrt{d} \text{ where}$$
$d \in \mathbb{R}^+$

→ When $b = 0$, $z = a + i \cdot 0 = a$. Since the complex number does not have a part containing i, it reduces to a real number. Similarly, when $a = 0$, $z = 0 + ib = ib$. Since the complex number has only a part containing the imaginary unit i, it is called a purely imaginary number.

Example 6

Find the real and imaginary parts of these complex numbers.

a $z = 3 + 2i$

b $z = 5i - 4$

c $z = -\dfrac{2}{3} + \sqrt{3}i$

d $z = \dfrac{\sqrt[3]{11}i - \sqrt{11}}{\pi}$

Answers

a $z = 3 + 2i \Rightarrow \begin{cases} \mathrm{Re}(z) = 3 \\ \mathrm{Im}(z) = 2 \end{cases}$

b $z = 5i - 4 \Rightarrow \begin{cases} \mathrm{Re}(z) = -4 \\ \mathrm{Im}(z) = 5 \end{cases}$

c $z = -\dfrac{2}{3} + \sqrt{3}i \Rightarrow \begin{cases} \mathrm{Re}(z) = -\dfrac{2}{3} \\ \mathrm{Im}(z) = \sqrt{3} \end{cases}$

d $z = \dfrac{\sqrt[3]{11}i + \sqrt{23}}{\pi} \Rightarrow \begin{cases} \mathrm{Re}(z) = \dfrac{\sqrt{23}}{\pi} \\ \mathrm{Im}(z) = \dfrac{\sqrt[3]{11}}{\pi} \end{cases}$

Geometric approach

Real numbers can be visualised on the number line that was introduced by John Wallis. Each point on the line represents one real number. In order to have numbers other than real numbers, we need to expand the line into the second dimension, which results in the complex plane.

> The first person to set up the plane model of complex numbers was **Jean-Robert Argand** (1768–1822). **Carl Friedrich Gauss** (1777–1855) independently developed and refined the plane model and therefore the geometrical visualization of complex numbers in a plane is known as an **Argand diagram** or **Gaussian plane**.

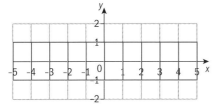

The complex plane is a two-dimensional coordinate plane where the usual coordinate axes x and y are now called the real and imaginary axes respectively. Each complex number $z = x + iy$ is represented by a point P(x, y) in the plane where the coordinates are the real and imaginary parts of the complex number itself.

Example 7

Plot these complex numbers in the Argand diagram.
$3 + 2i$, $2 - i$, $-3 - 3i$, $-4 + i$, $3i$ and -2.

Answer

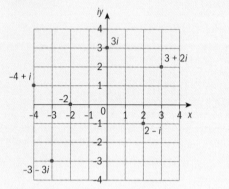

> The real part is measured along the real axis (horizontal axis) and the imagniary part along the imaginary axis (vertical axis).

Modulus of a complex number

You saw in Chapter 2 that the modulus, or absolute value, of a real number was algebraically defined as $|x| = \begin{cases} x, & x \geq 0 \\ -x, & x < 0 \end{cases}$. Geometrically it represents the distance from the number x on the number line to the origin 0. You can extend this idea to complex numbers: the modulus of a complex number $|z|$ is the distance from the point $P(x, y)$ (which represents the complex number $z = x + iy$) to the origin $(0, 0)$ in the complex plane.

To find the distance between two points in a coordinate plane use Pythagoras' theorem.

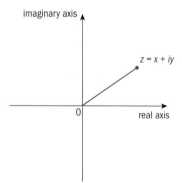

$$|z| = \sqrt{(x-0)^2 + (y-0)^2} = \sqrt{x^2 + y^2} = \sqrt{\mathrm{Re}^2(z) + \mathrm{Im}^2(z)}$$

The geometric interpretation will be discussed further in Chapter 12.

$$\rightarrow |z| = |x + iy| = \sqrt{x^2 + y^2}$$

Example 8

Find the modulus of these complex numbers.

a $3 - 4i$ **b** $-7 + \sqrt{11}i$ **c** $\dfrac{-5 - 12i}{13}$

Answers

a $|3 - 4i| = \sqrt{3^2 + (-4)^2} = \sqrt{9 + 16} = \sqrt{25} = 5$

b $\left|-7 + \sqrt{11}i\right| = \sqrt{(-7)^2 + \left(\sqrt{11}\right)^2} = \sqrt{49 + 11} = \sqrt{60} = 2\sqrt{15}$

c $\left|\dfrac{-5 - 12i}{13}\right| = \sqrt{\left(-\dfrac{5}{13}\right)^2 + \left(-\dfrac{12}{13}\right)^2} = \sqrt{\dfrac{25 + 144}{169}} = \sqrt{\dfrac{169}{169}} = 1$

Exercise 3D

1 Find the real and imaginary parts of these complex numbers.

a $z = 3i$ **b** $z = -7$ **c** $z = \dfrac{18 - 12i}{8}$

d $z = \dfrac{11}{4} + i\dfrac{\sqrt{7}}{5}$ **e** $z = \dfrac{4i - 2}{3\pi^2}$

2 Find the modulus of these complex numbers.

a $12 + 5i$ **b** $-24 - 7i$ **c** $2\sqrt{2} + i\sqrt{5}$

d $\dfrac{-21 + 20i}{29}$ **e** $\dfrac{-3 + 4i}{\pi}$

3.2 Operations with complex numbers

Two complex numbers are **equal** if, and only if, their **real** and **imaginary parts** are **equal**.

So given that $z_1 = a_1 + ib_1$, $z_2 = a_2 + ib_2$ and $a_1, b_1, a_2, b_2 \in \mathbb{R}$

$(z_1 = z_2) \Leftrightarrow (a_1 = a_2 \text{ and } b_1 = b_2)$

or

$(z_1 = z_2) \Leftrightarrow (\text{Re}(z_1) = \text{Re}(z_2) \text{ and } \text{Im}(z_1) = \text{Im}(z_2))$

Why is it not possible to define inequality relations (<, >) on complex numbers? Find reasons to declare the following statements false:

- $i > 0$
- $i < 0$

Addition and subtraction of complex numbers

The addition of complex numbers is defined in a very natural way:

$\rightarrow z_1 + z_2 = (a_1 + ib_1) + (a_2 + ib_2) = (a_1 + a_2) + i(b_1 + b_2)$

Likewise,

$\rightarrow z_1 - z_2 = (a_1 + ib_1) - (a_2 + ib_2) = (a_1 - a_2) + i(b_1 - b_2)$

Multiplication of complex numbers by a real number

To multiply a complex number by a real number use the distributive property.

> → $\lambda z = \lambda(a + ib) = (\lambda a) + i(\lambda b)$, $a, b, \lambda \in \mathbb{R}$

Example 9

If $z_1 = 2 + 3i$ and $z_2 = 3 - 4i$, calculate these and check your answers with a GDC.

a $z_1 + z_2$ **b** $5z_1 - \dfrac{1}{2}z_2$

Answers

a $z_1 + z_2 = 2 + 3i + 3 - 4i$

$\qquad = (2 + 3) + (3 - 4)i = 5 - i$

b $5z_1 - \dfrac{1}{2}z_2 = 5(2 + 3i) - \dfrac{1}{2}(3 - 4i)$

$\qquad = 10 + 15i - \dfrac{3}{2} + 2i$

$\qquad = \dfrac{17}{2} + 17i$

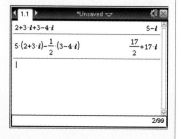

GDC help on CD:
Alternative demonstrations for the TI-84 Plus and Casio fx-9860GII GDCs are on the CD.

Multiplication of complex numbers

Use the distributive property and the fact that $i^2 = -1$ to multiply two complex numbers.

> → $z_1 \cdot z_2 = (a_1 + ib_1) \cdot (a_2 + ib_2) = a_1a_2 + ib_1a_2 + a_1ib_2 + \underset{-1}{i^2}b_1b_2$
> $\qquad = (a_1a_2 - b_1b_2) + i(a_1b_2 + a_2b_1)$

> This formula is not simple to memorize. In practice it is easier to apply the distributive property each time when multiplying complex numbers

Example 10

Given that $z_1 = 2 + 3i$, $z_2 = 3 - 4i$ and $z_3 = 1 - i$, calculate these and check your answers with a GDC.

a $z_1 \cdot z_2$

b $z_1 \cdot z_3 - 3z_2$

Answers

a $z_1 \cdot z_2 = (2 + 3i) \cdot (3 - 4i)$

$\qquad = 6 + 9i - 8i - 12i^2 = 6 + i - 12 \cdot -1$

$\qquad = 18 + i$

▶ Continued on next page

b $z_1 \cdot z_3 - 3z_2 = (2 + 3i) \cdot (1 - i) - 3(3 - 4i)$

$= 2 - 2i + 3i - 3i - 9 + 12i = -7 + 13i - 3 \cdot -1 = -4 + 13i$

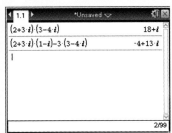

GDC help on CD:
Alternative demonstrations for the TI-84 Plus and Casio fx-9860GII GDCs are on the CD.

Exercise 3E

1 Given that $z_1 = 2 + 3i$, $z_2 = \dfrac{3}{2} - 4i$, $z_3 = 1 - 5i$ and $z_4 = \dfrac{3 + 4i}{5}$,

calculate these and check your answers with a GDC.

a $z_1 + z_3$ **b** $z_1 - 2z_2$ **c** $z_2 + z_4$ **d** $5z_4 - 2z_2$

e $3z_1 + 4z_2 - z_3 - 5z_4$ **f** $z_1 \cdot z_2 - z_3 \cdot z_4$ **g** $z_3^2 - \dfrac{2}{3} z_2 \cdot z_4$

Example 11

Find a complex number z that satisfies $(4 - 2i) \cdot z = 3z + 2 - 5i$.

> Solve this problem by using the equality of two complex numbers.

Answer

Let $z = a + ib$

$\Rightarrow (4 - 2i) \cdot (a + ib) = 3(a + ib) + 2 - 5i$ *Expand.*

$\Rightarrow 4a - 2ai + 4bi + 2b = 3a + 3bi + 2 - 5i$ *Collect the real and imaginary parts.*

$\Rightarrow (4a + 2b) + (-2a + 4b)i = (3a + 2) + (3b - 5)i$ *The real and imaginary parts are equal so set up a pair of simultaneous equations.*

$\Rightarrow \begin{cases} 4a + 2b = 3a + 2 \\ -2a + 4b = 3b - 5 \end{cases} \Rightarrow \begin{cases} a + 2b = 2 \\ -2a + b = -5 \end{cases}$ *Solve the simultaneous equations.*

$\Rightarrow \begin{cases} a = 2 - 2b \\ -2(2 - 2b) + b = -5 \end{cases}$ *Apply the method of substitution.*

$\Rightarrow \begin{cases} a = 2 - 2b \\ -4 + 4b + b = -5 \end{cases} \Rightarrow \begin{cases} a = 2 - 2b \\ 5b = -1 \end{cases}$

$\Rightarrow \begin{cases} a = 2 - 2 \cdot \left(-\dfrac{1}{5}\right) \\ b = -\dfrac{1}{5} \end{cases} \Rightarrow \begin{cases} a = \dfrac{12}{5} \\ b = -\dfrac{1}{5} \end{cases} \Rightarrow z = \dfrac{12}{5} - \dfrac{1}{5}i$

$a = 2 - 2b$

$= 2 + \dfrac{2}{5} = \dfrac{12}{5}$

> Remember to write down the final answer in the form asked for in the question, especially when solving long questions involving many different parts.

Conjugate complex numbers

Two complex numbers are said to be a conjugate pair if they have equal real parts and opposite sign imaginary parts.

If $z = a + ib$ then its conjugate is $z^* = a - ib$

> The conjugate of the number z is denoted by z^*.

Example 12

Given the complex number $z = a + ib$, find:

a $z + z^*$ **b** $z - z^*$ **c** $z \cdot z^*$

Answers

a $z + z^* = a + ib + a - ib = 2a$

b $z - z^* = (a + ib) - (a - ib)$
$\qquad\quad = a + ib - a + ib = 2ib$

c $z \cdot z^* = (a + ib) \cdot (a - ib)$ *Apply the formula*
$\qquad\quad = (a)^2 - (ib)^2$ $(A + B) \cdot (A - B) = A^2 - B^2$
$\qquad\quad = a^2 - \underset{-1}{i^2}\, b^2 = a^2 + b^2$

> $z + z^* = 2a \in \mathbb{R}$ and $z \cdot z^* \in \mathbb{R}, a^2 + b^2 \geq 0$.

Conjugate complex numbers have these properties:

i $(z^*)^* = z$

ii $(z_1 + z_2)^* = z_1{}^* + z_2{}^*$

iii $(z_1 \cdot z_2)^* = z_1{}^* \cdot z_2{}^*$

iv $z \cdot z^* = |z|^2$

v $(z^n)^* = (z^*)^n, n \in \mathbb{Z}$

> The first four properties can be easily proved. You are asked to do this in Exercise 3F. The fifth property can be proved using repeated application of property **iii**. In Chapter 12 you will see a simpler way of finding powers of complex numbers.

Division of complex numbers

You can divide complex numbers using several of the properties that you have learnt so far.

$$\frac{z_1}{z_2} = \frac{a_1 + ib_1}{a_2 + ib_2} \cdot \frac{a_2 - ib_2}{a_2 - ib_2}$$

Multiply the numerator and denominator by the conjugate of the denominator.

$$= \frac{a_1 a_2 + ib_1 a_2 - a_1 ib_2 - \overset{-1}{i^2} b_1 b_2}{a_2^2 + b_2^2}$$

Multiply the numerators and notice that the denominator becomes a positive real number.

$$= \frac{(a_1 a_2 + b_1 b_2) + i(a_2 b_1 - a_1 b_2)}{a_2^2 + b_2^2}$$

Separate the real and imaginary parts.

$$= \frac{a_1 a_2 + b_1 b_2}{a_2^2 + b_2^2} + i \frac{a_2 b_1 - a_1 b_2}{a_2^2 + b_2^2}$$

Collect like parts in the numerator.

Again notice that this formula is not very simple. In practise it is easier to apply this method each time when dividing complex numbers.

→ The division formula can be written in the form $\dfrac{z_1}{z_2} = \dfrac{z_1 \cdot z_2{}^*}{|z_2|^2}$

Example 13

Given that $z_1 = 5 + 5i$, $z_2 = 1 + 2i$ and $z_3 = 3 - 2i$, calculate these and check your answers with a GDC.

a $\dfrac{z_1}{z_2}$ **b** $\dfrac{z_1^2}{z_2 \cdot z_3^*}$

Answers

a $\dfrac{z_1}{z_2} = \dfrac{5+5i}{1+2i} \cdot \dfrac{1-2i}{1-2i}$

Multiply the numerator and denominator by the conjugate of the denominator.

$= \dfrac{5 + 5i - 10i - 10i^2}{1^2 + 2^2}$

Expand the numerator. Expand the denominator by using the difference of two squares.

$= \dfrac{15 - 5i}{5} = 3 - i$

Simplify.

b $\dfrac{z_1^2}{z_2 \cdot z_2^*} = \dfrac{(5+5i)^2}{(1+2i)\cdot(3+2i)}$

Expand the numerator and denominator.

$= \dfrac{25 + 50i + 25i^2}{3 + 2i + 6i + 4i^2}$

Multiply the numerator and denominator by the conjugate of the denominator.

$= \dfrac{50i}{-1+8i} \cdot \dfrac{-1-8i}{-1-8i}$

Expand the numerator and denominator.

$= \dfrac{50(-i - 8i^2)}{(-1)^2 + 8^2}$

$= \dfrac{50(8-i)}{65} = \dfrac{10(8-i)}{13}$

Simplify.

$= \dfrac{80}{13} - \dfrac{10}{13}i$

GDC help on CD:
Alternative demonstrations for the TI-84 Plus and Casio fx-9860GII GDCs are on the CD.

Once you know how to divide two complex numbers you can solve linear equations in complex numbers.

Example 14

Find the complex number z that satisfies $\dfrac{z+1}{3+i} = \dfrac{z-5i}{2i-1}$

Answer

$\dfrac{z+1}{3+i} = \dfrac{z-5i}{2i-1} \Rightarrow (z+1)(2i-1) = (z-5i)(3+i)$

$\Rightarrow z(2i - 1) + (2i - 1) = z(3 + i) - 5i(3 + i)$

$\Rightarrow z(2i - 1) - z(3 + i) = -2i + 1 - 15i + 5$

$\Rightarrow z(2i - 1 - 3 - i) = -17i + 6$

$\Rightarrow z(-4 + i) = 6 - 17i$

$z = \dfrac{6-17i}{-4+i} \cdot \dfrac{-4-i}{-4-i} \Rightarrow z = \dfrac{-24 + 68i - 6i - 17}{16 + 1} = \dfrac{-41 + 62i}{17}$

Exercise 3F

1 Given that $z_1 = 1 + 4i$, $z_2 = 2 - i$, $z_3 = \dfrac{1}{2} - \dfrac{5}{2}i$ and $z_4 = \dfrac{2i - 1}{3}$,

   Calculate these quotients and check your answers with a GDC.

   a $\dfrac{z_1}{z_2}$ **b** $\dfrac{z_1^*}{z_1}$ **c** $\dfrac{z_2 \cdot z_4}{z_3}$ **d** $\dfrac{3z_1 - 2z_3}{z_2 + 3z_4}$ **e** $\dfrac{z_1^2}{\left(z_2^*\right)^2}$

2 Find the real numbers a and b that satisfy these equations.

   a $(2 + i)(a + ib) = 11 - 2i$ **b** $\dfrac{a + ib}{2 - 5i} = -3 + 2i$

   c $(3i - 2)(a + ib) = 3 + 28i$ **d** $\left(\dfrac{1}{2} + \dfrac{3}{4}i\right)(a + ib) = -3 + 2i$

3 Find the real and imaginary parts of these numbers.

   a $\dfrac{3 - 2i}{4}$ **b** $\dfrac{5i - 2}{3i}$ **c** $\dfrac{1}{3i} + \dfrac{2}{1+i}$ **d** $\dfrac{2 - 3i}{2 + 3i} - \dfrac{2 + 3i}{2 - 3i}$

4 Given the numbers $z_1 = 1 + 3i$ and $z_2 = 3 - i$, find:

   a $z_1 \cdot z_2 + z_1 \cdot z_2^*$ **b** $z_1 \cdot z_2 - z_1^* \cdot z_2$ **c** $z_1 \cdot z_2 + (z_1 \cdot z_2)^*$

5 Find the complex number z that satisfies these equations.

   a $(z + 1)i = (z + 2i)(3 + 2i)$ **b** $(2z - 1)(1 + i) = (z - 1)(2 + 3i)$

   c $\dfrac{z - 3i + 2}{4 + 3i} = \dfrac{z - 1}{1 + i}$ **d** $\dfrac{3z - 2i}{2 + i} = \dfrac{2z + 5}{10 + 15i}$

6 What conditions must the real and imaginary parts of a complex number z satisfy so that $\dfrac{z}{2 - 7i} \in \mathbb{R}$?

7 What conditions must the real and imaginary parts of a complex number z satisfy so that $\dfrac{3 - 5i}{z^*}$ is purely imaginary?

8 Solve for $z \in \mathbb{C}$:

   a $|z| - z = 4 + 3i$ **b** $|z| + iz = 2 - i$ **c** $z^2 - z^* = 0$

9 Prove these properties of the modulus of a complex number.

   a $|z_1 \cdot z_2| = |z_1| \cdot |z_2|$ **b** $\left|\dfrac{z_1}{z_2}\right| = \dfrac{|z_1|}{|z_2|}$

   c $|z^n| = |z|^n$ **d** $|z_1 + z_2| \le |z_1| + |z_2|$

10 Prove these properties of conjugate complex numbers.

   a $(z^*)^* = z$ **b** $(z_1 + z_2)^* = z_1^* + z_2^*$ **c** $(z_1 \cdot z_2)^* = z_1^* \cdot z_2^*$

   d $z \cdot z^* = |z|^2$ **e** $|z| = |z^*|$

This table lists the fundamental properties or **axioms** of the operations on complex numbers. Other properties can be derived from these properties. The first four axioms refer to addition and the next four to multiplication, while the final axiom refers to both operations. 0 and 1 are real numbers but can be seen as complex, that is; $0 = 0 + 0i$ and $1 = 1 + 0i$.

→ **Axioms of complex numbers**

A1 For all complex numbers z_1 and z_2 then $z_1 + z_2$ is a complex number (Closure)

A2 For all complex numbers z_1 and z_2 then $z_1 + z_2 = z_2 + z_1$ (Commutativity)

A3 For all complex numbers z_1, z_2 and z_3 then $(z_1 + z_2) + z_3 = z_1 + (z_2 + z_3)$ (Associativity)

A4 There exists a complex number $0 = 0 + 0i$ such that for every complex number z, $0 + z = z + 0 = z$ (Additive identity)

A5 For every complex number z there exists a complex number $-z$ such that $z + -z = -z + z = 0$ (Additive inverse)

A6 For all complex numbers z_1 and z_2 then $z_1 \cdot z_2$ is a complex number (Closure)

A7 For all complex numbers z_1 and z_2 then $z_1 \cdot z_2 = z_2 \cdot z_1$ (Commutativity)

A8 For all complex numbers z_1, z_2 and z_3 then $(z_1 \cdot z_2) \cdot z_3 = z_1 \cdot (z_2 \cdot z_3)$ (Associativity)

A9 There exists a complex number $1 = 1 + 0i$ such that for every complex numbers z, $1 \cdot z = z \cdot 1 = z$ (Multiplicative identity)

A10 For all complex numbers z, $z \neq 0$, there exists a complex numbers z^{-1} such that $z \cdot z^{-1} = z^{-1} \cdot z = 0$ (Multiplicative inverse)

A11 For all complex numbers z_1, z_2 and z_3 then $z_1 \cdot (z_2 + z_3) = z_1 \cdot z_2 + z_1 \cdot z_3$ (Distributivity of multiplication over addition)

A structure in which addition and multiplication are defined and satisfy certain rules (shown left) is called the **field of complex numbers**. Since all real numbers can also be seen as complex and they satisfy the axioms, there is also a structure called the **field of real numbers**.

Investigation – axioms of a field

Decide if these sets of numbers satisfy the axioms of a field **A1-A11** given above.

a The integers, $\mathbb{Z}$

b The rational fractions, $\mathbb{Q}$

c The reals, $\mathbb{R}$

d Numbers of the form $p + q\sqrt{2}$ where p and q are rational fractions.

Powers and roots of complex numbers

To find powers and roots of complex numbers, you use the binomial theorem and powers of the imaginary unit, i.

Use the 'Σ-notation for a sum. Similarly, there is a product notation.

$$\prod_{k=1}^{n} i^k = i^1 \cdot i^2 \cdot \ldots \cdot i^n$$

For verification of the general rule for i^n, refer to the summary at the end of the chapter.

Example 15

Given the complex number $z = 1 - 2i$, find: **a** z^3 **b** $\left(z^5\right)^*$ **c** $\left(z^*\right)^5$
Check your answers using a GDC.

- -

Answers

a $z^3 = (1 - 2i)^3$
$= 1^3 - 3 \cdot 1^2 \cdot 2i + 3 \cdot 1 \cdot (2i)^2 - (2i)^3$
$= 1 - 6i - 12 + 8i$
$= -11 + 2i$

Use the binomial theorem.
Use $i^2 = -1$ and i^3
$= -1$

The binomial theorem states that $(a + x)^n \equiv$

$$\sum_{r=0}^{n} \binom{n}{r} a^{n-r} x^r$$

▶ Continued on next page

b $z^5 = z^2 \cdot z^3$
$$= (1 - 2i)^2 \cdot (-11 + 2i)$$
$$= (1 - 4i - 4) \cdot (-11 + 2i)$$
$$= (-3 - 4i) \cdot (-11 + 2i)$$
$$= 33 - 6i + 44i + 8 = 41 + 38i$$
$$\Rightarrow (z^5)^* = 41 - 38i$$

*Simplify the calculation by finding z^5 using the answer to part **a** and then finding its conjugate. Use the square of a difference.*

GDC help on CD:
Alternative demonstrations for the TI-84 Plus and Casio fx-9860GII GDCs are on the CD.

c $(z^*)^5 = (1 + 2i)^5$
$$= 1^5 + 5 \cdot 1^4 \cdot 2i + 10 \cdot 1^3 \cdot (2i)^2 +$$
$$10 \cdot 1^3 \cdot (2i)^3 + 5 \cdot 1 \cdot (2i)^4 + (2i)^5$$
$$= 1 + 10i - 40 - 80i + 80 + 32i$$
$$= 41 - 38i$$

*Use the binomial theorem.
Use $i^2 = -1$, $i^3 = -i$, $i^4 = 1$ and $i^5 = i$.*

Notice that the results in **b** and **c** are equal, that is, $(z^*)^n = (z^n)$, $n \in \mathbb{Z}$, as stated in the properties of conjugate complex numbers, that is, $(z^*)^n = (z^n)^*$, $n \in \mathbb{Z}$.

You can find the square roots of a complex number, z, by first squaring z so that you can work with the real and complex parts of $\sqrt{z}$ separately.

Example 16

Evaluate $\sqrt{8 - 6i}$.

Answer

Let $z = x + yi$, $x, y \in \mathbb{R}$ such that
$$z = \sqrt{8 - 6i} \Rightarrow z^2 = 8 - 6i$$
$$(x + yi)^2 = 8 - 6i$$
$$x^2 + 2xyi - y^2 = 8 - 6i$$

*Expand z and use $i^2 = -1$
Equate the real and imaginary parts.*

$$\begin{cases} x^2 - y^2 = 8 \\ 2xy = -6 \end{cases} \Rightarrow \begin{cases} x^2 - \left(-\dfrac{3}{x}\right)^2 = 8 \\ y = -\dfrac{3}{x} \end{cases}$$

Solve the simultaneous equations by using substitution.

$$\Rightarrow \begin{cases} x^2 - \dfrac{9}{x^2} = 8 \\ y = -\dfrac{3}{x} \end{cases} \Rightarrow \begin{cases} x^4 - 9 = 8x^2 \\ y = -\dfrac{3}{x} \end{cases}$$

$$\Rightarrow \begin{cases} x^4 - 8x^2 - 9 = 0 \\ y = -\dfrac{3}{x} \end{cases} \Rightarrow \begin{cases} (x^2 - 9)(x^2 + 1) = 0 \\ y = -\dfrac{3}{x} \end{cases}$$

Factorize the equation and apply the zero product property.

$$\Rightarrow \begin{cases} x = \pm 3 \\ y = -\dfrac{3}{\pm 3} \end{cases} \Rightarrow \begin{cases} x = \pm 3 \\ y = \mp 1 \end{cases}$$

Notice that ± 3 are the only real solutions for x.

$$\Rightarrow z_1 = 3 - i \text{ and } z_2 = -3 + i$$

A GDC will always give just one solution, but you need to be aware is that there another solution which is the negative of the number on the GDC.

GDC help on CD:
Alternative demonstrations for the TI-84 Plus and Casio fx-9860GII GDCs are on the CD.

The method shown in Example 16 for square roots is not an easy one and, for higher roots, algebraic skills are needed. In Chapter 12 you will learn a different method for finding roots of complex numbers.

Exercise 3G

1 Calculate:

a $i^5 + i^8 + i^{14} + i^{19}$

b $i^{123} + i^{172} + i^{256} + i^{375}$

c $(2 - i^{53}) \cdot (3 + 2i^{89})$

d $\dfrac{4i^{2010} - 3i^{2011}}{2i^{2012} + 5i^{2013}}$

e $\dfrac{i + i^2 + i^3 + \ldots + i^{2011}}{i \cdot i^2 \cdot i^3 \cdot \ldots \cdot i^{2011}}$

f $\dfrac{i^2 + i^4 + i^6 + \ldots + i^{2010}}{i^2 \cdot i^4 \cdot i^6 \cdot \ldots \cdot i^{2010}}$

2 Calculate these and check your answers with a GDC.

a $(2 + 3i)^2 + (1 - 4i)^2$

b $(3 + 2i)^2 + (3 - 2i)^2$

c $(3 + 2i)^3 + (3 - 2i)^3$

d $(1 + i)^4 + (1 - i)^4$

3 Evaluate these and check your answers with a GDC.

a $\sqrt{3 + 4i}$

b $\sqrt{12i - 5}$

c $\sqrt{\dfrac{5}{4} + 3i}$

d $\sqrt{\dfrac{55}{144} - \dfrac{1}{3}i}$

e $\sqrt{i}$

f $\sqrt{-i}$

4 Show that:

a $(1 + i)^{2n} = (2i)^n$, $n \in \mathbb{Z}$

b $(1 + i)^{2n+1} = (1 + i)(2i)^n$, $n \in \mathbb{Z}$

> For question 5, it may help to plot z, z^2, z^3, … on an Argand diagram and look for a pattern.

5 Given that $z = 1 - i$, find the values of $n \in \mathbb{N}$ such that:

a z^n is real

b z^n is purely imaginary.

3.3 Polynomial functions: graphs and operations

Historically, complex numbers appeared as roots of polynomial equations which means that they can be seen as zeros of polynomial functions. In this section we are going to study polynomial functions, their graphs and operations with the expressions that define them.

Polynomial functions and their graphs

The diagrams show the graphs of $f(x) = x^n$, $n = 0, 1, 2, 3, \ldots$ where n is a natural number.

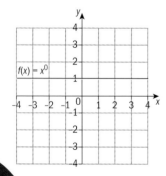

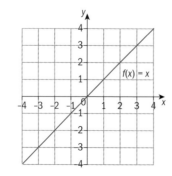

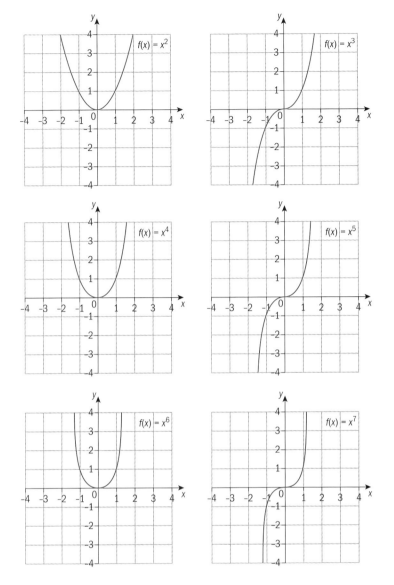

Apart from the first two powers of n, $n = 0$ and $n = 1$, notice that for:

- even powers $n = 2, 4, 6 \ldots$ the graphs a 'U' shape
- odd powers $n = 3, 5, 7 \ldots$ the graphs have a 'flex' shape.

The 'U' shape graph has a **local** minimum or maximum, while the 'flex' shape graph has a horizontal **inflexion**.

A linear combination of powers of x, for example $3 \cdot x^5 - 2 \cdot x^2 + 8x - 11$, is called a **polynomial**.

> → **A linear combination** of two functions f and g is an expression of the form $a \cdot f(x) + b \cdot g(x)$, where a and b are real numbers.
> A linear combination of n functions is an expression of the form $\sum_{k=1}^{n} a_k \cdot f_k(x)$, where f_k are functions and $a_k \in \mathbb{R}$.

In general, polynomials can be seen as a linear combination of the power functions $\{1, x, x^2, x^3, x^4, x^5, \ldots\}$

Polynomials are real functions of the real variable $f: \mathbb{R} \to \mathbb{R}$ of the form $f(x) = a_n x^n + a_{n-1} x^{n-1} + \ldots + a_1 x + a_0$, where $a_k \in \mathbb{R}$, $k = 0, \ldots, n$ are called the coefficients. The highest power of the variable x^n is called the degree of the polynomial, $\deg(f) = n$.

> The word 'polynomial' means 'many terms'. A polynomial of one term is called a monomial, of two terms a binomial, and of three terms a trinomial.

Polynomials of degree 0, 1, 2 and 3

- **Constant function** $f(x) = c$, $c \in \mathbb{R}$. The graph is a horizontal line. The degree of a constant polynomial is zero.

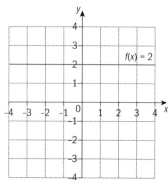 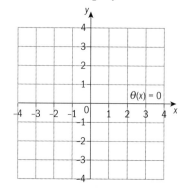

> You use the notation $\theta(x) = 0$ for the zero polynomial to distinguish it from other polynomials. The zero polynomial also has an important property as an additive identity element for polynomials, that is,
> $$f(x) + \theta(x) = \theta(x) + f(x) = f(x)$$
> for all polynomials f.

- **Zero polynomial** $\theta(x) = 0$. The graph is again a horizontal line but this time it is the x-axis itself.

- **Linear function** $f(x) = mx + c$, $m \neq 0$. This is a polynomial of the first degree. The graph is a straight line. By changing the parameters m and c, you change the steepness and the position of the line.

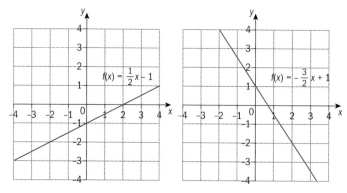

- **Quadratic function** $f(x) = ax^2 + bx + c$, $a \neq 0$. This is a polynomial of the second degree. The graph is a parabola, 'U' shaped, whose axis of symmetry is a vertical line. By changing the parameters a, b and c, you change the shape (wide or narrow), concavity (opens upwards or downwards) and position of the parabola.

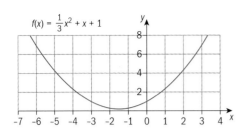

$f(x) = \frac{1}{3}x^2 + x + 1$

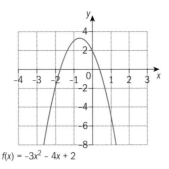

$f(x) = -3x^2 - 4x + 2$

Investigation – parameters of parabolas

Given a quadratic function $f(x) + ax^2 + b(x) + c,\ a \neq 0$, investigate the effect of the parameters a, b and c on the shape and the position of the parabola in the coordinate system. In Chapter 2 you were investigating the form $f(x) + a(x - h)^2 + k,\ a \neq 0$, where h and k were horizontal and vertical translations, respectively. Use this to find the effect of the parameter b.

- **Cubic function**, $f(x) = ax^3 + bx^2 + cx + d,\ a \neq 0$. This is a polynomial of the third degree.

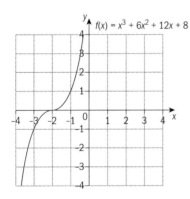

$f(x) = x^3 + 6x^2 + 12x + 8$

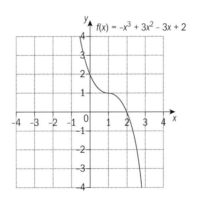

$f(x) = -x^3 + 3x^2 - 3x + 2$

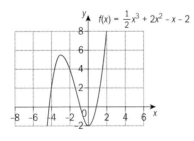

$f(x) = \frac{1}{2}x^3 + 2x^2 - x - 2$

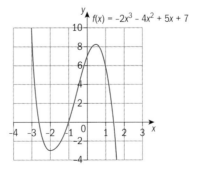

$f(x) = -2x^3 - 4x^2 + 5x + 7$

Cubic graphs have two different shapes. One shape looks like a 'flex' shape. The second shape is a combination of two 'U' shapes opening in opposite directions.

Investigation – parameters of cubics

Given a cubic function $f(x) = ax^3 + bx^2 + cx + d$, $a \neq 0$, investigate the effect of the parameters a, b, c and d on the shape and the position of the graphs. Start your investigation by taking two parameters at a time, for example a and b, c and c, a and d.

One interesting feature of polynomials of the same degree and with the same leading coefficient is that even though locally the graphs look very different if you change the scale on the axes they look very similar. For example, for a polynomial in x^3, $f(x)$ increases rapidly for large values of x.

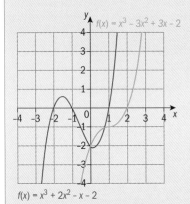

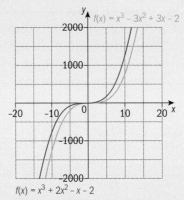

$f(x) = x^3 + 2x^2 - x - 2$

The functions $f(x) = x^3 - 3x^2 + 3x - 2$ and $f(x) = x^3 + 2x^2 - x - 2$ behave like polynomials that have only the leading term, x^3, since for extremely large values of x, both positive and negative, the other terms are insignificant to the total value and can be neglected. This is the so-called 'end behavior' property of polynomials.

> The end behavior of a polynomial function is determined by its degree and by the sign of its leading coefficients.

Polynomials are continuous functions, which means that you can draw their graphs without lifting the pen from the paper. You have to proceed in one direction (usually from left to right). Their graphs are also smooth curves with no sharp points.

Polynomials of degree 4 are called **quartic functions** and polynomials of degree 5 are called **quintic functions**. Special names are not usually used for polynomials of degree greater than 5.

In a graph of a polynomial of a higher degree you can see different types of 'U' and 'flex' shapes.

The graph shows a quintic polynomial.

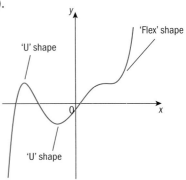

Investigation – higher-degree polynomials

1 Describe all the possible shapes of the graphs of polynomials of fourth degree. How many of each of the shapes of lower degree (2 and 3) can you have in those polynomials?

2 Describe all the possible shapes of the graphs of polynomials of fifth degree. How many of each of the shapes of lower degree (2, 3 and 4) can you have in those polynomials?

3 How many of each of the shapes of lower degree can you have in the polynomials of nth degree?

Operations with polynomials

Two polynomials $f(x) = a_n x^n + a_{n-1} x^{n-1} + \ldots + a_2 x^2 + a_1 x + a_0$ and $g(x) = b_m x^m + b_{m-1} x^{m-1} + \ldots + b_2 x^2 + b_1 x + b_0$ are **equal** if and only if:

i they have the **same degree**, $n = m$

ii all the **corresponding coefficients** are **equal**, $a_k = b_k$ for all $k = 0, 1, \ldots, n$.

Addition and multiplication of polynomials and multiplication by a real number follow the same rules for algebraic expressions that you met in the "Before you start" section.

> → The degree of a linear combination of two polynomials is not larger than the maximum of the degrees of either polynomial.
> $\deg(\lambda \cdot f(x) + \mu \cdot g(x)) \le \max\{\deg(f(x)), \deg(g(x))\}$

Example 17

Given the polynomials $f(x) = 4x^4 + 3x^3 - 2x^2 + 6x - 2$ and $g(x) = 2x^3 - 5x^2 + x - 3$, find $f(x) \cdot g(x)$.

Answer

The standard algebraic method is difficult to follow because so many terms arise. It is usually easier to use the 'grid method'.

	$4x^4$	$3x^3$	$-2x^2$	$6x$	-2
$2x^3$	$8x^7$	$6x^6$	$-4x^5$	$12x^4$	$-4x^5$
$-5x^2$	$-20x^6$	$-15x^5$	$10x^4$	$-30x^3$	$10x^2$
x	$4x^5$	$3x^4$	$-2x^3$	$6x^2$	$-2x$
-3	$-12x^4$	$-9x^3$	$6x^2$	$-18x$	6

The 'grid method' makes it easier to simplify the like terms.

$f(x) \cdot g(x) = 8x^7 - 14x^6 - 15x^5 + 13x^4 - 45x^3 + 22x^2 - 20x + 6$

The grid method for multiplication is also known as the 'box method'.

> → The degree of the product of two polynomials is the sum of the degrees of the factor polynomials:
> $\deg(f(x) \cdot g(x)) = \deg(f(x)) + \deg(g(x))$

Example 18

Given the polynomials $f(x) = x^2 + ax - 3$ and $g(x) = x^2 - 4x + b$, find the values of the real parameters a and b such that
$f(x) \cdot g(x) = x^4 - 22x^2 + 9$

Answer

$f(x) \cdot g(x) = (x^2 + ax - 3) \cdot (x^2 - 4x + b)$ $\qquad = x^4 + (a-4)x^3 + (a + 12 + b)x^2$ $\qquad \quad + (ab - 12)x - 3b$ $\qquad = x^4 - 22x^2 + 9$	*Use distribution.* *Simplify.*
$\Rightarrow \begin{cases} a - 4 = 0 \\ -4a - 3 + b = -22 \\ ab - 12 = 0 \\ -3b = 9 \end{cases}$	*Make the corresponding coefficients equal.*
$\Rightarrow \begin{cases} a = 4 \\ b = -3 \end{cases}$	*Check that the values of a and b satisfy all the equations.*

Exercise 3H

1 The polynomials $f(x) = 2x^2 + 3x + 1$ and $g(x) = 3x^2 - 2x - 5$ are given. Find the real parameters λ and μ such that:

   a $\lambda \cdot f(x) + \mu \cdot g(x) = 13x + 13$

   b $\lambda \cdot f(x) + \mu \cdot g(x) = 26x^2 + 26x$

2 Use the 'grid method' to find the product of the polynomials f and g given that:

   a $f(x) = x^3 - 2x$ and $g(x) = x^2 + 2$

   b $f(x) = 27x^3 - 36x^2 + 48x - 64$ and $g(x) = 3x^2 + 7x + 4$

3 Given the polynomials $f(x) = ax^2 - 3x + 5$ and $g(x) = 7x^2 + bx - 3$, find the values of the real parameters a and b such that
$f(x) \cdot g(x) = 14x^4 - 17x^3 + 23x^2 + 19x - 15$

4 Given the polynomials $f(x) = x^3 + ax^2 - x + 2$ and $g(x) = 2x^2 + bx + c$, find the values of the real parameters a, b and c such that
$f(x) \cdot g(x) = 2x^5 - 5x^4 + 3x^3 + 5x^2 - 8x + 4$

5 Given that a polynomial $f(x) = x^4 + 6x^3 + ax^2 + bx + 4$ can be written in the form $f(x) = (x^2 + px + q)^2$, find the values of a and b and the polynomial in the required form.

6 Find the polynomial g such that $g(x) = f(x-2)$, where
$f(x) = x^3 + 12x^2 + 6x + 3$

7 Find the polynomial f such that $f(2x-1) = 16x^4 - 32x^3 + 12x^2$

8 All the coefficients of the polynomial $f(x) = ax^4 + bx^3 + cx^2 + dx + e$ are positive integers smaller than 10. Find the polynomial given that $f(0) = 4$ and $f(10) = 32\,584$

Division of polynomials

You divide two polynomials using long division.

Example 19

Use long division to divide
$2x^4 + 4x^3 + 3x^2 + 2x - 7$
by
$x^2 + x + 2$

Answer

$$\begin{array}{r} 2x^2 + 2x - 3 \\ x^2+x+2\overline{)2x^4+4x^3+3x^2+2x-7} \\ -(2x^4+2x^3+4x^2) \\ \hline 2x^3 - x^2 + 2x \\ -(2x^3-2x^2+4x) \\ \hline -3x^2-2x-7 \\ -(-3x^2-3x-6) \\ \hline x-1 \end{array}$$

$2x^4 + 4x^3 + 3x^2 + 2x - 7$
$= (x^2+x+2) \cdot (2x^2+2x-3) + (x-1)$

Divide x^2 into $2x^4$

Multiply divisor $2x^2$

Divide x^2 into $2x^3$

Multiply divisor by $2x$

Divide x^2 into $-3x^3$

Multiply divisor by -3

Remainder is $x - 1$

Stop when the degree of the remainder is smaller than the degree of the divisor.

The same algorithm is used to divide numbers. Consider $657 \div 21$

$$\begin{array}{r} 31 \\ 21\overline{)657} \\ -63 \\ \hline 27 \\ -21 \\ \hline 6 \end{array}$$

So $657 = 21.31 + 6$

→ **Theorem**
For any two polynomials f and g there are unique polynomials q and r such that $f(x) = g(x) \cdot q(x) + r(x)$, for all real values of x.

dividend = divisior · quotient + remainder

The polynomial q is called the quotient and the polynomial r is called the remainder. The degree of the polynomial r is smaller than the degree of the polynomial g.

The proof of this theorem uses the Euclidian algorithm that is part of the Discrete option.

Example 20

Use long division to find the quotient and remainder when dividing
$f(x) = 2x^4 - 7x^3 - 7x^2 + 14x + 5$ by $g(x) = 2x + 3$

Answer

$$
\begin{array}{r}
x^3 - 5x^2 + 4x + 1 \\
2x+3\overline{)2x^4 - 7x^3 - 7x^2 + 14x + 5}
\end{array}
$$

$$
\begin{array}{r}
-(2x^4 + 3x^3) \\
\hline
-10x^3 - 7x^2 + 14x + 5 \\
-(-10x^3 - 15x^2) \\
\hline
8x^2 + 14x + 5 \\
-(8x^2 + 12x) \\
\hline
2x + 5 \\
-(2x + 3) \\
\hline
2 \leftarrow \text{remainder}
\end{array}
$$

So the quotient is $q(x) = x^3 - 5x^2 + 4x + 1$ and the remainder is $r(x) = 2$
Therefore,
$2x^4 - 7x^3 - 7x^2 + 14x + 5 = (2x + 3) \cdot (x^3 - 5x^2 + 4x + 1) + 2$

Exercise 3I

1 Use long division to divide f by g if:
  a $f(x) = x^4 + 5x^3 + 8x^2 + 3x - 2$ and $g(x) = x + 2$
  b $f(x) = x^5 + 3x^4 + x^3 - 4x^2 - 2x + 1$ and $g(x) = x^2 - 1$
  c $f(x) = 2x^5 - 3x^4 + x^3 - 2x^2 + 3x - 1$ and $g(x) = x^2 + x + 1$

2 Use long division to find the quotient and remainder when
  f is divided by g given that:
  a $f(x) = 2x^4 + 5x^3 + 4x^2 + 4x + 3$ and $g(x) = x + 1$
  b $f(x) = 3x^4 + 4x^3 + 6x^2 - 2x + 6$ and $g(x) = x^2 + 2x + 3$
  c $f(x) = x^6 + x - 1$ and $g(x) = x^2 + x + 1$

Polynomial remainder theorem

→ Given a polynomial

$$f(x) = a_n x^n + a_{n-1} x^{n-1} + \ldots + a_2 x^2 + a_1 x + a_0, \, a_k \in \mathbb{R}, \, k = 0, 1, \ldots, n, \, a_n \neq 0$$

and a real number p, then the remainder when $f(x)$ is divided by a
linear expression $(x - p)$ is $f(p)$.

Proof:
In the unique decomposition of the polynomial $f(x) = (x - p) \cdot q(x) + r$,
where the remainder r is a constant (one degree less than the
divisor) we input $x = p \Rightarrow f(p) = \underbrace{(p - p)}_{0} q(p) + r \Rightarrow f(p) = r$ QED

Extension material on CD:
Worksheet 3

The polynomial
remainder theorem
is also known
as 'Bézout's
little theorem'.
Étienne Bézout
(1730–1837) was
inspired by the work
of Euler and so
decided to become
a mathematician.
In 1763 he was
appointed examiner
of the Gardes de
la Marine (French
Naval Academy)
with the special
task of composing
a textbook
for teaching
mathematics to
the students.

Factor theorem

> → A polynomial
> $$f(x) = a_n x^n + a_{n-1} x^{n-1} + \dots + a_2 x^2 + a_1 x + a_0, \ a_k \in \mathbb{R}, \ k = 0, 1, 2, \dots, n, \ a_n \neq 0$$
> has a factor $(x - p)$, $p \in \mathbb{R}$ if and only if $f(p) = 0$.

This theorem is a direct consequence of the remainder theorem. Its proof is left as an exercise for you.

To evaluate a polynomial for a certain value of the variable x, **William George Horner** (1786–1837) discovered an algorithm that can be used in many different cases.

If you want to find the value of $f(x) = 3x^3 - 2x^2 - 5x - 1$ when $x = 2$, select the coefficients of all terms, including missing terms, and organize them in a tabular form:

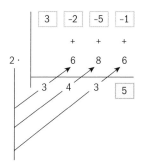

$$f(x) = ((3 \cdot x - 2) \cdot x - 5) \cdot x - 1$$
$$f(2) = ((3 \cdot 2 - 2) \cdot 2 - 5) \cdot 2 - 1$$
$$= ((6 - 2) \cdot 2 - 5) \cdot 2 - 1$$
$$= (4 \cdot 2 - 5) \cdot 2 - 1$$
$$= (8 - 5) \cdot 2 - 1$$
$$= 3 \cdot 2 - 1$$
$$= 6 - 1$$
$$= 5$$

Example 21

Use Horner's algorithm to find the remainder when dividing
$f(x) = 5x^3 + 13x^2 - 11x + 7$ by $g(x) = x + 3$

Answer

$x + 3 = x - (-3) \Rightarrow r = f(-3)$

Use the remainder theorem.

5	13	-11	7
	+	+	+

-3 ·   -15   6   15

| 5 | -2 | -5 | 22 |

Use Horner's algorithm.
$$f(-3) = ((5 \cdot -3 + 13) \cdot -3 - 11) \cdot$$
$$-3 + 7$$
$$= ((-15 + 13) \cdot -3 - 11) \cdot -3 + 7$$
$$= (-2 \cdot -3 - 11) \cdot -3 + 7$$
$$= (6 - 11) \cdot -3 + 7$$
$$= -5 \cdot -3 + 7$$
$$= 15 + 7$$
$$= 22$$

$r = f(-3) = 22$

> When you use Horner's algorithm, apart from getting the remainder (in the last row) you also obtain the coefficients of the quotient polynomial,
> $q(x) = 5x^2 - 2x - 5$
> This is the reason why this algorithm is also known as synthetic division.

> Investigate Horner's algorithm. Prove the general form of the algorithm and find in which other cases it can be used.

Since the algorithm also gives the quotient you can use successive division to search for factors of a polynomial.

Example 22

Show that $(x - 2)$ and $(x + 5)$ are factors of the polynomial
$f(x) = 2x^3 + 13x^2 + x - 70$

Answer

$x - 2 \Rightarrow r = f(2)$

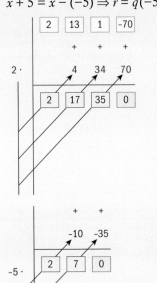

Use the remainder theorem and note that the remainders must be zeros for factors.

Use Horner's algorithm.
$$f(2) = ((2 \cdot 2 + 13) \cdot 2 + 1) \cdot 2 - 70$$
$$= ((4 + 13) \cdot 2 + 1) \cdot 2 - 70$$
$$= (17 \cdot 2 + 1) \cdot 2 - 70$$
$$= (34 + 1) \cdot 2 - 70$$
$$= 35 \cdot 2 - 70$$
$$= 70 - 70$$

$x + 5 = x - (-5) \Rightarrow r = q(-5)$

Use Horner's algorithm.
$$q(x) = 2x^2 + 17x + 35$$
$$q(-5) = (2 \cdot -5 + 17) \cdot -5 + 35$$
$$= (-10 + 17) \cdot -5 + 35$$
$$= 7 \cdot -5 + 35$$
$$= -35 + 35$$
$$= 0$$

Since the remainders are zeros you could proceed with the quotient polynomial as the quotient polynomial contains $x + 5$ as a factor. Notice that the last factor is $2x + 7$.

Exercise 3J

1 Use synthetic division to find the quotient and remainder when polynomial f is divided by g given that:
   a $f(x) = x^3 - x^2 - 4x - 5$ and $g(x) = x - 3$
   b $f(x) = 2x^3 + 5x^2 + 4x + 3$ and $g(x) = x + 1$
   c $f(x) = x^5 - 3x^3 - 2x + 1$ and $g(x) = x + 2$
   d $f(x) = 3x^6 - 2x^4 + 5x^2 - 2$ and $g(x) = x - 1$

2 Show that $(x - 2)$ and $(x + 3)$ are factors of
   $f(x) = 4x^4 - 27x^2 + 25x - 6$.

Corollary

Given a polynomial

$$f(x) = a_n x^n + a_{n-1} x^{n-1} + \ldots + a_2 x^2 + a_1 x + a_0, a_i \in \mathbb{R}, i = 0, 1, \ldots, n, a_n \neq 0$$

and real numbers a and b, $a \neq 0$, then the remainder when $f(x)$ is divided by a linear expression $(ax - b)$ is $f\left(\dfrac{b}{a}\right)$.

> A theorem easily derived from another theorem is a corollary of that theorem.

The proof can be conducted in a similar way to that of the theorem proof. The proof is left as an exercise for you.

In order to use synthetic division when dividing by a linear expression $(ax - b)$ you have to modify the algorithm.

$$f(x) = (ax - b) \cdot q(x) + r$$

$$\Rightarrow f(x) = a\left(x - \frac{b}{a}\right) \cdot q(x) + r$$

$$\Rightarrow f(x) = \left(x - \frac{b}{a}\right) \cdot (a \cdot q(x)) + r$$

Example 23

Use the synthetic division to find the quotient and remainder when dividing $f(x) = 2x^4 - 7x^3 - 7x^2 + 14x + 5$ by $g(x) = 2x + 3$

Answer

$$g(x) = 2x + 3 = 2\left(x - \left(-\frac{3}{2}\right)\right)$$

Use the remainder theorem.

Use synthetic division.

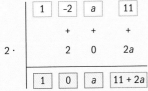

So the quotient is $q(x) = x^3 - 5x^2 + 4x + 1$ and the remainder is $r(x) = 2$.

> The coefficients of the quotient polynomial were multiplied by 2, so you need to divide them by 2 to obtain the quotient.

Example 24

When polynomial $f(x) = x^3 - 2x^2 + ax + 11$ is divided by $(x - 2)$ the remainder is 1. Find the value of a.

Answer

	1	-2	a	11
		+	+	+
$2 \cdot$		2	0	$2a$
	1	0	a	$11 + 2a$

Use synthetic division.

$11 + 2a = 1 \Rightarrow 2a = -10 \Rightarrow a = -5$

Use the remainder $r = 1$

Example 25

Find the remainder when polynomial $f(x) = x^{2011} - 3x^2 + 2x - 2$ is divided by $x^2 - 1$.

Answer

$f(x) = (x^2 - 1) \cdot q(x) + \underbrace{ax + b}_{r(x)}$

Use the theorem on unique decomposition. Note that the remainder is linear.

$x = 1 \Rightarrow f(1) = 1^{2011} - 3 \cdot 1^2 + 2 \cdot 1 - 2 = -2$
$x = -1 \Rightarrow f(-1) = (-1)^{2011} - 3 \cdot (-1)^2 + 2 \cdot -1 - 2 = -8$

Calculate the value of the polynomial at the zeros of the divisor.

$\begin{cases} f(1) = (1^2 - 1) \cdot q(1) + a \cdot 1 + b = -2 \\ f(-1) = ((-1)^2 - 1) \cdot q(-1) + a \cdot (-1) + b = -8 \end{cases}$

Substitute f (1) = −2 and f (−1) = −8 in the unique decomposition.

$\begin{cases} a + b = -2 \\ -a + b = -8 \end{cases} +$

Solve the simultaneous equations by elimination.

$2b = -10 \Rightarrow b = -5$
$a - 5 = -2 \Rightarrow a = 3$
Therefore, the remainder is $r(x) = 3x - 5$

Exercise 3K

1 Use synthetic division to find the quotient and remainder when polynomial f is divided by g given that:
   a $f(x) = 2x^5 - 3x^4 + 3x^3 + 3x^2 - 3$ and $g(x) = 2x - 1$
   b $f(x) = 3x^4 + 4x^3 + 4x^2 - 2x + 6$ and $g(x) = 3x + 1$

2 When you divide the polynomial f by the polynomial $g(x) = x^2 + 2x - 1$ you obtain the quotient $q(x) = 3x - 4$ and the remainder $q(x) = x + 2$. Find the polynomial f.

3 Polynomial $f(x) = x^5 - 4x^4 + 3x^3 + 2x^2 - 3x + a$ is divisible by $(x - 3)$. Find the value of a.

4 Polynomial $f(x) = x^5 - 2x^4 + 2x^3 + bx - 1$ is divisible by $(x - 1)$. Find the value of b.

EXAM-STYLE QUESTIONS

5 Polynomial $f(x) = 4x^3 + 5x^2 + ax + b$ is divisible by $(x + 2)$, and when divided by $(x - 1)$ there is a remainder of 6. Find the values of a and b.

6 When polynomial f is divided by $(x - 3)$ the remainder is 2, and when divided by $(x + 1)$ the remainder is −4. Find the remainder when polynomial f is divided by $(x^2 - 2x - 3)$.

7 Find the remainder when $f(x) = x^{2011} + x^{2010} + \ldots + x + 1$ is divided by $(x + 1)$.

8 Show that the polynomial $f(x) = (x + 1)^{2n} + (x + 2)^n - 1$ is divisible by $(x^2 + 3x + 2)$ for all $n \in \mathbb{Z}^+$.

9 Given a polynomial

$$f(x) = a_n x^n + a_{n-1} x^{n-1} + \ldots + a_2 x^2 + a_1 x + a_0, \, a_i \in \mathbb{R}, i = 1, 2, \ldots, n, a_n \neq 0$$

and real numbers a and b, $a \neq 0$, show that when $f(x)$ is divided

by a linear expression $(ax - b)$ the remainder is $f\left(\dfrac{b}{a}\right)$.

3.4 Polynomial functions: zeros, sum and product

The fundamental theorem of algebra

The fundamental theorem of algebra is one of the most important theorems in mathematics. It establishes the existence of the complex zeros of a polynomial (points at which the value of the function is zero). There are many theorems and corollaries that derive from this theorem which help in algebraic manipulation of equations and polynomial functions.

→ **The fundamental theorem of algebra (FTA)**
A polynomial $f(x) = a_n x^n + a_{n-1} x^{n-1} + \ldots + a_2 x^2 + a_1 x + a_0$ with real or complex coefficients $(a_n \neq 0)$ has at least one zero.
There is an $\omega \in \mathbb{C}$ such that $f(\omega) = 0$

This theorem was proved by Gauss, but is beyond the scope of this textbook.

→ **Corollary**
Each polynomial $f(x) = a_n x^n + a_{n-1} x^{n-1} + \ldots + a_2 x^2 + a_1 x + a_0$ with real or complex coefficients can be written in a factored form $f(x) = a_n (x - \omega_1)(x - \omega_2) \ldots (x - \omega_n)$ such that $\omega_k \in \mathbb{C}$, $k = 1, \ldots, n$

If a certain factor appears more than once, we say that the factor has a **multiplicity**. So, if there are fewer than n different zeros of the given polynomial, the sum of their multiplicities will add up to n.

$$f(x) = a_n \left(x - \omega_1\right)^{p_1} \left(x - \omega_2\right)^{p_2} \ldots \left(x - \omega_k\right)^{p_k}, k < n, \sum_{r=1}^{k} p_r = n$$

Example 26

Factorize the polynomial $f(x) = x^4 - 6x^3 + 11x^2 - 6x$, and check your answer with a GDC.

GDC help on CD:
Alternative demonstrations for the TI-84 Plus and Casio fx-9860GII GDCs are on the CD.

Answer

$f(x) = x^4 - 6x^3 + 11x^2 - 6x$

$\quad = x(x^3 - 6x^2 + 11x - 6)$

1	-6	11	-6
	+	+	+

$1 \cdot \quad\quad 1 \quad -5 \quad 6$

1	-5	6	0

Apply Horner's algorithm for $x = 1$ since the sum of the coefficients is equal to zero.

$\quad = x(x-1)(x^2 - 5x + 6)$

Apply the FTA to factorize the polynomial.

$\quad = x(x-1)(x-2)(x-3)$

Factorize the quadratic expression.

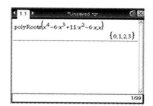

On a GDC you obtain zeros, but for the factor form of the polynomial you need to use the FTA.

Example 27

Given that 2 is a zero of the polynomial
$f(x) = x^5 - 4x^4 - 3x^3 + 34x^2 - 52x + 24$
and has a multiplicity of 3, factorize $f(x)$ fully and check your answer with a GDC.

GDC help on CD:
Alternative demonstrations for the TI-84 Plus and Casio fx-9860GII GDCs are on the CD.

Answer

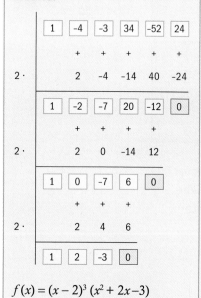

1	-4	-3	34	-52	24
	+	+	+	+	+

$2 \cdot \quad\quad 2 \quad -4 \quad -14 \quad 40 \quad -24$

1	-2	-7	20	-12	0
	+	+	+	+	

$2 \cdot \quad\quad 2 \quad 0 \quad -14 \quad 12$

1	0	-7	6	0
	+	+	+	

$2 \cdot \quad\quad 2 \quad 4 \quad 6$

1	2	-3	0

Successively apply Horner's algorithm with respect to the multiplicity of the given zero.

$f(x) = (x-2)^3(x^2 + 2x - 3)$

$\quad\quad = (x-2)^3(x+1)(x+3)$

Apply the FTA to factorize the polynomial.
Factorize the quadratic expression.

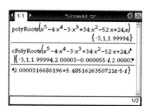

Due to the imperfection of the calculator's algorithm you obtain the approximation of the multiple zero (2), without its multiplicity. When using a complex roots finder you will find which zero has multiplicity, but an approximated value is given.

Exercise 3L

1 Given that k is a zero of multiplicity n of the polynomial f, factorize it fully and check your answers with a GDC.

   a $k = -2$, $n = 2$, $f(x) = 2x^4 + 3x^3 - 10x^2 - 12x + 8$

   b $k = \dfrac{1}{2}$, $n = 2$, $f(x) = 12x^3 - 32x^2 + 23x - 5$

2 Find a polynomial of the smallest degree, with integer coefficients, whose zeros are:

   a   1, 3 and 5

   b $-2, -1, 0$ and 1

   c $-\dfrac{2}{3}$, 1, 2 and 3

3 Find a polynomial of the smallest degree, with integer coefficients, whose zeros are:

   a $-\sqrt{2}$ and $\sqrt{3}$

   b $-\dfrac{1}{2}, \dfrac{3}{4}$ and $\sqrt{5}$

   c $-\dfrac{3}{5}$, $1 - \sqrt{2}$ and $\sqrt[3]{3}$

4 Factorize these polynomials and check your answers with a GDC.

   a $f(x) = x^3 - 2x^2 - 5x + 6$

   b $f(x) = 2x^3 - x^2 - 7x + 6$

   c $f(x) = 5x^4 - 12x^3 - 14x^2 + 12x + 9$

Conjugate root theorem

Given a polynomial

$f(x) = a_n x^n + a_{n-1} x^{n-1} + \ldots + a_2 x^2 + a_1 x + a_0$, $a_k \in \mathbb{R}$, $k = 0, 1, \ldots, n$, and $a_n \neq 0$, that has a complex zero z, then its conjugate z^* is also a zero of the polynomial f.

Proof:

Using the properties of conjugate numbers (see page 112):

$$f(z) = 0 \Rightarrow f(z^*) = a_n(z^*)^n + a_{n-1}(z^*)^{n-1} + \ldots + a_2(z^*)^2 + a_1(z^*) + a_0$$

$$= a_n(z^n)^* + a_{n-1}(z^{n-1})^* + \ldots + a_2(z^2)^* + a_1(z^*) + a_0 \qquad \textit{Conjugate of a power.}$$

$$= (a_n z^n)^* + (a_{n-1} z^{n-1})^* + \ldots + (a_2 z^2)^* + (a_1 z)^* + (a_0)^* \qquad \textit{Conjugate of a product.}$$

$$= (a_n z^n + a_{n-1} z^{n-1} + \ldots + a_2 z^2 + a_1 z + a_0)^* \qquad \textit{Conjugate of a sum.}$$

$$= (f(z))^* = 0^* = 0 \qquad \text{QED}$$

Example 28

Given that $4 + 5i$ is a complex zero of the polynomial $f(x) = x^3 - 6x^2 + 25x + 82$, find all the remaining zeros and check your answers with a GDC.

Answer

$x_1 = 4 + 5i \Rightarrow x_2 = 4 - 5i$

Use the conjugate zero theorem.

Method 1

$(x - (4 + 5i))(x - (4 - 5i)) = x^2 - 8x + 41$

Find the quadratic factor.

$$
\begin{array}{r}
x + 2 \\
x^2 - 8x + 41 \overline{\smash{)}\ x^3 - 6x^2 + 25x + 82} \\
-\underline{(x^3 - 8x^2 + 41x)} \\
2x^2 - 16x + 82 \\
-\underline{(2x^2 - 16x + 82)} \\
0
\end{array}
$$

Use long division to find the last linear factor.

$f(x) = (x - (4 + 5i))(x - (4 - 5i))(x + 2)$
$x - 2 = 0$
$\Rightarrow x_3 = 2$

GDC help on CD:
Alternative demonstrations for the TI-84 Plus and Casio fx-9860GII GDCs are on the CD.

Method 2

	1	-6	25	82
		+	+	+
$(4 + 5i) \cdot$		$4 + 5i$	$-33 + 10i$	-82
	1	$-2 + 5i$	$-8 + 10i$	0
		+	+	
$(4 - 5i) \cdot$		$4 - 5i$	$8 - 10i$	
	1	2	0	

Successively apply Horner's algorithm to the complex numbers $4 - 5i$ and $4 + 5i$.

$f(x) = (x - (4 + 5i))(x - (4 - 5i))(x + 2)$
$x + 2 = 0$
$\Rightarrow x_3 = -2$

Fully factorize the polynomial.
Find the last remaining zero.

To check with a GDC use the feature called 'Complex Roots of Polynomials' within the polynomial tools.

Example 29

Given that i is a complex zero of the polynomial $f(x) = x^4 - 2x^3 + 6x^2 + ax + 5$, $a \in \mathbb{R}$, find the value of a. Hence, find all the remaining zeros and check your answers with a GDC.

Answer

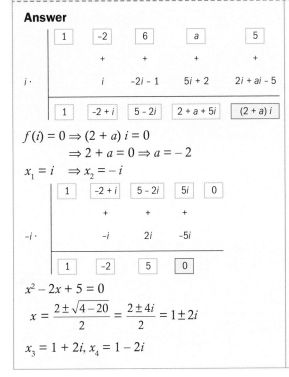

$f(i) = 0 \Rightarrow (2 + a) i = 0$

$\qquad \Rightarrow 2 + a = 0 \Rightarrow a = -2$

$x_1 = i \Rightarrow x_2 = -i$

$x^2 - 2x + 5 = 0$

$x = \dfrac{2 \pm \sqrt{4 - 20}}{2} = \dfrac{2 \pm 4i}{2} = 1 \pm 2i$

$x_3 = 1 + 2i,\ x_4 = 1 - 2i$

Apply Horner's algorithm for $x_1 = i$

Apply the remainder theorem.

Use the conjugate zero theorem.

Use $a = -2$ and continue to apply Horner's algorithm for $x_2 = -i$ to obtain the quotient.

Find the zeros of the quotient polynomial.

Apply the quadratic formula.

GDC help on CD:
Alternative demonstrations for the TI-84 Plus and Casio fx-9860GII GDCs are on the CD.

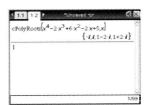

Exercise 3M

1 Given a polynomial f and the zero z, find all the remaining zeros.

a $f(x) = x^3 + 3x^2 + 4x + 12$, $z = 2i$

b $f(x) = x^3 - 6x^2 + 13x - 20$, $z = 1 - 2i$

c $f(x) = 5x^3 + 17x^2 + 21x + 6$, $z = -\dfrac{3}{2} + \dfrac{\sqrt{3}}{2}i$

d $f(x) = x^4 - 4x^3 + 5x^2 - 4x + 4$, $z = -i$

e $f(x) = 2x^4 + 3x^3 + 17x^2 - 12x - 10$, $z = -1 - 3i$

f $f(x) = 2x^4 + 9x^3 + 11x^2 - 7x - 15$, $z = -2 + i$

g $f(x) = 6x^4 + 26x^3 + 35x^2 + 36x + 9$, $z = -\dfrac{1}{2} + \dfrac{\sqrt{5}}{2}i$

h $f(x) = 3x^4 - 2x^3 + 4x^2 - 2x + 1$, $z = \dfrac{1}{3} + \dfrac{\sqrt{2}}{3}i$

2 Given that z is a complex zero of the polynomial f, find the missing coefficients. Hence, find all the remaining zeros and check your answers with a GDC.

a $z = -1$, $f(x) = x^3 - 13x + a$, $a \in \mathbb{R}$

b $z = 3$, $f(x) = x^3 - 7x^2 + ax - 15$, $a \in \mathbb{R}$

c $z = -1 - i$, $f(x) = x^4 + 2x^3 - 2x^2 - 8x + a$, $a \in \mathbb{R}$

d $z = -2i$, $f(x) = x^4 - 4x^3 + 9x^2 + ax + b$, $a, b \in \mathbb{R}$

Sum and product of polynomial roots

François Viète developed formulae that connect the zeros and the coefficients of a polynomial. Viète was the first to investigate this connection for positive real zeros. Albert Girard was the first to extend that to complex zeros.

Albert Girard
(1595–1632) introduced the abbreviations sin, cos and tan for trigonometric functions. He enrolled at the University of Leiden at the age of 22. Before that he was a professional musician, playing the lute.

Polynomials of the third degree

> **→ Theorem**
>
> Given a cubic equation $ax^3 + bx^2 + cx + d = 0$, $a, b, c, d \in \mathbb{R}$, $a \neq 0$
>
> and solutions x_1, x_2 and x_3 then
> $$\begin{cases} x_1 + x_2 + x_3 = -\dfrac{b}{a} \\[2mm] x_1 \cdot x_2 + x_1 \cdot x_3 + x_2 \cdot x_3 = \dfrac{c}{a} \\[2mm] x_1 \cdot x_2 \cdot x_3 = -\dfrac{d}{a} \end{cases}$$

These are Viète's formulae for cubic equations.

Proof

Using the previous results factorize the cubic polynomial
$$f(x) = ax^3 + bx^2 + cx + d = a(x - x_1)(x - x_2)(x - x_3)$$
$$\Rightarrow x^3 + \frac{b}{a}x^2 + \frac{c}{a}x + \frac{d}{a} = (x - x_1)(x - x_2)(x - x_3)$$

Expand the right-hand side of the equation and equate the corresponding coefficients:
$$(x - x_1)(x - x_2)(x - x_3)$$
$$= (x^2 - (x_1 + x_2)x + x_1x_2)(x - x_3)$$
$$= x^3 - (x_1 + x_2)x^2 + x_1x_2 \cdot x - x^2x_3 + (x_1 + x_2)x \cdot x_3 - x_1x_2x_3$$
$$= x^3 - (x_1 + x_2 + x_3)x^2 + (x_1x_2 + x_1x_3 + x_2x_3)x - x_1x_2x_3$$

$$\Rightarrow \begin{cases} -(x_1 + x_2 + x_3) = \dfrac{b}{a} \\[2mm] x_1 \cdot x_2 + x_1 \cdot x_3 + x_2 \cdot x_3 = \dfrac{c}{a} \\[2mm] -(x_1 \cdot x_2 \cdot x_3) = \dfrac{d}{a} \end{cases} \Rightarrow \begin{cases} x_1 + x_2 + x_3 = -\dfrac{b}{a} \\[2mm] x_1 \cdot x_2 + x_1 \cdot x_3 + x_2 \cdot x_3 = \dfrac{c}{a} \\[2mm] x_1 \cdot x_2 \cdot x_3 = -\dfrac{d}{a} \end{cases}$$

Investigation – Coefficients of a quartic polynomial

Viète's formulae connect the zeros and the coefficients of a cubic polynomial.

Find similar formulae that satisfy the relationship between the zeros and the coefficients of a quartic polynomial
$$f(x) = ax^4 + bx^3 + cx^2 + dx + e,\ a, b, c, d, e \in \mathbb{R},\ a \neq 0$$

Example 30

Given that the roots of a cubic equation $2x^3 + 4x^2 - 7x + 5 = 0$ are x_1, x_2 and x_3, without solving the equation, find:

a $x_1 + x_2 + x_3$ **b** $x_1 \cdot x_2 \cdot x_3$ **c** $x_1 \cdot x_2 + x_1 \cdot x_3 + x_2 \cdot x_3$

d $\dfrac{1}{x_1} + \dfrac{1}{x_2} + \dfrac{1}{x_3}$ **e** $x_1^2 + x_2^2 + x_3^2$

Answers

$a = 2,\ b = 4,\ c = -7,\ d = 5$ *Identify the coefficients of the cubic polynomial.*

a $x_1 + x_2 + x_3 = -\dfrac{4}{2} = -2$ *Use $x_1 + x_2 + x_3 = -\dfrac{b}{a}$.*

b $x_1 \cdot x_2 \cdot x_3 = -\dfrac{5}{2}$ *Use $x_1 \cdot x_2 \cdot x_3 = -\dfrac{d}{a}$.*

c $x_1 \cdot x_2 + x_1 \cdot x_3 + x_2 \cdot x_3 = -\dfrac{7}{2}$ *Use $x_1 \cdot x_2 + x_1 \cdot x_3 + x_2 \cdot x_3 = \dfrac{c}{a}$.*

d $\dfrac{1}{x_1} + \dfrac{1}{x_2} + \dfrac{1}{x_3} = \dfrac{x_2 x_3 + x_1 x_3 + x_1 x_2}{x_1 x_2 x_3}$ *Use the results found in parts **a** and **b**.*

$$= \dfrac{-\dfrac{7}{2}}{-\dfrac{5}{2}} = \dfrac{7}{5}$$

e $x_1^2 + x_2^2 + x_3^2$ *Use the formula $(x + y + z)^2$*
$$= (x_1 + x_2 + x_3)^2 - 2x_1 x_2 - 2x_1 x_3 - 2x_2 x_3$$ *$= x^2 + y^2 + z^2 + 2xy + 2xz + 2yz$.*

$$= (-2)^2 - 2\left(-\dfrac{7}{2}\right) = 4 + 7 = 11$$ *Use the results found in parts **a** and **c**.*

Theorem

Given a polynomial $f(x) = a_n x^n + a_{n-1} x^{n-1} + \ldots + a_2 x^2 + a_1 x + a_0$ with real or complex coefficients ($a_n \neq 0$) and zeros $x_1, x_2, \ldots, x_n$ then

$$\begin{cases} x_1 + x_2 + x_3 + \ldots + x_n = -\dfrac{a_{n-1}}{a_n} \\[2mm] x_1 x_2 + x_1 x_3 + \ldots + x_1 x_n + x_2 x_3 + x_2 x_4 + \ldots + x_2 x_n + \ldots + x_{n-1} x_n = \dfrac{a_{n-2}}{a_n} \\[2mm] \vdots \\[2mm] x_1 x_2 x_3 \cdot \ldots \cdot x_n = (-1)^n \dfrac{a_0}{a_n} \end{cases}$$

These are Viète's formulae for an 'equation' of the nth degree.

→ As a general system:
$$\sum_{1 \leq i_1 < i_2 < \ldots < i_k \leq n} \left(x_{i_1} \cdot x_{i_2} \cdot \ldots \cdot x_{i_k} \right) = (-1)^k \dfrac{a_{n-k}}{a_n},\ 1 \leq k \leq n$$

Proof

The proof is a direct consequence of the ability to write polynomials in factorized form:

$$f(x) = a_n(x - x_1)(x - x_2) \ldots (x - x_n).$$

By expanding the right-hand side and comparing coeffecients you obtain the given formulae.

Example 31

Find the sum and product of the zeros of these polynomials.

a $f(x) = x^4 - 3x^3 + 11x^2 + 17x - 4$

b $f(x) = 3x^5 + 11x^4 - 4x^3 + 5x^2 - 13x + 9$

c $f(x) = 17x^{13} + 4x^{12} + 122x^2 - 14x - 17$

d $f(x) = 3x^{2012} + 7x^{370} - 4x^{25} - 15x + 2$

Answers

a $f(x) = x^4 - 3x^3 + 11x^2 + 17x - 4$

$n = 4,\ a_4 = 1,\ a_3 = -3,\ a_2 = 11,\ a_1 = 17,\ a_0 = -4$

Identify n and the coefficients of the polynomial.

$$x_1 + x_2 + x_3 + x_4 = -\frac{a_3}{a_4} \Rightarrow x_1 + x_2 + x_3 + x_4 = -\frac{-3}{1} = 3$$

Use $x_1 + x_2 + \ldots + x_n = -\dfrac{a_{n-1}}{a_n}$

$$x_1 x_2 x_3 x_4 = (-1)^4 \frac{a_0}{a_4} \Rightarrow x_1 x_2 x_3 x_4 = \frac{-4}{1} = -4$$

Use $x_1 x_2 x_3 \cdots x_n = (-1)^n \dfrac{a_0}{a_n}$

b $f(x) = 3x^5 + 11x^4 - 4x^3 + 5x^2 - 13x + 9$

$n = 5,\ a_5 = 3,\ a_4 = 11,\ a_0 = 9$

For a polynomial of degree 5 you need a_5, a_4 and a_0

$$\sum_{r=1}^{5} x_r = -\frac{a_4}{a_5} \Rightarrow \sum_{r=1}^{5} x_r = -\frac{11}{3}$$

$$\prod_{r=1}^{5} x_r = (-1)^5 \frac{a_0}{a_5} \Rightarrow \prod_{r=1}^{5} x_r = (-1)^5 \frac{9}{3} = -3$$

c $f(x) = 17x^{13} + 4x^{12} + 122x^2 - 14x - 17$

$n = 13,\ a_{13} = 17,\ a_{12} = 4,\ a_0 = -17$

For a polynomial of degree 13 you need a_{13}, a_{12} and a_0

$$\sum_{r=1}^{13} x_r = -\frac{a_{12}}{a_{13}} \Rightarrow \sum_{r=1}^{5} x_r = -\frac{4}{17}$$

$$\prod_{r=1}^{13} x_r = (-1)^5 \frac{a_0}{a_{13}} \Rightarrow \prod_{r=1}^{13} x_r = (-1)^{17} \frac{-17}{17} = 1$$

d $f(x) = 3x^{2012} + 7x^{370} - 4x^{25} - 15x + 2$

$n = 2012,\ a_{2012} = 3,\ a_{2011} = 0,\ a_0 = 2$

$$\sum_{r=1}^{2012} x_r = -\frac{a_{2011}}{a_{2012}} \Rightarrow \sum_{r=1}^{2012} x_r = -\frac{0}{3} = 0$$

$$\prod_{r=1}^{2012} x_r = (-1)^{2012} \frac{a_0}{a_{2012}} \Rightarrow \prod_{r=1}^{2012} x_r = (-1)^{2012} \frac{2}{3} = \frac{2}{3}$$

In part **a** all the coefficients were listed but they are not all needed for the formulae.

For a polynomial of degree n you need a_n, a_{n-1} and a_0

Note: You can check the results for Example 31 using a GDC.

a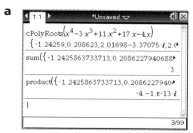

b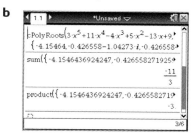

c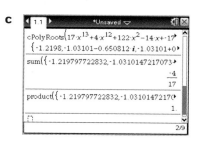

d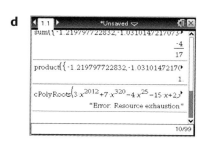

GDC help on CD:
Alternative demonstrations for the TI-84 Plus and Casio fx-9860GII GDCs are on the CD.

In part **c** you have to adjust the accuracy when converting to a decimal, while in part **d** the degree of the polynomial was too large for the algorithm for finding complex zeros. Note that for the sum and product of the solutions given in the form of a list you use the Math List menu.

Exercise 3N

1 The roots of a cubic equation $3x^3 - 2x^2 - 5x - 4 = 0$ are x_1, x_2 and x_3. Without solving the equation, find:

a $x_1 + x_2 + x_3$

b $x_1 \cdot x_2 \cdot x_3$

c $x_1 \cdot x_2 + x_1 \cdot x_3 + x_2 \cdot x_3$

d $\dfrac{6}{x_1} + \dfrac{6}{x_2} + \dfrac{6}{x_3}$

e $9x_1^2 + 9x_2^2 + 9x_3^2$

Check your results using a GDC.

2 The roots of a quartic equation
$x^4 - 3x^3 + 2x^2 - 4x - 6 = 0$ are x_1, x_2, x_3 and x_4.
Without solving the equation, find:

a $x_1 + x_2 + x_3 + x_4$

b $x_1 \cdot x_2 \cdot x_3 \cdot x_4$

c $x_1 \cdot x_2 + x_1 \cdot x_3 + x_1 \cdot x_4 + x_2 \cdot x_3 + x_2 \cdot x_4 + x_3 \cdot x_4$

d $x_1 \cdot x_2 \cdot x_3 + x_1 \cdot x_2 \cdot x_4 + x_1 \cdot x_3 \cdot x_4 + x_2 \cdot x_3 \cdot x_4$

e $\dfrac{3}{x_1} + \dfrac{3}{x_2} + \dfrac{3}{x_3} + \dfrac{3}{x_4}$

f $\dfrac{x_1^2}{5} + \dfrac{x_2^2}{5} + \dfrac{x_3^2}{5} + \dfrac{x_4^2}{5}$

Check your results using a GDC.

3 Find the sum and product of the zeros of these polynomials.

  a $f(x) = x^4 + 2x^3 - 3x^2 + 4x + 5$

  b $f(x) = 4x^6 + x^5 + 7x^4 - 3x^3 + 2x$

  c $f(x) = 11x^{10} - \dfrac{3}{7}x^7 + \sqrt{5} \cdot x^3 - \pi x + 22$

  d $f(x) = 5x^{7007} - 4x^{7006} + 2x^{231} + 10x + 8$

3.5 Polynomial equations and inequalities

Some useful theorems

Factorization is a common method used to solve polynomial equations. Descartes' rule of sign, the integer zero theorem and the rational zero theorem are valid for polynomials of all degrees and are an aid to finding factors.

Before factorizing, it is useful to know how many real zeros to expect for a given polynomial. **René Descartes** (1596–1650), in his work *La Géométrie,* noticed the following property.

Descartes' rule of signs

The number of positive real roots of a polynomial f is equal to the number of sign changes (from + to – or from – to +) of its coefficients, or an even number less. Also the number of negative real roots of a polynomial f is equal to the number of sign changes of the coefficients of $f(-x)$, or an even number less.

For example, the polynomial $f(x) = x^3 - 7x^2 - 9x + 18$ has the following sequence of signs: +, –, –, +. Here there are two sign changes so there are two or zero (an even number less) positive real roots. Now look at $f(-x) = (-x)^3 - 7(-x)^2 - 9(-x) + 18 = -x^3 - 7x^2 + 9x + 18$, which has the sequence of signs –, –, +, +. In this sequence there is only one sign change, so the polynomial f can have only one negative real root.

The following theorems are valid for polynomials with integer coefficients.

Integer zero theorem

> → Given a polynomial
> $$f(x) = a_n x^n + a_{n-1} x^{n-1} + \ldots + a_2 x^2 + a_1 x + a_0, \; a_k \in \mathbb{Z}, \; a_n \neq 0$$
> and an integer p such that $f(p) = 0$, then p is a factor of a_0.

Proof

$$f(p) = a_n p^n + a_{n-1} p^{n-1} + \ldots + a_2 p^2 + a_1 p + a_0 = 0$$
$$\Rightarrow a_n p^n + a_{n-1} p^{n-1} + \ldots + a_2 p^2 + a_1 p = -a_0$$
$$\Rightarrow p(a_n p^{n-1} + a_{n-1} p^{n-2} + \ldots + a_2 p + a_1) = -a_0$$

Therefore, p is a factor of a_0.   QED

Some cases of cubic equations were solved by the Babylonians (2000–1600 BCE). They used tables with perfect squares, perfect cubes and their sums. They were able to solve equations of the form $ax^3 + bx = c$. Later, in the 13th and 14th centuries, a group of Italian mathematicians, dal Ferro, Tartaglia and Cardano, developed a formula for solving a general cubic equation.

140   The long journey of mathematics

Sometimes, when the coefficient a_0 is not prime there are many possible factors. For example, if $a_0 = 18 \Rightarrow p \in \{\pm1, \pm2, \pm3, \pm6, \pm9, \pm18\}$.

In these cases the search for all possible zeros would take a long time. To speed up the process there is a corollary that reduces the set of possible zeros.

Corollary 1

Given a polynomial
$$f(x) = a_n x^n + a_{n-1} x^{n-1} + \ldots + a_2 x^2 + a_1 x + a_0, \; a_k \in \mathbb{Z}, \; a_n \neq 0$$
and an integer value p such that $f(p) = 0$, then for any integer value q, $p - q$ is a factor of $f(q)$.

Proof

$$f(p) = a_n p^n + a_{n-1} p^{n-1} + \ldots + a_2 p^2 + a_1 p + a_0 = 0 \qquad (1)$$
$$f(q) = a_n q^n + a_{n-1} q^{n-1} + \ldots + a_2 q^2 + a_1 q + a_0 \qquad (2)$$

Equation (1) – equation (2)

$$\underbrace{f(p)}_{0} - f(q) = a_n \left(p^n - q^n \right) + a_{n-1} \left(p^{n-1} - q^{n-1} \right) + \ldots + a_2 \left(p^2 - q^2 \right) + a_1 \left(p - q \right).$$

The terms on the right-hand side of the equation are grouped in such a way that every term containing $(p^r - q^r)$, $r = 1, 2, \ldots, n$, has a factor of $p - q$, so $p - q$ is a factor of $f(q)$. QED

This corollary is useful when there are many possible factors for integer zeros as you can eliminate some and simplify the search.

> $a^n - b^n$ is divisible by $a - b$ for all positive integers n. The formula $a^n - b^n = (a - b)(a^{n-1} + a^{n-2} b + \ldots + ab^{n-2} + b^{n-1})$ was proved in Chapter 1.

Example 32

Find all the possible integer zeros of the polynomial
$f(x) = x^3 - 7x^2 - 9x + 18$

Answer

$p \in \{\pm 1, \pm 2, \pm 3, \pm 6, \pm 9, \pm 18\}$	*List all the possible zeros, i.e. factors of 18, by using the integer zero theorem.*
$f(1) = 1^3 - 7 \times 1^2 - 9 \times 1 + 18 = 3$	*Use q = 1 to reduce the set of possible factors by using the corollary.*
$p - 1 \in \{\pm1, \pm3\}$	*p – 1 is a factor of 3.*
$\Rightarrow p \in \{-2, 0, 2, 4\}$	
$p \in \{-2, 2\}$	*This is the intersection of both sets.*

Note: By using the Corollary 1, you only need to inspect two values (instead of all twelve possible values).

By using Descartes' rule of signs you can eliminate the positive solution (+2) since there must either be two or no positive roots (there cannot be only one).

Apply synthetic division for both values (2 and −2) to check this conclusion.

Since the remainder when $f(x)$ is divided by $(x - 2)$ is −20, the polynomial is not divisible by $(x - 2)$, so Descartes' rule works well. The remainder when divided by $(x + 2)$ is zero.

$$x^3 - 7x^2 - 9x + 18 = (x + 2)(x^2 - 9x + 9)$$

Notice that by examining both possible integer zeros you can conclude that the only integer zero is −2, and you did not need to factorize the quadratic quotient.

The next theorem is a generalization from integers to rational zeros and the proof is similar.

Rational zero theorem

> → Given a polynomial
>
> $$f(x) = a_n x^n + a_{n-1} x^{n-1} + \ldots + a_2 x^2 + a_1 x + a_0, \ a_i \in \mathbb{Z}, \ a_n \neq 0 \text{ and a}$$
> rational number $\dfrac{p}{q}$, where $\gcd(p, q) = 1$ that is $\left(\dfrac{p}{q}\right)$ is in its
>
> simplest form, such that $f\left(\dfrac{p}{q}\right) = 0$, then p is a factor of a_0 and q is a factor of a_n.

$\gcd(p, q)$ means greatest common divisor of p and q.

Proof

$$f\left(\frac{p}{q}\right) = a_n \left(\frac{p}{q}\right)^n + a_{n-1}\left(\frac{p}{q}\right)^{n-1} + \ldots + a_2 \left(\frac{p}{q}\right)^2 + a_1 \frac{p}{q} + a_0 = 0 \quad (1)$$

You can also say that p and q are co-prime.

Multiply equation (1) by q^n

$$a_n p^n + a_{n-1} p^{n-1} q + \ldots + a_2 p^2 q^{n-2} + a_1 p q^{n-1} + a_0 q^n = 0 \quad (2)$$

Rearrange the equation (2)

$$p(a_n p^{n-1} + a_{n-1} p^{n-2} q + \ldots + a_2 p q^{n-2} + a_1 q^{n-1}) = -a_0 q^n \quad (3)$$

Since the left-hand side of the equation (3) has a factor p and $\gcd(p, q) = 1$, then p must be a factor of a_0.

In a similar way, we can rearrange equation (2) to obtain

$$a_n p^n = -q(a_{n-1}p^{n-1} + \ldots + a_2 p^2 q^{n-3} + a_1 p\, q^{n-2} + a_0\, q^{n-1}) \quad (4)$$

Again, since the right-hand side of the equation (4) has a factor q and $\gcd(p, q) = 1$,

we can conclude that q is a factor of a_n.　　　QED

Corollary 2

Given a polynomial $f(x) = a_n x^n + a_{n-1} x^{n-1} + \ldots + a_2 x^2 + a_1 x + a_0$,

$a_i \in Z$, $a_n \neq 0$ and a rational number $\dfrac{p}{q}$, where $\gcd(p, q) = 1$ such

that $f\left(\dfrac{p}{q}\right) = 0$, then for any real value k, $(p - qk)$ is a factor of $f(k)$.

Example 33

Given that the polynomial $f(x) = 2x^3 - 11x^2 - 11x + 15$ has no integer zeros, find its only rational zero.

Answer

$\dfrac{p}{q} \in \left\{ \pm\dfrac{1}{2}, \pm\dfrac{3}{2}, \pm\dfrac{5}{2}, \pm\dfrac{15}{2} \right\}$

List all the possible rational (non-integer) zeros by using the rational zero theorem.

$f(1) = 2 \times 1^3 - 11 \times 1^2 - 11 \times 1 + 15$
$\quad = -5$

Use corollary 2 with k=1 to reduce the set of possible zeros.

$p - q \in \{\pm 1, \pm 5\}$

p − 1 · q is a factor of −5.

$\dfrac{p}{q} \in \left\{ -\dfrac{3}{2}, \dfrac{1}{2}, \dfrac{3}{2} \right\}$

This is the intersection of both sets.

Sequence of signs for $f(x)$: $+, -, -, +$
$\Rightarrow$ 2 or 0 positive roots
Sequence of signs for $f(-x)$:
$-, -, +, +$
$\Rightarrow$ only 1 negative root

Apply Descartes' rule of signs.

Since there is only one rational zero it can only be the negative one $\left(-\dfrac{3}{2}\right)$, as complex zeros come in conjugate pairs.

	2	-11	-11	15
$-\dfrac{3}{2} \cdot$		+	+	+
		-3	21	-15
	2	-14	10	0

Use synthetic division.

$2x^3 - 11x^2 - 11x + 15$

$= \left(x + \dfrac{3}{2}\right)\underbrace{(2x^2 - 14x + 10)}_{2(x^2 - 7x + 5)}$

Factorize the quotient to simplify the divisor.

$= (2x + 3)(x^2 - 7x + 5)$
So again, the only rational

zero is $-\dfrac{3}{2}$

The quotient is a quadratic expression $x^2 - 7x + 5$ whose discriminant is 29. Therefore, the remaining two solutions are irrational.

Using Corollary 2, you only need to inspect three values (instead of all eight possible rational zeros).

Exercise 30

1   Solve these equations in the set of real numbers and check your
    answers with a GDC.

    a $x^3 - 6x^2 + 11x - 6 = 0$ **b** $x^3 + 2x^2 - 7x + 4 = 0$

    c $x^3 + 3x^2 - 4x - 12 = 0$ **d** $2x^3 - 5x^2 - 18x + 45 = 0$

Solving polynomial equations

In this section you will solve polynomial equations using an
algebraic method and then use a GDC to verify the solutions
(graphical method).

To solve a polynomial equation by a graphical method you need to
find the points of intersection of the graph with the x-axis. At these
points the value of the function is zero ($y = 0$), so these points are
called the *zeros of the function* or the *roots*.

Example 34

Solve these equations and check your answers by using a graphical method.

a $x^3 + 2x^2 - 5x - 6 = 0$

b $6x^4 + 17x^3 + 10x^2 - 7x - 6 = 0$

Answers

a Algebraic method:

$x^3 + 2x^2 - 5x - 6 = \underline{x^3 + 2x^2 + x} - 6x - 6$

$\qquad\qquad = x(x^2 + 2x + 1) - 6(x + 1)$

$\qquad\qquad = x(x + 1)^2 - 6(x + 1)$

$\qquad\qquad = (x + 1)(x(x + 1) - 6)$

$\qquad\qquad = (x + 1)(x^2 + x - 6)$

$\qquad\qquad = (x + 1)(x - 2)(x + 3)$

$\qquad\qquad \Rightarrow (x + 1)(x - 2)(x + 3) = 0$

$\qquad\qquad \Rightarrow x_1 = -1, x_2 = 2, x_3 = -3$

Split the linear term (5x) for a common factor.

Notice the perfect square $(x + 1)^2$
Use distribution with the common factor $(x + 1)$.
Factorize the quadratic factor $x^2 + x - 6$

Use the zero product theorem.

Graphical method:

$x^3 + 2x^2 - 5x - 6 = 0 \Rightarrow f(x) = x^3 + 2x^2 - 5x - 6$

The solutions are $x_1 = -3, x_2 = -1$ and $x_3 = 2$

b $6x^4 + 17x^3 + 10x^2 - 7x - 6 = 0$

Algebraic method:

$\dfrac{p}{q} \in \left\{ \pm 1, \pm 2, \pm 3, \pm 6, \pm \dfrac{1}{2}, \pm \dfrac{1}{3}, \pm \dfrac{2}{3}, \pm \dfrac{1}{6} \right\}$

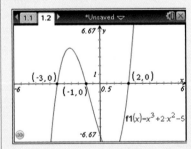

List all the possible rational zeros by using the rational zero theorem.

GDC help on CD:
Alternative demonstrations for the TI-84 Plus and Casio fx-9860GII GDCs are on the CD.

▶ Continued on next page

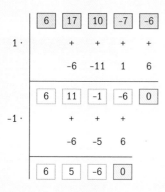

Use synthetic division.

$6x^4 + 17x^3 + 10x^2 - 7x - 6$
$= (x + 1)^2 (6x^2 + 5x - 6)$
$= (x + 1)^2 (2x + 3) (3x - 2)$
$\Rightarrow (x + 1)^2 (2x + 3) (3x - 2) = 0$

$\Rightarrow x_{1,2} = -1, \; x_3 = -\dfrac{3}{2}, \; x_4 = \dfrac{2}{3}$

Factorize the quadratic expression.

Use the zero product theorem.

Graphical method:
The solutions are

$x_1 = -\dfrac{3}{2}, \; x_{2,3} = -1$ and $x_4 = \dfrac{2}{3}$

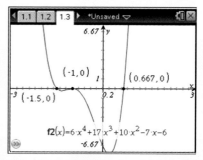

Note that at the point (−1, 0) the graph is just touching, that is tangent to, the x-axis. In this case (−1, 0) is a double, or repeated, zero of the function.

A polynomial of degree n can have up to n roots on the real number line. It is useful to be able to restrict any search for roots to a finite window. The next theorem provides such a search window.

Theorem

> → All the possible zeros of the polynomial
> $f(x) = a_n x^n + a_{n-1} x^{n-1} + \dots + a_1 x + a_0$ are in the interval
> $\left[-\left(\dfrac{M}{|a_n|} + 1 \right), \dfrac{M}{|a_n|} + 1 \right]$ where $M = \max \{ |a_n|, |a_{n-1}|, \dots |a_1|, |a_0| \}$

For the equations in Example 34:

a $M = 6, |a_3| = 1 \Rightarrow \left[-\left(\dfrac{6}{1} + 1 \right), \dfrac{6}{1} + 1 \right] = [-7, 7]$

b $M = 17, |a_3| = 6 \Rightarrow \left[-\left(\dfrac{17}{6} + 1 \right), \dfrac{17}{6} + 1 \right] = \left[-\dfrac{23}{6}, \dfrac{23}{6} \right]$

Notice that the zeros satisfy the conditions of the theorem.

You could solve both equations on a GDC using the *Polynomial tools* feature.

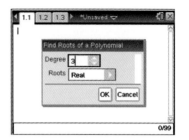 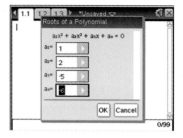 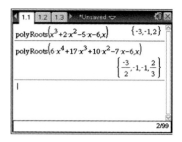

Notice that in the solution to part **b** the GDC writes the multiplicity of the zero by repeating the value of the zero (−1) twice.

GDC help on CD:
Alternative demonstrations for the TI-84 Plus and Casio fx-9860GII GDCs are on the CD.

Exercise 3P

1 Solve these equations in the set of real numbers and check your answers with a GDC.

  a $12x^3 + 17x^2 + 2x - 3 = 0$ **b** $x^3 - 4x^2 - 5x + 14 = 0$

  c $3x^3 - 13x^2 + 11x + 14 = 0$ **d** $x^4 - x^3 - 11x^2 + 9x + 18 = 0$

You will graph more polynomials in Chapter 4.

2 One of the the roots of the equation $x^3 + ax^2 - x - 3 = 0$ is equal to −3.

  a Find the value of a.

  b Find the other two roots.

3 The equation $ax^3 - 7x^2 + bx + 4 = 0$ has one double root which is equal to 2.

  a Find the values of a and b.

  b Find the remaining root.

4 Show that the polynomial $f(x) = x^3 + 5x + p$ does not have an integer zero when p is a prime number.

5 Two of the zeros of the polynomial $f(x) = x^3 + ax^2 + bx + c$, $a, b, c \in \mathbb{R}$ are opposite numbers.

  a Show that $ab = c$.

  b Find the third zero.

Solving polynomial inequalities

To solve polynomial inequalities by an algebraic method you factorize the polynomial and investigate the signs of the factors in a sign table. Then you find the values of x for which the inequality is true.

Quadratic inequalities are discussed in Chapter 14, section 2.12.

To solve polynomial inequalities by a graphical method use a GDC to graph the polynomial and identify the values of x for which the inequality is true.

Example 35

Solve these inequalities.

a $x^3 - x^2 - 10x - 8 \geq 0$

b $2x^3 - 5x^2 - 6x + 4 < 0$

c $1 - 4x^2 < 5x^3 + 4x$

Verify your solutions by using a graphical method.

Answers

a Algebraic method:

$x^3 - x^2 - 10x - 8 = 0$

$p \in \{\pm 1, \pm 2, \pm 4, \pm 8\}$

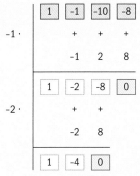

$x^3 - x^2 - 10x - 8 = (x + 2)(x + 1)(x - 4)$

x	$]-\infty, -2[$	-2	$]-2, -1[$	-1	$]-1\ 4[$	4	$]4\ \infty[$
$x + 2$	$-$	0	$+$	$+$	$+$	$+$	$+$
$x + 1$	$-$	$-$	$-$	0	$+$	$+$	$+$
$x - 4$	$-$	$-$	$-$	$-$	$-$	0	$+$
$x^3 - x^2 - 10x - 8$	$-$	0	$+$	0	$-$	0	$+$

$x \in [-2, -1] \cup [4, \infty[$

Graphical method:

$f(x) = x^3 - x^2 - 10x - 8$

$x \in [-2, -1] \cup [4, \infty[$

GDC help on CD:
Alternative demonstrations for the TI-84 Plus and Casio fx-9860GII GDCs are on the CD.

List all the possible zeros.

Use synthetic division.

Fully factorize the polynomial.

Construct the sign table.
Find the product of the signs and zeros.

Use a GDC to draw the graph of the polynomial and identify the parts of the graph that are above the x-axis.

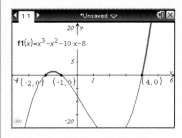

Identify the values of x that satisfy the inequality.

Include zeros since it is not a strict inequality.

▶ Continued on next page

b Algebraic method:

$2x^3 - 5x^2 - 6x + 4 = 0$

$\dfrac{p}{q} \in \left\{\pm 1, \pm 2, \pm 4, \pm\dfrac{1}{2}\right\}$

$k = -1 \Rightarrow f(-1) = 3$

$p + q \in \{\pm 1, \pm 3\}$

$\dfrac{p}{q} \in \left\{2, -4, \pm\dfrac{1}{2}\right\}$

$$
\begin{array}{c|ccc}
 & 2 & -5 & -6 & 4 \\
\frac{1}{2}\cdot & & + & + & + \\
 & & 1 & -2 & -4 \\
\hline
 & 2 & -4 & -8 & 0
\end{array}
$$

$2x^3 - 5x^2 - 6x + 4 = \left(x - \dfrac{1}{2}\right)\underbrace{\left(2x^2 - 4x - 8\right)}_{2\left(x^2 - 2x - 4\right)}$

$\qquad = (2x - 1)(x^2 - 2x - 4)$

$x^2 - 2x - 4 = 0 \Rightarrow x = \dfrac{2 \pm \sqrt{4 + 16}}{2}$

$\qquad\qquad = \dfrac{2 \pm 2\sqrt{5}}{2} = 1 \pm \sqrt{5}$

$2x^3 - 5x^2 - 6x + 4 = (2x - 1)\left(x - 1 + \sqrt{5}\right)\left(x - 1 - \sqrt{5}\right)$

x	$]-\infty, 1, -\sqrt{5}\,[$	$1-\sqrt{5}$	$]1-\sqrt{5}, \frac{1}{2}[$	$\frac{1}{2}$	$]\frac{1}{2}, 1+\sqrt{5}[$	$1+\sqrt{5}$	$]1+\sqrt{5}, \infty[$
$x - 1 + \sqrt{5}$	$-$	0	$+$	$+$	$+$	$+$	$+$
$2x - 1$	$-$	$-$	$-$	0	$+$	$+$	$+$
$x - 1 - \sqrt{5}$	$-$	$-$	$-$	$-$	$-$	0	$+$
Result	$-$	0	$+$	0	$-$	0	$+$

$x \in \left]-\infty, 1 - \sqrt{5}\left[\; \cup \;\right]\dfrac{1}{2}, 1 + \sqrt{5}\right[$

List all the possible rational zeros.

Use k = −1 to reduce the set of possible zeros.

p − (−1) · q must be a factor of 3.

Find the intersection of the two conditions.
Use synthetic division.

Solve the quadratic equation.

Construct the sign table.
Do not include the zeros since the inequality was strict.

▶ Continued on next page

Graphical method:

$f(x) = 2x^3 - 5x^2 - 6x + 4$

GDC help on CD: Alternative demonstrations for the TI-84 Plus and Casio fx-9860GII GDCs are on the CD.

Use a GDC to draw the graph of the polynomial and identify the parts of the graph that are below the x-axis.

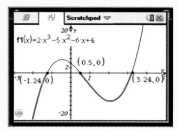

$x \in \left]-\infty, -1.24\right[\cup \left]0.5, 3.24\right[$

Identify the values of x that satisfy the inequality.

> $1 - \sqrt{5} = -1.24 \, (3\,sf)$ and
> $1 + \sqrt{5} = 3.24 \, (3\,sf).$

c Algebraic method:

$1 - 4x^2 < 5x^3 + 4x$

$0 < 5x^3 + 4x^2 + 4x - 1$

$5x^3 + 4x^2 + 4x - 1 = 0$

Rewrite the inequality so that the leading coefficient is positive.

$\dfrac{p}{q} \in \left\{ \pm 1, \pm \dfrac{1}{5} \right\}$

List all the possible rational zeros.

Use synthetic division.

	5	4	4	-1	0
$\frac{1}{5} \cdot$		+	+	+	
		1	1	1	
	5	5	5	0	

$5x^3 + 4x^2 + 4x - 1 = \left(x - \dfrac{1}{5} \right)\underbrace{\left(5x^2 + 5x + 5 \right)}_{5\left(x^2 + x + 1\right)}$

$= (5x - 1)(x^2 + x + 1)$

Solve the quadratic equation.

$x^2 + x + 1 = 0 \Rightarrow x = \dfrac{-1 \pm \sqrt{1-4}}{2} \notin \mathbb{R}$

So, the only real zero is: $x = \dfrac{1}{5}$

The quadratic equation has no real solution so the quadratic expression is irreducible on the set of real numbers.

x	$\left]-\infty, \dfrac{1}{5}\right[$	$\dfrac{1}{5}$	$\left]\dfrac{1}{5}, \infty\right[$
$5x - 1$	$-$	0	$+$
$x^2 + x + 1$	$+$	$+$	$+$
$5x^4 + 39x^3 + 32x^2 + 27x - 7$	$-$	0	$+$

Construct the sign table. Do not include the zeros since the inequality was strict.

$x \in \left]\dfrac{1}{5}, \infty\right[$

▶ Continued on next page

Graphical method:

$f(x) = 5x^3 + 4x^2 + 4x - 1$

$x\ ]0.2, \infty[$

GDC help on CD:
Alternative demonstrations for the TI-84 Plus and Casio fx-9860GII GDCs are on the CD.

Use a GDC to draw the graph of the polynomial and identify the parts of the graph that are above the x-axis.

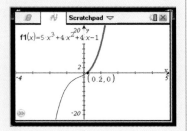

Identify the values of x that satisfy the inequality.

Example 36

Given the polynomials $f(x) = 4x^3 - 17x^2 + 30x + 5$ and $g(x) = -2x^3 + 8x^2 + 9x - 5$
find all the values of x such that $f(x) \leq g(x)$.
Verify your solution by using a graphical method on a GDC.

Answer

Algebraic method:

$4x^3 - 17x^2 + 30x + 5 \leq -2x^3 + 8x^2 + 9x - 5$

$6x^3 - 25x^2 + 21x + 10 \leq 0$

Rewrite the inequality.

$\dfrac{p}{q} \in \left\{\pm 1, \pm 2, \pm 5, \pm 10, \pm\dfrac{1}{2}, \pm\dfrac{5}{2}, \pm\dfrac{1}{3}, \pm\dfrac{2}{3}, \pm\dfrac{5}{3}, \pm\dfrac{1}{6}, \pm\dfrac{5}{6}\right\}$

List all the possible rational zeros.

$k = 1 \Rightarrow f(1) = 12$

Use k = 1 to reduce the set of possible zeros.

$p - q \in \{\pm 1, \pm 2, \pm 3, \pm 4, \pm 6, \pm 12\}$

p − 1 · q must be a factor of 12.

$\dfrac{p}{q} \in \left\{-1, \pm 2, \pm 5, \pm\dfrac{1}{2}, \dfrac{5}{2}, \pm\dfrac{1}{3}, \dfrac{2}{3}, \dfrac{5}{3}, \dfrac{5}{6}\right\}$

Find the intersection of both conditions.

Let $h(x) = 6x^3 - 25x^2 + 21x + 10$.
Sequence of signs for $h(x)$: +, −, +, +
Sequence of signs for $h(-x)$: −, −, −, +
There can be only one negative root and two or zero positive real roots.

Apply Descartes' rule of signs.

Use synthetic division.

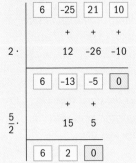

	6	-25	21	10
		+	+	+
2·		12	-26	-10
	6	-13	-5	0
		+	+	
$\frac{5}{2}$·		15	5	
	6	2	0	

▶ Continued on next page

$$6x^3 - 25x^2 + 21x + 10 = (x-2)\left(x - \frac{5}{2}\right)\underbrace{(6x+2)}_{2(3x+1)}$$

$$6x^3 - 25x^2 + 21x + 10 = (x-2)(2x-5)(3x+1)$$

x	$\left]-\infty, -\dfrac{1}{3}\right[$	$-\dfrac{1}{3}$	$\left]-\dfrac{1}{3}, 2\right[$	2	$\left]2, \dfrac{5}{2}\right[$	$\dfrac{5}{2}$	$\left]\dfrac{5}{2}, \infty\right[$
$x - 2$	$-$	$-$	$-$	0	$+$	$+$	$+$
$2x - 5$	$-$	$-$	$-$	$-$	$-$	0	$+$
$3x + 1$	$-$	0	$+$	$+$	$+$	$+$	$+$
$6x^3 - 25x^2 + 21x + 10$	$-$	0	$+$	0	$-$	0	$+$

$$x \in \left]-\infty, -\frac{1}{3}\right] \cup \left[2, \frac{5}{2}\right]$$

Graphical method 1:

$f(x) = 4x^3 - 17x^2 + 30x + 5$

$g(x) = -2x^3 + 8x^2 + 9x - 5$

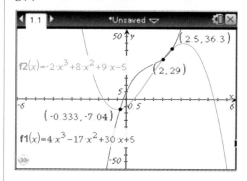

$x \in \,]{-\infty}, -0.333] \cup [2, 2.5]$

Graphical method 2:

Let $h(x) = f(x) - g(x) \le 0$

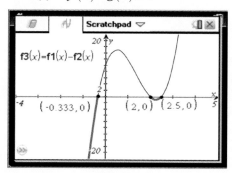

GDC help on CD:
Alternative demonstrations for the TI-84 Plus and Casio fx-9860GII GDCs are on the CD.

$x \in \,]{-\infty}, -0.333] \cup [2, 2.5]$

Fully factorize the polynomial.

Construct the sign table.

Use a GDC to draw the graphs of both polynomials and identify where the graph of f is below the graph of g.

Identify the values of x that satisfy the inequality.

Rewrite the inequality and call the new function $h(x)$.

Use a GDC to draw the graphs of the new polynomial and identify the parts where the graph is below the x-axis.

Identify the values of x that satisfy the inequality.

In Example 36 we used two different graphical methods.
When using method 1 you may need to examine different windows
to find the points of intersection between the graphs. Moreover, it
can be difficult to read which function is upper and which is lower,
particulary on calculators with poor resolution. Method 2 is more
suitable because you don't need to think about the size of the window
since the zeros appear along the x-axis. This saves having to explore
the different windows if the intersections between the graphs are
not visible in the original window.

> For the algebraic
> solution of $g(x) \geq f(x)$
> the syllabus restricts
> polynomials to degree
> 3 or below. However,
> graphical methods on
> a GDC can be used
> for solving polynomial
> equations and
> inequalities of degree
> 4 or higher.

Example 37

Use a GDC to solve the inequality $x^{11} - 3x^6 + 2 \leq 0$

Answer

$f(x) = x^{11} - 3x^6 + 2$

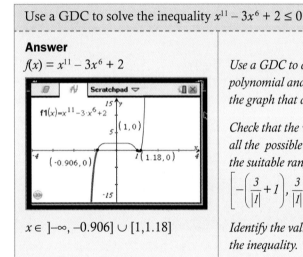

$x \in \ ]-\infty, -0.906] \cup [1, 1.18]$

*Use a GDC to draw the graph of the
polynomial and identify the parts of
the graph that are below the x-axis.*

*Check that the window shows
all the possible zeros by finding
the suitable range of x-values:*

$$\left[-\left(\frac{3}{|1|}+1\right), \frac{3}{|1|}+1\right] = [-4, 4].$$

*Identify the values of x that satisfy
the inequality.*

GDC help on CD:
*Alternative demonstrations
for the TI-84 Plus and Casio
fx-9860GII GDCs are on the
CD.*

Sometimes, when equations or inequalities can be easily split into simpler
polynomial curves, you can sketch these and find the solution by inspection.

Example 38

Use simple polynomial graphs to solve the inequality $x^3 - 3x + 2 \leq 0$

Answer

$\underbrace{x^3 + 2}_{f(x)} \leq \underbrace{3x}_{g(x)}$

Split the inequality into a cubic and linear function.

*Sketch the graphs of both the cubic and linear
functions. Identify the values of x for which the cubic
graph is below the linear graph.*

Note: You could split the inequality: $\underbrace{x^3}_{f(x)} \leq \underbrace{3x - 2}_{g(x)}$

*In this case the graphs would be exactly the same
shape but shifted 2 units down.*

$x \in \ ]-\infty, -6] \cup \{3\}$

Exercise 3Q

1 Solve these inequalities in the set of real numbers and check your answers with a GDC.

   a $x^3 - 6x^2 + 11x - 6 \geq 0$ **b** $x^3 + 2x^2 - 7x + 4 \leq 0$

   c $x^3 + 3x^2 - 4x - 12 < 0$ **d** $2x^3 - 5x^2 - 18x + 45 > 0$

   e $12x^3 + 17x^2 + 2x - 3 \leq 0$ **f** $x^3 - 4x^2 - 5x + 14 > 0$

   g $3x^3 - 13x^2 + 11x + 14 < 0$ **h** $x^4 - x^3 - 11x^2 + 9x + 18 \geq 0$

2 Given the polynomials $f(x) = 4x^3 - 17x^2 + 30x + 5$ and $g(x) = -2x^3 + 8x^2 + 9x - 5$, find all the values of x such that $f(x) > g(x)$.

   Verify your solution by using a graphical method on a GDC.

3 Use a GDC to solve these inequalities.

   a $x^7 - 2x^3 - 1 \geq 0$ **b** $x^9 - 2x^8 + 2x^5 + x \leq 0$

4 Use simple polynomial graphs to find the solutions of these inequalities.

   a $x^3 + x - 2 > 0$ **b** $-2x^3 + 3x + 1 \geq 0$ **c** $x^4 + 2x + 1 \leq 0$

3.6 Solving systems of equations

Systems of two linear equations with two unknowns with complex coefficients

When solving simultaneous equations with complex coefficients the methods of elimination and substitution can be very demanding. The method shown here will lead to general formulae for the solutions.

> For more on solving systems of two linear equations with two unknowns with real coefficients, see Chapter 14 section 2.5.

$\begin{cases} ax + by = e \\ cx + dy = f \end{cases}$

Multiply first equation by d and second equation by b to obtain equal coefficients for the variable y.

$\Rightarrow \begin{cases} adx + bdy = ed \\ bcx + bdy = fb \end{cases} -$

Subtract the equations to eliminate the variable y.

$\Rightarrow adx - bcx = ed - fb$

$\Rightarrow x(ad - bc) = ed - fb$

Factorize the left hand side.

$\Rightarrow x = \dfrac{ed - fb}{ad - bc}, \ ad - bc \neq 0$

$a \times \dfrac{ed - fb}{ad - bc} + by = e$

Substitute the value of x in the first equation to find the value of y.

$\Rightarrow by = e - a \times \dfrac{ed - fb}{ad - bc}$

$$\Rightarrow by = \frac{e(ad-bc)-a(ed-fb)}{ad-bc}$$

$$= \frac{ead-ebc-aed+afb}{ad-bc}$$

$$= \frac{afb-ebc}{ad-bc} \Rightarrow y = \frac{b(af-ec)}{ad-bc} \times \frac{1}{b}$$

$$= \frac{af-ec}{ad-bc}, \ ad-bc \neq 0.$$

So the general form of the solution is

$$(x,y) = \left(\frac{ed-fb}{ad-bc}, \frac{af-ec}{ad-bc}\right), \ ad-bc \neq 0$$

Notice that we could have substituted the value of x in the second equation to find the value of y.

These formulae are very efficient when the coefficients of the simultaneous linear equations are complex numbers.

Example 39

Solve the simultaneous equations
$$\begin{cases} 2x+(3-i)\,y = 3 \\ ix+(1+2i)\,y = 2i \end{cases}$$

Answers

$a = 2, b = 3-i, c = i, d = 1+2i, e = 3, \ f = 2i$ — *Identify the coefficients.*

$2 \times (1+2i) - (3-i) \times i = 2+4i-3i-1 = 1+i$ — *Find the denominator $ad - bc$*

$3 \times (1+2i) - (3-i) \times 2i = 3+6i-6i-2 = 1$ — *Find $ed - fb$, the numerator for x*

$2 \times 2i - 3 \times i = 4i - 3i = i$ — *Find $af - ec$, the numerator for y*

$x = \dfrac{1}{1+i} \times \dfrac{1-i}{1-i} = \dfrac{1-i}{2} = \dfrac{1}{2} - \dfrac{1}{2}i$ — *Apply the formulae for x and y.*

$y = \dfrac{i}{1+i} \times \dfrac{1-i}{1-i} = \dfrac{i+1}{2} = \dfrac{1}{2} + \dfrac{1}{2}i$

$(x,y) = \left(\dfrac{1}{2} - \dfrac{1}{2}i, \dfrac{1}{2} + \dfrac{1}{2}i\right)$

The same result can be obtained on a GDC, using Solve Systems of Linear Equations.

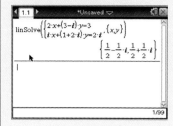

GDC help on CD:
Alternative demonstrations for the TI-84 Plus and Casio fx-9860GII GDCs are on the CD.

Systems of three linear equations with three unknowns

You can use the methods of substitution and elimination to reduce a system of three equations with three unknowns to a system of two equations with two unknowns.

Example 40

Solve the system of equations: $\begin{cases} 2x+4y+z=5 \\ 3x-5y-z=4 \\ x+y-z=6 \end{cases}$

Use:

a the method of substitution

b the method of elimination.

Answers

a $x+y-6=z \Rightarrow \begin{cases} 2x+4y+(x+y-6)=5 \\ 3x-5y-(x+y-6)=4 \end{cases}$

$\Rightarrow \begin{cases} 3x+5y=11 \\ 2x-6y=-2 \end{cases}$

$\Rightarrow \begin{cases} 3(3y-1)+5y=11 \\ x=3y-1 \end{cases}$

$\Rightarrow \begin{cases} 9y-3+5y=11 \\ x=3y-1 \end{cases}$

$\Rightarrow \begin{cases} 14y=14 \\ x=3y-1 \end{cases}$

$\Rightarrow \begin{cases} y=1 \\ x=3\cdot1-1=2 \end{cases}$

$\Rightarrow z = 2 + 1 - 6 = -3$

$\Rightarrow (x, y, z) = (2, 1, -3)$

Use the third equation to express z in terms of x and y and substitute for z in the first two equations. Use z because the coefficients of z are simpler.

Use $2x - 6y = -2$ to express x in terms of y and substitute for x in $3x + 5y = 11$

> If you use substitution to obtain a system of two equations with two unknowns, you don't have to use the same method to solve for the unknowns in this new system – you can use elimination.

b $\left.\begin{array}{l} 2x+4y+z=5 \\ 3x-5y-z=4 \end{array}\right\}+ \\ \left.\begin{array}{l} 2x+4y+z=5 \\ x+y-z=6 \end{array}\right\}+ \Rightarrow \begin{cases} 5x-y=9 \\ 3x+5y=11 \end{cases}$

$\Rightarrow \begin{cases} 25x-5y=45 \\ 3x+5y=11 \end{cases}+$

$\Rightarrow 28x = 56$

$\Rightarrow x = 2$

$\Rightarrow 5 \cdot 2 - y = 9 \Rightarrow 1 = y$

$\Rightarrow 2 + 1 - z = 6 \Rightarrow -3 = z$

$\Rightarrow (x, y, z) = (2, 1, -3)$

To eliminate z, add the first and second equations and the first and third equations.
You must eliminate the same unknown from both pairs of equations. Eliminate z because z has the simplest coeffiecients.

To eliminate y, multiply $5x - y = 9$ by 5 and then add $3x + 5y = 11$.

To find y substitute x = 2 in $5x - y = 9$
To find z substitute x = 2 and y = 1 in $x + y - z = 6$

Linear systems with three unknowns can also be solved using a GDC.

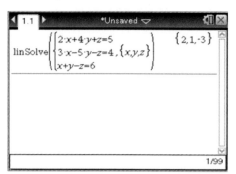

The geometrical interpretation of a linear equation in three variables as a plane is developed in Chapter 11.

GDC help on CD:
Alternative demonstrations for the TI-84 Plus and Casio fx-9860GII GDCs are on the CD.

When solving systems of three simultaneous linear equations with three unknowns there are again three possible types of solution:

i A unique triplet of numbers (the three variables) that satisfies all three equations.

ii No triplet of real numbers that satisfies all the equations.

iii Infinitely many triplets of real numbers that satisfy all the equations.

To solve three linear equations in three unknowns a special method of elimination was invented by **Carl Friedrich Gauss** (1777–1855). The method, called the Gaussian method, is more suitable to use when the coefficients of the system are in matrix form. It involves eliminating variables in order until you reach the last variable. Consequently you find the variables in reverse order to the order of elimination.

Example 41

Use the Gaussian method to solve the simultaneous equations $\begin{cases} x+3y-2z=3 \\ 2x-4y+3z=5 \\ 4x+y-z=6 \end{cases}$	

Answer

$\begin{cases} x+3y-2z=3 & (1) \\ 2x-4y+3z=5 & (2) \\ 4x+y-z=6 & (3) \end{cases}$

Eliminate x from equations (2) and (3)

To obtain equation (4) subtract (2) from $2 \cdot (1)$

$\begin{cases} x+3y-2z=3 & (1) \\ 10y-7z=1 & (4) \\ 11y-7z=6 & (5) \end{cases}$

To obtain equation (5) subtract (3) from $4 \cdot (1)$

$\begin{cases} x+3y-2z=3 & (1) \\ 10y-7z=1 & (4) \\ -\dfrac{7}{10}z=-\dfrac{49}{10} & (7) \\ z=7 \end{cases}$

To obtain equation (7) subtract (5) from $\dfrac{11}{10} \cdot (4)$

Use equation (7) to find z

▶ Continued on next page

$$\begin{cases} x+3y-2z=3 \\ 10y-7\cdot 7=1 \\ \quad\quad y=5 \end{cases}$$

$$\begin{cases} x+3\cdot5-2.7=3 \\ \quad\quad x+1=3 \\ \quad\quad\quad x=2 \end{cases}$$

The solution is $(x,\ y,\ z)=(2,\ 5,\ 7)$

| | *Substitute $z=7$ in equation (4) to find y* |
| | *Substitute $y=5$ and $z=7$ in equation (1) to find the value of x* |

Example 42

Discuss all the possible types of solution of this system of equations with respect to the real parameter a

$$\begin{cases} ax+y+z=3 \\ x+y+z=1 \\ x+2y-z=2 \end{cases}$$

Answer

$$\begin{cases} ax+y+z=3 & (1) \\ x+y+z=1 & (2) \\ x+2y-z=2 & (3) \end{cases} \Rightarrow \begin{cases} (a-1)x=2 & (4) \\ 2x+3y=3 & (5) \end{cases}$$

	Eliminate z
	To obtain equation (4) subtract (2) from (1).
	To obtain equation (5) add (2) and (3).

$$\Rightarrow \begin{cases} x=\dfrac{2}{a-1} \\ 2\cdot\dfrac{2}{a-1}+3y=3 \end{cases} \Rightarrow \begin{cases} x=\dfrac{2}{a-1} \\ y=\dfrac{3a-7}{3a-3} \end{cases}$$

| | *To find a unique solution assume that $a\neq 1$.* |
| | *To find y substitute for x in (5).* |

$$\Rightarrow \begin{cases} x=\dfrac{2}{a-1} \\ y=\dfrac{3a-7}{3a-3} \\ \dfrac{2}{a-1}+\dfrac{3a-7}{3a-3}+z=1 \end{cases}$$

| | *To find z substitute for x and y in (2).* |

$$\Rightarrow (x,\ y,\ z)=\left(\dfrac{2}{a-1},\dfrac{3a-7}{3a-3},\dfrac{-2}{3a-3}\right),\ a\neq 1$$

If $a=1 \Rightarrow 0\cdot x=2 \Rightarrow 0=2$

$\Rightarrow (x,\ y,\ z)\in\varnothing$

| | *The unique solution when $a\neq 1$* |
| | *Equation (4) gives a false statement therefore there is no solution when $a=1$.* |

Exercise 3R

1 Solve the following simultaneous equations and check your answers with a GDC.

a $\begin{cases} 2ix+(2+3i)\,y=1 \\ (1+i)x+2y=3 \end{cases}$

b $\begin{cases} (1+i)x+3iy=2+6i \\ (2-i)x-(4+3i)y=4i-3 \end{cases}$

2 Solve these systems of equations.

a $\begin{cases} x+y=-1 \\ x+z=4 \\ y+z=1 \end{cases}$
  b $\begin{cases} x-5y+3z=-1 \\ 3x-y+2z=4 \\ 2x+y-z=2 \end{cases}$

c $\begin{cases} 2x+y+2z=0 \\ 6x-4y-5z=-2 \\ 4x+y-3z=2 \end{cases}$
  d $\begin{cases} 3x-4y+3z=-2 \\ x+2y+6z=6 \\ 2x-6y-3z=-8 \end{cases}$

e $\begin{cases} x+2y+z=4 \\ 2x+y+2z=5 \\ 3x+2y+3z=12 \end{cases}$
  f $\begin{cases} 2x-3y+5z=-1 \\ 9x-7y+16z=0 \\ x-2y+3z=9 \end{cases}$

3 Find the value(s) of a real parameter k so that each system of equations has no unique solution.

a $\begin{cases} x+2y+z=0 \\ 2x+y+2z=1 \\ x+2y+kz=2 \end{cases}$
  b $\begin{cases} x+y+z=1 \\ 2x+ky+3z=-2 \\ 3x+5y+kz=-1 \end{cases}$

4 Find the value(s) of a real parameter k so that each system of equations has infinitely many solutions. Find the solutions.

a $\begin{cases} x+2y+3z=1 \\ kx+4y+3z=2 \\ 3x+6y-2z=3 \end{cases}$

b $\begin{cases} x+y+z=1 \\ 2x+ky+3z=-2 \\ 3x+5y+kz=-1 \end{cases}$

5 Find the values of a real parameter m so that the system of equations has a unique solution.

$\begin{cases} x+y+z=m \\ x+my+z=2m \\ x+y+mz=-1 \end{cases}$

Hence, find the solution in terms of m.

Review exercise

1 When the polynomial $f(x) = x^4 - 3x^3 + ax^2 - 4x + 7$ is divided by $(x + 2)$ the remainder is 7. Find the value of a.

2 Solve the simultaneous equations:
$$\begin{cases} 3x - 2y = i - 2 \\ 4y - (1-i)x = 3 + 3i \end{cases}$$

3 Find the value of m in the quadratic function
$f(x) = m - 2 + (2m + 1)x + mx^2$ if $f(x) \le 0$ for all real x.

4 Given that $1 - 2i$ is a complex root of the equation
$z^4 - 2z^3 + 14z^2 - 18z + 45 = 0$, find the remaining roots.

5 Find the value of m such that this system of equations has no unique solution.
$$\begin{cases} mx + 2y = 1 \\ 4x + (m+2)y = 4 \end{cases}$$

6 Find the value of a such that the roots α and β of the quadratic equation $x^2 + ax + a + 1 = 0$ satisfy $\alpha^3 + \beta^3 = 9$

7 Given that $z = \dfrac{1+i}{2}$, use mathematical induction to show that
$z^{2n} = \dfrac{i^n}{2^n}, \, n \in \mathbb{Z}^+$

8 Show that the imaginary part of the number $\left(\dfrac{1+i}{1-i}\right)^{2011}$ is -1.

9 The cubic equation $x^3 - 5x^2 + 6x - 3 = 0$ has solutions α, β and γ. Find the value of $\dfrac{1}{\alpha^2} + \dfrac{1}{\beta^2} + \dfrac{1}{\gamma^2}$

10 a Show that $\sqrt[3]{7 - \sqrt{50}} + \sqrt[3]{7 + \sqrt{50}}$ satisfies the equation
$x^3 + 3x - 14 = 0$
b Factorize the polynomial $f(z) = z^3 + 3z - 14$, $z \in \mathbb{C}$ and find all the possible zeros.
c Hence, find the value of $\sqrt[3]{7 - \sqrt{50}} + \sqrt[3]{7 + \sqrt{50}}$.

Review exercise

1 Solve the inequality $x^3 + 5x^2 + 2x - 22 \geq 0$

2 Find all the values of the real parameter m for which the equation $(mx)^2 + 3x + 1 - m = 0$ has no real solution.

3 Solve these simultaneous equations and write your answers as fractions.

$$\begin{cases} 2x + 14y + 9z = -7 \\ 4x - 3z = 4 + 7y \\ 10x - 28y = 5 + 6z \end{cases}$$

4 Given that α, β and γ are solutions of the equation $3x^3 + 2x = 5x^2 + 4$, find the value of $\alpha^3 + \beta^3 + \gamma^3$

5 Find the smallest zero of the polynomial
$f(x) = x^7 + 35x^6 - 97x^5 + 33x^2 + 4$

CHAPTER 3 SUMMARY

Zero factor property

$a \cdot b = 0 \Rightarrow a = 0$ or $b = 0$

Quadratic formula

$$ax^2 + bx + c = 0 \Rightarrow x = \frac{-b \pm \sqrt{b^2 - 4ac}}{2a}$$

Discriminant

$\Delta = b^2 - 4ac$

i If $\Delta > 0$ there are two distinct real roots.

ii If $\Delta = 0$ there is one repeated real root.

iii If $\Delta < 0$ there are no real roots (conjugate complex pair of solutions.)

Continued on next page

Operations with complex numbers

Given that $z_1 = a_1 + ib_1$, $z_2 = a_2 + ib_2$, a_1, b_1, a_2, $b_2 \in \mathbb{R}$

$(z_1 = z_2) \Leftrightarrow (a_1 = a_2 \text{ and } b_1 = b_2)$

$z_1 \pm z_2 = (a_1 \pm a_2) + i(b_1 \pm b_2)$

$\lambda z = \lambda(a + ib) = (\lambda a) + i(\lambda b), \lambda \in \mathbb{R}$

$z_1 \cdot z_2 = (a_1 a_2 - b_1 b_2) + i(a_1 b_2 + a_2 b_1)$

$|z| = |a + ib| = a^2 + b^2$

$$\frac{z_1}{z_2} = \frac{(a_1 a_2 + b_1 b_2) + i(a_2 b_1 - a_1 b_2)}{a_2^2 + b_2^2} = \frac{z_1 z_2^*}{|z_2|}$$

Axioms of complex numbers

A1 For all complex numbers z_1 and z_2 then $z_1 + z_2$ is a complex number (Closure)

A2 For all complex numbers z_1 and z_2 then $z_1 + z_2 = z_2 + z_1$ (Commutativity)

A3 For all complex numbers z_1, z_2 and z_3 then $(z_1 + z_2) + z_3 = z_1 + (z_2 + z_3)$ (Associativity)

A4 There exists a complex number $0 = 0 + 0i$ such that for every complex number z, $0 + z = z + 0 = z$ (Additive identity)

A5 For every complex number z there exists a complex number $-z$ such that $z + -z = -z + z = 0$ (Additive inverse)

A6 For all complex numbers z_1 and z_2 then $z_1 \cdot z_2$ is a complex number (Closure)

A7 For all complex numbers z_1 and z_2 then $z_1 \cdot z_2 = z_2 \cdot z_1$ (Commutativity)

A8 For all complex numbers z_1, z_2 and z_3 then $(z_1 \cdot z_2) \cdot z_3 = z_1 \cdot (z_2 \cdot z_3)$ (Associativity)

A9 There exists a complex numbers $1 = 1 + 0i$ such that for every complex numbers z, $1 \cdot z = z \cdot 1 = z$ (Multiplicative identity)

A10 For all complex numbers z, $z \neq 0$, there exists a complex numbers z^{-1} such that $z \cdot z^{-1} = z^{-1} \cdot z = 0$ (Multiplicative inverse)

A11 For all complex numbers z_1, z_2 and z_3 then $z_1 \cdot (z_2 + z_3) = z_1 \cdot z_2 + z_1 \cdot z_3$ (Distributivity of multiplication over addition)

Continued on next page

Viète's formulae for quadratic equations

$$ax^2 + bx + c = 0 \Rightarrow x_1 + x_2 = -\frac{b}{a} \text{ and } x_1 \cdot x_2 = \frac{c}{a}$$

Viète's formulae for cubic equations

$$ax^3 + bx^2 + cx + d = 0 \Rightarrow \begin{cases} x_1 + x_2 + x_3 = -\dfrac{b}{a} \\ x_1 \cdot x_2 + x_1 \cdot x_3 + x_2 \cdot x_3 = \dfrac{c}{a} \\ x_1 \cdot x_2 \cdot x_3 = -\dfrac{d}{a} \end{cases}$$

Viète's formula for equations of the nth degree

$$a_n x^n + a_{n-1} x^{n-1} + \dots + a_2 x^2 + a_1 x + a_0 = 0 \Rightarrow$$

$$\sum_{1 \le i_1 < i_2 < \dots < i_k \le n} \left(x_{i_1} \cdot x_{i_2} \cdot \dots \cdot x_{i_k} \right) = (-1)^k \frac{a_{n-k}}{a_n}, \ 1 \le k \le n$$

Degree of polynomials

The degree of a polynomial, $f(x) = a_n x^n + a^{n-1} x^{n-1} + \dots + a_1 x + a_0$, is the largest power of x appearing: $\deg(f) = n$.

For a linear combination of two polynomials, $af(x) + bg(x)$ with $a, b \in \mathbb{R}$, or the product of two polynomials, $f(x) \cdot g(x)$, the degree is given by

$\deg(af + bg) = \max\{\deg(f), \deg(g)\}$
$\deg(f \cdot g) = \deg(f) + \deg(g)$

Unique decomposition

For any two polynomials f and g there are unique polynomials q and r such that $f(x) = g(x) \cdot q(x) + r(x)$, for all real values of x.

Remainder theorem

Given a polynomial
$f(x) = a_n x^n + a_{n-1} x^{n-1} + \dots + a_2 x^2 + a_1 x + a_0$, $a_k \in \mathbb{R}$,
$k = 0, 1, 2, \dots, n$, $a_k \ne 0$ and a real number p, then the remainder when $f(x)$ is divided by a linear expression $(x - p)$ is $f(p)$.

Continued on next page

Factor theorem

A polynomial $f(x) = a_n x^n + a_{n-1} x^{n-1} + \ldots + a_2 x^2 + a_1 x + a_0$ with real coefficients $(a_n \neq 0)$ has a factor $(x - p)$, $p \in \mathbb{R}$, if and only if $f(p) = 0$.

Fundamental theorem of algebra (FTA)

A polynomial $f(x) = a_n x^n + a_{n-1} x^{n-1} + \ldots + a_2 x^2 + a_1 x + a_0$ with real or complex coefficients $(a_n \neq 0)$ has at least one zero.

Each polynomial $f(x) = a_n x^n + a_{n-1} x^{n-1} + \ldots + a_2 x^2 + a_1 x + a_0$ with real coefficients can be written in a factor form $f(x) = a_n(x - \omega_1)(x - \omega_2) \ldots (x - \omega_n)$ such that $\omega_k \in \mathbb{C}$, $k, \ldots, n$.

Integer zero theorem

Given a polynomial $f(x) = a_n x^n + a_{n-1} x^{n-1} + \ldots + a_2 x^2 + a_1 x + a_0$, $a_k \in \mathbb{Z}$, $a_n \neq 0$ and an integer p such that $f(p) = 0$, then p is a factor of a_0.

Rational zero theorem

Given a polynomial $f(x) = a_n x^n + a_{n-1} x^{n-1} + \ldots + a_2 x^2 + a_1 x + a_0$, $a \in \mathbb{Z}$ $a_n \neq 0$ and a rational number $\frac{p}{q}$, where $\gcd(p, q) = 1$ that

is $\left(\frac{p}{q} \right)$ is in its simplest form, such that $f\left(\frac{p}{q} \right) = 0$, then p is a

factor of a_0 and q is a factor of a_n.

All the possible zeros of the polynomial

$f(x) = a_n x^n + a_{n-1} x^{n-1} + \ldots + a_1 x + a_0$ are in the interval

$\left[-\left(\frac{M}{|a_n|} + 1 \right), \frac{M}{|a_n|} + 1 \right]$ where $M = \max \{ |a_n|, |a_{n-1}|, \ldots |a_1|, |a_0| \}$

Is mathematics invented or discovered?

Number history

At different points in its development the real number system was sometimes considered to be an invention, and sometimes to reflect the reality of the universe. The Pythagoreans did not acknowledge the existence of irrational numbers. Negative numbers were first thought to be meaningless entities.

Complex numbers were originally called 'imaginary numbers', but as work on polynomial equations developed and ultimately led to the Fundamental Theorem of Algebra, imaginary numbers were seen in a new light, and eventually they became known as complex numbers.

- Explore how the history of number highlights the central themes in the discussion of whether mathematics is invention or discovery, considering:

▶ The real number system as a historical artifact

▶ The 'existence' of numbers other than our real number system

▶ The definition of number operations on the non-real numbers

▶ The intrinsic connection between number and geometry, e.g. the Argand diagram

▶ The relationship between number and dimension

▶ Higher dimensional numbers, e.g. Quaternions and Octonions, and order of operations

▶ The problem of rotations and higher dimensional numbers.

- Have we invented/discovered all the kinds of numbers that there are?

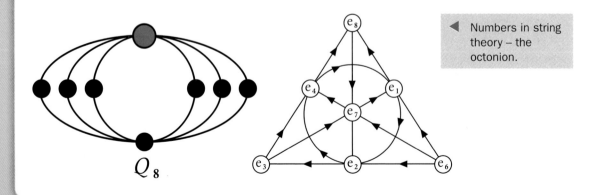

◀ Numbers in string theory – the octonion.

"We used to think that if we knew one, we knew two, because one and one are two. We are finding that we must learn a great deal more about 'and'."

Sir Arthur Eddington, British astrophysicist (1882–1944)

The meaning of 1 + 1 = 2

1 + 1 = 2 is probably one of the first arithmetic facts you ever learned.

- Is the concept behind the numbers always the same? What does 1 mean in group theory? What does 1 + 1 mean in group theory?

- Is the representation of the concept of the numbers always the same? How is it expressed in different number bases?

- Are these considerations and their answers cross-cultural?

Gödel's Incompleteness Theorem

In 1931, Czech-born mathematician Kurt Gödel (1906–78) published his famous Incompleteness Theorem which, put simply, stated that in any axiomatic system certain statements cannot be either proved or disproved. Or in other words, mathematical truth is greater than proof.

Mathematicians have still not recovered from the shock waves that Gödel's theorem created in the mathematical world.

- Research how Gödel's theorem challenged the existing notions of mathematics.

US theoretical physicist and Nobel laureate Frank Wilczek (1951–) has said that mathematics is both invented and discovered, but he thinks "it's mostly discovered". Argentine–American mathematician and philosopher Gregory Chaitin (1947–) on the other hand, believes that mathematics is empirical: "we invent it as we go".

The question 'Is mathematics invented or discovered?' gives rise to other questions.

- What is the meaning of the words 'mathematics', 'invention' and 'discovery'?

- If mathematics is invented, what does this say about its ability to describe reality?

- If mathematics is discovered, what does this say about its ability to describe reality?

- Are some parts of mathematics invented, and others discovered? If so, which are which?

4

Modeling the real world

CHAPTER OBJECTIVES:

6.1 Informal ideas of limit, continuity, and convergence; definition of the derivative from first principles; the derivative interpreted as a gradient function and as a rate of change; finding equations of tangents and normals; identifying increasing and decreasing functions; the second derivative; higher derivatives

6.2 Derivative of x^n; differentiation of sums and multiples of functions; the product and quotient rules; the chain rule for composite functions; related rates of change; implicit differentiation

6.3 Local maximum and minimum values; optimization problems; points of inflexion with zero and non-zero gradients; graphical behavior of functions including the relationship between the graphs of f, f', and f''

6.6 Kinematic problems involving displacement s, velocity v and acceleration a; total distance traveled

Before you start

You should know how to:

1 Draw graphs of rational functions.

e.g., sketch the graph of $y = \dfrac{1}{x-1}$, clearly showing any asymptotes as dotted lines.

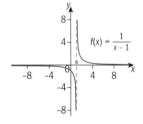

2 Find the sum of infinite geometric series.

e.g., since, $|r| < 1$, $\displaystyle\sum_{r=0}^{\infty} \left(\frac{1}{2}\right)^r = \dfrac{1}{1-\dfrac{1}{2}} = 2$.

Skills check

1 Sketch the graph of $f(x) = \dfrac{1}{x-3}$, clearly labeling all intercepts and asymptotes.

2 Find $\displaystyle\sum_{n=0}^{\infty} 5\left(\frac{1}{2}\right)^n$

From abstract models to real-world applications

A mathematical model uses mathematical language and systems of functions to describe, explain, interpret, and predict real-world phenomena. Climate scientists and meteorologists have collected vast amounts of data about weather systems and CO_2 concentrations in the atmosphere over many years. They have created mathematical models that fit the historical data and that they can now use to predict future climate changes.

Mathematical models are used today in all areas of human endeavor, from the natural sciences to the creative arts. In this chapter you will learn how to work with functions that may be derived from real-world situations, such as mechanics and economics.

The global financial crisis of 2008, was mainly due to a mathematical model created by economist David X. Li, to manage financial risk. His model was used in financial institutions throughout the world to assist in the calculation of risk factors in certain investment strategies. Was it a flaw in the model or in its interpretation that caused the crisis?

Mathematical modeling has many beneficial applications. However, what are the possible pitfalls of modeling real-life phenomena? What are the limits of mathematical modeling?

4.1 Limits, continuity and convergence

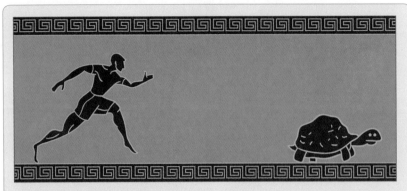

Zeno of Elea, a philosopher and logician, posed this problem about 2500 years ago. Achilles and a tortoise were engaged in a footrace. Achilles allowed the tortoise a head start of 100 metres. Both started running at a constant speed. Who won the race?

Zeno analysed the problem as follows. After a short time into the race, Achilles arrives at the tortoise's starting point of 100 m. In that time, the tortoise advances further. It then takes more time for Achilles to run this extra distance, in which time the tortoise advances even further. So, whenever Achilles reaches some point that the tortoise has already been at, he still has further to go. Since Achilles has an infinite number of points to cover before he reaches where the tortoise was, Achilles is still trying to win this race today!

It has taken several millennia for mathematicians to arrive at the language and concepts needed to solve this paradox satisfactorily. In this section you will learn some of the mathematics developed by 17th, 18th and 19th-century mathematicians in an attempt to deal with the concepts of time and infinity.

An informal treatment of limits

You can think of a limit as a way of describing the output of a function as the input gets close to a certain value.

The rules for finding limits are quite straightforward, and can be algebraic, graphical, numerical, or a combination of these methods.

As an example, consider the rational function $y = \dfrac{x^2 - 1}{x - 1}, x \neq 1$

This function is not defined at $x = 1$ and its domain is $\{x \mid x \in \mathbb{R}, x \neq 1\}$.

Now, with your GDC, trace along the graph of this function. You will notice that as x gets closer to 1 from the left, the value of the function gets closer to 2. Trace along the graph from the right, and notice that the value of the function likewise gets closer to 2.

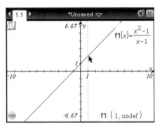

This table shows these results.

Approaching $x = 1$ from the left.   Approaching $x = 1$ from the right.

x	0.6	0.7	0.8	0.9	1	1.1	1.2	1.3	1.4
$y = \dfrac{x^2 - 1}{x - 1}$	1.6	1.7	1.8	1.9	undef.	2.1	2.2	2.3	2.4

Approaching 1 in steps or **increments** of 0.1.

You can write this result using this notation: $\lim_{x \to 1} \dfrac{x^{2}-1}{x-1} = 2$

This means that the limit of the functions as x approaches 1 *both* from the left *and* from the right is 2.

Example 1

a Sketch the graph of $y = \dfrac{2^{x}-1}{x}$, $x \neq 0$

b Find $\lim_{x \to 0} \dfrac{2^{x}-1}{x}$, giving your answer to 2 decimal places.

x

In Example 1, the limits were the same whether approaching from the left or from the right. In Example 2, the limits are different when approached from the left and from the right. Therefore we say that in Example 2, the function has no limit.

For the limit of a function to exist as x approaches a particular value, the function does not need to be defined at the value but the value of the limit as the function approaches from the left and from the right must be the same.

The double arrow is read 'if and only if'. An 'if and only if' definition or theorem has the form: if p then q and if q then p, where p and q are statements. This means that the two parts of the definition or theorem are equivalent. To prove an 'if and only if' theorem it is necessary to prove both, if p then q and also prove if q then p.

> → The notation used to say that the limit, L, of a function f exists as x approaches a real value c is
> $$\left(\lim_{x \to c} f(x) = L\right) \Leftrightarrow \left(\lim_{x \to c^+} f(x) = L \text{ and } \lim_{x \to c^-} f(x) = L\right) \text{ for } L \in \mathbb{R}.$$

Exercise 4A

Using a GDC, sketch the graph of each function and find the limit, if it exists.

1 $\lim\limits_{x \to -1} \dfrac{x^2 - 1}{x + 1}$

2 $\lim\limits_{x \to 1} \dfrac{x^3 - 1}{x - 1}$

3 $\lim\limits_{x \to 2} \begin{cases} 3x - 1, & x < 2 \\ \dfrac{1}{x^2 - 1}, & x \geq 2 \end{cases}$

4 $\lim\limits_{x \to 0} \dfrac{|x|}{x}$

5 $\lim\limits_{x \to 6} (x - 6)^{\frac{2}{3}}$

6 $\lim\limits_{x \to 3} \lfloor x \rfloor$

$y = [x]$ or $y = \text{int}(x)$ is the floor function. It is defined as the 'largest integer less than or equal to x'. This function will be defined for you in an examination.

Asymptotes and continuity

Does $\lim\limits_{x \to 0} \dfrac{1}{x^2}$ exist? Here is the graph of the function $\dfrac{1}{x^2}$
You can see that as x approaches 0 from the left and from the right, the values of the function increase without bound, and approach positive infinity. The limit therefore does not exist, since the limit is not a real number.

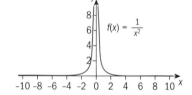

The line $x = 0$ is the vertical asymptote of this function. We can now define the vertical asymptote of a function.

You met vertical asymptotes in Chapter 2.

> → The line $x = c$ is a **vertical asymptote** of the graph of a function $y = f(x)$ if either $\lim\limits_{x \to c+} f(x) = \pm\infty$ or $\lim\limits_{x \to c-} f(x) = \pm\infty$.

On page 168 you saw that the graph of $y = \dfrac{x^2 - 1}{x - 1}$, $x \neq 1$, is linear.

Simplifying, gives $y = \dfrac{x^2 - 1}{x - 1} = x + 1$

However, since $x \neq 1$, there is a gap or hole in the function at $x = 1$. For the function $y = x + 1$, however, there is no gap at $x = 1$. Hence, both functions have a limit of 2 as x approaches 1, but the graph of $y = \dfrac{x^2 - 1}{x - 1}$ is discontinuous at $x = 1$. The graph of $y = x + 1$ has no holes anywhere in its domain, so $y = x + 1$ is continuous.

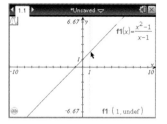

> → A function $y = f(x)$ is **continuous** at $x = c$, if $\lim\limits_{x \to c} f(x) = f(c)$.
>
> The three necessary conditions for f to be continuous at $x = c$ are:
>
> **1** f is defined at c, i.e., c is an element of the domain of f.
>
> **2** the limit of f at c exists.
>
> **3** the limit of f at c is equal to the value of the function at c.
>
> A function that is not continuous at a point $x = c$ is said to be **discontinuous** at $x = c$.

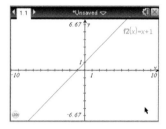

A function is said to be continuous on an open interval I if it is continuous at every point in the interval.

A function is said to be continuous if it is continuous at *every* point in its domain.

A function that is not continuous is said to be **discontinuous**.

GDC help on CD:
Alternative demonstrations for the TI-84 Plus and Casio fx-9860GII GDCs are on the CD.

Example 3

a Sketch the graph of $f(x) = \begin{cases} 1, & x \le -1 \\ -x, & -1 < x < 0 \\ 1, & x = 0 \\ -x, & 0 < x < 1 \\ 1, & x \ge 1 \end{cases}$

b Find the limits, if they exist, as x approaches -1, 0 and 1.

c Determine if f is continuous at $x = -1$, $x = 0$, and $x = 1$.

A polynomial function such as $3x^2 + 2x - 4$ is continuous at every point in its domain.

Answers

a

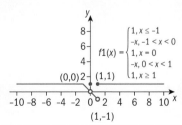

▶ Continued on next page

b $\displaystyle\lim_{x \to -1} f(x) = 1$	*As x approaches −1 from the left and from the right, f approaches 1.*
$\displaystyle\lim_{x \to 0} f(x) = 0$	*As x approaches 0 from the left and from the right, f approaches 0.*
$\displaystyle\lim_{x \to 1^-} f(x) = -1, \lim_{x \to 1^+} f(x) = 1$ $\Rightarrow \displaystyle\lim_{x \to 1} f(x) = undefined.$	*As x approaches 1 from the left, f approaches −1, and as x approaches 1 from the right, f approaches 1, that is, the limit as x approaches 1 does not exist.*
c $\displaystyle\lim_{x \to -1} f(x) = 1$ and $f(-1) = 1$, hence at $x = 1$ f is continuous. $\displaystyle\lim_{x \to 0} f(x) = 0$, but $f(0) = 1$, hence at $x = 0$, f is discontinuous. $\displaystyle\lim_{x \to 1} f(x) = $ is undefined, and $f(1) = 1$, hence at $x = 1$, f is discontinuous.	

Example 4

$$f(x) = \begin{cases} \dfrac{x^3 - 3x^2 + 4}{x+1}, & x \neq -1 \\ k, & x = -1 \end{cases}.$$

Determine the value of k in order that $f(x)$ be continuous at $x = -1$.

Answer

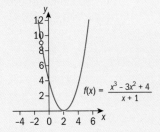

$f(x) = \dfrac{x^3 - 3x^2 + 4}{x+1}$

$\dfrac{x^3 - 3x^2 + 4}{x+1}$ is equivalent to $(x-2)^2$ in their respective domains. When $x = -1$, $(x-2)^2 = 9$ Hence, when $k = 9$, $f(-1) = \displaystyle\lim_{x \to -1} \dfrac{x^3 - 3x^2 + 4}{x+1} = 9$ so f is continuous at $x = -1$	$x^3 - 3x^2 + 4 = \dfrac{(x-2)^2(x+1)}{x+1}$ $= (x-2)^2$ *For f to be continuous, f (−1) must equal 9.*

Exercise 4B

1 $f(x) = \begin{cases} 3-x, & x \geq 1 \\ (x-1)^2, & x < 1 \end{cases}$. Determine if f is continuous at $x = 1$.

2 $f(x) = \begin{cases} x^2 + 4x + 5, & x \leq -2 \\ 2x + 5, & x > -2 \end{cases}$. Determine if f is continuous at $x = -2$.

3 $f(x) = \begin{cases} \dfrac{x-1}{|x-1|}, & x \neq 1 \\ 0, & x = 1 \end{cases}$. Determine if f is continuous at $x = 1$.

EXAM-STYLE QUESTIONS

4 Find a value for k such that $f(x) = \begin{cases} x^2 - 1, & x < 3 \\ 2kx, & x \geq 3 \end{cases}$ is continuous at 3.

5 Find the value of a such that $f(x) = \begin{cases} ax^2 - a, & x \geq 3 \\ 4, & x < 3 \end{cases}$ is continuous for all values of x.

6 Determine if these functions are continuous on the set of real numbers. If they are not continuous for all real x, state the values of x for which the function is discontinuous.

a $f(x) = \dfrac{x^2 + 1}{x^2 - 1}$ **b** $f(x) = \dfrac{x+1}{4 - x^2}$ **c** $f(x) = \dfrac{x}{x^2 + 1}$

d $f(x) = \dfrac{x^2 + 3x + 5}{x^2 + 3x - 4}$ **e** $f(x) = \dfrac{x^2 + 1}{x^3 - 1}$ **f** $f(x) = \dfrac{x+1}{\sqrt{x^2 + 1}}$

Limits to infinity

Infinity is not a number. It lies beyond all finite bounds. Hence, when discussing the behavior of a function as x approaches positive or negative infinity, written $\pm\infty$, we look for the value that the function approaches as x increases and decreases without bound.

For example, consider the behavior of the function

$y = \dfrac{1}{x}$ as x approaches $\pm\infty$.

The equation of the vertical asymptote is $x = 0$. The value of the function approaches 0 as x approaches $\pm\infty$, but is never equal to 0. There is no real number x such that $\dfrac{1}{x} = 0$.

The line $y = 0$ is the horizontal asymptote.

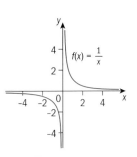

→ The line $y = k$, $k \in \mathbb{R}$, is the horizontal asymptote of $f(x)$ if either $\lim\limits_{x \to \infty} = k$ or $\lim\limits_{x \to -\infty} = k$.

Example 5

Sketch the graph of $y = \dfrac{x}{\sqrt{x^2+1}}$ for $-20 \leq x \leq 20$, show clearly any asymptote(s).

Check your answer on a GDC.

Answer

x	−20	−10	−5	−2	−1	0
$\dfrac{x}{\sqrt{x^2+1}}$	−0.999	−0.995	−0.981	−0.894	−0.707	0
x	1	2	5	10	20	
$\dfrac{x}{\sqrt{x^2+1}}$	0.707	0.894	0.981	0.995	0.999	

As x increases in the positive direction y approaches 1.

As x decreases in the negative direction y approaches −1.

Points on the graph are the values from the table.

GDC help on CD:
Alternative demonstrations for the TI-84 Plus and Casio fx-9860GII GDCs are on the CD.

$\lim\limits_{x\to\infty} f(x) = 1$ and $\lim\limits_{x\to-\infty} f(x) = -1$

The horizontal asymptotes are $y = 1$ and $y = -1$

Example 6

Sketch the graph of $f(x) = \dfrac{x+2}{x^2+1}$ for $-20 \leq x \leq 20$, show clearly any asymptote(s).

Check your answer on a GDC.

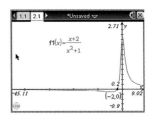

Answer

x	−20	−10	−5	−2	−1	0
$\dfrac{x+2}{x^2+1}$	−0.004	−0.084	−0.199	0	0.5	2
x	1	2	5	10	20	
$\dfrac{x+2}{x^2+1}$	1.5	0.8	0.269	0.119	0.005	

Notice that when $x = -2$, $y = 0$

Points on the graph are the values from the table.

GDC help on CD:
Alternative demonstrations for the TI-84 Plus and Casio fx-9860GII GDCs are on the CD.

$\lim\limits_{x\to\infty} f(x) = 0$ and $\lim\limits_{x\to-\infty} f(x) = 0$

The horizontal asymptote is $y = 0$, the x-axis.

The horizontal asymptote tells you the behavior of the function for very large values of x. However unlike the vertical asymptote, the function can assume the value of the horizontal asymptote for small values of x as happened in Example 6 at $x = 2$.

Other asymptotes

While not explicitly on the syllabus, it is useful to know that some asymptotes are neither vertical nor horizontal. For example, consider the graph of $f(x) = \dfrac{x^2 + 2x + 1}{x} = 2 + x + \dfrac{1}{x}$

The point (1, 4) is a local minimum of the function. The point (–1, 0) is a local maximum.

You will notice that there is a slant, or oblique, asymptote which passes between the local minimum and maximum points. As x approaches $\pm\infty$ the function resembles ever more closely the straight line $y = x + 2$

On the graph, the line $y = 2 + x$ is an asymptote to the function

$$f(x) = 2 + x + \frac{1}{x}$$

For very large values of x the value of $\dfrac{1}{x}$ is very small

As is clear from the graph, the difference, $\dfrac{1}{x}$, between the full function, $\dfrac{x^2 + 2x + 1}{x}$ and its slant asymptote, $x + 2$, becomes vanishingly small as $x \to \pm\infty$.

The word asymptote comes from the Greek *asymptotos*, meaning 'not falling together'.

Now consider the graph of $f(x) = 2 + x + x^2 + \dfrac{1}{x}$

We know that there is a vertical asymptote at $x = 0$. The limit of this function as x approaches ∞ is $\lim\limits_{x \to \infty}\left(2 + x + x^2 + \dfrac{1}{x}\right) = 2 + x + x^2$.

The curve $2 + x + x^2$ is an asymptote to the function.

Hence, an asymptote can be defined more generally as a line tangent to a curve at infinity.

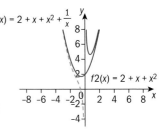

$f1(x) = 2 + x + x^2 + \dfrac{1}{x}$

$f2(x) = 2 + x + x^2$

Finding limits algebraically

Up to now we have been finding limits graphically and confirming our results numerically. We can find some limits algebraically using these properties of limits.

You will only be required to use informal methods to find limits in the exam

→ **Properties of limits as** $x \to \pm\infty$

Let L_1, L_2, and k be real numbers and $\lim\limits_{x \to \pm\infty} f(x) = L_1$ and $\lim\limits_{x \to \pm\infty} g(x) = L_2$. Then,

1 $\lim\limits_{x \to \pm\infty}(f(x) \pm g(x)) = \lim\limits_{x \to \pm\infty} f(x) \pm \lim\limits_{x \to \pm\infty} g(x) = L_1 \pm L_2$

2 $\lim\limits_{x \to \infty}(f(x) \cdot g(x)) = \lim\limits_{x \to \infty} f(x) \cdot \lim\limits_{x \to \infty} g(x) = L_1 \cdot L_2$

3 $\lim\limits_{x \to \pm\infty}(f(x) \div g(x)) = \lim\limits_{x \to \pm\infty} f(x) \div \lim\limits_{x \to \pm\infty} g(x) = L_1 \div L_2$,

provided $L_2 \neq 0$.

4 $\lim\limits_{x \to \pm\infty} kf(x) = k \lim\limits_{x \to \pm\infty} f(x) = kL_1$

5 $\lim\limits_{x \to \pm\infty}[f(x)]^{\frac{a}{b}} = L_1^{\frac{a}{b}}, \dfrac{a}{b} \in \mathbb{Q}$ (in simplest form),

provided $L_1^{\frac{a}{b}}$ is real.

These properties also hold when finding the limit as $x \to c$, $c \in \mathbb{R}$.

Example 7

Find the horizontal asymptote of $y = \dfrac{x+2}{2x+3}$

Answer

$\lim\limits_{x \to \infty} \dfrac{x+2}{2x+3} = \lim\limits_{x \to \infty} \dfrac{1 + \dfrac{2}{x}}{2 + \dfrac{3}{x}}$

Divide numerator and denominator by largest power of x.

$\lim\limits_{x \to \infty} \dfrac{1 + \dfrac{2}{x}}{2 + \dfrac{3}{x}} = \lim\limits_{x \to \infty}\left(1 + \dfrac{2}{x}\right) \div \lim\limits_{x \to \infty}\left(2 + \dfrac{3}{x}\right)$

Apply limit property 3: the limit of a quotient is the quotient of the limits.

$\lim\limits_{x \to \infty}\left(1 + \dfrac{2}{x}\right) = \lim\limits_{x \to \infty} 1 + \lim\limits_{x \to \infty} \dfrac{2}{x} = 1 + 0 = 1$

Apply limit property 1: the limit of a sum is the sum of the limits.

$\lim\limits_{x \to \infty}\left(2 + \dfrac{3}{x}\right) = \lim\limits_{x \to \infty} 2 + \lim\limits_{x \to \infty} \dfrac{3}{x} = 2 + 0 = 2$

Hence, $\lim\limits_{x \to \infty} \dfrac{x+2}{2x+3} = \dfrac{1}{2}$, and the horizontal asymptote is $y = \dfrac{1}{2}$.

Remember that the line $y = k$ is an horizontal asymptote if $\lim\limits_{x \to \infty} f(x) = k$.

▶ Continued on next page

Graphing the function confirms the limit graphically and numerically:

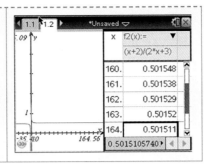

GDC help on CD:
Alternative demonstrations for the TI-84 Plus and Casio fx-9860GII GDCs are on the CD.

As shown in Example 7, when finding limits of rational algebraic expressions, it is often useful to divide the numerator and denominator by the largest power of x. For example, when finding $\lim\limits_{x \to \infty} \dfrac{x+3}{x^2+1}$, dividing both numerator and denominator by x^2 gives

$$\lim_{x \to \infty} \frac{\dfrac{1}{x}+\dfrac{3}{x^2}}{1+\dfrac{1}{x^2}}.$$

Using the properties of limits it is easy to verify that $\lim\limits_{x \to \infty} \dfrac{x+3}{x^2+1} = 0$

Similarly, $\lim\limits_{x \to \infty} \dfrac{-2x^3+2}{x^3-x} = \lim\limits_{x \to \infty} \dfrac{-2+\dfrac{2}{x^3}}{1-\dfrac{1}{x^2}} = -2$

Hence, the horizontal asymptote is $y = -2$

> You may wish to confirm this result using your GDC.

Investigation – graphs of $x^n + y^n = 1$

Graph the equation $x^2 + y^2 = 1$ using a graphing program. How would you enter the same equation in your GDC in order to see the same shape as the software produces?

Now graph $x^4 + y^4 = 1$. How does it compare with the graph of $x^2 + y^2 = 1$? Experiment with different even values of n for $x^n + y^n = 1$. What do you notice? From your observations, conjecture the shape of the graph of $x^n + y^n = 1$, when n is an even number, and n approaches infinity.

Investigation – graphs of polynomials

Graph functions of the type $\dfrac{P_n(x)}{Q_m(x)}$, such that n and m are positive integers, that represent the degree of the polynomial function.

Investigate the limit of the polynomial functions as x approaches $\pm\infty$ when

a $n < m$ **b** $n = m$ **c** $n > m$

Make a conjecture regarding the horizontal asymptotes of your functions, and justify your conjecture for the different cases in **a**, **b** and **c**.

> Possible examples are, $P_2(x) = x^2 + 3$
> $Q_4(x) = 3x^4 + x^2 + 1$

Exercise 4C

1 Find the required limit algebraically, if it exists.

a $\lim\limits_{x \to 4}\left(\dfrac{x+3}{x-3}\right)$

b $\lim\limits_{x \to -2}\left(\dfrac{x^2+x-2}{x+2}\right)$

c $\lim\limits_{x \to -2}\left(\dfrac{x^6-64}{x^3-8}\right)$

d $\lim\limits_{x \to 0}\dfrac{x^2-1}{x^2-x}$

e $\lim\limits_{x \to 1}\dfrac{x^2-1}{x^2-x}$

f $\lim\limits_{x \to 1}\dfrac{1}{1+\dfrac{1}{1-x}}$

g $\lim\limits_{x \to 0}\dfrac{(2+3x)^2-4(1+x)^2}{6x}$

h $\lim\limits_{x \to a}\dfrac{x^2-a^2}{x-a}$

2 Find the limit of $f(x)$ algebraically as x approaches $+\infty$, if it exists.

a $\dfrac{2x}{x+2}$

b $\dfrac{3x^2}{x^2-1}$

c $\dfrac{2x^2+x-1}{3x^2+5x-1}$

d $\dfrac{5x^2}{4x^3+2}$

e $\dfrac{x-1}{x^2-3x+5}$

f $\dfrac{\sqrt{4x+3}+2\sqrt{1+x}}{\sqrt{x}}$

3 Find, algebraically, any horizontal asymptotes of these functions.

a $\dfrac{3x^2-x+5}{x^2-4}$

b $\dfrac{2x}{4x-1}$

c $\dfrac{2x^2-3x+1}{x^3}$

d $\dfrac{x^2+1}{1-x^2}$

e $\dfrac{2x^3-3x}{2x^2+1}$

Convergence of sequences

The concept of limits can be used to describe the value that a sequence approaches as its index approaches a certain value.

> Sequences were introduced in Chapter 1

You know that $\lim\limits_{x \to \infty}\dfrac{1}{x}=0$. Now consider $\lim\limits_{n \to \infty} a_n$, $n \in \mathbb{Z}^+$, $a_n = \dfrac{1}{n}$

Write out the terms of this sequence:

$$1,\frac{1}{2},\frac{1}{3},...,\frac{1}{n},...$$

If m and n are positive integers, then when $m < n$, $\dfrac{1}{m} > \dfrac{1}{n}$. Hence, as the number of terms in the sequence increases, the value of the expression $\dfrac{1}{n}$ decreases, until it is very close to 0.

Hence, $\lim\limits_{n \to \infty}\dfrac{1}{n} = 0$, and we say that the sequence converges to 0.

You can investigate this graphically and numerically with the GDC.

If the sequence has a finite limit, then the sequence is said to be **convergent**, otherwise it is **divergent**.

The properties of limits of sequences are the same as those for limits of functions.

Example 8

Find $\lim\limits_{n \to \infty} \dfrac{n^2 + 3n}{2n^2 + 1}$, $n \in \mathbb{Z}^+$

GDC help on CD:
*Alternative demonstrations
for the TI-84 Plus and Casio
fx-9860GII GDCs are on the
CD.*

Answer

$$\lim\limits_{n \to \infty} \dfrac{n^2 + 3n}{2n^2 + 1} = \lim\limits_{n \to \infty} \dfrac{1 + \dfrac{3}{n}}{2 + \dfrac{1}{n^2}} = \dfrac{1}{2}$$

Hence, the sequence converges to $\dfrac{1}{2}$.

*Divide both numerator and
denominator by n^2, and use the
properties of limits.*

*Put your GDC in sequence mode.
Confirm this result graphically and
numerically on the GDC.*

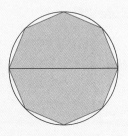

Investigation – inscribed polygons

Consider a polygon inscribed in a circle.
You can form a sequence of rational numbers by taking
the ratio of the perimeter of a regular polygon to its diameter.

Begin with an equilateral triangle in a circle. Calculate its perimeter
and write the ratio of its perimeter to its diameter. Do the same for
regular polygons of up to 10 sides. Formulate a conjecture.
Test your conjecture by calculating the same ratio for polygons with
many sides, e.g., 60, 80, 100, etc.

Determine the limit to infinity of your sequence, and justify your answer.

Convergence of series

In Chapter 1 you learned that if a geometric series has a finite sum,
it **converges** to its sum.

Recall the formula for finding the sum of a finite geometric series,

$$S_n = \dfrac{u_1(1 - r^n)}{1 - r}$$

→ For a geometric series, $\displaystyle\sum_{n=0}^{\infty} u_1 r^n = \lim\limits_{n \to \infty} \dfrac{u_1(1 - r^n)}{1 - r}$

When $-1 < r < 1$, $\lim\limits_{n \to \infty} r^n = 0$ and the series converges to $S = \dfrac{u_1}{1 - r}$

Consider the geometric series $\displaystyle\sum_{n=1}^{\infty} \left(\dfrac{1}{2}\right)^n$. Writing out this series, you

obtain $\dfrac{1}{2} + \dfrac{1}{4} + \dfrac{1}{8} + ...$ Since $|r| < 1$, this infinite geometric series has a

finite sum, $S = \dfrac{\dfrac{1}{2}}{1 - \dfrac{1}{2}} = 1$. The series converges to 1.

If the series does not have a finite sum, the series **diverges**.

Example 9

Determine whether the series $\displaystyle\sum_{n=0}^{\infty}\frac{5^n+4^n}{6^n}$ converges.

Answer

$\displaystyle\sum_{n=0}^{\infty}\frac{5^n+4^n}{6^n}=\sum_{n=0}^{\infty}\frac{5^n}{6^n}+\sum_{n=0}^{\infty}\frac{4^n}{6^n}=\sum_{n=0}^{\infty}\left(\frac{5}{6}\right)^n+\sum_{n=0}^{\infty}\left(\frac{4}{6}\right)^n$	*This is the sum of two geometric series.*
$\displaystyle\sum_{n=0}^{\infty}\left(\frac{5}{6}\right)^n=\frac{1}{1-\dfrac{5}{6}}=6$ and $\displaystyle\sum_{n=0}^{\infty}\left(\frac{4}{6}\right)^n=\frac{1}{1-\dfrac{4}{6}}=3$	*Find limits separately.*
Hence, $\displaystyle\sum_{n=0}^{\infty}\frac{5^n+4^n}{6^n}=9$, so the series converges to 9.	*Using limit property 1, the limit of a sum is the sum of the limits.*

Exercise 4D

1 Determine whether these sequences converge.

a $\displaystyle\lim_{n\to\infty}\frac{n+1}{n}$
b $\displaystyle\lim_{n\to\infty}\frac{n+1}{2n+1}$
c $\displaystyle\lim_{n\to\infty}\frac{n^2-n}{2n^2+\sqrt{n}}$

d $\displaystyle\lim_{n\to\infty}\frac{1-n^3}{n^2+1}$
e $\displaystyle\lim_{n\to\infty}\frac{n^2+1}{1-n^3}$

2 Determine whether each series converges. If it converges, determine its sum.

a $\displaystyle\sum_{n=0}^{\infty}\frac{(-1)^n}{2^n}$
b $\displaystyle\sum_{n=1}^{\infty}\left(\frac{\pi}{3.14}\right)^n$
c $\displaystyle\sum_{n=1}^{\infty}5\left(\frac{1}{3}\right)^n$

d $\displaystyle\sum_{n=1}^{\infty}\frac{3}{10^n}$
e $\displaystyle\sum_{n=1}^{\infty}\frac{2^n-3^n}{7^n}$
f $\displaystyle\sum_{n=1}^{\infty}4(-0.6)^{n-1}$

EXAM-STYLE QUESTIONS

3 A geometric series has $u_1=35$ and $r=2^x$.
  a Find the values of x for which the series is convergent.
  b Find the value of x for which the series converges to 40.

4 Find the set of values of x for which the series $\displaystyle\sum_{n=0}^{\infty}\left(\frac{3x}{x+1}\right)^n$ converges.

4.2 The derivative of a function

In mathematics, the derivative is the rate at which one quantity changes with respect to another. The process of finding the derivative is called **differentiation**. These ideas are central to the area of mathematics called the calculus.

Calculus was the result of centuries of work and debate. **Isaac Newton** (1642–1726) said "If I have seen further it is by standing on the shoulders of giants". There is evidence that both Newton and **Leibniz** (1646–1716) developed calculus independently within the same ten-year period, approximately 1665 to 1675.

Average rates of change

The graph shows the exchange rate of the euro to the US$ over the indicated time period in 2010. What was the average daily drop in exchange rate from September 29 to October 25?

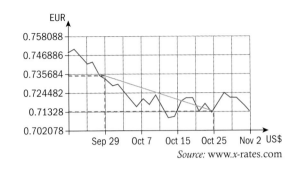

Source: www.x-rates.com

The average daily drop in the exchange rate

$$= \frac{\text{total change in the rate}}{\text{change in the time period}}$$

The change in the exchange rate over the indicated time period is approximately $0.735684 - 0.71328$, or -0.02356. The number of days between October 25 and September 29 is 26. Hence, the average daily exchange rate drop, $-0.2356 \div 26$, is about 0.001.

Graphically, the average rate of change between two points is the gradient of the line joining the two points.

Example 10

This graph shows the growth of internet domains on the World Wide Web since 1994. Estimate the average yearly growth between January 2001 and January 2010.

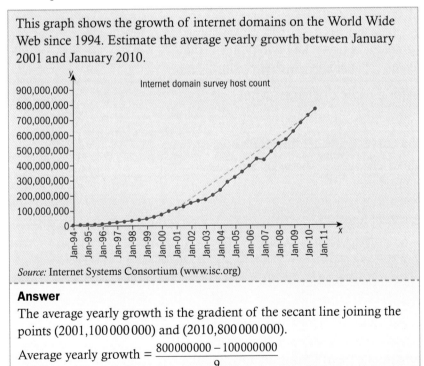

Source: Internet Systems Consortium (www.isc.org)

Answer

The average yearly growth is the gradient of the secant line joining the points $(2001, 100\,000\,000)$ and $(2010, 800\,000\,000)$.

$$\text{Average yearly growth} = \frac{800\,000\,000 - 100\,000\,000}{9}$$

$$\approx 77.8 \text{ million domains yearly}$$

A secant line joins two points on a curve.

Example 11

A ball rolling toward the edge of a ping-pong table is d cm from the edge at any time t seconds, $t > 1$, and $d = -t^2 + t + 6$. Find the average speed of the ball between the first and third second.

Answer

Average speed = total distance ÷ total time

$$\frac{\left[-(3)^2 + 3 + 6\right] - \left[-(1)^2 + 1 + 6\right]}{3 - 1} = -3 \text{ cm}\,s^{-1}$$

The speed of the ball is $3\,\text{cm}\,s^{-1}$.

> → In general, the average rate of a function f between two input values x_1 and x_2 is given by
> $$\frac{f(x_2) - f(x_1)}{x_2 - x_1}, \text{ or } \frac{\Delta y}{\Delta x}$$
> (read as 'the change in y divided by the change in x' where Δ is the Greek letter delta.)

The rate of change, $\dfrac{\Delta y}{\Delta x}$, at a point, is the gradient of the graph at the point.

If a function is linear, the gradient between any two points is the same, hence the rate of change, $\dfrac{\Delta y}{\Delta x}$, between any two points is the same, and will be the same for the rate of change at any particular point.

This changes, however, for a curve. Consider the graph of the function $y = x^2$, and the rate of change, $\dfrac{\Delta y}{\Delta x}$, between any point on the curve and the point $(1, 1)$.

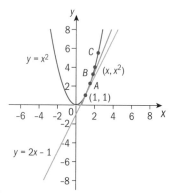

The gradient of any of the secant lines is $\dfrac{\Delta y}{\Delta x} = \dfrac{x^2 - 1}{x - 1} = x + 1$, $x \neq 1$

What is the gradient at $x = 1$, since according to this definition, x cannot equal 1? You can see geometrically that as the points move closer to $(1,1)$ the secant lines approach a line which is a tangent to the curve at $(1,1)$.

Now take a point on the curve arbitrarily close to the point $(1, 1)$, whose x coordinate is $1 + h$, where h is a very small quantity, $h \neq 0$. The corresponding y-coordinate is $(1 + h)^2$. You can now find the gradient between the two points $(1, 1)$ and $(1 + h, (1 + h)^2)$:

$$\frac{\Delta y}{\Delta x} = \frac{(1+h)^2 - 1^2}{(1+h) - 1} = \frac{1^2 + 2h + h^2 - 1^2}{h} = \frac{2h + h^2}{h} = 2 + h$$

The limit of this expression as h approaches 0 is $\lim\limits_{h \to 0}(2 + h) = 2$.

Hence, the gradient of the tangent at the point $(1, 1)$ is 2.

Since speed refers to how fast an object is moving, it is always positive. Velocity refers to the rate at which an object changes its position, hence it can be positive or negative. For example, if I move forward at a rate of 2 km h^{-1} and then return at the same rate, my speed is always the same, but the direction in which I'm moving has changed. Moving forward I have a positive velocity, whereas returning I have a negative velocity.

> → The gradient of a curve $y = f(x)$ at the point $(a, f(a))$ is
> $\lim\limits_{h \to 0} \dfrac{f(a+h) - f(a)}{h}$, provided this limit exists.

Example 12

Find the gradient of the curve $y = x^2$ at the point $x = -2$.

Answer

$\dfrac{\Delta y}{\Delta x} = \dfrac{(-2+h)^2 - (-2)^2}{(-2+h) - (-2)} = \dfrac{4 - 4h + h^2 - 4}{h}$ *Use the definition for gradient of a curve at a point.*

$= \dfrac{-4h + h^2}{h} = \dfrac{h(-4+h)}{h} = -4 + h$ *Simplify.*

$\lim\limits_{h \to 0} (-4 + h) = -4$ *Evaluate the limit.*

Example 13

Find the points on the curve $y = \dfrac{1}{x}$ such that the gradient at these points is $-\dfrac{1}{9}$.

Answer

Consider the point $\left(a, \dfrac{1}{a}\right)$ and a   *Use the definition for gradient of a curve at a point, then simplify.*

neighboring point $\left(a+h, \dfrac{1}{a+h}\right)$

$\dfrac{\Delta y}{\Delta x} = \dfrac{\dfrac{1}{a+h} - \dfrac{1}{a}}{(a+h) - a} = \dfrac{\dfrac{a - (a+h)}{a(a+h)}}{h} = \dfrac{-h}{a^2 + ah}$

$= \dfrac{-1}{a^2 + ah}$ *Evaluate the limit.*

$\lim\limits_{h \to 0} \dfrac{-1}{a^2 + ah} = -\dfrac{1}{a^2}$

$-\dfrac{1}{a^2} = -\dfrac{1}{9}$, hence $a = \pm 3$ *Set the expression equal to the gradient, and solve for a.*

The points are $(3, \dfrac{1}{3})$ and $(-3, -\dfrac{1}{3})$.

In geometrical terms, a curve is a set of points under a specific condition. For example, a circle is a set of points equidistant from a fixed point called its center. Recall from Chapter 2 that other geometrical curves, such as the parabola, ellipse and hyperbola are obtained when a plane intersects a cone at different angles. Many curves have been either discovered, or invented for the solution of special problems, for example, in mechanics.

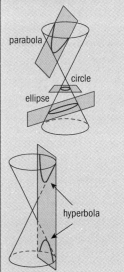

parabola
circle
ellipse
hyperbola

Find examples of real-life problems where conic sections are used, and model them.

Exercise 4E

1 Find the gradient of the curve at the given value of x.

a $y = 2x^2 - 1$ at $x = 1$ 　　**b** $y = \dfrac{2}{x}$ at $x = -2$

c $y = x^3$ at $x = 1$ 　　**d** $y = -x^2$ at $x = 1$

e $y = \dfrac{x}{x+1}$ at $x = 0$ 　　**f** $y = \dfrac{1}{x^2}$ at $x = 2$

2 Find the point on the curve $y = \dfrac{1}{x^2}$ such that the gradient at the point is 2.

3 Find the point on the curve $y = 2x^2 + \dfrac{1}{x}$ and then the point on the curve whose gradient is 3.

Investigation – gradients

a Find the gradient to $y = x^n$, n a positive integer, at different points along the curve. You have already found two such values for $y = x^2$, at $x = 1$ and $x = -2$.

b Conjecture a rule to find the gradient of the tangent to $y = x^n$ at any point on its curve.

You have developed the definition of the gradient of a point on a curve, and looked at some examples. In the Investigation above you derived a rule for the gradient function of the given curve for all points on the curve. The derivative of a function at each point along its curve can now be found.

→ The **derivative**, or **gradient function**, of a function f with respect to x is the function $f'(x) = \lim\limits_{h \to 0} \dfrac{f(x+h) - f(x)}{h}$, provided this limit exists.

If f' exists, then f has a derivative at x, or is **differentiable** at x. ($f'(x)$ is read 'f dash', or 'f prime', of x.) Another notation for the derivative is $\dfrac{dy}{dx}$, the derivative of the function $y = f(x)$ with respect to x.

A function is differentiable if the derivative exists for all x in the domain of f.

Example 14

Find $f'(x)$ given that $f(x) = 2x^2 + x$, and hence find the gradient of the function at $x = -3$.

Answer

$f'(x) = \lim_{h \to 0} \dfrac{2(x+h)^2 + (x+h) - (2x^2 + x)}{h}$	*Use the definition of the derivative,*
$\quad = \lim_{h \to 0}(4x + 1 + 2h)$	*then simplify,*
$\quad = 4x + 1$	*then evaluate the limit.*
$f'(-3) = 4(-3) + 1 = -11$	*Evaluate f' at $x = -3$.*

Example 15

If $f(x) = \sqrt{x}$, find $f'(x)$, and then find the gradient to the curve at $x = 4$.

Answer

$f'(x) = \lim_{h \to 0} \dfrac{\sqrt{x+h} - \sqrt{x}}{h}$	*To simplify use the difference of two squares.*
$\quad = \lim_{h \to 0} \dfrac{\sqrt{x+h} - \sqrt{x}}{h} \cdot \dfrac{\sqrt{x+h} + \sqrt{x}}{\sqrt{x+h} + \sqrt{x}}$	*Multiply by $\dfrac{\sqrt{x+h} + \sqrt{x}}{\sqrt{x+h} + \sqrt{x}}$*
$\quad = \lim_{h \to 0} \dfrac{\frac{(x+h) - x}{h}}{\sqrt{x+h} + \sqrt{x}}$	*Place h as shown so as to arrive at the next result.*
$\quad = \lim_{h \to 0} \dfrac{1}{\sqrt{x+h} + \sqrt{x}}$	
$\quad = \dfrac{1}{2\sqrt{x}}$	
When $x = 4$, the gradient to the curve is $\dfrac{1}{2\sqrt{4}} = \dfrac{1}{4}$	

Example 16

A particle moves in a straight line so that its position from its starting point at any time t, in seconds, is given by $s = 4t^2$, where s is in metres. The particle passes through a point P when $t = a$ and then sometime later it passes through point Q when $t = a + h$. Find the average velocity as the particle travels from point P to point Q, and deduce its velocity at the instant it passes through P.

Answer

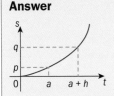

$P(a, 4a^2)$ and $Q(a + h, 4(a + h)^2)$

$\text{Average velocity} = \dfrac{4(a+h)^2 - 4a^2}{(a+h) - a}$

$= \dfrac{4(a^2 + 2ah + h^2) - 4a^2}{h}$

$= \dfrac{4h^2 + 8ah}{h}$

$= h\dfrac{4h + 8a}{h}$

$= 4h + 8a\,\text{m s}^{-1}$

Velocity at $P = 8a\,\text{m s}^{-1}$

Sketch a graph.

Average velocity =
total distance traveled

total traveling time

To find velocity at P find

$\lim\limits_{h \to 0}(4h + 8a)$.

The velocity of the particle when $t = a$ is the **instantaneous velocity**.

Some functions do not have a derivative at every point in their domain. You can easily prove that if a function is differentiable at $x = c$, then it will be continuous at $x = c$. In other words, **differentiability at a point implies continuity at the point**.

Let f be differentiable at $x = c$. You want to show that $\lim\limits_{x \to c} f(x) = f(c)$.

Since f is differentiable at $x = c$, and the point $x = c$ is excluded from the limit $x \to c$,

$$\lim_{x \to c}[f(x) - f(c)] = \lim_{x \to c}\left[\frac{f(x) - f(c)}{x - c} \cdot x - c\right] = \lim_{x \to c}\frac{f(x) - f(c)}{x - c} \cdot \lim_{x \to c}(x - c) = f'(c) \cdot 0 = 0.$$

Since $\lim\limits_{x \to c}[f(x) - f(c)] = \lim\limits_{x \to c} f(x) - \lim\limits_{x \to c} f(c) = \lim\limits_{x \to c} f(x) - f(c)$ it follows that $\lim\limits_{x \to c} f(x) = \lim\limits_{x \to c} f(c)$, hence f is continuous at c.

Now consider the converse, i.e., if a function is continuous at $x = c$, it is differentiable at $x = c$. To find a counter - example you need to find a function that is continuous at $x = c$, but whose left and right limits as x approaches c are either not equal or do not exist.

One such function that you are familiar with is $y = |x|$. You know that for all $x < 0$, the function is equivalent to $y = -x$, hence the gradient of all points to the left of $x = 0$ is -1. For all $x > 0$, the function is equivalent to $y = x$, hence the gradient of all points to the right of $x = 0$ is 1. Using mathematical notation,

$$\lim_{h \to 0^+} \frac{|x+h|-|x|}{h} = -1 \text{ and } \lim_{h \to 0^-} \frac{|x+h|-|x|}{h} = 1$$

Since the left and right limits differ, the function does not have a derivative at $x = 0$.

> → If a function is differentiable at c, it is continuous at c.
>
> A function that is continuous at c may not be differentiable at c.

Linearity

A visual approach to deciding if a function is differentiable at the point is local linearity at the point. If you zoom in with your GDC at a point on a function that is differentiable, for example, x^2 at $x = 0$, the function seems to 'flatten' at this point. The more you zoom in, the more linear it appears at $x = 0$.

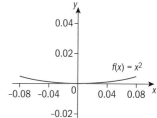

Test this visual approach on several functions at points that have a derivative.

When you perform the zoom test on $y = |x|$ at $x = 0$, it will remain unchanged regardless of how closely you zoom in.

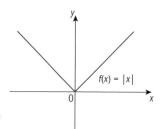

Exercise 4F

1 Find the gradient function of the given curve, and then the value of the gradient to the curve at the given point.

  a $y = x^2 + 2x + 1$ at $x = 0$
  b $y = x^3 - 1$ at $x = 1$

  c $y = \dfrac{2}{x}$ at $x = 3$
  d $y = \sqrt{x-1}$ at $x = 2$

  e $y = \sqrt{x+3}$ at $x = 1$
  f $y = \dfrac{1}{\sqrt{x}}$ at $x = 4$

EXAM-STYLE QUESTION

2 A particle moves in a straight line so that its position from its starting point after t seconds is $12 - 5t^2$. If the particle passes through point A when $t = a$, and point B when $t = a + h$, find
  a the average velocity of the object as it moves from A to B
  b the velocity as it passes through point A.

Equations of tangents and normals

If m is the gradient at a point (x_1, y_1) on a curve the equation
of the tangent at that point is

$(y - y_1) = m(x - x_1)$

Gradient-point formula:
$$y - y_1 = m(x - x_1)$$

You can also find the normal to the function at a particular point,
since the gradient of the normal is the negative reciprocal of the
gradient of the tangent.

Example 17

Given $f(x) = 3x^2 - 2$, find

a the gradient to the curve at $x = 1$

b the equation of the tangent to the curve at $x = 1$

c the equation of the normal to the tangent at $x = 1$

Answers

a $f'(x) = \lim_{h \to 0} \dfrac{\left[3(x+h)^2 - 2\right] - (3x^2 - 2)}{h}$ *Find the gradient function.*

$= \lim_{h \to 0} \dfrac{3x^2 + 6xh + 3h^2 - 2 - 3x^2 + 2}{h}$

$= \lim_{h \to 0} \dfrac{6xh + 3h^2}{h} = \lim_{h \to 0} 6x + 3h = 6x$

Hence, at $x = 1$, the gradient is 6.   *Evaluate the gradient function at $x = 1$.*

Equation of tangent is

$y - 1 = 6(x - 1)$ or $y = 6x - 5$ *The gradient of the tangent is 6 and it goes through the point (1, 1).*

c Equation of normal is

$y - 1 = -\dfrac{1}{6}(x - 1)$ *The gradient of the normal to the tangent is $-\dfrac{1}{6}$ and the normal goes*

or $y = -\dfrac{x}{6} + \dfrac{7}{6}$ *through the point (1, 1).*

GDC help on CD:
*Alternative demonstrations
for the TI-84 Plus and Casio
fx-9860GII GDCs are on the
CD.*

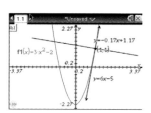

Exercise 4G

1 Given $f(x) = 9 - x^2$, find

   a the gradient of the curve at $x = -1$

   b the equation of the tangent to the curve at $x = -1$

   c the equation of the normal to the curve at $x = -1$

2 Find the points on the curve $y = \dfrac{1}{x-1}$ whose gradient is -1, and find the equations of the tangents through these points.

3 Find any points on the following curves that have horizontal tangents, i.e., tangents parallel to the x-axis.

   a $y = 4 - 3x - 3x^2$ **b** $y = x^3 + 1$ **c** $y = \dfrac{1}{x}$

   d $y = x^2 - 3x$ **e** $y = \sqrt{x}$

4 Find the equations of the tangent and normal to the curve $y = x + \dfrac{1}{x}$ at $x = 1$

4.3 Differentiation rules

Derivative of a constant function

The graph of $f(x) = c$, $c \in \mathbb{R}$, is a straight line whose equation is $y = c$ and it is parallel to the x-axis. Its gradient is therefore 0 for all x. Hence,

> $\rightarrow$ If $f(x) = c$, and $c \in \mathbb{R}$, then $f'(x) = 0$

Positive integer powers of x

From the Investigation on the derivative of $y = x^2$, you will probably have conjectured that its gradient function is $2x$. What happens with higher powers of x? Here is a rule developed from first principles.

Using the definition of the derivative

$$f'(x) = \lim_{h \to 0} \frac{(x+h)^n - x^n}{h}$$

$$f'(x) = \lim_{h \to 0} \frac{h[(x+h)^{n-1} + (x+h)^{n-2}x + \dots + (x+h)x^{n-2} + x^{n-1}]}{h}$$

$$= \lim_{h \to 0} ((x+h)^{n-1} + (x+h)^{n-2}x + \dots + (x+h)x^{n-2} + x^{n-1}) = nx^{n-1}$$

> $\rightarrow$ If n is a positive integer, and $f(x) = x^n$, then $f'(x) = nx^{n-1}$

You have already seen this result in the investigation in the previous section, that is, if $f(x) = x^2$, then $f'(x) = 2x$

The power rule holds for all n, where n is a real number, and this result will be used without proof.

Use the algebraic identity (see Chapter 1).
$a^n - b^n = (a-b)$
$(a^{n-1} + n^{n-2}b + \dots$
$+ ab^{n-2} + b^{n-1}), n \in \mathbb{Z}^+$
with $a = x + h$ and $b = x$.

There are n terms, each having the limit x^{n-1} as h approaches 0.

This is the power rule.

Constant multiple of a function

→ For $c \in \mathbb{R}$, $(cf)'(x) = cf'(x)$ provided $f'(x)$ exists.

The sum and difference of functions

Let $f(x)$ be the sum of two functions in x whose derivatives exist, i.e., $f(x) = u(x) + v(x)$. Then,

$$f'(x) = \lim_{h \to 0} \frac{[u(x+h) + v(x+h)] - [u(x) + v(x)]}{h}$$

$$= \lim_{h \to 0} \frac{u(x+h) - u(x)}{h} + \frac{v(x+h) - v(x)}{h}$$

$$= \lim_{h \to 0} \frac{u(x+h) - u(x)}{h} + \lim_{h \to 0} \frac{v(x+h) - v(x)}{h}$$

$$= u'(x) + v'(x)$$

→ If $f(x) = u(x) \pm v(x)$, then $f'(x) = u'(x) \pm v'(x)$

The proof for the difference of two functions is left as an exercise for you.

The proof for negative integer powers of x is left for you as an exercise. This result can be extended to all real number powers of x.

Example 18

Differentiate $y = \dfrac{1}{4}x^5 - x^3 + 5x^2 - \dfrac{1}{2}x + 3$ with respect to x.

Answer	
$\dfrac{dy}{dx} = \dfrac{5}{4}x^4 - 3x^2 + 10x - \dfrac{1}{2}$	*The derivative of a sum is the sum of the derivatives. The first four terms use the power and constant multiple rules, and the last term uses the constant function rule.*

When the function is given as y, write the derivative as $\dfrac{dy}{dx}$

Example 19

Find $f'(x)$ if $f(x) = \dfrac{2x^4 - 3x^3 + 1}{x^2}$, $x \neq 0$

Answer	
$f(x) = \dfrac{2x^4}{x^2} - \dfrac{3x^3}{x^2} + \dfrac{1}{x^2}$	*Write as a sum.*
$= 2x^2 - 3x + x^{-2}$	*Simplify.*
$f'(x) = 4x - 3 - 2x^{-3} = 4x - 3 - \dfrac{2}{x^3}$	*Differentiate each term.*

When the function is given as $f(x)$, write the derivative as $f'(x)$.

Example 20

Find the equation of the normal to the curve $f(x) = -2x^3 + x - 1$ at $x = 0$.

Answer

At $x = 0$, $f(0) = -1$, so the point on the curve is $(0, -1)$.	*Differentiate $f'(x)$ to get the gradient function.*
$f'(x) = -6x^2 + 1$	
$f'(0) = 1$	*Evaluate $f'(x)$ at $x = 0$*
Gradient of normal $= -\dfrac{1}{1} = -1$	*Find gradient of normal.*
$y - 1 = -1(x - 0)$	*Use $y - y_1 = m(x - x_1)$*
$\quad y = -x + 1$	*Simplify.*

Exercise 4H

1 Find $\dfrac{dy}{dx}$ for each function.

a $y = 4 - x - 3x^2$

b $y = 2x^4 - 3x + 1$

c $y = 4x^3 - \dfrac{1}{x^3} + 2x^2 + \dfrac{2}{3x^2}$

d $y = \dfrac{2 - 3x^2 + 5x^4}{x}$

2 Find the equation of the tangent to the curve $y = 2(3x^2 - 2x)$ at $x = 1$

3 Find the equation of the normal to the curve $y = \dfrac{x-3}{x}$ at the point $x = -1$

The chain rule

The function $y = (2x - 1)^3$ is a polynomial so it is differentiable for all x. To differentiate this function, expand it, and then use the sum and difference rules.

$y = (2x - 1)^3$ is a composite function where $y = f(g(x))$, $g(x) = (2x - 1)$, and $f(x) = x^3$

Let $u = g(x)$, then $y = u^3$

Consider the relationship $\dfrac{dy}{dx} = \dfrac{dy}{du} \cdot \dfrac{du}{dx}$

$\dfrac{dy}{du} = 3u^2$ and $\dfrac{du}{dx} = 2$ (since $u = 2x - 1$)

then using the relationship above

$\dfrac{dy}{dx} = 3u^2 \cdot 2 = 6u^2$

Replacing u

$\dfrac{dy}{dx} = 6(2x - 1)^2$

This is as example of the chain rule.

This function,
$y = (2x - 1)^3$
$= 8x^3 - 4x^2 + 6x - 1$
Hence
$\dfrac{dy}{dx} = 24x^2 - 4x + 6$

$\dfrac{dy}{du}$ and $\dfrac{du}{dx}$ are not fractions, hence this relationship is not arrived at by cancelling du. Since, however, these are rates of change, we can intuitively see that if, for example, y changes twice as fast as u and u changes three times as fast as x, then y would change 6 times as fast as x.

> → If f is differentiable at the point $u = g(x)$, and g is differentiable at x, then the composite function $(f \circ g)(x)$ is differentiable at x. Furthermore, if $y = f(u)$ and $u = g(x)$, then $\dfrac{dy}{dx} = \dfrac{dy}{du} \cdot \dfrac{du}{dx}$
>
> Another definition for the chain rule is
>
> $(f \circ g)'(x) = f'(g(x)) \cdot g'(x)$

Example 21

Differentiate $y = (1 - 3x)^7$ with respect to x.

Answer

Let $u = 1 - 3x$, then $\dfrac{du}{dx} = -3$	Define u and find $\dfrac{du}{dx}$
Hence, $y = u^7$ and $\dfrac{dy}{du} = 7u^6$	Write y in term of u. Find $\dfrac{dy}{du}$
$\dfrac{dy}{dx} = \dfrac{dy}{du} \cdot \dfrac{du}{dx} = 7u^6 \cdot (-3)$	Use the chain rule.
$\dfrac{dy}{dx} = 7(1 - 3x)^6 (-3) = -21(1 - 3x)^6$	Substitute for u and simplify.

Example 22

Differentiate $\sqrt{3x^2 - 4}$

Answer

$(f \circ g)(x) = \sqrt{3x^2 - 4}$ for $f(x) = \sqrt{x}$ and $g(x) = 3x^2 - 4$	Find f and g for the composite function.
$f'(g(x)) = \dfrac{1}{2}(g(x))^{-\frac{1}{2}} = \dfrac{1}{2}(3x^2 - 4)^{-\frac{1}{2}}$	Differentiate $f(g)$
$g'(x) = 6x$	Differentiate $g(x)$
$(f \circ g)'(x) = \dfrac{1}{2}(3x^2 - 4)^{-\frac{1}{2}} \cdot 6x$	Apply the chain rule.
$= \dfrac{3x}{\sqrt{3x^2 - 4}}$	Simplify.

You can use the chain rule to show that the derivative of an odd function is an even function. Recall the definition of an odd function, i.e., if f is odd, then $f(-x) = -f(x)$

Hence, $-f'(x) = f'(-x)(-1)$, and it follows that $f'(-x) = f'(x)$. f' is therefore an even function.

> Odd and even functions are discussed in Section 2.2

Similarly, if f is an even function, then $f(-x) = f(x)$. Hence, $f'(-x)(-1) = f'(x)$, and it follows that $f'(-x) = -f'(x)$, and therefore f' is an odd function.

Exercise 4I

1 Find $\dfrac{dy}{dx}$ for each function.

   a $y = (2x + 3)^5$

   b $y = \sqrt{2 - 3x}$

   c $y = \dfrac{2 - 3x^2 + 5x^4}{x}$

   d $y = \dfrac{-3}{\sqrt{5x^2 + 1}}$

   e $y = \left(\dfrac{x}{1 - \sqrt{x}}\right)^3$

2 Find the equation of the tangent to the curve $y = \sqrt{3x^2 - 2x}$ at $x = 1$

3 Find the equation of the normal to the curve $\dfrac{x - 3}{x}$ at the point $x = 1$

4 Find the point of the curve $\dfrac{1}{3x^2 - 6x + 1}$ where the tangent to the curve is parallel to the x-axis.

5 Find the derivative, with respect to x, of the function $y = \sqrt{1 - \sqrt{x}}$

Product rule

You use the chain rule to differentiate composite functions. To find the derivative of a product of functions, you use the product rule.

The derivative of $y = x^2$ is $2x$
Rewrite x^2 as $x \cdot x$. The derivative of x is 1

Thus the product of the derivatives of the component functions is $1 \times 1 = 1 \neq 2x$. Thus, in general, the derivative of a product of functions is not equal to the product of the derivatives of the functions.

You can derive the product rule from first principles.

Let $f(x) = u(x)\,v(x)$, where $u(x)$ and $v(x)$ are differentiable functions.

Then

$$f'(x) = \lim_{h \to 0} \frac{u(x + h)v(x + h) - u(x)v(x)}{h}$$

$$= \lim_{h \to 0} \frac{u(x + h)v(x + h) - [u(x + h)v(x) - u(x + h)v(x)] - u(x)v(x)}{h}$$

$$= \lim_{h \to \infty} \left[u(x + h)\frac{v(x + h) - v(x)}{h} + v(x)\frac{u(x + h) - u(x)}{h} \right]$$

$$= \lim_{h \to 0} u(x + h).\lim_{h \to 0} \frac{v(x + h) - v(x)}{h} + \lim_{h \to 0} v(x).\lim_{h \to 0} \frac{u(x + h) - u(x)}{h}$$

$$= u(x)v'(x) + v(x)u'(x)$$

> Insert the expression in square brackets – this is equal to 0

> Factorize, and then rearrange.

Hence, if $f(x) = u(x)v(x)$, where $u(x)$ and $v(x)$ are differentiable functions then

$f'(x) = u(x)v'(x) + v(x)u'(x)$

> → If $y = uv$ then $\dfrac{dy}{dx} = u\dfrac{dv}{dx} + v\dfrac{du}{dx}$
>
> where u and v are functions of x and differentiable.
>
> Another way of writing this is:
>
> If $f(x) = u(x)v(x)$, where $u(x)$ and $v(x)$ are differentiable
> functions then $f'(x) = u(x)v'(x) + v(x)u'(x)$.

Example 23

Find $f'(x)$ if $f(x) = (2x + 3)(4 - 3x)$

Answer

Let $u(x) = 2x + 3$, then $u'(x) = 2$	*Define u and v.*
Let $v(x) = 4 - 3x$, then $v'(x) = -3$	*Find $\dfrac{du}{dx}$ and $\dfrac{dv}{dx}$*
$f'(x) = -3(2x + 3) + 2(4 - 3x)$	*By the product rule:*
$\qquad = -1 - 12x$	$f'(x) = u(x)v'(x) + v(x)u'(x)$

Example 24

Find the equation of the tangent to the curve $y = \dfrac{x^2 + 1}{x + 1}$, $x \neq -1$ at $(0, 1)$.

Answer

$y = (x^2 + 1)(x + 1)^{-1}$	*Change quotient to a product.*
Let $u = x^2 + 1$, then $\dfrac{du}{dx} = 2x$	
Let $v = (x + 1)^{-1}$, then	
$\dfrac{dv}{dx}(x) = -(x + 1)^{-2}$	
$\dfrac{dy}{dx} = 2x(x + 1)^{-1} - (x^2 + 1)(x + 1)^{-2}$	*Use product rule:* $\dfrac{d(uv)}{dx} = u\dfrac{dv}{dx} + v\dfrac{du}{dx}$
$\qquad = (x + 1)^{-2}[2x(x + 1) - (x^2 + 1)]$	*Factorize.*
$\dfrac{2x^2 + 2x - x^2 - 1}{(x + 1)^2} = \dfrac{x^2 + 2x - 1}{(x + 1)^2}$	*Evaluate $f'(x)$ at $x = 0$*
$f'(0) = -1$	*Use gradient point formula:*
$y = -x + 1$	$y - y_1 = m(x - x_1)$

Exercise 4J

Differentiate these functions, with respect to x.

1 $y = (x - 1)(x + 3)^3$ 　　　　　 **2** $y = (2x - 3)^2(4x + 1)^3$

3 $y = \dfrac{x + 1}{x - 1}$ 　　　　　 **4** $y = x\sqrt{1 - 2x}$

5 $y = \dfrac{1}{x^4 - 3x + 1}$ 　　　　　 **6** $y = (x - 1)^4(3x - 2)^{\frac{2}{3}}$

7 Find the equations of the tangent and normal to the curve
$f(x) = (x^2 + 1)(x^2 + 3)$ at the point $(-1, 4)$.

Changing a quotient to a product in order to differentiate, is not always straightforward, so you need the quotient rule.

The quotient rule

If $y = \dfrac{u}{v}$, where u and v are both differentiable functions in x, then

$$f'(x) = \lim_{h \to 0} \frac{\dfrac{u(x+h)}{v(x+h)} - \dfrac{u(x)}{v(x)}}{h}$$

$$= \lim_{h \to 0} \frac{v(x)u(x+h) - u(x)v(x+h)}{hv(x+h)v(x)}$$

$$= \lim_{h \to 0} \frac{v(x)u(x+h) - v(x)u(x) + v(x)u(x) - u(x)v(x+h)}{h}$$

> Add and subtract $v(x)\,u(x)$

$$= \lim_{h \to 0} \frac{v(x)\dfrac{u(x+h)-u(x)}{h} - u(x)\dfrac{v(x+h)-v(x)}{h}}{v(x+h)v(x)}$$

> Factorize

$$= \frac{\displaystyle\lim_{h \to 0} v(x) \cdot \lim_{h \to 0} \frac{u(x+h)-u(x)}{h} - \lim_{h \to 0} u(x) \cdot \lim_{h \to 0} \frac{v(x+h)-v(x)}{h}}{\displaystyle\lim_{h \to 0}\left[v(x+h)v(x)\right]}$$

> Take the limits in both numerator and denominator.

$$= \frac{v(x)u'(x) - u(x)v'(x)}{(v(x))^2}$$

Hence, if $u(x)$ and $v(x)$ are differentiable functions, and

$f'(x) = \dfrac{u(x)}{v(x)}, v(x) \neq 0$ then $f'(x) = \dfrac{v(x)u'(x) - u(x)v'(x)}{(v(x))^2}$

→ If $y = \dfrac{u}{v}$ then $\dfrac{dy}{dx} = \dfrac{v\dfrac{du}{dx} - u\dfrac{dv}{dx}}{v^2}$

where u and v are differentiable functions of x.

An alternative way of writing this is:

if $u(x)$ and $v(x)$ are differentiable functions, and

$f(x) = \dfrac{u(x)}{v(x)}, v(x) \neq 0$ then $f'(x) = \dfrac{v(x)u'(x) - u(x)v'(x)}{(v(x))^2}$

Example 25

Using the quotient rule, differentiate $y = \dfrac{x^2+1}{x+1}$ $x \neq -1$

Answer

Let $u = x^2 + 1$, then $\dfrac{du}{dx} = 2x$

Let $v = (x + 1)$, then $\dfrac{dv}{dx} = 1$

Hence,

$\dfrac{dy}{dx} = \dfrac{2x(x+1)-(x^2+1)}{(x+1)^2} = \dfrac{x^2+2x-1}{(x+1)^2}$

This is the same function as in Example 24 – and answer, using the quotient rule is the same.

Example 26

Differentiate $y = \dfrac{x^2+1}{x^2-1}$, $(x \neq \pm 1)$, with respect to x, and hence find the derivative at $x = 2$

Answer

Let $u = x^2 + 1$, then $\dfrac{du}{dx} = 2x$

Let $v = x^2 - 1$, then $\dfrac{dv}{dx} = 2x$

$\dfrac{dy}{dx} = \dfrac{2x(x^2-1)-2x(x^2+1)}{(x^2-1)^2}$

$\quad = \dfrac{-4x}{(x^2-1)^2}$

At $x = 2$, $\dfrac{dy}{dx} = -\dfrac{8}{9}$

Find $\dfrac{dv}{dx}$ and $\dfrac{du}{dx}$

Use the quotient rule.

Check your answer using the GDC.

GDC help on CD:
Alternative demonstrations for the TI-84 Plus and Casio fx-9860GII GDCs are on the CD.

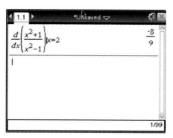

Exercise 4K

1 Differentiate these functions, with respect to x:

a $y = \dfrac{x^2-7}{x^3}$

b $y = \dfrac{x}{\sqrt{x^2+1}}$

c $y = \dfrac{1}{x^4-3x+1}$

d $y = \dfrac{1+\sqrt{x}}{1-\sqrt{x}}$

e $y = \sqrt{\sqrt{x}-x}$

f $y = \dfrac{1}{x^3\sqrt{1-2x+3x^2}}$

2 Find the gradient of the curve known as 'Newton's Serpentine', $y = \dfrac{4x}{x^2+1}$, at $x = -1$

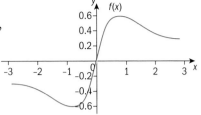

3 Find the equation of the normal to the curve known as 'The Witch of Agnesi', $y = \dfrac{8}{4+x^2}$, at $x = 1$

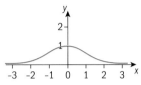

4 Find $f'(x)$ if $f(x) = \sqrt[3]{\left(1 - \dfrac{1}{2+x}\right)^2}$

> Throughout history famous curves have often been given special names, such as the two in questions 2 and 3. What is the significance of the name given to a curve? What properties of curves, if any, do the names highlight? Find applications of these famous curves, or real-life situations that the curves model.

Higher derivatives

If f is a differentiable function, then $f'(x)$ is the derivative of $f(x)$. Similarly, if $f'(x)$ is a differentiable function, $f''(x)$ is the derivative of $f'(x)$. Since multiple dash or prime notation begins to lose its efficiency after about the third derivative, for higher derivatives we write $f^{(n)}(x)$.

> f'' is 'f double dash', or 'f double prime', of x.
> $f^{(n)}(x)$ is the nth derivative of f with respect to x.

Using $\dfrac{\mathrm{d}y}{\mathrm{d}x}$ notation, we write

$$f'(x) = \frac{\mathrm{d}y}{\mathrm{d}x} \quad f''(x) = \frac{\mathrm{d}^2 y}{\mathrm{d}x^2} \quad f'''(x) = \frac{\mathrm{d}^3 y}{\mathrm{d}x^3} \quad f^{(n)}(x) = \frac{\mathrm{d}^n y}{\mathrm{d}x^n} \quad n = 4, 5, \ldots$$

Example 27

Find the first five derivatives of $f(x) = x^4 - 3x^2 + 2x - 1$

Answer	
$f'(x) = 4x^3 - 6x + 2$	*When using a superscript for a derivative the numbers are placed in brackets, as shown.*
$f''(x) = 12x^2 - 6$	
$f'''(x) = 24x$	
$f^{(4)}(x) = 24$	
$f^{(5)}(x) = 0$	*For $n \geq 5$, $f^{(n)}(x) = 0$*

Example 28

A particle moves in a straight line so that its position from a fixed point after t seconds is given by $s(t) = 3t + 5t^2 - t^3$, s in cm.

a Find the velocity of the particle at $t = 2$.

b If the acceleration is the derivative of the velocity, find the acceleration of the particle at $t = 2$.

▶ Continued on next page

Answers

a $s'(t) = 3 + 10t - 3t^2$	*Differentiate s(t).*
$s'(2) = 11 \text{ cm s}^{-1}$	*Evaluate $v(t) = s'(t)$ at $t = 2$*
b $s''(t) = 10 - 6t$	*Differentiate $s'(t)$.*
$s''(2) = -2 \text{ cm s}^{-2}$	*Evaluate $a(t) = s''(t)$ at $t = 2$*

Exercise 4L

1 If $f(x) = 4x + 1 + \dfrac{1}{x}$, find $f''(x)$.

2 If $f(x) = x^4 - 2x - 1$, find $f'(0)$ and $f''(-1)$.

3 If $f(x) = x^4 - 4x^3 + 16x - 16$, find x such that
$f(x) = f'(x) = f''(x) = 0$

4 $f(x) = x^4 + rx^2 + sx + t$ passes through the point $(-1, 16)$.
At this point, $f''(x) = -f'(x) = 16$. Find the values of r, s, and t.

5 A particle moves in a straight line such that its position at any
time t is $s(t) = (t - 4)^3(3 - 2t)^2$ metres. Find

  a the velocity after 4 seconds

  b the acceleration after 4 seconds

  c the jerk of the particle after 1 second.

> The derivative of the acceleration is called the 'jerk'.

6 Given $f(x) = \dfrac{1}{x}$, find f', f'', f''', $f^{(4)}$, $f^{(5)}$ and hence find an
expression for $f^{(n)}(x)$. Prove your result using the method of
mathematical induction.

Investigation – Leibniz's formula

$f(x) = uv$ is the product of two functions in x. You can find $f'(x)$ using
the product rule,
$f'(x) = u'v + uv'$
You can find $f''(x)$, using the product rule.

$$f''(x) = (u'v)' + (uv')'$$
$$= (u''v + u'v') + (u'v' + uv'')$$
$$= u''v + 2u'v' + uv''$$

Now find the 3rd derivative and 4th derivatives. Note the similarity
to the binomial formula, and conjecture a formula for $f^{(n)}(x)$. Use
this formula to find the 5th derivative of $f(x)$.
The general case for $f^{(n)}(x)$ is called **Leibniz's formula**.

Graphical meaning of the derivative

Local maximum and minimum points

Look at a quadratic function whose leading coefficient is

i   positive ($a > 0$) or   **ii**   negative ($a < 0$).

What is the gradient of the parabola at its vertex?

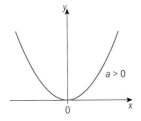 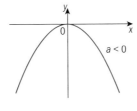

In both cases the tangent to the vertex is parallel to the x-axis. This means that the gradient of the targent to the vertex of a quadratic function is 0.

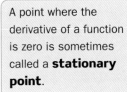

A point where the derivative of a function is zero is sometimes called a **stationary point**.

i   For $a > 0$ the vertex is a **minimum point** (the curve is concave upwards). What are the signs of the gradients of the points on the left and right of this vertex?

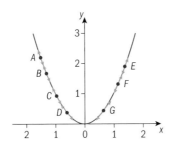

To the left of the minimum point the gradients of the tangents to points A, B, C and D are all negative. Also, the function is **decreasing** in the interval where the **gradients are negative**.

To the right, the gradients of the tangents to points E, F and G are all positive. The function is **increasing** in the interval where the **gradients are positive**.

→   Hence, the gradients of the points change from negative to positive in going through the minimum point.

ii For $a < 0$ the vertex is a **maximum point** (the curve is concave downwards).

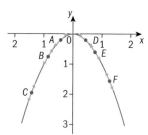

To the left of the **maximum point** the gradients of the tangent to points A, B, and C are all positive. In the interval where the **gradients are positive**, the function is **increasing**.

To the right, the gradients of the tangent to points D, E, and F are all negative. In the interval where the **gradients are negative**, the function is **decreasing**.

→ The gradients of the points change from positive to negative in going through the maximum point.

Example 29

Find the maximum and minimum points on the curve $y = 2x^4 - 4x^2 + 1$

Answer

$\dfrac{dy}{dx} = 8x^3 - 8x$

Set the first derivative equal to 0 and solve.

$8x^3 - 8x = 0$, hence $8x(x^2 - 1) = 0$

and $x = 0, \pm 1$

Test values to the left and right of these, using a sign diagram:

values of x	$x < -1$	$-1 < x < 0$	$1 < x < 1$	$x > 1$
sign of $\dfrac{dy}{dx}$	–	+	–	+

Since the gradients go from negative, through zero, to positive at $x = \pm 1$ the function has local minimum points. At $x = 0$ the gradient goes from positive, through zero, to negative and is therefore a maximum.

GDC help on CD:
Alternative demonstrations for the TI-84 Plus and Casio fx-9860GII GDCs are on the CD.

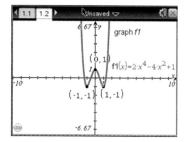

Exercise 4M

1 Find any maximum and minimum points of these functions and classify them as such.

a $y = x^2 - 3x + 1$
b $y = -2x^3 + 6x^2 - 3$
c $y = 3x^4 - 2x^3 - 3x^2 + 4$
d $y = x^4 - 4x^3$

Points of inflexion

In question **1d** of Exercise 4M, there was a point on the curve whose gradient was 0, but it was neither a maximum nor a minimum point. Unlike at a maximum, or minimum, the sign of the gradient did not change when going through the point. If you look at its graph, you will see that the graph changes from concave up to concave down at this point, $x = 0$.

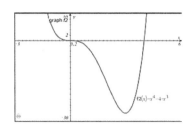

Now look at the graph of the function $y = x^3$

At the point $(0, 0)$, the gradient is 0, and this is neither a maximum nor a minimum of the function. At this point the curve changes from being concave downwards to being concave upwards. The point where the concavity of a curve changes is called a **point of inflexion**. At a point of inflexion the tangent line at the point crosses the curve. A horizontal point of inflexion has a gradient of 0.

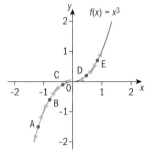

> → A point whose gradient is equal to 0 is either a maximum, minimum, or horizontal point of inflexion.

No sign change in gradients going from the left to the right of the point of inflexion.

First derivative test

On both sides of the point of inflexion in the graph of $y = x^3$ the gradients are **positive**, i.e., there is **no change in sign**. Since the gradients are positive for all x, the function is **increasing** throughout its domain ($x \neq 0$).

Now look at the graph of $y = -x^3$

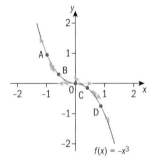

On both sides of the point of inflextion the gradients are negative, i.e., there is **no change in sign**. Since the gradients are **negative** for all x, the function is **decreasing** throughout its domain ($x \neq 0$).

No sign change in gradients going from the left to the right of the point of inflexion.

> → Consider the function $f(x)$ and suppose that $f'(c) = 0$. To determine if the point $x = c$ is a maximum, minimum or horizontal point of inflexion, make a sign table and test values of $f(x)$ to the left and right of c.
>
> • If the signs of gradients change from negative to positive, then f has a minimum at $x = c$.
> • If the signs of the gradients change from positive to negative, then f has a maximum at $x = c$.
> • If there is no sign change, then f has a horizontal point of inflexion at $x = c$.

> → Let $f(x)$ be continuous on $[a, b]$ and differentiable on $[a, b]$.
>
> • If $f' > 0$ for all $x \in]a, b[$, then f increases on $[a, b]$
> • If $f' < 0$ for all $x \in]a, b[$, then f decreases on $[a, b]$

Example 30

a Find and classify any maxima, minima or points of inflexion of the function $f(x) = x^3 - 3x + 1$

b State the intervals where f is increasing and where f is decreasing.

c Sketch the graph of the function.

Answers

a $f'(x) = 3x^2 - 3$

Set $f'(x) = 0$

$3x^2 - 3 = 0 \Rightarrow x = \pm 1$

Differentiate, set $f'(x) = 0$ and solve for x.

Make a sign diagram.

x	$x < -1$	$x = -1$	$-1 < x < 1$	$x = 1$	$x > 1$
$f'(x)$	+	0	−	0	+
f	increasing	stationary	decreasing	stationary	increasing

$f(-1) = 3$ and $f(1) = -1$

Hence, the cubic has a maximum at $(-1, 3)$ and a minimum $(1, -1)$

b f is increasing at $]-\infty, -1[\cup]1, \infty[$

f is decreasing at $]-1, 1[$

c

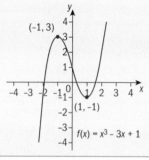

Sketch the curve.
Label the maximum and minimum points.
Set f(x) = 0 to find where the curve crosses the x-axis.
Find f(0), where the curve crosses the y-axis.

> The cubic has a positive leading coefficient so as $x \to -\infty, f(x) \to -\infty$ and as $x \to +\infty$, $f(x) \to +\infty$. Given the general shape of a cubic this means that the only possibilities are a maximum followed by a minimum or a single point of inflexion.

Example 31

Find and classify any maxima, minima or points of inflexion of $y = x^4 + 2x^3$, and the intervals where the function is increasing or decreasing.

Answers

a $\dfrac{dy}{dx} = 4x^3 + 6x^2$

$4x^3 + 6x^2 = 0 \Rightarrow 2x^2(2x + 3) = 0 \Rightarrow x = 0, x = -\dfrac{3}{2}$

Differentiate.

Set $\dfrac{dy}{dx} = 0$

x	$x < -\dfrac{3}{2}$	$x = -\dfrac{3}{2}$	$-\dfrac{3}{2} < x < 0$	$x = 0$	$x > 0$
$\dfrac{dy}{dx}$	−	0	+	0	+
f	decreasing	stationary	increasing	stationary	increasing

▶ Continued on next page

Hence, $f(x)$ has a minimum at $\left(-\dfrac{3}{2}, -\dfrac{27}{16}\right)$ and a horizontal point of inflexion at $(0, 0)$.

The function is increasing at $\left]-\dfrac{3}{2}, 0\right[\cup \left]0, \infty\right[$

and is decreasing at $\left]-\infty, -\dfrac{3}{2}\right[$

Sketch the graph of the quartic function to confirm these results.

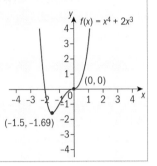

Exercise 4N

1 Use the graphs to estimate where f' is
  i 0   **ii** positive   **iii** negative.

a
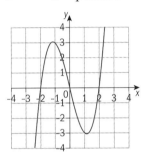

b
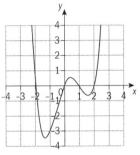

2 Use these graphs of gradient functions for a function f to determine
  i where f has any maxima, minima or points of inflexion
  ii intervals where f is increasing
  iii intervals where f is decreasing.

a
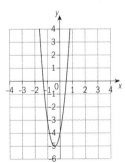

b
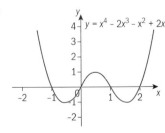

3 For each function, find:
  i any stationary points, and justify your results
  ii any intervals where f is increasing
  iii any intervals where f is decreasing.

  a $y = -3x^2 + 6x - 1$
  b $y = x\sqrt{2-x^2}$
  c $y = \dfrac{x}{x^2+1}$

  d $y = x^{\frac{1}{3}}(x-2)$
  e $y = x^2\sqrt{2-x^2}$

Second derivative test for maximum and minimum points

This is another test to determine the nature of maximum and minimum points. Here are the graphs of $y = x^2$ and its derivative.

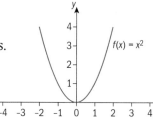

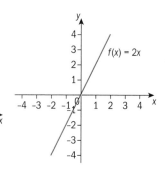

The gradients go from negative to positive at the minimum point, the second derivative is positive, and $f''(x) = 2$. When the graph of f is concave up its second derivative is positive.

Here are the graphs of $y = -x^2$ and its derivative.

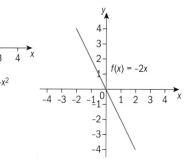

The gradients go from positive to negative at the maximum point, the second derivative is negative, and $f''(x) = -2$. When the graph of f is concave down, the second derivative is negative.

> → • If $f'(c) = 0$ and $f''(c) < 0$, then $f(x)$ has a local maximum at $x = c$
> • If $f'(c) = 0$ and $f''(c) > 0$, then $f(x)$ has a local minimum at $x = c$

If both $f'(c) = 0$ and $f''(c) = 0$ the test is inconclusive. You shall study this later on in this section.

Example 32

Find and classify all maxima, minima and horizontal points of inflexion of the function $y = 3 + x + \dfrac{1}{x}$

Confirm your findings with a sketch graph of the function.

- -

Answer

$\dfrac{dy}{dx} = 1 - \dfrac{1}{x^2} = \dfrac{x^2 - 1}{x^2}$

$x^2 - 1 = 0 \Rightarrow x = \pm 1$

$\dfrac{d^2y}{dx^2} = \dfrac{2}{x^3}$

$f''(1) = 2 > 0 \Rightarrow f$ has a minimum at $x = 1$

$f''(-1) = -1 < 0 \Rightarrow f$ has a maximum at $x = -1$

Hence, the stationary points are: minimum $(1, 5)$ and maximum $(-1, 1)$.

Set $\dfrac{dy}{dx} = 0$ and

solve for x

Using the second derivative test.

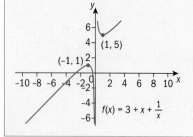

The graph of the function confirms the result.

Exercise 40

1 Find and classify any maxima, minima and horizontal points of inflexion of these functions.

> Do this exercise analytically and confirm your results on a GDC.

  a $y = 2x^3 + 3x^2 - 12x - 3$ **b** $y = -x^4 + 2x - 1$

  c $y = x^5 - 5x$ **d** $y = \dfrac{12}{x^2 + 2x - 3}$

  e $y = \dfrac{3x + 3}{x(3 - x)}$

EXAM-STYLE QUESTION

2 For each of these functions

  i   Find the coordinates of any maxima, minima or horizontal points of inflexion, and state their type. Justify your answers.

  ii   Indicate the intervals where f is increasing, and the intervals where f is decreasing.

  iii   Sketch the curve, showing clearly the points you have found in **i**, as well as the intercepts and any asymptotes.

  a $y = x^5 - 5x^4$ **b** $y = \dfrac{1 - x}{x^2 + 8}$

4.4 Exploring relationships between f, f' and f''

Points of inflexion and concavity

Here are the graphs of the cubic $f(x) = x^3 - 3x + 1$, its first derivative $f'(x) = 3x^2 - 3$ and second derivative $f''(x) = 6x$.

Notice that there is a point on f which corresponds to a stationary point on f' at $x = 0$. At this point, f' has a gradient of 0. This is evident from the graph of f''.

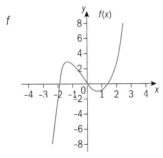

f

$f(x)$ has a point of inflexion at $x = 0$

This is a non-horizontal point of inflexion, since the first derivative at this point is not equal to 0.
At this point f changes concavity from concave down to concave up.

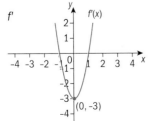

f'

Where f has a point of inflexion at $x = 0$, f' has a minimum.

> See the next page for the graph of f''.

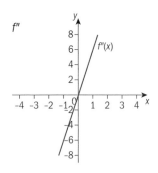

f''

Where f has a point of inflexion at $x = 0$, f'' is negative on the left and positive on the right.

→ • f is concave down in an open interval if for all x in the interval, $f''(x) < 0$
  • f is concave up in an open interval if for all x in the interval, $f''(x) > 0$

→ • If a curve has a point of inflexion its second derivative will be 0 at this point.

However if a curve has a point whose second derivative is 0 that point is not always a point of inflexion.

For example, consider $f(x) = x^4$

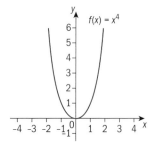

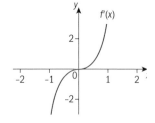

Here, $f'(x) = 4x^3$, and $f'(x) = 0$ for $x = 0$.
At $(0, 0)$ $y = x^4$ has a minimum. $f''(x) = 12x^2$, and $f''(x) = 0$ for $x = 0$, but $(0, 0)$ is clearly not a point of inflexion.

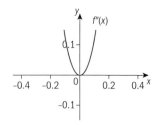

Example 33

Consider the function $f(x) = 2x^4 - 4x^2 + 1$

a Find any maxima, minima or horizontal points of inflexion.

b Find the intervals where the function is   i decreasing   ii increasing.

c Find the intervals where the function is   i concave up   ii concave down.

d Sketch the function, indicating any maxima, minima and points of inflexion.

Answers

a $f'(x) = 8x^3 - 8x$

  $8x^3 - 8x = 0$, hence $8x(x^2 - 1) = 0$, and $x = 0$, or $x = \pm 1$.

  $f''(x) = 24x^2 - 8$

  $f''(0) = -8$, $f''(0) < 0$,   hence at $x = 0$, f has a maximum.

  $f''(-1) = 16$, $f''(-1) > 0$,   hence at $x = -1$ f has a minimum.

  $f''(1) = 16$, $f''(1) > 0$,   hence at $x = 1$, f has a mimimum.

  The stationary points are therefore $(0, 1)$, $(-1, -1)$, and $(1, -1)$.

Set $f'(x) = 0$ to find any stationary points.

Using the second derivative test.

b Sign diagram

x	$x < -1$	$-1 < x < 0$	$0 < x < 1$	$x > 1$
sign of f'	−	+	−	+
behavior of f	decreasing	increasing	decreasing	increasing

  i f is decreasing in the intervals $]-\infty, 1[\cup]0, 1[$

  ii f is increasing in the intervals $]-1, 0[\cup]1, \infty[$

c $24x^2 - 8 = 0$, hence $x = \pm\dfrac{1}{\sqrt{3}}$

x	$x < -\dfrac{1}{\sqrt{3}}$	$-\dfrac{1}{\sqrt{3}} < x < \dfrac{1}{\sqrt{3}}$	$x > \dfrac{1}{\sqrt{3}}$
sign of f''	+	−	+
concavity of f	concave up	concave down	concave up

To find any points of inflexion, set $f''(x) = 0$

$f''(x) < 0$ means concave down.

$f''(x) > 0$ means concave up.

  i f is concave up on the interval $\left]-\infty, -\dfrac{1}{\sqrt{3}}\right[\cup \left]\dfrac{1}{\sqrt{3}}, \infty\right[$

  ii f is concave down on the interval $\left]-\dfrac{1}{\sqrt{3}}, \dfrac{1}{\sqrt{3}}\right[$

d

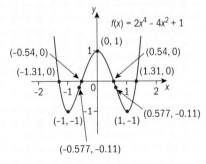

Although the question does not ask for zeros in the sketch, you should indicate on the graph where they are.

$f(x) = 2x^4 - 4x^2 + 1$
$f(0) = 1$

Exercise 4P

1 For each function
  i find any points of inflexion
  ii determine the intervals where the function is concave up or concave down.
  Justify your answers.

  a $y = x^3 - x$ **b** $y = x^4 - 3x + 2$ **c** $y = \sqrt{4x - x^2}$

  d $y = (x-1)^{\frac{2}{3}}$ **e** $y = \dfrac{3x^2}{x-1}$

EXAM-STYLE QUESTION

2 Here is the graph of f' for a function f.
  From the graph, indicate
  a the x-coordinates of any points where the gradient of $f(x)$ is zero and determine the nature of the points
  b the intervals where f is
    i increasing   **ii** decreasing
  c the intervals where f is
    i concave up   **ii** concave down.
  d Sketch f on a copy of the graph.

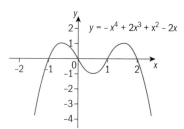

Investigation – cubic polynomials

For the cubic polynomial $y = ax^3 + bx^2 + cx + d$, define the conditions for a, b, and c such that the cubic has stationary points, and show that the cubic always has a point of inflexion.

4.5 Applications of differential calculus: kinematics

Kinematics is the study of how objects move. A particle moving in a straight line is the simplest type of motion. To describe simple linear motion you need a starting point, a direction, and a distance.

- The path is the set of points between the start and end location.

- The displacement describes the difference in the particle's position between its start and end points.

- The distance is the length of the path between two points on the path the particle has traveled.

Although Leibniz and Newton discovered calculus at about the same time, Newton was led to calculus when studying motion. In fact, he used the term 'fluxions' for derivatives. Leibniz's notation for the derivative $\dfrac{dy}{dx}$ is the one we use today. Newton's notation $\dot{y}$, with a dot over the dependent variable, was less popular.

For example, if you jog once around a standard track the distance you will cover is 400 m, but your displacement is 0, since your start and end positions are the same.

Displacement and distance are related. In physics terms, displacement is a vector quantity that measures the change in position between a start and an end point, and distance is a scalar. In the diagram below, the path of a particle follows the squiggly line from A to B. The length in metres of this path is the distance traveled. The displacement is given by the vector $\overrightarrow{AB}$. Hence, the displacement is always less than or equal to the distance traveled.

Instantaneous velocity and acceleration

- Velocity at a particular instant is the derivative of the position function, $s(t)$ with respect to time.

$$v(t) = \frac{ds}{dt} \text{ or } s'(t)$$

- Acceleration is the derivative of the velocity with respect to time, $a(t)$. It is therefore the second derivative of the position function. If a particle's velocity at time t is $v(t)$, then

$$a(t) = \frac{dv}{dt} = \frac{d^2s}{dt^2} \text{ or } s''(t)$$

- When v and a have the same sign, the particle is speeding up (accelerating).
- When v and a have opposite signs, the particle is slowing down (decelerating).

> The derivative of the acceleration is called the jerk. It is exactly what you experience when there is a sudden change in acceleration, i.e., you feel your body being moved abruptly. For example, if you are in an airplane when there is a sudden change in acceleration, your drink might spill, or items might fall off trays. It is interesting that the travel sickness that many people experience is often due to jerk, even jerk that is not noticeable. Note that the jerk as the derivative of acceleration is not explicitly on the HL syllabus but you may be asked to calculate a higher derivative.

Example 34

A particle moves in a horizontal line so that its position from a fixed point after t seconds is s metres, where $s = 5t^2 - t^4$
Find the velocity and acceleration of the particle after 1 second.
Is the particle speeding up or slowing down at $t = 1$?

Answer

$v = \dfrac{ds}{dt} = 10t - 4t^3$	*Differentiate s to find velocity.* *Evaluate at t = 1*
At $t = 1$, $v = 6 \, ms^{-1}$	
$a = \dfrac{dv}{dt} = \dfrac{d^2s}{dt^2} = 10 - 12t^2$	*Differentiate v to find acceleration.* *Evaluate at t = 1*
At $t = 1$, $a = -2 \, ms^{-1}$	*The signs of v and a are different*
At $t = 1$ the particle is slowing down.	*which means the particle is slowing down.*

Example 35

A particle moves along a line so that its position at any time t, in seconds, is
$$s(t) = t^3 - 7t^2 + 11t - 2.5$$
a Find the velocity and acceleration at any time t.
b Find the times when the particle is at rest, and the times when it is speeding up and slowing down. Justify your answers.
c Find the values of t when the particle changes direction.
d Find the total distance traveled in the first 3 seconds.

Answers

a $v(t) = s'(t) = 3t^2 - 14t + 11$
$a(t) = v'(t) = 6t - 14$

b $3t^2 - 14t + 11 = 0$, $t = 1$ or $t = \dfrac{11}{3}$

$6t - 14 = 0$, $t = \dfrac{7}{3}$

When the particle is at rest $v(t) = 0$

When the particle has constant velocity, $a(t) = 0$

t	$0 < t < 1$	$1 < t < \dfrac{11}{3}$	$t > \dfrac{11}{3}$
sign of v	$+$	$-$	$+$

When $a = 0$, $t = 2\dfrac{1}{3}$

t	$t < \dfrac{7}{3}$	$t > \dfrac{7}{3}$
sign of a	$-$	$+$

v and a have different signs for $0 < t < 1$ and $\dfrac{7}{3} < t < \dfrac{11}{3}$
In these time intervals the particle is slowing down. This makes sense since these are the times just before the particle comes to rest.
v and a have the same signs for $1 < t < \dfrac{7}{3}$ and $t > \dfrac{11}{3}$
At these time intervals the particle is speeding up.

From the sign diagram for $v(= s')$, s is increasing on the intervals $0 < t < 1$ and $t > \dfrac{11}{3}$, hence the particle is moving to the right. During the interval $1 < t < \dfrac{11}{3}$, s is decreasing, hence the particle is moving to the left.

c $t = 1$ and $t = \dfrac{11}{3}$

d The total distance traveled in the first 3 seconds is
$[s(1) - s(0)] + [s(3) - s(1)]$.
$s(1) - s(0) = 2.5 - (-2.5) = 5$
In the first second the particle moves 5 m to the right.
$s(3) - s(1) = -5.5 - 2.5 = -8$
Between the first and third seconds, the particle moves 8 m to the left.
Hence the particle has covered a total distance of 13 m.

Since the particle changes direction at $t = 1$, find the distance traveled from $t = 0$ to $t = 1$, and from $t = 1$ to $t = 3$, separately.

Exercise 4Q

1 At any time t, in seconds, a diver's position after diving off a board can be modeled by the function $s(t) = -5t^2 + 5t + 10$, where s is the height, in metres, above the water. Find
   a the height of the diving board
   b how long it takes the diver to hit the water
   c the velocity and acceleration of the diver at impact. Interpret your answers.

2 A detonation in the earth propels a rock straight up. Its height at any time t can be modeled by the function $s(t) = 50t - 15t^2$ where s is in metres. Find
   a the maximum height of the rock
   b the velocity and speed when the rock is 20 m above the ground, and interpret your answers
   c the acceleration of the rock at any time t
   d the time taken for the rock to hit the ground again.

3 A particle moves in a straight line such that its displacement t seconds later is s metres, where $s(t) = 7t + 5t^2 - 2t^3$
   a Find the initial velocity and acceleration, and interpret your answer.
   b Find the velocity and acceleration after 2 seconds, and interpret your answer.

4 A particle moves in a straight line such that its displacement from a fixed point after t seconds is s metres, where $s(t) = 10t^2 - t^3$
   a Find the average velocity in the first 3 seconds.
   b Find the velocity at $t = 3$ and acceleration at $t = 3$.
   c Determine if the particle is speeding up or slowing down at $t = 3$.
   d Find the value of t when the direction of the particle changes.

EXAM-STYLE QUESTION

5 The position of a particle at any time t seconds after it starts moving is given by the function $s(t) = \frac{1}{3}t^3 - 3t^2 + 8t$ metres. Find
   a the velocity and acceleration at any time t
   b the times when the particle is
      i at rest   **ii** is speeding up   **iii** is slowing down
   c the acceleration when the particle's velocity is 0, and interpret your answer
   d the times when the particle changes direction
   e the total distance traveled in the first 5 seconds.

4.6 Applications of differential calculus: economics

Calculus is applied in basic economic theory in marginal analysis. Economists analyze how small changes, for example, increasing the production of a product by a single unit, affect profits and costs. Marginal analysis quantifies the benefits of performing such an action against the costs. When the benefits, or profits, exceed the cost of the action, you can proceed on this course until this balance changes. The break-even point occurs when the production costs and the total revenues, the amount of income generated before any deductions are made, are the same.

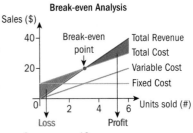

Break-even Analysis

Source: www.12manage.com

There are three basic terms in marginal analysis.

- Marginal profit is the rate of change of profit with respect to the number of units produced or sold.
- Marginal revenue is the rate of change of revenue with respect to the number of units sold.
- Marginal cost is the rate of change of cost with respect to the number of units sold.

Here is a summary of the basic terms and corresponding formulae.

x = number of units produced (or sold)

$r(x)$ = total revenue from selling x amount of units

$c(x)$ = total cost of producing x amount of units

$p(x)$ = profit in selling x amount of units

$r'(x)$ = marginal revenue, the extra revenue for selling one extra unit

$c'(x)$ = marginal cost, the extra cost for selling one extra unit

$p'(x)$ = marginal profit, the extra profit for selling one additional unit

> The variables are connected by the formula $p(x) = r(x) - c(x)$

Example 36

The profit, in euros, obtained from selling x pairs of shoes can be modeled by $p(x) = 0.00025x^3 + 10x$

a Find the marginal profit for a production of 50 pairs of shoes.

b Find the actual gain in profit obtained by raising the production level from 50 to 51 pairs of shoes.

c Comment on your answers to **a** and **b**.

- -

Answers

a $p'(x) = 0.00075\,x^2 + 10$
$p'(50) = 0.00075\,(50)^2 + 10 = 11.88$ euros

b $p(50) = 0.00025\,(50)^3 + 10(50) = 531.25$ euros
$p(51) = 0.00025\,(51)^3 + 10(51) = 543.16$ euros
The additional profit is therefore $543.16 - 531.25 = 11.91$ euros

c The marginal profit is the extra profit for selling one additional pair of shoes. This approximates very well the actual profit, according to the profit formula, for selling an extra pair of shoes.

In the real world, however, profit does not always work according to the profit model. It is more likely that a company will be able to maintain sales only by lowering prices at some stage. So other models, such as a demand function, need to be introduced in order to make a more accurate marginal analysis.

What are the variables affecting maximum profit, and when does maximum profit occur?

Profit is the difference between revenues and costs, i.e., $p(x) = r(x) - c(x)$. Differentiating, $p'(x) = r'(x) - c'(x)$. Profits will be maximized when $p'(x) = 0$, i.e., $0 = r'(x) - c'(x) \Rightarrow r'(x) = c'(x)$

Maximum profit occurs when the marginal revenue and marginal cost are equal.

Of course, minimum profit can also occur when $p'(x) = 0$. Profit will at any rate occur when marginal revenues and marginal costs are equal.

Example 37

The cost of manufacturing fishing poles, in thousands of units, is modeled by $c(x) = x^3 - 10x^2 + 20x$. The revenue is modeled by $r(x) = 7x + 3$
Find a production level that maximizes profits.

Answer

$c'(x) = 3x^2 - 20x + 20$ $r'(x) = 7$ $c'(x) = r'(x)$ $\Rightarrow 3x^2 - 20x + 20 = 7$ $\Rightarrow 3x^2 - 20x + 13 = 0$ $\Rightarrow x_1 \approx 0.730, x_2 \approx 5.94$ Therefore, maximum profit occurs at a production level of 5 940 fishing poles. Minimum profit occurs at a production level of approximately 730 fishing poles.	*Differentiate both c(x) and r(x), and set them equal.* *Since p(x) = r(x) − c(x) is a cubic function, if it has a maximum it will also have a minimum.*

The ideal production level is the one that minimizes costs.
If the cost of producing x units is $c(x)$, then the average cost per unit is $\dfrac{c(x)}{x}$. If the cost of producing x units can be minimized, then it will occur when $\dfrac{d}{dx}\left(\dfrac{c(x)}{x}\right) = 0$

$$\frac{d}{dx}\left(\frac{c(x)}{x}\right) = \frac{xc'(x) - c(x)}{x^2} = 0$$

Use quotient rule.

$$\Rightarrow xc'(x) = c(x)$$

$$\Rightarrow c'(x) = \frac{c(x)}{x}$$

If the production cost of x units can be minimized, it will occur when the marginal cost of producing one extra unit is the same as the average cost of producing x units.

Example 38

The cost of manufacturing fishing poles, in thousands of units, is modeled by $c(x) = x^3 - 10x^2 + 20x$. Find a production level, if it exists, that minimizes average cost.

Answer

$c'(x) = 3x^2 - 20x + 20$

$\dfrac{c(x)}{x} = x^2 - 10x + 20$

For minimum cost,
$c'(x) = \dfrac{c(x)}{x}$

Hence, $3x^2 - 20x + 20 = x^2 - 10x + 20$

$\Rightarrow 2x^2 - 10x = 0 \Rightarrow 2x(x - 5) = 0$

$\Rightarrow x = 0, x = 5$

$\dfrac{d}{dx}\left(\dfrac{c(x)}{x}\right) = 2x - 10$

Use the second derivative test.

$\dfrac{d}{dx}(2x - 10) = 2$

The only production level that would possibly minimize average cost is $x = 5$.

Since $2 > 0$, the second derivative is positive for all x, so $x = 5$ is a minimum.

Since x is in thousands, 5000 fishing poles is the production level necessary to minimize costs.

In Examples 37 and 38 the maximum profit is for a production level of almost 6000 fishing poles, but to minimize average cost, the production level is 5000 poles.

Hence, these results would have to be analyzed and a decision taken as to which production level the company should aim for.

Ultimately, every company's goal is to maximize profits, which means maximizing revenues. Obviously, a company with minimal production costs, but no revenues, will not make a profit.

Exercise 4R

1 A company manufactures oil tanks for reservoirs. The total weekly cost in euros of producing the tanks can be modeled by

$c(x) = 20\,000 + 180x - 0.1x^2$

   a Find the marginal cost function.

   b Find the marginal cost of producing 100 tanks per week.

   c Find the cost of producing 101 tanks, and compare your answer with **b**.

2 A company does market research before producing a new type of memory stick. Initial overheads and fixed costs of production, in euros, for x memory sticks can be modeled by $c(x) = 500 + 3x$. The projected selling price is modeled by $p(x) = 7 - 0.002x$

  a Find

    i the domain of the price function

    ii the marginal cost function, and interpret its meaning

    iii the revenue function, and its domain.

  b Graph the revenue and cost functions, and find the break-even points, and interpret what they mean.

  State the memory stick production levels in the form $a < x < b$, (a and b are integers), that must be met in order for the company to make a profit.

3 A company specializes in making units from rare metals for nuclear plants. The total cost, in dollars, is modeled by $c(x) = 500\,x^2 + 1000$, where x represents hundreds of units. Find the number of units the company should make in order to minimize costs.

4 The cost, in euros, for producing x number of jackets is $c(x) = 400 + 20x - 0.2x^2 + 0.0004\,x^3$

  a Find the number of jackets that should be produced to minimize costs.

  b Find the number of jackets that should be produced to maximize profits if the revenue function can be modeled by $r(x) = 35x - 3$

  c Interpret your answers to **a** and **b**.

4.7 Optimization and modeling

The primary purpose of applied mathematics is to describe, investigate, explain and solve real-world problems. This process is called mathematical modeling. The steps are:

- Identify all variables, parameters, limits and constraints.
- Translate the real-world problem into a mathematical system.
- Solve the mathematical system.
- Interpret the reasonableness of the solution in light of the real-world problem.

Optimization problems deal with finding the most effective solutions to real-world problems, for example, how to minimize the surface area of a container for a required volume.

Example 39

An open box is made from cutting congruent squares from the corners of a 4 m by 4 m cardboard sheet. How large should the squares be so that the box has a maximum volume? What is the maximum capacity of the box?

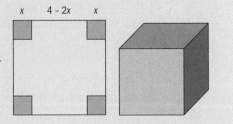

Answer

Let x be a side of the square in cm

$L = W = 4 - 2x$

$H = x$

$0 < x < 2$

$V = x(4 - 2x)^2$

$V' = (4 - 2x)^2 + 2x(4 - 2x)(-2)$

$\quad = (4 - 2x)^2 - 4x(4 - 2x)$

$V' = 12x^2 - 32x + 16$

Set $V' = 0$, $12x^2 - 32x + 16 = 0$

$4(3x^2 - 8x + 4) = 0$

$4(3x - 2)(x - 2) = 0$

$x = \dfrac{2}{3}$ or $x = 2$

Since $x < 2$, reject $x = 2$

$V'' = 24x - 32$

When $x = \dfrac{2}{3}$, $V'' = -16 < 0$ hence

at $x = \dfrac{2}{3}$ m, V has a maximum.

Hence, $V = \dfrac{2}{3}(4 - \dfrac{4}{3})^2 = 4.74 \text{ m}^3$

Identify any variables.
Express the dimensions of the box in terms of the side of the square, x.

Identify constraints.

Write the function for volume.
Differentiate the function.

Solve for x.

Use the second derivative test to check $x = \dfrac{2}{3}$ gives a maximum.

Check your answer on a GDC.

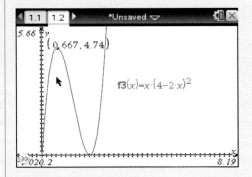

GDC help on CD:
Alternative demonstrations for the TI-84 Plus and Casio fx-9860GII GDCs are on the CD.

Example 40

You have been asked to design a cylindrical can to hold 1 litre of car oil, with the minimum surface area in order to minimize costs. Find the dimensions of the can.

Answer

$V = 1$ litre $= 1000 \, \text{cm}^3$ *Identify constraints.*

$r =$ radius of base   *Identify variables.*

$h =$ height of can

$V = \pi r^2 h$

Surface area $A = 2(\pi r^2) + 2\pi rh$ *Identify function to be optimized.*

$1000 = \pi r^2 h$, hence $h = \dfrac{1000}{\pi r^2}$ *Use the constraint to reduce the function to be maximized to one*

$A = 2\pi r^2 + 2\pi r \cdot \dfrac{1000}{\pi r^2}$ *unknown.*

$\quad = 2\pi r^2 + \dfrac{2000}{r}$

$A' = 4\pi r - \dfrac{2000}{r^2}$ *Differentiate.*

$4\pi r - \dfrac{2000}{r^2} = 0$

$4\pi r^3 - 2000 = 0$

$r = \sqrt[3]{\dfrac{500}{\pi}} \approx 5.42, \; h = 10.8 \, \text{cm}$ *Substitute r and h into A.*

$A = 554 \, \text{cm}^2$ *Check your answer on the GDC.*

GDC help on CD:
Alternative demonstrations for the TI-84 Plus and Casio fx-9860GII GDCs are on the CD.

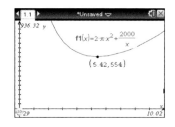

Exercise 4S

1 Find the dimensions of the rectangle with largest area, if its base is on the x-axis, and its upper corners are on the parabola $10 - x^2$.

2 A rectangular plot of land is bounded on one side by a river, and by a fence on the other three sides. Find the largest area that can be enclosed using 800 m of fencing.

3 A stained glass window is to be designed and entered in an annual competition in the UK. The window must be in the shape of a semicircle over a rectangle, such that the diameter is on a side of the rectangle. If the perimeter of the window is to be 4 m, find the dimensions that will result in the rectangular part having the largest possible area.

4 You wish to make an open rectangular box from a 24 cm by 45 cm piece of cardboard, by cutting out congruent squares from its corners and folding up the sides. Find the dimensions of the box of largest volume you can make this way, and find the volume.

5 Find the volume of the largest right circular cone that can be inscribed in a sphere whose radius is 10 cm.

6 A rectangular sheet of tin whose dimensions are l cm by w cm and whose perimeter is 36 cm will be rolled to create a cylinder.
 a Find the values of the length and width that will give the greatest volume.
 b The same sheet of tin will be revolved about one of its sides to create a cylindrical figure. Find the values of l and w that will give the greatest volume.

4.8 Differentiation of implicit functions

The functions you have studied so far have been explicitly defined, i.e., the dependent variable is defined in terms of the independent variable. The equation of the circle however, whose center is at the origin is $x^2 + y^2 = r^2$, where r is the radius of the circle. In this case, both x and y are implicitly defined.

You can often change an implicitly defined function into an explicit function by expressing one variable in terms of the other. However, this is not always easy, or possible, and so to analyze such functions you need to differentiate them implicitly.

For example, differentiate $y^2 = 4x$ implicitly with respect to x.

Differentiate the left-hand side with respect to x using the chain rule:
$$\frac{d(y^2)}{dx} = \frac{d(y^2)}{dy} \cdot \frac{dy}{dx} = 2y\frac{dy}{dx}$$

Differentiate the right-hand side: $\dfrac{d(4x)}{dx} = 4$

Hence: $2y\dfrac{dy}{dx} = 4 \Rightarrow \dfrac{dy}{dx} = \dfrac{4}{2y} = \dfrac{2}{y}$

Therefore: $\dfrac{dy}{dx} = \dfrac{2}{y}$

In order to check the result, you can (if possible) write the function explicitly, and differentiate.

In this case $y^2 = 4x \Rightarrow y = \pm\sqrt{4x} = \pm 2\sqrt{x}$

Taking $y = 2\sqrt{x}$
$$\frac{dy}{dx} = \frac{d(2x^{\frac{1}{2}})}{dx} = 2\frac{d(x^{\frac{1}{2}})}{dx} = 2 \cdot \frac{1}{2}x^{-\frac{1}{2}} = \frac{1}{\sqrt{x}} = \frac{2}{2\sqrt{x}} = \frac{2}{y}$$

Likewise, taking $y = -2\sqrt{x}$
$$\frac{dy}{dx} = -\frac{1}{\sqrt{x}} = \frac{2}{-2\sqrt{x}} = \frac{2}{y}$$

Thus you obtain the same result by either differentiating implicitly or explicitly.

Example 41

> Find the gradient of the circle $x^2 + y^2 = 1$ at the point $(0, 1)$.
>
> **Answer**
>
> | | |
> |---|---|
> | $2x + 2y\dfrac{dy}{dx} = 0$ | *Differentiate implicitly with respect to x.* |
> | $\dfrac{dy}{dx} = -\dfrac{x}{y}$ | *Rearrange.* |
> | Substitute $(0, 1)$ for x and y $\dfrac{dy}{dx} = 0$ | *Again, you can check by first writing the equation explicitly and then differentiating.* |

Extension material on CD:
Worksheet 4

$y = x^2 + 2$ is an explicitly defined function. $x^3y + xy^3 = 3$ is an implicitly defined function.

As in these examples, it is sometimes possible to check the result of implicit differentiation by defining the function explicitly, and then differentiating. For example, consider the Folium of Descartes whose equation is $x^3 + y^3 = 9xy$, and is graphed here. This is obviously not a function, but consists of three distinct piecewise functions joining to form the graph.

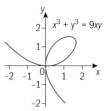

▲ The Folium of Descartes

Example 42

Differentiate the folium of Descartes, $x^3 + y^3 - 9xy = 0$

Answer

$3x^2 + 3y^2\dfrac{dy}{dx} - 9\left(y + x\dfrac{dy}{dx}\right) = 0$	*Use the product rule to differentiate* xy.
$3x^2 + 3y^2\dfrac{dy}{dx} - 9y - 9x\dfrac{dy}{dx} = 0$	*Expand.*
$\dfrac{dy}{dx}(3y^2 - 9x) = 9y - 3x^2$	*Factorize* $\dfrac{dy}{dx}$
$\dfrac{dy}{dx} = \dfrac{9y - 3x^2}{3y^2 - 9x} = \dfrac{3y - x^2}{y^2 - 3x}$	*Solve for* $\dfrac{dy}{dx}$

Example 43

The point $P(2, m)$, where $m < 0$, lies on the curve $2x^2y + 3y^2 = 16$
a Calculate the value of m.
b Find the gradient of the normal to the tangent at P.

Answers

a $2(2)^2(m) + 3m^2 = 16$	*Substitute (2, m) into the implicit function for (x, y).*
$8m + 3m^2 = 16$	
$3m^2 + 8m - 16 = 0$	*Solve the quadratic for m.*
$(3m - 4)(m + 4) = 0$	
$3m - 4 = 0$ or $m + 4 = 0$	
$m = \dfrac{4}{3}$ or $m = -4$	
Since $m < 0$, $m = -4$	*Reject positive value of m.*
b $2\left(2xy + x^2\dfrac{dy}{dx}\right) + 6y\dfrac{dy}{dx} = 0$	*Differentiate implicitly.*
$4xy + 2x^2\dfrac{dy}{dx} + 6y\dfrac{dy}{dx} = 0$	
$\dfrac{dy}{dx} = \dfrac{-4xy}{2x^2 + 6y}$	
At $(2, -4)$, $\dfrac{dy}{dx} = \dfrac{1}{2}$	*Substitute your solution for x and y.*
Hence, the gradient of the normal at P is -2.	

Exercise 4T

1 Find $\dfrac{dy}{dx}$ by differentiating implicitly with respect to x.

a $3y^2 + x^2 = 4$

b $y^4 = x^3 + 1$

c $x^2 + y^2 - 3x + 4y = 2$

d $2x^2 - 3x^2 y^2 + y^2 = 9$

e $(x + y)^2 = 5 - 2x$

f $x^2 = \dfrac{x - y}{x + y}$

2 Find the equation of the tangent to the curve $x^2 - y^2 = 9$ at the point $(5, 4)$.

3 Find the equation of the normal to the curve $y^2 = 3x + 1$ at the point $(1, -2)$.

4 Find the equations of the tangent and the normal to the curve $x^2 - \sqrt{3}\,xy + 2y^2 = 5$ at the point $(\sqrt{3}, 2)$.

5 Find the coordinates of the where the gradient is zero points on the curve $x^2 + y^2 - 6x - 8y = 0$

6 Given the curve $3x^2 + 2xy + y^2 = 3$, find $\dfrac{dy}{dx}$ and $\dfrac{d^2 y}{dx^2}$ at the point $(1, -2)$.

7 Given the curve $x^2 + xy + y^2 = 3$, find the x-intercepts and show that the tangents to the curve at the x-intercepts are parallel.

8 A rectangular tank with square base x m and height y m is designed so that the top of the tank is at ground level. The purpose of the tank is to store excess water that runs off from the ground, which has a low porous index. The proposed volume of the tank is 125 m³. The costs for such a design is modeled by the function $C(x) = 3(x^2 + 2xy) + 8xy$ in local currency. Find the dimensions of the tank that will minimize the costs.

⋮ **EXAM-STYLE QUESTION**

⋮ **9** Given the curve $x + y = x^2 - 2xy + y^2$

⋮ a find $\dfrac{dy}{dx}$

⋮ b show that $1 - \dfrac{dy}{dx} = \dfrac{2}{2x - 2y + 1}$

⋮ c show that $\dfrac{d^2 y}{dx^2} = \left(1 - \dfrac{dy}{dx}\right)^3$

4.9 Related rates

Related rates problems look at the effect that a change in a particular rate has on another rate. For example, when a balloon is filled with air at a certain rate, at what rate is its surface area increasing? Suppose the balloon then begins to lose air at a certain rate, at what rate is its surface area decreasing?

The next example shows the necessary steps to solving related rates problems.

Example 44

A 10 m long industrial ladder is leaning against a wall on a building site. It starts to slip down the wall at a rate of 0.5 ms^{-1}.
How fast is the foot of the ladder moving along the ground when it is 6 m from the wall?

Answer

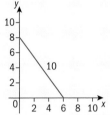

	Sketch a diagram of the problem, naming the variables.
When $x = 6$ and $\dfrac{dy}{dt} = -0.5$, find $\dfrac{dx}{dt}$	*Write down the given information, and what you are asked to find.*
$x^2 + y^2 = 100$ $2x\dfrac{dx}{dt} + 2y\dfrac{dy}{dt} = 0$	*Write down an equation relating the variables.*
When $x = 6$, $y = 8$ $2(6)\dfrac{dx}{dt} + 2(8)(-0.5) = 0$ $\dfrac{dx}{dt} = \dfrac{2}{3}$	*Differentiate with respect to time.* *Find any missing values necessary to solve for the required rate.* *Substitute and solve.*
The ladder is moving along the ground at a rate of approximately 0.667 m s^{-1}	*Interpret the answer in the context of the given problem using appropriate units.*

Example 45

Water is poured into a conical tank at the rate of 3 m³ min⁻¹. The tank stands with the point downward. How fast is the water level in the tank rising when the depth is 2 m and the radius of the water surface is 1.5 m?

Answer

At time t, r is the radius of the water surface.

h is the depth of water in the tank.

Find $\dfrac{dh}{dt}$ when $\dfrac{dV}{dt} = 3$, $h = 2$, $r = 1.5$

$V = \dfrac{1}{3}\pi r^2 h$

$\dfrac{dV}{dt} = \dfrac{1}{3}\pi\left(2hr\dfrac{dr}{dt} + r^2\dfrac{dh}{dt}\right)$

$\dfrac{r}{h} = \dfrac{1.5}{2} \Rightarrow 2r = 1.5h \Rightarrow r = 0.75h$

$\dfrac{dr}{dt} = 0.75\dfrac{dh}{dt}$ so $r\dfrac{dr}{dt} = 1.5 \cdot 0.75\dfrac{dh}{dt}$

$3 = \dfrac{1}{3}\pi[(2)(2)(1.125)\left(\dfrac{dh}{dt}\right) +$

$\qquad (2.25)\dfrac{dh}{dt}]$

$\dfrac{dh}{dt} = 0.424$

When the depth is 2m, the water level is rising at a rate of 0.424m min⁻¹ (3 sf).

Sketch a diagram.
Define the variables.

Write down the given information and what you have to find.

Write down a formula connecting the variables.
Differentiate with respect to t.

Similar triangles.

Differentiate to get $\dfrac{dr}{dt}$ in terms of $\dfrac{dh}{dt}$

Substitute values in $\dfrac{dV}{dt}$ and solve.

Interpret the answer in the context of the problem.

Alternatively,
$\dfrac{dr}{dh} = \dfrac{dr}{dh} \times \dfrac{dh}{dt} = 0.75\dfrac{dh}{dt}$

Exercise 4U

1   The area of circle and its radius are related by the formula $A = \pi r^2$. Write an equation relating the rate of change of the area to the rate of change of its radius.

2   The formula for the surface area of closed cylinder is $A = 2\pi r^2 + 2\pi rh$. Write an expression relating the rate of change of the area to the rates of change of both the radius and the height of the cylinder.

3 If l, w, and h are respectively the length, width and height of a rectangular box, express the rate of change of the diagonal of the box in terms of the rates of change of dimensions of the box.

4 The length of a rectangle is increasing at a rate of $2\,\mathrm{cm\,s^{-1}}$ while its width is decreasing at a rate of $2\,\mathrm{cm\,s^{-1}}$. When the length and width of the rectangle are 12 cm and 5 cm respectively, find the rate of change of

   a the area   **b** the perimeter   **c** the diagonal.

5 A cube is increasing in volume at a rate of $1.5\,\mathrm{m^3\,s^{-1}}$. Find the rate at which the surface area of the cube is changing when the cube has a volume of 81 m³.

6 A ladder 5 m long is leaning against the side of a building. Its base begins to slide away from the wall, and when it is 3 m from the wall, its slides at a rate of $0.5\,\mathrm{m\,s^{-1}}$. At this point find

   a how quickly the top of the ladder is sliding down the wall

   b the rate of change of the area between the ladder, the wall and the ground.

7 A spark from a fire burns a hole in a paper napkin. The hole initially has a radius of 1 cm and its area is increasing at a rate of $2\,\mathrm{cm^2\,s^{-1}}$. Find the rate of change of the radius when the radius is 5 cm.

8 An airplane is flying at an altitude of 8 miles and passes over a radar station. When the airplane is 12 miles from the base of the station, the radar detects that its horizontal distance is changing at a rate of 320 mph. Find how fast the airplane is flying at this point in time.

9 Kim is flying her kite at a height of approximately 10 m. The wind is blowing horizontally at the kite at a rate of $1\,\mathrm{m\,s^{-1}}$. How quickly must Kim let out the string when the kite is 20 m away from her?

EXAM-STYLE QUESTION

10 Two concentric circles are expanding in size. At time t the radius of the outer circle is 9 m and it is expanding at the rate of $1.2\,\mathrm{m\,s^{-1}}$. The radius of the inner circle is 1 m and it is expanding at the rate of $1.5\,\mathrm{m\,s^{-1}}$. Find the rate of change of the area of the ring between the circles, at time t.

11 Consider a ramp modeled by the function $y = \dfrac{1}{x}$, $x > 0$. A ball slides down the ramp so that the x-coordinate of its position at any time t seconds is increasing by a rate of $f(x)$ units per second. If its y-coordinate is decreasing at a constant rate of 1 unit per second, find $f(x)$.

12 A conical tank with vertex pointed downward has a radius of 10 m at its top and is 24 m high. Water flows out of the tank at a rate of $20\,\text{m}^3\,\text{min}^{-1}$. How fast is the depth of the water in the tank decreasing when it reaches a depth of 16m?

Review exercise

1 Find the limits, if they exist, of the following.

   a $\displaystyle\lim_{x\to 1}\frac{x^2-3}{x+1}$ **b** $\displaystyle\lim_{x\to\infty}\frac{\sqrt{x^2-1}}{x}$ **c** $\displaystyle\lim_{x\to 2}\frac{3^x-1}{x}$

   d $\displaystyle\lim_{x\to 0}\frac{3x^2+x^2}{x^2}$ **e** $\displaystyle\lim_{x\to\infty}\frac{5x^2}{2x^3+1}$ **f** $\displaystyle\lim_{x\to-\infty}\frac{7}{x^3+1}$

2 Determine if $y = \begin{cases} x^2+2x,\ x\le 2 \\ x^3-6x,\ x>2 \end{cases}$ is continuous at $x = 2$.

3 Determine if the sequence $a_n = \dfrac{2n^2-3}{n^3-2}$ converges as n tends to $+\infty$.

4 Determine if the series $\displaystyle\sum_{n=0}^{\infty} 3\left(\frac{(-1)^n}{5^n}\right)$ converges, and if it does, find its sum.

5 Find the values of a for which the series $a^2 + \dfrac{a^2}{1+a^2} + \dfrac{a^2}{(1+a^2)^2} + \ldots$ is convergent, and find its sum.

6 Given $y = \dfrac{x^3-2x^2+5}{x^2-x^3}$, find

   a its horizontal asymptote

   b the points where the curve intersects its horizontal asymptote, for small values of x.

7 Find the equation of the tangent and normal to the curve $y = \dfrac{2x+1}{x^2+1}$ at $x = 0$

8 Let f be an even function with domain $(-a, a)$, $a > 0$. f is differentiable throughout its domain. Show that the tangent to the graph of f at $x=0$ is parallel to the x-axis.

9 Find any points on the curve $y = x\sqrt{x+1}$ those tangents are parallel to the line $x + y = -3$

10 The normal to the curve $y = \dfrac{1}{2}(2x^4 - 5x^3 - 5x^2 + 3x)$ at the point where $x = 1$ meets the curve again at point P. Find the coordinates of P.

11 If f is a function such that $f(x) = [g(x)]^3$, $g(0) = -\dfrac{1}{2}$, $g'(0) = \dfrac{8}{3}$, find the equation of the tangent to $f(x)$ at $x = 0$.

12 Differentiate y with respect to x.

    a $y = (1 - 3x)^7(3x + 5)^3$ **b** $y = \sqrt{(4x^2 - 3x + 1)^5}$

    c $y = \dfrac{x^2 - 3}{\sqrt{x + 1}}, x \neq -1$ **d** $y = \sqrt{x + \sqrt{x^2 + 1}}$

    e $(x + 2 + (x - 3)^8)^3$

13 Consider the polynomial function $f(x) = ax^3 + 6x^2 - bx$. Determine the values of a and b if f has a minimum at $x = -1$, and a point of inflexion at $x = 1$.

14 Consider the function $y = x - \sqrt[3]{x}$

    a Find the intercepts of the function.

    b Find any stationary points and distinguish between them.

    c Find any points of inflexion.

    d Determine the intervals where

      i the function increases   **ii** the function decreases.

15 Consider the function $y = \dfrac{2x}{x^2 - 1}$

    a Find the vertical and horizontal asymptotes.

    b Show that the function is an odd function.

    c Show that $\dfrac{dy}{dx} < 0$ for all x in the domain.

    d Sketch the function.

16 Consider the function $f(x) = \dfrac{(x - 3)^2}{x^2 - 3}$

    a Find any zeros, intercepts, and asymptotes of f.

    b Find any stationary points, and justify your answers.

    c Find any points of inflexion.

    d Find the intervals where f is

      i increasing,   **ii** decreasing.

    e Sketch the function showing all features found.

17 Given $x = y^5 - y$, find $\dfrac{dy}{dx}$, if it exists, at the points where $x = 0$

Review exercise

1 Find the shortest distance between the point $(1.5, 0)$ and the curve $y = \sqrt{x}$

2 A piece of wire 80 cm in length is cut into three parts: two equal circles and a square. Find the radius of the circles if the sum of the three areas is to be minimized.

3 The radius of a right circular cylinder is increasing at a rate of 3 cm min^{-1} and the height is decreasing at a rate of 4 cm min^{-1}. Find the rate at which the volume is changing when the radius is 9 cm and the height is 12 cm, and determine if the volume is increasing or decreasing.

4 A poster has a total area of 180 cm^2 with a 1 cm margin at the bottom and sides, and a 2 cm margin at the top. Find the dimensions that will give the largest printing area.

5 A particle travels along the x-axis. Its velocity at any point x is $\dfrac{dx}{dt} = \dfrac{1}{1+2x}$. Find the particle's acceleration at $x = 2$ in terms of x.

CHAPTER 4 SUMMARY

Continuous functions

- A function $y = f(x)$ is **continuous** at $x = c$, if $\lim\limits_{x \to c} f(x) = f(c)$. The three necessary conditions for f to be continuous at $x = c$ are:

  1 f is defined at c, i.e., c is an element of the domain of f.

  2 the limit of f at c exists.

  3 the limit of f at c is equal to the value of the function at c.

  A function that is not continuous at a point $x = c$ is said to be **discontinuous** at $x = c$.

Properties of limits

- **Properties of limits as $x \to \pm\infty$**

  Let L_1, L_2, and k be real numbers and $\lim\limits_{x \to \pm\infty} f(x) = L_1$ and $\lim\limits_{x \to \pm\infty} g(x) = L_2$. Then,

  1 $\lim\limits_{x \to \pm\infty} (f(x) \pm g(x)) = \lim\limits_{x \to \pm\infty} f(x) \pm \lim\limits_{x \to \pm\infty} g(x) = L_1 \pm L_2$

Continued on next page

2 $\lim\limits_{x \to \infty}(f(x) \cdot g(x)) = \lim\limits_{x \to \infty} f(x) \cdot \lim\limits_{x \to \infty} g(x) = L_1 \cdot L_2$

3 $\lim\limits_{x \to \pm\infty}(f(x) \div g(x)) = \lim\limits_{x \to \pm\infty} f(x) \div \lim\limits_{x \to \pm\infty} g(x) = L_1 \div L_2$,

provided $L_2 \neq 0$.

4 $\lim\limits_{x \to \pm\infty} kf(x) = k \lim\limits_{x \to \pm\infty} f(x) = kL_1$

5 $\lim\limits_{x \to \pm\infty}[f(x)]^{\frac{a}{b}} = L^{\frac{a}{b}}_1, \dfrac{a}{b} \in \mathbb{Q}$ (in simplest form),

provided $L^{\frac{a}{b}}_1$ is real.

Convergence of series

- For a geometric series, $\displaystyle\sum_{n=0}^{\infty} u_1 r^n = \lim_{n \to \infty} \frac{u_1(1-r^n)}{1-r}$

When $-1 < r < 1$, $\lim\limits_{n \to \infty} r^n = 0$ and the series converges to $S = \dfrac{u_1}{1-r}$

The derivative of a function

- The **derivative**, or **gradient function**, of a function f with respect to x is the function $f'(x) = \lim\limits_{h \to 0} \dfrac{f(x+h)-f(x)}{h}$, provided this limit exists.
- If f' exists, then f has a derivative at x, or is **differentiable** at x. ($f'(x)$ is read 'f dash', or 'f prime', of x.) Another notation for the derivative is $\dfrac{dy}{dx}$, the derivative of the function $y = f(x)$ with respect to x.
- A function is differentiable if the derivative exists for all x in the domain of f.

Basic Differentiation rules

- If $f(x) = c$, and $c \in \mathbb{R}$, then $f'(x) = 0$
- If n is a positive integer, and $f(x) = x^n$, then $f'(x) = nx^{n-1}$
- For $c \in \mathbb{R}$, $(cf)'(x) = cf'(x)$ provided $f'(x)$ exists.
- If $f(x) = u(x) \pm v(x)$, then $f'(x) = u'(x) \pm v'(x)$

The chain rule

- If f is differentiable at the point $u = g(x)$, and g is differentiable at x, then the composite function $(f \circ g)(x)$ is differentiable at x. Furthermore, if $y = f(u)$ and $u = g(x)$, then $\dfrac{dy}{dx} = \dfrac{dy}{du} \cdot \dfrac{du}{dx}$

Another definition for the chain rule is $(f \circ g)'(x) = f'(g(x)) \cdot g'(x)$

Continued on next page

The product rule

- If $y = uv$ then $\dfrac{dy}{dx} = u\dfrac{dv}{dx} + v\dfrac{du}{dx}$

  where u and v are functions of x and differentiable.

  Another way of writing this is:

  If $f(x) = u(x)v(x)$, where $u(x)$ and $v(x)$ are differentiable functions then $f'(x) = u(x)v'(x) + v(x)u'(x)$.

The quotient rule

- If $y = \dfrac{u}{v}$ then $\dfrac{dy}{dx} = \dfrac{v\dfrac{du}{dx} - u\dfrac{dv}{dx}}{v^2}$

  where u and v are differentiable functions of x.

  An alternative way of writing this is:
  if $u(x)$ and $v(x)$ are differentiable functions, and
  $f(x) = \dfrac{u(x)}{v(x)}$, $v(x) \neq 0$ then $f'(x) = \dfrac{v(x)u'(x) - u(x)v'(x)}{(v(x))^2}$

Higher derivatives

- $f'(x) = \dfrac{dy}{dx}, f''(x) = \dfrac{d^2 y}{dx^2}, f'''(x) = \dfrac{d^3 y}{dx^3}, f^n(x) = \dfrac{d^n y}{dx^n}$ $n = 4, 5, \ldots$

Maxima, minima and horizontal points of inflexion

- A point whose gradient is equal to 0 is either a maximum, minimum, or horizontal point of inflexion.

First derivative test

- Consider the function $f(x)$ and suppose that $f'(c) = 0$.
  To determine if the point $x = c$ is a maximum, minimum or horizontal point of inflexion, make a sign table and test values of $f(x)$ to the left and right of c.
  - If the signs of gradients change from negative to positive, then f has a minimum at $x = c$.
  - If the signs of the gradients change from positive to negative, then f has a maximum at $x = c$.
  - If there is no sign change, then f has a horizontal point of inflexion at $x = c$.

Second derivative test

- If $f'(c) = 0$ and $f''(c) < 0$, then $f(x)$ has a local maximum at $x = c$
- If $f'(c) = 0$ and $f''(c) > 0$, then $f(x)$ has a local minimum at $x = c$

Continued on next page

Kinematics

- Velocity at a particular instant is the derivative of the position function, $s(t)$ with respect to time.

  $v(t) = \dfrac{ds}{dt}$ or $s'(t)$

- Acceleration is the derivative of the velocity with respect to time, $a(t)$. It is therefore the second derivative of the position function. If a particle's velocity at time t is $v(t)$, then

  $a(t) = \dfrac{dv}{dt} = \dfrac{d^2 s}{dt^2}$ or $s''(t)$

- When v and a have the same sign, the particle is speeding up (accelerating).
- When v and a have opposite signs, the particle is slowing down (decelerating).

Economics

- Marginal profit is the rate of change of profit with respect to the number of units produced or sold.
- Marginal revenue is the rate of change of revenue with respect to the number of units sold.
- Marginal cost is the rate of change of cost with respect to the number of units sold.

If x is the number of units produced a sold:
- $r(x)$ = total revenue from selling x amount of units
- $c(x)$ = total cost of producing x amount of units
- $p(x)$ = profit in selling x amount of units
- $r'(x)$ = marginal revenue, the extra revenue for selling one extra unit
- $c'(x)$ = marginal cost, the extra cost for selling one extra unit
- $p'(x)$ = marginal profit, the extra profit for selling one additional unit

Differentiation of implicit functions

To differentiate $y^2 = 4x$ multiply with respect of x:

LHS: $\dfrac{d(y^2)}{dx} = \dfrac{d(y^2)}{dy} \cdot \dfrac{dy}{dx} = 2y \dfrac{dy}{dx}$

RHS: $\dfrac{d}{dx}(4x) = 4$

$2y \dfrac{dy}{dx} = 4 \Rightarrow \dfrac{dy}{dx} = \dfrac{4}{2y} = \dfrac{2}{y}$

An infinity of ideas

Solving the paradox

This chapter began with Zeno's paradox about an endless day at the races with Achilles and the tortoise. Intuition and common sense tell us that there is, of course, a solution to this paradox. We know that an infinite series can have a finite sum.

You could solve Zeno's paradox by translating it into an infinite geometric series whose ratio is less than 1, and then finding its sum. Developing this mathematics took several centuries of long and laborious work.

> *"How wonderful that we have met a paradox.*
> *Now we have some hope of making progress"*
>
> Niels Bohr, Danish physicist (1885–1962), winner of Nobel Prize for Physics 1922

Finite or infinite?

In reality, the finite and infinite are closely related. Using the formula for an infinite geometric series, you can prove that $0.999999\ldots = 1$. It is not approximately 1, but is exactly 1.

◀ A circle can be thought of as a polygon with an infinite number of sides.

- Find other examples in which both the finite and the infinite define the same concept or object.

- How does our perception define it in one way, or the other?

> *"Even as the finite encloses an infinite series*
> *And in the unlimited limits appear,*
> *So the soul of immensity dwells in minuta*
> *And in the narrowest limits, no limits in here*
> *What joy to discern the minute in infinity!*
> *The vast to perceive in the small, what Divinity!"*
>
> Jakob Bernouilli, *Ars Conjectandi* (*The Art of Conjecture*), 1713

- Where did the symbol for infinity come from?
- Who was the first to use it?

To infinity ...

Some research into the history of mathematics will give you insight into the journey of the mathematics of infinity. Are there different kinds of infinities? An article in *Scientific American* claimed that there exist an infinite number of infinities, some infinitely small, others infinitely large. What can this mean?

Scientific American Special Report: Parallel Universes, 2010

Calculus

The discovery, or invention, of calculus is regarded as a great innovation.

The development of calculus was a culmination of centuries of work by mathematicians all over the world. English scientist Isaac Newton and German mathematician Gottfried Wilhelm Leibniz, both approached calculus in an intrinsically different manner, but yet arrived at the same set of concepts.

The argument over which of them invented or discovered calculus first, and whether there was any plagiarism, is one of the most famous conflicts in mathematical history.

- Was calculus discovered or invented?
- What specific concepts of calculus had immediate applications to the problems being considered in the 17th century, when Newton and Leibniz were working on it?

Applications of pure mathematics

Practical applications are often found for pure mathematics – sometimes many years after the first ideas were formulated.

1 George Boole, an English mathematician, developed his Boolean Logic system in the 1850s. This system was later used in digital electronics.

2 In physics, elementary particles were discovered through arguments involving the beauty, symmetry or elegance of the underlying mathematics.

Perhaps all 'pure' mathematics will be used to model some aspect of real life one day.

- What other concepts were discovered, or invented, through studying pure mathematics?
- How were they later used to solve specific real world problems?
- If mathematicians did not discover, or invent, pure mathematical theories, would our world progress at the rate that it has?
- Should knowledge always be immediately applicable? Is there merit in knowledge for knowledge's sake?
- What is the difference between pure and applied mathematics? How does it compare to the difference between fine arts and commercial arts?

5 Aesthetics in mathematics

Before you start

You should know how to:

1 Find an inverse function.

e.g. $f : x \mapsto \dfrac{x+1}{x-3}$, $x \neq 3$

$y = \dfrac{x+1}{x-3} \Rightarrow x = \dfrac{y+1}{y-3}$

$\Rightarrow x(y-3) = y+1$

$\Rightarrow y = \dfrac{3x+1}{x-1}$

$\therefore f^{-1} : x \mapsto \dfrac{3x+1}{x-1}$, $x \neq 1$

2 Work with composite functions.

e.g. $f(x) = 2x - 3$ and $g(x) = \dfrac{1}{2}x^2$

$(f \circ g)(x) = f\left(\dfrac{1}{2}x^2\right)$

$= 2\left(\dfrac{1}{2}x^2\right) - 3 = x^2 - 3$

Skills check

1 Show that if $f(x) = \dfrac{x}{x-1}$, $x \neq 1$, then $f^{-1}(x) = f(x)$.

Use a sketch of $f(x)$ to justify this result.

2 Show that if $f(x) = ax - b$ and $g(x) = \dfrac{x+b}{a}$, $(f \circ g)(x) = (g \circ f)(x)$ for all values of x.

A journey through number

The golden ratio

"Perhaps the most surprising place in which the golden ratio crops up is in the physics of black holes, a discovery made by Paul Davies of the University of Adelaide in 1989. Black holes and other self-gravitating bodies such as the sun have a 'negative specific heat'. This means they get hotter as they lose heat. Basically, the loss of heat robs the gas of a body, such as the sun, of internal pressure, enabling gravity to squeeze it into a smaller volume. The gas then heats up, for the same reason that the air in a bicycle pump gets hot when it is squeezed. However, things are not so simple for a spinning black hole, since there is an outward 'centrifugal force' acting to prevent any shrinkage of the hole. The force depends on how fast the hole is spinning. It turns out that, at a critical value of the spin, a black hole flips from negative to positive specific heat – that is, from growing hotter as it loses heat to growing colder. What determines the critical value? The mass of the black hole and the golden ratio!"

Marcus Chown,
Guardian (16 January 2003)

"Researchers from the Helmholtz-Zentrum Berlin für Materialien und Energie (HZB), in cooperation with colleagues from Oxford and Bristol Universities, as well as the Rutherford Appleton Laboratory, UK, have for the first time observed a nanoscale symmetry hidden in solid state matter. They have measured the signatures of a symmetry showing the same attributes as the golden ratio famous from art and architecture. The magnetic field is used to tune the chains of spins to a quantum critical state. The resonant modes ('notes') are detected by scattering neutrons. These scatter with the characteristic frequencies of the spin chains. The first two notes show a perfect relationship with each other. Their frequencies (pitch) are in the ratio of 1.618..., which is the golden ratio famous from art and architecture."

Press release, Helmholtz Association of German Research Centres (7 January 2010)

The first and second movements of Mozart's Piano Sonata No. 1 in C Major, which he composed when he was 18 years old, contain approximations of the golden ratio. The first movement consists of 100 measures. These measures are divided into two parts: the first part contains 38 measures and the second part 62. The ratio 38 : 62 is a close approximation of the golden ratio. The second movement of this sonata is also divided into parts whose ratio approximates the golden ratio; however, the same cannot be said of the third part. Studies show that the golden ratio appears in a number of Mozart's works.

So from the infinitely large to the infinitely small the golden ratio is present! The golden ratio can be defined as follows:

→ If a line segment AB is divided into two segments by a point P such that $\dfrac{AB}{AP} = \dfrac{AP}{PB}$, we say that the line is divided in the golden ratio. The golden ratio is denoted by φ.

5.1 Recursive functions

One of the most fascinating aspects of the golden ratio is that it seems to appear in unexpected ways.

Investigation – random numerical patterns and their behavior

1 Think of two positive numbers a_0 and a_1. Let $a_n = a_{n-1} + a_{n-2}$
   - Use a spreadsheet to generate the first 1000 terms of this sequence.
   - Add a column in your spreadsheet to represent $r = \dfrac{a_n}{a_{n-1}}$
   - What do you observe?

2 Think of a number k_0. Let $k_n = 1 + \dfrac{1}{k_{n-1}}$
   - Use a spreadsheet to generate the first 1000 terms of this sequence.
   - What do you observe? Write down your conjecture and test it further.

Recursive formulae were first introduced in Chapter 1. You saw that factorials can be represented by a sequence using the recursive rule:

$u_0 = 1$

$u_n = n \times u_{n-1}, n \in \mathbb{Z}^+$

Similarly, the triangular numbers, also first seen in Chapter 1, can be represented by the recursive rule:

$T_1 = 1$

$T_n = T_{n-1} + n, n \in \mathbb{Z}^+$

> → A recursive function is one in which initial values $u_0, u_1, \ldots$ are given together with a rule $u_n = F(u_{n-1}, u_{n-2}, \ldots)$ which explains how a term is obtained from the previous terms.

In the first part of the Investigation on the previous page, the general term is written in terms of the previous two terms, $u_n = F(u_{n-1}, u_{n-2})$, and two initial values, u_0 and u_1, are required. In the second part of the Investigation, the general term is given in terms of the previous term, $u_n = F(u_{n-1})$, and only one initial value, u_0, is required.

Example 1

Find the first four terms of the triangular number sequence given by

$u_0 = 0$

$u_n = n + u_{n-1}$

Hence, show that $u_n = \dfrac{n}{2}(n + 1)$

Answer

$u_0 = 0$

$u_1 = 1 + u_0 = 1$

$u_2 = 2 + u_1 = 2 + 1 = 3$

$u_3 = 3 + u_2 = 3 + 2 + 1 = 6$

$\therefore u_n = n + (n-1) + (n-2) + \ldots + 3 + 2 + 1$ *Sum of first n positive integers.*

$\quad = \dfrac{n}{2}(n + 1)$

See the example involving Gauss (when he was 11 years old) in Chapter 1 on page 12.

Example 2

a Find the first six terms of the sequence given by:

$$\begin{cases} u_0 = 1 \\ u_1 = 1 \\ u_n = 2u_{n-1} + 3u_{n-2} \end{cases}$$

b Prove by mathematical induction that $u_n = \dfrac{3^n + (-1)^n}{2}$

Answers

a $u_0 = 1$

$u_1 = 1$

$u_2 = 2 + 3 = 5$

$u_3 = 10 + 3 = 13$

$u_4 = 26 + 15 = 41$

$u_5 = 82 + 39 = 121$

b Let $P(n)$ be the statement $u_n = \dfrac{3^n + (-1)^n}{2}$

$P(0): u_0 = \dfrac{3^0 + (-1)^0}{2} = 1$

$P(1): u_1 = \dfrac{3^1 + (-1)^1}{2} = 1$

For the first step in the proof you need to show that this statement is true for the first two possible values since the recursive function uses two previous terms.

Assume that $P(k-1)$ and $P(k)$ are true for some $k \in \mathbb{Z}^+$

i.e. $u_{k-1} = \dfrac{3^{k-1} + (-1)^{k-1}}{2}$ and $u_k = \dfrac{3^k + (-1)^k}{2}$

Prove that $u_{k+1} = \dfrac{3^{k+1} + (-1)^{k+1}}{2}$

Proof: $u_{k+1} = 2u_k + 3u_{k-1}$

Using the recursive definition.

$= 2\left(\dfrac{3^k + (-1)^k}{2}\right) + 3\left(\dfrac{3^{k-1} + (-1)^{k-1}}{2}\right)$

Using the assumption.

$= \dfrac{2(3^k) + 2(-1)^k + 3^k + 3(-1)^{k-1}}{2}$

Expanding brackets and taking a common denominator.

$= \dfrac{3(3^k) - 2(-1)(-1)^k + 3(-1)^2(-1)^{k-1}}{2}$

Adding the terms in 3^k, and using identities:
$2(-1)^k \equiv -2(-1)(-1)^k$
$(-1)^2 \equiv 1$
and $3(-1)^k \equiv 3(-1)^2(-1)^{k-1} \equiv 3(-1)^{k+1}$

$= \dfrac{3^{k+1} - 2(-1)^{k+1} + 3(-1)^{k+1}}{2}$

$= \dfrac{3^{k+1} + (-1)^{k+1}}{2}$

Simplifying terms.

Since it was shown that $P(0)$ and $P(1)$ are true and it was also proved that if the statement is true for $n = k$ and $n = k-1$, $k \in \mathbb{Z}^+$, $k \geq 1$, it is also true for $n = k+1$, it follows by the principle of mathematical induction that the statement is true for all values of $n \in \mathbb{N}$.

Exercise 5A

1 Find the first five terms of the sequence given by the recursive

function $\begin{cases} u_1 = 1 \\ u_n = \dfrac{u_{n-1}}{1 + u_{n-1}}, n \in \mathbb{Z}^+ \end{cases}$ Hence, write down u_n in terms of n.

2 Find the first six terms of the sequence given by the recursive

function $\begin{cases} u_1 = 2 \\ u_n = \dfrac{u_{n-1}}{1 - u_{n-1}}, n \in \mathbb{Z}^+, n \geq 2 \end{cases}$

Hence, write down u_n in terms of n.

3 a Find the first six terms of the sequence given by the recursive

function $\begin{cases} u_0 = 2 \\ u_n = u_{n-1} - \dfrac{1}{2^n}, n \in \mathbb{Z}^+ \end{cases}$

  b Conjecture a formula for u_n in terms of n.

  c Prove your conjecture using mathematical induction.

EXAM-STYLE QUESTION

4 a Find the first five terms of the sequence given by the recursive

function $\begin{cases} u_1 = 1 \\ u_n = u_{n-1} + 2n - 3, n \in \mathbb{Z}^+ \end{cases}$

  b Prove by mathematical induction that $u_n = n^2 - 2n + 2$

Investigation – sequences

1 The Fibonacci sequence is defined by the recursive function $\begin{cases} F_0 = 0 \\ F_1 = 1 \\ F_n = u_{n-1} + u_{n-2} \end{cases}$

Find the first eight terms of the sequence.

2 If a line segment AB is divided into two segments by a point P such that $\dfrac{AB}{AP} = \dfrac{AP}{PB}$,

we say that the line is divided in the golden ratio. The golden ratio is denoted by φ.

Use this definition of the golden ratio and the diagram to show

that the exact value of the golden ratio is given by $\dfrac{1 + \sqrt{5}}{2}$.

x units   1 units

A   P   B

3 Consider the geometric sequence $\phi, \phi^2, \phi^3, \ldots$

Use your result from question **2** to simplify ϕ, and hence show that $\phi^2 = F_2 \phi + F_1$

Also simplify ϕ^3 and ϕ^4.

Make a conjecture connecting ϕ^n, F_n and F_{n-1}. Prove your conjecture.

Continued on next page

4 Now consider the sequence $\left(-\dfrac{1}{\phi}\right), \left(-\dfrac{1}{\phi}\right)^2, \left(-\dfrac{1}{\phi}\right)^3, \ldots$

Show that $\left(-\dfrac{1}{\phi}\right) = \dfrac{1-\sqrt{5}}{2}$

Hence, show that $\left(-\dfrac{1}{\phi}\right)^2 = F_2\left(-\dfrac{1}{\phi}\right) + F_1$

Write similar equations for $\left(-\dfrac{1}{\phi}\right)^3$ and $\left(-\dfrac{1}{\phi}\right)^4$

Make a conjecture connecting $\left(-\dfrac{1}{\phi}\right)^n$, F_n and F_{n-1}. Prove your conjecture.

5 Use your results from questions **3** and **4** to find a formula for F_n in terms of n.

5.2 Properties of exponents and logarithms

By definition, if $m \in \mathbb{Z}^+$, a^m denotes the product of m factors each equal to a. m is called the exponent, index or power of a and a is called the base.

> Explore the connections between the Fibonacci numbers, the Lucas numbers and Pascal's triangle.

Using this definition, you can deduce the three fundamental properties for the combination of indices.

In all the cases below $m, n \in \mathbb{Z}^+$, $m \geq n$

1 $\quad \rightarrow a^m \times a^n = a^{m+n}$

Using the definition:

$$\text{LHS} = \underbrace{a \times a \times \ldots \times a}_{m} \times \underbrace{a \times a \times \ldots \times a}_{n} = \underbrace{a \times a \times \ldots \times a}_{m+n} = a^{m+n}$$

$$= \text{RHS}$$

2 $\quad \rightarrow a^m \div a^n = a^{m-n}$

Using the definition:

$$\text{LHS} = \dfrac{a^m}{a^n} = \dfrac{\overbrace{a \times a \times \ldots \times a}^{m}}{\underbrace{a \times a \times \ldots \times a}_{n}} = \underbrace{a \times a \times \ldots \times a}_{m-n} = a^{m-n}$$

$$= \text{RHS}$$

> n of m factors in the numerator are cancelled by the denominator, leaving $m - n$ factors of a.

3 $\quad \rightarrow (a^m)^n = a^{mn}$

Using the definition:

$$\text{LHS} = \underbrace{a^m \times a^m \times \ldots \times a^m}_{n} = \underbrace{\overbrace{a \times a \times \ldots \times a}^{m} \times \overbrace{a \times a \times \ldots \times a}^{m} \times \ldots \times \overbrace{a \times a \times \ldots \times a}^{m}}_{n} = \underbrace{a \times a \times \ldots \times a}_{mn}$$

$$= a^{m \cdot n}$$

$$= \text{RHS}$$

A special case of property 2 gives this result.

4 $\quad\boxed{\;\rightarrow\; a^0 = 1\;}$

$$1 = a^m \div a^m = a^{m-m}$$
$$= a^0$$

The properties of indices can be consistently extended to negative powers using this definition.

5 $\quad\boxed{\;\rightarrow\; a^{-n} = \dfrac{1}{a^n} \qquad a \neq 0\;}$

Similarly the properties of indices can be consistently extended to rational powers using these definitions.

6 $\quad\boxed{\;\rightarrow\; a^{\frac{1}{m}} = \sqrt[m]{a} \qquad a > 0\;}$

The restriction $a > 0$ is necessary to avoid possible inconsistencies.
For example
$$(-8)^{\frac{1}{3}} = \sqrt[3]{-8} = -2$$
$$\text{or} \quad = (-8)^{\frac{2}{6}} = \sqrt[6]{(-8)^2}$$
$$= \sqrt[6]{64} = 2$$

Example 3

Show that: **a** $\quad a^{\frac{m}{n}} = \sqrt[n]{a^m}$ $\qquad$ **b** $\quad a^{\frac{m}{n}} = \left(\sqrt[n]{a}\right)^m$

where $a > 0$, and $m, n \in \mathbb{Z}^+$

Answer

a $\quad a^{\frac{m}{n}} = a^{m \times \frac{1}{n}} = \left(a^m\right)^{\frac{1}{n}} = \sqrt[n]{(a^m)}$ $\qquad$ **b** $\quad a^{\frac{m}{n}} = a^{\frac{1}{n} \times m} = \left(a^{\frac{1}{n}}\right)^m = \left(\sqrt[n]{a}\right)^m$

$$\boxed{\;\rightarrow\; a^{\frac{m}{n}} = \sqrt[n]{a^m} = \left(\sqrt[n]{a}\right)^m \qquad a > 0\;}$$

Example 4

Evaluate:

a $\quad\left(\dfrac{27}{8}\right)^{\frac{2}{3}}$ $\qquad$ **b** $\quad 144^{-\frac{1}{2}}$ $\qquad$ **c** $\quad -125^{-\frac{1}{3}}$

Answers

a $\quad\left(\dfrac{27}{8}\right)^{\frac{2}{3}} = \left(\sqrt[3]{\dfrac{27}{8}}\right)^2$ $\qquad$ **b** $\quad 144^{-\frac{1}{2}} = \dfrac{1}{144^{\frac{1}{2}}}$ $\qquad$ **c** $\quad -125^{-\frac{1}{3}} = \dfrac{1}{-125^{\frac{1}{3}}}$

$$= \left(\dfrac{3}{2}\right)^2 \qquad\qquad\qquad = \dfrac{1}{\sqrt{144}} \qquad\qquad\qquad = \dfrac{1}{\sqrt[3]{-125}}$$

$$= \dfrac{9}{4} \qquad\qquad\qquad\quad = \dfrac{1}{12} \qquad\qquad\qquad\quad = -\dfrac{1}{5}$$

Example 5

Simplify the expression $(20 \times 9^{2n+1}) \div (5 \times 3^{4n+1})$

Answer

$(20 \times 9^{2n+1}) \div (5 \times 3^{4n+1})$

$= \dfrac{20 \times (3^2)^{2n+1}}{5 \times 3 \times 3^{4n}} = \dfrac{20 \times 3^{4n+2}}{15 \times 3^{4n}} = \dfrac{20 \times 9 \times 3^{4n}}{15 \times 3^{4n}} = 12$

Example 6

Solve the equation $3^x + 3^{1-x} = 4$

Answer

$3^x + 3^{1-x} = 4$

$\Rightarrow 3^x + \dfrac{3}{3^x} = 4$

$\Rightarrow 3^{2x} - 4 \times 3^x + 3 = 0$

$\Rightarrow (3^x)^2 - 4(3^x) + 3 = 0$

$3^{1-x} = 3^1 \div 3^x$

Multiply throughout by 3^x and rearrange.

Let $3^x = y$

$\Rightarrow y^2 - 4y + 3 = 0$

$\Rightarrow (y-3)(y-1) = 0$

$\quad y = 3 \quad$ or $\quad y = 1$

$\Rightarrow 3^x = 3 \quad$ or $\quad \Rightarrow 3^x = 1$

$\quad \therefore x = 1 \quad$ or $\quad \therefore x = 0$

Exercise 5B

1 Evaluate:

   a $(64)^{\frac{2}{3}}$ **b** $\left(\dfrac{8}{27}\right)^{\frac{1}{3}}$ **c** $\left(\dfrac{81}{16}\right)^{-\frac{3}{4}}$

2 Show that:

   a $\left(\dfrac{b^{-3}x^{-2}}{8x}\right)^{-\frac{2}{3}} = 4b^2x^2$

   b $\dfrac{a^{-1} - a^{-2}}{a^{-3}} = a(a-1)$

   c $\dfrac{x^3 \times x^{-7}}{x^{-4}} = 1$

3 Simplify $\sqrt{y^3} \div \sqrt[3]{y^2}$ and use your answer to show that

   when $y = 64$, $\sqrt{y^3} \div \sqrt[3]{y^2} = 32$

4 Show that $\dfrac{(x^4yz^{-3})^2 \times \sqrt{x^{-5}y^2z}}{(xz)^{\frac{7}{2}}} \equiv \dfrac{x^2y^3}{z^9}$

5 Simplify $5 \times 4^{3n+1} - 20 \times 8^{2n}$

6 Solve the equation $4^x + 2 = 3 \times 2^x$

Investigation – music and indices

The diagram below shows keys on a piano keyboard. If you play the A below the middle C the piano string vibrates at 220 hertz.

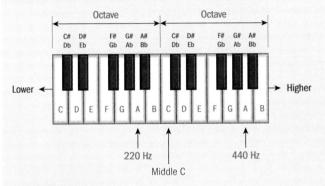

220 Hz

440 Hz

Middle C

Non-musicians might find it helpful to ask a piano player for an explanation of 'middle C'.

This table shows the frequencies for some notes above A.

Note	A#	B	Middle C	C#
Frequency (Hz)	$220 \times 2^{\frac{1}{12}}$	$220 \times 2^{\frac{2}{12}}$	$220 \times 2^{\frac{3}{12}}$	$220 \times 2^{\frac{4}{12}}$

- What is the frequency of G# below middle C?
- What is the frequency of G# above middle C?
- The interval between these two notes is called an octave.
  Explain the significance of this term.
- What is the frequency of the C in the next octave above middle C?
- The diagram shows the keyboard of a grand piano. How would you use the results above to find the lowest frequency and the highest frequency that can be played on a grand piano?

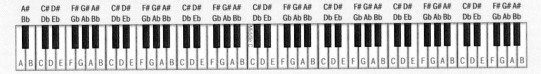

Financial matters and indices

This shows how the laws of indices are used to solve financial problems.

Example 7

When the first Android phone was launched it sold at $600.
Six months later its selling price was $360. Use the laws of indices to
calculate the average percentage depreciation per month.

Answer

Assuming that the depreciation rate is constant, we can argue that the
value of the Android when launched was A_0 and, if the depreciation
rate per month is r, the selling price after one month was:

$A_1 = A_0(1 - r)$

Over the second month this value again depreciated by r, giving:

$A_2 = A_1(1 - r) = A_0(1 - r)^2$

After 6 months:

$A_6 = A_0(1 - r)^6 \Rightarrow 360 = 600(1 - r)^6 \Rightarrow (1 - r)^6 = \dfrac{360}{600} = 0.6$

$\therefore 1 - r = \sqrt[6]{0.6}$

$\Rightarrow r = 1 - \sqrt[6]{0.6} = 0.082$

So, the average percentage depreciation rate per month is 8.2%.

Exercise 5C

1 A house bought for €250 000 in 1990 was sold for €450 000 in
   2010. What was the annual rate of appreciation of the house?
   Give your answer to the nearest percent.

See Chapter 1 for
more on compound
interest and
population growth.

2 Tensions in the Middle East cause oil prices to rise. The table
   shows the cost price for a barrel of oil at various points in time.

December 1999	$17.48
December 2006	$61.08
December 2010	$77.45
February 2010	$72.99
February 2011	$97.87

a What was the annual average percentage rise in oil price between
   1999 and 2006?
b What was the annual average percentage rise between 2006 and 2010?
c Calculate the average percentage rise per month between February
   2010 and February 2011.

3 Samira invests $1000 at Better Bank which offers 8% interest
   compounded quarterly. Her brother Hemanth invests $500 at Good
   Bank which offers 8% interest compounded annually and another
   $500 at Superior Bank which offers 8.4% interest compounded
   monthly. Calculate the value of each investment after 15 years.

4 Giuseppe borrows €15000 to buy a new car. The bank charges 5% interest compounded yearly. Giuseppe agrees to pay two-thirds of the amount owing at the end of each year. How much would Giuseppe have paid back at the end of 5 years?

5.3 Euler's number and exponential functions

Investigation – compound interest

In Chapter 1 you looked at financial matters and compound interest. If €1000 was invested at a compound interest of 2% compounded annually, you could find out how much it is worth after n years by calculating $1000(1.02)^n$.

If the interest is compounded every six months the formula becomes $1000\left(1+\dfrac{0.02}{2}\right)^{2n}$

And if the interest is compounded monthly the formula becomes $1000\left(1+\dfrac{0.02}{12}\right)^{12n}$

If you calculate the value after one year in each of these cases you get:

Interest compounded annually: €1020

Interest compounded every six months: €1020.10

Interest compounded monthly: €1020.18

Look at the growth of €1 invested for one year at 100% interest compounded at different intervals of n.

- Write down the general formula to obtain the growth of this investment after one year.
- Use technology to draw up a table with the value of the investment for different values of n.
- Plot these values and comment on your results.

In the previous Investigation the amount starts growing quickly but then as the value of n gets larger there is hardly any increase in the outcome. The graph becomes less and less steep, seemingly approaching a certain value.

In fact your results should indicate that $\lim_{n \to \infty}\left(1+\dfrac{1}{n}\right)^n \approx 2.718$

In order to establish that this limit exists and to find its value, expand $\left(1+\dfrac{1}{n}\right)^n$, $n \in \mathbb{Z}^+$ using the binomial theorem:

$$\left(1+\frac{1}{n}\right)^n = \sum_{r=0}^{n}\binom{n}{r}\left(\frac{1}{n}\right)^r$$

$$= \frac{n!}{0!(n-0)!}\left(\frac{1}{n}\right)^0 + \frac{n!}{1!(n-1)!}\left(\frac{1}{n}\right) + \frac{n!}{2!(n-2)!}\left(\frac{1}{n}\right)^2 + \ldots + \frac{n!}{r!(n-r)!}\left(\frac{1}{n}\right)^r$$

$$+ \ldots + \frac{n!}{n!(n-n)!}\left(\frac{1}{n}\right)^n$$

$$= 1 + \frac{n}{1!}\left(\frac{1}{n}\right) + \frac{n(n-1)}{2!}\left(\frac{1}{n^2}\right) + \ldots + \frac{n(n-1)(n-2)\ldots(n-r+1)}{r!}\left(\frac{1}{n^r}\right)$$

$$+ \ldots + \frac{n(n-1)(n-2)\ldots(n-n+1)}{n!}\left(\frac{1}{n^n}\right)$$

$$= 1 + \frac{1}{1!} + \frac{\frac{n}{n}\left(\frac{n-1}{n}\right)}{2!} + \ldots + \frac{\left(\frac{n}{n}\right)\left(\frac{n-1}{n}\right)\left(\frac{n-2}{n}\right)\ldots\left(\frac{n-r+1}{n}\right)}{r!}$$

$$+ \ldots + \frac{\left(\frac{n}{n}\right)\left(\frac{n-1}{n}\right)\left(\frac{n-2}{n}\right)\ldots\left(\frac{n-n+1}{n}\right)}{n!}$$

$$= 1 + \frac{1}{1!} + \frac{1\left(1-\frac{1}{n}\right)}{2!} + \ldots + \frac{1\left(1-\frac{1}{n}\right)\left(1-\frac{2}{n}\right)\ldots\left(1-\frac{r-1}{n}\right)}{r!}$$

$$+ \ldots + \frac{1\left(1-\frac{1}{n}\right)\left(1-\frac{2}{n}\right)\ldots\left(1-\frac{n-1}{n}\right)}{n!}$$

$$= 1 + \frac{1}{1!} + \frac{1}{2!}\left(1-\frac{1}{n}\right) + \ldots + \frac{1}{r!}\left(1-\frac{1}{n}\right)\left(1-\frac{2}{n}\right)\ldots\left(1-\frac{r-1}{n}\right)$$

$$+ \ldots + \frac{1}{n!}\left(1-\frac{1}{n}\right)\left(1-\frac{2}{n}\right)\ldots\left(1-\frac{n-1}{n}\right)$$

Since every term in this expansion is positive it is clear that any limit must be greater than 2. Looking at the general term of the expansion you can draw two conclusions.

$$\frac{1}{r!}\left(1-\frac{1}{n}\right)\left(1-\frac{2}{n}\right)\left(1-\frac{3}{n}\right)\left(1-\frac{4}{n}\right)\ldots\left(1-\frac{r+1}{n}\right) < \frac{1}{r!} = \frac{1}{1\cdot2\cdot3\cdots r} \le \frac{1}{1\cdot2\cdot2\cdots2} = \frac{1}{2^{r-1}}$$

The first inequality follows because each term in brackets is less than one. The second inequality follows because 2, 3, ..., $r \ge 2$. Using these bounds on individual terms you can place a bound on the whole expression:

$$\left(1+\frac{1}{n}\right)^n < 1 + 1 + \frac{1}{2} + \frac{1}{4} + \frac{1}{8} + \ldots + \frac{1}{2^{n-1}} < 2 + 1 = 3$$

Thus any limit must also be less than 3.

Now consider the general term with r fixed and allow n to increase. As n increases each term in brackets also increases, $1 - \dfrac{1}{n} \to 1$ etc., therefore so must the general term. Thus, for example, the third term in the expansion for n is less than the third term in the expansion for $n+1$. As n increases so does the number of terms in the expansion. Since each individual term gets larger and there are more terms as n increases you can conclude that

$\left(1 + \dfrac{1}{n}\right)^{n}$ is an increasing function of n. This confirms the result in the numerical investigation.

In summary this shows that $\left(1 + \dfrac{1}{n}\right)^{n}$ is an increasing function of n whose value lies between 2 and 3. While it is beyond the scope of this course to prove this result, you can use the result that any increasing sequence which is bounded above must have a limit. In this case

$$\lim_{n \to \infty} \left(1 + \frac{1}{n}\right)^{n} = 1 + \frac{1}{1!} + \frac{1}{2!} + \frac{1}{3!} + \ldots + \frac{1}{r!} + \ldots$$

→ This limit is called the Euler number and is denoted by
   $e \approx 2.718\,281\,828\,459\,045\ldots$

e is a transcendental number.

Negative numbers satisfy equations such as $x + 2 = 0$. Rational numbers satisfy equations such as $3x - 2 = 0$. An irrational number like $\sqrt{2}$, which cannot be written exactly as an integer fraction, still satisfies the equation $x^2 - 2 = 0$. Together these numbers, which are the solutions of a polynomial equation with integer coefficients, are called algebraic numbers. That is, if $f(x) = a_n x^n + a_{n-1} x^{n-1} + \ldots + a_2 x^2 + a_1 x + a_0$ is a polynomial function of degree n, with $a_n \in \mathbb{Z}$ for all values of n, and k is a number such that $f(k) = 0$ then k is an algebraic number.

A number which is not algebraic and, therefore necessarily, irrational is called **transcendental**. It was not until 1844 that **Joseph Liouville** (1809–82) showed that an actual transcendental number existed. Now we know that there are many more transcendental numbers than algebraic numbers but finding examples is notoriously difficult.

In 1873 **Charles Hermite** proved that e is transcendental and in 1882 **Ferdinand Von Lindemann** proved that π is transcendental.

Exponential functions and their properties

This section looks at functions of the form $f(x) = a^x$. This time the independent variable is the power and the base is a fixed number a.

Start with one particular function $f(x) = 2^x$. Sketch the function.

Note that $y = a^x$ is called an exponential function but $y = e^x$ is **the** exponential function.

The graph
- has no stationary points
- is always positive
- is always increasing
- has the y-axis as a horizontal asymptote
- has no vertical asymptotes
- is injective or one-to-one.

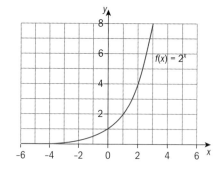

A function is injective or one-to-one if it maps every point in its domain to a unique point in its range. A function such as $f(x) = x^2$ is not injective as $f(2) = f(-2) = 4$; it would fail the horizontal line test.

Investigation – properties of exponential functions

Use a GDC to answer these questions about the properties of exponential functions.

- What do the graphs of $f(x) = a^x$, $a > 1$ have in common?
- Why do we refer to these graphs as representations of exponential growth?
- How are the graphs of $f(x) = a^x$, $0 < a < 1$ different? Why do they represent exponential decay?
- Why do we put the following restrictions on a, $a > 0$, $a \neq 1$?
- Compare the graphs $f(x) = a^x$ and $g(x) = x^a$ for $a \in \{2, 3, 5, 10\}$ and explain why the following statement is true: 'Exponential growth is bigger and faster than polynomial growth of any degree.'

Example 8

The table shows values for a function of the form $y = ba^x$, $a > 0$.
Find the values of a and b and sketch the graph.

x	0	2	4	6
y	3	12	48	192

Answer

$3 = ba^0 \Rightarrow b = 3$

$3a^2 = 12$

$\Rightarrow a^2 = 4$

$\therefore a = 2$

$f(x) = 3 \cdot 2^x$

The point (0, 3) is on the graph.

The point (2, 12) is on the graph.

Only the positive solution for a is required since the values of y are all positive.

Note: Only two points were required to find the solution.

Example 9

For well-tempered tuning of an instrument, the interval between the
first tone at $440\,\text{Hz}$ and the thirteenth tone at $880\,\text{Hz}$ is divided so that
the ratio of two consecutive tones is a constant. What is this ratio?
Find a formula for obtaining all the tones of the interval.

Answer

$440 \cdot a^{12} = 880$

$\therefore a^{12} = 2 \Rightarrow a = 2^{\frac{1}{12}}$

Use 12 since $880\,\text{Hz}$ is the twelfth note after $440\,\text{Hz}$.

So, the formula for obtaining all
the tones is given by

$u_n = 440\left(2^{\frac{1}{12}}\right)^n, \; n \in \{1,2,...,11\}$

Example 10

The diagram shows the graph of $f(x) = 2^x$
and $g(x)$, the reflection of $f(x)$ in the
y-axis. Find $g(x)$.

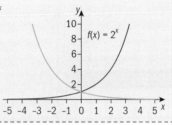

Answer

$g(x) = f(-x)$

$\therefore g(x) = 2^{-x} = (2^{-1})^x = 0.5^x$

*$g(x)$ is a reflection of $f(x)$ in the
y-axis.*

Exercise 5D

1 The diagram shows graphs of exponential
   functions of the form $y = a^x$. For each
   of the functions, find the value of a.

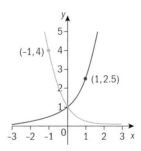

2 Given that $f(x) = 2^x$, show that $f(x+1) = 2f(x)$
   Express $f(x+a)$ in terms of $f(x)$.

3 Sketch the graphs of $f(x) = 2^x$ and $g(x) = 2^{-x}$ on the same set of
   axes. Hence, sketch the graph of $h(x) = 2^{|x|}$ on a separate diagram.

   See Chapter 2 for
   help with graph
   transformations and
   sketching.

4 Solve the equation $e^x + 1 = e^{x+1}$.

5 Find the geometric transformations of the graph of $f(x) = e^x$
   which give these functions.

   a $y = e^{-x}$　　**b** $y = -e^x$　　**c** $y = -e^{-x}$

   Hence, sketch all the functions on the same set of axes.

   Learn these useful
   results.

5.4 Invariance and the exponential function – a different approach to Euler's number

Investigation – the graph of $y = a^x$

Use a GDC to complete a table like this by graphing the function a^x for the given values of a and using your calculator to find the derivative at $x = a$.

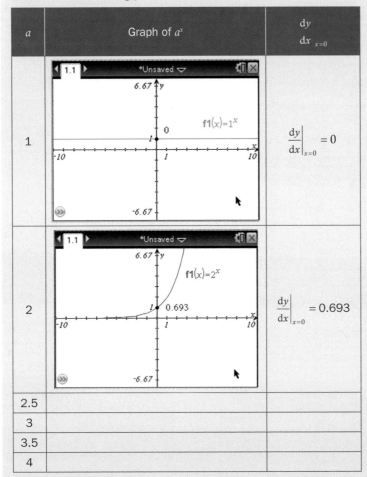

a	Graph of a^x	$\dfrac{dy}{dx}\Big|_{x=0}$
1	$f1(x) = 1^x$	$\dfrac{dy}{dx}\Big|_{x=0} = 0$
2	$f1(x) = 2^x$ 0.693	$\dfrac{dy}{dx}\Big|_{x=0} = 0.693$
2.5		
3		
3.5		
4		

GDC help on CD:
Alternative demonstrations for the TI-84 Plus and Casio fx-9860GII GDCs are on the CD.

The notation $\dfrac{dy}{dx}\Big|_{x=a}$ stands for the derivative of y with respect to x evaluated at $x = a$.

Leave enough space for a graph in each row.

You will have noticed that the graph of function $y = a^x$ passes through $(0,1)$ and that the gradient of the function $y = a^x$ at this point increases as the value of a increases. However, betwen $a = 2$ and $a = 3$, the gradient goes from a value below 1 to a value greater than 1.

Is it possible to find a value of a such that the value of the gradient of $\dfrac{da^x}{dx}\Big|_{x=0} = 1$?

Here is a simplified way of showing that the value of a is e.
For an exponential function $y = a^x$, attempting to find the derivative from first principles gives:

$$\frac{dy}{dx} = \lim_{h \to 0} \frac{a^{x+h} - a^x}{h} = \lim_{h \to 0} \frac{a^x(a^h - 1)}{h} = a^x \lim_{h \to 0} \frac{(a^h - 1)}{h}$$

Once again it is beyond the scope of this course to show that

$\lim_{h \to 0} \dfrac{(a^h - 1)}{h}$ exists, so you can accept that it does and call it k.

This leads to the result that:

$$y = a^x \Rightarrow \frac{dy}{dx} = ka^x$$

In other words, the derivative of any exponential function is proportional to the function itself and $\left. \dfrac{da^x}{dx} \right|_{x=0} = k$

So the number a such that the value of the gradient of $\left. \dfrac{da^x}{dx} \right|_{x=0} = 1$ is actually the number for which $k = 1$.

In other words, you want to find a value for a such that $\displaystyle\lim_{h\to 0} \frac{a^h - 1}{h} = 1$

Suppose $\displaystyle\lim_{h\to 0} \frac{a^h - 1}{h} = 1$

Then for small h,

$$\frac{a^h - 1}{h} \simeq 1$$
$$\Rightarrow a^h \simeq 1 + h$$
$$\Rightarrow a \simeq (1+h)^{\frac{1}{h}}$$

Now let $h = \dfrac{1}{m}$, so that small h corresponds to large m

$$\lim_{h\to 0} \frac{a^h - 1}{h} = 1 \quad\longleftrightarrow\quad a = \lim_{m\to\infty}\left(1 + \frac{1}{m}\right)^m = e$$

> You may wish to explore more invariants in mathematics. Why are they important?

$$\rightarrow \quad y = e^x \Rightarrow \frac{dy}{dx} = e^x$$

The exponential function is **invariant** under differentiation.

5.5 Logarithms and bases

Napier and Briggs and logarithm tables

You can use the properties of indices to calculate products of large numbers. The table shows powers of 3 from 3^{-5} to 3^9. To multiply two number that are powers of 3, look for the corresponding exponents in the table, add them and then use the table by looking at the number whose exponent is the answer.

For example: 243×81
$$5 + 4 = 9$$
$$\therefore 243 \times 81 = 19\,683$$

Similarly, to divide two numbers subtract the exponents.

For example: $6561 \div 729$
$$8 - 6 = 2$$
$$\therefore 6561 \div 729 = 9$$

n	3^n
−5	$\frac{1}{243}$
−4	$\frac{1}{81}$
−3	$\frac{1}{27}$
−2	$\frac{1}{9}$
−1	$\frac{1}{3}$
0	1
1	3
2	9
3	27
4	81
5	243
6	729
7	2187
8	6561
9	19683

This concept gave the Scottish mathematician John Napier the inspiration to invent logarithms. He argued that if you can write any number as a^x, where a is a constant and x varies for different numbers, calculations would be much easier since you would only need to add or subtract powers of x. He needed to find an appropriate number for the base a which was close enough to 1 so that its powers did not grow too quickly. He chose $a = 1 - 10^{-7} = 0.9999999$, and spent the next 20 years painstakingly calculating the value of x, which he called the logarithm, for the numbers 10^{-7} to 4998609. Napier published his results in 1614.

Henry Briggs, a contemporary English mathematician, was fascinated by this work but realized that the tables would be easier to use with a different choice of a. He visited Napier in Scotland and suggested that by choosing $a = 10$ the work would be easier.

▲ **John Napier**
(1550–1617)

Here is what Briggs had in mind:
Let $x = 10^y$

The power to which 10 must be raised to give x is y. This is represented mathematically as:

$$x = 10^y \Rightarrow \log_{10} x = y$$

Napier gave Briggs the job of rewriting the tables using 10 as the base because he considered himself too old to undertake this task. Briggs published his first tables in 1624. Until the introduction of electronic handheld calculators in 1967, students still used logarithm tables for all their calculations.

Logarithms

Logarithms are defined as:

> → $a = b^x \Leftrightarrow x = \log_b a \quad a,b \in \mathbb{R}^+$ and $b \neq 1$

The restrictions on a and b are required in order to obtain sensible results. You should always check that they are satisfied.

Example 11

Write $2^3 = 8$ in logarithmic form.

Answer

$2^3 = 8 \Rightarrow \log_2 8 = 3$ | *The base is 2 and the power is 3.* |
| | *So, the logarithm to base 2 of 8 is 3.* |

Example 12

Write $\log_3 81 = 4$ in exponent form.

Answer

$\log_3 81 = 4 \Rightarrow 3^4 = 81$

Before the appearance of logarithms and calculators people used an abacus or counting frame to help them do calculations. An abacus was first used around 2500 BCE in Mesopotamia and are still being used today. You can use an abacus to do addition, subtraction, multiplication, division and even find square and cube roots very quickly.

Example 13

Solve these equations for x.

a $x = \log_{16} 4$ **b** $\log_7 x = 2$

c $\log_b 1 = x$ **d** $\log_x 32 = \dfrac{5}{2}$

- -

Answers

a $\qquad x = \log_{16} 4$

$\Rightarrow 16^x = 4$

$\therefore x = \dfrac{1}{2}$

Use the definition: $x = \log_b a \Leftrightarrow b^x = a$

$16^{\frac{1}{2}} = 4$

b $\log_7 x = 2 \qquad \Rightarrow x = 7^2 = 49$

c $\log_b 1 = x \qquad \Rightarrow b^x = 1$

$\therefore x = 0$

$b^0 = 1$ *for any value of b.*

d $\log_x 32 = \dfrac{5}{2} \qquad \Rightarrow x^{\frac{5}{2}} = 32$

$\Rightarrow x = 32^{\frac{2}{5}} = 4$

Exercise 5E

1 Write each of these in logarithmic form.

a $5^3 = 125$ **b** $10^3 = 1000$ **c** $27^{\frac{1}{3}} = 3$

d $10^{-3} = 0.001$ **e** $m = n^2$ **f** $a^b = 2$

2 Write each of these in index (exponent) form.

a $\log_3 9 = 2$ **b** $\log_{10} 1\,000\,000 = 6$ **c** $\log_{49} 7 = \dfrac{1}{2}$

d $\log_a 1 = 0$ **e** $\log_4 a = b$ **f** $\log_p q = r$

3 Evaluate:

a $\log_8 64$ **b** $\log_9 3$ **c** $\log_{10} 0.01$

d $\log_{144} 12$ **e** $\log_{37} 1$ **f** $\log_a \sqrt[3]{a}$

4 Solve for x:

a $\log_x 81 = 2$ **b** $\log_3 x = 4$ **c** $\log_{11} 121 = x$

d $\log_x 5 = \dfrac{1}{3}$ **e** $\log_x 16 = \dfrac{2}{3}$ **f** $\log_x 32 = -5$

Properties of logarithms

You derive the properties of logarithms from the properties of indices.

1 $\boxed{\rightarrow a^m \times a^n = a^{m+n} \Rightarrow \log_a x + \log_a y = \log_a xy}$

For which values of a, x and y are these properties valid?

Let $\log_a x = m \Rightarrow a^m = x$

Let $\log_a y = n \Rightarrow a^n = y$

$\therefore xy = a^m \times a^n$

$\qquad = a^{m+n}$

$\log_a xy = m + n$

$\log_a xy = \log_a x + \log_a y$

Substitute $\log_a x$ for m and $\log_a y$ for n.

2 $\rightarrow a^m \div a^n = a^{m-n} \Rightarrow \log_a x - \log_a y = \log_a\left(\dfrac{x}{y}\right)$

> The derivation of this rule is very similar to that of property 1 and is left as an exercise for you.

3 $\rightarrow \log_a x^n = n\log_a x$

$$\log_a x^n = \log_a (\overbrace{x \times x \times ... \times x}^{n}) = \overbrace{\log_a x}^{n} + \log_a x + ... + \log_a x$$
$$= n\log_a x$$

> Repeatedly using property 1

4 $\rightarrow a^0 = 1 \Rightarrow \log_a 1 = 0$

5 $\rightarrow a^1 = a \Rightarrow \log_a a = 1$

6 $\rightarrow -\log_a x = \log_a\left(\dfrac{1}{x}\right)$

$$\log_a\left(\dfrac{1}{x}\right) = \log_a 1 - \log_a x$$
$$= -\log_a x$$

> Using properties 2 and 4

Example 14

Express as a single logarithm:
a $\log_a 5 + 2\log_a 7 - \log_a 35$
b $\log_a p + 2\log_a q - 3\log_a r$
c $1 - \log_a ab$

Answers

a $\log_a 5 + 2\log_a 7 - \log_a 35$

$= \log_a 5 + \log_a 7^2 - \log_a 35$

$= \log_a \dfrac{5 \times 49}{35}$

$= \log_a 7$

> Use property 3
> Use properties 1 and 2.

> It is important to ensure all logarithms use the same base.

b $\log_a p + 2\log_a q - 3\log_a r$

$= \log_a p + \log_a q^2 - \log_a r^3$

$= \log_a \dfrac{pq^2}{r^3}$

c $1 - \log_a ab$

$= 1 - \log_a a - \log_a b$

$= -\log_a b$

> Use property 1
> Use property 5

> When the domain restrictions are not given, you should assume that you are working on the largest possible domain.

Example 15

Evaluate $2(\log 5 + \log 2) - 1$.	
Answer	
$2(\log 5 + \log 2) - 1$ $= 2\log 10 - \log 10 = \log 10 = 1$	$\log 5 + \log 2 = \log 10$ and $1 = \log 10$

> Since base 10 is used so often it is common to just write $\log x$ for $\log_{10} x$

Exercise 5F

1 Express in terms of $\log_a p$ and $\log_a q$:

a $\log_a \dfrac{p^2}{q}$

b $\log_a \sqrt[3]{\dfrac{p}{q^2}}$

2 Express as a single logarithm:

a $\log 4 + 2\log 3 - \log 6$

b $\dfrac{1}{2}\log_a p + \dfrac{1}{4}\log_a q^2$

c $2 - \log 5$

3 Express as a rational number:

a $\log 5 + \log 8 - \log 4$

b $\log_2 48 - \dfrac{1}{3}\log_2 27$

c $2 + \log_5 10 - \log_5 2$

4 Express y in terms of x:

a $3\log y = 2\log x$

b $\log y = \log x + \log 2$

c $\log y - 3\log x = \log 2$

d $\log y = 2 + 3x$

Changing the base of logarithms

Before calculators became commonly available people often used tables of logarithms to do calculations. The tables usually used base 10 and so it was sometimes necessary to know how to change the base.

You can show that

> $\rightarrow \log_a x = \dfrac{\log_b x}{\log_b a}$

Let $y = \log_a x$

$\Rightarrow a^y = x$

$\Rightarrow y \log_b a = \log_b x$

$\Rightarrow y = \dfrac{\log_b x}{\log_b a}$

$\Rightarrow \log_a x = \dfrac{\log_b x}{\log_b a}$

> Take logarithms to base b of both sides.

> Substitute $\log_a x$ for y.

> Why can we take logarithms of both sides?

Example 16

Show that $\log_a b = \dfrac{1}{\log_b a}$

Answer	
$\log_a b = \dfrac{\log_b b}{\log_b a}$	*Change to base b*
$\therefore \log_a b = \dfrac{1}{\log_b a}$	$\log_b b = 1$

Learn this useful result.

Example 17

Evaluate:

a $\log_3 5 \times \log_5 3$ **b** $\log_2 3 \times \log_3 32$

Answers	
a $\log_3 5 \times \log_5 3$	
$= \log_3 5 \times \dfrac{\log_3 3}{\log_3 5} = 1$	
b $\log_2 3 \times \log_3 32$	$32 = 2^5 \Rightarrow \log_5 2^5 = 5\log_2 2$
$= \log_2 3 \times \dfrac{\log_2 32}{\log_2 3} = 5\log_2 2 = 5$	

Exercise 5G

1 Evaluate:

  a $\log_3 2 \times \log_2 81$ **b** $\log_6 10 \times \log 6$

  c $\log_{125} 8 \times \log_8 5$ **d** $\dfrac{1}{\log_2 6} + \dfrac{1}{\log_3 6}$

  e $\dfrac{1}{\log_4 6} + \dfrac{1}{\log_9 6}$ **f** $\log_5 40 - \dfrac{1}{\log_8 5}$

2 Show that:

  a $a^{\log b} = b^{\log a}$ **b** $\dfrac{1}{\log_a ab} + \dfrac{1}{\log_b ab} = 1$

EXAM-STYLE QUESTION

3 Let $p = \log_a x$ and $q = \log_a y$. Express $\log_x a$ in terms of p and $\log_y a$ in terms of q. Hence, show that:

  a $\log_{xy} a = \dfrac{1}{p+q}$ **b** $\log_{\frac{x}{y}} a = \dfrac{1}{p-q}$

Logarithmic and exponential equations

You can use the properties of exponents and logarithms to solve equations as shown in these examples.

Example 18

Solve for x:

a $4^x = 9$ **b** $3^{x-1} = 8$ **c** $3(2^x) = 8$

Answers

a $4^x = 9$

$\Rightarrow x\log 4 = \log 9$

$\therefore x = \dfrac{\log 9}{\log 4} = 1.58$

Take logarithms of both sides.
Use the rule $\log x^n = n \log x$

Using GDC.

b $3^{x-1} = 8$

$\Rightarrow (x - 1) \log 3 = \log 8$

$\Rightarrow x - 1 = \dfrac{\log 8}{\log 3}$

$\Rightarrow x = 1 + \dfrac{\log 8}{\log 3} = 2.89$

Using GDC.

c $3(2^x) = 8$

$\Rightarrow 2^x = \dfrac{8}{3}$

$\Rightarrow x \log 2 = \log\left(\dfrac{8}{3}\right)$

$\Rightarrow x = \dfrac{\log\left(\dfrac{8}{3}\right)}{\log 2} = 1.42$

Using GDC.

GDC help on CD:
Alternative demonstrations for the TI-84 Plus and Casio fx-9860GII GDCs are on the CD.

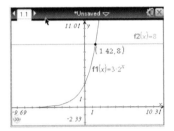

Example 19

Solve the equation $3^{2-x} = 9^{2x}$

Answer

$3^{2-x} = 9^{2x}$

$\Rightarrow 3^{2-x} = (3^2)^{2x}$

$\Rightarrow 3^{2-x} = 3^{4x}$

$\Rightarrow 2 - x = 4x$

$\Rightarrow x = \dfrac{2}{5}$

$3^2 = 9$
Use the rule: $(a^m)^n = a^{mn}$
Equate the exponents

Example 20

Solve the equation $2\sqrt{8} = 4^{2x}$

Answer

$2\sqrt{8} = 4^{2x}$

$\Rightarrow 2 \times 2^{\frac{3}{2}} = (2^2)^{2x}$

$\Rightarrow 2^{\frac{5}{2}} = 2^{4x}$

$\Rightarrow \dfrac{5}{2} = 4x$

$\therefore x = \dfrac{5}{8}$

$\sqrt{8} = 4^{\frac{1}{2}} = (2^2)^{\frac{1}{2}} = 2^{\frac{3}{2}}$

Equate the exponents.

Exercise 5H

1 Solve these equations.

    a $5^x = 7$ **b** $4^{2x-1} = 3$

2 Solve $(2^x)(5^x) = 0.01$

The next set of examples shows how exponents and logarithms are used in more complicated equations.

Example 21

Solve the equation $\log_6 x + \log_6 (x - 5) = 1$

Answer	
$\log_6 x + \log_6 (x - 5) = 1$	
$\Rightarrow \log_6 x(x - 5) = 1$	
$\Rightarrow x(x - 5) = 6$	
$\Rightarrow x^2 - 5x - 6 = 0 \Rightarrow$	
$(x - 6)(x + 1) = 0$	
$\therefore x = 6$ or $x = -1$	*Check whether the results are actually*
But $x \neq -1$; therefore, $x = 6$	*solutions of the equation. $\log_6(-1)$*
	does not make sense so $x \neq -1$

Example 22

Solve the equation $\log_3 x - 4\log_x 3 + 3 = 0$

Answer	
$\log_3 x - 4\log_x 3 + 3 = 0$	$\log_x 3 = \dfrac{1}{\log_3 x}$
$\Rightarrow \log_3 x - \dfrac{4}{\log_3 x} + 3 = 0$	
Let $y = \log_3 x$	
$y - \dfrac{4}{y} + 3 = 0$	
$\Rightarrow \quad y^2 + 3y - 4 = 0$	
$\Rightarrow (y + 4)(y - 1) = 0$	
$\Rightarrow y = 1$ or $y = -4$	
$\Rightarrow \log_3 x = 1$ or $\log_3 x = -4$	
$\Rightarrow x = 3$ or $x = 3^{-4} = \dfrac{1}{81}$	

Example 23

Solve the equation $3^{2x+1} + 3^x - 2 = 0$

Answer

$3^{2x+1} + 3^x - 2 = 0$

$\Rightarrow 3(3^x)^2 + (3^x) - 2 = 0$

Let $(3^x) = y$

$\Rightarrow \quad 3y^2 + y - 2 = 0$

$\Rightarrow (3y - 2)(y + 1) = 0$

$\Rightarrow y = \dfrac{2}{3}$ *$y = \dfrac{2}{3}$ or $y = -1$*

$\Rightarrow 3^x = \dfrac{2}{3}$ *Since $3^x > 0$ for all $x \in \mathbb{R}$, $3^x \neq -1$.*

$\Rightarrow x \log 3 = \log \dfrac{2}{3}$

$\Rightarrow x = \dfrac{\log \dfrac{2}{3}}{\log 3} = -0.369$ *Using GDC.*

Example 24

Solve these simultaneous equations.

$\log_3 x + 4\log_9 y = 2$

$2\log_4 x + \log_2 y = 1$

Answer

$\log_3 x + 4\log_9 y = 2$ *Change to base 3.*

$\Rightarrow \log_3 x + 4\dfrac{\log_3 y}{\log_3 9} = 2$ *$\log_3 9 = 2$*

$\Rightarrow \log_3 x + \dfrac{4\log_3 y}{2} = 2$

$\Rightarrow \log_3 xy^2 = 2 \quad \Rightarrow xy^2 = 9$

$2\log_4 x + \log_2 y = 1$ *Change to base 2.*

$\Rightarrow 2\dfrac{\log_2 x}{\log_2 4} + \log_2 y = 1$ *$\log_2 4 = 2$*

$\Rightarrow \log_2 x + \log_2 y = 1 \quad \Rightarrow xy = 2$

Combining the two results:

$\dfrac{xy^2}{xy} = \dfrac{9}{2} \Rightarrow y = \dfrac{9}{2}$

$x = \dfrac{2}{y} \Rightarrow x = \dfrac{4}{9}$ *Using $xy = 2$ and substituting $y = \dfrac{9}{2}$*

Exercise 5I

1 Solve these equations.
   a $2^{3x} = 5$ **b** $3^x(3^{x-1}) = 10$

2 Solve these equations.
   a $4\log_3 x = \log_x 3$ **b** $3\log_2 x + \log_2 27 = 3$

3 Solve the equation $9^x - 6(3^x) - 16 = 0$

4 Solve the equation $\log_4 x + 12\log_x 4 - 7 = 0$

EXAM-STYLE QUESTIONS

5 Solve the equation $5^{x+1} + \dfrac{4}{5^x} - 21 = 0$

6 Solve the equation $\log_3 x + \log_x 9 - 3 = 0$

7 Solve the equation $3 \times 9^x - 2 \times 4^x = 5 \times 6^x$ giving your answer to three significant figures.

8 Solve these simultaneous equations.
   $6\log_2 x + 6\log_8 y = 7$ $4\log_4 x + 4\log_2 y = 9$

9 Solve these simultaneous equations.
   $2\log_x y = 1$ $xy = 125$

10 Solve these simultaneous equations.
   $y\log_2 8 = x$ $2^x + 8^y = 64$

11 Solve these simultaneous equations.
   a $\log_5 x = y = \log_{25}(2x - 1)$
   b $\log(x + y) = 0$ $2\log x = \log(y + 5)$

Investigation – e^{π} or π^e – which is the greater?

Without using a calculator determine which is greater e^{π} or π^e

- Write the equation of the tangent line to $y = \ln x$ at a general point $(a, \ln a)$
- Using a graph, explain why $\ln x < \dfrac{x}{e}$ for all x
- What is the relationship between $\ln(x^e)$ and x for all $x > 0$, $x \neq e$?
- Make your conclusions.

5.6 Logarithmic functions and their behavior

In Chapter 2 you studied functions and their properties as well as the conditions which allow you to find the inverse of a function, if it exists.

Inverse of an exponential function

Since the exponential function is one-to-one, continuous and defined for all values of $x \in \mathbb{R}$, its inverse exists. Also you know that the inverse of a function is given by the image of the function reflected in the line $y = x$, so you would expect the inverse of an exponential function to pass through the point $(1, 0)$ and to have the y-axis as a vertical asymptote.

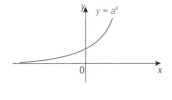 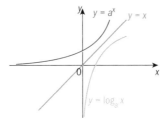

$y = a^x \Rightarrow x = \log_a y$

$\therefore f(x) = a^x$

$\Rightarrow f^{-1}(x) = \log_a x$

Note that the y-axis is a vertical asymptote.
The domain of the function is $x \in \mathbb{R}$, $x > 0$.

> → The inverse of the exponential function $y = e^x$ is given by $y = \ln x$, which is also called the natural logarithm function. Similarly, the inverse of $y = 10^x$ is given by $y = \log x$.

Notation: $\log_e x = \ln x$

Example 25

Show that $\log x = \log e \times \ln x$.

Answer

$\ln x = \dfrac{\log x}{\log e}$ *Change from base e to base 10.*

$\therefore \log x = \log e \times \ln x$

Example 26

The diagram shows the graph of
$y = \log_a x + k$.
Find k in terms of x.

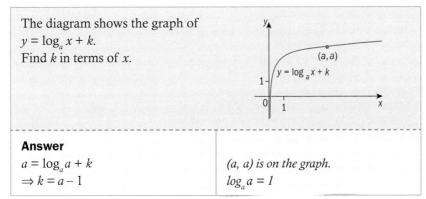

Answer

$a = \log_a a + k$ *(a, a) is on the graph.*

$\Rightarrow k = a - 1$ $\log_a a = 1$

Example 27

Identify the domain, any asymptotes and intercepts of these graphs, and hence sketch the functions.

a $y = \log_2(x - 1) + 1$ **b** $y = \ln(2 - x) - 3$

Answers

a Asymptote: $x = 1$ is a vertical asymptote

Domain: $\{x \mid x \in \mathbb{R}, x > 1\}$

For x-intercept:

$y = 0 \Rightarrow \log_2(x - 1) + 1 = 0$

$\therefore \log_2(x - 1) = -1$

$\Rightarrow x - 1 = 2^{-1}$

$\Rightarrow x = 1 + \dfrac{1}{2} = \dfrac{3}{2}$

x-intercept is $\left(\dfrac{3}{2}, 0\right)$

$\displaystyle\lim_{x \to 1^+} \log_2(x - 1) = -\infty$

$\log_2(x - 1)$ is not defined for $x - 1 \le 0$

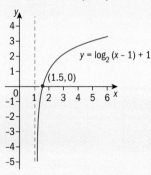

b Asymptote: $x = 2$ is a vertical asymptote.

Domain: $\{x \mid x \in \mathbb{R}, x < 2\}$

For x-intercept:

$y = 0 \Rightarrow \ln(2 - x) - 3 = 0$

$\therefore \ln(2 - x) = 3$

$\Rightarrow 2 - x = e^3$

$\therefore x = 2 - e^3$

x-intercept is $(2 - e^3, 0)$

For y-intercept:

$x = 0 \Rightarrow y = \ln 2 - 3$

y-intercept is $(0, \ln 2 - 3)$

$\displaystyle\lim_{x \to 2^-} \ln(x - 2) = -\infty$

$\ln(2 - x)$ is not defined for $2 - x \le 0$

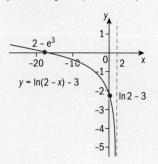

Exercise 5J

1 On the same set of axes, sketch the graphs of $f(x) = e^x$ and $f^{-1}(x) = \ln x$. Hence, state the domain and range of each function.

2 Given that $f(x) = a^x$, find $f^{-1}(x)$ and $f \circ f^{-1}(x)$ and use your answers to show that $a^{\log_a x} = x$

3 On the same set of axes, sketch the graphs of $f(x) = -\ln x$ and $g(x) = |\ln x|$

4 On the same set of axes, sketch the graphs of $f(x) = |\ln x|$ and $g(x) = \ln|x|$

5 On separate sets of axes, draw the graphs of
$y = \log_3(x - 3)$ and $y = (\log_3 x) - 3$
Identify the two geometric transformations of $y = \log_3 x$ which produce these functions.

6 Identify the domain, any asymptotes and intercepts of these graphs, and hence sketch the functions.
  a $y = \ln(x - 1) - 1$ **b** $y = \log_3(9 - 3x) + 2$

> The answers to question 1 are important and a key fact.

Extension material on CD:
Worksheet 5

5.7 Derivatives of exponential and logarithmic functions

In section 5.4 you saw that the exponential function is invariant under differentiation, that is $\dfrac{d}{dx}(e^x) = e^x$

To differentiate the composite function $y = e^{f(x)}$, use the chain rule.

Let $u = f(x)$
Then $y = e^u$ and $\dfrac{dy}{dx} = \dfrac{dy}{du} \times \dfrac{du}{dx}$

Now, $\dfrac{dy}{du} = e^u$ and $\dfrac{du}{dx} = f'(x)$ $\Rightarrow \dfrac{dy}{dx} = e^u \times f'(x)$

So

> → $y = e^{f(x)} \Rightarrow \dfrac{dy}{dx} = f'(x)\, e^{f(x)}$

These examples illustrate the techniques used to find the derivatives of exponential functions.

- Investigate why and how a logarithmic scale is used to measure quantities such as acidity (pH), earthquake magnitude, brightness of stars and loudness of sound.
- What is the Weber–Fechner law in physiology?

Example 28

Differentiate these with respect to x.

a $y = e^{7x^2}$ **b** $y = e^{-4x}$ **c** $y = \dfrac{1}{e^{(x^2+1)}}$

Answers

a $\dfrac{dy}{dx} = f'(x)e^{f(x)}$ so

$\dfrac{d(e^{7x^2})}{dx} = 14x\,e^{7x^2}$

Alternatively, use the chain rule.

$u = 7x^2 \qquad y = e^u$

$\dfrac{du}{dx} = 14x \qquad \dfrac{dy}{du} = e^u$

$\therefore \dfrac{dy}{dx} = 14x\,e^{7x^2}$

b $\dfrac{dy}{dx} = -4e^{-4x}$

$\dfrac{dy}{dx} = f'(x)e^{f(x)}$

c $y = \dfrac{1}{e^{(x^2+1)}} = e^{-(x^2+1)}$

$\therefore \dfrac{dy}{dx} = -2xe^{-(x^2+1)} = \dfrac{-2x}{e^{(x^2+1)}}$

Example 29

Given that $y = x^2e^{5x}$, find $\dfrac{dy}{dx}$. Hence, show that $\dfrac{dy}{dx} - 5y = \dfrac{2y}{x}$

Answer

$y = x^2\,e^{5x}$

$\dfrac{dy}{dx} = 5x^2e^{5x} + 2xe^{5x}$

$\dfrac{dy}{dx} = 5\left(x^2e^{5x}\right) + \dfrac{2}{x}\left(x^2e^{5x}\right)$

$\dfrac{dy}{dx} = 5y + \dfrac{2}{x}y \quad \therefore \dfrac{dy}{dx} - 5y = \dfrac{2y}{x}$

Use the product rule.

$\dfrac{d(uv)}{dx} = u\dfrac{dv}{dx} + v\dfrac{du}{dx}$

$u = x^2 \Rightarrow \dfrac{du}{dx} = 2x$

$v = e^{5x} \Rightarrow \dfrac{dv}{dx} = 5e^{5x}$

Exercise 5K

1 Differentiate these functions with respect to x.

a $y = \dfrac{3}{2}e^{x^2}$ **b** $y = -\dfrac{5}{e^{3x-1}}$ **c** $y = e^{4x-1} + 4$

d $y = e^x + \dfrac{1}{e^x}$ **e** $y = e^{-(1-3x)}$ **f** $y = 2e^{\sqrt{x}}$

2 Use the product rule to find the derivatives of these functions with respect to x.

a $y = xe^x$ **b** $y = \dfrac{x^2}{e^x}$ **c** $y = \dfrac{e^{2x}}{\sqrt{x}}$ **d** $\sqrt{x}e^{\sqrt{x}}$

3 Use the quotient rule to find the derivatives of these functions with respect to x.

a $y = \dfrac{e^{2x}}{\sqrt{x}}$ **b** $y = \dfrac{1-x^2}{e^x}$ **c** $y = \dfrac{e^{3x}}{1+x}$ **d** $y = \dfrac{1+e^x}{1-e^x}$

4 Differentiate these functions with respect to x.

a $y = \dfrac{xe^x}{1+e^x}$ **b** $y = (1 + e^x)^2$ **c** $y = \sqrt{1+e^{-x}}$

d $y = \dfrac{x+e^x}{e^{-x}}$ **e** $y = \dfrac{e^x + e^{-x}}{e^x - e^{-x}}$

5 $f(x) = xe^x$ for $-3 \le x \le 3$. Show that $f(x)$ has one stationary point and one point of inflexion.

> Maxima, minima and points of inflexion were studied in Chapter 4.

  a Identify the nature of the stationary point and its position.
  b Find the coordinates of the point of inflexion, and hence sketch the graph of $f(x)$.
  c Find the equation of the tangent at the point of inflexion.
  d Find the point where this tangent meets the x-axis.
  e Calculate the area bounded by this tangent and the x- and y-axes.

The derivative of $y = \ln x$ and $y = a^x$

By definition:

$$y = \ln x \implies x = e^y$$

Differentiate implicitly with respect to x:

$$1 = e^y \frac{dy}{dx} \implies \frac{dy}{dx} = \frac{1}{e^y} = \frac{1}{x}$$

> Another way of obtaining this result is by using the property $\dfrac{dy}{dx} = \dfrac{1}{\frac{dx}{dy}}$. This is left as an exercise for you.

Therefore

> $\dfrac{d}{dx}(\ln x) = \dfrac{1}{x}$

In section 5.4 you used differentiation from first principles to show that $\dfrac{d}{dx}(a^x) = ka^x$, where $k = \lim\limits_{h \to 0} \dfrac{a^h - 1}{h}$

Now use the result $\dfrac{d}{dx}(\ln x) = \dfrac{1}{x}$ to evaluate k.

Let $y = a^x$

Then $\ln y = x \ln a$

Differentiating implicitly with respect to x gives:

$$\frac{1}{y}\frac{dy}{dx} = \ln a \implies \frac{dy}{dx} = y \ln a = (\ln a)a^x$$

So

> $\dfrac{d}{dx}(a^x) = (\ln a)a^x$

But since $\dfrac{d}{dx}(a^x) = ka^x$, where $k = \lim\limits_{h \to 0} \dfrac{a^h - 1}{h}$, you can now say that

$$\lim_{h \to 0} \frac{a^h - 1}{h} = \ln a$$

Also note that $\dfrac{d}{dx}(a^x) = (\ln a)a^x$

$$\Rightarrow \dfrac{d}{dx}\left(a^x\right)\Big|_{x=0} = \ln a$$

For an exponential function $y = a^x$, the gradient at the y-intercept is always equal to $\ln a$.

Note that this result leads to a previously obtained result in this chapter, that is.

$$\dfrac{d}{dx}e^x\Big|_{x=0} = \ln e = 1$$

Example 30

Find the derivatives of these functions with respect to x.

a $y = 4^x$ **b** $y = 4^{2x}$ **c** $y = x \cdot 4^x$

Answers

a $y = 4^x \Rightarrow \dfrac{dy}{dx} = (\ln 4)4^x$ $\qquad$ $\dfrac{d}{dx}(a^x) = (\ln a)\,a^x$

b $y = 4^{2x} \Rightarrow \dfrac{dy}{dx} = (\ln 4)4^{2x} \times 2$ $\qquad$ *Use the chain rule.*

$\therefore \dfrac{dy}{dx} = (2\ln 4)4^{2x}$ $\qquad$ *You could also say that $\dfrac{dy}{dx} = (\ln 16)4^{2x}$*

c $y = x \cdot 4^x$ $\qquad$ *Use the product rule.*

$\Rightarrow \dfrac{dy}{dx} = x(\ln 4)4^x + 4^x$

$\Rightarrow \dfrac{dy}{dx} = 4^x(x(\ln 4) + 1)$

Example 31

Differentiate these functions with respect to x.

a $y = \ln(4x)$ **b** $y = \ln(1 + x^2)$ **c** $y = \log_a x$ **d** $y = \ln(f(x))$

Answers

a $y = \ln 4x$ $\qquad$ *Use the chain rule.*

$\Rightarrow \dfrac{dy}{dx} = 4 \times \dfrac{1}{4x} = \dfrac{1}{x}$ $\qquad$ *Note:*

b $y = \ln(1 + x^2)$ $\qquad$ $\dfrac{d}{dx}\ln(kx) = \dfrac{d}{dx}(\ln k + \ln x) = \dfrac{1}{x},$

$\Rightarrow \dfrac{dy}{dx} = \dfrac{1}{1 + x^2} \times 2x = \dfrac{2x}{1 + x^2}$ $\qquad$ *where k is a constant.*

c $y = \log_a x$ $\qquad$ *Change the base.*

$\Rightarrow y = \dfrac{\ln x}{\ln a}$ $\qquad$ $\dfrac{\ln x}{\ln a} = \dfrac{1}{\ln a} \times \ln x,$ *where $\dfrac{1}{\ln a}$ is a*

$\Rightarrow \dfrac{dy}{dx} = \dfrac{1}{x\ln a}$ $\qquad$ *constant.*

d $y = \ln(f(x))$ $\qquad$ *Use the chain rule.*

$\Rightarrow \dfrac{dy}{dx} = \dfrac{1}{f(x)} \times f'(x) = \dfrac{f'(x)}{f(x)}$

$\rightarrow \Rightarrow \dfrac{dy}{dx}(\log_a x) = \dfrac{1}{x\ln a}$

Learn this important result:
if $y = \ln(f(x))$

$$\dfrac{dy}{dx} = \dfrac{f'(x)}{f(x)}$$

Exercise 5L

1 Find the derivatives of these functions with respect to x.

   a $y = 5^{3x}$ **b** $y = \ln(4x + 1)$

2 Differentiate these functions with respect to x.

   a $y = 1 + 2\ln x$ **b** $y = \dfrac{1}{\ln x}$

Example 32

Find the derivative with respect to x of $y = x^x$

Answer

$y = x^x \Rightarrow \ln y = x\ln x$	*Differentiate implicitly with respect to x.*
$\dfrac{1}{y}\dfrac{dy}{dx} = x \times \dfrac{1}{x} + \ln x$	*Use the product rule to differentiate $x\ln x$.*
$\Rightarrow \dfrac{dy}{dx} = y(1 + \ln x) = x^x(1 + \ln x)$	

Example 33

The equation of a curve is given by $y = e^{e^x}$.
a Show that the curve has no maxima or minima for $x \in \mathbb{R}$.
b Find the equation of the tangent to the curve at the point where $x = 0$.
c Find the equation of the normal to the curve at the point where $x = 0$.
d Sketch the two lines and hence find the area of the region bounded by the two lines and the x-axis.

Answers

a $y = e^{e^x}$	*To locate maxima or minima find for what values the first derivative is zero.*
$\Rightarrow \dfrac{dy}{dx} = e^x e^{e^x} \neq 0$ for $x \in \mathbb{R}$	
Therefore, there are no maxima or minima on the curve.	
$\dfrac{d^2y}{dx^2} = e^x \times e^x e^{e^x} + e^x \times e^{e^x}$	*Use the product rule.*
$\Rightarrow \dfrac{d^2y}{dx^2} = e^x e^{e^x}(e^x + 1) \neq 0$, for $x \in \mathbb{R}$	*To locate points of inflexion find for what values the second derivative is zero.*
Therefore, there are no points of inflexion on the curve.	
b $y = e^{e^x} \Rightarrow y = e$ when $x = 0$	
$\Rightarrow \dfrac{dy}{dx} = e^x e^{e^x} = e$ when $x = 0$	
The equation of the tangent at $(0, e)$ is	
$y = ex + e$	

▶ Continued on next page

c The gradient of the normal at $(0, e)$ is $-\dfrac{1}{e}$.

So, the equation of the normal at $(0, e)$ is

$$y = -\frac{x}{e} + e$$

The two lines are shown on the graph.

d

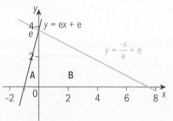

Total area $= \dfrac{1}{2} \times (1 + e^3) \times e = \dfrac{e}{2}(1 + e^2)$

$y = -\dfrac{x}{e} + e$ *intersects the x-axis at* e^2.

Exercise 5M

1 Find the derivatives of these functions with respect to x.

    a $y = x^2 \ln x$ **b** $y = x a^x$

2 Differentiate these functions with respect to x.

    a $y = \ln\left(\dfrac{1}{x}\right)$ **b** $y = \ln x^2$ **c** $y = \dfrac{\ln x}{x}$

3 Differentiate these equations with respect to x.

    a $x^y = e^x$ **b** $y = x^{2x}$

4 The equation of a curve is given by $y = e^x(x - 1)$

    a Show that this curve has only one stationary point.

    b Identify this stationary point and find its coordinates.

    c Find the coordinates of the point where this curve crosses the x-axis.

    d Sketch the curve.

EXAM-STYLE QUESTION

5 A curve has equation $f(x) = \ln(1 + x^2)$

    a For what values of x is this function defined?

    b Explain why the y-axis is a line of symmetry for this function.

    c Find $f'(x)$ and $f''(x)$ and use your results to show that the curve has one minimum and two points of inflexion.

    d Sketch the curve.

    e Show that the normals and tangents at the two points of inflexion form a square and find its area.

5.8 Angles, arcs and areas

An angle is formed when two rays meet at a point. The point is called the vertex and the two rays are called the sides of the angle.

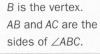

> B is the vertex.
> AB and AC are the sides of ∠ABC.

An angle is a measure of rotation about the vertex for one ray to coincide with the other. Then, if two rays are coincident, both of these statements are true:

- The angle they make is zero.
- The angle they make is one revolution (a full turn).

A circle and two radii are shown in the diagram. Start with OB on OA. The angle is zero. As OB turns anticlockwise about the center, O, the angle AOB gets bigger until B once more reaches A, when it has gone through one revolution. In fact, if OB continues to rotate about O, each time it goes through another revolution OB will coincide with OA. When B first reaches C, the angle is half a revolution.

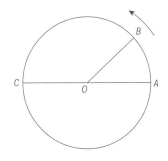

> It is thought that the Sumerians were responsible for dividing a full turn into 360 degrees (360°). The Sumerians (Babylonians) had a calendar of 12 months, each month having exactly 30 days. This was based on their observation that it took the sun about 360 days to complete its circular track across the sky. The Sumerians also used base 60 in their number system.

Another unit for measuring angles is the **radian**. To measure the angle $A\hat{O}B = \theta$ draw a circle, radius r, at O and measure the length of the arc $AB = S$. The angle θ in radius is defined as

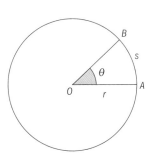

$$\rightarrow \quad \theta = \frac{s}{r} \iff s = \theta r$$

1 radian is the angle which is subtended by an arc of length equal to the radius of the circle.

> The definition does not depend on the size of the circle since all circles are similar.

Since the circumference of a circle $= 2\pi r$, one full turn $= 360° = 2\pi$ radians.

$$\rightarrow \quad 1 \text{ radian} = \frac{180°}{\pi} \approx 57° \qquad 1° = \frac{\pi}{180} \text{ radians}$$

Investigation – converting between degrees and radians

The first diagram shows a unit circle, that is, a circle whose radius is 1 unit long. The radius *OB* is rotated by one radian six consecutive times and the angle measure at each stage, in radians, is marked on the circumference.

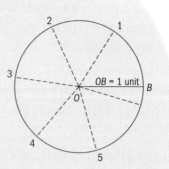

- What is the radian measure if the radius is rotated through one whole revolution? Give reasons for your answer based on the definition of a radian.
- The radius *OB* is now rotated anticlockwise about *O* by half a revolution, as shown in the second diagram. What is the angle measure in radians?
- What is the angle measure in radians if *OB* is rotated clockwise about *O* by half a revolution? How would you distinguish between clockwise and anticlockwise rotation?
- What is the angle measure if *OB* is rotated anticlockwise about *O* by
  i a quarter of a revolution **ii** three quarters of a revolution?
- Use your results to complete the conversion table.

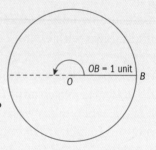

Angle measure conversion table	
Degrees (°)	Radians
30	
	$\dfrac{\pi}{4}$
60	
	$\dfrac{\pi}{2}$
75	
	$\dfrac{2\pi}{3}$
135	
150	
	π
210	
	$\dfrac{5\pi}{4}$
240	
	$\dfrac{3\pi}{2}$
285	
	$\dfrac{5\pi}{3}$
315	
	$\dfrac{11\pi}{6}$
360	

2π radians = 360°

So 1 radian = $\dfrac{180}{\pi}$

Slices of pi: areas and perimeters of sectors

Two radii of a circle divide it into two sectors. The one with the larger area is called the major sector and the smaller one is called the minor sector.

Investigation – areas and perimeters of sectors

A diameter divides a circle into two congruent sectors, and two perpendicular diameters divide a circle into four congruent sectors. Copy and complete this table.

1	Number of diameters	Number of congruent sectors	Angle subtended by minor arc (radians)	Area of one sector	Length of minor arc
	1	2	π	$\frac{1}{2}\left(\pi r^{2}\right)=\left(\frac{\pi}{2}\right)r^{2}$	$\frac{1}{2}\left(2\pi r\right)=\pi r$
	2	4	$\frac{\pi}{2}$	$\frac{1}{4}\left(\pi r^{2}\right)=\left(\frac{\pi}{4}\right)r^{2}$	$\frac{1}{4}\left(2\pi r\right)=\frac{\pi}{2}r$

- Use your results to make a conjecture for the area of a sector in terms of the angle.
- Use your results to make a conjecture for the length of an arc in terms of the angle.

Your results from the Investigation should have led to:

> → Area of sector = $\frac{1}{2}\theta r^2$
>
> Length of arc = θr, where θ is measured in radians.

You can derive similar formulae when θ is measured in degrees.

Example 34

A sector of a circle has radius 4 cm and central angle 0.25 radians. Find its arc length and area.

Answer

Area = $\frac{1}{2} \times 0.25 \times 16 = 2\,cm^2$	*Area of sector* = $\frac{1}{2}\theta r^2$
Arc length = $0.25 \times 4 = 1\,cm$	*Length of arc* = θr

Example 35

A planet is in opposition when it is directly opposite the Sun from our viewpoint on Earth. During opposition in March 2012 Mars is closer to the Earth than at any other time at a distance of approximately 9.98×10^8 km. Mars has an approximate diameter of 6.8×10^3 km. What is the angle subtended by Mars when viewed during closest approach?

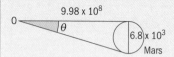

Answer

$\theta = \dfrac{6.8 \times 10^3}{9.98 \times 10^8} = 6.8 \times 10^{-6}$ radians (2 sf)	*Arc length* = θr *So* $\theta = \dfrac{arc\ length}{r}$

Example 36

The diagram shows two arcs which subtend the same angle.
Given that $OA = 3$ cm, $OP = 5$ cm and $\angle AOB = 0.8$ radians, find the area which is **not** shaded and its perimeter.

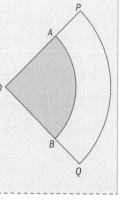

▶ Continued on next page

Answer

Area required = area of sector OPQ – area of sector OAB $= \dfrac{1}{2}\,\theta(OP)^2 - \dfrac{1}{2}\,\theta(OA)^2$ $= \dfrac{1}{2}\,\theta[(OP)^2 - (OA)]^2$ $= \dfrac{1}{2} \times 0.8 \times (5^2 - 3^2) = 6.4 \text{ cm}^2$ Perimeter = $2AP$ + arc length AB $\qquad\qquad$ + arc length PQ $\qquad\qquad = 2 \times (5-3) + 0.8 \times 3 + 0.8 \times 5$ $\qquad\qquad = 10.4 \text{ cm}$	*Area of sector* $= \dfrac{1}{2}\,\theta r^2$ *Length of arc* $= \theta r$

Example 37

a The diagram shows a square $ABCD$ with sides a cm long which just fits inside a circle. Find the area of the shaded segment.

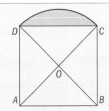

b On each of the sides of the square, semicircles are drawn to form four crescents. Show that the area shaded in dark grey is equal to the area of the square .

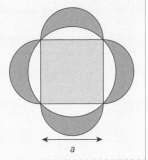

Answers

a Length of diagonal = $\sqrt{2}\,a$ $\qquad \therefore$ radius of sector $= \dfrac{\sqrt{2}}{2}\,a$ Area of triangle $DOC = \dfrac{a^2}{4}$ $\angle DOC = \dfrac{\pi}{2}$ area of sector $= \dfrac{1}{2} \times \dfrac{\pi}{2} \times \left(\dfrac{\sqrt{2}}{2}\,a\right)^2 = \dfrac{\pi a^2}{8}$ Area of segment = area of sector – area of triangle $\qquad\qquad = \dfrac{\pi a^2}{8} - \dfrac{a^2}{4} = \dfrac{a^2}{4}\left(\dfrac{\pi}{2} - 1\right)$	*Use Pythagoras' theorem* *Diagonals divide a square into four congruent triangles.*

▶ Continued on next page

b Area of crescent = area of semicircle – area of segment

$$= \frac{1}{2}\pi\left(\frac{a}{2}\right)^2 - \frac{a^2}{4}\left(\frac{\pi}{2}-1\right) = \frac{a^2}{4}$$

Area of four crescents $= 4 \times \dfrac{a^2}{4} = a^2 =$ area of square

Area of semicircle is half the area of a circle with radius $\dfrac{a}{2}$.

Exercise 5N

1 For each figure, find the length of the arcs and the area of the sector.

a

5 cm
$\dfrac{\pi}{4}$

b
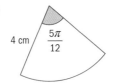
4 cm
$\dfrac{5\pi}{12}$

c
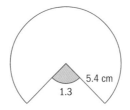
5.4 cm
1.3

2 The diagram shows an aerial view of a swimming pool, *ABQP*, formed by two sectors of a circle. The angle *AOB* is 0.8 radians. Find the surface area and the perimeter of the pool.

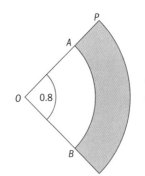

$OA = 5\,\mathrm{m}$
$OP = 9\,\mathrm{m}$

EXAM-STYLE QUESTION

3 In a special offer, three cans of cat food are sold for the price of two. The cans are wrapped by a plastic foil as shown in the diagram. Each can has a diameter of 7.5 cm.
Find the length of plastic foil required to hold the cans together. (Assume that no overlapping is required.)

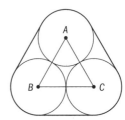

4 The Greek mathematician Eratosthenes noticed that at noon on a summer solstice the rays of the sun shone straight down a deep well in Syene (now known as Aswan). At the same time, in Alexandria, the rays made an angle of 7.2° with the vertical, as shown in the diagram.
The distance between Alexandria and Syenne was 5000 stadia. One stadion measures approximately 185 m. Eratosthenes used this information to calculate the circumference of the Earth.

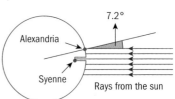

a What result did Eratosthenes obtain?

b The circumference of the Earth is 40 008 km to 5 sf.
What was Eratosthenes' percentage error?

5 The diagram shows a right-angled triangle ABC with the right angle at B. Two crescents are formed by constructing semicircles with AC, AB and BC as diameters.
Find the sum of the areas of the segments $APBA$ and $BQCB$.
Hence, show that the sum of the areas of the two crescents is equal to the area of the triangle.

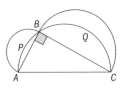

Review exercise

1 Simplify the expression $\dfrac{9^{2n+2} \times 6^{2n-3}}{3^{5n} \times 6 \times 4^{n-2}}$

2 Evaluate $\dfrac{8^{\frac{2}{3}} + 4^{\frac{3}{2}}}{16^{\frac{3}{4}}}$

3 Solve these equations.
   a $9^x - 12(3^x) + 27 = 0$
   b $3^x - \dfrac{9}{3^x} = 8$

4 Solve these logarithmic equations.
   a $\log_a x + \log_a 3 - \log_a 7 = \log_a 12$ **b** $\log_4 x - \log_4 5 = \dfrac{5}{2}$
   c $\log_3 x - \dfrac{6}{\log_3 x} = 1$ **d** $\log_7 x + 2\log_x 7 = 3$

5 Solve these simultaneous equations.
   a $xy = 81$
    $3 \log_x y = 1$
   b $y \log_2 4 = x$
    $2^x + 4^y = 512$
   c $\ln 8 + \ln(x - 6) = 2 \ln y$
    $2y - x = 2$

6 Simplify these expressions.
   a $\log y + \log \dfrac{1}{y}$
   b $\dfrac{\log x^5 - \log x^2}{3\log x + \log \sqrt{x}}$
   c $\ln(\ln x^2) - \ln(\ln x)$

7 Find the value of x given that
$\log_2 x + \log_2 x^2 + \log_2 x^3 + \ldots + \log_2 x^m = 3m(m + 1)$

8 Given that $y = 5e^{2x} + 8e^{-2x}$, show that $\dfrac{d^2 y}{dx^2} = 4y$

9 Given that $y = e^{3x}(2 + 5x)$, show that $\dfrac{d^2 y}{dx^2} - 6\dfrac{dy}{dx} + 9y = 0$

10 Find the value of x which satisfies the equation $e^x - e^{-x} = 4$
Hence, show that for this value of x
$e^x + e^{-x} = 2\sqrt{5}$

11 Show that the function $f(x) = \dfrac{4e^x}{(e^x + 1)^2}$ has a maximum point at $(0, 1)$. Find the x-coordinates of the two points of inflexion.

12 a For what values of x is the function $f(x) = (\ln x)^2$ defined?

b Evaluate $f'(x)$ and $f''(x)$.

c Use your results to show that the function has a minimum and a point of inflexion and find their coordinates.

d Find the equation of the tangent which passes through the point of inflexion.

e Find the area of the region enclosed by this tangent, the y-axis and the line $y = 1$

13 Figure 1 shows an equilateral triangle ABC with sides of length $2a$. An arc of a circle with center A and tangent BC is drawn as shown. Calculate the shaded area and call it S_1.

Another segment is formed as shown in Figure 2. Find the area of this second segment S_2.

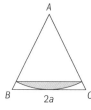

Fig.1

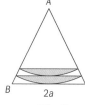

Fig.2

The process is repeated and a third segment is formed as shown in Figure 3. Calculate the area of the third segment. Hence, show that the areas S_1, S_2 and S_3 form a geometric sequence and find the common ratio.

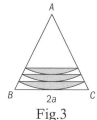

Fig.3

The process of constructing segments can continue indefinitely. What would be the total area of segments formed?

14 Seven circular glass coasters, each with diameter 8 cm, are placed on a table so that they are touching each other as shown in the diagram. Find the area of the space between the coasters.

CHAPTER 5 SUMMARY

If a line segment AB is divided into two segments by a point P such that $\dfrac{AB}{AP} = \dfrac{AP}{PB}$, we say that the line is divided in the golden ratio. The golden ratio is denoted by φ.

Recursive functions

An initial value u_0 is given and a function $u_n = F(u_{n-1})$ is defined to give a sequence.
E.g. $u_0 = 1$, $u_n = 3u_{n-1} + 5$

Continued on next page

Exponents and their properties

- $a^m \times a^n = a^{m+n}$
- $(a^m)^n = a^{mn}$
- $a^{\frac{1}{m}} = \sqrt[m]{a}$
- $a^{\frac{m}{n}} = \sqrt[n]{a^m} = \left(\sqrt[n]{a}\right)^m,\ a > 0$

- $a^m \div a^n = a^{m-n}$
- $a^0 = 1$
- $a^{-n} = \dfrac{1}{a^n},\ a \neq 0$
- $a^{-\frac{m}{n}} = \dfrac{1}{\sqrt[n]{a^m}} = \dfrac{1}{\left(\sqrt[n]{a}\right)^m}$

$m, n \in \mathbb{R}$

$a \in \mathbb{R}^+$

Logarithms and their properties

Definition: $x = \log_b a \Leftrightarrow a = b^x$

Notation: $\log a = \log_{10} a$

$\quad\quad\quad\quad \ln a = \log_e a$

$a, b \in \mathbb{R}^+, b \neq 1$

Properties:

- $\log_a xy = \log_a x + \log_a y$
- $\log_a \dfrac{x}{y} = \log_a x - \log_a y$
- $\log_a x^n = n\log_a x$
- $\log_a 1 = 0$
- $\log_a a = 1$
- $\log_a \dfrac{1}{x} = -\log_a x$

$a, x, y \in \mathbb{R}^+, a \neq 1$

Change of base formula: $\log_a x = \dfrac{\log_b x}{\log_b a}$

$a, b, x \in \mathbb{R}^+, a, b \neq 1$

Euler number: $e \approx 2.718\,281\,828\,459\,045\ldots$

Exponential and logarithmic functions

Exponential functions: $y = a^x$

The exponential function and its inverse:

$\quad f(x) = e^x \Leftrightarrow f^{-1}(x) = \ln x$

$x > 0$

Derivatives of exponential and logarithmic functions:

- $y = e^x \Rightarrow \dfrac{dy}{dx} = e^x$

- $y = e^{f(x)} \Rightarrow \dfrac{dy}{dx} = f'(x)\,e^{f(x)}$

- $y = \ln x \Rightarrow \dfrac{dy}{dx} = \dfrac{1}{x}$

Remember to consider domain restrictions where appropriate.

- $y = a^x \Rightarrow \dfrac{dy}{dx} = (\ln a)a^x$

- $y = \log_a x \Rightarrow \dfrac{dy}{dx} = \dfrac{1}{x \ln a}$

Angles, arcs and areas

Degree to radian measure: $360° = 2\pi$ $\quad$ or $\quad$ $180° = \pi$

Length of arc of circle $= \theta r$ $\quad$ $1° = \dfrac{\pi}{180}$ radians $\quad$ 1 radian $= \dfrac{180°}{\pi}$

Area of sector of circle $= \dfrac{1}{2}\theta r^2$

Beauty in mathematics

> *"The greatest mathematics has the simplicity and inevitableness of supreme poetry and music, standing on the borderland of all that is wonderful in science, and all that is beautiful in art."*
>
> Herbert Westren Turnbull (1885–1961)
> *The Great Mathematicians*, 1929

Following the pattern

The Swiss mathematician Leonhard Euler (1707–83) introduced the concept of a function and used the notation $f(x)$, as well as the letter e for the transcendental number 2.71828.......

He also calculated the values of a set of continuing fractions, like the ones shown below.

$$1 + \frac{1}{1+1} \approx$$

$$1 + \frac{1}{1+\frac{1}{2}} \approx$$

$$1 + \frac{1}{1+\frac{1}{2+\frac{1}{3}}} \approx$$

$$1 + \frac{1}{1+\frac{1}{2+\frac{2}{3+\frac{3}{4}}}} \approx$$

$$1 + \frac{1}{1+\frac{1}{2+\frac{2}{3+\frac{3}{4+\frac{4}{5}}}}} \approx$$

- Continue the pattern to build up the fraction and evaluate your results. What do you notice?

- Would you call this a beautiful fraction?

- Are there other similar continued fractions that lead to other curious numbers?

- Euler first came up with an equation using this continued fraction, which included e. Use your results to deduce Euler's equation.

- Rigor was restored to mathematics at the beginning of the 19th century, a century after Euler's demise. Why was Euler able to make so many advances in mathematics before formalization?

> *"Pure mathematics is, in its way, the poetry of logical ideas."*
>
> Albert Einstein (1879–1955)

The Four-Color Problem

In 1852, South African mathematician and botanist Francis Guthrie (1831–99) proposed that to color any planar map so that no two regions sharing a common border have the same color, you need at most four colors.

■ How could you color this map using four colors instead of five?

This simple-sounding conjecture was difficult to prove.

It was finally proved to be true in 1976 by Kenneth Appel and Wolfgang Haken at the University of Illinois. The technique they used came as a shock to most mathematicians. They had divided all planar maps into certain types and then used a computer to analyze each different type separately with all of its possibilities, verifying that all possible maps could be colored using just four colors. Thus the theorem was proved.

■ The four-color theorem was proved by exhaustion using computers. Is this a valid proof? What makes a proof valid?

■ Is there beauty in the proof of the four-color theorem?

■ How does technology influence knowledge claims?

■ In 2000 the Clay Mathematics Institute offered a $1 million prize for a proof of the Riemann hypothesis. The prize can only be claimed provided that the proof is not computer generated. Is this condition fair in today's world, which is so dependent on technology?

▼ A maximum of four colors ensures no two adjoining regions are the same color.

The art of mathematics

M.C. Escher (1898–1972) was a Dutch graphic artist known for his tessellations, impossible structures and other works of art inspired by mathematics.

■ Research how Escher's works based upon symmetry, impossible objects and hyperbolic planes are linked to mathematics.

■ Was Escher a mathematician as well as an artist?

■ Is there place for aesthetic beauty in mathematics?

You met Sierpinski's triangle in this chapter. By coloring different sets of triangles within it you can produce different patterns.

■ Use Sierpinski's triangle to generate different patterns. Is this art?

▶ Sierpinski's triangle

6

Exploring randomness

CHAPTER OBJECTIVES:

5.1 Concepts of population, sample, random sample and frequency distribution of discrete and continuous data; grouped data: mid-interval values, interval width, upper and lower interval boundaries; mean, variance, standard deviation

5.2 Concepts of trial, outcome, equally likely outcomes, sample space (U) and event; the probability of an event A as $P(A) = \dfrac{n(A)}{n(U)}$; the complementary events A and A' (not A); use of Venn diagrams, tree diagrams, counting principles and tables of outcomes to solve problems

5.3 Combined events, the formula for $P(A \cup B)$; mutually exclusive events

5.4 Conditional probability; the definition: $P(A|B) = \dfrac{P(A \cap B)}{P(B)}$; independent events; the definition: $P(A|B) = P(A) = P(A|B')$; use of Bayes' theorem for a maximum of three events

Before you start

You should know how to:

1 Calculate simple statistics from discrete data, including mean, median, mode, quartiles, range, interquartile range.
e.g. For the data set:
6, 7, 8, 9, 11, 12, 14, 15, 15, 19, 20

median (middle) $\dfrac{n+1}{2}$ = 6th value, 12
mode is 15
mean is 12.4 (3 sf)

2 Use counting techniques. e.g. A student must answer 7 out of 10 questions. The order does not matter. How many ways are there to answer the questions?
Number of ways =
$\begin{pmatrix} 10 \\ 7 \end{pmatrix} = \dfrac{10!}{7!3!} = \dfrac{10 \cdot 9 \cdot 8}{3 \cdot 2 \cdot 1} = 120$

Skills check

1 The masses, in kg, of the 12 members of a basketball team are: 94, 110, 88, 103, 97, 85, 91, 95, 107, 103, 114, 96
Find the
a median   **b** mode
c mean   **d** range
e quartiles **f** interquartile range.

2 In a group of 20 children, 8 have blue eyes and 12 have brown eyes. 3 children are selected from the group. How many ways are there to select:
a 3 children with blue eyes
b at least one child with blue eyes?

Statistics and probability

We are bombarded with data in everyday life, in news reports, sports results, education, business – even the weather.

Statistics allows you to collect, analyze, interpret and present data in a format that is easy to understand. These basic skills, covered in this chapter, ensure that you can then properly understand data that is presented to you, and turn it into information that you can use to support decisions or even make predictions.

Probability allows you to calculate the likelihood of a given outcome of an event. On the whole, people's intuition tends to let them down when considering probabilities. How else do you explain why people buy lottery tickets, when the probability of winning is usually around $\frac{1}{14\,000\,000}$?

In this chapter you can do most of the calculations on your GDC, but if you know how to do them by hand too it will help your understanding. The emphasis is on understanding and interpreting the results you obtain, in context. Statistical tables are not allowed in examinations – you will need to use your GDC.

Discussion activity – what should we do with our test scores?

32 students took a test scored out of 10. Their results were:

0, 1, 1, 2, 2, 2, 3, 3, 4, 4, 4, 5, 5, 5, 5, 5, 6, 6, 6, 6, 7, 7, 7,
7, 7, 7, 8, 8, 8, 8, 9, 10

What should the teacher do with this data?

How could you organize the data to give a better picture of the scores?

6.1 Classification and representation of statistical data

The study of statistics is important because it allows us to make predictions.

In this section you will be looking at:

- collecting data
- organizing the data
- analyzing the data
- drawing conclusions and making predictions.

The data you collect can be **qualitative** or **quantitative**.

Qualitative data	Quantitative data
Qualitative data is described in words and is sometimes called categorical data.	Quantitative data is numerical: it can be counted or measured.
Questions that give qualitative data include:	Questions that give quantitative data include:
What is your favorite pen color?	How many pens do you own?
How do you travel to school?	How long does it take you to get to school?
What brand of computer do you own?	How many computers have you owned?

Is the data from the test scores in the discussion activity qualitative or quantitative?

Quantitative data, usually called a 'statistical variable', can be split up into two categories: **discrete** and **continuous**.

> → A discrete variable has exact numerical values.

The number of CDs that you have or the number of children in your family are two example of discrete variables.

▲ Discrete. How many pairs of shoes do you own?

> → A continuous variable can be measured. Its accuracy depends on the accuracy of the measuring instrument used.

Length, weight and time, are examples of continuous variables which might have fraction or decimal values.

▲ Continuous.
What is the speed of the train?

What is the difference between a population and a sample?

When we think of the term **population**, we usually think of people in our town, region, state or country.

> → In statistics, the term **population** includes all members of a defined group that you are collecting data from.

> → A part of the population is called a **sample**, i.e. it is a selection of members or elements from a subset of the population.

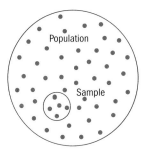

If the study is of a small set you can collect data from every element and the analysis will be an accurate study of the whole population. For example, if you want to analyze the grades obtained in Maths HL, SL and Studies in a particular school you can use the grades of every student. The students who took IB mathematics are the population.

Very often, however, the population is too large for you to be able to study every element. In this case you use a statistical sample that represents the whole population. For example, to predict the outcome of an election data is collected from a **random** sample.

Random samples must have two characteristics:

1 Every element has an equal opportunity of being selected.
2 The sample has essentially the same characteristics as the population.

Once you have collected the data you can draw a graph. Diagrams are much easier to interpret than tables.

Example 1

A student counted how many cars passed his house in one-minute intervals for 30 minutes. His results were:

23, 22, 22, 22, 24, 22, 21, 21, 23, 23, 27, 21, 21, 22, 23, 25, 27, 26, 23, 23, 22, 27, 26, 25, 28, 26, 22, 20, 21, 20.

Display this data in a frequency table.

Draw a bar chart for this data.

A bar chart is sometimes called a column graph.

Answer

Number of cars per minute	Tally	Frequency
20	||	2
21	卌	5
22	卌 ||	7
23	卌 |	6
24	|	1
25	||	2
26	|||	3
27	|||	3
28	|	1

Tally each data item in the correct row. Write the totals in the frequency column.
The number 21 appears 5 times in the data.

Use the vertical scale for the frequency and the horizontal scale for number of cars per minute.

A bar chart is suitable for discrete data or qualitative data. It should have gaps between the bars and all of the bars should be the same width.

→ When you have a lot of data, you can organize it into groups in a **grouped frequency table**.
If the data are continuous, you can draw a **histogram**.

A histogram is like a bar chart, but there are no gaps between the bars and the *areas* of the bars are proportional to the frequencies of the classes.

If all the classes have the same width then the *heights* of the bars are proportional to the frequencies.

Example 2

The ages of 200 members of a tennis club are:
20, 22, 23, 24, 25, 25, 25, 26, 26, 26, 26, 28, 28, 29, 29, 29, 30, 30, 30, 30,
30, 30, 30, 32, 32, 33, 33, 33, 34, 34, 34, 34, 34, 34, 34, 34, 35, 35, 35, 35,
36, 36, 36, 36, 36, 37, 37, 37, 38, 38, 38, 39, 39, 39, 40, 40, 40, 41, 41, 41, 42, 42, 42, 42, 42, 42, 42,
42, 43, 43, 43, 43, 43, 43, 44, 44, 44, 44, 44, 44, 45, 45, 45, 45, 45, 45, 45, 45, 46, 46, 46, 46, 46, 46,
46, 46, 47, 47, 47, 47, 47, 47, 47, 47, 47, 48, 48, 48, 48, 48, 48, 48, 48, 48, 49, 49, 49, 49, 49, 49, 49,
49, 50, 50, 50, 50, 50, 50, 51, 51, 51, 51, 51, 51, 51, 52, 52, 52, 52, 52, 53, 53, 53, 53, 53, 53, 53, 53,
53, 54, 54, 54, 54, 55, 55, 55, 55, 55, 56, 56, 56, 57, 57, 57, 57, 57, 57, 57, 57, 57, 57, 58, 58, 58, 59,
59, 59, 60, 60, 60, 60, 60, 61, 61, 61, 62, 62, 62, 63, 63, 63, 63, 64, 64, 64, 64, 65, 65, 68, 69.
Draw a grouped frequency table and a histogram for the data.

> Is age data continuous or discrete?

Answer

Age	Tally	Frequency
$20 \le$ age < 25	||||	4
$25 \le$ age < 30	ⅼⱧ ⅼⱧ ||	12
$30 \le$ age < 35	ⅼⱧ ⅼⱧ ⅼⱧ ⅼⱧ	20
$35 \le$ age < 40	ⅼⱧ ⅼⱧ ⅼⱧ |||	18
$40 \le$ age < 45	ⅼⱧ ⅼⱧ ⅼⱧ ⅼⱧ ⅼⱧ |	26
$45 \le$ age < 50	ⅼⱧ ⅼⱧ ⅼⱧ ⅼⱧ ⅼⱧ ⅼⱧ ⅼⱧ ⅼⱧ ||	42
$50 \le$ age < 55	ⅼⱧ ⅼⱧ ⅼⱧ ⅼⱧ ⅼⱧ ⅼⱧ |	31
$55 \le$ age < 60	ⅼⱧ ⅼⱧ ⅼⱧ ⅼⱧ ||||	24
$60 \le$ age < 65	ⅼⱧ ⅼⱧ ⅼⱧ ||||	19
$65 \le$ age < 70	||||	4

Equal class intervals of 5 years

25 is in the class $25 \le$ age < 30

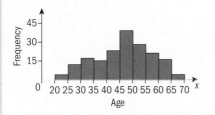

Numbers go on the edges of the bars on an x-axis like scale.

No gaps between the bars

Another type of graphical representation that you can use to represent grouped data is a **frequency polygon**.

In a frequency polygon plot the midpoint of each class against the frequency of that class on a graph. Then join the points with straight lines. (If the class intervals are not of equal width, **frequency density** is used instead. You will learn more about this later in the chapter.)

You can draw a frequency polygon for the tennis club in Example 2.

Make a new table with three columns headed Age, Frequency and Midpoint.

The midpoint (or classmark) is the mid-value of the class. For a graph you say 'midpoint' but for a table you say 'mid-value'.

Age	Frequency	Midpoint
$20 \leq$ age < 25	4	22.5
$25 \leq$ age < 30	12	27.5
$30 \leq$ age < 35	20	32.5
$35 \leq$ age < 40	18	37.5
$40 \leq$ age < 45	26	42.5
$45 \leq$ age < 50	42	47.5
$50 \leq$ age < 55	31	52.5
$55 \leq$ age < 60	24	57.5
$60 \leq$ age < 65	19	62.5
$65 \leq$ age < 70	4	67.5

Here we have treated age as a continuous variable. Why is the midpoint of the first class 22.5?

Plot frequency on the vertical axis against midpoint on the horizontal axis.

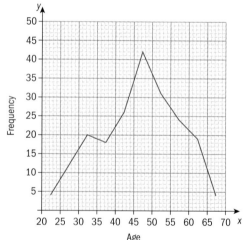

Sometimes you want to find a running total or **cumulative frequency**.

For example, how many members of the tennis club in Example 2 are under 40?

There are 4 members under 20, 16 members (4 + 12) under 30 and 36 members (4 + 12 + 20) under 40.

To find the cumulative frequency of a class add the frequencies of all the classes up to and including that class.

Example 3

Here are the exam scores for a group of 24 students. The maximum possible mark was 90.

47 54 63 77 23 15 66 32 56 83 16 49

52 67 44 9 62 46 38 58 37 25 55 46

Construct a frequency distribution table with mark intervals of 0–9, 10–19, and so on, and find cumulative frequencies for each mark interval.

Answer

Mark	Midpoint	Frequency	Cumulative frequency
0–9	4.5	1	1
10–19	14.5	2	3
20–29	24.5	2	5
30–39	34.5	3	8
40–49	44.5	5	13
50–59	54.5	5	18
60–69	64.5	4	22
70–79	74.5	1	23
80–90	85	1	24

You can draw a histogram to represent the data in Example 3, using the frequency column.

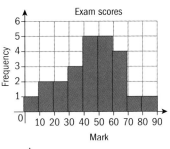

You can draw a **cumulative frequency diagram**, using the cumulative frequency column.

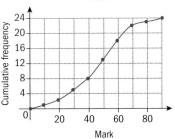

> When drawing a histogram for discrete variables, the boundaries are shifted by half a unit to the left and to the right. In this histogram, there are boundaries at –0.5, 9.5, 19.5 etc.

In the next example the distribution table does not have equal-width intervals. You have to calculate the **frequency density** of each class in order to draw the histogram. You use frequency density instead of frequency so that the area of each bar in the histogram corresponds to the frequency of the class it represents. Frequency density is calculated using the formula:

> → frequency density $= \dfrac{\text{frequency}}{\text{interval width}}$

> Notice that the cumulative frequency diagram starts from the minimum observation with a cumulative frequency of 0. Each cumulative frequency corresponds to the end-point of the interval, not the midpoint as in the histogram.

Example 4

These are the yearly salaries of the employees in a small start-up company.
€12 350, €19 820, €13 540, €8 440, €11 950, €11 320, €7 840, €8 450,
€14 550, €18 740, €12 360, €14 620, €22 380, €32 420, €36 780

a Copy and complete the table.

Salary (in thousands €)	Frequency	Interval width	Density
$7.5 \leq x < 10$			
$10 \leq x < 15$			
$15 \leq x < 25$			
$25 \leq x < 40$			

b Draw the histogram representing the salaries in the company.

Answers

a

Salary (in thousands €)	Frequency	Interval width	Frequency density
$7.5 \leq x < 10$	3	2.5	1.2
$10 \leq x < 15$	7	5	1.4
$15 \leq x < 25$	3	10	0.3
$25 \leq x < 40$	2	15	0.133

Use the formula:

$$Frequency\ density = \frac{frequency}{interval\ width}$$

b

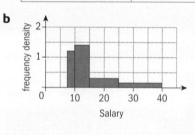

Histograms provide a clear picture of the distribution of the data. The histogram in Example 2 is a typical symmetrical histogram showing that the data has a normal distribution. The histogram in Example 4 is skewed showing that there were more employees with salaries at the lower end of the scale.

Exercise 6A

1 All of the IB students in a school were asked how many minutes a day they spent studying mathematics. The results are given in the table.

Time spent studying mathematics (minutes)	$0 \leq t < 15$	$15 \leq t < 30$	$30 \leq t < 45$	$45 \leq t < 60$	$60 \leq t < 75$	$75 \leq t < 90$
Number of students	21	32	35	41	27	11

a Is this data continuous or discrete?

b Use your GDC to help you draw a fully labeled histogram to represent this data.

2 The table shows the age distribution of mathematics teachers who work at Caring High School.

Age	Number of teachers
$20 \le x < 30$	5
$30 \le x < 40$	4
$40 \le x < 50$	3
$50 \le x < 60$	2
$60 \le x < 70$	3

 a Is the data discrete or continuous?
 b How many mathematics teachers work at Caring High School?
 c Use your GDC to help you draw a fully labeled histogram to represent this data.

3 The histogram shows data on frozen chickens in a supermarket. The masses in kilograms are grouped such that $1 \le w < 2$, $2 \le w < 3$ and so on.

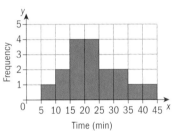

 a Is the mass of the chickens discrete or continuous data?
 b Draw the grouped frequency table for this histogram.
 c How many frozen chickens are there in the supermarket?

4 The histogram shows how many minutes it takes for students to return home after school.
 a Is the data discrete or continuous?
 b Represent the data in a grouped frequency table.
 c What is the shortest time that a student takes to get home?

EXAM-STYLE QUESTION

5 These histograms show four data sets A, B, C and D with the same number of values and the same range.

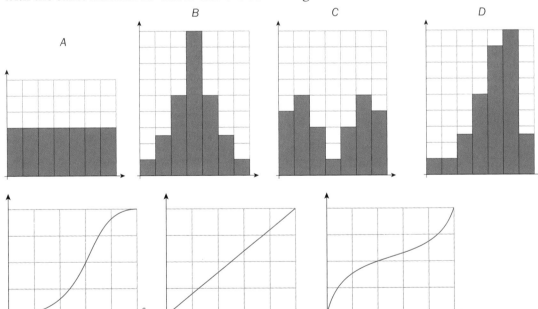

 a Decide which data set A, B, C, or D goes with each of these cumulative frequency diagrams.
 b Sketch a cumulative frequency diagram for the remaining data set.

6.2 Measures of central tendency

A measure of central tendency tells you where the middle of a set of data lies. The three most common measures of central tendency are the **mode**, the **mean** and the **median**.

> Another word for measure of central tendency is 'average'. This is commonly used to mean 'the mean'.

The mode

> → The **mode** is the value that occurs most frequently in a set of data.

> There can be more than one mode – or no mode at all.

In a frequency table, the mode is the group with the highest frequency.

Continuous data is grouped in intervals and to find the mode you have to use a histogram. If all the intervals have the same width then the modal interval is the one with the highest frequency. If the intervals have different widths, then the modal interval is the interval with the highest frequency density.

> Notice that if several intervals have the same highest frequency or frequency density then all of them are modal intervals.

Exercise 6B

1 Find the mode of the data in each frequency table.

a

Goals	Frequency
0	4
1	7
2	3
3	3
4	1

b

Height	Frequency
$140 \leq h < 150$	6
$150 \leq h < 160$	6
$160 \leq h < 170$	5
$170 \leq h < 180$	10
$180 \leq h < 190$	8

> A set of data is **bimodal** if it has two modes.

2 The table shows the waiting times, in minutes, of 50 customers in a bank.

2.5	1.3	2.2	1.4	5.2	3.0	7.1	4.2	1.0	0.5
3.2	2.0	5.3	3.1	1.2	1.8	4.1	2.2	1.2	1.8
3.1	2.7	0.2	6.4	2.0	3.1	1.1	4.2	4.3	0.5
1.2	1.4	2.1	5.4	3.1	4.3	2.5	4.2	5.2	0.5
1.4	0.3	4.2	2.2	2.4	0.6	3.2	4.2	0.8	0.5

a Construct a frequency distribution table with time intervals $0 \leq t < 1.0$, $1.0 \leq t < 2.0$,... and find cumulative frequencies for each time interval.

b Draw the cumulative frequency diagram and estimate the percentage of customers who waited longer than 5 minutes.

c Find the mean, median and mode of the data.

3 The histogram shows the heights of some trees in the park.

a Find the corresponding frequencies for each class and construct the frequency distribution table.

b Find the cumulative frequencies and draw the cumulative frequency diagram.

c Estimate the percentage of trees higher than 18 metres.

d Find the mean, median and mode of the data.

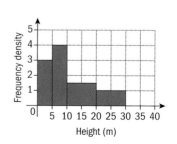

4 A data set contains these values

$$2, 3, 3, 3, 6, 6, 7$$

a Write down the values of the mode and median.

b Suppose that a new data value a is added to the set. Find the value of a that would make the mean and the median of the new data set the same. Hence, state the effect of this new data value on the mode of the set.

5 Consider the sequence $\ln a, \ln\sqrt{a}, \ln\sqrt[4]{a}, \ln\sqrt[8]{a},\ldots$

where $a > 1$

a Show that the series $\ln a + \ln\sqrt{a}, + \ln\sqrt[4]{a}, + \ln\sqrt[8]{a} +\ldots$ converges and find its sum.

b A data set consists of the first n terms of the sequence in **a**. Find an expression for its mean in terms of n and $\ln a$.

c Hence find the minimum value of n for which the mean of this data set is less than 1% of the first term of the sequence.

The mean

The arithmetic **mean** is usually just called the mean, and is the most common measure of central tendency.

μ is pronounced 'mu'.

> → Arithmetic mean $\mu = \dfrac{\sum_{i=1}^{k} f_i x_i}{n}$

$\Sigma f_i x_i$ is the sum of the data values and n is the number of data values in the population.
To calculate the mean of grouped data use the mid-value of each class with its frequency.

The median

> → The **median** is the value in the middle when the data are arranged in order of size. If the number of data values is even, then the median is the mean of the two middle values.

To find the median of grouped data you use a cumulative frequency curve. Take the horizontal line that passes through the midpoint of the range of the variable. Then look at the point of intersection with the cumulative frequency curve to make an estimation of the median from the horizontal axis.

The next example shows you how to find the measures of central tendency for data grouped in classes of different widths.

Example 5

The continuous variable X has a frequency distribution as shown in the table.

Find the:

a mode **b** median **c** mean.

X	Frequency
$0 \le x < 100$	8
$100 \le x < 200$	11
$200 \le x < 400$	24
$400 \le x < 600$	15
$600 \le x < 1000$	14

Answers

a

X	Frequency	Interval width	Frequency density
$0 \le x < 100$	8	100	0.08
$100 \le x < 200$	10	100	0.1
$200 \le x < 400$	22	200	0.11
$400 \le x < 600$	14	200	0.07
$600 \le x < 1000$	16	400	0.04

The modal interval $= 200 \le x < 400$

The intervals don't have equal widths so find the frequency densities.

b

X	Frequency	Cumulative frequency
$0 \le x < 100$	8	8
$100 \le x < 200$	10	18
$200 \le x < 400$	22	40
$400 \le x < 600$	14	54
$600 \le x < 1000$	16	70

Make a cumulative frequency table and draw a cumulative frequency graph.

The median = 350

Draw a horizontal line through 35, the midpoint of the range. Estimate the value of the median on the horizontal axis.

c

Mean = 404

Find the midpoints of the intervals and use them as the values of the variable X. On the GDC use the mean value feature from the list menu.

Investigation – what happens to the measures of central tendency when the data values are adjusted?

State your conjecture, test them and then prove them.

	Data	Mean	Mode	Median
Data set	6, 7, 8, 10, 12, 14, 14, 15, 16, 20			
Add 4 to each piece of data in the set.				
Multiply each piece of data by 2.				

Now copy and complete these sentences to explain what happens to the mean, mode and median of the original data set.
a   If you add 4 to each data value..
b   If you multiply each data value by 2......................................

6.3 Measures of dispersion

Measures of dispersion describe the spread of the data around a central value.

Look at these sets of data.

$A = \{1, 2, 3, 3, 4, 4, 4, 5, 5, 6, 7\}$ and $B = \{3, 3, 4, 4, 4, 4, 4, 4, 4, 5, 5\}$

Notice that they both have the same mode, median and mean, but they have other features that are different.

> → The **range** is the difference between the largest value and the smallest value.

The range of set $A = 7 - 1 = 6$
The range of set $B = 5 - 3 = 2$

The range is the easiest measure of dispersion to calculate but it can be affected by extreme values. It doesn't tell you how the data are distributed within the range.

Quartiles

When the data is arranged in order of size the median separates the data into two halves.

Quartiles separate the data into quarters.

- One-quarter, or 25%, of the values are less than or equal to the **lower quartile**, Q_1.
- Three-quarters, or 75%, of the values are less than or equal to the **upper quartile**, Q_3.

> → The **interquartile range** (IQR) is the difference between the upper and lower quartiles.
>
> $$IQR = Q_3 - Q_1$$

For set $A = \{1, \quad 2, \quad 3, \quad 3, \quad 4, \quad 4, \quad 4, \quad 5, \quad 5, \quad 6, \quad 7\}$

	Q_1		Q_2		Q_3	
	lower quartile		median		upper quartile	

$$IQR = Q_3 - Q_1 = 5 - 3 = 2$$

For set $B = \{3, \quad 3, \quad 4, \quad 4, \quad 4, \quad 4, \quad 4, \quad 4, \quad 4, \quad 5, \quad 5\}$

	Q_1		Q_2		Q_3	
	lower quartile		median		upper quartile	

$$IQR = Q_3 - Q_1 = 4 - 4 = 0$$

The interquartile range tells you the spread of the middle 50% of the data and is useful because it is not affected by a small number of extreme values.

For grouped data you use cumulative frequencies to find the interquartile range.

Example 6

The table shows the frequency distribution from Example 4.
a Find the cumulative frequencies and draw a cumulative frequency diagram.
b Use the cumulative frequency diagram to estimate:
  i the upper quartile   **ii** the lower quartile
  iii the interquartile range (IQR).

Answers

a

Salary (in thousands €)	Frequency	Cumulative frequency
$7.5 \leq x < 10$	3	3
$10 \leq x < 15$	7	10
$15 \leq x < 25$	3	13
$25 \leq x < 40$	2	15

▶ Continued on next page

b

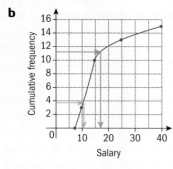

Salary

	Total frequency = 15
	$15 \div 4 = 3.75$, *so* Q_1 *is at 3.75 and* Q_3 *is at*
	$3 \times 3.75 = 11.25$
i $Q_1 = 10.5$	*Draw a horizontal line through 3.75 to the*
	cumulative frequency polygon to estimate Q_1
ii $Q_3 = 17.5$	*Draw a horizontal line through 11.25 to the*
iii IQR $= 17.5 - 10.5 = 7$	*cumulative frequency polygon to estimate* Q_3

Exercise 6C

EXAM-STYLE QUESTION

1 The depths of snow at a ski resort are measured on 31st January every year for 12 years. All data is in centimetres.
30, 75, 125, 55, 60, 75, 65, 65, 45, 120, 70, 110
Find the **a** range **b** median **c** lower quartile **d** upper quartile
e interquartile range.

2 The cumulative frequency diagram shows the reach in cm of 100 boxers.
  a Estimate the median reach of a boxer.
  b What is the interquartile range?
  c What does the interquartile range tell you?

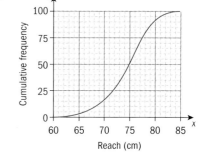

3 The table shows the length of 40 flash drives in a computer store.
Show this data on a cumulative frequency diagram.

Length (mm)	f	Upper class boundary	Length (/mm)	Cumulative frequency
6 – 10	0	10.5	$l \leq 10.5$	0
11 – 15	2	15.5	$l \leq 15.5$	2
16 – 20	4	20.5	$l \leq 20.5$	6
21 – 25	8	25.5	$l \leq 25.5$	14
26 – 30	14	30.5	$l \leq 30.5$	28
31 – 35	6	35.5	$l \leq 35.5$	34
36 – 40	4	40.5	$l \leq 40.5$	38
41 – 45	2	45.5	$l \leq 45.5$	40

Plot the points at the upper class boundary, usually the midpoint between classes.

4 a The table shows the cumulative frequency distribution for the times taken by 100 students to eat lunch.

Time (min)	Number of students
2 and under	0
4 and under	6
6 and under	18
8 and under	24
10 and under	40
12 and under	60
14 and under	78
16 and under	92
18 and under	100

Using a scale of 1 cm for 10 students on the vertical axis and 1 cm for 2 minutes on the horizontal axis, plot and draw a cumulative frequency diagram.
Use your graph to estimate
  i the median **ii** the interquartile range.
b The data in **a** can be represented in the form of a table. Find the values of p and q.

Time	$2 < t \leq 8$	$8 < t \leq 12$	$12 < t \leq 16$	$16 < t \leq 20$
Frequency	24	36	p	q

5 An IB exam marked out of 120 is taken by 4200 students. Here is a cumulative frequency graph of the marks.

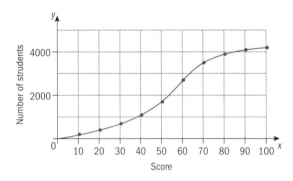

a Estimate the number of students who scored 40 marks or fewer on the test.
b The middle 50% of test results lie between marks a and b, where $a < b$. Estimate the values of a and b.
c If 80 marks is the minimum score to be awarded a grade 7, estimate the percentage of students in the group who achieved this grade.

6 The graph shows the time that students listen to music during school.

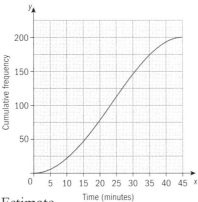

Estimate

a the median time that students listen to music

b the interquartile range

c the time a student must spend listening to music to be in the top 10% of listening times.

Variance and standard deviation

Variance and **standard deviation** are measures of dispersion.

The symbol for variance is σ^2.

To find the variance of a set of data:

- calculate the mean, μ
- calculate the difference, $x_i - \mu$, between each data value, x_i, and the mean, μ
- find the sum of the squares of the differences
- divide this sum by n, the number of data values

The formula for the variance is:

$$\rightarrow \sigma^2 = \frac{\sum_{i=1}^{k} f_i(x_i - \mu)^2}{n}, \text{ where } n = \sum_{i=1}^{k} f_i$$

σ is pronounced 'sigma'.

You can use algebraic manipulations to deduce a simplified form of this formula:

$$\sigma^2 = \frac{\sum_{i=1}^{n}(x_i - \mu)^2}{n} = \frac{\sum_{i=1}^{n}\left(x_i^2 - 2x_i\mu + \mu^2\right)}{n}$$

Expand the square of the difference.

$$= \frac{\sum_{i=1}^{n} x_i^2}{n} - 2\mu\frac{\sum_{i=1}^{n} x_i}{n} + \mu^2\frac{\sum_{i=1}^{n} 1}{n}$$

Use distributive properties.

$$= \frac{\sum_{i=1}^{n} x_i^2}{n} - 2\mu \times \mu + \mu^2 \times \frac{n}{n}$$

Use the definition of the mean and simplify the third sum.

$$= \frac{\sum_{i=1}^{n} x_i^2}{n} - 2\mu^2 + \mu^2 = \frac{\sum_{i=1}^{n} x_i^2}{n} - \mu^2$$

$$\rightarrow \sigma^2 = \frac{\sum_{i=1}^{k} f_i (x_i - \mu)^2}{n} = \frac{\sum_{i=1}^{k} f_i x_i^2}{n} - \mu^2$$

Standard deviation, σ, is the square root of the variance.

$$\rightarrow \sigma = \sqrt{\sigma^2} = \sqrt{\frac{\sum_{i=1}^{n} (x_i - \mu)^2}{n}} = \sqrt{\frac{\sum_{i=1}^{n} x_i^2}{n} - \mu^2}$$

Although you will usually calculate the variance on the GDC the next example shows you how to calculate the variance step by step using the formula.

Example 7

The table shows the distribution of broken eggs in 40 boxes, each with 10 eggs.
Find the variance and standard deviation of the number of broken eggs per box.

x_i	f_i
0	22
1	12
2	4
3	2

Answer

x_i	f_i	$x_i \times f_i$	$x_i^2 \times f_i$
0	22	0	0
1	12	12	12
2	4	8	16
3	2	6	18
$n = \sum f_i = 40$		$\mu = \dfrac{\sum x_i f_i}{40} = \dfrac{26}{40} = 0.65$	$\dfrac{\sum x_i^2 f_i}{40} = \dfrac{46}{40} = 1.15$

$\sigma^2 = 1.15 - 0.65^2 = 0.7275 \qquad \sigma = \sqrt{\sigma^2} = \sqrt{0.7275} = 0.853$

The GDC uses $n - 1$ instead of n in the formulae for estimating population variance and standard deviation from a sample. This correction was introduced by Friedrich Wilhelm Bessel (1784–1846), to correct bias in the estimations. On the GDC, in One Variable Statistics, the standard deviation is denoted by σ_x.

The standard deviation shows how much variation there is from the mean and gives an idea of the shape of the distribution.

- A low standard deviation shows that most of the data are close to the mean.
- A high standard deviation indicates that the data is spread out over a large range of values.

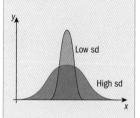

The next example shows you how to use your GDC to find the variance and standard deviation of a data set.

Example 8

Find the variance and the standard deviation of the continuous data in the table.

X	Frequency
$0 \le x < 100$	8
$100 \le x < 200$	11
$200 \le x < 400$	24
$400 \le x < 600$	15
$600 \le x < 1000$	14

GDC help on CD:
Alternative demonstrations for the TI-84 Plus and Casio fx-9860GII GDCs are on the CD.

Answer

x_{mi}	f_i
50	8
150	11
300	24
500	15
800	14

$\sigma = 245 \qquad \sigma^2 = 64\,600$

When data is grouped, use the midpoints of the interval. Now use the GDC where you have already stored the frequency distribution and use "One Variable Statistics".

Notice that both answers are given correct to 3 significant figures.

Exercise 6D

1 The mean value of six numbers $a, b, 2, 3, 5, 5$ is 3 and the variance is $\dfrac{7}{3}$. Find the values of a and b given that $a < b$.

2 a For the set of data $\{a - 1, a, a + 2, a + 3\}$, $a \in \mathbb{R}$:

find the mean and the variance in terms of a.

b Each number in the set is now decreased by 3.
Find the mean and the variance of the new set in terms of a.

3 A small employment agency is processing applications.
In the table are the numbers of applications processed in June.

Number of applications	Number of days
8	7
9	12
10	6
11	4
15	1

a Find the mean and standard deviation of the number of job applications

b Find the interquartile range of the number of job applications.

4 Two dice are rolled four times and the sums of their scores are: 2, 3, 6 and 9. Then the dice are rolled twice more. If the new mean score sum was 6 and the standard deviation of the score sum was $\sqrt{10}$ find the values of the two last score sums. Hence find the range and interquartile range of the distribution of the score sums.

5 The data set $A = \{4k - 2, k, k + 1, 2k + 4, 3k\}$, where $k \in \mathbb{R}$

  a Find the mean of set A in terms of k.

  b Hence, find an expression for the variance of set A in the form of $ak^2 + bk + c$ where $a, b, c \in \mathbb{R}$

Each number in the set A is now decreased by 2.

  c Find the mean of this new data set in terms in k.

  d Explain the effect of this change on the variance of set A.

6 The data set $B = \{a, 3a, 5a, ..., (2n - 1)\}a$ where $a \in \mathbb{Z}^+$

  a Show that the mean of this data set is given by an.

  b Given that

  $$\sum_{k=1}^{n} k^2 = \frac{n(n+1)(2n+1)}{6} \text{ for all } n \in \mathbb{Z}^+$$

  show that $\sum_{k=1}^{n} (2k-1)^2 = \frac{n(4n^2 - 1)}{3}$ for all $n \in \mathbb{Z}^+$

  c Write down the data set obtained when $a = 1$. Use the result obtained in **c** to find an expression for the variance of this data set in the form $pn^2 + qn$.

Investigation – variance and constants

What is the effect on the variance and standard deviation of a set of data when:

a a constant is added to all the values.

b all the values are multiplied by a constant?

The investigation leads to the general rule:

> → If you **add** a constant value k to all the numbers in a set, the mean increases by k but the standard deviation **remains the same**.
>
> If you **multiply** all the numbers in a set by a positive value k, both the mean and the standard deviation are **multiplied** by k.
>
> If k is negative, the mean is multiplied by k, but the standard deviation is multiplied by $-k$.

> k can be positive or negative, but not zero.

> k can be greater or less than 1.

6.4 Theoretical probability

Investigation – a dice problem

During the mid-1600s, mathematicians Blaise Pascal, Pierre de Fermat and Antoine Gombaud puzzled over this simple gambling problem:

Which is more likely–rolling a 'six' on four throws of one dice, or rolling a 'double six' on 24 throws with two dice?

Which option do you think is the most likely? Why?

Probability gives a numerical value that represents the chance or likelihood of a certain **event** occurring.

→ An **experiment** is the process by which you obtain an observation.
A **trial** is an experiment that you conduct a number of times under the same conditions.
An **event** is an outcome or outcomes from a trial.
A **random experiment** is one where there is uncertainty over which event may occur.

A random experiment can have a number of different equally likely outcomes.
One outcome or several outcomes form an event.

Some examples of random experiments are:

- rolling a dice three times
- tossing a coin once
- picking two cards from a pack of 52 playing cards
- recording the number of cars that pass the school gate in a 5-minute period.

Write $P(A)$ to represent the probability of an event A occurring, where $0 \leq P(A) \leq 1$.

A **theoretical probability** can be calculated from previous knowledge, for example, when a coin is thrown

$$P(\text{head}) = \frac{1}{2}$$

A set of all possible outcomes is called the **sample space**, U. For example, when you roll a dice, the sample space can be $\{1, 2, 3, 4, 5, 6\}$. The notation $n(U) = 6$ shows that there are six members of the sample space.

If event A is 'the number 6' then $n(A) = 1$ and

$$P(A) = \frac{n(A)}{n(U)} = \frac{1}{6}$$

Even though the foundations of probability were laid down in the correspondence between **Blaise Pascal** (1623–62) and **Pierre de Fermat** (1601–65), the first published work on probability was made by **Christiaan Huygens** (1629–95). He was encouraged by Pascal to publish his work. Huygens was also known for his work in astronomy (discovering Titan, one of Saturn's moons) and horology (inventing a pendulum clock).

Probabilities are measured on a scale from 0 to 1.

$$P(A) = \frac{\text{number of outcomes that are 6}}{\text{total number of outcomes}}$$

Example 9

Two unbiased dice are rolled. Event A is that the numbers on the upper face are equal and event B is that both numbers are odd.
Find these probabilities:

a $P(A)$

b $P(B)$

c $P(A \cap B)$

d $P(A \cup B)$

Answers

	1	2	3	4	5	6
1	(1, 1)	(1, 2)	(1, 3)	(1, 4)	(1, 5)	(1, 6)
2	(2, 1)	(2, 2)	(2, 3)	(2, 4)	(2, 5)	(2, 6)
3	(3, 1)	(3, 2)	(3, 3)	(3, 4)	(3, 5)	(3, 6)
4	(4, 1)	(4, 2)	(4, 3)	(4, 4)	(4, 5)	(4, 6)
5	(5, 1)	(5, 2)	(5, 3)	(5, 4)	(5, 5)	(5, 6)
6	(6, 1)	(6, 2)	(6, 3)	(6, 4)	(6, 5)	(6, 6)

Draw a sample space diagam.

Notice that there are 36 possible outcomes, 6 on the first die and 6 on the second dice.

a $P(A) = \dfrac{6}{36} = \dfrac{1}{6}$

There are 6 favorable outcomes {(1, 1), (2, 2),... (6, 6)}.

b $P(B) = \dfrac{9}{36} = \dfrac{1}{4}$

There are 9 favorable outcomes {(1, 1), (1, 3),... (5, 5)}.

c $P(A \cap B) = \dfrac{3}{36} = \dfrac{1}{12}$

There are 3 favorable outcomes {(1, 1), (3, 3),(5, 5)} that saytisfy both conditions.

d $P(A \cup B) = \dfrac{1}{6} + \dfrac{1}{4} - \dfrac{1}{12} = \dfrac{2+3-1}{12} = \dfrac{4}{12} = \dfrac{1}{3}$

Apply the formula
$P(A \cup B) = P(A) + P(B) - P(A \cap B)$

→ The **theoretical probability** of an event A is $P(A) = \dfrac{n(A)}{n(U)}$ where $n(A)$ is the number of outcomes that give event A and $n(U)$ is the total number of equally likely possible outcomes.

This formula was introduced by **Pierre-Simon Laplace** (1749–1827). Even though he was primarily a physicist who proved the stability of the solar system, Laplace also put the theory of probability on a sound footing in his second edition of *Théorie Analytique des Probabilités* published in 1814.

'Probability a priori' (before the experience) is another way of saying 'theoretical probability'.

Example 10

A fair coin is tossed three times.
a Find the sample space *U*.
b Hence write down the probability of obtaining exactly one tail.

Answers

a $U = \{HHH, HHT, HTH, HTT,$ | *List all the outcomes*
$\quad THH, THT, TTH, TTT\}$ |
b P(tail once) = P(THH, HTH, | $\dfrac{Number\ of\ favorable\ outcomes}{Total\ number\ of\ outcomes}$
$\quad HHT) = \dfrac{3}{8}$ |

Exercise 6E

1 A fair octahedral (eight-sided) dice is thrown. The faces are numbered 1 to 8. What is the probability that the number thrown is:

  a an even number

  b a multiple of 3

  c a multiple of 4

  d not a multiple of 4

  e less than 4?

2 A used car dealer has 150 used cars on his lot. The dealer knows that 30 of the cars are defective. One of the 150 cars is selected at random. What is the probability that it is defective?

> 'At random' means that any car has an equal chance of being selected. One of the 30 defective cars is as likely to be chosen as one of the cars that is not defective.

3 A fair coin was tossed and a dice rolled. The table shows all possible outcomes of this experiment

		1	2	3	4	5	6
Head	H	(1, H)	(2, H)	(3, H)	(4, H)	(5, H)	(6, H)
Tail	T	(1, T)	(2, T)	(3, T)	(4, T)	(5, T)	(6, T)
		1	2	3	4	5	6

Rolling a die

Let *A* be the event 'a tail was obtained' and *B* be the event 'the dice score was at least 3'.

Find

  a P(*A*)   **b** P(*B*)   **c** P(*A* ∪ *B*)

  d P(*A* ∩ *B*)   **e** P(*A'* ∪ *B*).

4 In a certain road $\frac{1}{3}$ of the houses have no newspapers delivered. If $\frac{1}{4}$ have a national paper delivered and $\frac{3}{5}$ have a local paper delivered, what is the probability that a house chosen at random has both?

5 Two fair dice are rolled. Let A be the event the product of the scores is even and B the event the sum of the scores is odd. Find

   a $P(A)$

   b $P(B)$

   c $P(A \cup B)$

   d $P(A \cap B)$

   e $P(A' \cup B')$.

Look back at the sample space diagram for two dice in Example 9.

Venn diagrams

To solve probability problems you need to use diagrams. Venn diagrams are useful in solving theoretical probability problems involving two or more events.

In a Venn diagram, the rectangle always represents the sample space U, which is the set of all of the possible outcomes of an experiment. If you use a circle to represent an event A, the part of the rectangle outside the circle represents A', the complement of this event, i.e., the set of the elements of the sample space that are not elements of A.

Example 11

In a group of 50 students, 10 have blue eyes and blond hair and 12 have neither blue eyes nor blond hair. If the total number of students with blond hair is 34, find the total number of students with blue eyes.

Answer

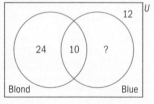

Draw a Venn diagram.

The number of students with blue eyes $= 50 - (24 + 12) = 14$

John Venn was born in Hull, England in 1834. His father and grandfather were priests and John was also encouraged to follow in their footsteps. In 1853 he went to Gonville and Caius College, Cambridge and graduated in 1857, becoming a fellow of the college. For the next five years he went into the priesthood and returned to Cambridge in 1862 to teach logic and probability theory. John Venn developed a graphical way to look at sets. This graph became known as a Venn diagram.

Example 12

At New Blue Bay International School, 15% of the Diploma students take Mathematics HL, 20% of these students also take Physics HL and 21% take Physics HL but not Mathematics HL.

If you select a Diploma student from this school, what is the probability that the student takes

a both Mathematics HL and Physics HL

b neither Mathematics HL nor Physics HL?

Answers

a $P(M \cap P) = 0.2 \times 0.15 = 0.03$

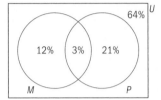

Find 20% of 15%.

Use a Venn diagram to display all the information.

b $P(M' \cap P')$
$= 1 - (0.12 + 0.03 + 0.21)$
$= 0.64$

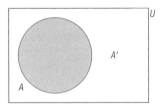

Clearly, $P(A') = 1 - P(A)$

Suppose that you are asked to pick a card from a pack of 52. Let A be the event 'red card is obtained' and B the event 'club card is obtained'. A and B are examples of mutually exclusive events as a club is a black card and therefore you cannot pick a red card and a club at the same time.

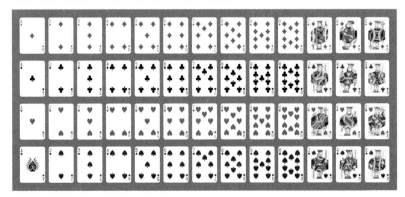

Two events, A and B, are **mutually exclusive** if whenever A occurs it is impossible for B to occur and, similarly, whenever B occurs it is impossible for A to occur.

> For example, in tossing a coin, the events 'a head' and 'a tail' are also mutually exclusive.

Events A and A' are the most obvious example of mutually exclusive events – either one or the other must occur, but A and A' cannot occur at the same time.

Here is the Venn diagram for mutually exclusive events A and B.

The two sets do not overlap, $A \cap B = \varnothing$.

→ Events A and B are mutually exclusive if and only if $P(A \cap B) = 0$

Exercise 6F

1 In a class of 25 students, 15 study French, 13 study Malay and 5 study neither language.
   One of these students is chosen at random. What is the probability that student studies both French and Malay?

2 Of the 32 students in a class, 18 play golf, 16 play the piano and 7 play both. How many play neither? One person is chosen at random. Find the probability that:
   a the student plays golf but not the piano,
   b the student plays the piano but not golf.

3 In a town, 40% of the population read newspaper 'A', 30% read newspaper 'B', 10% read newspaper 'C'. It is found that 5% read both 'A' and 'B', 4% read both 'A' and 'C' and 3% read both 'B' and 'C'. Also, 2% of the people read all three newspapers. Find the probability that a person chosen at random from the town
   a reads only 'A'   **b** reads only 'B'
   c reads none of the three newspapers.

> For this question you will need to use three circles in the Venn diagram – one to represent each newspaper.
>
>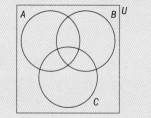

4 If X and Y are two events such that $P(X) = \dfrac{1}{4}$ and $P(Y) = \dfrac{1}{8}$ and $P(X \cap Y) = \dfrac{1}{8}$, find
   a $P(X \cup Y)$ **b** $P(X \cup Y)'$.

> Use a Venn diagram to solve these problems. Write the probability of each set in the diagram instead of the number of elements.

5 If $P(A) = 0.2$ and $P(B) = 0.5$ and $P(A \cap B) = 0.1$, find
   a $P(A \cup B)$ **b** $P(A \cup B)'$ **c** $P(A' \cup B)$.

> Recall that
> $(A \cup B)' = A' \cap B'$

Counting techniques and probability

In Chapter 1 you looked at systematic ways of counting and learnt about the number of permutations and combinations of r objects selected from a group of n. Now you are going to use these techniques to solve probability problems involving large numbers of possible events.

The next example is typical of a probability problem where you need to count the number of possible subsets of a given set meeting specific requirements.

Example 13

There are 14 girls and 11 boys in a class. Two students are selected to represent the class in the school assembly.

a In how many different ways can the two students be selected?
b Find the probability that the selected students are:

  i both girls **ii** both boys **iii** of different genders.

Answers

a $\dbinom{25}{2} = \dfrac{25 \times 24}{1 \times 2} = 300$

There are 14 + 11 = 25 students in the class out of which we choose 2.

b i $P(B_i) = \dfrac{\dbinom{14}{2}}{300} = \dfrac{\frac{14 \times 13}{1 \times 2}}{300} = \dfrac{91}{300}$

There are 14 girls and we choose 2 over the number of all possible outcomes.

ii $P(B_{ii}) = \dfrac{\dbinom{11}{2}}{300} = \dfrac{\frac{11 \times 10}{1 \times 2}}{300} = \dfrac{55}{300}$

There are 11 boys and we choose 2 over the number of all possible outcomes.

iii $P(B_{iii}) = \dfrac{14 \times 11}{300} = \dfrac{154}{300}$

Choose 1 girl and 1 boy out of 14 girls and 11 boys over the number of all possible outcomes.

Exercise 6G

1 Four students from a group of ten are selected randomly to form a team. What is the probability that both the students Sophie and Jerome are in the chosen team?

2 There are 5 lemons and 3 limes in a fruit bowl. Two pieces of fruit are selected at random. Find the probability that the selected fruit are:

  a two limes **b** two different pieces of fruit.

3 There are 7 red and 5 yellow fish in an aquarium. Three fish are randomly caught in a net. Find the probability that the fish caught were:

  a all red **b** not all of the same color.

4 There are 4 green, 5 orange and 6 purple marbles in a jar. Three marbles are randomly drawn from the jar. Find the probability that the marbles selected are:

  a all orange **b** all of a different color
  c at least one green.

5 Two archers Bill and Bob shoot a target in turns. The game finishes when one of them scores the center. Bill is a better archer and he starts first. The probability that Bill scores center is 0.3, while the probability that Bob scores centre is 0.25

   a Find the probability that:
   i Bob scores on his first shoot
   ii Bill scores on his third shoot
   iii Bob scores on his nth shoot.
   b If p is the probability that Bill wins the game, show that $p = 0.3 + 0.525\,p$
   c Hence or otherwise find the probability that Bob wins the game.

6.5 Probability properties

In this section you learn some useful probability properties and use Venn diagrams to illustrate them.

> → Two events A and $B \subseteq U$ have these properties:
>
> **i** $0 \le P(A) \le 1$
> **ii** $P(U) = 1$
> **iii** $P(A \cup B) = P(A) + P(B),\ A \cap B = \varnothing$
> **iv** $P(\varnothing) = 0$
> **v** $P(A') = 1 - P(A)$
> **vi** If $A \subseteq B$ then $P(B \backslash A) = P(B) - P(A)$

An event that has a probability of 0 is an **impossible event**, while an event that has a probability of 1 is a **certain event**.

In probability theory, the first three properties (with a slight modification of property **iii**) are called **probability axioms**. All the other properties can be proved by using the first three.

$B \backslash A$ means the set $B - A$ which contains all the elements of B that are not in $B \cap A$. $A \subseteq B$ means A is a subset of B.

Andrey Nikolaevich Kolmogorov (1903–87) published his work in 1933 in which he built up probability theory in a rigorous way using fundamental axioms, comparable to Euclid's treatment of geometry.

There are just three probability axioms. Research other axiomatic theories and discuss their characteristics. What characteristics do they have in common?

Example 14

Use the first three properties of the probability function to show these properties:

a $P(\varnothing) = 0$ **b** $P(A') = 1 - P(A)$.

Answers

a $P(A) = P(A \cup \varnothing)$
  $\Rightarrow P(A) = P(A) + P(\varnothing)$
  $\Rightarrow 0 = P(\varnothing)$

A set can be written as the union of the empty set and itself.
*A set is disjoint with the empty set so apply property **iii**.*
Subtract P(A) from both sides.

b $P(U) = P(A \cup A')$
  $\Rightarrow 1 = P(A) + P(A')$
  $\Rightarrow 1 - P(A) = P(A')$

Universal set can be written as the union of a set and its complement.
*A set is disjoint with its complement so apply properties **ii** and **iii**.*
Subtract P(A) from both sides.

This property is very useful: $P(A \cup B) = P(A) + P(B) - P(A \cap B)$

Proof

If $A \cap B = \varnothing$ then apply property **iv** to obtain the exact property **iii**.

$$P(A \cup B) = P(A) + P(B) - P(A \cap B) = P(A) + P(B) - P(\varnothing)$$
$$= P(A) + P(B)$$

Now examine the case when $A \cap B \neq \varnothing$.
To visualize the events use a Venn diagram.

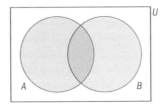

Rewrite the union: $A \cup B = A \cup (B \backslash (A \cap B))$

Notice that the two sets are disjoint, therefore you can apply the probability properties.

> Disjoint sets have no member in common.

$P(A \cup B) = P(A \cup (B \backslash (A \cap B)))$ *Apply property **iii***

$\quad\quad = P(A) + P(B \backslash (A \cap B))$ *$A \cap B$ is a complete subset of*

$\quad\quad = P(A) + P(B) - P(A \cap B)$ *B, so we can apply property **iv**.*

Paul Erdős (1913–96) was a Hungarian mathematician, who published more papers than any other mathematician in history. He wrote a total of 1525 mathematical articles during his lifetime, working with 511 different collaborators. After having been awarded a doctorate in mathematics at the age of just 21, Erdős worked on problems from a wide range of fields, including combinatorics, number theory, and classical analysis.

Paul Erdős is known for his 'legendarily eccentric' personality. Possessions meant very little to him – awards and other earnings were generally donated to people in need and other worthy causes. Most of his life was spent traveling between scientific conferences and the homes of colleagues all over the world. He would typically show up on a colleague's doorstep and announce "my brain is open", staying long enough to collaborate on a few papers, before moving on to his next challenge a few days later.

Extension material on CD:
Worksheet 6

Exercise 6H

1 The events A and B are such that $P(A) = 0.4$, $P(B) = 0.6$ and $P(A \cup B) = 0.7$. Find these probabilities:

   a $P(A \cap B)$ **b** $P(A \cap B')$ **c** $P(A' \cup B')$

2 Of the students in a class, 80% are taller than 160 cm and 75% are not taller than 180 cm. If a student is randomly selected from the class, find the probability that his height is a value between 160 cm and 180 cm.

3 If $P(A) = 0.6$, $P(B) = 0.55$ and $P(A \cap B) = 0.2$ find

   a $P(A \cup B)$

   b $P(A' \cap B)$

   c $P((A \cup B) \backslash (A \cap B))$

4 If $P(A \cap B') = 0.5$, $P(A \cap B) = 0.2$ and $P(A \cup B) = 0.85$ find

   a $P(A)$

   b $P(B)$

   c $P(A' \cap B)$

5 Show that for any two events A and B that are not impossible

$$P(A \cap B) \times P(A \cup B) \leq P(A) \times P(B)$$

6 Prove this formula for three events A, B and C.

$$P(A \cup B \cup C)$$
$$= P(A) + P(B) + P(C) - P(A \cap B) - P(A \cap C) - P(B \cap C) + P(A \cap B \cap C)$$

7 Use the properties of the probability function to show the last property: $A \subseteq B \Rightarrow P(B \backslash A) = P(B) - P(A)$

6.6 Experimental probability

You can also find a probability by doing an experiment.

For example, each member of a class of 20 students tossed a coin 10 times.

There are only two possible outcomes of the experiment, head or tail.

The sample space $U = \{$number of heads$\} = \{0, 1, 2, ..., 8, 9, 10\}$

The numbers of heads the students obtained were:

4, 5, 5, 3, 4, 4, 6, 8, 7, 3, 5, 4, 7, 8, 3, 4, 5, 4, 6, 7

Construct a frequency table.

Number of heads	3	4	5	6	7	8
Frequency	3	6	4	2	3	2

The range is $8 - 3 = 5$ which is quite large, but half of the results are either 4 or 5 heads.

When you are estimating a probability by experiment you need to perform the experiment a large number of times to get a reliable answer.

For example, to find the probability of getting a head the result will be more reliable if you toss the coin 100 times rather than 10 times. This can be time consuming so you can use your GDC to simulate the results.

The **relative frequency** of the number of heads is the proportion of heads in the total number of throws.

> → Relative frequency = $\dfrac{\text{Frequency of occurrence of event } A}{\text{Number of trials}}$

To simulate 20 students each tossing a coin 100 times use the random integer feature on the GDC with 1 to represent 'head' and 0 to represent 'tail'. Then add all the 1's in the list to find the number of heads.
Here is a typical result.

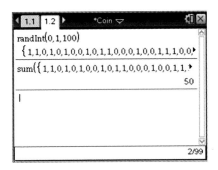

GDC help on CD:
Alternative demonstrations for the TI-84 Plus and Casio fx-9860GII GDCs are on the CD.

The number of heads obtained was:
50, 41, 50, 50, 51, 52, 51, 45, 51, 43,
50, 48, 44, 53, 65, 50, 45, 54, 49, 44

Notice that the relative frequency (number of heads ÷ number of trials) ranges from 0.41 to 0.65, which is smaller than the range obtained (0.3 to 0.8) when the coin was tossed only 10 times.

The mean relative frequency = $\dfrac{\text{total number of heads}}{\text{total number of trials}} = \dfrac{986}{2000} = 0.493$

Next use the GDC to simulate 20 students each tossing a coin 1000 times. Suppose the results are:

x_i	50	41	50	50	51	52	51	45	51	43	50	48	44	53	65	50	45	54	49	44
r_i	0.5	0.41	0.5	0.5	0.51	0.52	0.51	0.45	0.51	0.43	0.5	0.48	0.44	0.53	0.65	0.5	0.45	0.54	0.49	0.44

The results can be analyzed on the GDC.

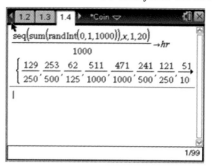

GDC help on CD:
Alternative demonstrations for the TI-84 Plus and Casio fx-9860GII GDCs are on the CD.

Is mathematics an experimental science? Is probability mathematics? How can we deal with two definitions of probability, experimental and theoretical? Are they consistent?

Notice that the range of relative frequencies is now even narrower, from 0.471 to 0.517, and the mean relative frequency is 0.4999.
It seems that the more times you repeat the experiment the more the relative frequencies tend towards 0.5. This result is an **experimental probability**.

'Probability a posteriori' (after the experience) is another way of saying 'experimental probability'.

> → You can use relative frequency as an estimate of probability. The larger the number of trials, the closer the relative frequency is to the probability.

The next example shows a useful application of experimental probability – deciding whether or not a dice is fair.

The US National Weather Service used this method to find the probability of being struck by lightning, using

$$\frac{\text{No of people struck}}{\text{No of people in population}}$$

Example 15

Use an Excel spreadsheet to simulate an experiment in which 100 people roll an unbiased six-sided dice 1000 times each.
a   For each person, find the relative frequency of obtaining a score of 6 on the upper face.
b   Find the total number of times the score 6 was obtained by the whole group of 100 people, i.e. the average of all the relative frequencies.
c   Comment on your answer in **b** and compare it with the theoretical value for the probability of getting a score of 6 when rolling a fair dice.

Answers

In the first cell A1 put Rand Between(1, 6). Copy the cell in column A to obtain 1000 rolls.
At the end of the column in cell A1001 use function Count If(A1:A1000;6)
In the cell A1002 calculate relative frequency of '6' occurring in 1000 rolls.
Extend the number of columns to find results for 100 people.
a   The relative frequencies range from 0.137 to 0.195.

This gives a random integer number between 1 and 6 (outcome of one roll).
This counts the number of sixes in the column.

Relative frequencies are in the yellow cells.

▶ Continued on next page

b Average relative frequency = 0.16688

c This is the experimental probability of getting a '6' when 100 people do 1000 rolls each.

$P(\frac{1}{6}) = 0.166667$ (3 sf) so the result is close to the theoretical probability.

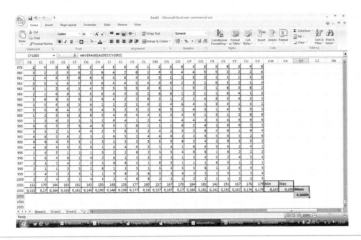

Exercise 6I

1 The table shows the relative frequencies of the ages of the students at a high school.

  a A student is randomly selected. Find the probability that the student is:

   i 15 years old

   ii 16 years of age or older.

  There are 1200 students at this school.

  b Calculate the number of 15-year-old students.

Age (in years)	Relative frequency
13	0.15
14	0.31
15	0.21
16	0.19
17	0.14
Total	1

2 The sides of a 6-sided spinner are numbered from 1 to 6. The table shows the results for 100 spins.

Number on spinner	1	2	3	4	5	6
Frequency	27	18	17	15	16	7

  a What is the relative frequency of getting a 1?

  b Do you think the spinner is fair? Give a reason for your answer.

  c The spinner is spun 3000 times. Estimate the number of times the result will be a 4.

3 Two dice are thrown 500 times. For each throw, the sum of the two numbers shown on the dice is written down.

The table shows the result.

Sum	2	3	4	5	6	7	8	9	10	11	12
Frequencies	6	8	21	34	65	80	63	77	68	36	42

Using these frequencies, calculate the probability of:
a the sum being exactly divisible by 5
b the sum being an even number
c the sum being exactly divisible by 5 or being an even number.

4 A 10-sided dice, with faces 1 to 10, is rolled. Calculate the probability that the number scored is:
a a prime number
b either a prime number or a multiple of 4
c either a multiple of 4 or a multiple of 3.

6.7 Conditional probability

The Monty Hall dilemma

This is a famous probability puzzle based on the American television game show 'Let's Make a Deal'.
The name comes from the show's original host, Monty Hall.

Contestants on the game show are given the choice of three doors.
Behind one door is the main prize (a car) and behind the other two doors there are unwanted prizes. The car and the unwanted prizes are placed randomly behind the doors before the show.

The rules of the game are as follows: after the contestants have chosen a door, the door remains closed for the time being. Monty Hall, who knows what is behind the doors, then opens one of the two remaining doors and always reveals an unwanted prize. After he has opened one of the doors, Monty Hall asks the participants whether they want to stay with their first choice or to switch to the last remaining door.

What should they do?

a Stick with their first choice.
b Switch to the other remaining closed door
c It does not matter. Chances are even.

> We will revisit this problem later in the chapter.

Here is a Venn diagram showing students who do archery and badminton.

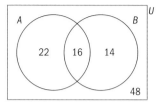

If you know that a particular student does badminton how does this affect the probability that the student also does archery?

Altogether 30 students do badminton; of these 16 also do archery.

You write the probability that a student does archery, given that they do badminton, as $P(A|B)$.

Note that
$$P(A|B) = \frac{n(A \cap B)}{n(B)} = \frac{16}{30} = \frac{8}{15}$$

This is known as **conditional probability**.

It also follows that $P(A|B) = \dfrac{P(A \cap B)}{P(B)} = \dfrac{\frac{16}{100}}{\frac{30}{100}}$

$$= \frac{16}{30} = \frac{8}{15}$$

> → For two events A and B the probability of A occurring given that B has occurred can be found using
>
> $$P(A|B) = \frac{P(A \cap B)}{P(B)}$$
>
> Rearranging the formula gives $P(A \cap B) = P(A|B) \times P(B)$
>
> This is known as the **multiplicative probability law**.

When solving probability problem, a tree diagram can help organize your working. We will see more of these later on.

Example 16

Shuyi rolls a fair dice. Let A be the event 'obtain a score of at least 4' and B the event 'obtain an even score'.

a Write down $P(A)$ and $P(B)$.

b Draw the tree diagram.
Use it to show that
$$P(A \cap B) = P(A|B) \times P(B)$$

dice rolled

$\frac{1}{2}$ B (2, 4, 6)

$\frac{2}{3}$ A (4 or 6)

not A (2)

not B (1, 3, 5)

▶ Continued on next page

Answers

a $P(A) = \dfrac{n(\{4,5,6\})}{n(\{1,2,3,4,5,6\})} = \dfrac{3}{6} = \dfrac{1}{2}$ $P(A) = \dfrac{n(A)}{n(U)}$

and $P(B) = \dfrac{n(\{2,4,6\})}{n(\{1,2,3,4,5,6\})} = \dfrac{3}{6} = \dfrac{1}{2}$ $P(B) = \dfrac{n(A)}{n(U)}$

b $P(A \cap B) = \dfrac{2}{6} = \dfrac{1}{3}$ *For $P(A \cap B)$ move*
along the top branch

$P(B) = \dfrac{1}{2}$ and $P(A|B) = \dfrac{2}{3}$ *Here the sample space*
is reduced to B

So $P(A|B) \times P(B) = \dfrac{2}{3} \times \dfrac{1}{2} = \dfrac{1}{3} = P(A \cap B)$

The next example shows how to use a Venn diagram to organize the information and find the missing values needed to calculate the probabilities.

Example 17

Of the 53 staff at a school, 36 drink tea, 18 drink coffee, and 10 drink neither tea nor coffee.
a How many staff drink both tea and coffee?
One member of staff is chosen at random. Find the probability that:
b this teacher drinks tea but not coffee
c the teacher is a tea drinker who also drinks coffee
d the teacher is a tea drinker who does not drink coffee.

Answers

a

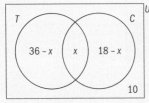

Let $n(T \cap C) = x$ $n(T \cap C)$ *This is the number who*
so   *drink both tea and coffee.*
$36 - x + x + 18 - x + 10 = 53$ *53 is the total number of staff on the*
$64 - x = 53$ *Venn diagram.*
    $x = 11$ *Solve for x.*
11 staff drink both tea and coffee.

Therefore $P(T \cap C) = \dfrac{11}{53}$ *Since x = 11 and total = 53.*

▶ Continued on next page

b $P(T \cap C') = \dfrac{25}{53}$ *36 − 11 = 25*

Probability that staff member drinks tea but not coffee is $\dfrac{25}{53}$

c $P(C \mid T) = \dfrac{P(C \cap T)}{P(T)} = \dfrac{\frac{11}{53}}{\frac{36}{53}}$

$\quad = \dfrac{11}{\cancel{53}} \times \dfrac{\cancel{53}}{56} = \dfrac{11}{36}$

Probability that staff member is a tea drinker who drinks coffee is $\dfrac{11}{36}$

d $P(C' \mid T) = \dfrac{P(C' \cap T)}{P(T)} = \dfrac{\frac{25}{53}}{\frac{36}{53}}$ *P(C′ ∩ T) = P(T ∩ C′)*

$\quad = \dfrac{25}{\cancel{53}} \times \dfrac{\cancel{53}}{36} = \dfrac{25}{36}$

Exercise 6J

1 There are 27 students in a class. 15 take Art and 20 take Theater. Four do neither subject. How many students do both subjects? One person is chosen at random. Find the probability that
  a he or she takes Theater but not Art
  b he or she takes at least one of the two subjects
  c he or she takes Theater, given that they take Art.

2 For events A and B it is known that: $P(A' \cap B') = 0.35$, $P(A) = 0.25$ and $P(B) = 0.6$ Find
  a $P(A \cap B)$ **b** $P(A \mid B)$ **c** $P(B' \mid A')$

3 48% of all teenagers own a skateboard and 39% of all teenagers own a skateboard and roller blades. What is the probability that a teenager owns roller blades given that the teenager owns a skateboard?

4 A number is chosen at random from this list of eight numbers:
      1   2   4   7   11   16   22   29

Find:
  a P(it is even | it is not a multiple of 4)
  b P(it is less than 15 | it is greater than 5)
  c P(it is less than 5 | it is less than 15)
  d P(it lies between 10 and 20 | it lies between 5 and 25)

5 In my town 95% of all households have a desktop computer. 61% of all households have a desktop computer and a laptop computer. What is the probability that a household has a laptop computer given that it has a desktop computer?

6 The probability that a student takes Technology and Spanish is 0.1. The probability that a student takes Technology is 0.6. What is the probability that a student takes Spanish given that the student is taking Technology?

7 U and V are mutually exclusive events. $P(U) = 0.26$ and $P(V) = 0.37$. Find:
   a $P(U \text{ and } V)$
   b $P(U \mid V)$
   c $P(U \text{ or } V)$

8 A teacher gave her class an IB Paper 1 and an IB Paper 2. 35% of the class passed both tests and 52% of the class passed the first test. What percent of those who passed the first test also passed the second test?

9 A jar contains black and white marbles. Two marbles are chosen without replacement. The probability of selecting a black marble and then a white marble is 0.34, and the probability of selecting a black marble on the first draw is 0.47. What is the probability of selecting a white marble on the second draw, given that the first marble drawn was black?

10 The table shows the number of left- and right-handed table-tennis players in a sample of 50 males and females.

	Left-handed	Right-handed	Total
Male	5	32	37
Female	2	11	13
Total	7	43	50

A table-tennis player was selected at random from the group. Find the probability that the player is:
   a male and left-handed
   b right-handed
   c right-handed, given that the player selected is female.

11 Your neighbor has two children. You learn that he has a son, Sam. What is the probability that Sam's sibling is a brother?

This is not as obvious as it might seem!

The Monty Hall problem revisited!

Take a typical situation in the game. Suppose the contestant has chosen Door 3 and Monty Hall reveals that there is an unwanted prize behind Door 2. What is the conditional probability that the car is behind Door 1?
Let A stand for the condition that there is a car behind Door 1 and the contestant has chosen Door 3. Let B stand for the condition that Monty Hall has revealed that there is an unwanted prize behind Door 2 given that the contestant has chosen Door 3.

Analysis of the Monty Hall problem using conditional probability.

The probability of A and B ($P(A \cap B)$) is just $\frac{1}{3} \times \frac{1}{3} = \frac{1}{9}$ because if the car is behind Door 1 and the contestant has chosen Door 3 Monty Hall has to show what is behind Door 2.

The problem is the computation of the probability of being shown an unwanted prize behind Door 2 given that the choice was Door 3. This situation can arise in two ways:

1 when the car is behind Door 1

2 when the car is behind Door 3.

The first way has a probability of $\frac{1}{9}$, as shown above.

In the second way, the host could reveal either what is behind Door 1 or Door 2. If he is equally likely to choose either of these doors then the probability of showing what is behind Door 2 is $\frac{1}{2} \times \frac{1}{9} = \frac{1}{18}$.

Therefore the probability of there being revealed an unwanted prize behind Door 2 when the contestant has chosen Door 3 is
$$\frac{1}{9} + \frac{1}{2} \times \frac{1}{9} = \frac{3}{18}$$
This is $P(B)$, the probability of B.

We want the conditional probability, $P(A \mid B)$. This is given by

$$P(A \mid B) = \frac{P(A \cap B)}{P(B)} = \frac{\frac{1}{9}}{\frac{3}{18}} = \frac{2}{3}$$

This means that the conditional probability that the car is behind Door 3 given that the contestant has chosen Door 3 and has been shown that there is an unwanted prize behind Door 2 is only $\frac{1}{3}$.
Therefore it is worthwhile to switch!

6.8 Independent events

When two events are not influenced by each other, they are called **independent events**.

Consider this example:

Luka is playing darts and Hannah is swimming in a pool.
Events A and B are:
A: 'Luka scores a bullseye'
B: 'Hannah swims 50 metres in under 30 seconds'

The corresponding probabilities are:
$P(A) = 0.45$, $P(B) = 0.72$

The occurrence of one of the events will not influence the occurrence of the other. So,

$P(A \mid B) = 0.45$ will remain the same as $P(A)$,

while $P(B \mid A) = 0.72$ will remain the same as $P(B)$.
It follows that

> → $P(A|B) = P(A) \Rightarrow P(A|B') = P(A)$
> $P(A|B) = P(A) \Rightarrow P(A'|B) = P(A')$
> $P(A|B) = P(A) \Rightarrow P(A'|B') = P(A')$

These three formulae mean that the probability that A or its complement A' occurs is **independent** of B or its complement B' occuring.

Using the multiplicative probability law, you can obtain another very important probability formula that is only valid for **independent events:**

> → For independent events A and B
> $P(A \cap B) = P(B) \times P(A|B) \Rightarrow P(A \cap B) = P(A) \times P(B)$

This formulae can be generalized for three more three or more events for example, if A, B and C are independent events.

$P(A \cap B \cap C)$
$= P(A) \times P(B) \times P(C)$
if A, B, C are independent.

Example 18

For these events:
A: 'Luka scores a bullseye'
B: 'Hannah swims 50 metres in under 30 seconds'
Assign the corresponding probabilities: $P(A) = 0.45$, $P(B) = 0.72$
   a Find the probability that Luka scores a bullseye and Hannah is going to swim 50 metres in under 30 seconds.
   b Find the probability that exactly one of the events happens.

Answers

a $P(A \cap B) = 0.45 \times 0.72$
$= 0.324$

Apply the formula $P(A \cap B)$ $= P(A) \times P(B)$

b $P((A \cap B') \cup (A' \cap B))$
$= 0.45 \times (1-0.72) + (1-0.45) \times 0.72$
$= 0.522$

*Apply the multiplicative probability law for independent events and probability function property **iii**.*

> Luka scores and Hannah doesn't swim her distance in under 30 seconds or Luka doesn't score and Hannah finishes in under 30 seconds.

The next example shows you how to combine the properties studied so far to find probabilities.

Example 19

Independent events A and B are such that $P(B) = 0.4$ and $P(A \cup B) = 0.75$. Find these probabilities:
   a $P(A)$
   b $P(A \cap B')$

Answers

a $P(A \cup B) = P(A) + P(B) - P(A) \times P(B)$

$0.75 = P(A) + 0.4 - P(A) \times 0.4$

$0.35 = 0.6 \times P(A)$

$P(A) = 0.583$

Apply the additional and multiplicative probability laws for independent events.

b $P(A \cap B') = P(A) \times P(B')$

$P(A \cap B') = 0.583 \times (1 - 0.4) = 0.233$

Given that A and B are independent events then A and B' are independent too.

The next example is about an event that takes place n times with each outcome independent of the previous ones.

Example 20

Maika goes on a photo safari. The probability that Maika takes a photo of a cheetah on any day is 0.3. Find how many days the safari must last so that the probability that Maika takes at least one photo of a cheetah will exceed 95%.

Answer

If the safari is for n days

$1 - 0.7^n \geq 0.95$

$0.05 \geq 0.7^n$

$n \log(0.7) \leq \log(0.05)$

$n \geq \dfrac{\log(0.05)}{\log(0.7)} = 8.40$

The safari should last at least 9 days.

Use the complementary event, 'Maika did not take a photo of cheetah in n days'.

Take logarithms on both sides. Both logs are negative.

The inequality can be solved immediately by a GDC.

GDC help on CD:
Alternative demonstrations for the TI-84 Plus and Casio fx-9860GII GDCs are on the CD.

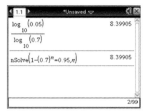

Exercise 6K

1 If $P(A) = 0.4$, $P(B) = 0.6$, $P(C) = 0.3$,

$P(A \cap B) = 0.24$, $P(B \cap C) = 0.15$ and $P(A \cup C) = 0.82$,
which of the events A, B and C are independent?
Give reasons for your answers.

2 A card is drawn from a standard deck of 52 cards.
Of these events:
- A 'the card is a Queen',
- B 'the color of the card is red'
- C 'the card is a face card'
  which are independent? Explain your answer.

3 Given that A and B are independent events, show that:
a A and B' are independent
b A' and B are independent
c A' and B' are independent.

4 Given that $P(A) = \dfrac{1}{3}$, $P(A \cup B) = \dfrac{5}{6}$ and $P(B|A) = \dfrac{3}{4}$, find $P(B)$.
Are the events A and B independent?

5 Independent events A and B are such that $P(A) = 0.45$ and $P(A \cap B) = 0.18$. Find these probabilities:
a $P(B)$ **b** $P(A \cup B)$ **c** $P(A' \cap B')$

6 Independent events A and B are such that

$P(A) = a$, $P(B) = 2a$ and $P(A \cup B) = \dfrac{5}{8}$.

Find $P(A)$ and $P(B)$.

7 In a game the players flip a coin and roll a dice. Draw a tree diagram to represent the game, and find the probabilities of these events:

   a 'a tail on the coin and a 6 on the dice'

   b 'a head on the coin and an even number on the dice'

   c 'no tail on the coin and a multiple of 3 on the dice.'

8 A combination lock on a suitcase consists of four digits. Notice that the leading digits can be zeros. Norbu is setting up the combination. What is the probability that his combination will be divisible by:

   a 2   **b** 5   **c** 4?

9 A set contains a large number of integers. How many integers must you select so that the probability of selecting at least one odd integer is at least 0.92?

10 Julia plays tennis. She can score a point from the base line with a probability 0.55. How many times does she need to hit the ball so that the probability of Julia scoring a point exceeds 0.999?

6.9 Probability tree diagrams

"Solving problems is a practical art, like swimming, or skiing, or playing the piano . . . if you wish to learn swimming you have to go in the water, and if you wish to become a problem solver you have to solve problems." – George Polya, *Mathematical Discovery* (1981).

George Polya (1887–1985) was born and educated in Hungary. He obtained a PhD in mathematics from Budapest and taught in Switzerland and at Brown, Smith and Stanford Universities in the United States. He was granted numerous honors and awards in mathematics, and taught and lectured in virtually every country of the world. He encouraged the use of problem solving techniques in learning mathematics and is best known for his outstanding book, the classic *How to Solve It* (1945).

Research Polya's steps to problem solving, which he outlines in *How To Solve It*.

Tree diagrams are useful tools for tackling probability problems involving two or more events. The next set of examples will show you how to use them.

Example 21 is a problem 'with replacement', i.e. the conditions for the second event are exactly the same as they were for the first event.

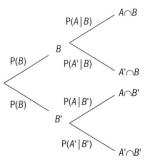

Example 21

The probability that Samuel, a keen member of the school Archery Club, hits the bullseye is 0.8. Samuel takes two shots. Assume that success with each shot is independent from the previous shot.

Represent this information on a tree diagram.

Find the probability that Samuel

a hits two bullseyes

b hits only one bullseye

c hits at least one bullseye.

Answers

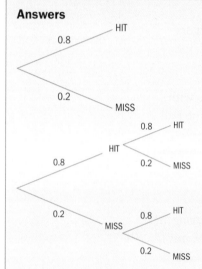

The first section of the tree diagram represents Samuel's first shot. He will either hit the bullseye or miss it. The probability that he misses is
$1 - 0.8 = 0.2$

The second shot will also either hit or miss the bullseye.

There are therefore four possible outcomes of this 'experiment':
- *a hit followed by a hit (H and H)*
- *a hit followed by a miss (H and M)*
- *a miss followed by a hit (M and H)*
- *a miss followed by a miss (M and M).*

a Probability of 2 bullseyes
= P(*H* and *H*).
So P(*H* and *H*) = 0.8 × 0.8
= 0.64

Since a hit with the first shot is independent of getting a hit with the second shot, multiply the probabilities together (the product rule).

b P(*H* and *M*) + P(*M* and *H*)
= (0.8 × 0.2) + (0.2 × 0.8)
= 0.32

Just one bullseye could be either a hit on the first or a hit on the second and missing the other one.
These two events, (H and M) and (M and H) are mutually exclusive: they can't both happen at the same time.
Multiply along each branch (as again events are independent) and then add (as the 2 outcomes are mutually exclusive).

c = 1 − (0.2 × 0.2)
= 1 − 0.04
= 0.96

P(at least one bullseye)
= 1 − P(miss the bullseye both times)
= 1 − P(M and M)

> Write the probabilities of the given events on the branches of the tree and then multiply the probabilities along the branches to obtain the probability of the event at the end of the branches.

In the next example the events are not independent.

Example 22

There are 3 white and 7 black balls in a box. One ball is taken from the box, its color noted and then it is left outside the box. Another ball is taken from the box and its color noted.
a Draw the probability tree diagram that represents this information.
b Find these probabilities:
    i both balls are white
    ii both balls are black.

Answers

a

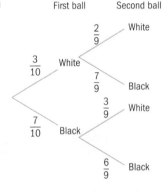

First ball   Second ball

b i $P(WW) = \dfrac{3}{10} \times \dfrac{2}{9} = \dfrac{1}{15}$

Use conditional probability:
$P(W \cap W) = P(W \mid W) \times P(W)$

  ii $P(BB) = \dfrac{7}{10} \times \dfrac{6}{9} = \dfrac{7}{15}$

or **i** $P(WW) = \dfrac{\dbinom{3}{2}}{\dbinom{10}{2}} = \dfrac{1}{15}$

Notice that to find these probabilities you can also use combinations, e.g.

$P(WW) = \dfrac{\dbinom{3}{2}}{\dbinom{10}{2}}$

  ii $P(BB) = \dfrac{\dbinom{7}{2}}{\dbinom{10}{2}} = \dfrac{7}{15}$

The multiplicative probability law can be extended to three or even more events.

$$P(A \cap B \cap C) = P(A) \times P(B \mid A) \times P(C \mid (A \cap B)).$$

Event A occurs first, then event B given that A has already occurred, and finally event C given that both A and B have occurred.
The next example shows this formula in action.

Example 23

A shelf has 12 toys on it. There are 7 cars and 5 trucks. A child can take 3 toys from the shelf.

a Draw a tree diagram to represent this information.
b Find the probability that:
  i all the toys are cars
  ii the child picks at least one car.

Answers

a

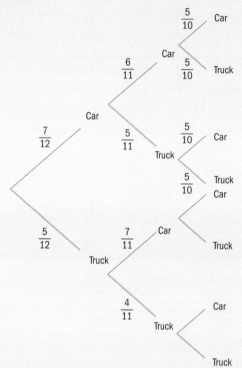

b i $P(CCC) = \dfrac{7}{12} \times \dfrac{6}{11} \times \dfrac{5}{10} = \dfrac{7}{44}$

ii $1 - P(TTT) = 1 - \dfrac{5}{12} \times \dfrac{4}{11} \times \dfrac{3}{10}$

$= 1 - \dfrac{1}{22} = \dfrac{21}{22}$

The complementary event of 'at least one car' is 'all three toys are trucks'.

or

i $P(CCC) = \dfrac{\dbinom{7}{3}}{\dbinom{12}{3}} = \dfrac{7}{44}$

To find these probabilities we can also use combinations.

ii $P(CCC) = 1 - \dfrac{\dbinom{5}{3}}{\dbinom{12}{3}} = \dfrac{21}{22}$

GDC help on CD:
Alternative demonstrations for the TI-84 Plus and Casio fx-9860GII GDCs are on the CD.

In the next example some branches of the tree diagram are shorter than others. Subsequent events depend on the outcomes of previous ones.

Example 24

Toby is a rising star of the school tennis club. He has found that when he gets his first serve in the probability that he wins that point is 0.75. When he uses his second serve there is a 0.45 chance of him winning the point. He is successful at getting his first serve in on 3 out of 5 occasions and his second serve in on 3 out of 4 occasions.

a Find the probability that the next time it is Toby's turn to serve he wins the point.

b Given that Toby wins the point, what is the probability that he got his first serve in?

Answers

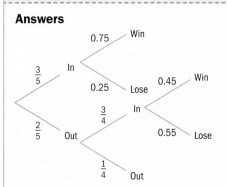

On this tree diagram, it is not necessary to continue the branches once the point has been won.

a P (win) = (get first serve in and win) + (miss first serve, get second serve in and win)

$\left(\frac{3}{5} \times 0.75\right) + \left(\frac{2}{5} \times \frac{3}{4} \times 0.45\right)$

= 0.45 + 0.135

= 0.585

Multiply along the branches.

b P(first serve in | win)
P(first serve in and win point | win point)

$= \dfrac{\left(\frac{3}{5} \times 0.75\right)}{0.585} = 0.769 \text{ (3 sf)}$

*Both of these values have been found in part **a**.*

Exercise 6L

1 The probability that it will rain on a certain day is 0.2. The probability that Shikma arrives late for work if it's raining is 0.4, while the probability that she will be late for work when it's not raining is 0.1. Find the probability that on a given day it will rain and Shikma will not be late for work.

2 Shankari performs blood tests on a virus. The probability that the test gives a positive result is 0.85. The probability that the person does have the virus given that the test was positive is 0.98, and the probability that the person doesn't have the virus given that the test was negative is 0.12. What is the probability that Shankari's diagnosis is correct?

3 Jan shoots two free throws in a basketball game. The probability that he scores the first shot is 0.75. The probability that he misses the second shot given that he scored the first shot is 0.15. The probability that he is going to score the second shot given that he missed the first shot is 0.8. What is the probability that Jan scores only one shot?

4 Given that $P(A) = \frac{1}{3}, P(B|A) = \frac{3}{5}$ and $P(B|A') = \frac{1}{2}$ find

   a $P(B')$

   b $P(A' \cup B')$

5 There are 12 purple and 18 orange chips in a bag. We draw three chips from the bag without replacing them. Find the probability that:

   a all the chips are orange

   b there is at least one purple chip

   c there are more orange chips.

6 Sam draws three cards from a standard deck of 52 cards without replacing them. Find the probability that all three cards are:

   a red

   b hearts

   c of the same suit

   d faces cards in the same suit.

> Face cards are King, Queen, Jack.

6.10 Bayes' theorem

As you will see, Bayes' theorem enables you to solve more difficult probability problems where you need to analyze different possibilities. To visualize the situation you can use a Venn diagram.

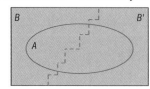

The Venn diagram shows that event A may occur when event B occurs or A may occur when B does not occur. These are mutually exclusive situations. You can use this fact to write $P(A)$ as the sum of two expressions which correspond to the two situations.

$$P(A) = P((A \cap B) \cup (A \cap B')) = P(B) \times P(A|B) + P(B') \times P(A|B')$$

> This formula is called the **total probability formula**.

→ The events B and B' are called **hypotheses** since we don't know whether they have occurred or not.

The next example uses the total probability formula in a real-life context.

Example 25

In a computer factory each mouse is checked by one of two controllers. The first controller checks 55% of the mouses and finds 4% of them to be defective. The second controller finds that 3% of his mouses are defective. What is the probability that a mouse will be found to be defective by the controllers?

Answer

Let the events 'mouse checked by first controller' and 'mouse checked by second controller' be called F and S respectively. Let D be the event 'mouse is defective'. Use a probability tree diagram.

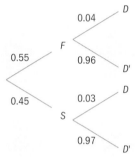

$$P(D) = P(F) \times P(D|F) + P(S) \times P(D|S)$$
$$= 0.55 \times 0.04 + 0.45 \times 0.03 = 0.0355$$

Sometimes instead of only two hypotheses we have three or even more hypotheses. In that case you use a similar calculation.

Let H_1, H_2 and H_3 be events such that:

i $H_1 \cap H_2 = \varnothing, H_1 \cap H_3 = \varnothing$ and
$H_2 \cap H_3 = \varnothing$

ii $H_1 \cup H_2 \cup H_3 = U$

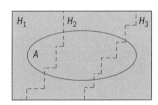

$$P(A) = P\big((A \cap H_1) \cup (A \cap H_2) \cup (A \cap H_3)\big)$$
$$= P(H_1) \times P(A|H_1) + P(H_2) \times P(A|H_2) + P(H_3) \times P(A|H_3)$$

In exams you will not be asked questions that involve more than three hypotheses.

Example 26

There are two jars containing green and purple marbles. In jar A there are 4 green and 6 purple marbles, while jar B contains 1 green and 2 purple marbles. We transfer 2 marbles from jar A to jar B and then draw a marble from jar B. What is the probability that the marble drawn from jar B is green?

Answer

Let H_1 be the event 'we transfer two green marbles', H_2 be the event 'we transfer one green and one purple marble', H_3 be the event 'we transfer two purple marbles'.

There are three different hypotheses and the content of jar B is different with respect to each of them.

$$P(H_1) = \frac{\binom{4}{2}}{\binom{10}{2}} = \frac{\frac{4 \times 3}{1 \times 2}}{\frac{10 \times 9}{1 \times 2}} = \frac{2}{15}$$

There are 10 marbles in total and we choose 2. Out of 4 green marbles we choose 2 from jar A.

$$P(H_2) = \frac{4 \times 6}{\binom{10}{2}} = \frac{4 \times 6}{\frac{10 \times 9}{1 \times 2}} = \frac{8}{15}$$

Out of 4 green and 6 purple marbles we choose one of each color from jar A.

$$P(H_3) = \frac{\binom{6}{2}}{\binom{10}{2}} = \frac{\frac{6 \times 5}{1 \times 2}}{\frac{10 \times 9}{1 \times 2}} = \frac{1}{3}$$

Out of 6 purple marbles we choose 2 from jar A.

> Notice that the sum of the three probabilities is 1. Give a reason for this.

Draw the probability tree diagram.

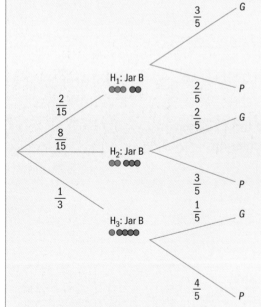

$$P(G) = P(H_1) \times P(G|H_1) + P(H_2) \times P(G|H_2) + P(H_3) \times P(G|H_3)$$

$$= \frac{2}{15} \times \frac{3}{5} + \frac{8}{15} \times \frac{2}{5} + \frac{1}{3} \times \frac{1}{5}$$

$$= \frac{6 + 16 + 5}{75} = \frac{27}{75} = \frac{9}{25}$$

Here is Bayes' theorem, which allows you to calculate conditional probabilities:

→ **Bayes' theorem**

$$P(B|A) = \frac{P(B \cap A)}{P(A)} = \frac{P(B) \times P(A|B)}{P(B) \times P(A|B) + P(B') \times P(A|B')}$$

Thomas Bayes (1702–61) was an English mathematician. He studied logic and theology at the University of Edinburgh since, as a non-conformist, he was not allowed to study at either Oxford or Cambridge. He set out his probability theory in the *Essays towards solving a problem in the doctrine of chances* which were published in 1764, after his death.

Bayes' theorem has made remarkable contributions to history. It has been used to search for nuclear weapons, create actuarial tables, improve low-resolution computer images, and to help determine who really wrote the Federalist papers (a series of articles promoting the ratification of the American constitution). It was used by Alan Turing and others during World War II in performing cypto-analytic work, and most recently, is helping to determine the false positive rate to mammograms.

Example 27

At a university there are 65% female and 35% male students. Only 25% of the female students are taking statistics, whist 60% of the male students are taking the course. A randomly selected student from the university is taking statistics. Find the probability that the student is female.

Answer

Let *F* denote that the randomly selected student is female, and *M* that the student is male. Let *S* denote that the student is taking statistics.

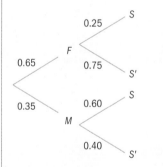

$$P(F|S) = \frac{P(F \cap S)}{P(S)}$$

Apply the conditional probability formula.

$$= \frac{P(F) \times P(S|F)}{P(F) \times P(S|F) + P(M) \times P(S|M)}$$

Apply the total probability formula.

$$= \frac{0.65 \times 0.25}{0.65 \times 0.25 + 0.35 \times 0.60}$$

$$= \frac{0.1625}{0.1625 + 0.21} = 0.436 \text{ (3 sf)}$$

Notice that we were taking the favorable branch divided by the sum of all the possible branches for the given event.

Now look at Example 26 again and calculate the probability that an event had occurred first, given that we know what happened next.

Example 28

There are two jars with purple and green marbles. In jar A there are 4 green and 6 purple marbles. Jar B contains 1 green and 2 purple marbles. We transfer 2 marbles from jar A to jar B and then draw a marble from jar B. A randomly drawn marble from the jar B is green. What is the probability that 2 marbles of different colors were transferred from jar A?

Answer

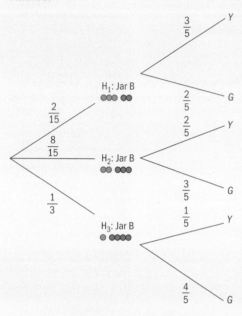

Draw the probability tree diagram

$$P(H_2|G) = \frac{P(H_2 \cap G)}{P(G)}$$

Apply the conditional probability formula.

$$= \frac{P(H_2) \times P(G|H_2)}{P(H_1) \times P(G|H_1) + P(H_2) \times P(G|H_2) + P(H_3) \times P(G|H_3)}$$

Apply the total probability formula.

$$= \frac{\dfrac{8}{15} \times \dfrac{2}{5}}{\dfrac{2}{15} \times \dfrac{3}{5} + \dfrac{8}{15} \times \dfrac{2}{5} + \dfrac{1}{3} \times \dfrac{1}{5}} = \frac{\dfrac{16}{75}}{\dfrac{27}{75}} = \frac{16}{27}$$

Exercise 6M

EXAM-STYLE QUESTION

1   Two boxes contain cards. In the first box there are 9 cards with numbers 1 to 9 written on them, and in the second box there are 5 cards with numbers 4 to 8 written on them. A box is randomly selected and a card is drawn from it.

   a What is the probability that the card will show an even number?

   b What is the probability that an even card was drawn from the first box?

2 In a factory two machines produce bolts and the bolts are stored in a warehouse. The first machine produces 60% of all the bolts but 5% of them are defective while 2% of the bolts produced by the second machine are defective. A bolt is randomly selected from the warehouse.

   a What is the probability that the bolt will be defective?

   b What is the probability that the bolt was produced on the first machine given that it was defective?

3 40% of the children attending a summer camp are girls and 60% are boys. The children vote for their favorite sport. 75% of the girls and 35% of the boys vote for beach volleyball as their favorite sport. A child is randomly selected.

   a What is the probability the child voted for beach volleyball as their favorite sport?

   b What is the probability that beach volleyball was not their favorite sport given that the child was a girl?

4 In one box there are 14 white and 16 black balls, while in a second box there are 7 white and 12 black balls. A ball is drawn from the first box and placed in the second box, and then two balls are drawn from the second box.

   a What is the probability that both balls are black?

   b What is the probability that the ball drawn from the first box was white, given that both balls drawn from the second box were white?

5 A sample space U contains the events A and B. These probabilities are given: $P(B) = \dfrac{2}{3}$, $P(A|B) = \dfrac{5}{6}$ and $P(A'|B') = \dfrac{1}{4}$

   a Draw a probability tree diagram representing this information

   b Find: **i** $P(A)$ **ii** $P(B|A)$ **iii** $P(B'|A')$.

6 A pair of dice is in a non-transparent bag. One dice is biased and the probability of obtaining a 6 on that dice is $\dfrac{2}{3}$, while the other dice is unbiased. A dice is taken from the bag at random and roll it.

   a What is the probability of obtaining a 6?

   b What is the probability that the unbiased dice was taken given that you did not obtain a 6?

7 A survey of a large group of adults shows that 18% have lung problems. Of these it is found that 70% are heavy smokers, 20% smoke occasionally and 10% are non-smokers. Of those who don't have lung problems it is found that 5% are heavy smokers, 15% smoke occasionally and 80% are non-smokers. An adult is selected at random from the group.

   a Find the probability that the selected person is a non-smoker.

   b Find the probability that the selected person has lung problems given that they are a heavy smoker.

8 There are three urns labelled A, B and C. In urn A there are 2 white and 4 red cubes, in urn B there are 5 white and 3 red cubes and urn C contains 4 white and 6 red cubes. An urn is selected at random and a cube is drawn from it.

   a Find the probability that a red cube is drawn.

   b Given that a red cube is drawn, find the probability that it is drawn from urn C.

9 On her way to school in the morning Anne can take three different routes, A, B and C. The probabilities that she takes routes A, B and C are 0.45, 0.20 and 0.35 respectively. The probability that she will get to school on time if she takes route A is 0.95. If she takes route B it is 0.90 and the probability if she takes route C is 0.80.

   a Find the probability that Anne will get to school on time.

   b Find the probability that Anne took route A, given that she got to school on time.

   c Find the probability that Anne took route B, given that she did not get to school on time.

10 There are two jars containing pink and brown marbles. In the first jar there are 5 pink and 10 brown marbles while in the second jar there are 4 pink and 5 brown marbles. We transfer 2 marbles from the first jar to the second jar and then draw a marble from the second jar.

   a What is the probability that the marble drawn from the second jar is brown?

   b What is the probability that we transferred 2 pink marbles from the first jar, given that a pink marble is drawn from the second jar?

   c What is the probability that we transferred 2 brown marbles from the first jar, given that a pink marble is drawn from the second jar?

11 In a company employees are organized in the following sections: management, production and marketing. 10% of all the employees are in management and 25% are in marketing. The percentages of female employees in management, production and marketing are 40%, 30% and 70% respectively.

   a Find the probability that a randomly selected employee from this company is male.

   b Given that a randomly selected employee is male find the probability that he works in the management section.

   c Given that a randomly selected employee is female find the probability that she works in the marketing section.

12 Three machines produce nuts. The first machine produces 50% of all the nuts, the second machine produces 35% and the third machine produces 15%. Of the nuts produced by the first machine 4% are defective, versus 3% from the second machine and 6% from the third machine. Given that a randomly selected nut is not defective, find the probability that the nut was produced by the second machine.

13 There are 20 laptops in a classroom. 12 have a hard disk with a capacity of 160 GB and 8 with a capacity of 320 GB. A teacher randomly takes two laptops away to be tested. A student then takes a laptop for her project. What is the probability that the teacher takes both laptops with 320 GB given that the student takes one with 160 GB?

14 In a football team there are 11 players who can perform a penalty kick. 4 players are excellent and they score with a probability of 0.9, 4 have medium ability and they score with a probability of 0.6, whilst the remaining 3 are poor and they score with a probability of 0.2. A randomly selected player shoots the penalty kick. What is the probability that the player will score?

15 Two letters are drawn from the set $\{a, b, c, d, e, f, o\}$ without replacement. Then another letter is drawn. What is the probability that the third letter drawn is a vowel?

Review exercise

1 The mode, median and mean of a set of positive integers are 6, 7 and 8 respectively. Find the smallest possible set of positive integers with the smallest variance that satisfies these conditions.

2 A and B are two independent events. Given that $P(B|A)=\dfrac{1}{3}$ and $P(A\cup B)=\dfrac{11}{12}$ find $P(A)$.

3 The weight, in kilograms, of students in a class is measured and the cumulative frequency diagram is shown.
   a Estimate the median weight of the students.
   b Estimate the middle 50% of the weight of the students.
   c How many students are in the class?
   d Construct the frequency distribution table and find the modal weight of the students in the group.

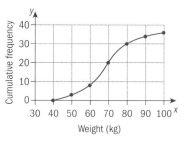

4 There are 7 boys and 5 girls in an environmental activity group. A committee of three members is selected from the group.
   a How many different committees can be selected?
   b Find the probability that Agatha and Jacob (two students from the group) are not both on the committee.
   c What is the probability that there will be more girls than boys on the committee?

5 Bassel invests in different companies. The probability that he invests in company X is $\dfrac{1}{3}$ and in company Y is $\dfrac{5}{9}$. The probability that he doesn't invest is $\dfrac{1}{9}$. The probability that the investments in companies X and Y yield a dividend are $\dfrac{3}{7}$ and $\dfrac{3}{5}$ respectively. The performances of the companies are independent.
   a Draw a probability tree diagram with the corresponding probabilities of all possible outcomes.
   b What is the probability that Bassel receives a dividend?
   c Find the probability that the dividend Bassel receives is from company Y.

6 There are 3 yellow, 4 blue and 5 green marbles in a bag. We randomly draw a marble from the bag and, without replacement, draw another marble. Given that the second marble drawn is green, what is the probability that the first marble was green too?

7 A pair of unbiased dice is rolled and the product of the numbers showing on the upper faces is noted. Find the probabilities that the product is:

a A prime number

b An even number

c A number divisible by 3

d A number divisible by 6 given that it is an even number.

8 A class of 30 students takes a test on statistics and their results are such that: $\sum_{i=1}^{30} m_i = 540$ and $\sum_{i=1}^{30} m_i^2 = 9990$ where m_i represents the mark of an individual student. Calculate:

a **i** the mean mark in the test

   ii the standard deviation of the marks in the test;

b Given that the pass mark for the test is 8, do you expect a student to fail the test?

9 Each odd number from 1 to $3n$, where n is odd, is written on a card and the cards are placed in a bag.

a How many cards are placed in the bag?

b What is the probability, in terms of n, that a card drawn randomly from the bag has a number divisible by 3?

 # Review exercise

1 There are four groups at a science conference. The heights of participants in each group are measured and the following mean heights of the groups are obtained.

Subject	Number of participants	Mean height(cm)
Biology	23	168
Chemistry	17	171
Environmental systems	8	163
Physics	20	177

Find the mean height of all the participants at the science conference.

2 **a** From the word STATISTICS how many arrangements of all the letters can be formed?

b What is the probability that the arrangement will start with the letter S?

c What is the probability that the arrangement will not end with a vowel?

3 A code lock consists of three digits. What is the probability that the randomly chosen code is:

a an even three-digit number

b a number divisible by 7

c a perfect square?

4 The probability that it rains on a particular day during July on the island of Hvar is 0.07. The hotel owner will give a 50% discount for a room on a rainy day in the month of July. David rents a room in the hotel for two days. Given that the rate for a room in the hotel is €85 per day, find the probability that for those two days David will pay less than €170.

5 Electric car batteries are tested and the distances obtained on a single charge are given in the table.

Distance (km)	Frequency
$0 \leq d < 100$	2
$100 \leq d < 200$	5
$200 \leq d < 300$	7
$300 \leq d < 400$	12
$400 \leq d < 500$	10
$500 \leq d < 600$	4

a Find an estimate of the mean distance traveled on the batteries.

b Find estimate of the standard deviation of the distance traveled on the batteries. Estimate the maximum distance the manufacturer can claim with 95% certainty that the car would travel on a single charge.

6 A group of 140 competitors were each given a puzzle to complete. The times taken to do this were recorded. The diagram is the histogram of the scores.

a Construct a frequency distribution table.

b Two competitors from the group were selected at random. Find the probability that:

i both managed to assemble the puzzle within 20 seconds

ii no competitor was able to do it in less than 10 seconds.

c Find the estimations of the mean and standard deviation of times taken to assemble the puzzle.

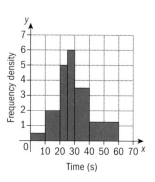

7 How many times would you need to roll a pair of dice so that the probability of obtaining a sum of 7 at least once is greater than 0.95?

8 The red blood cell (RBC) count is a blood test which determines the number of red blood cells per micro-litre in a blood sample. The normal range for children is typically between 3.8 and 5.5 million per micro-litre. The results of the blood test for 120 children are given in the table.

RBC	Frequency
$3.4 < n \le 3.8$	7
$3.8 < n \le 4.2$	15
$4.2 < n \le 4.6$	36
$4.6 < n \le 5.0$	22
$5.0 < n \le 5.4$	27
$5.4 < n \le 5.8$	13

  a Find the mean and the standard deviation of the RBC count.
  b Construct the cumulative frequency distribution and draw the cumulative frequency diagram to estimate the median result of the RBC count.
  c It is known that a higher RBC count is found in people who live at high altitudes. Given that all the children are healthy use your graph in **b** to estimate how many tested children live at high altitudes.

9 A school buys 20 new books. There are 6 books on statistics, 7 books on calculus, 4 books on geometry and 3 books on discrete mathematics. Given that the books are randomly put on the shelf in a classroom, find the probability that
  a all the books on statistics will be at the beginning of the shelf
  b all the books on calculus will be together.

10 The local basketball league consists of 12 teams. Team A is ranked fifth and the probability it will win against any of the higher ranked teams is 0.4. The probability it will win against the teams ranked in sixth to eighth place is 0.55 and of winning against the teams ranked ninth to twelfth is 0.75. Keith doesn't know which team team A will play against in the next game, but he would like to bet on A winning.
  a What is the probability that Keith will win the bet?
  b Given that Keith has lost the bet, what is the probability that team A played against a team with a better rank?

CHAPTER 6 SUMMARY

Classification and representation of statistical data

- A discrete variable has exact numerical values.
- A continuous variable can be measured. Its accuracy depends on the accuracy of the measuring instrument used.
- In statistics, the term **population** includes all members of a defined group that you are collecting data from.
- A part of the population is called a **sample**, i.e. it is a selection of members or elements from a subset of the population.
- When you have a lot of data, you can organize it into groups in a **grouped frequency table**.
  If the data are continuous, you can draw a **histogram**.
- frequency density $= \dfrac{\text{frequency}}{\text{interval width}}$

Measures of central tendency

- The **mode** is the value that occurs most frequently in a set of data.
- Arithmetic mean $\mu = \dfrac{\sum\limits_{i=1}^{k} f_i x_i}{n}$
- The **median** is the value in the middle when the data are arranged in order of size. If the number of data values is even, then the median is the mean of the two middle values.

Measures of dispersion

- The **range** is the difference between the largest value and the smallest value.
- The **interquartile range** (IQR) is the difference between the upper and lower quartiles.
  IQR $= Q_3 - Q_1$

- $\sigma^2 = \dfrac{\sum\limits_{i=1}^{k} f_i (x_i - \mu)^2}{n} = \dfrac{\sum\limits_{i=1}^{k} f_i x_i^2}{n} - \mu^2$

- $\sigma = \sqrt{\sigma^2} = \sqrt{\dfrac{\sum\limits_{i=1}^{n}(x_i - \mu)^2}{n}} = \sqrt{\dfrac{\sum\limits_{i=1}^{n} x_i^2}{n} - \mu^2}$

- If you **add** a constant value k to all the numbers in a set, the mean increases by k but the standard deviation **remains the same**.
- If you **multiply** all the numbers in a set by a positive value k, both the mean and the standard deviation are **multiplied** by k.
- If k is negative, the mean is multiplied by k, but the standard deviation is multiplied by $-k$.

Continued on next page

Theoretical probability

- An **experiment** is the process by which you obtain an observation.
  A **trial** is an experiment that you conduct a number of times under the same conditions.
  An **event** is an outcome or outcomes from a trial.
  A **random experiment** is one where there is uncertainty over which event may occur.
- The **theoretical probability** of an event A is $P(A) = \dfrac{n(A)}{n(U)}$

  where $n(A)$ is the number of outcomes that give event A
  and $n(U)$ is the total number of equally likely possible outcomes.
- Events A and B are mutually exclusive if and only if $P(A \cap B) = 0$

Probability properties

- Two events A and $B \subseteq U$ have these properties:
  i $0 \le P(A) \le 1$
  ii $P(U) = 1$
  iii $P(A \cup B) = P(A) + P(B), A \cap B = \varnothing$
  iv $P(\varnothing) = 0$
  v $P(A') = 1 - P(A)$
  vi   If $A \subseteq B$ then $P(B \backslash A) = P(B) - P(A)$

Conditional probability

- For two events A and B the probability of A occurring given

  that B has occurred can be found using $P(A \mid B) = \dfrac{P(A \cap B)}{P(B)}$

  Rearranging the formula gives $P(A \cap B) = P(A \mid B) \times P(B)$
  This is known as the **multiplicative probability law**.

Independent events

- $P(A|B) = P(A) \Rightarrow P(A|B') = P(A)$
  $P(A|B) = P(A) \Rightarrow P(A'|B) = P(A')$
  $P(A|B) = P(A) \Rightarrow P(A'|B') = P(A')$
- For independent events A and B
  $P(A \cap B) = P(B) \times P(A|B) \Rightarrow P(A \cap B) = P(A) \times P(B)$

Bayes' theorem

- **Total probability theorem**:
  $$P(A) = P\big((A \cap B) \cup (A \cap B')\big) = P(B) \times P(A|B) + P(B') \times P(A|B')$$
- The events B and B′ are called **hypotheses** since we don't know whether they have occurred or not.

- $P(B|A) = \dfrac{P(B \cap A)}{P(A)} = \dfrac{P(B) \times P(A|B)}{P(B) \times P(A|B) + P(B') \times P(A|B')}$

Statistics and probability – but is it mathematics?

Is statistics part of mathematics?

Statistics is about interpreting data. Calculating mean, standard deviation, etc. involves mathematics, but is interpreting what they mean in the context of the problem mathematics?

Statistics uses a lot of mathematics, and so it is a mathematical science. But just because it uses mathematics, this doesn't make it mathematics! Tax accountants use mathematics too, but we don't consider accountancy to be a branch of mathematics.

Mathematical statistics is a separate discipline that overlaps with statistics and focuses on the theoretical basis of statistics, using tools such as probability theory and decision theory.

- Mathematics is an axiomatic system. Is statistics an axiomatic system too?

> An axiom is a statement which is assumed to be true without proof, used as a basis for developing an argument.

- Is probability mathematics?
- How is mathematical language useful in statistics and probability?

Can you use statistics to prove anything?

Results of statistical experiments or analyses are often given within a confidence interval of, say, 95%.

- Discuss these statements:

"Statistics is a systematic method for finding a wrong answer with a 95% confidence interval." Anonymous

"Statistics may be defined as a body of methods for making wise decisions in the face of uncertainty." W. Allen Wallis, American economist and statistician, (1912-88)

"I could prove God statistically. Take the human body alone – the chances that all the functions of an individual would just happen is a statistical monstrosity." George Gallup, American statistician and father of the 'Gallup Poll', (1901-84)

"An unsophisticated forecaster uses statistics as a drunken man uses lamp-posts, for support rather than illumination." Andrew Lang, Scottish poet, (1844-1912)

- Are there aspects of mathematics you can choose whether or not to believe?

Information and data

Find the dictionary definitions of 'information' and 'data'.

- Do we derive data from information, or information from data?
- Is data always true?
- Is information always true?

> *"There are Five kinds of lies: Lies, Damned Lies, Statistics, Politicians quoting Statistics and Novelists quoting Politicians on Statistics."*
>
> Stephen K. Tagg, University of Strathclyde Business School

How are statistics and probability related?

When you roll a dice, probability theory says the probability of a 'six' on a fair dice is $\frac{1}{6}$. When you do a probability experiment, such as rolling the dice 60 times, and collecting data on the number of sixes, you are using statistics.

Suppose you roll the dice 60 times and get 8 sixes, is the dice fair? Statistics uses probability theory (that $P(6) = \frac{1}{6}$) to help answer this question. A statistician could decide whether the dice is fair and give confidence limits for this decision.

You cannot carry out statistics without the theoretical base of probability, but you can study probability without using data. However, for many situations the best estimate of probability is the relative frequency of the event, based on real data.

- Think of real-life situations where
  - ▶ the best estimate of probability has to come from real data
  - ▶ the best estimate of probability is based on probability theory.

Probability and intuition

Most of the time formal definitions of probability seem to agree with our intuitive understanding. If we pick a ball from a bag with with eight red and two green counters, we understand that we are more likely to pick a red one, but a green one is unlikely, not impossible.

Sometimes our intuition leads us to wrong conclusions.

The birthday problem

In a class of 25 people, what is the probability that two people share the same birthday?

- What do you think? 1%? Maybe 5%, or even as much as 10%?
- Find the solution to this classic problem – either work it out yourself or find one online or in a book.

In probability theory, independence is defined as:

$P(A \cap B) = P(A) \times P(A)$

- Can we understand this independence intuitively?

7 The evolution of calculus

Before you start

You should know how to:

1 Find the derivatives of linear, polynomial, rational, exponential and logarithmic functions. e. g. Differentiate
$$y = e^{2x+3} \ln(1-x)^2$$
$$\frac{dy}{dx} = 2e^{2x+3}\ln(1-x)^2 - e^{2x+3}\frac{2(1-x)}{(1-x)^2}$$
Using the product and chain rules
$$\frac{dy}{dx} = \frac{2e^{2x+3}((x-1)\cdot\ln((x-1)^2)+1)}{x-1}$$

2 Find points of intersection between the graphs of two functions. e. g. find the point where the graphs of the functions $y = e^{-\frac{x}{2}}$ and $y = e^{x+1}$ intersect.

At this point $e^{-\frac{x}{2}} = e^{x+1}$ therefore $-\frac{x}{2}$
$$= x + 1$$
$$x = -\frac{2}{3}, y = e^{\frac{1}{3}}$$
so point of intersection is $(-\frac{2}{3}, e^{\frac{1}{3}})$

3 Find the velocity and acceleration given the displacement. e. g. For a displacement function $s(t)$, velocity is the first derivative and acceleration is the second derivative.

Skills check

1 Find the derivatives of these functions.

a $y = x\ln(x)$

b $y = \dfrac{e^{2x-3}}{\sqrt{2-x}}$

> Use the differentiation rules from Chapter 4.

c $y = x^4 - \dfrac{1}{x^4}$

2 Find the point(s) of intersection of the graphs of these functions.

a $y = 3x - 2$ and $y = x^2 - 2x + 4$

b $y = 1 - x$ and $y = \sqrt{2x+1}$

c $y = \dfrac{6}{x} + 3x$ and $y = x^3 - 5x$

3 A particle moves along a line so that its displacement at any time t is $s(t) = 3t^4 - t^3 + t$. Find expressions for the velocity and acceleration of the particle at any time t.

Integral calculus

How to calculate surface areas and volumes of regular shapes such as rectangles and cylinders has been common knowledge for thousands of years; but how do architects and engineers calculate areas and volumes of curved spaces, such as the aquarium in Valencia, Spain?

About 2000 years ago, **Archimedes** was one of the first mathematicians to attempt to find the area between a parabola and a chord. His method was to fill the area with shapes whose areas were known, for example triangles. He did this until the space not covered was so small as to be negligible, or in the words of Newton and Leibniz, infinitesimally small. Modern-day mathematicians call this the method of 'exhaustion'.

This chapter looks at integral calculus. In section 7.3 you will see how integration is related to areas under curves.

> Although we study the derivative first, some of the concepts of integration were known long before differentiation. These ideas were important in the beginnings of fair trade, which depended in part on knowing how to work out areas of regular and irregular shapes.

Method of exhaustion

Take a circle and start filling it with isosceles triangles from its center. The sides of the triangles are radii. The altitude, $CD = h$, is shorter than the radius, CB. If we create n such triangles, then the sum of the areas of the triangles approximates the area of the circle, $A \approx \sum_{i=1}^{n} \frac{1}{2} b_i h_i$

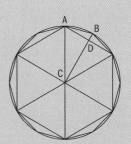

As we increase the number of triangles, the altitudes of the triangles get closer to the length of the radius, and the sum of the bases of the triangles approaches the actual circumference of the circle, so we can write

$A \approx \sum_{i=1}^{n} \frac{1}{2} r(b_i) \approx \frac{1}{2} r \cdot (2\pi r) \approx \pi r^2$. We can see that as we increase the number of triangles, the sum of their areas gets closer to the actual area of the circle, until $\lim_{n \to \infty} \sum_{i=1}^{n} \frac{1}{2} r(b_i) = \pi r^2$

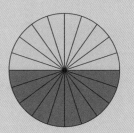

This is an example of the method of exhaustion.
Archimedes figured out the area between a parabola and a chord. How did he do it?

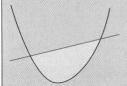

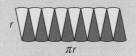

Choose a convenient shape whose area you know how to find, and fill the space between the chord and the parabola with these shapes, to 'exhaustion'!
Perhaps this prompted Leibniz to say, "He who understands Archimedes and Apollonius will admire less the achievements of the foremost men of later times."

> Archimedes did not use coordinate axes – this system was invented by **René Descartes** in the 17th century.

7.1 Integration as anti-differentiation

The process of finding a function $f(x)$ whose derivative is $f'(x)$ is called anti-differentiation, which relates to **integration**.

For example, you know that the derivative of x^2 with respect to x is $2x$, hence when you anti-differentiate $2x$ with respect to x you obtain x^2. This, however, is not the only answer, since, for example, $y = x^2 + 3$ also has a derivative of $2x$.

You can easily see that $2x$ is the derivative for any function of the form $y = x^2 + c$, where c is any real number. This set, or family, of all anti-derivatives of a function is called the **indefinite integral** of the function, and c is called the constant of integration.

This can be written using symbols as $\int 2x \, dx = x^2 + c, c \in \mathbb{R}$

> Mathematical models provide solutions to real-world problems. Analyze mathematical models used to approximate areas and volumes of irregular shapes. Discuss how welll these models approximate the actual areas and volumes of the shapes found through calculus methods.

> The integration symbol is an elongated S, and was first used by Leibniz.

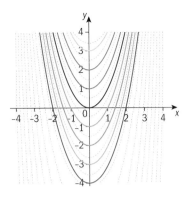

▲ Graph of family of curves of $y = x^2 + c$ for different values of c

Notice the lines in the background of the graphs. They form a slope field for the family of curves $y = x^2 + c$, i.e., they show the direction of the tangent lines at the different values of x.

The tangent lines are parallel for corresponding values of x.

Slope fields is a topic in the calculus option.

In general terms

→ $$\int f(x)\,dx = F(x) + c,\ c \in \mathbb{R}$$

$f(x)$ is called the **integrand**, and x is the variable of integration.

Differentiating x^n:

$x^n \rightarrow$ | multiply the coefficient of x by n | $\rightarrow$ | decrease the power of n by 1 | $\rightarrow nx^{n-1}$

Reversing the process:

$nx^{n-1} \rightarrow$ | increase the exponent by 1 | $\rightarrow$ | divide by the new exponent | $\rightarrow$ | add a constant of integration, c | $\rightarrow x^n + c$

In general

→ $$\int x^n\,dx = \frac{x^{n+1}}{n+1},\ n \neq -1$$

Recall from chapter 4 the constant multiple rule for differentiation,

for c a real number, $f'(cx) = cf'(x)$ provided $f'(x)$ exists.

The reverse is also true, i.e., $\int cf(x)\,dx = c\int f(x)\,dx$.

Example 1

a Find the indefinite integral of $-4x^3$ **b** Find $\int -\dfrac{5}{x^7}\,dx$

Answers

a $$\int -4x^3\,dx = -4\int x^3\,dx$$
$$= -4\frac{x^4}{4} + c$$
$$= -x^4 + c$$

Differentiate your answer mentally to check your result, differentiating $-x^4$ gives $-4x^3$. Don't forget the constant of integration.

▶ Continued on next page

b $\displaystyle\int -\frac{5}{x^7}\,dx = -5\int x^{-7}\,dx$

$\displaystyle\quad\quad\quad\quad = -5\,\frac{x^{-6}}{-6}+c$

$\displaystyle\quad\quad\quad\quad = \frac{5}{6x^6}+c$

Again, differentiate your answer to check your result.

Example 2

Integrate $f(x)=\sqrt[3]{x^2}$

Answer

$\displaystyle\int \sqrt[3]{x^2}\,dx = \int x^{\frac{2}{3}}\,dx = \frac{x^{\frac{5}{3}}}{\frac{5}{3}}+c = \frac{3}{5}x^{\frac{5}{3}}+c$

Change the radical to a rational exponent and use the power rule.

Exercise 7A

Find these indefinite integrals, with respect to x.

1 $-2x$ **2** $3x^8$

3 $-5x^4$ **4** $\dfrac{1}{x^5}$

5 $\sqrt{x^3}$ **6** $\dfrac{1}{\sqrt{x^3}}$

7 $\dfrac{2x}{\sqrt{x}}$ **8** $-\dfrac{\sqrt[4]{x^5}}{7x^3}$

There is another rule that is useful in integrating functions. It is the reverse of the sum and difference differentiation rule.

$$\rightarrow\quad \int [f(x)\pm g(x)]\,dx = \int f(x)\,dx \pm \int g(x)\,dx$$

Example 3

Integrate $1-\sqrt[4]{x}$ with respect to x.

Answer

$1-\sqrt[4]{x} = 1- x^{\frac{1}{4}}$

$\displaystyle\int\left(1-x^{\frac{1}{4}}\right)dx = x - \frac{x^{\frac{5}{4}}}{\frac{5}{4}}+c$

$\displaystyle\quad\quad\quad\quad\quad = x - \frac{4}{5}x^{\frac{5}{4}}+c$

Change radicals to exponents.
Integrate term by term.
Note that
$\displaystyle\int 1\,dx = \int 1\times x^0\,dx = \frac{x^1}{1} = x+c$

From the family of curves, you can find a specific curve that passes through a given point.

Example 4

If $\dfrac{dy}{dx} = \left(1 - \dfrac{1}{x^2}\right)^2$

find y given that the graph of the function passes through the point $(1,0)$.

Answer

$\left(1 - \dfrac{1}{x^2}\right)^2 = 1 - \dfrac{2}{x^2} + \dfrac{1}{x^4}$	*Expand.*
$\qquad\qquad = 1 - 2x^{-2} + x^{-4}$	
$\dfrac{dy}{dx} = 1 - 2x^{-2} + x^{-4} + c$	*Use properties of indices, and integrate term by term. Don't forget the constant of integration.*
$\Rightarrow y = x + \dfrac{2}{x} - \dfrac{1}{3x^3} + c$	
At $(1,0)$, $0 = 1 + 2 - \dfrac{1}{3} + c$	*Substitute (1, 0) into the equation for y, and find c.*
hence $c = -2\dfrac{2}{3}$	
$\therefore y = x + \dfrac{2}{x} - \dfrac{1}{3x^3} - 2\dfrac{2}{3}$	*Rewrite y with the value of c.*

Exercise 7B

1 Integrate these with respect to x.

a $5x^2 - \dfrac{1}{5x^2}$　　　**b** $(x + 3)(2x - 1)$　　　**c** $\dfrac{x^2 - 1}{x^4}$

d $\left(x + \dfrac{1}{x}\right)^2$　　　**e** $\dfrac{(x+3)(x-4)}{x^5}$　　　**f** $\sqrt{x} - \dfrac{5}{\sqrt[3]{x}}$

2 If $\dfrac{dy}{dx} = (3x^2 - 4)$, find y given that the function passes through the point $(2, -1)$.

3 If $f'(t) = t + 3 - \dfrac{1}{t^2}$, find f given that the curve goes through the point $\left(1, -\dfrac{1}{2}\right)$.

4 If $\dfrac{dy}{dx} = (2x + 3)^3$, find y if $y = 2$ when $x = -1$.

5 Find A in terms of x if $\dfrac{dA}{dx} = (2x + 1)(x^2 - 1)$, and $A = 0$ when $x = 1$.

6 Find s in terms of t if $\dfrac{ds}{dt} = 3t - \dfrac{8}{t^2}$, and $s = 1.5$ when $t = 1$.

7 Find y in terms of x given that $\dfrac{d^2 y}{dx^2} = 6x - 1$, and when $x = 2$, $\dfrac{dy}{dx} = 4$ and $y = 0$.

EXAM-STYLE QUESTION

8 A particle moves in a straight line such that at time t seconds, its acceleration $a(t) = 6t + 1$. When $t = 0$, the velocity is $2\ \text{m s}^{-1}$, and its displacement from the origin is $1\ \text{m}$. Find expressions for the velocity and the displacement.

In question 4 of exercise 7B, you found the integral of $(2x + 3)^3$ by first expanding and then integrating each term. It would be more efficient to find a method of integration without needing to expand the expression, especially if the power is large.

Investigation – integrating $(ax + b)^n$

Integrate different expressions of the form $(ax + b)^n$, where a, b and n are real numbers, and $a \neq 0$, $n \neq -1$. Predict the integral of all expressions of this form. Prove your conjecture by differentiating your result.

Use the form from the investigation to integrate question 4 of exercise 7B, and then apply your prediction from the investigation.

In order to integrate $(2x + 3)^3$, let $u = 2x + 3$, and hence $\dfrac{du}{dx} = 2$, so $dx = \dfrac{du}{2}$. You can therefore write

$$\int (2x + 3)^3 \, dx = \int u^3 \frac{du}{2} = \frac{1}{2} \int u^3 \, du$$

> Although, strictly speaking $\dfrac{du}{dx}$ is not a fraction, it conveniently behaves as one, See chapter 4, the chain rule.

> The variable in the integrand must be the same as the variable of integration, i.e. here you have u^3 and du.

Integrating u^3 with respect to u,

$$\frac{1}{2} \int u^3 \, du = \frac{1}{2} \cdot \frac{u^4}{4} + c = \frac{u^4}{8} + c = \frac{(2x+3)^4}{8} + c$$

Substitute the original expression for u.

The result obtained from the investigation is called the compound formula.

$$\rightarrow \int (ax + b)^n \, dx = \frac{1}{a(n+1)} (ax + b)^{n+1} + c, \; a \neq 0$$

> The compound formula can be used for linear functions only.

Example 5

Integrate $\sqrt{1 - 2x}$ with respect to x.

Answer

Solution 1:

Let $u = 1 - 2x$, then $\dfrac{du}{dx} = -2$, and $dx = \dfrac{du}{-2}$. Hence

By substitution

$$\int \sqrt{1 - 2x} \, dx = -\frac{1}{2} \int u^{\frac{1}{2}} \, du = -\frac{1}{2} \frac{u^{\frac{3}{2}}}{\frac{3}{2}} + c = -\frac{1}{3} u^{\frac{3}{2}} + c$$

$$\int \sqrt{1 - 2x} \, dx = -\frac{1}{3}(1 - 2x)^{\frac{3}{2}} + c$$

Solution 2:

$$\int (1 - 2x)^{\frac{1}{2}} dx = \frac{1}{-2\left(\frac{3}{2}\right)} (1 - 2x)^{\frac{3}{2}} + c$$

$$= -\frac{1}{3}(1 - 2x)^{\frac{3}{2}} + c$$

Use the compound formula.

> There is a more advanced integration by substitution method in chapter 9.

Using the compound formula is quicker and easier than using the method of substitution.

Example 6

Find $\displaystyle\int \frac{3}{\sqrt[3]{4-5x}}\,dx$.

Answer

$\displaystyle\int \frac{3}{\sqrt[3]{4-5x}}\,dx = 3\int (4-5x)^{-\frac{1}{3}}\,dx$ *Apply compound formula.*

$\displaystyle = \frac{3}{-5\cdot\frac{2}{3}}(4-5x)^{\frac{2}{3}} + c$

$\displaystyle = -\frac{9}{10}(4-5x)^{\frac{2}{3}} + c$

Exercise 7C

Integrate these with respect to x.

1 $(3x-1)^7$ **2** $-2\sqrt{2x+1}$ **3** $\dfrac{1}{(4x-1)^5}$

4 $\dfrac{2}{\sqrt[4]{3-x}}$ **5** $\dfrac{2}{(2-5x)^{\frac{1}{3}}} + \sqrt[3]{1-x}$ **6** $4\sqrt{2-3x} - 6(3x+2)^{\frac{2}{3}}$

Integration of exponential functions

In Chapter 5 you learned how to differentiate exponential functions. In particular, $\dfrac{d}{dx}(e^x) = e^x$.

$y = e^x$ is the only function whose gradient function is equal to the function itself for all x in the domain.

Therefore

$\rightarrow \displaystyle\int e^x\,dx = e^x + c$

Furthermore, it is easy to confirm that

$\rightarrow \displaystyle\int e^{ax+b}\,dx = \frac{1}{a}e^{ax+b} + c$

Use substitution:

let $u = ax + b$, then $\dfrac{du}{dx} = a$, or $dx = \dfrac{du}{a}$

Hence $\displaystyle\int e^{ax+b}\,dx = \int e^u\,\frac{du}{a} = \frac{1}{a}\int e^u\,du = \frac{1}{a}e^u + c = \frac{1}{a}e^{ax+b} + c$

Example 7

Find $\int 4e^{-2x}\,dx$

Answer

$$\int 4e^{-2x}\,dx = 4\int e^{-2x}\,dx = \frac{4}{-2}\,e^{-2x} + c = -2e^{-2x} + c$$

Example 8

Integrate $\sqrt{e^{5-3x}}$ with respect to x.

Answer	
$\sqrt{e^{5-3x}} = (e^{5-3x})^{\frac{1}{2}} = e^{\frac{5}{2}-\frac{3}{2}x}$	*Write $\sqrt{e^{5-3x}}$ using exponents.*
$\int e^{\frac{5}{2}-\frac{3}{2}x}\,dx = -\frac{2}{3}e^{\frac{5}{2}-\frac{3}{2}x} + c$	

Recall also that $\dfrac{d}{dx}(2^x) = 2^x \ln(2)$

Hence, $\int 2^x \ln(2)\,dx = \ln(2)\int 2^x\,dx = 2^x + c$

If you now want to integrate 2^x, you need to divide by $\ln(2)$, since $\ln(2)$ is not part of the integral. That is, $\int 2^x\,dx = \dfrac{1}{\ln(2)}2^x + c$

If you now differentiate the result, you obtain 2^x.

Using the compound formula, you can also integrate 2^{3x-1} with respect to x. In particular

→ $\int m^{ax+b}\,dx = \dfrac{1}{a\ln(m)}m^{ax+b} + c$, where m is a positive real number, $a \neq 0$.

Example 9

Find $\int 2^{3x-1}\,dx$.

Answer

$$\int 2^{3x-1}\,dx = \frac{1}{3\ln(2)}2^{3x-1} + c$$

Exercise 7D

In questions 1 to 6, integrate with respect to x.

1 $-5e^{-2x}$ **2** $\dfrac{1}{e^{3x+2}}$ **3** $\sqrt[3]{e^x} - \dfrac{2}{e\sqrt{e^{2x}}}$

4 3^x **5** $\dfrac{1}{3^{2x}}$ **6** 4^{1-x}

7 Use the method of substitution to derive the compound rule for exponential functions, to show that for a real positive number m,

for $a \neq 0$ $\displaystyle\int m^{ax+b}\,dx = \dfrac{1}{a\ln(m)}m^{ax+b} + c$

Integration and logarithmic functions

In Chapter 5 you differentiated logarithmic functions.

For $x > 0$, $\dfrac{d}{dx}(\ln x) = \dfrac{1}{x}$ so for $x > 0$, $\displaystyle\int \dfrac{1}{x}\,dx = \ln x + c$

For $x < 0$, $\dfrac{d}{dx}\ln(-x) = \dfrac{1}{-x}(-1) = \dfrac{1}{x}$ so for $x < 0$, $\displaystyle\int \dfrac{1}{x}\,dx = \ln(-x) + c$

The two statements above can be combined into

$$\rightarrow \int \dfrac{1}{x}\,dx = \ln|x| + c$$

Similarly, using the compound formula,

$$\rightarrow \int \dfrac{1}{(ax+b)}\,dx = \dfrac{1}{a}\ln|ax+b| + c,\ a \neq 0$$

You can confirm this result by differentiation.

Example 10

Find $\displaystyle\int \dfrac{3}{1-2x}\,dx$

Answer

$\displaystyle\int \dfrac{3}{1-2x}\,dx = 3\int \dfrac{1}{1-2x}\,dx = -\dfrac{3}{2}\ln|1-2x| + c$

Exercise 7E

Integrate with respect to x, $x \neq 0$.

1 $\dfrac{1}{3x}$ **2** $-\dfrac{6}{x}$ **3** $\dfrac{1}{2-3x}$

4 $\dfrac{5}{3-5x}$ **5** $-2(4+3x)^{-1}$

7.2 Definite integration

As you have seen in the previous section, the result of indefinite integration is a family of functions. The process of **definite integration**, however, results in a numerical answer.

In Chapter 4 you worked on kinematic problems. Since velocity is the rate of change of the displacement with respect to time, to obtain the velocity you differentiate the displacement function. Hence, to obtain the displacement from the velocity function, you reverse the process, and anti-differentiate, or integrate the velocity function.

Consider an example. The velocity of a particle at any time t, in seconds, is given by $3t^2 + t\ \text{m s}^{-1}$. Find the total distance traveled from $t = 1\,\text{s}$ to $t = 2\,\text{s}$.

In order to find the total distance traveled, see if the particle changed direction anywhere in the interval [1, 2]. The graph of the function $f(t) = 3t^2 + t$ shows that the velocity is positive throughout this interval, so the particle did not change direction.

You can evaluate the displacement at $t = 1$ and $t = 2$, and the distance traveled will be the difference of these two values.

Integrate the velocity function to get the displacement function:

$$\frac{ds}{dt} = 3t^2 + t \Rightarrow s = t^3 + \frac{t^2}{2} + c$$

Evaluate the displacement at $t = 1$ and $t = 2$:

When $t = 1$, $s = 1.5 + c$, and when $t = 2$, $s = 10 + c$

Subtracting these two values for s gives $8.5\,\text{m}$ as the total distance traveled between $t = 1$ and $t = 2$.
The constant of integration cancels out when subtracting.

There is a special notation for evaluating a definite integral in this manner.

upper limit

$$\int_1^2 (3t^2 + t)\, dt = \left[t^3 + \frac{t^2}{2} \right]_1^2 = \left(8 + \frac{2^2}{2} \right) - \left(1 + \frac{1}{2} \right) = 8.5$$

lower limit

evaluate at upper limit

evaluate at lower limit

If a function f is continuous on an interval $[a, b]$, then its definite integral exists over this interval. Here are some properties of definite integrals.

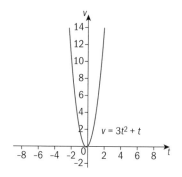
$v = 3t^2 + t$

All applications of the definite integral used later in this chapter require the numerical evaluation of an integral.

See Chapter 4, Example 36.

$10 + c - (1.5 + c)$
$= 8.5$

You will study more applications of definite integration later in the chapter.

Write the integral in square brackets, with upper and lower limits as shown. Since c always cancels out, you don't need to write it.

The proofs of some of these properties are beyond the scope of this course.

Properties of definite integrals

If the integral of f with respect to x in the interval $[a, b]$ exists, then

→ **1** $\displaystyle\int_a^b f(x)\,dx = -\int_b^a f(x)\,dx$

2 $\displaystyle\int_a^a f(x)\,dx = 0$

3 $\displaystyle\int_a^b kf(x)\,dx = k\int_a^b f(x)\,dx$

4 $\displaystyle\int_a^b (f(x) \pm g(x))\,dx = \int_a^b f(x)\,dx \pm \int_a^b g(x)\,dx$

5 $\displaystyle\int_a^b f(x)\,dx + \int_b^c f(x)\,dx = \int_a^c f(x)\,dx$

You can test these properties using the particle example. For example, testing the first property,

$$\int_2^1 (3t^2 + t)\,dt = \left[t^3 + \frac{t^2}{2} \right]_2^1 = \left(1 + \frac{1}{2}\right) - \left(2^3 + 2\right) = -8.5$$

> Use the particle example to test properties 2 to 5 of definite integrals.

Example 11

Evaluate $\displaystyle\int_0^1 (x^2 + 4x + 2)\,dx$

Answer

$\displaystyle\int_0^1 (x^2 + 4x + 2)\,dx = \left[\dfrac{x^3}{3} + 2x^2 + 2x \right]_0^1$	*Use property 4.*
$\qquad = \left(\dfrac{1}{3} + 2 + 2\right) - 0$	
$\qquad = 4\dfrac{1}{3}$	

> **GDC help on CD:**
> *Alternative demonstrations for the TI-84 Plus and Casio fx-9860GII GDCs are on the CD.*

Example 12

Evaluate $\displaystyle\int_{-1}^0 \frac{3}{1-2x}\,dx$

Answer

$\displaystyle\int_{-1}^0 \frac{3}{1-2x}\,dx = -\frac{3}{2}\left[\ln|1-2x| \right]_{-1}^0$	*Take out $-\dfrac{3}{2}$ as a factor.*
$\qquad = -\dfrac{3}{2}[\ln 1 - \ln 3]$	*Use property 3.*
$\qquad = \dfrac{3}{2}\ln(3)$	*ln(1) = 0*

> You can confirm the results of Examples 11 and 12 on a GDC:

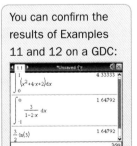

Exercise 7F

Evaluate these definite integrals. Check your results using a GDC.

1 $\displaystyle\int_1^3 (3x + \frac{1}{x^2})dx,\ x \neq 0$

2 $\displaystyle\int_0^2 3\sqrt{4x+1}\,dx$

3 $\displaystyle\int_{-1}^2 -2e^{1-3x}\,dx$

4 $\displaystyle\int_1^3 3(2^{x+1})dx$

5 $\displaystyle\int_{-2}^0 2(1-3x)^5\,dx$

6 $\displaystyle\int_1^4 \frac{1-\sqrt{x}}{\sqrt{x}}dx,\ x \neq 0$

The properties of the definite integral are based on the assumption that the integral exists within the specific bounds of integration. Before integrating you need to check if f is continuous in the given interval.

Example 13

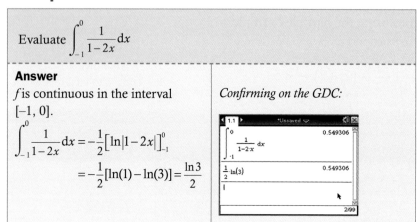

Evaluate $\displaystyle\int_{-1}^0 \frac{1}{1-2x}dx$

Answer

f is continuous in the interval $[-1, 0]$.

$$\int_{-1}^0 \frac{1}{1-2x}dx = -\frac{1}{2}\Big[\ln|1-2x|\Big]_{-1}^0$$
$$= -\frac{1}{2}[\ln(1)-\ln(3)] = \frac{\ln 3}{2}$$

Confirming on the GDC:

GDC help on CD:
Alternative demonstrations for the TI-84 Plus and Casio fx-9860GII GDCs are on the CD.

If f is not continuous in the interval of integration, it is possible to obtain a numerical answer, but this answer is invalid.

Example 14

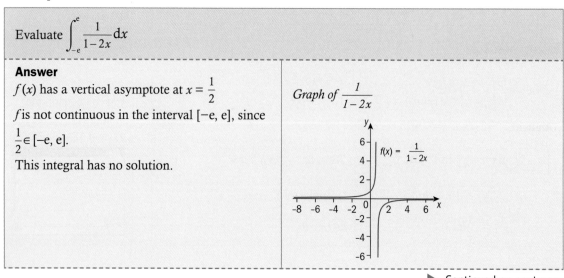

Evaluate $\displaystyle\int_{-e}^e \frac{1}{1-2x}dx$

Answer

$f(x)$ has a vertical asymptote at $x = \dfrac{1}{2}$

f is not continuous in the interval $[-e, e]$, since

$\dfrac{1}{2} \in [-e, e]$.

This integral has no solution.

Graph of $\dfrac{1}{1-2x}$

▶ Continued on next page

$$\int_{-e}^{e} \frac{1}{1-2x}\,dx = -\frac{1}{2}\left[\ln|1-2x|\right]_{-e}^{e}$$

$$= -\frac{1}{2}\left(\ln|1-2e| - \ln|1+2e|\right)$$

$$= -\frac{1}{2}\ln\frac{|1-2e|}{|1+2e|}$$

This result, however, is meaningless since the basic condition necessary is not met, namely, continuity throughout the integrating interval. The GDC integrates it numerically, so the GDC has made a mistake! It does state though that the accuracy is questionable. Some GDCs may give a 'divide by zero' error here.

Although the integral has no solution, you could still proceed and integrate and get a number.

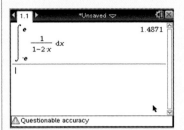

GDC help on CD:
Alternative demonstrations for the TI-84 Plus and Casio fx-9860GII GDCs are on the CD.

Exercise 7G

Evaluate these integrals, if possible.

1 $\displaystyle\int_{-1}^{0} (2r-1)^4\,dr$

2 $\displaystyle\int_{0}^{4} \frac{1-\sqrt{s}}{\sqrt{s}}\,ds$

3 $\displaystyle\int_{0}^{2} \frac{x+1}{x^2-1}\,dx$

4 $\displaystyle\int_{0}^{1} \frac{dx}{(2x+1)^3}$

5 $\displaystyle\int_{-2}^{-1} \frac{1}{x+1}\,dx$

6 $\displaystyle\int_{0}^{1} \left(\frac{3}{3x+4} - \frac{2}{x+1}\right)dx$

7 $\displaystyle\int_{-1}^{1} \frac{e^x+4}{e^x}\,dx$

8 $\displaystyle\int_{0}^{2} 10^x\,dx$

7.3 Geometric significance of the definite integral

Areas between graphs of functions and the axes

Consider a rectangle in the first quadrant formed by the lines $y = h$, $x = b$, and the points (b, h) and the x- and y-axes.

The area of the rectangle is bh. The definite integral of $y = h$ between $x = 0$ and $x = b$ is

$$\int_{0}^{b} h\,dx = h[x]_{0}^{b} = hb$$

Integration gives the area under the line $y = h$ between $x = 0$ and $x = b$

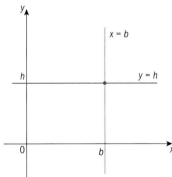

Now, consider a right-angled triangle in the first quadrant formed by the lines $y = \dfrac{h}{b}x$, $x = b$, and the points (b, h) and the x-axis.

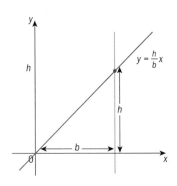

The geometric formula for the area is $\dfrac{1}{2}bh$.

The definite integral of y between $x = 0$ and $x = b$ is

$$\int_0^b \frac{h}{b}x\,dx = \frac{h}{b}\int_0^b x\,dx = \frac{h}{b}\left[\frac{x^2}{2}\right]_0^b = \frac{h}{b}\left[\frac{b^2}{2} - 0\right] = \frac{1}{2}bh$$

Integration gives the area of the triangle.

Consider $\triangle OBC$ formed by the line $y = 2x$, the x-axis, and the line $x = 5$.
Find the area enclosed by the lines $x = 5$ and $x = 2$.

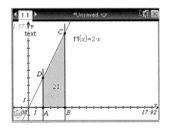

Geometrically it is clear that the area of the shaded part is the difference between the areas of $\triangle OBC$ and $\triangle OAD$.

Area of $\triangle OBC$ is $\dfrac{1}{2}(5 \times 10) = 25$

Area of $\triangle OAD$ is $\dfrac{1}{2}(2 \times 4) = 4$

The difference of the areas is $25 - 4 = 21$ square units.

Use integration:
$$\int_2^5 2x\,dx = \left[x^2\right]_2^5 = 5^2 - 2^2 = 21$$

Consider the area under the curve of the graph of $y = \sqrt{4 - x^2}$
You may recognize this as the equation of a semicircle whose center is the origin, and whose radius is 2.

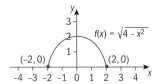

Using the formula for the area of a semicircle,

$$A = \frac{\pi r^2}{2}, \text{ then } A = \frac{4\pi}{2} = 2\pi$$

Now, compare this to the result of $\displaystyle\int_{-2}^{2} \sqrt{4 - x^2}\,dx$, using the GDC.

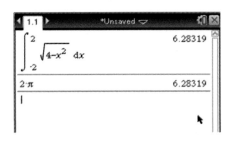

The examples show the relationship between the definite integral and areas of familiar shapes.

In Chapter 9 you will learn how to integrate integrals of this kind analytically.

GDC help on CD:
Alternative demonstrations for the TI-84 Plus and Casio fx-9860GII GDCs are on the CD.

You are now ready to formalize one of the most astonishing and important results of Newton's and Leibniz's work: the connection between differentiation and integration. The theorem justifies the procedures for evaluating definite integrals, and is still regarded as one of the most significant developments of modern-day mathematics.

> **→ The fundamental theorem of calculus**
>
> If f is continuous in $[a, b]$ and if F is any anti-derivative of f on $[a, b]$ then $\int_a^b f(x)\,dx = F(b) - F(a)$

Both Newton and Leibniz approached calculus intuitively. The fundamental theorem of calculus, however, was actually formalized and proved by Augustin-Louis Cauchy (1789–1857). His proof elegantly joined the two branches of differential and integral calculus. Cauchy's last words before he died were indeed self-prophetic, "Men pass away, but their deeds abide".

▲ **Augustin-Lewis Cauchy** (1789–1857) formalized the fundamental theorem of calculus.

Areas of irregular shapes

Look at the area under the curve $y = x^2$ from $x = 0$ to $x = 1$ in the diagrams. On the left is the actual area and on the right is an approximation of this area, using rectangles of base 0.125 and height x^2. Notice that the error in the approximation is the total area of the white space between the curve and the rectangles. You can use the method of exhaustion to fill the space with more rectangles of smaller width.

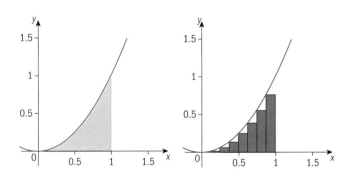

Newton approached the problem of finding areas by viewing the area function as the inverse of the tangent, i.e., the area function depended on the ratio of the difference of the y-values to the x-values, $\dfrac{dy}{dx}$, and employed the use of infinite series. Leibniz, on the other hand, approached the problem by summing the of areas of infinitely thin rectangles, hence the use of an elongated S, the integral symbol.

Using graphing software, it is easy to change n, the number of rectangles under the curve.
Using 15 similar rectangles, the approximation of the area under the curve is 0.3 square units.

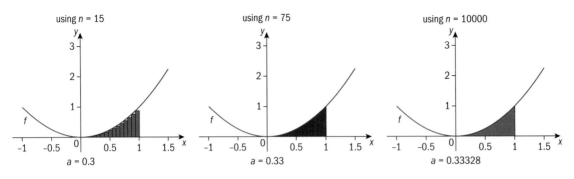

using $n = 15$

$a = 0.3$

using $n = 75$

$a = 0.33$

using $n = 10000$

$a = 0.33328$

We get a better approximation when $n = 75$.

When $n = 10\,000$, the area is about 0.333 sq. units.

You have considered rectangles below the curve, the so-called lower bound sum. You can also approximate the area by drawing rectangles above the curve, the upper bound sum. This time, the error in the approximation is the sum of the areas of the purple spaces above the curve.

Again, consider the upper bound sum with 15, 75 and then 10000 rectangles:

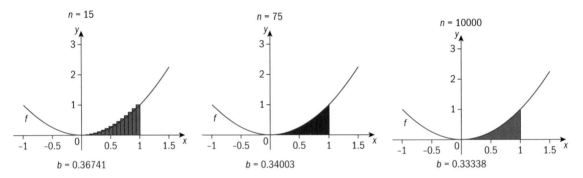

$n = 15$

$b = 0.36741$

$n = 75$

$b = 0.34003$

$n = 10000$

$b = 0.33338$

When $n = 15$, the area is approximately 0.367 sq. units.

When $n = 75$, the area is approximately 0.340 sq. units.

When $n = 10\,000$, the area is approximately 0.333 sq. units.

As the number of rectangles increases, the approximate area approaches the actual area.

This method of approximating the area under a curve is called Riemann sums, after the German mathematician **Georg Friedrich Bernhard Riemann** (1826–66).

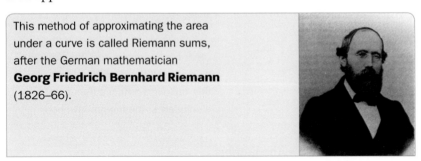

Mathematicians developed different methods to approximate the area under the curve of a graph. Explore some of these methods, and analyze the error of the approximations of the areas that these methods produce.

If f is continuous in the interval $[a, b]$, to find the area under the curve of $f(x)$ from $x = a$ to $x = b$, you can divide $[a, b]$ into n sub-intervals of equal length, $\dfrac{(b-a)}{n}$, and call this $\triangle x$. In each sub-interval, select the height of the rectangle such that a corner of the rectangle is on the curve, and call this $f(c)$.

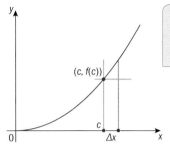

$\triangle$ is the Greek upper case delta. $\triangle x$ is "delta x".

Then, the area under the curve of i such sub-intervals is approximated by $\displaystyle\sum_{i=1}^{n} f(c_i)\triangle x_i$

As $\triangle x$ approaches 0, the number of rectangles n approaches infinity and the approximate area approaches the actual area. You can now ready to define the area under a curve as a definite integral.

> → If the integral of f exists in the interval $[a, b]$, and f is non-negative in this interval, then the area A under the curve $y = f(x)$ from a to b is $A = \displaystyle\int_a^b f(x)\,dx$

Example 15

Find the area bounded by the graph of $y = x^3$, $x = 0$, $x = 2$, and the x-axis.

Answer

Since $y = x^3$ is non-negative in the interval $[0, 2]$

$$A = \int_0^2 x^3 dx = \left[\frac{x^4}{4}\right]_0^2 = \frac{2^4}{4} = \frac{16}{4} = 4 \text{ sq. units}$$

Area $= 4$ sq. units

Confirm on the GDC. The area is entirely below the x-axis.

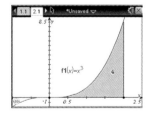

GDC help on CD:
Alternative demonstrations for the TI-84 Plus and Casio fx-9860GII GDCs are on the CD.

Example 16

Find the area of a triangle formed by $f(x) = \begin{cases} x+4, & -4 \le x \le -1 \\ 2-x, & -1 \le x \le 2 \end{cases}$

a using the formula for the area of a triangle
b by integration.

Answers

a

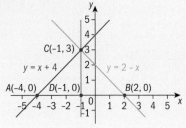

$A = \dfrac{1}{2} \times 6 \times 3 = 9$ sq. units

Sketch the graph.

Area $= \dfrac{1}{2}bh$

b Since both functions are non-negative in the interval $[-4, 2]$
Area of $\triangle ACD$

$$= \int_{-4}^{-1} (x+4)\mathrm{d}x = \left[\frac{x^2}{2} + 4x\right]_{-4}^{-1} = \left(\frac{(-1)^2}{2} + 4(-1)\right) - \left(\frac{(-4)^2}{2} + 4(-4)\right)$$

$$= -3.5 + 8 = 4.5 \text{ sq. units}$$

Divide the triangle into two smaller triangles.

Area of $\triangle BCD$

$$= \int_{-1}^{2} (2-x)\mathrm{d}x = \left[2x - \frac{x^2}{2}\right]_{-1}^{2} = \left(2(2) - \frac{(2)^2}{2}\right)_{-1}^{2} - \left(2(-1) - \frac{(-1)^2}{2}\right)_{-1}^{2}$$

$$= 2 + 2.5 = 4.5 \text{ sq. units}$$

Integrate to find the area of each triangle.

Hence, area of $\triangle ABC = 4.5 + 4.5 = 9$ sq. units

Add the areas.

Alternative solution

$\triangle ACD \equiv \triangle BCD$ (RHS), so area of $\triangle ABC = 2 \times$ Area of $\triangle ACD$
Area of $\triangle ACD = 4.5$ sq. units
Area of $\triangle ABC = 2 \times 4.5 = 9$ sq. units

Notice that the triangle is symmetrical about the line CD.

Now look at areas below the x-axis,
for example, the area above the graph of
$y = x^3$, between $x = -1$ and $x = 0$.
Calculating the integral

$$A = \int_{-1}^{0} x^3 \mathrm{d}x = \left[\frac{x^4}{4}\right]_{-1}^{0} = -\frac{1}{4}$$

Since area is positive, take the absolute

value: $A = \dfrac{1}{4}$ sq. unit

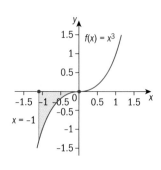

> → When f is negative for all $x \in [a, b]$, then the area bounded by
> the curve and the lines $x = a$ and $x = b$ is $\left| \int_a^b f(x)dx \right|$

For the area below the x-axis for $y = x^3$

$$A = \left| \int_{-1}^0 x^3 dx \right| = \left| \left[\frac{x^4}{4} \right]_{-1}^0 \right| = |{-0.25}| = 0.25$$

Confirming this result on the GDC:

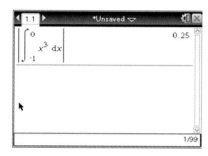

This confirms numerically using the absolute value of the function.

Now look at the area bounded by the graph of $y = x^3$,
$x = -1$, $x = 1$, and the x-axis.

Since the area is partly above and partly below the x-axis,
you have to integrate the functions in the two intervals separately.

$$A = \left| \int_{-1}^0 x^3 dx \right| + \int_0^1 x^3 dx = |{-0.25}| + 0.25 = 0.5$$

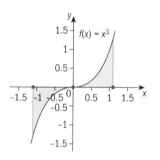

You can also evaluate this area graphically on the GDC by graphing
$y = |x^3|$. To evaluate the integral numerically on the GDC, enter the
integral of the absolute value of the function. This eliminates the
need for separating the integrals.

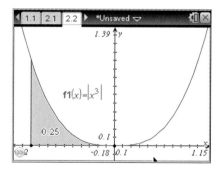

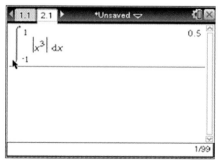

GDC help on CD:
*Alternative demonstrations
for the TI-84 Plus and Casio
fx-9860GII GDCs are on the
CD.*

Example 17

Find the area of the region bounded by the graph of the function
$y = \dfrac{1}{2}(x-1)(x+2)(x-3)$ and the x-axis and confirm your answer graphically on the GDC.

Answer

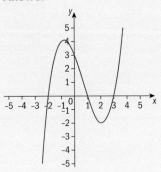

Graph the function on your GDC. Since part of the graph lies below the x-axis, integrate the function separately in the intervals where it is above and below the x-axis.

$$A = \left| \int_{-2}^{1} \left(\frac{1}{2}(x-1)(x+2)(x-3)\right) dx \right. + \left. \int_{1}^{3} \left(\frac{1}{2}(x-1)(x+2)(x-3)\right) dx \right|$$

$$A = \int_{-2}^{1} \left(\frac{1}{2}x^3 - x^2 - \frac{5}{2}x + 3\right) dx \ + \left| \int_{1}^{3} \left(\frac{1}{2}x^3 - x^2 - \frac{5}{2}x + 3\right) dx \right|$$

$$= \left[\frac{x^4}{8} - \frac{x^3}{3} - \frac{5x^2}{4} + 3x\right]_{-2}^{1} + \left| \left[\frac{x^4}{8} - \frac{x^3}{3} - \frac{5x^2}{4} + 3x\right]_{1}^{3} \right|$$

$$= \frac{63}{8} + \left|-\frac{8}{3}\right| = \frac{253}{24} = 10.5 \text{ sq. units to 3 sf.}$$

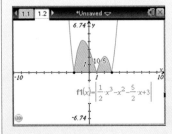

On the GDC, enter the absolute value of the function and the interval itself as the lower and upper bound.

GDC help on CD:
Alternative demonstrations for the TI-84 Plus and Casio fx-9860GII GDCs are on the CD.

$A = 10.5$ sq. units

→ The total area of $f(x)$ in an interval $[a, b]$, where its graph is partly above and partly below the x-axis is $A = \displaystyle\int_{a}^{b} \left| f(x) \right| dx$.

Investigation – odd and even functions

In Example 15 you looked at $f(x) = x^3$ which is an odd function.
Choose different odd functions continuous in an interval $[a, b]$.

For your examples, find $\displaystyle\int_{-a}^{a} f(x)\,dx$

Make a conjecture and justify it.
Does your conjecture hold when applying this definite integral to areas? Explain.
Do the same for even functions continuous in an interval $[a, b]$.

Exercise 7H

Find the area of the region bounded by the graph of the function, the x-axis, and the given lines.

1 $y = x^4 - x$, $x = -1$ and $x = 1$

2 $y = x^2 - 2x - 3$, $x = -1$ and $x = -3$

3 $y = x^2 - 2x - 3$, $x = -3$ and $x = 1$

> Graph the functions on your GDC. Find the areas by integration. Then check your answer on your GDC.

In questions 4 –11, find the area of the region bounded by the graph of the function, the x-axis, and the given lines.

4 $y = e^x - 3$, $x = 0$, $x = 3$

5 $y = x^4 + 3x^3 - 3x^2 - 7x + 6$, $x = -3$; $x = 1$

6 $y = \sqrt{4 - x}$, $x = 0$, $x = 4$

7 $y = \dfrac{1}{x^2} + 1$, $x = \dfrac{1}{2}$, $x = 5$

8 $y = 2^x$, $x = 1$, $x = 2$

9 $y = 2e^{-x+1} - 1$, $x = 0$, $x = 3$

10 $y = \dfrac{1}{x+2}$, $x = -1$, $x = 2$

11 $y = \dfrac{2}{3 - 4x}$, $x = 1$, $x = 3$

12 Find the area of the region bounded by the graph of $y = -x^3 + 6x^2 + x - 30$, its x-intercepts, and the x-axis.

13 Find the area of the region enclosed by $y = \begin{cases} x^2, & 0 \le x < 1 \\ 2 - x, & 1 \le x \le 2 \end{cases}$ and the x-axis.

14 Find the area of the region enclosed by $y = \begin{cases} \sqrt{x}, & 0 \le x < 1 \\ x^2, & 1 \le x \le 2 \end{cases}$ and the x-axis.

The graph shows the region bounded by the graph of the function $y = e^x$; the y-axis, and the line $y = e$.

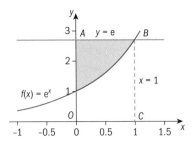

You can find this area by first finding the area of the region below the curve bounded by the graph of the function, the x-axis, and the lines $x = 0$ and $x = 1$.
Then subtract this area from that of the rectangle $OABC$, which is e sq. units.
Hence, the area of the desired region is

$$e - \int_0^1 e^x dx = e - \left[e^x \right]_0^1 = e - (e - 1) = 1 \text{ sq. unit}$$

You can also obtain the result by rearranging to make x the subject and then integrating with respect to y, from $y = 1$ to $y = e$.
If $y = e^x$ then $x = \ln(y)$, and, $A = \int_1^e \ln(y) dy = 1$ sq. unit

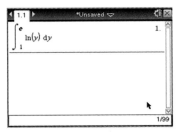

Since you don't yet know how to integrate $\ln(y)$ analytically (this will be covered in chapter 9), use the GDC to confirm the result.

Example 18

Find the area of the region bounded by the graph of the function $y = \dfrac{1}{x^2}$ and the lines $y = 1$ and $y = 4$.

Answer

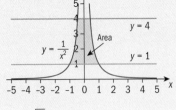

$x = \sqrt{\dfrac{1}{y}}$

$A = 2 \int_1^4 \sqrt{\dfrac{1}{y}} dy$

$A = 2 \left[2y^{\frac{1}{2}} \right]_1^4 = 4 \left(4^{\frac{1}{2}} - 1 \right)$

$= 4$ sq. units

Graph the function.

Make x the subject.

Integrate with respect to y.

$\int \sqrt{\dfrac{1}{y}} dy$ *gives the area to the right of the y-axis.*

By symmetry, A is double the area on the right of the y-axis.

Confirm on the GDC.

GDC help on CD:
Alternative demonstrations for the TI-84 Plus and Casio fx-9860GII GDCs are on the CD.

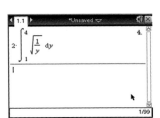

Example 19

Find the area of the region bounded by the graph of the function
$y = x^3 + 1$, the y-axis, and the lines $y = 1$ and $y = 9$.

Answer

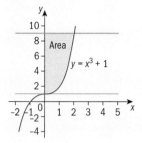

Graph the function on your GDC and identify the area.

$x = \sqrt[3]{y-1}$

Make x the subject.

$$A = \int_1^9 (y-1)^{\frac{1}{3}}\,dy$$

Integrate with respect to y.

$$= \left[\frac{3(y-1)^{\frac{4}{3}}}{4} \right]_{-1}^9$$

$$= \frac{3(8)^{\frac{4}{3}}}{4} - 0 = 12 \text{ sq. units}$$

Confirm on the GDC.

Alternative solution

Area of rectangle
OABC $= 9 \times 2 = 18$

Area above curve

$$= 18 - \int_0^2 (x^3 + 1)\,dx$$

$$= 18 - \left[\frac{x^4}{4} + x \right]_0^2 = 18 - 6$$

$$= 12 \text{ sq. units}$$

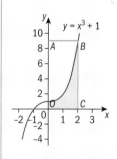

Why does the alternative method used here not work for Example 18?

Exercise 7I

Find the areas of the regions bounded by the function, the y-axis, and the given lines.

1 $y = x^2 + 1,\quad y = 1, y = 10$

2 $y = \sqrt{x},\quad y = 0, y = 4$

3 $y = \sqrt{4-x},\quad y = 0, y = 2$

4 $y = 4 - x^2, y = 3, y = 4$

5 $y = \dfrac{1}{\sqrt{-x+4}}, y = \dfrac{1}{2}, y = 2$

Areas of regions between curves

The graph shows two curves $f(x)$ and $g(x)$.

The regions bounded by the two curves are shaded.

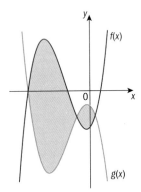

Translate both graphs vertically so that both areas are above the x-axis.

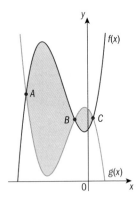

A translation of both graphs by the same amount in the same direction preserves the original area.

The area between points A and B is the difference of the areas under the curves $f(x)$ and $g(x)$ from A to B.

> → If functions f and g are continuous in the interval $[a, b]$, and $f(x) \geq g(x)$ for all $x \in [a, b]$, then the area between the graphs of f and g is
>
> $$A = \int_a^b f(x)\, dx - \int_a^b g(x)\, dx = \int_a^b (f(x) - g(x))\, dx$$

Similarly the area between points B and C is the difference of the areas under the curves $g(x)$ and $f(x)$ between points B and C,

$$A = \int_b^c g(x)\, dx - \int_b^c f(x)\, dx = \int_b^c (g(x) - f(x))\, dx$$

To find the total area between A and C, add the areas of the two regions.

Example 20

Find the area enclosed by the graphs of the curves

$$f(x) = \frac{1}{2}x^3 + 2x^2 + 2x - \frac{1}{2} \text{ and } g(x) = -\frac{1}{2} + 3x + 2x^2 - \frac{1}{2}x^3$$

Answer

$$\frac{1}{2}x^3 + 2x^2 + 2x - \frac{1}{2} = -\frac{1}{2} + 3x + 2x^2 - \frac{1}{2}x^3$$

$$x^3 - x = 0 \Rightarrow x(x+1)(x-1) = 0$$

$$x = 0, \pm 1$$

$$A = \int_{-1}^{0} [f(x) - g(x)] \, dx + \int_{0}^{1} [g(x) - f(x) \, dx]$$

$$A = \int_{-1}^{0} (x^3 - x) \, dx + \int_{0}^{1} (x - x^3) \, dx$$

$$A = \left[\frac{x^4}{4} - \frac{x^2}{2} \right]_{-1}^{0} + \left[\frac{x^2}{2} - \frac{x^4}{4} \right]_{0}^{1}$$

$$A = \frac{1}{4} + \frac{1}{4} = \frac{1}{2} \text{ sq. unit}$$

Let f(x) = g(x) to find the points of intersection of the two curves.

Since the leading coefficient of f(x) is positive and the leading coefficient of g(x) is negative, we know that in the interval [–1, 0], f(x) > g(x) and in the interval [0, 1], g(x) > f(x).

If we are not sure which function is greater in the given interval, it is sufficient to place the integrals in an absolute value sign.

Check your answers on the GDC.

GDC help on CD:
Alternative demonstrations for the TI-84 Plus and Casio fx-9860GII GDCs are on the CD.

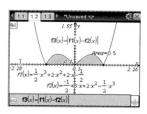

The total area of the regions enclosed by the graphs of two functions *f* and *g* that intersect at $x = a$, $x = b$ and $x = c$, $a < b < c$ is

$$A = \int_{a}^{c} \left| f(x) - g(x) \, dx \right|$$

In area problems, a region may be not be entirely enclosed between two functions. The next example highlights this case.

Example 21

Find the area of the region in the first quadrant that is enclosed by $y = \sqrt{x}$, the x-axis, and the line $y = x - 2$.

Answer

$A = R_1 + R_2$

$R_1 = \displaystyle\int_0^2 \sqrt{x}\ dx = \left[\dfrac{2x^{\frac{3}{2}}}{3}\right]_0^2 = \dfrac{4\sqrt{2}}{3}$

Sketch the graph.

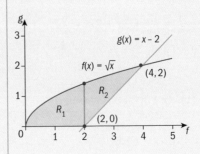

$R_2 = \displaystyle\int_2^4 (\sqrt{x} - (x-2))\, dx$

$= \left[\dfrac{2x^{\frac{3}{2}}}{3} - \dfrac{x^2}{2} + 2x\right]_2^4$

$= \dfrac{2(4)^{\frac{3}{2}}}{3} - \dfrac{4^2}{2} + 2(4) - \left(\dfrac{2(2)^{\frac{3}{2}}}{3} - \dfrac{2^2}{2} + 2(2)\right)$

$= \dfrac{16}{3} - \left(\dfrac{4\sqrt{2}}{3} + 2\right) = \dfrac{10 - 4\sqrt{2}}{3}$

$\therefore A = \dfrac{4\sqrt{2}}{3} + \dfrac{10 - 4\sqrt{2}}{3}$

$= \dfrac{10}{3}$

$= 3.33$ sq. units to 3 sf

Check on a GDC.

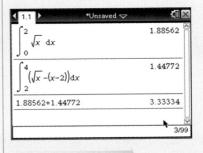

GDC help on CD:
Alternative demonstrations for the TI-84 Plus and Casio fx-9860GII GDCs are on the CD.

Exercise 7J

In questions 1–11, find the area of the region enclosed by the graphs of the curves.

 Do not use a GDC for questions 1–6.

1 $y = 2 - x^2$ and $y + x = 0$

2 $y = x^3$ and $y = x^2$

3 $y = 4 - x^2$ and $y = 2 - x$

4 $y = |x|$ and $y = x^{\frac{2}{3}}$

5 $y = 16 - x^2$ and $y = x^2 - 4x$

6 $y = x^4 - 2x^2$ and $y = 2x^2$

7 $y = 2x^3 + 5x^2 + x - 2$ and $y = 8 - 4x - 20x^2 - 8x^3$

8 $y = x^4 - 4$ and $y = \dfrac{1}{1+x}$, for $x > 0$

9 $y = e^{1-x} - 1$; $y = \sqrt{x}$; $x = 4$

EXAM-STYLE QUESTION

10 In this graph, the regions bounded by the curve $y = x^2$ and the lines $y = 4$ and $y = a$ is equal to the region bounded by the curve $y = x^2$ and $y = a$. Find the value of a.

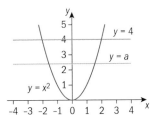

In questions 11–13, find the area of the region whose boundary is defined by the functions or lines.

11 $y = 2 - x$ and $y = x^2$

12 $y = e^x$, $y = e^{-x}$, $x = \pm 1$

13 $y = \dfrac{1}{x}$, $y = x^{\frac{2}{3}}$, x-axis and $x = 3$

Areas and kinematics

At the beginning of section 7.2, you found the total distance traveled by a particle in a given time interval by integrating the velocity function, evaluating the displacement at the end points of the interval, and then subtracting these results. The velocity in this case was positive throughout this interval.

Consider a similar problem where the velocity changes direction within the given interval. The velocity function will be partly above and partly below the t-axis.

Example 22

A particle moves in a straight line such that its velocity at any time t can be modeled by $v(t) = t - t^3$ ms^{-1}.

Find the total distance traveled by the particle in the time interval $[1, 2]$

Answer

Sketch the function to see if it is entirely above or below the t-axis, or if part of the graph is below and part above the t-axis.

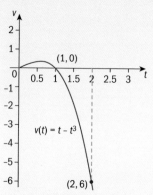

$$d(t) = \int_0^1 (t - t^3)\,dt + \left| \int_1^2 (t - t^3)\,dt \right|$$

Integrate the parts separately above and below the x-axis

$$= \left[\frac{t^2}{2} - \frac{t^4}{4} \right]_0^1 + \left| \left[\frac{t^2}{2} - \frac{t^4}{4} \right]_1^2 \right|$$

$$= \frac{1}{4} + \left| -2 - \left(\frac{1}{2} - \frac{1}{4} \right) \right| = 2.5 \, \text{m}$$

Graphically:

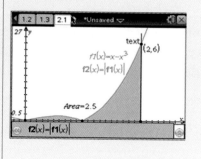

Numerically:

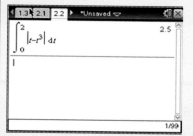

The total distance is the integral of the absolute value of the function on the interval [0, 2].

GDC help on CD:
Alternative demonstrations for the TI-84 Plus and Casio fx-9860GII GDCs are on the CD.

From Example 22, you can see that:

→ If v is a velocity function in terms of t, then the total distance traveled between times t_1 and $t_2 = \displaystyle\int_{t_1}^{t_2} |v|\,dt$

Exercise 7K

1 A particle starts from rest and moves in a straight line.
Its velocity at any time t seconds is given by $v(t) = t(t-4)\ \text{m s}^{-1}$
Find the distance traveled between the two times when the
particle is at rest.

2 A particle moves in a straight line so that after t seconds its
velocity is $v(t) = 5 + 4t - t^2$ m.
Find the total distance traveled by the particle
  a in the first second
  b between the first second and the sixth second.

3 A particle starts from rest and its acceleration, in m s^{-2}, can be
modeled by $a(t) = 1 - e^{-2t}$, $0 \le t \le 3$.
Find the distance traveled in the first 3 seconds.

EXAM-STYLE QUESTION
4 The velocity of a particle moving in a straight line is given by
$v(t) = 10 + 5e^{-0.5t}\ \text{m s}^{-1}$
  a Show that the acceleration of the particle at any time t is
always negative.
  b Find the total distance covered in the first 2 seconds.

Volumes of solids of revolution

A lathe is a machine that rotates material on its axis to make objects
with circular cross-sections and curved sides, such as vases.
A variety of materials, such as metal or plastic, may be used.

In mathematics, objects like those made with a lathe are called
solids of revolution. A solid figure with curved sides is obtained by
rotating the curve through 360° about a line; for example, the x-axis.

Here is the graph of $y = 2x$ between $x = 0$ and $x = 2$. Rotating the
line $y = 2x$ about the x-axis gives a cone.

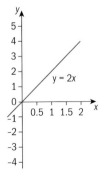

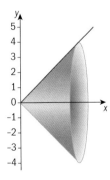

To find the volume of the cone in an interval $[a, b]$, take cross-sectional slices, as with the area and inscribed rectangles. These slices are 3-D cylinders each with radius y, and height tending to $\mathrm{d}x$ so each has volume $\pi y^2 \mathrm{d}x$. Then, to find the volume of the cone, add the volume of all the cylinders, i.e., $\sum \pi y^2 \mathrm{d}x$. When $\mathrm{d}x$ is infinitesimally small,

then $v = \int_a^b \pi y^2 \mathrm{d}x = \pi \int_a^b y^2 \mathrm{d}x$

$\rightarrow \mathrm{d}x \leftarrow$

> → The volume of a solid formed when a function $y = f(x)$, continuous in the interval $[a, b]$, is rotated 2π radians about the
> x-axis is $V = \pi \int_a^b y^2 \mathrm{d}x$

2π radians $= 360°$

The volume of the cone obtained by rotating the line $y = 2x$ in the interval $[0, 2]$ through 2π radians about the x-axis is

$$V = \pi \int_0^2 (2x)^2 \mathrm{d}x = 4\pi \int_0^2 x^2 \mathrm{d}x = 4\pi \left[\frac{x^3}{3} \right]_0^2 = 4\pi \left[\frac{8}{3} \right] = \frac{32\pi}{3} \text{ cubic units}$$

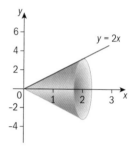

> Compare this to the result obtained using the formula
> for the volume of a cone, $V = \frac{1}{3}\pi r^2 h$
> $V = \frac{\pi}{3}(4^2)(2) = \frac{32\pi}{3}$ cu. units

Similarly, you can find the volume of the cone formed when the line $y = 2x$ is rotated 2π radians about the y-axis in the same interval.

The cylinders have radius x and height $\mathrm{d}y$.

> → The volume of a solid of revolution formed when $y = f(x)$ in
> the interval $y = c$ to $y = d$ is rotated 2π radians about the y-axis
> is $V = \pi \int_c^d x^2 \mathrm{d}y$ $f(a) = c$ $f(b) = d$

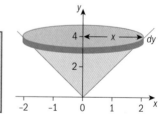

To find the volume of the cone formed by rotating the line $y = 2x$ about the y-axis, rearrange to give $x = \frac{y}{2}$. The interval $[0, 2]$ on the x-axis corresponds to $[0, 4]$ on the y axis. Then

$$V = \pi \int_0^4 \left(\frac{y}{2} \right)^2 \mathrm{d}y = \frac{\pi}{4} \left[\frac{y^3}{3} \right]_0^4 = \frac{\pi}{4} \left(\frac{4^3}{3} \right) = \frac{16\pi}{3} \text{ cu. units.}$$

Example 23

Find the volume of the solid formed when the graph of the curve $y = \sqrt{x}$ in the interval $[1, 4]$ is rotated 2π radians about **a** the x-axis **b** the y-axis.

Answers

a $V = \pi \int_1^4 \left(\sqrt{x}\right)^2 dx = \pi\left[\dfrac{x^2}{2}\right]_1^4 = \pi\left(\dfrac{16}{2} - \dfrac{1}{2}\right) = \dfrac{15\pi}{2}$ cu. units

Use $V = \pi \displaystyle\int_a^b y^2 dx$

b $y = \sqrt{x} \Rightarrow x = y^2$; when $x = 1$, $y = 1$; when $x = 4$, $y = 2$

$V = \pi \int_1^2 y^4 dy = \pi\left[\dfrac{y^5}{5}\right]_1^2 = \pi\left(\dfrac{32}{3} - \dfrac{1}{5}\right) = \dfrac{31\pi}{5}$ cu.units

Rearrange to make x the subject and find the values of y when $x = 1$ and $x = 4$

Use $V = \pi \displaystyle\int_c^d y^2 dy$

Example 24

Find the volume of the solid formed when the graph of the curve $y = e^{1-x}$ is rotated 2π radians about the x-axis between $x = 0$ and $x = 1$.

Answer

Sketch the graph.

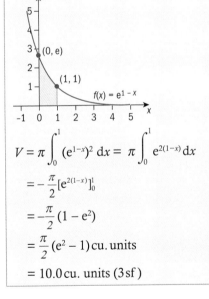

$f(x) = e^{1-x}$

$V = \pi \int_0^1 (e^{1-x})^2 dx = \pi \int_0^1 e^{2(1-x)} dx$

$= -\dfrac{\pi}{2}[e^{2(1-x)}]_0^1$

$= -\dfrac{\pi}{2}(1 - e^2)$

$= \dfrac{\pi}{2}(e^2 - 1)$ cu. units

$= 10.0$ cu. units ($3\,sf$)

Exercise 7L

In questions 1 and 2, find the volume of the solid formed by rotating the region enclosed by the graph of the function and the x-axis, through 2π radians about the x-axis, in the given interval.

1 $y = (x - 1)^2 - 1$, $[0, 1]$

2 $y = 1 + \sqrt{x}$, $[0, 2]$

Sangaku are Japanese geometrical puzzles in Euclidean geometry on wooden tablets. They were placed as offerings at Shinto shrines or Buddhist temples during the Edo period (1603–1867) as offerings to the gods .The tablets were created using only Japanese mathematics. For example, the connection between an integral and its derivative (the fundamental theorem of calculus) was unknown, so Sangaku problems on areas and volumes were solved by expansions in infinite series and term-by-term calculation. You may wish to select a Sangaku puzzle, and through research investigate the method of calculating areas and volumes.

3 When the graph of the function $y = \dfrac{x^2}{2}$ is revolved 2π radians about the y-axis, it models the shape of a bowl.

Find the volume of the bowl between $y = 0$ and $y = 2$.

4 A paperweight is modeled by the graph of the function $y = \sqrt{2x - x^2}$ when it is revolved 2π radians about the x-axis between $x = 1$ and $x = 2$.

Find the volume of the paperweight.

5 Find the volume of the solid of revolution formed when the graph of the function $y = x^{\frac{3}{2}}$ is revolved about the y-axis between $y = 1$ and $y = 3$.

6 A wine bottle stopper is modeled by the function $y = \dfrac{x}{12}\sqrt{36 - x^2}$

Find the volume of the stopper when it is rotated 2π radians about the x-axis between $x = 0$ and $x = 6$.

> There are several methods to find the volume of a solid of revolution. Investigate the different methods, such as disc, shell and washer methods, and explore the conditions under which the various methods are employed.

Now look at the volume of a solid formed by the region between two curves. The graph shows the region formed between the curves $y = \sqrt{\dfrac{x}{2}}$ and $y = \dfrac{x^2}{4}$

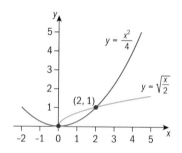

Geometrically, the volume of the region between the two curves rotated 2π radians about the x-axis is the difference in the volumes of the solids formed by the curve and the x-axis.

> → Hence, if $f(x) \geq g(x)$ for all x in the interval $[a, b]$, then the volume of revolution formed when rotating the region between the two curves 2π radians about the x-axis in the interval $[a, b]$ is
>
> $$V = \pi \int_a^b (f(x))^2 \, dx - \pi \int_a^b (g(x))^2 \, dx, \text{ or}$$
>
> $$V = \pi \int_a^b ([f(x)]^2 - [g(x)]^2) \, dx$$

> It is easy enough to find the points of intersection analytically by setting the two equations equal to each other, and solving for x.

For the two curves $y = \sqrt{\dfrac{x}{2}}$ and $y = \dfrac{x^2}{4}$,

$$V = \pi \int_0^2 \left(\sqrt{\dfrac{x}{2}} \right)^2 dx - \pi \int_0^2 \left(\dfrac{x^2}{4} \right)^2 dx$$

$$= \pi \int_0^2 \dfrac{x}{2} \, dx - \pi \int_0^2 \dfrac{x^4}{16} \, dx = \dfrac{\pi}{2}\left[\dfrac{x^2}{2} \right]_0^2 - \dfrac{\pi}{16}\left[\dfrac{x^5}{5} \right]_0^2$$

$$= \dfrac{\pi}{2}(2) - \dfrac{\pi}{16} \cdot \left(\dfrac{32}{5} \right) = \pi - \dfrac{2\pi}{5} = \dfrac{3\pi}{5} \text{ cu. units.}$$

Now rotate the region about the *y*-axis in the same interval.
Rearrange both equations to make *x* the subject:

$x = 2y^2$ and $x = 2\sqrt{y}$

For both functions, when $x = 0$, $y = 0$ and when $x = 2$, $y = 1$, so the curves intersect at $(0, 0)$ and $(2, 1)$.

Hence, $V = \pi \int_0^1 (2\sqrt{y})^2 \mathrm{d}y - \pi \int_0^1 (2y^2)^2 \mathrm{d}y$

$$V = 4\pi \int_0^1 y\,\mathrm{d}y - 4\pi \int_0^1 y^4\,\mathrm{d}y = 4\pi \left[\frac{y^2}{2}\right]_0^1 - 4\pi \left[\frac{y^5}{5}\right]_0^1$$

$$= \pi \frac{4}{2} - \pi \frac{4}{5} = \frac{6\pi}{5} \text{ cu. units}$$

> → If x_1 and x_2 are relations in *y* such that $x_1 \geq x_2$ for all
> *y* in the interval $[c, d]$, then the volume formed
> when rotating the region between the two curves 2π
> radians about the *y*-axis in the interval $[c, d]$ is
>
> $$V = \pi \int_d^c x_1^2 \mathrm{d}y - \pi \int_d^c x_2^2 \mathrm{d}y$$
>
> or $V = \pi \int_c^d (x_1^2 - x_2^2)\mathrm{d}y$

The astronomer **Johann Kepler** (1571–1630) expanded upon Archimedes' work on finding volumes of irregular shapes. Legend has it that at his wedding, Kepler was distracted by the problem of how much wine was in the barrels his guests were being served from. The problem so fascinated him that he dedicated an entire book to its solution. The book, published in 1615, was entitled *Nova stereometria doliorum vinariorum* or *New volume measurements of wine barrels*.

Example 25

The graphs of $x = \dfrac{y^4}{4} - \dfrac{y^2}{2}$ and $x = \dfrac{y^2}{2}$ completely enclose a region. Find the volume of the solid formed when this region is rotated 2π radians about the *y*-axis in the interval $[c, d]$, $c, d \geq 0$.

Answer

$\dfrac{y^4}{4} - \dfrac{y^2}{2} = \dfrac{y^2}{2} \Rightarrow \dfrac{y^4}{4} - y^2 = 0 \Rightarrow y^2\left(\dfrac{y^2}{4} - 1\right) = 0 \Rightarrow y = 0, \ y = \pm 2$

Without a graph it is safer to use the absolute value in the interval.

$$V = \pi \left| \int_0^2 \left(\frac{y^4}{4} - \frac{y^2}{2}\right)^2 - \left(\frac{y^2}{2}\right)^2 \mathrm{d}y \right| = \pi \left| \int_0^2 \left(\frac{y^8}{16} - \frac{y^6}{4}\right) \mathrm{d}y \right|$$

$$V = \pi \left| \left[\frac{y^9}{144} - \frac{y^7}{28}\right]_0^2 \right| = \pi \left|\frac{2^9}{144} - \frac{2^7}{28}\right| = \frac{64\pi}{63}$$

The volume formed from $y = -2$ to $y = 0$ is twice the volume from $y = 0$ to $y = 2$, hence the total volume is

$2 \cdot \dfrac{64\pi}{63} = \dfrac{128\pi}{63} = 6.38 \text{ cu. units}$

Confirm on a GDC:

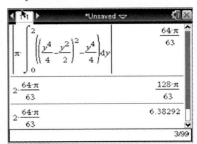

GDC help on CD:
Alternative demonstrations for the TI-84 Plus and Casio fx-9860GII GDCs are on the CD.

Exercise 7M

1 Find the volume of the solid formed when the region between the graphs of the functions $y = x$ and $y = \dfrac{x}{2}$ is rotated through 2π radians about the x-axis between $x = 2$ and $x = 5$.

2 Find the volume of the solid formed when the region between the graphs of the functions $y = x - 4$ and $y = x^2 - 4x$ is revolved 2π radians about the x-axis.

3 Find the volume of the solid formed when the region between the graphs of $y = x$ and $y^2 = 2x$ is revolved 2π radians about the y-axis.

4 Find the volume of the solid formed when the region between the graphs of the functions $y = 2x - 1$, $y = x^{\frac{1}{2}}$, and $x = 0$ is revolved 2π radians about the y-axis.

Extension material on CD:
Worksheet 7

Review exercise

1 The gradient function of a curve is $\dfrac{dy}{dx} = ax + \dfrac{b}{x^2}$. The curve passes through the point $(-1, 2)$, and has a point whose gradient is 0 at $(-2, 0)$. Find the equation of the curve.

2 Calculate the area enclosed by the graphs of $y = x^2$ and $y^2 = x$

3 The region enclosed by $y = 1 + 3x - x^2$ and $y = \dfrac{2}{x}$ for $x > 0$ is rotated 2π radians about the x-axis. Find the volume of the solid formed.

4 Evaluate

a $\displaystyle\int_1^2 \left(x + \dfrac{1}{x^2} - \dfrac{1}{x^4} \right) dx$

b $\displaystyle\int_1^4 \dfrac{5x^2 - 4}{\sqrt{x}}\, dx$

c $\displaystyle\int_1^2 \dfrac{1}{x-3}\, dx$

d $\displaystyle\int_1^e \dfrac{1}{1-4x}\, dx$

Review exercise

1. A particle moves in a straight line so that its velocity after t seconds is $v(t) = t^3 - 4t \text{ m s}^{-1}$
   Find the total distance traveled in the first 3 seconds.

EXAM-STYLE QUESTION

2. The velocity of a particle moving in a straight line is
   $v(t) = t^3 - 3t^2 + 2 \text{ m s}^{-1}$
   Find the total distance traveled between the maximum and minimum velocities.

3. Find the total area of the region enclosed by the graph of
   $y = x^2 - 4 + \dfrac{3}{x^2}$ and the x-axis.

4. Integrate these where possible with respect to x.

   a $\dfrac{3x^4 + 6}{x^2}$

   b $\left(x + \dfrac{1}{x}\right)\left(x - \dfrac{1}{x}\right)$

   c $\dfrac{1}{2 - 3x}$

   d $\dfrac{2}{\sqrt{1 - 4x}}$

   e $2e^{-3x} + \sqrt[3]{e^x}$

5. Find the quotient when $2x^2 + 3x$ is divided by $2x - 1$.
   Hence, evaluate $\displaystyle\int_1^2 \left(\dfrac{2x^2 + 3x}{2x - 1}\right) dx$

6. Find the area enclosed by the graph of $y = \dfrac{1}{(x+1)}$, the y-axis, and the line $y = 5$.

7. Find the area enclosed by the graph of $y = \sqrt{x+1}$, and the x- and y-axes.

EXAM-STYLE QUESTION

8. The area enclosed by the curve $y = 3x(a - x)$ and the x-axis is 4 units^2. Find the value of a.

9. The region between the graphs of $y = 3^x$, $y = 3^{-x}$, and the lines $x = -1$ and $x = 1$ is rotated 2π radians about the x-axis. Find the volume of the solid formed.

CHAPTER 7 SUMMARY
Integration

- $\int f(x)dx = F(x) + c, c \in \mathbb{R}$

- $\int x^n \, dx = \dfrac{x^{n+1}}{n+1}, n \neq -1$

- $\int [f(x) \pm g(x)]dx = \int f(x)dx \pm \int g(x)dx$

- $\int (ax + b)^n \, dx = \dfrac{1}{a(n+1)}(ax + b)^{n+1} + c, a \neq 0$

- $\int e^x dx = e^x + c$

- $\int e^{ax+b} dx = \dfrac{1}{a}e^{ax+b} + c, a \neq 0$

- $\int m^{ax+b} \, dx = \dfrac{1}{a\ln(m)} m^{ax+b} + c$, where m is a positive real number, $a \neq 0$.

- $\int \dfrac{1}{x}dx = \ln|x| + c$

- $\int \dfrac{1}{(ax+b)}dx = \dfrac{1}{a}\ln|ax+b| + c, a \neq 0$

Definite integration

- $\displaystyle\int_a^b f(x)\,dx = -\int_b^a f(x)\,dx$

- $\displaystyle\int_b^a f(x)\,dx = 0$

- $\displaystyle\int_a^b kf(x)\,dx = k\int_a^b f(x)\,dx$

- $\displaystyle\int_a^b (f(x) \pm g(x)\,dx = \int_a^b f(x)\,dx \pm \int_a^b g(x)\,dx$

- $\displaystyle\int_a^b f(x)dx + \int_b^c f(x)dx = \int_b^c f(x)dx$

The fundamental theorem of calculus

- If f is continuous in $[a, b]$ and if F is any anti-derivative of f on $[a, b]$ then $\displaystyle\int_a^b f(x)\,dx = F(b) - F(a)$

Continued on next page

Areas between graphs of functions and the axes

- If the integral of f exists in the interval $[a, b]$, and f is non-negative in this interval, then the area A under the curve $y = f(x)$ from a to b is

  $$A = \int_a^b f(x) \, dx.$$

- When f is negative for all $x \in [a, b]$, then the area bounded by the curve and the lines $x = a$ and $x = b$ is $\left| \int_a^b f(x) \, dx \right|$

- If functions f and g are continuous in the interval $[a, b]$, and $f(x) \geq g(x)$ for all $x \in [a, b]$, then the area between the graphs

  of f and g is $A = \int_a^b (f(x) - g(x)) \, dx$

Kinematics

- If v is a velocity function in terms of t, then the total distance traveled between times t_1 and t_2 is $\int_{t_1}^{t_2} |v| \, dt$

Volumes of revolution

- The volume of a solid formed when a function $y = f(x)$, continuous in the interval $[a, b]$, is rotated 2π radians about the x-axis is $V = \pi \int_a^b y^2 \, dx$.

- The volume of a solid of revolution formed when $x = f(y)$ in the interval $y = c$ to $y = d$ is rotated 2π radians about the y-axis

  is $V = \pi \int_c^d x^2 \, dy$

- If $f(x) \geq g(x)$ for all x in the interval $[a, b]$, then the volume formed when rotating the region between the two curves 2π radians about the x-axis in the interval $[a, b]$ is

  $$V = \pi \int_a^b (f(x))^2 \, dx - \pi \int_a^b (g(x))^2 \, dx, \text{ or } V = \pi \int_a^b ([f(x)]^2 - [g(x)]^2) \, dx.$$

- If x_1 and x_2 are relations in y such that $x_1 \geq x_2$ for all y in the interval $[c, d]$, then the volume formed when rotating the region between the two curves 2π radians about the y-axis in

  the interval $[c, d]$ is $V = \pi \int_d^c x_1^2 \, dy - \pi \int_d^c x_2^2 \, dy$

  or $V = \pi \int_c^d (x_1^2 - x_2^2) \, dy$

The evolution of calculus

Calculus wars

The two 'fathers of calculus' were the British mathematician Isaac Newton and the German mathematician Gottfried Leibniz. The original dispute over who first invented, or discovered, calculus has been settled in favor of both men. The evidence shows that they came upon calculus independently of each other, and within the same ten-year period.

This dispute, however, gravely affected them and also the further development of mathematics in the early 18th century. It also highlighted how influential the Royal Society was in Britain and throughout Europe.

■ What role did the Swiss mathematician Nicolas Fatio de Duillier play in this dispute? Were his actions ethical?

■ What was the Royal Society? What role did it take in resolving the dispute? Did any organization in Germany support Leibniz? Were their actions ethical?

■ What was Isaac Newton's role in the Royal Society's commission? Were there any ethical issues with this?

Although several centuries would pass before Leibniz was finally acquitted of plagiarizing Newton's work, it was one of Newton's supporters, Samuel Clarke, who together with Leibniz paved the way for Einstein's Theory of Relativity. Perhaps this was history's way of vindicating Leibniz for his unjust treatment during the calculus dispute.

Isaac Newton · Gottfried Leibniz

Expert opinion

In our search for knowledge, we often rely on the 'experts' and their experience and opinions.

■ How can we judge the experts' expertise?

■ What obligations do experts in the same field have to each other?

■ What are experts' ethical responsibilities in the dissemination of their knowledge?

From calculus to chaos

Before calculus, mathematicians could use geometry to analyze 2-D and 3-D shapes such as circles, ellipses, cones and spheres. Calculus provided a powerful tool for analyzing any type of smooth curve, and scientists used it to model the world and the universe in fields such as oceanography, astronomy, materials science, physics and engineering. The two great 20th century scientific theories – relativity and quantum mechanics – were based on analytical calculus.

But calculus only works on smooth curves – and in the real world not all curves are smooth. Chaos theory is a way of solving physical problems that cannot be solved by calculus. Although many mathematicians had been working on chaos theory since the late 19th century, it did not develop more fully until later in the 20th century when computers became available, as chaos mathematics often involves repeatedly applying mathematical formulae.

▶ Rocks are fractals. You cannot tell if you are looking at a close-up of a small boulder, or a mountain

Fractals

Fractals are an example of chaos in space. They are geometric shapes that do not become simpler the more closely you look at them or analyze them. They look the same in close up as they do from far away.

Sierpinski's triangle, which you saw in Chapter 1, is an example of a fractal.

■ Where else do fractals occur in nature?

■ How are fractals used to create realistic 'natural' environments for animated films, computer games and screensavers?

◀ A fern leaf is a fractal. Under a microscope its structure is more complex than you see with the naked eye.

■ What is the 'butterfly effect'? How is this an example of time-chaos?

"Chaos is the rediscovery that calculus does not have infinite power."

Michel Baranger, Professor of Physics Emeritus at the Massachusetts Institute of Technology

8 Ancient mathematics and modern methods

CHAPTER OBJECTIVES:

3.2 Definition of $\cos\theta$, $\sin\theta$ and $\tan\theta$ in terms of the unit circle;
exact values of sin, cos and tan of $0, \frac{\pi}{6}, \frac{\pi}{4}, \frac{\pi}{3}, \frac{\pi}{2}$ and their multiples;
definition of the reciprocal trigonometric ratios $\sec\theta$, $\csc\theta$ and $\cot\theta$;
Pythagorean identities: $\cos^2\theta + \sin^2\theta = 1$; $1 + \tan^2\theta = \sec^2\theta$; $1 + \cot^2\theta = \csc^2\theta$

3.3 Compound angle identities; double angle identities

3.4 The circular functions $\sin x$, $\cos x$ and $\tan x$; their domains and ranges;
their periodic nature; their graphs; composite functions of the form
$f(x) = a\sin[b(x + c) + d]$; applications

3.5 The inverse functions $x \rightarrow \arcsin x$, $x \rightarrow \arccos x$, $x \rightarrow \arctan x$; their domains
and ranges; their graphs

3.6 Algebraic and graphical methods of solving trigonometric equations in a finite
interval including the use of trigonometric identities and factorization

3.7 The cosine rule; the sine rule including the ambiguous case; area of a triangle
as $\frac{1}{2}ab\sin C$; applications in two and three dimensions

Before you start

You should know how to:

1 Work with similar triangles,
e.g. for the shape ABCD, use similar
triangles to find the height of C above AD.
Construct a perpendicular from C to
AD and draw a horizontal line through B.
In triangle ABD, $BD^2 = 6^2 + 8^2 \Rightarrow BD = 10$
$\angle ADB = \angle DBL$ and $\angle DBL = \angle BCL$
Therefore, triangles BCL and BAD
are similar.

$$\Rightarrow \frac{CL}{AD} = \frac{CB}{BD} \Rightarrow \frac{CL}{8} = \frac{5}{10}$$

So, CL = 4.
    LM = BA = 6.
Therefore, CM = 6 + 4 = 10.

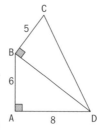

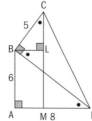

Skills check

1 Three vertical
poles, AB, CD and
EF, form part of a
hanging bridge. AC
is horizontal. Given
that AB = 6.5 m and
CD = 4 m, use similar
triangles to calculate
the height EF.

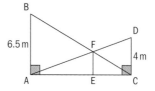

Trigonometry and its multiple applications

The word *trigonometry* comes from the Greek words *trigonon* meaning 'triangle' and *metria* meaning 'measure'.

'How did the Egyptians plan the construction of the pyramids without the aid of trigonometry?'

The answer is that they must have had some knowledge of the properties of similar triangles, which lead to the development of trigonometric ratios in right-angled triangles. The Ancient Greeks calculated the diameter of the Earth by measuring the distance of the horizon, and Thales determined the height of the Cheops pyramid by comparing the length of its shadow to that of the shadow of a rod of known length.

Cartography and navigation in the early 15th century are said to have flourished because of trigonometry and methods of triangulation. One could argue that we no longer need these antiquated methods for map making and navigation because in the 21st century a click of a button does it all

for us. We can take a photograph with our camera and when we view it we not only get its image but its geographical coordinates and position on a map. It can also be argued that having paper maps in cars is no longer necessary due to the increase in use of GPS (Global Positioning System) receivers. The GPS consists of 27 satellites orbiting the Earth so that at any point on the Earth's

Pedro Nunes, (1502–78) Portuguese mathematician Discovered a method to find the latitude of a point by the height of the sun and how to indicate this as a chart. This helped Portuguese discoverers become some of the best technical navigators of their era.

surface at least four of them are visible, and if one is equipped with a GPS receiver a vehicle's location can be instantly calculated. So is trigonometry obsolete? Not at all, since a GPS satellite has to be accurately positioned in its orbit using trigonometrical calculations.

In this chapter you will approach trigonometry from two different perspectives. The first considers functions of angles where trigonometric functions will be defined in terms of the ratios of the sides of right-angled triangles. The second considers functions of real numbers where trigonometric functions will be defined in terms of real numbers.

8.1 The right-angled triangle and trigonometric ratios

The trigonometric ratios for acute angle θ in triangle ABC are:

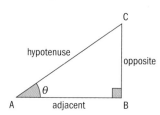

> We refer to the sides of the triangle as opposite, adjacent and hypotenuse because we need the sides and their relative position to the acute angle θ.

$$\rightarrow \quad \sin\theta = \frac{BC}{AC} = \frac{\text{opposite}}{\text{hypotenuse}}$$

$$\cos\theta = \frac{AB}{AC} = \frac{\text{adjacent}}{\text{hypotenuse}}$$

$$\tan\theta = \frac{BC}{AB} = \frac{\text{opposite}}{\text{adjacent}}$$

You can use the properties of similar triangles to show that the above trigonometric ratios hold for a right-angled triangle of any size.

Consider the triangle ABC with a right angle at B and vertex A placed at the origin as shown.

Triangles ABC and AQP are similar since they have equal angles – each has a right angle and an angle θ. It follows therefore that $\dfrac{PQ}{AP} = \dfrac{BC}{AC} = \sin\theta$ by our definition above.

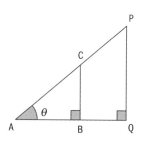

Similarly, $\dfrac{AQ}{AP} = \dfrac{AB}{AC} = \cos\theta$ and $\dfrac{PQ}{AQ} = \dfrac{BC}{AB} = \tan\theta$.

With these definitions, together with a calculator, you are now equipped to solve right-angled triangles as follows:

The GDC can calculate trigonometric ratios of any angle measured in either degrees or radians. You must ensure that the mode setting for angle measurement is set correctly.

- Given any angle (other than the right angle) and the length of one side, you can find the third angle and the lengths of the other two sides.
- Given any two sides of the triangle, you can find the length of the third side and the sizes of the two unknown angles.

If you want to find the angle, θ, whose trigonometric ratio we know e.g. $\sin\theta = \dfrac{5}{7}$, we use the notation $\theta = \arcsin\dfrac{5}{7}$. On the GDC this is denoted by $\sin^{-1}$.

Therefore, $\theta = \sin^{-1}\left(\dfrac{p}{q}\right) = \arcsin\left(\dfrac{p}{q}\right) \neq \dfrac{1}{\sin\theta}$.

Example 1

For each triangle, solve for unknown angles and sides.

a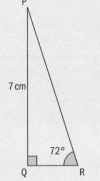

b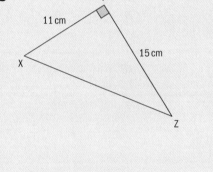

Answers

a In triangle PQR:

$Q\hat{P}R = 90° - 72° = 18°$

$\dfrac{PQ}{PR} = \sin72° \Rightarrow PR = \dfrac{7}{\sin72°}$

$= 7.36\,\text{cm}$

$QR = \sqrt{(7.36)^2 - 7^2} = 2.27\,\text{cm}$

Use the sine ratio to find PR given PR = hyp and PQ = opp.

Use Pythagoras' theorem to find QR.

b In triangle XYZ:

$XZ = \sqrt{11^2 + 15^2} = 18.6\,\text{cm}$

$\tan Y\hat{X}Z = \dfrac{15}{11} \Rightarrow$

$Y\hat{X}Z = \arctan\left(\dfrac{15}{11}\right) = 53.7°$

$Y\hat{Z}X = 90° - 53.7° = 36.3°$

Use Pythagoras' theorem to find XZ.

Use the tangent ratio to find $Y\hat{X}Z$ given YZ = opp and YX = adj.

When using a previous answer in a new calculation, make sure you don't use the approximate answer to 3 sf.

Example 2

The photo shows the river Rhine and the building known as Langer Eugen in Bonn which houses the United Nations. The building is 114 m high.

The angles α and β from the top of the tower to the edges of the river are measured. Given that $\alpha = 75°$ and $\beta = 19°$, calculate the width of the river, DC, giving your answer to the nearest metre.

Answer

$B\widehat{A}C = 90°$, AB = 114 m

$\dfrac{AD}{AB} = \tan \beta$

$\Rightarrow AD = 114 \tan 19° \approx 39.25$ m

In triangle ABD, use the tangent ratio to find AD given
AD = opp and AB = adj.

$\dfrac{AC}{AB} = \tan \alpha$

$\Rightarrow AC = 114 \tan 75° \approx 425.25$ m

In triangle ABC, use the tangent ratio to find AC given
AC = opp and AB = adj.

DC = 425.45 − 39.25 = 386.2

Therefore, the river has a width of 386 m.

The earliest mathematician to use trigonometry was a Greek mathematician, **Hipparchus**, in the second century BCE. He produced Chord Tables to aid in the finding of the heights and distances of inaccessible objects. **Ptolemy** cited and used this concept about 300 years later, but it was **Aryabhatta** who first defined trigonometric ratios in terms of right-angled triangles.

Aryabhatta's book, written around 499 CE in India, contains mathematical rules of arithmetic, algebra and trigonometry. He talks of the '*jya*' meaning half chord. This was translated later into Latin as '*sinus*', which means cove or bay. This became the sine ratio you use today.

Example 3

The solid below is made up of a cone and a hemisphere.

a Find the height of the solid.

b Show that the surface area of the hemispherical base is equal to the curved surface area of the cone.

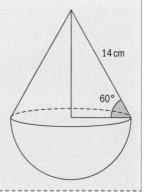

Answers

a Let the height of the cone be h and the radius of the hemisphere be r.

$$\frac{h}{14} = \sin 60° \Rightarrow h$$
$$= 14\sin 60°$$
$$= 12.12\,\text{cm}$$

Use the sine ratio to find h given opp = h and hyp = 14.

$$\frac{r}{14} = \cos 60° \Rightarrow r$$
$$= 14\cos 60° = 7\,\text{cm}$$

Use the cosine ratio to find r given adj = r and hyp = 14.

Total height of solid $= 12.12 + 7$
$$= 19.12\,\text{cm}$$

b Surface area of solid = curved surface area of hemisphere + area of sector.

Length of arc $L = 2\pi r$
$$= 43.98\,\text{cm}$$

A cone opens up into a sector of a circle when cut along a slanting edge. The length of the arc is equal to the circumference (L) of the base of the cone.

But $L = \theta R \Rightarrow \theta = \dfrac{L}{R}$

$$= \frac{43.98}{14} = 3.14\ \text{radians}$$

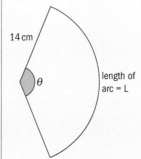

The formula for arc length, $l = \theta r$ and area of a sector, $A = \dfrac{1}{2}\theta r^2$, are only true when θ is in radians.

See Section 5.8.

Area of sector
$$= \frac{1}{2}\theta R^2 = 307.72\,\text{cm}^2$$

Curved surface area of hemisphere
$$= 2\pi r^2 = 307.87\,\text{cm}^2$$

Exercise 8A

1 For each triangle, solve for unknown angles and sides:

a

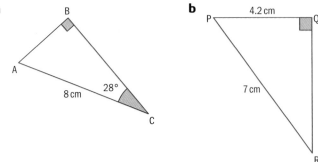

b

2 The image below illustrates how triangulation was used to measure the width of a river.
The diagram is a simplified representation.

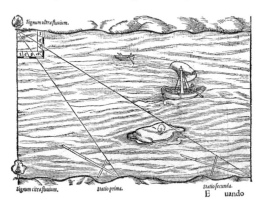

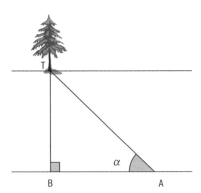

The distance AB = 30 m and the angle $\alpha = 52.3°$.
Find the width of the river.

3 The cone in the diagram below is made from the sector shown on the right.

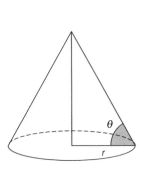

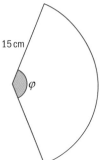

Given that $\theta = 1.23$ radians, find φ.

4 Milk is transported in stainless steel cylindrical containers mounted on trucks. The diagram below shows the cross-section of such a cylinder containing milk. The radius of the cylinder is 1.5 m and the height of milk inside the cylinder is 1.8 m.

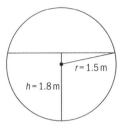

Given that the cylinder has a length of 3 m, calculate the quantity of milk, in cubic metres, that is being transported.

Great circles are those circles on a sphere which have the same radius and center as the sphere. On a globe the Earth is divided in longitudes and latitudes. All longitudes are great circles but the equator is the only latitude which is a great circle. You may wish to research how trigonometry and great circles are used to solve navigational problems, such as calculating the amount of fuel needed for a flight from London to São Paulo.

8.2 The unit circle and trigonometric ratios

Consider the **unit circle** shown below with centre at the origin $(0,0)$ and point P on the circumference. The length of OP is 1 and the coordinates of the point P are (x, y).

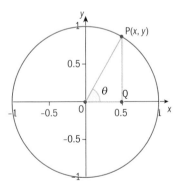

A unit circle is a circle with a radius of one unit and center at the origin $(0,0)$.

Applying the definition of the three trigonometric ratios given in the previous section:

$$\rightarrow \sin\theta = \frac{y}{\text{OP}} = \frac{y}{1} \qquad \cos\theta = \frac{x}{\text{OP}} = \frac{x}{1} \qquad \tan\theta = \frac{y}{x}, x \neq 0$$

Trigonometric ratios for angles which are not acute

In the previous section, trigonometric ratios were defined for the unit circle.

$$\sin \theta = \frac{y}{\text{OP}} = \frac{y}{1}$$

$$\cos \theta = \frac{x}{\text{OP}} = \frac{x}{1}$$

$$\tan \theta = \frac{y}{x}, \, x \neq 0$$

You can extend the definition of the trigonometric functions to angles larger than 90° by looking at the coordinates of the point P outside the first quadrant. When P is in the second quadrant, θ is obtuse, the x-coordinate is negative and the y-coordinate is positive, so all the ratios except for $\sin \theta$ are negative in this quadrant.

Consider the signs of the coordinates of P in each quadrant, and assign signs to the three ratios. This is summarized in the diagram below.

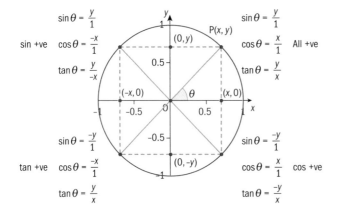

The hypotenuse OP is always equal to 1.

From the above diagram, we can also obtain the following identities for $0 \leq \theta \leq \frac{\pi}{2}$.

→ $\sin (\pi - \theta) = \sin \theta$ $\sin (\pi + \theta) = -\sin \theta$ $\sin (2\pi - \theta) = -\sin \theta$
    $\cos (\pi - \theta) = -\cos \theta$ $\cos (\pi + \theta) = -\cos \theta$ $\cos (2\pi - \theta) = \cos \theta$
    $\tan (\pi - \theta) = -\tan \theta$ $\tan (\pi + \theta) = \tan \theta$ $\tan (2\pi - \theta) = -\tan \theta$

180° = π radians
360° = 2π radians

These results suggest that sine, cosine and tangent are periodic. The periodicity of the trigonometric functions is discussed in detail in section 8.5.

For a review of radian measure see Chapter 5.

Reciprocal trigonometric ratios

The reciprocal trigonometric ratios are defined as:

$$\rightarrow \csc\theta = \frac{1}{\sin\theta}, \ \sec\theta = \frac{1}{\cos\theta}, \ \cot\theta = \frac{1}{\tan\theta}$$

It is very important to remember that $\csc\theta = \dfrac{1}{\sin\theta} = (\sin\theta)^{-1}$.

For example: $\sin\theta = 0.3246 \Rightarrow \csc\theta = \dfrac{1}{\sin\theta} = (\sin\theta)^{-1} = 3.081$

You should also note that by definition:

- $-1 \le \sin\theta \le 1 \Rightarrow |\csc\theta| \ge 1$, and
- $-1 \le \cos\theta \le 1 \Rightarrow |\sec\theta| \ge 1$

> Recall that $\sin^{-1}\theta$ is used to represent arcsin θ, the inverse of sin θ.

Example 4

Write each of these angles in terms of acute angles.

a $\sin 156°$ **b** $\tan 140°$ **c** $\cos 320°$ **d** $\cos\left(\dfrac{5\pi}{8}\right)$

Answers

a $\sin 156° = \sin(180 - 24)° = \sin 24°$ *Use sin (π − θ) = sin θ.*

b $\tan 140° = \tan(180 - 40)° = -\tan 40°$ *Use tan (π − θ) = −tan θ.*

c $\cos 320° = \cos(360 - 40)° = \cos 40°$ *Use cos (2π − θ) = cos θ.*

d $\cos\left(\dfrac{5\pi}{8}\right) = \cos\left(\pi - \dfrac{3\pi}{8}\right) = -\cos\dfrac{3\pi}{8}$ *Use cos (π − θ) = −cos θ.*

Example 5

Given that $\sin\theta = \dfrac{3}{5}$ and $0 < \theta < \dfrac{\pi}{2}$, find the exact values of $\cos\theta$ and $\tan\theta$.

Answer

Using Pythagoras' theorem:

AB (adjacent) $= \sqrt{5^2 - 3^2} = 4$

Therefore,

$\cos\theta = \dfrac{4}{5}$ and $\tan\theta = \dfrac{3}{4}$

Since $\sin\theta = \dfrac{3}{5}$ and θ is an acute angle we can use the triangle ABC to solve.

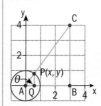

So AB = 4, BC = 3 and AC = 5.

Example 6

> Find the maximum value of $1 - 3\sin\theta$, and evaluate the smallest positive value of θ for which this occurs.

Answer

$-1 \le \sin\theta \le 1$

$\Rightarrow 3 \ge -3\sin\theta \ge -3$

$1 - 3\sin\theta = 1 + (-3\sin\theta)$ has a maximum value of 4.

This value occurs when

$-3\sin\theta = 3$

$\sin\theta = -1$

$\theta = \dfrac{3\pi}{2}$

Sine is positive in the 1st and 2nd quadrants.

Multiply the inequality by -3.

> Look back at p 390 where the 1st, 2nd, 3rd and 4th quadrants are defined.

Exercise 8B

1 Write each of these in terms of acute angles.

   a $\sin 144°$ **b** $\cos 210°$

   c $\tan 230°$ **d** $\sin\left(\dfrac{7\pi}{8}\right)$

   e $\tan\left(\dfrac{7\pi}{3}\right)$ **f** $\cos\left(\dfrac{7\pi}{6}\right)$

2 Given that $\sin\theta = \dfrac{5}{13}$ and $\dfrac{\pi}{2} < \theta < \pi$, find the exact values of $\cos\theta$ and $\tan\theta$.

3 Given that $\sec\theta = -\dfrac{5}{4}$ and $\pi < \theta < 2\pi$, find the exact values of $\cos\theta$, $\tan\theta$ and $\sin\theta$.

4 For each trigonometric expression below write down

   i the maximum value

   ii the minimum value.

State the smallest positive value of θ for which these values occur.

   a $2 + 4\cos\theta$

   b $5 - 3\sin\theta$

   c $2\sin\theta - 1$

   d $-2\cos\theta - 3$

Investigation – trigonometric identities

Use your GDC to copy and complete the following table.

> $\sin^2\theta$ is standard notation for $(\sin\theta)^2$.
> On a GDC you must enter $(\sin\theta)^2$.

$\theta°$	$\sin\theta$	$\cos\theta$	$\tan\theta$	$\sin(90-\theta)$	$\cos(90-\theta)$	$\tan(90-\theta)$	$\dfrac{\sin\theta}{\cos\theta}$	$\sin^2\theta + \cos^2\theta$
23								
37								
44								
56								
38								
87								
14								

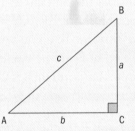

a Using the table above, suggest three relationships connecting the trigonometric ratios of θ and $(90 - \theta)$.

b Write a conjecture connecting the three trigonometric ratios.

c Write a conjecture connecting $\sin^2\theta$ and $\cos^2\theta$.

d Use the general right-angled triangle on the right to prove your conjectures.

e Use your results to find a relationship between $\tan^2\theta$ and $\sec^2\theta$.

f Find a relationship between $\cot^2\theta$ and $\csc^2\theta$.

Summary of the results of the investigation

> → **Co-function identities**
>
> $\sin\theta = \cos(90° - \theta)$ and $\cos\theta = \sin(90° - \theta)$

> → **Tangent identity**
>
> $\tan\theta = \dfrac{\sin\theta}{\cos\theta}$

> → **Pythagorean identities**
>
> $\sin^2\theta + \cos^2\theta = 1$
>
> $1 + \tan^2\theta = \sec^2\theta$
>
> $1 + \cot^2\theta = \csc^2\theta$

> Why do we call these identities? What is the difference between identities and equations?

Angles and radian measure

Until now in this chapter we have mainly used degrees for angle measure. You should recall that in Chapter 5, section 5.8, **radians** were introduced as another unit for angle measure.

Why do we use radians as an angle measure?

For an introduction to radian measure see Chapter 5.

> → For a circle of radius r, the angle at the centre, which subtends a part of the circumference of length r, is equal to one radian.
>
> As the circumference of the circle is $2\pi \times r$ then the total angle at the centre is 2π radians which is also 360° (a complete circle).

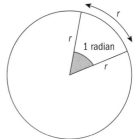

Hence

- 2π radians = 360° or π radians = 180°
- 1 radian = $\dfrac{180}{\pi}$ degrees and 1 degree = $\dfrac{\pi}{180}$ radians.

You can use this information to convert between degrees and radians.

Consider the unit circle and the real number line as shown in the diagram. Point B on the circle coincides with zero on the number line. Imagine that you wrap the number line around the circle, with point B on the circle remaining fixed. As the number line is wrapped around the circle, each real number on the number line is mapped onto a point (x, y) on the circle.

The point P on the number line coincides with P′ on the circle, the point Q on the number line coincides with Q′, and the points π and $-\pi$ both coincide with point A.

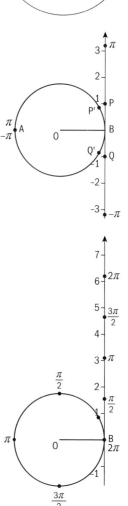

Notice that since the number line is infinite you could continue wrapping it around the circle and if you did this the points 2π and -2π would again coincide with point B on the circle. In fact all the points $\pm 2n\pi$, $n \in \mathbb{Z}$, on the number line will coincide with the point B on the circle. Any real number, x, on the number line will coincide with a unique point on the unit circle. However this point on the unit circle coincides with all the numbers $x \pm 2n\pi$, $n \in \mathbb{Z}$, on the number line.

In a similar way all the angles $\theta \pm n360$, $n \in \mathbb{Z}$, measured in degrees are equivalent to the same point on a circle.

It does not really matter whether you use degrees or radians when measuring angles; however, when using radians you will find that many formulae become simple.

Trigonometric ratios of some special angles

Returning to the definitions of the three trigonometric ratios and the diagram, you can evaluate the ratios for 0 and $\dfrac{\pi}{2}$.

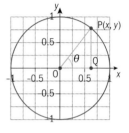

$\sin 0 = y = 0$ $\qquad\qquad \sin\left(\dfrac{\pi}{2}\right) = y = 1$

$\cos 0 = x = 1$ $\qquad\qquad \cos\left(\dfrac{\pi}{2}\right) = x = 0$

$\tan 0 = \dfrac{y}{x} = \dfrac{0}{1} = 0$ $\qquad \tan\left(\dfrac{\pi}{2}\right) = \dfrac{y}{x} = \dfrac{1}{0}$ is undefined

Since $\theta = \pm 2n\pi$ is equivalent to $\theta = 0$ and $\theta = \left(\dfrac{\pi}{2} \pm 2n\pi\right)$ is equivalent to $\theta = \dfrac{\pi}{2}$ on the unit circle, you can say that:

When $\theta = 0$ point P has coordinates (1, 0). When $\theta = \dfrac{\pi}{2}$ point P has coordinates (0, 1).

$\sin(\pm 2n\pi) = y = 0$ $\qquad\qquad \sin\left(\dfrac{\pi}{2} \pm 2n\pi\right) = y = 1$

$\cos(\pm 2n\pi) = x = 1$ $\qquad\qquad \cos\left(\dfrac{\pi}{2} \pm 2n\pi\right) = x = 0$

$\tan(\pm 2n\pi) = \dfrac{y}{x} = \dfrac{0}{1} = 0$ $\qquad \tan\left(\dfrac{\pi}{2} \pm 2n\pi\right) = \dfrac{y}{x} = \dfrac{1}{0}$ is undefined

$\theta = \pi \pm 2n\pi$ is equivalent to $\theta = \pi \pm (2n - 2)\pi$ which means all the odd multiples of π.

Similarly, $\theta = \pi \pm 2n\pi$ is equivalent to $\theta = \pi$ and $\theta = \dfrac{3\pi}{2} \pm 2n\pi$ is equivalent to $\theta = \dfrac{3\pi}{2}$ on the unit circle, which means that:

$\sin(\pi + 2n\pi) = y = 0$ $\qquad\qquad \sin\left(\dfrac{3\pi}{2} \pm 2n\pi\right) = y = -1$

$\cos(2n\pi) = x = -1$ $\qquad\qquad \cos\left(\dfrac{3\pi}{2} \pm 2n\pi\right) = x = 0$

$\tan(2n\pi) = \dfrac{y}{x} = \dfrac{0}{-1} = 0$ $\qquad \tan\left(\dfrac{3\pi}{2} \pm 2n\pi\right) = \dfrac{y}{x} = \dfrac{-1}{0}$ is undefined

When $\theta = \pi$ point P has coordinates (−1, 0). When $\theta = \dfrac{3\pi}{2}$ point P has coordinates (0, −1).

Investigation – exact values of sin, cos and tan

a The diagram on the right shows an isosceles right-angled triangle PQR. Choose any value for the length of PQ to evaluate the trigonometric ratios for $\theta = \dfrac{\pi}{4}$.

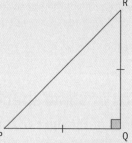

b The diagram on the right shows an equilateral triangle XYZ. A perpendicular from X to YZ has been constructed. Choose any value for the length of XY and evaluate the trigonometric ratios for $\theta = \dfrac{\pi}{3}$ and $\theta = \dfrac{\pi}{6}$.

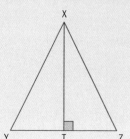

Summary of the results of the investigation

→ **Values of sin, cos and tan for common acute angles**

$$\sin\left(\frac{\pi}{6}\right) = \frac{1}{2} \qquad \cos\left(\frac{\pi}{6}\right) = \frac{\sqrt{3}}{2} \qquad \tan\left(\frac{\pi}{6}\right) = \frac{1}{\sqrt{3}}$$

$$\sin\left(\frac{\pi}{4}\right) = \frac{\sqrt{2}}{2} \qquad \cos\left(\frac{\pi}{4}\right) = \frac{\sqrt{2}}{2} \qquad \tan\left(\frac{\pi}{4}\right) = 1$$

$$\sin\left(\frac{\pi}{3}\right) = \frac{\sqrt{3}}{2} \qquad \cos\left(\frac{\pi}{3}\right) = \frac{1}{2} \qquad \tan\left(\frac{\pi}{3}\right) = \sqrt{3}$$

$\dfrac{\pi}{6} = 30°, \dfrac{\pi}{4} = 45°, \dfrac{\pi}{3} = 60°$

Example 7

Given that $\sin\theta = -\dfrac{1}{3}$ where $-\pi \le \theta \le -\dfrac{\pi}{2}$, find the values of $\cos\theta$ and $\tan\theta$.

Answer

$\cos\theta = \pm\sqrt{1 - \sin^2\theta} = \pm\sqrt{1 - \left(-\dfrac{1}{3}\right)^2}$

$\Rightarrow \cos\theta = -\sqrt{\dfrac{8}{9}} = -\dfrac{2\sqrt{2}}{3}$

$\tan\theta = \dfrac{\sin\theta}{\cos\theta} = \dfrac{-\dfrac{1}{3}}{-\dfrac{2\sqrt{2}}{3}} = \dfrac{1}{2\sqrt{2}}$

Use the identity $\sin^2\theta + \cos^2\theta = 1$
$\Rightarrow \cos^2\theta = 1 - \sin^2\theta$

Since $-\pi \le \theta \le -\dfrac{\pi}{2}$, θ is in the 3rd quadrant. Therefore, $\cos\theta$ is negative.

Use the identity $\tan\theta = \dfrac{\sin\theta}{\cos\theta}$.
$\tan\theta$ is positive in the 3rd quadrant.

$\tan\theta$ can also be found using the method shown in Example 8. It is left as an exercise for you to use this alternative method to verify the results obtained.

Example 8

Given that $\tan\theta = \dfrac{4}{3}$, find the possible values of $\sin\theta$ and $\cos\theta$.

Answer

$$\sec\theta = \pm\sqrt{1+\tan^2\theta} = \pm\sqrt{1+\left(\frac{4}{3}\right)^2}$$

$$\sec\theta = \pm\sqrt{\frac{25}{9}} = \pm\frac{5}{3}$$

Therefore, $\cos\theta = \pm\dfrac{3}{5}$

$$\sin\theta = \tan\theta\cos\theta = \left(\frac{4}{3}\right)\times\left(\pm\frac{3}{5}\right) = \pm\frac{4}{5}$$

Since tan θ > 0, θ can be in the 1st or 3rd quadrant.
Use the identity 1 + tan²θ = sec²θ.
(sec θ is positive in the 1st quadrant and negative in the 3rd quadrant.)

Use the identity tan θ = $\dfrac{\sin\theta}{\cos\theta}$.
(sin θ is positive in the 1st quadrant and negative in the 3rd quadrant.)

Example 9

Show that $\dfrac{1+\cos\theta}{\sin\theta} + \dfrac{\sin\theta}{1+\cos\theta} \equiv \dfrac{2}{\sin\theta}$ where $0 < \theta < \dfrac{\pi}{2}$.

Answer

$$\frac{1+\cos\theta}{\sin\theta} + \frac{\sin\theta}{1+\cos\theta}$$

$$= \frac{(1+\cos\theta)^2 + \sin^2\theta}{\sin\theta(1+\cos\theta)}$$

$$= \frac{1+2\cos\theta+\cos^2\theta+\sin^2\theta}{\sin\theta(1+\cos\theta)}$$

$$= \frac{2+2\cos\theta}{\sin\theta(1+\cos\theta)}$$

$$= \frac{2(1+\cos\theta)}{\sin\theta(1+\cos\theta)} = \frac{2}{\sin\theta}$$

Add the fractions.

Expand (1 + cosθ)².

Use the identity sin²θ + cos²θ = 1.
Note: with the condition $0 < \theta < \dfrac{\pi}{2}$, sinθ ≠ 0 and (1 + cosθ) ≠ 0.

> The equivalence symbol ≡ means that the two expressions either side of it are the same for all values of θ.

Exercise 8C

1 Given that $\sin\theta = \dfrac{1}{4}$ where $\dfrac{\pi}{2} \le \theta \le \pi$, find the values of $\cos\theta$ and $\tan\theta$.

2 Given that $\cos\theta = -\dfrac{12}{13}$ where $0 \le \theta \le \pi$, find the values of $\sin\theta$ and $\tan\theta$.

3 Show that $\sin\left(\arcsin\left(\dfrac{\sqrt{3}}{2}\right) - \arctan\left(\dfrac{1}{\sqrt{3}}\right)\right) = \dfrac{1}{2}$.

4 Show that $\dfrac{\sin\theta}{1-\cos\theta} + \dfrac{1-\cos\theta}{\sin\theta} \equiv \dfrac{2}{\sin\theta}$.

5 Prove the identity $\tan\theta + \cot\theta \equiv \sec\theta\csc\theta$ and hence show that $(\sin\theta + \cos\theta)(\tan\theta + \cot\theta) \equiv \sec\theta + \csc\theta$.

6 Show that $\cot^2\theta - \cos^2\theta \equiv \cos^4\theta\csc^2\theta$.

8.3 Compound angle identities

In this section you will derive useful identities, known as the compound angle identities, which express the trigonometric ratios of the sum or difference of two angles, A and B, in terms of the ratios of the separate angles.

Consider the diagram below which shows two right-angled triangles PQS and RQS. $P\hat{Q}S = A$ and $R\hat{Q}S = B$.

A perpendicular, RM, has been drawn from R to PQ, and a line parallel to PQ is drawn from S to meet RM at L.

From the diagram, you can see that $Q\hat{S}L = A = S\hat{R}L$.
In triangle RMQ,

$$\sin(A + B) = \frac{RM}{RQ} = \frac{LM+RL}{RQ}$$

$RM = LM + RL$

$$= \frac{LM}{RQ}+\frac{RL}{RQ}$$

$LM = PS$ since PMLS

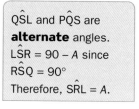

QŜL and PÔS are **alternate** angles.
LŜR = 90 − A since
RŜQ = 90°
Therefore, SR̂L = A.

$$= \frac{PS}{RQ} \times \frac{SQ}{SQ} + \frac{RL}{RQ} \times \frac{RS}{RS}$$

$\frac{RS}{RS} = 1 = \frac{SQ}{SQ}$

$$= \frac{PS}{SQ} \times \frac{SQ}{RQ} + \frac{RL}{RS} \times \frac{RS}{RQ}$$

Rearranging

$$= \sin A \cos B + \cos A \sin B$$

Therefore, $\sin(A + B) \equiv \sin A \cos B + \cos A \sin B$. This is the identity for the sine of the sum of two angles.

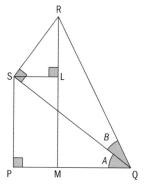

Investigation – cosines of compound angles

1 Use this diagram with $\angle PQR = A$ and $\angle SQR = B$ to prove that $\sin(A - B) \equiv \sin A \cos B - \cos A \sin B$

2 Use the identities for $\sin(A + B)$ and $\sin(A - B)$ to derive these identities:

   • $\cos(A + B) \equiv \cos A \cos B - \sin A \sin B$

   • $\cos(A - B) \equiv \cos A \cos B + \sin A \sin B$

3 Can you also derive these identities using the diagram?

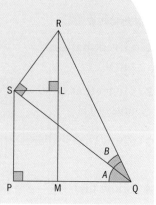

You can now use the tangent identity, $\tan\theta = \dfrac{\sin\theta}{\cos\theta}$, to derive the compound angle identities for tangent.

$$\tan(A+B) = \frac{\sin(A+B)}{\cos(A+B)}$$

Expand using the identities for the sine and cosine of the sum of two angles.

$$= \frac{\sin A\cos B + \cos A\sin B}{\cos A\cos B - \sin A\sin B}$$

Divide both the numerator and denominator by $\cos A\cos B$.

$$= \frac{\dfrac{(\sin A\cos B + \cos A\sin B)}{\cos A\cos B}}{\dfrac{(\cos A\cos B - \sin A\sin B)}{\cos A\cos B}}$$

Simplify.

$$= \frac{\dfrac{\sin A}{\cos A} + \dfrac{\sin B}{\cos B}}{1 - \dfrac{\sin A\sin B}{\cos A\cos B}} = \frac{\tan A + \tan B}{1 - \tan A\tan B}$$

Therefore, $\tan(A+B) = \dfrac{\tan A + \tan B}{1 - \tan A\tan B}$. This is the identity for the tangent of the sum of two angles.

It is left as an exercise for you to show that $\tan(A - B) = \dfrac{\tan A - \tan B}{1 + \tan A\tan B}$

This is the identity for the tangent of the difference of two angles.

→ **Compound angle identities**

$\sin(A+B) = \sin A\cos B + \cos A\sin B$ $\qquad \sin(A-B) = \sin A\cos B + \cos A\sin(-B)$

$\cos(A+B) = \cos A\cos B - \sin A\sin B$ $\qquad \cos(A-B) = \cos A\cos B + \sin A\sin B$

$\tan(A+B) = \dfrac{\tan A + \tan B}{1 - \tan A\tan B}$ $\qquad\qquad \tan(A-B) = \dfrac{\tan A - \tan B}{1 + \tan A\tan B}.$

Example 10

Show that $\cos\left(\dfrac{\pi}{12}\right) = \dfrac{\sqrt{3}+1}{2\sqrt{2}}$.

Answer

$$\cos\left(\frac{\pi}{12}\right) = \cos\left(\frac{\pi}{4} - \frac{\pi}{6}\right)$$

$$= \cos\left(\frac{\pi}{4}\right)\cos\left(\frac{\pi}{6}\right) + \sin\left(\frac{\pi}{4}\right)\sin\left(\frac{\pi}{6}\right)$$

Use the identity for the cosine of the difference of two angles.

$$= \frac{1}{\sqrt{2}} \times \frac{\sqrt{3}}{2} + \frac{1}{\sqrt{2}} \times \frac{1}{2} = \frac{\sqrt{3}+1}{2\sqrt{2}}$$

Example 11

Show that $\cos(A + B)\cos(A - B) \equiv \cos^2 A - \sin^2 B$.

Answer

$\cos(A + B)\cos(A - B)$	*Difference of two squares.*
$= (\cos A \cos B - \sin A \sin B) \times (\cos A \cos B + \sin A \sin B)$	$\cos^2 B = 1 - \sin^2 B$ *and*
$= \cos^2 A \cos^2 B - \sin^2 A \sin^2 B$	$\sin^2 B = 1 - \cos^2 B$ *using the identity*
$= \cos^2 A (1 - \sin^2 B) - \sin^2 A (1 - \cos^2 B)$	$\sin^2 \theta + \cos^2 \theta = 1.$
$= \cos^2 A - \cos^2 A \sin^2 B - \sin^2 A + \sin^2 A \cos^2 B$	
$= \cos^2 A - \sin^2 B$	*Expand and simplify.*

Example 12

Given that $\sin\theta = \dfrac{3}{5}$ and $\sin\phi = \dfrac{12}{13}$, where both θ and ϕ are acute angles,

find the exact value of $\cos(\theta - \phi)$.

Answer

$\sin\theta = \dfrac{3}{5} \Rightarrow \cos\theta = \sqrt{1 - \left(\dfrac{3}{5}\right)^2} = \dfrac{4}{5}$	*Use the identity* $\sin^2 \theta + \cos^2 \theta = 1$
$\sin\phi = \dfrac{12}{13} = \cos\phi = \sqrt{1 - \left(\dfrac{12}{13}\right)^2} = \dfrac{5}{13}$	$\Rightarrow \cos^2 \theta = 1 - \sin^2 \theta$ *(θ and ϕ are acute angles so they are in the 1st quadrant. So sin b and sin ϕ are positive.)*
$\cos(\theta - \phi) = \cos\theta \cos\phi + \sin\theta \sin\phi$	*Use the identity for the cosine of the difference of two angles.*
$\qquad = \dfrac{4}{5} \times \dfrac{5}{13} + \dfrac{3}{5} \times \dfrac{12}{13} = \dfrac{56}{65}$	

Exercise 8D

1 Find the exact value of:

   a $\sin 75°$ **b** $\tan 15°$ **c** $\sec 105°$

2 Evaluate these expressions:

   a $\cos 70° \cos 10° + \sin 70° \sin 10°$ **b** $\dfrac{\tan 75°}{\tan 15°}$

3 Given that $0 < \theta < \dfrac{\pi}{2}$ and $\sin\theta = \dfrac{24}{25}$, and $\dfrac{\pi}{2} < \phi < \pi$ and $\sin\phi = \dfrac{3}{5}$,

   find the exact value of $\tan(\theta + \phi)$.

4 Show that $\cot(A + B) \equiv \dfrac{\cot A \cot B - 1}{\cot A + \cot B}$.

5 Prove these identities:

   a $\dfrac{\sin(A + B)}{\cos A \cos B} \equiv \tan A + \tan B$ **b** $(\sin A + \cos A)(\sin B + \cos B) \equiv (\sin A + B) + \cos(A - B)$

EXAM-STYLE QUESTION

6 a Show that $\arctan\left(\dfrac{1}{4}\right) + \arctan\left(\dfrac{3}{5}\right) = \dfrac{\pi}{4}$

   b Hence, or otherwise, find the value of $\arctan(4) + \arctan\left(\dfrac{5}{3}\right)$

8.4 Double angle identities

By letting $A = B = \theta$ we can use the compound angle identities to obtain formulae for $\sin 2\theta$, $\cos 2\theta$ and $\tan 2\theta$, that is, double angle identities.

> → **Double angle identities**
>
> $\sin 2\theta = 2\sin\theta \cos\theta$ $\tan 2\theta = \dfrac{2\tan\theta}{1-\tan^2\theta}$
>
> $\cos 2\theta = \cos^2\theta - \sin^2\theta$
>
> $\quad\quad\quad = 1 - 2\sin^2\theta$
>
> $\quad\quad\quad = 2\cos^2\theta - 1$

This example shows how to use the double angle formulae to obtain ratios for multiples of an angle.

Example 13

Given that $\sin A = \dfrac{2}{5}$, use the double angle identities to evaluate $\sin 2A$ and $\cos 4A$.

Answer

$\sin A = \dfrac{4}{5} \Rightarrow \cos A = \sqrt{1 - \left(\dfrac{4}{5}\right)^2} = \dfrac{3}{5}$

$\sin 2A = 2\sin A \cos A$

$\quad\quad = 2 \times \dfrac{4}{5} \times \dfrac{3}{5} = \dfrac{24}{25}$

$\cos 4A = \cos 2(2A)$

$\quad\quad = 1 - 2\sin^2(2A)$

$\quad\quad = 1 - 2\left(\dfrac{24}{25}\right)^2 = -\dfrac{527}{625}$

Use the identity
$\sin^2 A + \cos^2 A = 1$
$\Rightarrow \cos^2 A = 1 - \sin^2 A$
Use the double angle identity for
$\sin 2A$.

In the next example compound and double angle formulae are used to prove identities.

Example 14

Show that $\sin 3A = 3\sin A - 4\sin^3 A$.

Answer

$\sin 3A = \sin A \cos 2A + \cos A \sin 2A$

$\quad\quad = \sin A(1 - 2\sin^2 A) + \cos A(2\sin A\cos A)$

$\quad\quad = \sin A - 2\sin^3 A + 2\sin A\cos^2 A$

$\quad\quad = \sin A - 2\sin^3 A + 2\sin A(1 - \sin^2 A)$

$\quad\quad = \sin A - 2\sin^3 A + 2\sin A - 2\sin^3 A$

$\sin 3A = 3\sin A - 4\sin^3 A$

Write $\sin 3A = \sin(A + 2A)$ and use the identity for the sine of the sum of two angles.
Use the double angle identity for $\sin 2A$. Since you want the answer in terms of $\sin A$, choose the appropriate double angle identity for $\cos 2A$, that is, $\cos 2A = 1 - 2\sin^2 A$.
Use the Pythagorean identity $\cos^2 A + \sin^2 A = 1$ to write $\cos^2 A$ in terms of $\sin^2 A$.

Example 15

Express $4 - 2\cos^2 A$ in terms of $\cos 2A$.

Answer

$$4 - 2\cos^2 A = 4 - (\cos 2A + 1)$$
$$= 3 - \cos 2A$$

The expression contains the term $\cos^2 A$ so use the double angle identity:
$\cos 2A = 2\cos^2 A - 1$
$\Rightarrow 2\cos^2 A = \cos 2A + 1$

Exercise 8E

1 Use the compound angle identities to derive the double angle identities:

  a $\sin 2\theta = 2\sin\theta\cos\theta$

  b $\cos 2\theta = \cos^2\theta - \sin^2\theta$

  c $\tan 2\theta = \dfrac{2\tan\theta}{1 - \tan^2\theta}$

> Let $A = B = \theta$, then
> $\sin 2\theta = \sin(\theta + \theta)$.

EXAM-STYLE QUESTIONS

2 Given that $\cos\alpha = \dfrac{4}{5}$ and $\cos\beta = \dfrac{7}{25}$, find the possible values of $\cos(\alpha + \beta)$.

3 Given that $\cos A = \dfrac{1}{3}$, find the value of $\cos 2A$ and $\cos 4A$.

4 Show that $\tan\left(\theta + \dfrac{\pi}{3}\right)\tan\left(\theta - \dfrac{\pi}{3}\right) = \dfrac{\tan^2\theta - 3}{1 - 3\tan^2\theta}$.

5 Express the following in terms of $\cos 2A$:

  a $2\cos^2 A + \sin^2 A$ **b** $\cos^4 A$ **c** $\sin^4 A$

6 Show that:

  a $(1 + \tan^2\theta)(1 - \cos 2\theta) = 2\tan^2\theta$

  b $(1 + \tan^2\theta)(1 + \cos 2\theta) = 2$

7 Prove these identities:

  a $\dfrac{1 - \cos 2A}{1 + \cos 2A} = \tan^2 A$

  b $\dfrac{1 - \tan^2 A}{1 + \tan^2 A} = \cos 2A$

  c $\dfrac{\sin 2A}{1 - \cos 2A} = \cot A$

  d $\cos 3A = 4\cos^3 A - 3\cos A$

8.5 Graphs of trigonometric functions

Consider again the definitions of the trigonometric ratios using the unit circle. The diagram to the right shows an angle θ moved through by the radius OP. P has coordinates (x, y) and

$$\sin\theta = \frac{y}{\text{OP}} = y$$

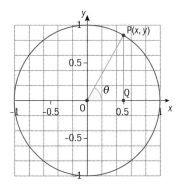

You learned that, when measuring angles in radians, their sizes correspond to real numbers which can be represented on a number line wrapped around the unit circle.

Now consider $\sin\theta$ for several angles on the circle. Since $\sin\theta$ is represented by the y-coordinate, imagine a number of vertical lines from points on the circumference of the circle to the x-axis. Open up the wrapped number line carrying these vertical lines to obtain:

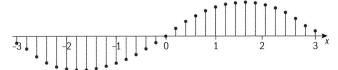

▲ The values on the x-axis correspond to the size of the angles and the bars correspond to the values of their sines.

To see an animation of the sine curve created by opening up the wrapped number line, go to http://clem.mscd.edu/~talmanl/HTML/SineCurve.html

Now consider $\cos\theta = \frac{x}{\text{OP}} = x$ and, in the same way, consider $\cos\theta$ for several angles on the unit circle.

Since $\cos\theta$ is represented by the x-coordinate, imagine a number of horizontal lines from points on the circumference of the circle to the y-axis. Open up the wrapped number line carrying these horizontal lines to obtain:

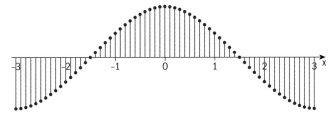

▲ The values on the x-axis correspond to the size of the angles and the bars correspond to the values of their cosines.

Consider the end points of these lines – these are the values for $\sin\theta$ and $\cos\theta$ for corresponding values of θ, that is, the functions $f(\theta) = \sin\theta$ and $g(\theta) = \cos\theta$. Since $\theta \in \mathbb{R}$ the domain of both functions will be all real values of θ. Also, if you wrap the whole number line around the unit circle, and repeat the process, the graphs for $f(\theta) = \sin\theta$ and $f(\theta) = \cos\theta$ will repeat themselves for every full turn. In other words, the sine and cosine functions are **periodic with period 2π**. Finally, the range of these functions is given by $-1 \leq f(\theta) \leq 1$.

Now do the same for $f(\theta) = \tan\theta$. Since $\tan\theta = \dfrac{y}{x}$, $x \neq 0$, you know that with the wrapped number line $\tan\theta$ is undefined whenever the line crosses the y-axis on the unit circle, that is, whenever $\theta = (2n+1)\dfrac{\pi}{2}$, $n \in \mathbb{Z}$. The graph of $f(\theta) = \tan\theta$ will have vertical asymptotes for all these values of θ. The domain of the tangent function is therefore $\theta \in \mathbb{R}$, $\theta \neq \dfrac{(2n+1)\pi}{2}$. The range of the tangent function will be $f(\theta) \in \mathbb{R}$.

You can verify that the range of the tangent function consists of all real numbers by considering values of $f(\theta) = \tan\theta$ as θ approaches either side of the vertical asymptotes.

In order to visualize the shape of the tangent function, consider this diagram which shows the unit circle and a tangent drawn at the point (1, 0). Triangles OPQ and OLM are similar and hence $\dfrac{\text{LM}}{\text{OM}} = \dfrac{\text{PQ}}{\text{OQ}} \Rightarrow \text{LM} = \dfrac{y}{x} = \tan\theta$. Therefore, the value of $\tan\theta$ is obtained by extending OP to meet this tangent line at L. The y-coordinate of the point L is then equal to $\tan\theta$. If you do this for several values of θ, you get this graph.

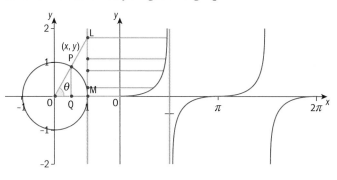
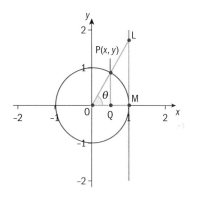

Note that for $f(\theta) = \tan\theta$ the **period of the function is π.**

You can summarize these results.

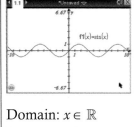
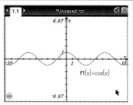
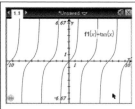

Domain: $x \in \mathbb{R}$	Domain: $x \in \mathbb{R}$	Domain: $x \in \mathbb{R}$
Range: $y \in \mathbb{R}$, $-1 \leq y \leq 1$	Range: $y \in \mathbb{R}$, $-1 \leq y \leq 1$	$x \neq \dfrac{(2n+1)\pi}{2}$
Amplitude = 1	Amplitude = 1	Range: $y \in \mathbb{R}$
Period = $2\pi \Rightarrow$	Period = $2\pi \Rightarrow$	Period = $\pi \Rightarrow$
$\sin(x + 2n\pi) = \sin(x)$	$\cos(x + 2n\pi) = \cos(x)$	$\tan(x + n\pi) = \tan(x)$
Odd function $\Rightarrow$	Even function $\Rightarrow$	Odd function $\Rightarrow$
$\sin(-x) = -\sin(x)$	$\cos(-x) = \cos(x)$	$\tan(-x) = -\tan(x)$

An odd function satisfies $f(-x) = -f(x)$. It is symmetric under a 180° rotation about the origin. An even function satisfies $f(-x) = f(x)$. It is symmetric under reflection in the y-axis. See Section 2.4.

Amplitude is defined as $\dfrac{1}{2}$ (max – min)

GDC help on CD: *Alternative demonstrations for the TI-84 Plus and Casio fx-9860GII GDCs are on the CD.*

Example 16

Use the graph of $f(\theta) = \cos\theta$ to deduce the graph and properties of $g(\theta) = \sec\theta$.

Answer

Since $g(\theta) = \dfrac{1}{f(\theta)}$ you can obtain the graph using this reasoning:

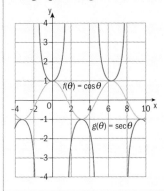

- $\sec\theta = 1$ when $\cos\theta = 1$ i.e. at $\theta = \pm 2n\pi$, $n \in \mathbb{Z}$
- $\sec\theta = -1$ when $\cos\theta = -1$ i.e. at $\theta = \pm(2n+1)\pi$, $n \in \mathbb{Z}$
- $g(\theta)$ is undefined when $f(x) = 0 \Rightarrow$ vertical asymptotes at $\theta = \pm(2n+1)\dfrac{\pi}{2}$
- $\sec\theta \to \infty$ as $\cos\theta \to 0^{+}$
  $\sec\theta \to -\infty$ as $\cos\theta \to 0^{-}$

Example 17

Use the graph of $y = \sin x$ to sketch the graph of $y = 3\sin 4x$.

Answer

Let $f(x) = \sin x$ and $g(x) = 3\sin 4x$. Then $g(x) = 3f(4x)$.

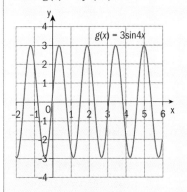

Stretch the function of $y = \sin x$ by a scale factor of 3 parallel to the y-axis. The amplitude of $g(x)$ is 3.

Stretch the function of $y = \sin x$ by a scale factor of $\dfrac{1}{4}$ parallel to the x-axis.

The period of the new function, $g(x)$, is therefore $\dfrac{2\pi}{4} = \dfrac{\pi}{2}$.

High-tech musical composers and developers of computer games need to use the basic rules of trigonometry. How is music related to trigonometric functions?

'Sketch' means give a general shape of the graph, showing any zeros, the y-intercept, any asymptotes and any maxima or minima.

Exercise 8F

1 Use the graph $f(\theta) = \sin\theta$ to deduce the graph and properties of $g(\theta) = \csc\theta$.

2 Use the graph of $y = \cos x$ to sketch each of these functions.

 a $y = 5\cos 2x$

 b $y = \cos\left(x + \dfrac{\pi}{2}\right)$

 c $y = -4\cos(\pi x)$

 d $y = 2\cos(4x + \pi) + 1$

For a reminder about transforming graphs of functions see Section 2.4.

3 Use the graph of $y = \sin x$ to sketch each of these functions.

a $y = 3\sin x$ **b** $y = \sin 3x$

c $y = 3\sin 3x$ **d** $y = 3(1 + \sin(3x))$

4 On the same set of axes, sketch the graphs of $f(x) = \cos 2x$ and $g(x) = 2\cos x$ for $0 \le x \le \pi$.

How many solutions are there to the equation $f(x) = g(x)$ in the interval $0 \le x \le \pi$?

5 For each of these functions, determine whether they are odd or even, give the period of each function and hence sketch each of the functions over the period $0 \le x \le 2\pi$.

a $f(x) = 4\sin x \cos x$

b $g(x) = 1 - 2\sin^2 x$

c $h(x) = x\sin x$

Investigation – properties of the sine function

Consider the function $f(x) = a\sin[b(x + c)] + d$, $b, c \in \mathbb{R}$ $b > 0$.

By considering different values of a, b, c and d show that

- the amplitude of the function is given by $|a|$
- the function has a period of $\dfrac{2\pi}{b}$
- the line $y = d$ is a line of symmetry
- $d = \dfrac{\text{maximum value} + \text{minimum value}}{2}$
- the function can be obtained by shifting the graph of $y = a\sin(bx)$ by c units to the left and d units vertically upwards when $c, d > 0$.

> The number c is called the phase shift of $f(x)$.

Steps for fitting data to a sine function

For the sine function $f(x) = a\sin[b(x + c)] + d$:

a Calculate the amplitude

$$a = \frac{\text{maximum value} - \text{minimum value}}{2}$$

b Calculate the vertical shift

$$d = \frac{\text{maximum value} + \text{minimum value}}{2}$$

c Find the value of

$$b = \frac{2\pi}{\text{period}}$$

d Calculate the horizontal shift by choosing given coordinates of a data point.

> Trigonometric functions of the form $f(x) = a\sin[b(x + c)] + d$ apply to wave motion and the motion of a Ferris wheel. The first Ferris wheel was built in 1893 to a Chicago exhibition. The current tallest Ferris wheel is 165 m (the Singapore Flyer), opened in 2008.

Example 18

Find the amplitude, period, phase shift and vertical shift of the function $f(x) = 2\sin(4x + \pi) + 5$. Hence, sketch the graph.

Answer

Amplitude = 2

Period = $\dfrac{2\pi}{4} = \dfrac{\pi}{2}$

Phase shift = $\dfrac{\pi}{4}$

Vertical shift = 5

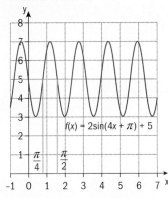

Rewrite the function as:

$$f(x) = 2\sin\left[4\left(x + \frac{\pi}{4}\right)\right] + 5$$

Compare with
$$f(x) = a\sin[b(x + c)] + d$$
where a = amplitude,

$$b = \frac{2\pi}{period}\left(\Rightarrow period = \frac{2\pi}{b}\right)$$

c = phase shift
d = vertical shift

Example 19

The number of hours of sunlight in Wellington, the southernmost city in New Zealand, on the shortest day is 9.18 hours and on the longest day 15.13 hours.

Given that the hours of sunlight over a year follows a function of the form $f(x) = a\sin[b(x + c)] + d$, find the values of a, b, c and d.

Use a graph of this function to find the number of hours of sunshine on 21 March 2012.

(Assume that this is the 60th day of the year and that there are 365 days in a year.)

> Assume that in the southern hemisphere 21 June is the shortest day and 21 December is the longest day.

Answer

For $f(x) = a\sin[b(x + c)] + d$;
The amplitude of the function is given by

$$\frac{15.13 - 9.18}{2} = 2.98 = a$$

The vertical shift is given by

$$\frac{15.13 + 9.18}{2} = 12.16 = d$$

The period of the function is

$$\frac{2\pi}{b} = 365 \Rightarrow b = \frac{2\pi}{365} \approx 0.0172$$

So, $f(x) = 2.98\sin\left[\dfrac{2\pi}{365}(x + c)\right] + 12.16$

The data repeats itself every 365 days, that is, period = 365.

▶ Continued on next page

$$15.13 = 2.98\sin\left[\frac{2\pi}{365}(355 + c)\right] + 12.16$$

On 21 December (longest day),
x = 365 − 10 = 355 and f(x) = 15.13
Substitute these values and solve for c.

$$2.97 = 2.98\sin\left[\frac{2\pi}{365}(355 + c)\right]$$

$$\frac{2.97}{2.98} = \sin\left[\frac{2\pi}{365}(355 + c)\right]$$

$$\left[\frac{2\pi}{365}(355 + c)\right] = \arcsin\left(\frac{2.97}{2.98}\right)$$
$$= 1.489$$

GDC help on CD:
*Alternative demonstrations
for the TI-84 Plus and Casio
fx-9860GII GDCs are on the
CD.*

$$355 + c = \frac{1.498 \times 365}{2\pi}$$

$$c = \frac{1.498 \times 365}{2\pi} - 355 = -268$$

Therefore,

$$f(x) = 2.98\sin\left[\frac{2\pi}{365}(x - 268)\right] + 12.16$$

From the graph of the function, there are 13.4 hours of sunshine on 21 March 2012.

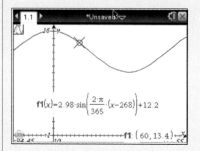

Exercise 8G

1 a Find the amplitude, period and phase shift of functions **i** and **ii** below.

   b Calculate the minimum and maximum values and sketch each function.

   i $f(x) = 7\sin\left[6\left(x + \frac{\pi}{12}\right)\right] + 3$

   ii $f(x) = -3\sin\left(2x + \frac{\pi}{2}\right) - 5$

2 The voltage, V, produced by an AC generator is given by $V(t) = 220\sin(120\pi t)$.

   a Find the maximum voltage produced.

   b Find the minimum voltage produced.

   c Write down the amplitude of the function V.

   d Write down the period of the function V.

   e Sketch the graph of V over two periods starting at $t = 0$.

3 Water tides can be modeled by the function

$$h(t) = a \sin [b(t + c)] + d$$

where $h(t)$ is the height of water at time t, measured in hours after midnight.

At Blue Harbor on Sunny Island the time between consecutive high tides is 12 hours. The height of the water at high tide is 14.4 m and the height of the water at low tide is 1.2 m.

On a particular day the first high tide occurs at 08:15.

a Use the information given to find the values of a, b, c and d.

b Plot the graph of the function and calculate the time of the first low tide.

A fishing boat is only allowed to leave or enter the harbor if the height of the water is at least 5 m.

c Find the time intervals during which a boat could enter or leave the harbor on that particular day.

4 In Miami, Florida, the sun shines for 12.75 hours on the 21 June and on the shortest day there are 10.65 hours of sunshine. Given that the hours of sunlight over a year follows a function of the form $f(x) = a \sin[b(x + c)] + d$, find the values of a, b, c and d.

Use a graph of this function to find the number of hours of sunshine on Independence Day, 4 July.

(Assume that there are 365 days in a year.)

8.6 The inverse trigonometric functions

Before discussing the nature of the inverse trigonometric functions, you can list the characteristics of a function and its inverse (as defined in Chapter 2).

Inverse functions are discussed in Section 2.3.

- The inverse $f^{-1}(x)$ of a function $f(x)$ exists if and only if $f(x)$ is a one-to-one function.
- $(f^{-1} \circ f)(x) = (f \circ f^{-1})(x) = x$
- The domain of $f(x)$ is the range of $f^{-1}(x)$ and the range of $f(x)$ is the domain of $f^{-1}(x)$.
- The graph of $f^{-1}(x)$ is the mirror image of $f(x)$ in the line $y = x$.

The inverse sine function

You know by definition that: $y = \sin x \Leftrightarrow x = \arcsin y$. The graph of the function $f(x) = \sin x$ is not one-to-one since for $-1 < f(x) < 1$ there are infinitely many values of x that give the same value for $f(x)$.

The horizontal line test shows that the function is not one-to-one.

However, if you restrict the domain of $f(x) = \sin x$ to the interval $-\dfrac{\pi}{2} \leq x \leq \dfrac{\pi}{2}$, the function becomes one-to-one and hence it will have an inverse, $f^{-1}(x) = \arcsin x$.

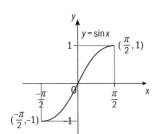

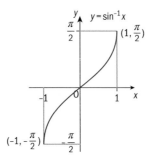

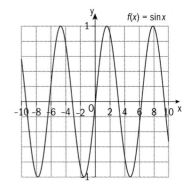

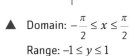

▲ Domain: $-\dfrac{\pi}{2} \leq x \leq \dfrac{\pi}{2}$

   Range: $-1 \leq y \leq 1$

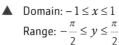

▲ Domain: $-1 \leq x \leq 1$

   Range: $-\dfrac{\pi}{2} \leq y \leq \dfrac{\pi}{2}$

The inverse cosine function

You can do the same for the function $f(x) = \cos x$ by restricting the domain to $0 \leq x \leq \pi$.

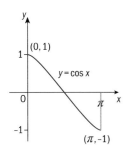

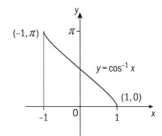

▲ Domain: $0 \leq x \leq \pi$

   Range: $-1 \leq y \leq 1$

▲ Domain: $-1 \leq x \leq 1$

   Range: $0 \leq y \leq \pi$

The inverse tangent function

The function $f(x) = \tan x$, for the domain $-\dfrac{\pi}{2} < x < \dfrac{\pi}{2}$ is one-to-one and therefore has an inverse.

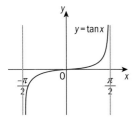

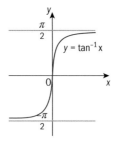

A GDC often uses a different notation for inverse trigonometric functions.

$\sin^{-1} x = \arcsin x$

$\cos^{-1} x = \arccos x$

$\tan^{-1} x = \arctan x$

▲ Domain: $-\dfrac{\pi}{2} \leq x \leq \dfrac{\pi}{2}$

   Range: all real numbers

▲ Domain: all real numbers

   Range: $-\dfrac{\pi}{2} \leq y \leq \dfrac{\pi}{2}$

The next examples show how inverse trigonometric functions are used in typical examination-style questions.

Example 20

Write $\sin(\arcsin a + \arccos b)$ in terms of a and b.

Answer

Let $\arcsin a = \theta$ and $\arccos b = \phi$.

$\sin(\arcsin a + \arccos b)$

$= \sin(\theta + \phi)$

$= \sin\theta\cos\phi + \cos\theta\sin\phi$

$= ab + \sqrt{1-a^2}\,\sqrt{1-b^2}$

$\arcsin a = \theta \Rightarrow \sin\theta = \alpha, -\dfrac{\pi}{2} < \theta < \dfrac{\pi}{2}$

Use the identity $\sin^2\theta + \cos^2\theta = 1 \Rightarrow$

$\cos\theta = \sqrt{1-\sin^2\theta} = \sqrt{1-a^2}, \cos\theta \geq 0$

$\arccos b = \phi \Rightarrow \cos\phi = b, \, 0 < \phi < \pi$

$\sin\phi = \sqrt{1-\cos^2\phi} = \sqrt{1-b^2}, \, \sin\phi \geq 0$

Example 21

Evaluate $\sin\left(\arctan\left(\dfrac{1}{\sqrt{3}}\right)\right) + \cos\left(\arcsin\left(\dfrac{1}{\sqrt{2}}\right)\right)$

Answer

$\arctan\left(\dfrac{1}{\sqrt{3}}\right) = \dfrac{\pi}{6}$

$\therefore \sin\left(\arctan\left(\dfrac{1}{\sqrt{3}}\right)\right) = \sin\dfrac{\pi}{6} = \dfrac{1}{2}$

$\arcsin\left(\dfrac{1}{\sqrt{2}}\right) = \dfrac{\pi}{4}$

$\therefore \cos\left(\arcsin\left(\dfrac{1}{\sqrt{2}}\right)\right) = \cos\dfrac{\pi}{4} = \left(\dfrac{1}{\sqrt{2}}\right)$

Therefore,

$\sin\left(\arctan\left(\dfrac{1}{\sqrt{3}}\right)\right) + \cos\left(\arcsin\left(\dfrac{1}{\sqrt{2}}\right)\right)$

$= \dfrac{1}{2} + \dfrac{1}{\sqrt{2}} = \dfrac{1+\sqrt{2}}{2}$

Let $\arctan\left(\dfrac{1}{\sqrt{3}}\right) = \theta \Rightarrow \tan\theta = \dfrac{1}{\sqrt{3}}$

$\Rightarrow \theta = \dfrac{\pi}{6}$

Let $\arcsin\left(\dfrac{1}{\sqrt{2}}\right) = \phi \Rightarrow \sin\phi = \dfrac{1}{\sqrt{2}}$

$\Rightarrow \phi = \dfrac{\pi}{4}$

Exercise 8H

1 Find the exact value of each of these expressions.

a $\cos\left(\arcsin\dfrac{\sqrt{2}}{2}\right)$

b $\sec\left(\arctan\dfrac{1}{2}\right)$

c $\cos\left(\arcsin\left(-\dfrac{\sqrt{3}}{2}\right)\right)$

d $\tan\left(\arctan\dfrac{5\pi}{6}\right)$

e $\arccos\left(\sin\left(\dfrac{3\pi}{4}\right)\right)$

f $\arcsin\left(\sin\left(-\dfrac{7\pi}{6}\right)\right)$

2 Find the exact value of these expressions.

a $\sin\left(\arcsin\dfrac{1}{2}+\arccos\dfrac{1}{2}\right)$

b $\cos\left(\arcsin\dfrac{3}{5}-\arccos\left(-\dfrac{4}{5}\right)\right)$

c $\tan\left(2\arctan\left(\dfrac{3}{4}\right)\right)$

3 a Show that $\tan(\arcsin a)=\dfrac{a}{\sqrt{1-a^2}}$

b Show that $\cos(\arcsin a + \arccos a) = 0$

c Show that $\tan(\arccos a)=\dfrac{\sqrt{1-a^2}}{a}$

8.7 Solving trigonometric equations

In this section you will look at the different forms of trigonometric equations and their solutions. Because of the periodic nature of trigonometric functions, general solutions to trigonometric equations will give an infinite number of solutions. However, you will only be looking at solutions over a finite interval.

Simple trigonometric equations

Example 22

Solve the equation $\sin\theta = -\dfrac{\sqrt{2}}{2}$ for $0 \le \theta \le 2\pi$.

- -

Answer

Method 1: analytic solution

$\theta = \pi+\dfrac{\pi}{4}=\dfrac{5\pi}{4}$ or

$\theta = 2\pi-\dfrac{\pi}{4}=\dfrac{7\pi}{4}$

Method 2: graphical solution

$\theta = 3.93$ radians or
$\theta = 5.50$ radians

*Since $\sin\theta < 0$, it follows that
θ is in the 3rd or 4th quadrant.*

$\sin\dfrac{\pi}{4}=\dfrac{\sqrt{2}}{2}.$

*Alternatively, use a graphical
method to find the solution.*

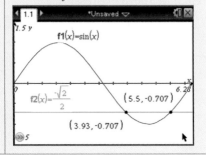

GDC help on CD:
*Alternative demonstrations
for the TI-84 Plus and Casio
fx-9860GII GDCs are on the
CD.*

Trigonometric equations which are quadratic

These are equations which can be written as $a\sin^2\theta + b\sin\theta + c = 0$
or $a\cos^2\theta + b\cos\theta + c = 0$

These examples show they are solved using trigonometric identities.

Example 23

Solve the equation $2\tan\theta = \cos\theta$ for $-\pi \le \theta \le \pi$.

Answer

$\dfrac{2\sin\theta}{\cos\theta} + \cos\theta = 0$

$\Rightarrow 2\sin\theta + \cos^2\theta = 0$

$\Rightarrow 2\sin\theta + 1 - \sin^2\theta = 0$

$\Rightarrow \sin^2\theta - 2\sin\theta - 1 = 0$

$\Rightarrow (\sin\theta - 1)^2 = 0$

$\Rightarrow \sin\theta = 1$

$\Rightarrow \theta = \dfrac{\pi}{2}$

Rewrite the term $2\tan\theta$ using the identity $\tan\theta = \dfrac{\sin\theta}{\cos\theta}$.

Multiply throughout by $\cos\theta$.

Rewrite the term $\cos^2\theta$ using the identity
$\sin^2\theta + \cos^2\theta = 1 \Rightarrow$
$\cos^2\theta = 1 - \sin^2\theta$.

Example 24

Solve the equation $3\sin x\cos x + \sin x - 9\cos x - 3 = 0$ for $0 \le x \le 2\pi$.

Answer

$3\sin x\cos x + \sin x - 9\cos x - 3 = 0$

$\Rightarrow \sin x(3\cos x + 1)$
$\quad - 3(3\cos x + 1) = 0$

$\Rightarrow (3\cos x + 1)(\sin x - 3) = 0$

$\Rightarrow 3\cos x + 1 = 0$ or $\sin x - 3 = 0$

$\Rightarrow \cos x = -\dfrac{1}{3}$ since $\sin x \ne 3$

$x = 1.91$ radians or $x = 4.37$ radians

Factorize.

Use a GDC to evaluate x.

Example 25

Solve the equation $e^{\frac{x}{2}}\cos 2x = e^{\sqrt{x}}\sin x$ for $0 \le x \le 2\pi$.

Answer

$x = 0.432$ radians

$x = 2.68$ radians

$x = 4.45$ radians

$x = 5.12$ radians

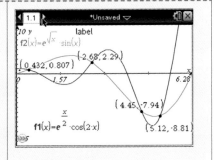

Using the GDC to sketch the graphs. Notice that there are four solutions in the given interval.

GDC help on CD:
Alternative demonstrations for the TI-84 Plus and Casio fx-9860GII GDCs are on the CD.

Example 26

Solve these equations for $-\pi \le x \le \pi$.

a $\sin 2x + \sin x = 0$ **b** $\cos 2x + \sin x = 0$

Answers

a $\sin 2x + \sin x = 0$

$\Rightarrow 2\sin x \cos x + \sin x = 0$

$\Rightarrow \sin x (2\cos x + 1) = 0$

$\Rightarrow \sin x = 0$ or $\cos x = -\dfrac{1}{2}$

$\Rightarrow x = \arcsin 0 = 0, \pm \pi$

or $x = \arccos\left(-\dfrac{1}{2}\right) = \pm\dfrac{2\pi}{3}$

Use the double angle identity for sine: $\sin 2x = 2\sin x \cos x$.
Factorize.

b $\cos 2x + \sin x = 0$

$\Rightarrow 1 - 2\sin^2 x + \sin x = 0$

$\Rightarrow 2\sin^2 x - \sin x - 1 = 0$

$\Rightarrow (2\sin x + 1)(\sin x - 1) = 0$

$\Rightarrow \sin x = -\dfrac{1}{2}$ or $\sin x = 1$

$\Rightarrow x = \arcsin\left(-\dfrac{1}{2}\right) = -\dfrac{\pi}{6}, -\dfrac{5\pi}{6}$

or $x = \arcsin 1 = \dfrac{\pi}{2}$

Use the double angle identity for cosine: $\cos 2x = 1 - 2\sin^2 x$.
Factorize.

> Since $\sin x$ can be equal to 0 we cannot divide throughout by $\sin x$, otherwise we would miss three possible solutions.

Exercise 8I

1 Solve $3\sin x = 2\tan x$ where $-\pi \le x \le \pi$.

2 The angle θ satisfies the equation $\cot \theta + \sin \theta = 6$.
   Find all the possible values of θ in the interval $[0, \pi]$.

3 Find all the values of θ in the interval $[-\pi, \pi]$ which satisfy the
   equation $3\cos 2\theta = 2\cos^2 \theta$.

EXAM-STYLE QUESTION

4 Given that $3\tan^2 \theta - \dfrac{14}{\cos \theta} + 18 = 0$, find the possible values for $\sec \theta$.

5 Solve $\sin x - \cos x = 1$ for $0 \le x \le \pi$.

EXAM-STYLE QUESTIONS

6 Solve the equation $\csc \theta + \sin \theta = 2$ for $-\pi \le \theta \le \pi$.

7 Given that $\dfrac{\sin x - 3\cos x}{\sin x - \cos x} = 7$ find the value of $\tan x$. Hence,
   find the exact values of:

   a $\tan 2x$ **b** $\tan\left(\dfrac{x}{2}\right)$

8 Find all the solutions to the equation $\frac{x}{2}\sin 2x = \sqrt{x}\sin x$ in the interval $0 \le x \le 2\pi$.

9 Find all the solutions to the equation $-5x^2 \cos 8x = \tan x$ in the interval $0 \le x \le \frac{\pi}{2}$.

8.8 The cosine rule

In the triangle on the right, a perpendicular has been drawn from B to AC. The height of the perpendicular is h and the lengths of the sides AB, AC and BC are c, b and a respectively.

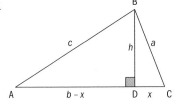

In $\triangle$ABD, $h^2 = c^2 - (b-x)^2$
In $\triangle$BDC, $h^2 = a^2 - x^2$ $\Big\}$ $c^2 - (b-x)^2 = a^2 - x^2$

Simplifying, you obtain:

$c^2 - (b^2 - 2bx + x^2) = a^2 - x^2 \Rightarrow$

$c^2 - b^2 + 2bx = a^2 \Rightarrow$

$c^2 = a^2 + b^2 - 2bx \Rightarrow$

$c^2 = a^2 + b^2 - 2ab\cos C$

In $\triangle$BDC, $\cos C = \frac{x}{a} \Rightarrow x = a\cos C$

Rearranging you find:

$$\cos C = \frac{a^2 + b^2 - c^2}{2ab}$$

> In this form, c is the side opposite the chosen angle, a and b are the sides adjacent to the chosen angle and C is the chosen angle.

→ The cosine rule states for any triangle ABC with corresponding sides a, b and c:

$c^2 = a^2 + b^2 - 2ab\cos C \Rightarrow \cos C = \frac{a^2 + b^2 - c^2}{2ab}$

> Can you derive the cosine rule if the triangle ABC looks like this?

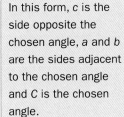

Two other forms of the cosine rule are:

$a^2 = b^2 + c^2 - 2bc\cos A \Rightarrow \cos A = \frac{b^2 + c^2 - a^2}{2bc}$

$b^2 = a^2 + c^2 - 2ac\cos B \Rightarrow \cos B = \frac{a^2 + c^2 - b^2}{2ac}$

The cosine rule is used to solve triangles when

- three sides are given, or
- two sides and the **included angle** are given.

Example 27

In triangle PQR, PQ = 9 cm, QR = 16 cm and PR = 11 cm.
Calculate the smallest angle in the triangle to the nearest degree.

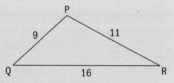

Answer

$$\cos R = \frac{p^2 + q^2 - r^2}{2pq}$$

$$\Rightarrow P\hat{R}Q = \arccos\left(\frac{16^2 + 11^2 - 9^2}{2 \times 16 \times 11}\right)$$

$$\Rightarrow P\hat{R}Q = 33°$$

The smallest angle is opposite the smallest side, that is, angle R.

Use the cosine rule.

Example 28

In triangle ABC, the lengths of the sides a, b and c are in the ratio 2 : 5 : 6 respectively.
Find the largest side of the triangle.

Answer

$$\cos Z = \frac{x^2 + y^2 - z^2}{2xy}$$

$$\Rightarrow X\hat{Z}Y = \arccos\left(\frac{2^2 + 5^2 - 6^2}{2 \times 2 \times 5}\right)$$

$$= 110.5° \text{ (to 1 dp)}$$

$a : b : c = 2 : 5 : 6$
This means that triangle ABC is similar to a triangle XYZ with sides 2, 5 and 6 units long.

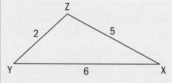

Since similar triangles are equi-angular you can solve for triangle PQR.
The largest angle is opposite the largest side.

Example 29

From a sailing boat, B, out at sea, two lighthouses, A and C, can be seen on the coastline. The boat is 6 km away from A and 9 km from C.
Given that the angle $A\hat{B}C$ is 70°, what is the distance between the two lighthouses?

Answer

$b^2 = a^2 + c^2 - 2ac \cos B$

$\Rightarrow b = \sqrt{9^2 + 6^2 - 2 \times 9 \times 6 \cos 70}$

$\qquad = 8.95 \,(\text{to 2 dp})$

So the two lighthouses are 8.95 km apart.

Draw a sketch to represent the information given in the question.

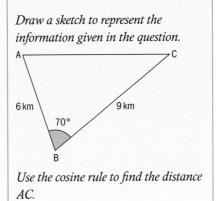

Use the cosine rule to find the distance AC.

Exercise 8J

1 Use the cosine rule to find the missing angles and sides of these triangles.

a

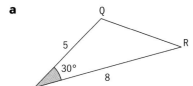

b

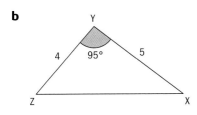

c

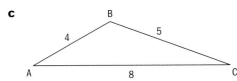

2 Find the largest angle in triangle ABC given that $a = 4.5$ cm, $b = 3.9$ cm and $c = 2.3$ cm.

3 In triangle PQR the sides PQ, QR and RP are in the ratio $3 : 2 : 4$. Find the smallest angle of the triangle.

4 Triangle ABC has sides of length 5, x and $(2x-1)$. Given that $\widehat{BAC} = 60°$, find x and hence calculate the other two angles of the triangle.

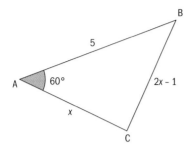

5 A parallelogram ABCD has sides AB and AD of length a and b respectively. The diagonals AC and BD have lengths p and q as shown in the diagram.

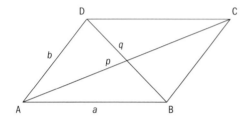

Show that $p^2 + q^2 = 2(a^2 + b^2)$.

8.9 The sine rule

In this triangle, a perpendicular has been drawn from B to AC. The height of the perpendicular is h and the lengths of the sides AB, AC and BC are c, b and a respectively.

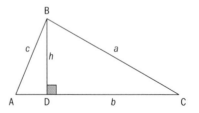

In $\triangle$ABD, $\sin A = \dfrac{h}{c} \Rightarrow h = c\sin A$

In $\triangle$CBD, $\sin C = \dfrac{h}{a} \Rightarrow h = a\sin C$

$\Rightarrow c\sin A = a\sin C \Rightarrow \dfrac{c}{\sin C} = \dfrac{a}{\sin A}$

Now consider the same triangle but this time with a perpendicular of height H drawn from C to AB.

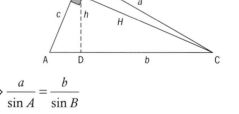

In $\triangle$BCE, $\sin B = \dfrac{H}{a} \Rightarrow H = a\sin B$

In $\triangle$ACE, $\sin A = \dfrac{H}{b} \Rightarrow H = b\sin A$

$\Rightarrow a\sin B = b\sin A \Rightarrow \dfrac{a}{\sin A} = \dfrac{b}{\sin B}$

Combining the two results you get: $\dfrac{a}{\sin A} = \dfrac{b}{\sin B} = \dfrac{c}{\sin C}$

Alternatively: $\dfrac{\sin A}{a} = \dfrac{\sin B}{b} = \dfrac{\sin C}{c}$

> → The sine rule states for any triangle ABC with corresponding
> sides a, b and c: $\dfrac{a}{\sin A} = \dfrac{b}{\sin B} = \dfrac{c}{\sin C}$

The sine rule is used to solve triangles when
- two angles and any side are given, or
- two sides and a **non-included angle** are given.

Example 30

This illustration shows the angles of elevation of the highest point of the Great Pyramid of Cheops measured from two observation points A and B. Given that A and B are 32 m apart, calculate the height of the pyramid h.

 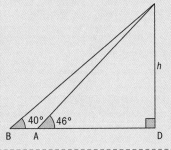

Answer

In triangle ABC:

$\dfrac{BC}{\sin 134°} = \dfrac{AB}{\sin 6°}$

$BC = \dfrac{32\sin 134°}{\sin 6°} \approx 220.22$ m

In triangle BCD:

$\sin 40° = \dfrac{h}{BC}$

$\Rightarrow h \approx 220.22 \sin 40°$

Therefore, $h \approx 141.6$ m.

$C\widehat{A}B = 180° - 46° = 134°$

$A\widehat{C}B = 180° - (134° + 40°) = 6°$

Apply the sine rule to triangle ABC given AB = 32.

Use the sine ratio given opp = h and hyp = BC ≈ 220.22.

The construction of the Great Pyramid of Cheops has amazed people through the ages. Did you know that the ratio of the side length of the square base to the height of the pyramid is equal to $\dfrac{\pi}{2}$?

Ambiguous case

Be careful when finding angles using the sine rule because there are two angles between 0° and 180° with a particular ratio for sine. In order to understand how this is possible, consider the same triangle used in the derivation of the sine rule above.

Given the lengths of two sides, BC and AB, and the measure of a non-included angle, C, you can actually construct two triangles, ABC and A'BC, as shown in this diagram. This is often referred to as the ambiguous case.

When given two sides and a non-included angle (the ambiguous case) it should not be assumed that there are always two solutions for a missing angle. You should always check that each answer makes sense. This is demonstrated in Examples 31 and 32.

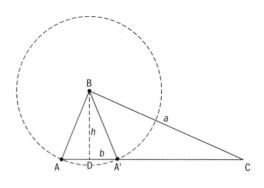

$$B\hat{A}C = 180° - B\hat{A}'C$$

Example 31

The diagram shows a river with a fence, AB, of length 5 m built at an angle of 34° to the riverside.
Farmer Brown wants to fence off an area in the shape of a triangle ABC (as shown in the diagram) for his three goats. He has 3 m of fencing left. Find the angles ACB and ABC.

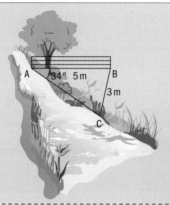

This solution can also be obtained using the cosine rule:
Let AC = x. Then use the cosine rule:
$x^2 + 5^2 - 10\,x\cos 34° = 9^2$.
This will give you two solutions for AC and hence the two possible solutions for triangle ABC.

Answer

For triangle ABC':

$$\frac{\sin C'}{5} = \frac{\sin 34°}{3} \Rightarrow \sin C' = \left(\frac{5\sin 34°}{3}\right)$$

$$C' = \arcsin\left(\frac{5\sin 34°}{3}\right) = 68.7°$$

For triangle ABC:
$C = \sin(180 - 68.7)° = 111.3°$
So, A$\hat{C}$B = 68.7° or 111.3°

$\triangle$ABC:
A$\hat{C}$B = 68.7°, A$\hat{B}$C = 180° - (34 + 68.7)
     = 77.3°

$\triangle$ABC: A$\hat{C}$B = 111.3°,
A$\hat{B}$C = 180° - (34 + 111.3)° = 34.7°

You are given two sides and the non-included angle so you may have two solutions.

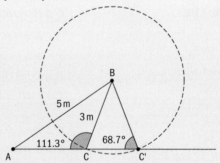

Solve for acute angle C'.
Use the sine rule in triangle ABC' given c = 5, a = 3 and A = 34°.
Use the fact that sin(180° − θ) = sin θ to solve for the obtuse angle C.

This will give you two solutions for AC and hence the two possible solutions for triangle ABC.

Example 32

Consider the same situation as in Example 31 but with AB = 8 m making an angle of 40° with the riverside as shown in the diagram. This time farmer Brown has 12 m of fencing left. Find the angles ACB and ABC.

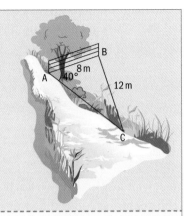

In Example 32 we have one possible triangle (one solution) and $a \geq c$.

Answer

For acute angle C in triangle ABC:

$$\frac{\sin C}{8} = \frac{\sin 40°}{12} \Rightarrow \sin C = \left(\frac{8\sin 40°}{12}\right)$$

$$C = \arcsin\left(\frac{8\sin 40°}{12}\right) = 25.4°$$

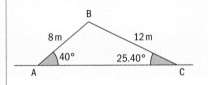

For obtuse angle:
$(180 - 25.4)° = 154.6°$
But this is not possible because
$40° + 154.6° > 180°$
So there is only one solution:
$A\hat{C}B = 25.4°$,
$A\hat{B}C = 180° - (40 + 25.4)$
$= 114.6°$

As with Example 31 consider the possibility of two solutions.
Use the sine rule given c = 8,
a = 12 and A = 40°.
Use sin(180° − θ) = sin θ to solve for the obtuse angle.

Exercise 8K

1 Find the unknown angles and sides in these triangles:

a

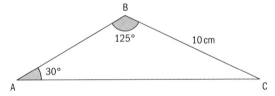

b

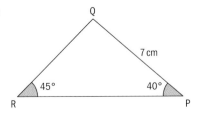

c

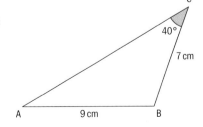

2 A plane flying from P to Q followed a course from P that had a 15° error, as shown in the diagram below. After traveling for 80 km, the pilot corrected the course by changing direction at point R and flew a further 150km to reach Q.

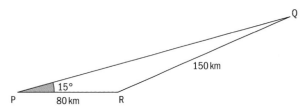

Assuming that the plane flew at a constant $400\,\text{km}\,\text{h}^{-1}$, calculate the amount of time (to the nearest second) that was lost due to the error.

3 From a hot air balloon the angles of depression to each end of a lake are 68° and 32°. Given that the balloon is 250 m above the ground, find the length of the lake. Give your answer to the nearest metre.

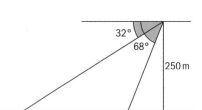

4 This diagram shows three points, A, B and C, on level ground. A vertical mast, MA, stands at A. The top of the mast is supported by wires fastened to the ground at B and C.
M$\hat{\text{B}}$A = 64° and M$\hat{\text{C}}$A = 23°

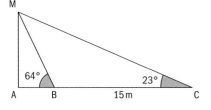

Given that B and C are 15 m apart, find the length of both wires and the height of the mast, MA.

5 Show that two triangles, ABC, can be drawn in which AB = 31 cm, AC = 27 cm and A$\hat{\text{B}}$C = 55°. Find the size of the angles of each triangle, giving your answers to the nearest degree.

Investigation – the sine rule and the triangle inequality

This diagram shows a triangle ABC.

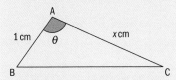

Copy and complete the table.

x (cm)	θ (°)	BC (cm)
0.7	26.8	0.49
0.8		0.64
0.9	50.1	
1.0		1.0
1.1	70.2	
1.2		1.44

Make a conjecture about your results for BC.

Investigate what happens to your results for θ as you change values of x and the corresponding values of BC according to your conjecture.

Answer these questions based on your results.

- What values can x take?
- Do your results indicate any limits to the values x can take?
- Write the triangle inequality for the case $0 < x < 1$, and hence solve for x
- Write the triangle inequality for the case $x \geq 1$, and solve for x.
- Combine the last two results into one inequality and comment on your findings.

> The triangle inequality states that the sum of the lengths of any two sides must be greater than the length of the remaining side. If the sum of the lengths of two sides is equal to the length of the third side then the three points are collinear.

8.10 Area of a triangle

Consider triangle ABC shown in the diagram on the right.

Area of triangle ABC $= \dfrac{1}{2}ah$

In triangle ABD, $\sin C = \dfrac{h}{b} \Rightarrow h = b\sin C$

Substituting for h we obtain:

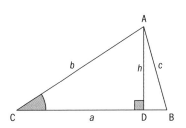

> → Area of $\triangle ABC = \dfrac{1}{2}ab\sin C$ where a and b are adjacent sides and C is the included angle.

> Can you derive the area formula if the triangle ABC looks like this?
>
>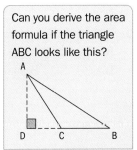

Example 33

Find the area of the quadrilateral ABCD.

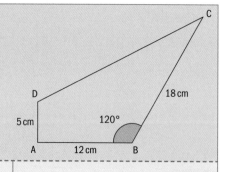

Answer

Area of triangle:

$$ABD = \frac{1}{2} \times 12 \times 5$$

$$= 30 \, cm^2$$

$$\theta = \arctan\left(\frac{5}{12}\right)$$

$$= 22.6°$$

$$\therefore \, D\hat{B}C = 120° - 22.6°$$

$$= 97.4°$$

$$BD = \sqrt{5^2 + 12^2}$$

$$= 13$$

Area of triangle:

$$BDC = \frac{1}{2} \times 13 \times 18 \times \sin 97.4°$$

$$= 116 \, cm^2$$

Hence, area of $ABCD = 30 + 116$

$$= 146 \, cm^2$$

Join B to D to obtain two triangles so area ABCD = area ABD + area BDC.

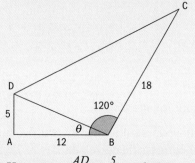

Use $\tan \theta = \dfrac{AD}{AB} = \dfrac{5}{12}$ to find θ.

Use Pythagoras' theorem to find BD.
For $\triangle BDC$, area of triangle
$$= \frac{1}{2} bc \sin A$$

Exercise 8L

1 Find the area of quadrilateral PQRS.

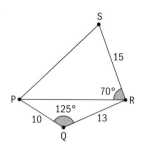

2 Find the difference between the areas of two possible triangles ABC in which
$B\hat{A}C = 20°$, $BC = 52 \, cm$ and $AC = 2BC$.

3 The diagram shows two chords XY and YZ drawn on a circle with center O and radius 5 cm. Given that XY = 3 cm and YZ = 7 cm, find the area of quadrilateral OXYZ.

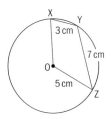

4 This diagram shows a mast AB of length 12 m. C and D are points on the ground such that the angle of elevation from C to B is 60° and the angle of elevation from D to B is 55°. Given that the distance between C and D is 15 m, calculate the angle CAD and hence find the area of the triangle CAD.

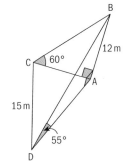

5 In the diagram, O is the center of a circle with radius r.

$$\widehat{POR} = \widehat{QOS} = \frac{\pi}{4} \text{ and } \widehat{ROS} = \frac{\pi}{6}$$

a Find the area of triangles POQ and ROS.

b Find the area of the minor segment formed by the chord PQ.

c Find the area of the minor segment formed by the chord RS.

d Show that the shaded area is equal to $\frac{r^2}{4}\left(\pi + 1 - \sqrt{3}\right)$.

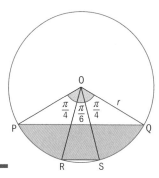

✗ **Review exercise**

1 Use the triangle on the right to show that if $\tan\left(\dfrac{\theta}{2}\right) = t$, then

$$\sin\theta = \frac{2t}{1+t^2} \text{ and } \cos\theta = \frac{1-t^2}{1+t^2}.$$

Hence, solve the equation:

$$\sqrt{3}\sin\theta + \cos\theta = 1 \text{ for } 0 \le \theta \le 2\pi$$

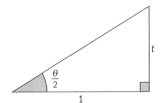

2 Find the exact value of these expressions:

a $\sin 165°$ **b** $\tan 105°$ **c** $\cos\left(\dfrac{5\pi}{12}\right)$ **d** $\tan\left(\dfrac{\pi}{8}\right)$

3 Prove these identities:

a $\dfrac{\cos\theta}{\cos\theta - \sin\theta} \equiv \dfrac{1}{1-\tan\theta}$ **b** $\dfrac{\cos(A-B)}{\cos A \cos B} \equiv 1 + \tan A \tan B$

c $\cos 3A - \sin 3A \equiv (\cos A + \sin A)(1 - 4\cos A \sin A)$

d $2\sin 2\theta\,(1 - 2\sin^2\theta) \equiv \sin 4\theta$

e $1 + 2\cos 2A + \cos 4A \equiv 4\cos^2 A \cos 2A$

EXAM-STYLE QUESTIONS

4 Find the value of each of these expressions:

a $\cos\left(\arcsin\dfrac{3}{5} - \arccos\dfrac{1}{2}\right)$

b $\sin\left[2\arccos\left(-\dfrac{3}{5}\right)\right]$

c $\sin\left[\arctan(-1) + \arccos\left(-\dfrac{4}{5}\right)\right]$

5 The graph below shows the function $f(x) = A\cos Bx + C$, for
$-\dfrac{\pi}{B} \le x \le \dfrac{\pi}{B}$

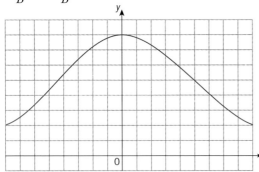

On the same set of axes, sketch the graph of
$g(x) = -\dfrac{A}{2}\cos(2Bx) + C$.

6 Given that arcsin x, arccos x and arcsin$(1 - x)$ are all acute angles, prove that $\sin[\arcsin x - \arccos x] \equiv 2x^2 - 1$.

Hence, show that if $\arcsin x - \arccos x = \arcsin(1 - x)$ then $x = \dfrac{1}{4}\left(\sqrt{17}-1\right)$.

7 Use the identity $\tan(A+B) \equiv \dfrac{\tan A + \tan B}{1 - \tan A \tan B}$ to show that if

$2x + y = \dfrac{\pi}{4}$, then $\tan y = \dfrac{1 - 2\tan x - \tan^2 x}{1 + 2\tan x - \tan^2 x}$

Review exercise

1 Show that $\cos(A - B) - \cos(A + B) \equiv 2 \sin A \sin B$.
Use this result to solve the equation $\sin 3x \sin x = -1$ for $0 \le x \le \pi$.

2 A system of equations is given by:
$$\sin y + \sin x = 1.1$$
$$\cos y + \sin 2x = 1.8$$
a Express y in terms of x for each of the equations.
b Hence, solve the system of equations for $0 < x < \dfrac{\pi}{2}$, $0 < y < \dfrac{\pi}{2}$.

EXAM-STYLE QUESTION

3 The diagram below shows a circle, center O, inscribed in a kite
ABCD. The sides of the kite are tangents to the circle.

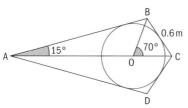

BC = 0.6 m
$B\widehat{A}O = 15°$
$B\widehat{O}A = 70°$
a Calculate the size of angle ABC.
b Find the length of AB.
c Hence, calculate the radius of the circle.

4 a Explain why the function
$$f(x) = \frac{2 + 3\sin x}{4 + 3\cos x}$$

$0 \le x \le 2\pi$
does not have any vertical asymptotes.
b Write down the y-intercept of $f(x)$.
c Write down the x-intercepts, p and q (where $p < q$).
d Sketch the graph of $f(x)$, labeling any stationary points p, q
and the y-intercept.
e Given that $g(x) = \cos 2x$, for what values of x is $f(x) > g(x)$?
f Hence or otherwise, calculate the maximum value of $f(x) - g(x)$.

5 The diagram below shows a right-angled rectangular prism ABCDEF.

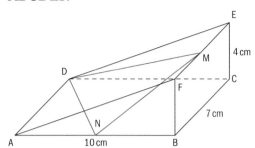

The base of the prism ABCD is a rectangle with AB = 10 cm and BC = 7 cm. The face BCEF is rectangular and is perpendicular to the base. The height of the prism CE = 4 cm. M is the midpoint of EF and N is the midpoint of AB.

Calculate

a the length of DN   **b** the length of DM

c the length of MN   **d** the angle DMN

e the area of triangle DMN.

EXAM-STYLE QUESTIONS

6 a Prove that in a triangle ABC, the length of the perpendicular h from C to AB is given by $\dfrac{ab}{c}\sin C$.

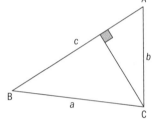

b This diagram shows the triangle ABC which lies on horizontal ground. $A\hat{C}B = 150°$. A mast CD stands vertically and is 10 m high.

Given that $D\hat{B}C = 30°$ and $D\hat{A}C = 45°$, find the lengths of the sides of triangle ABC and the length of the perpendicular from C to AB.

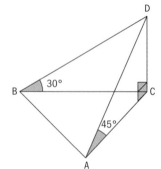

7 The drive wheel of an engine has a radius of 7 cm, and the pulley on the rotary pump has a radius of 3 cm. The shafts of the drive wheel and the pulley are 24 cm apart as shown in the diagram.

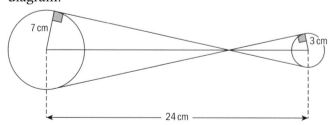

Calculate the length of belt required to join the wheel and pulley as shown in the diagram.

Extension material on CD:
Worksheet 8

CHAPTER 8 SUMMARY

Trigonometric ratios

$\sin \theta = \dfrac{\text{opposite}}{\text{hypotenuse}}$ $\qquad$ $\csc \theta = \dfrac{1}{\sin \theta}$

$\cos \theta = \dfrac{\text{adjacent}}{\text{hypotenuse}}$ $\qquad$ $\sec \theta = \dfrac{1}{\cos \theta}$

$\tan \theta = \dfrac{\text{opposite}}{\text{adjacent}}$ $\qquad$ $\cot \theta = \dfrac{1}{\tan \theta}$

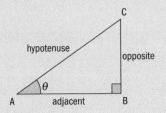

Values of sin, cos and tan for common acute angles

$\sin\left(\dfrac{\pi}{6}\right) = \dfrac{1}{2}$ $\qquad$ $\cos\left(\dfrac{\pi}{6}\right) = \dfrac{\sqrt{3}}{2}$ $\qquad$ $\tan\left(\dfrac{\pi}{6}\right) = \dfrac{1}{\sqrt{3}}$

$\sin\left(\dfrac{\pi}{4}\right) = \dfrac{\sqrt{2}}{2}$ $\qquad$ $\cos\left(\dfrac{\pi}{4}\right) = \dfrac{\sqrt{2}}{2}$ $\qquad$ $\tan\left(\dfrac{\pi}{4}\right) = 1$

$\sin\left(\dfrac{\pi}{3}\right) = \dfrac{\sqrt{3}}{2}$ $\qquad$ $\cos\left(\dfrac{\pi}{3}\right) = \dfrac{1}{2}$ $\qquad$ $\tan\left(\dfrac{\pi}{3}\right) = \sqrt{3}$

For the unit circle shown below:

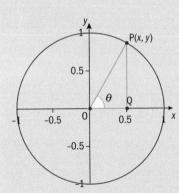

$\sin \theta = \dfrac{y}{\text{OP}} = \dfrac{y}{1}$

$\cos \theta = \dfrac{x}{\text{OP}} = \dfrac{x}{1}$

$\tan \theta = \dfrac{y}{x}, \; x \neq 0$

Odd/even function identities

$\sin(-\theta) = \sin \theta$ $\qquad$ $\cos(-\theta) = -\cos \theta$ $\qquad$ $\tan(-\theta) = -\tan \theta$

Co-function identities

$\sin \theta = \cos\left(\dfrac{\pi}{2} - \theta\right)$ $\qquad$ $\cos \theta = \sin\left(\dfrac{\pi}{2} - \theta\right)$

Tangent and cotangent identity

$\tan \theta = \dfrac{\sin \theta}{\cos \theta}$ $\qquad$ $\cot \theta = \dfrac{\cos \theta}{\sin \theta}$

Continued on next page

Pythagorean identities

$$\sin^2\theta + \cos^2\theta = 1 \quad 1 + \tan^2\theta = \sec^2\theta \quad 1 + \cot^2\theta = \csc^2\theta$$

Compound angle identities

$$\sin(A \pm B) = \sin A \cos B \pm \cos A \sin B$$

$$\cos(A - B) = \cos A \cos B \mp \sin A \sin B$$

$$\tan(A \pm B) = \frac{\tan A \pm \tan B}{1 \mp \tan A \tan B}$$

Double angle identities

$$\sin 2\theta = 2\sin\theta\cos\theta \qquad \tan 2\theta = \frac{2\tan\theta}{1 - \tan^2\theta}$$

$$\cos 2\theta = \cos^2\theta - \sin^2\theta$$

$$= 1 - 2\sin^2\theta = 2\cos^2\theta - 1$$

Graphs of trigonometric functions

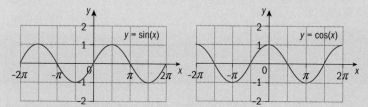

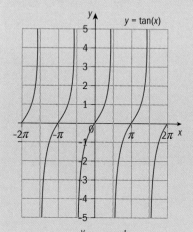

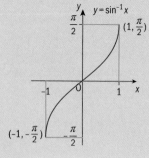

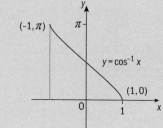

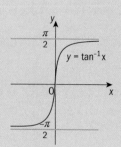

Continued on next page

The cosine rule

For any triangle ABC with corresponding sides a, b and c:

$$c^2 = a^2 + b^2 - 2ab\cos C \qquad \cos C = \frac{a^2 + b^2 - c^2}{2ab}$$

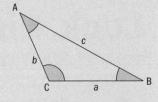

The sine rule

For any triangle ABC with corresponding sides a, b and c:

$$\frac{a}{\sin A} = \frac{b}{\sin B} = \frac{c}{\sin C}$$

Area of a triangle

$A = \dfrac{1}{2} ab \sin C$ where a and b are adjacent sides and C is the included angle.

Luck + intuition = ingenuity

To what extent does luck play a role in the development of mathematics?

The right place, at the right time

Eratosthenes of Cyrene (3rd century BCE) – a Libyan mathematician – lived in Alexandria, which is on the same line of longitude as the city of Syene.

On 21 June he measured the angle of elevation of the sun at noon in Alexandria. He knew that the angle of elevation of the sun at Syene on that day was 90° – there was a very deep well in the city, which only reflected the sun on 21 June. The difference between those two angles was 7.2°.

Eratosthenes also knew the distance between the two cities. Using this distance, the 7.2° difference and assuming that the Earth is round and the rays from the sun are essentially parallel, he calculated the circumference of the Earth.

- Have any other mathematicians been enlightened because of luck?
- Has luck played a part in any other mathematical discoveries?

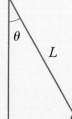

"All perceiving is also thinking, all reasoning is also intuition, all observation is also invention."

Rudolf Arnheim, German art theorist and psychologist (1904–2007)

Making assumptions

Eratosthenes made assumptions about the shape of the Earth and about the sun's rays when he calculated the circumference of the Earth.

When modeling physical phenomena, mathematicians often have to make assumptions.

The mathematical model for the motion of a simple pendulum is derived using Newton's laws of motion. It reduces to this second order differential equation:

$$\frac{d^2 \theta}{dt^2} = \frac{-g}{L} \sin \theta$$

where L represents the length of the string, g the acceleration due to gravity and θ the angle made by the string to the vertical line of suspension.

This differential equation is difficult to solve, because of the $\sin \theta$ term. But if we replace $\sin \theta$ by θ a solution can easily be found. The solution describes the motion of the pendulum by expressing θ as a function of time. However this is only true for small values of θ.

- Why can $\sin \theta$ be replaced by θ in the differential equation?
- How small should the angle be to justify this assumption?
- 'Pure mathematics is precise. Approximations are contrary to the nature of the subject.' Explain this statement.
- The applied mathematician claims that approximations are justified when results are matched to observations. Why does the argument 'the end justifies the means' defy the claim of absolute truth in mathematics?

Intuition = perception?

- How true is Gödel's realism about mathematics?
- How does he defend it?

> A pathological function is a function developed specifically to violate an almost universally valid property. Pathological problems can provide interesting examples of counter-intuitive behavior.

Counter-intuitive results – pathological functions

A function has to be continuous in order to be differentiable but this does not mean that any continuous function is differentiable. French mathematician Joseph Fourier (1768–1830) showed that any reasonable smooth function $f(\theta)$ in the interval $-\pi < \theta < \pi$, can be expanded as a Fourier series.

Look at the 'saw tooth' function shown in green.

The series $f(\theta) = \sum_{n=1}^{5}(-1)^{n+1}\frac{2}{n}\sin(n\theta)$ gives the black curve, which is an approximation of the saw tooth curve.

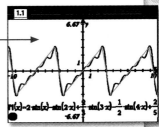

As the number of terms increases the function becomes a better approximation of the saw tooth function. When $n = 10$ the graph looks like this:

A Weierstrass function is a special type of Fourier series.
$$\omega(x) = \sum_{n=0}^{\infty}\frac{1}{2^n}\sin(2^nx) = \sin x + \frac{1}{2}\sin 2x + \frac{1}{4}\sin 4x + ...$$

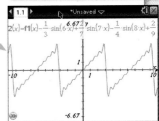

- Use your GDC to graph the Weierstrass function $\omega(x) = \sum_{n=0}^{3}\frac{1}{2^n}\sin(2^nx)$
- Then graph the function $\omega_1(x) = \sum_{n=0}^{5}\frac{1}{2^n}\sin(2^nx)$
- What happens to the function as the number of terms in the series increases?
- Is the function periodic?
- What happens to the function as you zoom into an interval which you keep making smaller?
- Why is this function continuous everywhere but nowhere differentiable?
- How does this function compare to a fractal?

◄ Visualization of the derivative of a Weierstrass function. Does this sense perception of the function help intuitive understanding of the function?

9 The power of calculus

CHAPTER OBJECTIVES:

6.1 Definition of a derivative from first principles

6.2 Derivative of $\sin x$, $\cos x$ and $\tan x$, $\sec x$, $\csc x$, $\cot x$, $\arcsin x$, $\arccos x$ and $\arctan x$.

6.4 Indefinite integral of $\sin x$ and $\cos x$; other indefinite integrals using the results from 6.2; the composites of any of these with a linear function

6.5 Areas of regions enclosed by curves; volumes of revolution about the x-axis or y-axis

6.6 Kinematic problems involving displacement s, velocity v and acceleration a

6.7 Integration by substitution; integration by parts

Before you start

You should know how to:

1 Transform trigonometric expressions.

e.g. prove $\sin 2\theta \equiv \dfrac{2\tan\theta}{1+\tan^2\theta}$

$$\text{RHS} = \dfrac{\dfrac{2\sin\theta}{\cos\theta}}{\dfrac{\cos^2\theta+\sin^2\theta}{\cos^2\theta}} = \dfrac{2\sin\theta\cos\theta}{(\cos^2\theta+\sin^2\theta)}$$

$$= 2\sin\theta\cos\theta = \sin 2\theta$$

2 Apply the product and quotient rules on x^n, e^x and $\ln x$, how to do implicit differentiation and the chain rule.

e.g. Differentiate $f(x) = e^{x^2}\ln(2x-1)$

$$f'(x) = e^{x^2}\cdot 2x \cdot \ln(2x-1) + e^{x^2}\dfrac{1}{2x-1}\cdot 2$$

$$= 2xe^{x^2}\ln(2x-1) + \dfrac{2e^{x^2}}{2x-1}$$

Skills check

1 Prove these identities.

a $\cos 2\theta \equiv \dfrac{1-\tan^2\theta}{1+\tan^2\theta}$

b $\tan 2\theta \equiv \dfrac{2\tan\theta}{1-\tan^2\theta}$

2 Find the derivative of:

a $f(x) = 3e^{2x} - 2x^2$

b $g(x) = (x+1)\cdot\ln(x^2+2x+1)$

c $h(x) = \dfrac{e^{x^2}}{x+1}$

Further calculus and applications

The geometrical name for a doughnut shape is a ring torus. It is a solid of revolution, created by rotating a circle about a vertical axis at the centre of the 'hole' in the torus.

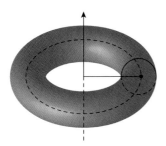

The torus was studied by a Greek geometer **Pappus** of Alexandria (290–350 CE).
There are three types of torus called ring, horn and spindle. Investigate the properties of these tori (the plural of torus).

You learned about solids of revolution in Chapter 7.

In this chapter you will learn more differentiation and integration techniques, and use these to model and analyze real-world problems. Integration can be used to find the volume of a torus – or the volume of dough needed to make a doughnut.

This chapter will also show you how to solve optimization problems – such as how to calculate the amount of dough needed to create the optimum number of doughnuts for a day's sales, with minimum wastage.

9.1 Derivatives of trigonometric functions

To find derivatives of trigonometric functions we are going to use the definition of the derivative, i.e., differentiate from first principles. First find the values of some trigonometric limits that will appear in the process.

Trigonometric limits

One of the most useful limits that involves trigonometric functions is $\lim_{h \to 0} \dfrac{\sin h}{h}$.

To determine its value, consider the unit circle.

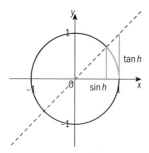

Notice that the arc whose length is denoted by h can be 'squeezed in' between two vertical line segments that represent $\sin h$ and $\tan h$ values. From the diagram:

$$\sin h \le h \le \tan h$$

$$1 \le \frac{h}{\sin h} \le \frac{1}{\cos h}$$

Divide by $\sin h$, $\sin h \ne 0$.

$$\cos h \le \frac{\sin h}{h} \le 1$$

Use reciprocal values.

$$\lim_{h \to 0^+} \cos h \le \lim_{h \to 0^+} \frac{\sin h}{h} \le \lim_{h \to 0^+} 1$$

Take a limit $h \to 0^+$

$$1 \le \lim_{h \to 0^+} \frac{\sin h}{h} \le 1$$

Use properties of limits and $\cos 0 = 1$.

$$\lim_{h \to 0^+} \frac{\sin h}{h} = 1$$

The graph of $\dfrac{\sin x}{x}$ is even, so the same from both sides.

Therefore since the limit from the left is equal to the limit from the right you can conclude that

$$\rightarrow \lim_{h \to 0} \frac{\sin h}{h} = 1$$

Another confirmation of the result can be obtained numerically by using a calculator.

GDC help on CD:
Alternative demonstrations for the TI-84 Plus and Casio fx-9860GII GDCs are on the CD.

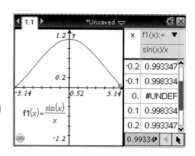

The graph also confirms that $\lim\limits_{h\to 0}\dfrac{\sin h}{h}=1$.

Another limit that will be useful is $\lim\limits_{h\to 0}\dfrac{\cos h-1}{h}$. To find its value use these trigonometric identities.

$$\sin^2\left(\frac{\theta}{2}\right)=\frac{1-\cos\theta}{2}\Rightarrow \cos\theta-1\equiv -2\sin^2\left(\frac{\theta}{2}\right)$$

Using this identity:

$$\lim_{h\to 0}\frac{\cos h-1}{h}=\lim_{h\to 0}\frac{-2\sin^2\left(\dfrac{h}{2}\right)}{h}$$

$$=-\lim_{h\to 0}\left(\frac{\sin\left(\dfrac{h}{2}\right)}{\dfrac{h}{2}}\times \sin\left(\frac{h}{2}\right)\right)$$

> Rewrite the square and fractions into a desirable form.

> Use the results of the known limits.

$$=-\lim_{\frac{h}{2}\to 0}\frac{\sin\left(\dfrac{h}{2}\right)}{\dfrac{h}{2}}\times \lim_{h\to 0}\left(\sin\left(\frac{h}{2}\right)\right)=-1\times 0=0$$

> Notice that when $h\to 0$ then $\dfrac{h}{2}\to 0$ too.

Again you can verify the value of the limit by looking at the graph of the function $g(x)=\dfrac{\cos x-1}{x}$, $x\neq 0$.

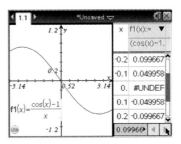

Derivatives of trigonometric functions

Now you can differentiate trigonometric functions from first principles.

Example 1

Find the derivative of $f(x)=\sin x$ from first principles.		Derivatives of trigonometric functions from first principles are not examinable. The derivative of sine is included here so that you can understand the result.
Answer **Solution 1**		
$f(x)=\lim\limits_{h\to 0}\dfrac{f(x+h)-f(x)}{h}$	*Use the definition of the derivative.*	
$=\lim\limits_{h\to 0}\dfrac{\sin(x+h)-\sin(x)}{h}$	*Use addition formula for sine.*	
$=\lim\limits_{h\to 0}\dfrac{\sin x\cos h+\cos x\sin h-\sin x}{h}$		
$=\lim\limits_{h\to 0}\left(\cos x\dfrac{\sin h}{h}+\sin x\dfrac{\cos h-1}{h}\right)$	*Rewrite the expression.*	
$=\cos x\cdot\lim\limits_{h\to 0}\dfrac{\sin h}{h}+\sin x\cdot\lim\limits_{h\to 0}\dfrac{\cos h-1}{h}$	*Use the properties of limits*	
$=\cos x\cdot 1+\sin x\cdot 0=\cos x$		

You can use the GDC to confirm the gradient function of $y = \sin x$. First, graph $f\,1(x) = \sin x$, then graph $f\,2(x) = nDeriv\,(\sin x, x = x)$. This is the graph of the derivative which calculates the value of a derivative at all the points in the window range.

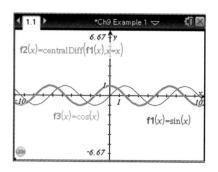

Graph $f\,3(x) = \cos(x)$ and change the graphing mode to dashed blue. This will trace over $f\,2(x)$, so the GDC confirms your analytical result.

This graphical method can be used to confirm all the results of differentiation. You need to input the function that you need to differentiate in $f\,1$ and your answer in $f\,3$.

In a similar way you can find that $\dfrac{d(\cos x)}{dx} = -\sin x$.

You can confirm this in Exercise 9A.

GDC help on CD:
Alternative demonstrations for the TI-84 Plus and Casio fx-9860GII GDCs are on the CD.

nDeriv and *centralDiff* are equivalent commands.

All of these results can be obtained by finding derivatives from first principles but some of the calculations are challenging.

Exercise 9A

1   Use a graphical method to confirm these results.

a $\dfrac{d(\cos x)}{dx} = -\sin x$

b $\dfrac{d\left(\sin\frac{x}{2}\right)}{dx} = \dfrac{1}{2}\cos\dfrac{x}{2}$

c $\dfrac{d(\cos 3x)}{dx} = -3\sin 3x$

d $\dfrac{d(\sin(2x-1))}{dx} = 2\cos(2x-1)$

e $\dfrac{d(\tan x)}{dx} = \sec^2 x$

f $\dfrac{d(\cot x)}{dx} = -\csc^2 x$

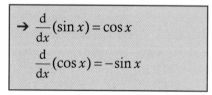

$\rightarrow \quad \dfrac{d}{dx}(\sin x) = \cos x$

$\dfrac{d}{dx}(\cos x) = -\sin x$

Since you know the derivatives of sine and cosine functions you can find the derivative of the tangent function by using the quotient rule.

Example 2

Find the derivative of $f(x) = \tan x$ by using the quotient rule.

Answer

$f(x) = \tan x = \dfrac{\sin x}{\cos x} \Rightarrow$	*Rewrite tangent as quotient of sine and cosine.*
$f'(x) = \dfrac{\dfrac{d(\sin x)}{dx} \cdot \cos x - \sin x \cdot \dfrac{d(\cos x)}{dx}}{(\cos x)^2}$	*Apply the quotient rule.*
$= \dfrac{\cos x \cdot \cos x - \sin x \cdot (-\sin x)}{\cos^2 x}$	*Use the derivatives of sin and cos and simplify the expression.*
$= \dfrac{\cos^2 x + \sin^2 x}{\cos^2 x}$	*Use the fundamental trigonometric identity $\cos^2 x + \sin^2 x = 1$.*
$= \dfrac{1}{\cos^2 x} = \sec^2 x$	

Example 3

Find the derivative of the function $f(x) = \sin x \cdot \cos x$.

Answer
Solution 1

$f'(x) = \dfrac{d(\sin x)}{dx} \cdot \cos x + \sin x \cdot \dfrac{d(\cos x)}{dx}$	*Use the product rule.*
$= \cos x \cdot \cos x + \sin x \cdot (-\sin x)$	*Use the derivatives of sin and cos.*
$= \cos^2 x - \sin^2 x$	*Simplify the expression.*

Solution 2

$f(x) = \sin x \cdot \cos x$	*Use $\sin 2\theta = 2 \cdot \sin\theta \cdot \cos\theta$ to rewrite the product.*
$= \dfrac{1}{2}\sin 2x \Rightarrow$	
$f'(x) = \dfrac{1}{2} \cdot \dfrac{d(\sin 2x)}{dx}$	*Use the chain rule.*
$= \dfrac{1}{2} \cdot \cos 2x \cdot 2 = \cos 2x$	*Simplify the expression.*

> These two results from the solutions are equivalent since the trigonometric formula for a cosine of double angle is $\cos 2\theta = \cos^2\theta - \sin^2\theta$.

This example looks at the derivative of a reciprocal trigonometric function.

Example 4

Find the derivative of $f(x) = \sec x$.

Answer

$f(x) = \sec x = (\cos x)^{-1} \Rightarrow$	*Secant is reciprocal cosine.*
$f'(x) = -1 \cdot (\cos x)^{-2} \cdot (-\sin x)$	*Use the chain rule.*
$= \dfrac{\sin x}{\cos^2 x} = \sec x \cdot \tan x$	*Simplify and rewrite.*

> Write the reciprocal functions as composite functions, and then apply the chain rule.

Exercise 9B

1 Differentiate with respect to x.

  a $y = \cot x$ **b** $y = \csc x$ **c** $y = \sin 3x$

  d $y = \tan(5x - 3)$ **e** $y = \cos(8 - 3x)$ **f** $y = \csc\left(\dfrac{x-3}{4}\right)$

  g $y = \cot\left(\dfrac{7-2x}{13}\right)$

2 Use the chain rule to find $\dfrac{dy}{dx}$.

  a $y = \sin(x^5 - 3)$ **b** $y = \cos(e^x)$

  c $y = \csc(x^2 + 11)$ **d** $y = \cot(4x^3 - 2x^2 + 7x + 17)$

  e $y = \tan(\ln(2x + 1))$ **f** $y = \sec\left(\sqrt{e^x + 1}\right)$

  g $y = \sin(\cos(\tan x))$

Now you can find derivatives of composite functions, products and quotients of trigonometric and other functions.

Example 5

Find the derivatives with respect to x of:

a $y = x^2 \sin 2x$ **b** $y = \dfrac{e^{3x-1}}{\cos x}$ **c** $y = \ln(x^2 + 1)\tan\dfrac{x}{2}$

Answers

a $y = x^2 \sin 2x$ *Use product rule.*

$\Rightarrow y' = 2x \sin 2x + x^2 \cos 2x \cdot 2$

$= 2x \sin x^2 + 2x^2 \cos 2x$ *Simplify the expression.*

$[= 2x(\sin 2x + x \cos 2x)]$

b $y = \dfrac{e^{3x-1}}{\cos x}$ *Use the quotient rule.*

$\Rightarrow y' = \dfrac{e^{3x-1} \cdot 3 \cdot \cos x - e^{3x-1}(-\sin x)}{\cos^2 x}$

$= \dfrac{3e^{3x-1}\cos x + e^{3x-1}\sin x}{\cos^2 x}$

c $y = \ln(x^2 + 1)\tan\dfrac{x}{2} \Rightarrow$

$y' = \dfrac{2x}{x^2 + 1}\tan\dfrac{x}{2} + \ln(x^2 + 1)\sec^2\dfrac{x}{2} \cdot \dfrac{1}{2}$ *Use product rule.*

$= \dfrac{2x \tan\dfrac{x}{2}}{x^2 + 1} + \dfrac{1}{2}\ln(x^2 + 1)\sec^2\dfrac{x}{2}$ *Simplify the expression.*

> Notice that sometimes you can leave answers in a factorized form, especially if you need to do further calculations on the derivatives.

> Some derivative expressions are very long and it may not be possible to simplify them.

440   The power of calculus

Gradients of curves

You can use the derivative to determine the gradient of a function at a given point.

See Chapter 4.

Example 6

Find the gradient of the curve $y = 3x\cos(2x)$ at the point $\left(\dfrac{5\pi}{6}, \dfrac{5\pi}{4}\right)$.

Answer

$y = 3x\cos(2x)$

$\Rightarrow y' = 3\cos(2x) + 3x(-\sin(2x) \cdot 2)$

$\quad = 3(\cos(2x) - 2x\sin(2x))$

$y'\left(\dfrac{5\pi}{6}\right) = 3\left(\cos\left(2 \times \dfrac{5\pi}{6}\right) - 2 \cdot \dfrac{5\pi}{6} \cdot \sin\left(2 \cdot \dfrac{5\pi}{6}\right)\right)$

$\quad = 3\left(\underbrace{\cos\left(\dfrac{5\pi}{3}\right)}_{\frac{1}{2}} - \dfrac{5\pi}{3} \cdot \underbrace{\sin\left(\dfrac{5\pi}{3}\right)}_{-\frac{\sqrt{3}}{2}}\right) = \dfrac{3}{2} + \dfrac{5\pi\sqrt{3}}{2}$

This result can be obtained from the GDC. Notice that the GDC gives a decimal form, so you need to verify our answer.

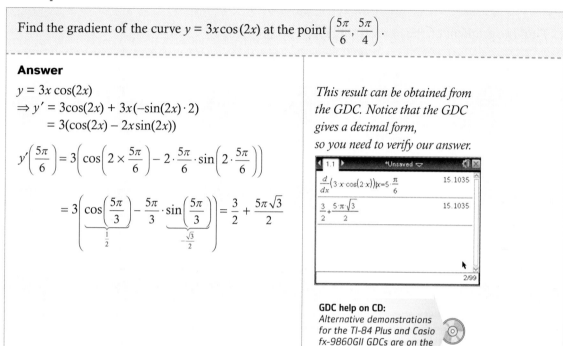

GDC help on CD:
Alternative demonstrations for the TI-84 Plus and Casio fx-9860GII GDCs are on the CD.

Sometimes it is easier to first rewrite and simplify the trigonometric expression and then to differentiate it.

Example 7

Find the derivative of $f(x) = (1 + \tan^2 x) \cdot (1 - \sin^2 x)$

Answer

Solution 1 - differentiate first

$f'(x) = \dfrac{d(1 + \tan^2 x)}{dx} \cdot (1 - \sin^2 x) + (1 + \tan^2 x) \cdot \dfrac{d(1 - \sin^2 x)}{dx}$

$\quad = (2\tan x \cdot \sec^2 x) \cdot (1 - \sin^2 x) + (1 + \tan^2 x) \cdot (-2\sin x \cdot \cos x)$

$\quad = 2\dfrac{\sin x}{\cos^3 x} \cdot \cos^2 x + \dfrac{1}{\cos^2 x} \cdot (-2\sin x \cdot \cos x)$

$\quad = 2\dfrac{\sin x}{\cos x} - 2\dfrac{\sin x}{\cos x} = 0$

Use the product rule.

Use trigonometric identities to simplify.

Solution 2 - simplify first

$f(x) = (1 + \tan^2 x) \cdot (1 - \sin^2 x)$

$\quad = \sec^2 x \cdot \cos^2 x = 1 \Rightarrow$

$f'(x) = \dfrac{d(1)}{dx} = 0$

Use trigonometric identities.

Differentiate the constant.

Exercise 9C

1 Use product and quotient rules to differentiate with respect to x.

a $y = (2x - 1)\cos x$ **b** $y = (3x - x^2)\sin 2x$ **c** $y = e^{1-x}\tan x$

d $y = \dfrac{\sin x}{x}$ **e** $y = \dfrac{2x + 3}{\sin 2x}$ **f** $y = \dfrac{\tan x}{\sqrt{2 - x}}$

2 Find the gradient of the curve at the given point.

a $y = \sin 2x$, at $x = \dfrac{\pi}{6}$ **b** $y = \cos 3x$, at $x = \dfrac{7\pi}{12}$

c $y = \tan(-x)$, at $x = \dfrac{5\pi}{4}$ **d** $y = (x - 2)\sin x$, at $x = 0$

e $y = -3x \cos x$, at $x = \dfrac{\pi}{2}$ **f** $y = x^2 \tan x$, at $x = \dfrac{3\pi}{4}$

g $y = e^x \sec x$, at $x = 0$

3 Find the derivatives of these expressions with respect to the variable indicated.

a $y = \sin^2 \alpha + \cos^2 \alpha$, α **b** $y = \dfrac{\tan \beta}{\sin \beta}$, β

c $y = \dfrac{2 \tan 2\theta}{1 - \tan^2 2\theta}$, θ **d** $y = \dfrac{\sin \rho + \sin 2\rho}{\cos \rho + \cos 2\rho}$, ρ

e $y = \dfrac{(\sin \varphi \sin 2\varphi - \cos \varphi)\sec \varphi}{\sin \varphi - \cos \varphi}$, φ

Derivatives of inverse trigonometric functions

To differentiate the inverse trigonometric functions, $y = \arcsin(x)$, $y = \arccos(x)$ and $y = \arctan(x)$ introduced in Chapter 8, you can proceed as follows:

Let $y = \arcsin x$ then $x = \sin y$ so $\dfrac{dx}{dy} = \cos y$

Using $\dfrac{dx}{dy} = \dfrac{1}{\dfrac{dx}{dy}}$ and $\sin^2 x + \cos^2 x = 1$ gives

$$\dfrac{dx}{dy} = \dfrac{1}{\sqrt{1 - \sin^2 y}} = \dfrac{1}{\sqrt{1 - x^2}}$$

Using this result and the chain rule you can find a general formula.

If $y = \arcsin\dfrac{x}{a}$ $\dfrac{x}{a}\dfrac{dx}{dy} = \dfrac{1}{\sqrt{1 - \left(\dfrac{x}{a}\right)^2}} \cdot \dfrac{1}{a} = \dfrac{1}{\sqrt{a^2 - x^2}}$

> → If $y = \arcsin x$ then $\dfrac{dy}{dx} = \dfrac{1}{\sqrt{1 - x^2}}$
>
> If $y = \arcsin \dfrac{x}{a}$ then $\dfrac{dy}{dx} = \dfrac{1}{\sqrt{a^2 - x^2}}$

> **arcsin x:**
> $$[-1, 1] \to \left[-\dfrac{\pi}{2}, \dfrac{\pi}{2}\right]$$
> In this interval, the cosine value will always be positive so don't consider the negative square root.

GDC help on CD:
Alternative demonstrations for the TI-84 Plus and Casio fx-9860GII GDCs are on the CD.

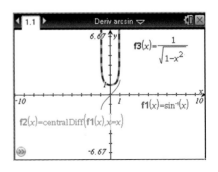

Example 8

Find the derivative of the function $g(x) = \arctan x$, $x \in \mathbb{R}$.

Answer

$\tan(g(x)) = x$	*Composition of a function and its inverse function gives the identity function.*
$\Rightarrow \sec^2(g(x)) \cdot g'(x) = 1$	*Differentiate with respect to x by using the chain rule.*
$\Rightarrow g'(x) = \cos^2(\arctan(x))$	*Rearrange to make g'(x) the subject. Use the trigonometric identity*
$= \dfrac{1}{1 + \tan^2(\arctan(x))}$	$\cos^2\theta = \dfrac{1}{1 + \tan^2\theta}$ *and simplify.*
$= \dfrac{1}{1 + x^2}$, $x \in \mathbb{R}$	

Exercise 9D

1 Find the derivatives of

   a $f(x) = \arccos x$ **b** $f(x) = \arcsin 3x$ **c** $f(x) = \arctan(2x + 1)$

2 Find $\dfrac{dy}{dx}$.

   a $y = 2x \arcsin x$ **b** $y = \dfrac{\arccos x}{x}$ **c** $y = (2x + 1)\arctan x$

   d $y = \sqrt{1 - x^2}\,\arcsin x$ **e** $y = (4x^2 + 1)\arctan 2x$

3 Show that these identities are valid and explain why:

   a $\dfrac{d(\arcsin x + \arccos x)}{dx} = 0$

   b $\dfrac{d(\arctan x + \arctan(-x))}{dx} = 0$

   c $\dfrac{d\left(2\arctan x - \arcsin\left(\dfrac{2x}{x^2 + 1}\right)\right)}{dx} = 0$

4 Differentiate with respect to x these implicitly defined functions.

   a $x = \sin y$ **b** $x + y = \tan y$

   c $x + \sin x = y + \cos y$ **d** $e^{\sin y} = x^2$

   e $\cos y = \dfrac{x}{y}$ **f** $\ln(xy) = \tan 2y$

> See Section 4.8

Tangents and normals

As discussed in Section 4.2, equations of a tangent and a normal to the curve $y = f(x)$ at the point (x_1, y_1) are given by

$y = f'(x_1)(x - x_1) + y_1$ and $y = -\dfrac{1}{f'(x_1)}(x - x_1) + y_1$ respectively.

Example 9

> Given the function $f(x) = 2\sin(3x) + 1$, $-\pi < x < \pi$, find the equation of:
> **a** the tangent **b** the normal at the point where the graph of the function meets the y-axis.

These results use the point-slope form of a straight line.

Answers

$x = 0 \Rightarrow f(0) = 2\sin(3 \cdot 0) + 1$
  $= 1 \Rightarrow P(0, 1)$

Calculate the y-coordinate of the point of intersection.

$f'(x) = 2\cos(3x) \cdot 3 = 6\cos(3x)$
$x = 0 \Rightarrow f'(0) = 6\cos(3 \cdot 0) = 6$

Calculate the gradient of the curve at the point.

a Tangent:
  $y = f'(0)(x - 0) + 1$
  $y = 6x + 1$

Apply the formula for the equation of a tangent.

b Normal:
  $y = -\dfrac{1}{f'(0)}(x - 0) + 1$

  $y = -\dfrac{1}{6}x + 1$

Apply the formula for the equation of a normal.
You can confirm our results on the GDC.

GDC help on CD:
Alternative demonstrations for the TI-84 Plus and Casio fx-9860GII GDCs are on the CD.

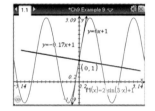

In this example you use **implicit** differentiation.

Example 10

> Find the equation of the normal to the curve $y + 2x = \cos(xy)$ at the point $P(0, 1)$ in the form $y = mx + c$.

This is a good example of an exam-style question.

Answer

$y + 2x = \cos(xy)$

$\dfrac{dy}{dx} + 2 = -\sin(xy)\left(y + x\dfrac{dy}{dx}\right)$

Differentiate the implicitly defined function with respect to x.

$m_1 + 2 = -\sin(0 \cdot 1)(1 + 0 \cdot m_1)$
$m_1 + 2 = 0 \Rightarrow m_1 = -2$

Find the slope of the curve at the given point.

$m_2 = \dfrac{1}{2} \Rightarrow N: y = \dfrac{1}{2}(x - 0) + 1$

Apply the formula for a normal.

$\Rightarrow N: y = \dfrac{1}{2}x + 1$

In this next example you need to use a GDC because the equation cannot be solved using the algebraic methods you have learned so far.

Example 11

Graphs of the functions $f(x) = \arctan(2x)$ and $g(x) = e^x - 3$ are given in the diagram. The point P is the point of intersection between the curves, the line T is the tangent to f at P, and N is the normal to g at P.

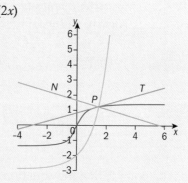

a Find the coordinates of P.

b Find the area of the triangle enclosed by the tangent T, normal N and the x-axis.

Answers

a P(1.44, 1.24), given correct to 3 sf.

b Area = 6.82

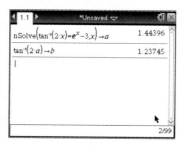

The calculator working is shown here.

(a, b) are the coordinates of P.

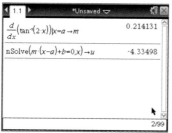

The slope of the tangent T is stored in m and the zero of the tangent T is stored in u.

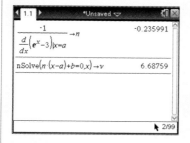

The slope of the normal N is stored in n and the zero of the normal N is stored in v.

GDC help on CD:
Alternative demonstrations for the TI-84 Plus and Casio fx-9860GII GDCs are on the CD.

▶ Continued on next page

The base of the triangle is calculated by adding the absolute value of u (since u < 0) and v, whilst the height of the triangle is the y-coordinate of the point P.

Exercise 9E

1 Given a function $f(x)$ and the point P, find the equation of the tangent.

a $f(x) = \tan(3x)$, $P(0,0)$

b $f(x) = \sin(2x) - 1$, $P\left(\dfrac{\pi}{3}, y\right)$

c $f(x) = 2\cos\left(\dfrac{x}{2}\right) - e^{2x}$, $P(0,1)$

d $f(x) = \ln\left(\tan\left(\dfrac{x}{3}\right)\right) + 2$, $P\left(\dfrac{3\pi}{4}, y\right)$

2 Given a function $f(x)$ and the point P find the equation of the normal.

> You may need to use your GDC for some of these equations.

a $f(x) = \cos(2x)$, $P(0,1)$

b $f(x) = \tan(4x)$, $P\left(\dfrac{\pi}{16}, y\right)$

c $f(x) = 2e^x \sin\left(\dfrac{x}{2}\right)$, $P(0, y)$

d $f(x) = x\cos(2x) - 3$, $P\left(\dfrac{\pi}{2}, y\right)$

EXAM-STYLE QUESTIONS

3 Given a curve $\ln(x) = \tan y$ find the equation of tangent at the point $P(0,1)$.

4 Given a curve $y + y^2 = \sin 2x$ find the equation of normal at the point $P(0,-1)$.

5 Consider the curves $y = \cos(x^2)$ and $y = e^{x^2} - 2$.

a Find the point of intersection between the curves that lies in the first quadrant.

b Find the equations of tangents to both curves at the point of intersection.

c Find the angle between the tangents in part **b**.

6 Find the area of a triangle enclosed by the y-axis, the tangent and the normal to the curve $e^y = \sin x + 1$ at the point $P(-\pi, 0)$.

Higher derivatives of trigonometric functions

Higher derivatives were discussed in Chapter 4. You can now investigate them for trigonometric functions.

Example 12

$y = x \tan x$

a Find $\dfrac{d^2 y}{dx^2}$

b Calculate the exact value of second derivative at $x = \dfrac{\pi}{3}$.

Answers

a $y = x \tan x \Rightarrow \dfrac{dy}{dx} = \tan x + x \times \sec^2 x$ *Find the first derivative using the product rule.*

$\Rightarrow \dfrac{d^2 y}{dx^2} = \sec^2 x + \sec^2 x + x \times 2 \sec x \times$ *Differentiate to find the second derivative.*

$\sec x \times \tan x$

$= 2\sec^2 x + 2x \sec^2 x \tan x$

$= 2\sec^2 x \, (1 + x \tan x)$ *Simplify.*

b $\dfrac{dy^2}{dx^2}\left(\dfrac{\pi}{3}\right) = 2\sec^2\left(\dfrac{\pi}{3}\right)\left(1 + \dfrac{\pi}{3}\tan\left(\dfrac{\pi}{3}\right)\right)$ *Substitute $x = \dfrac{\pi}{3}$*

$= 8 \times \left(1 + \dfrac{\pi\sqrt{3}}{3}\right) = 8 + \dfrac{8\pi\sqrt{3}}{3}$ *You can check this on a GDC*

GDC help on CD:
Alternative demonstrations for the TI-84 Plus and Casio fx-9860GII GDCs are on the CD.

This example shows an interesting connection between the trigonometric functions sine and cosine.

> You may have already noticed that their graphs are similar.

Example 13

Find the pattern that emerges in higher derivatives of the function $f(x) = \sin x$.

Answer

$f(x) = \sin x$

$\Rightarrow f'(x) = \cos x \quad \Rightarrow f^{(3)}(x) = -\cos x \quad \Rightarrow f^{(5)}(x) = \cos x$

$\Rightarrow f''(x) = -\sin x \quad \Rightarrow f^{(4)}(x) = \sin x$

Notice that you completed a cycle and began the same cycle again.

$$f^{(n)}(x) = \begin{cases} \cos x, n = 4k-3 \\ -\sin x, n = 4k-2 \\ -\cos x, n = 4k-1 \\ \sin x, n = 4k \end{cases} \quad k \in \mathbb{Z}^+$$

The graph of the cosine function is related to the graph of the sine function by a horizontal translation of $\dfrac{\pi}{2}$ units:

$$f^{(n)}(x) = \sin\left(x + \dfrac{n\pi}{2}\right), \, n = 0, 1, 2, \dots,$$

where the 0th derivative is the original function itself.

You can prove this formula using **mathematical induction**.

> This is very similar to the emerging pattern of the powers of the imaginary unit i. Later on in Chapter 12 you will use the polar form of a complex number to explain this emerging pattern.

Exercise 9F

1 Find the exact value of the second derivative for these functions at the given value of x.

a $f(x) = \tan x$, $x = \dfrac{\pi}{3}$

b $f(x) = x \sin x$, $x = 0$

c $f(x) = (x^2 + 1) \cos x$, $x = 0$

d $f(x) = \sqrt{x} \, \cos\dfrac{x}{2}$, $x = 1$

e $f(x) = e^x \sin 2x$, $x = \dfrac{\pi}{4}$

f $f(x) = 2x \sec x$, $x = \pi$

Check all your answers by using a calculator.

2 Describe any emerging patterns when successively differentiating these functions:

a $f(x) = \cos x$

b $g(x) = \sin 3x$

c $h(x) = \cos(ax + b)$, $a, b \in \mathbb{R}$, $a \neq 0$.

3 A function $f(x) = \sin 2x$ defines a sequence in such a way that the general term of the sequence is defined by the formula

$$a_n = f^{(n-1)}\left(\frac{\pi}{8}\right), \quad n = 1, 2, 3, \ldots,$$

where the 0th derivative is the original function itself.

a Write the first four terms of the sequence.

b Find the sum of the first 10 terms of the sequence.

EXAM-STYLE QUESTION

4 Prove the following statements by mathematical induction:

a $f(x) = \sin x \Rightarrow f^{(n)}(x) = \sin\left(x + \dfrac{n\pi}{2}\right)$, $n = 0, 1, 2, \ldots$

b $g(x) = \cos x \Rightarrow g^{(n)}(x) = \sin\left(x + \dfrac{(n+1)\pi}{2}\right)$, $n = 0, 1, 2, \ldots$

where the 0th derivative is the original function itself.

You should now be able to differentiate a variety of trigonometric functions. These results will be useful when doing further integrals.

Example 14

Find the derivatives of

a $f(x) = \ln\left(\dfrac{1+\sin x}{\cos x}\right) + c$

b $f(x) = \ln\left(\dfrac{\cos x}{1-\sin x}\right)$

c $f(x) = \ln(\tan x + \sec x)$

> Plot and compare the graphs of the three functions.
> What do you notice?

Answers

a $f'(x) = \dfrac{1}{\frac{1+\sin x}{\cos x}} \cdot \dfrac{-\cos x \cdot \cos x - (1+\sin x)\cdot(-\sin x)}{\cos^2 x}$

$\quad = \dfrac{\cos x}{1+\sin x} \times \dfrac{\cos^2 x + \sin x + \sin^2 x}{\cos^2 x}$

$\quad = \dfrac{1}{1+\sin x} \cdot \dfrac{1+\sin x}{\cos x}$

$\quad = \sec x$

Use $\dfrac{d}{dx}(\ln x) = \dfrac{1}{x}$ and product rule.

b $f'(x) = \dfrac{1}{\frac{\cos x}{1-\sin x}} \cdot \dfrac{-\sin x \cdot (1-\sin x) - \cos x \cdot (-\cos x)}{(1-\sin x)^2}$

$\quad = \dfrac{1-\sin x}{\cos x} \cdot \dfrac{-\sin x + \sin^2 x + \cos^2 x}{(1-\sin x)^2}$

$\quad = \dfrac{1}{\cos x} \cdot \dfrac{1-\sin x}{1-\sin x}$

$\quad = \sec x$

$\dfrac{1}{\cos x} = \sec x$

c $f'(x) = \dfrac{1}{\tan x + \sec x} \cdot \left(\sec^2 x + \sec x \tan x\right)$

$\quad = \dfrac{\sec x \cdot (\sec x + \tan x)}{\tan x + \sec x}$

$\quad = \sec x$

9.2 Related rates of change with trigonometric expressions

The derivative of a function, $y = f(x)$, measures the rate of change of the independent variable, y, with respect to a change in the dependent variable, x.

Example 15

A 10 m long industrial ladder is leaning against a wall on a building construction site. It starts to slip down the wall at a rate of 0.5 ms^{-1}. How fast is the angle between the ladder and the ground changing when the vertical height of the ladder is 8 m?

Answer

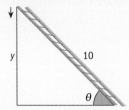

Sketch a diagram representing the given information.

$\dfrac{dy}{dt} = -0.5$ and $y = 8$

You need to find $\dfrac{d\theta}{dt}$.

Write down the given information, and what you are asked to find.

$\sin\theta = \dfrac{y}{10}$

Identify the relationship between the height of the ladder and the angle.

$\cos\theta \cdot \dfrac{d\theta}{dt} = \dfrac{1}{10}\dfrac{dy}{dt}$

Differentiate implicitly with respect to time.

$y = 8 \Rightarrow \sin\theta = \dfrac{8}{10} = \dfrac{4}{5}$

Evaluate sin θ at y = 8.

$\Rightarrow \cos\theta = \sqrt{1 - \left(\dfrac{4}{5}\right)^2} = \dfrac{3}{5}$

Use $\cos\theta = \sqrt{1 - \sin^2\theta}$.

$\dfrac{3}{5} \cdot \dfrac{d\theta}{dt} = \dfrac{1}{10} \cdot \left(-\dfrac{1}{2}\right) \Rightarrow \dfrac{d\theta}{dt}$

Substitute and solve.

$= -\dfrac{1}{12}\ {}^{c}s^{-1}$

So the angle is decreasing at a rate of $\dfrac{1}{12}\ {}^{c}s^{-1}$ or $4.77°\,s^{-1}$.

Notation ${}^{c}s^{-1}$ denotes radians per second.

Interpret your answer in the context of the problem.

$°s^{-1}$ denotes degrees per second

Example 16

There are two ships at sea, *Zadar* and *Rab*. At a given moment *Zadar* is 40 km south and 50 km east of *Rab*. *Zadar* sails north at a rate of 12 kmh^{-1}, whilst *Rab* sails east at a rate of 15 kmh^{-1}.

a How fast are the two ships approaching each other after 2 hours?

b How fast is the bearing of *Zadar* from *Rab* changing after 2 hours?

Answers

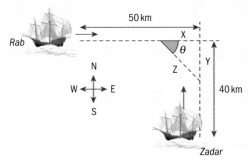

Sketch a diagram representing the given information.

a Given that $x = 50 - 15t \Rightarrow \dfrac{dx}{dt} = -15$

and $y = 40 - 12t \Rightarrow \dfrac{dy}{dt} = -12$

you need to find $\dfrac{dz}{dt}$.

Write down the given information, and what we are asked to find.

$z^2 = x^2 + y^2$

Identify the relationship between the variables using Pythagoras' theorem.

$2z\dfrac{dz}{dt} = 2x\dfrac{dx}{dt} + 2y\dfrac{dy}{dt}$

Differentiate as an implicit function with respect to time.

$z\dfrac{dz}{dt} = x\dfrac{dx}{dt} + y\dfrac{dy}{dt}$

Simplify.

$x = 50 - 15 \cdot 2 = 20$
$y = 40 - 12 \cdot 2 = 16$
$z = \sqrt{20^2 + 16^2} = 25.612\ldots$

Calculate x, y and z when t = 2.

$25.6\dfrac{dz}{dt} = 20 \cdot (-15) + 16 \cdot (-12)$

Substitute and solve.

$\dfrac{dz}{dt} = \dfrac{-492}{25.6} = -19.2$ kmh^{-1}, correct to 3 sf.

Interpret your answer in the context of the problem.

So the distance between the two ships is decreasing at a rate of 19.2 kmh^{-1}.

Notice that the bearing of Zadar from Rab is 90° + θ, therefore the bearing is changing at the same rate as the angle θ itself.

b Given that $\dfrac{dx}{dt} = -15$ and $\dfrac{dy}{dt} = -12$

you need to find $\dfrac{d\theta}{dt}$.

Write down the given information, and what you are asked to find.

▶ Continued on next page

$$\tan\theta = \frac{y}{x}$$

$$\sec^2\theta \cdot \frac{d\theta}{dt} = \frac{\frac{dy}{dt}\cdot x - y\cdot\frac{dx}{dt}}{x^2}$$

$$x = 50 - 15.2 = 20$$
$$y = 40 - 12.2 = 16$$

$$\tan\theta = \frac{16}{20} = \frac{4}{5} \Rightarrow \sec^2\theta = 1 + \left(\frac{4}{5}\right)^2 = \frac{41}{25}$$

$$\frac{41}{25}\cdot\frac{d\theta}{dt} = \frac{-12\cdot 20 - 16\cdot(-15)}{20^2}$$

$$\frac{d\theta}{dt} = 0°\,\mathrm{h}^{-1}$$

So the bearing is not changing at all.

| | |
Identify the relationship between the variables.

Differentiate as an implicit function with respect to time.
Calculate x, y and z when t = 2.

Substitute and solve.

Notice that the bearing from one ship to another is not changing since the ratio of the initial positions of the ships is equal to the ratio of their corresponding velocities.

Example 17

A reef 120 m from a straight shoreline is marked by a beacon which rotates six times per minute.

a How fast is the beam moving along the shoreline at the moment when the light beam and the shoreline are at right angles?

b How fast is that beam moving along the shoreline when the beam hits the shoreline 50 m from the point on the shoreline closest to the lighthouse?

c What is happening to the velocity of the light beam when the ray is parallel to the shoreline?

Answer

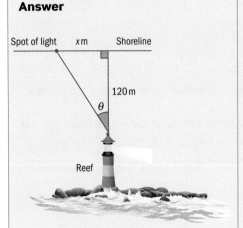

Sketch a diagram representing the given information.

Assume the beacon is at the same height as the shoreline.

▶ Continued on next page

$\dfrac{d\theta}{dt} = \dfrac{6 \cdot 2\pi}{60} = \dfrac{\pi}{5}\,^c\text{s}^{-1}$, from the speed of the beam and you need to find $\dfrac{dx}{dt}$.	*Write down the given information, and what we are asked to find.*
$\tan\theta = \dfrac{x}{120}$	*Identify the relationship between the variables.*
$\sec^2\theta\,\dfrac{d\theta}{dt} = \dfrac{1}{120} \cdot \dfrac{dx}{dt}$	*Differentiate as an implicit function with respect to time.*
$\theta = 0$	*When the beam is at 90° to the shoreline, $\theta = 0$.*
$\sec^2 0 \cdot \dfrac{\pi}{5} = \dfrac{1}{120} \cdot \dfrac{dx}{dt}$	*Substitute and solve.*
$\Rightarrow \dfrac{dx}{dt} = 24\pi = 75.4\text{ ms}^{-1}$, (3 sf)	
The beam is moving along the shoreline at 75.4 ms^{-1}	*Interpret your answer in the context of the problem.*
b $\sec^2\theta\,\dfrac{d\theta}{dt} = \dfrac{1}{120} \cdot \dfrac{dx}{dt}$	*Start with the derivative again.*
$\tan\theta = \dfrac{50}{120} = \dfrac{5}{12}$	
$\Rightarrow \sec^2\theta = 1 + \left(\dfrac{5}{12}\right)^2 = \dfrac{169}{144}$	*Use $\sec^2\theta = 1 + \tan^2\theta$*
$\dfrac{169}{144} \cdot \dfrac{\pi}{5} = \dfrac{1}{120} \cdot \dfrac{dx}{dt}$	*Substitute and solve.*
$\Rightarrow \dfrac{dx}{dt} = \dfrac{169}{6}\pi = 88.5\text{ ms}^{-1}$ (3 sf)	
The beam is moving along the shoreline at 88.5 ms^{-1}	*Interpret your answer in the context of the problem.*

c As the light ray approaches the position parallel to the shoreline, then angle

$\theta \to \dfrac{\pi}{2} \Rightarrow \sec^2\theta \to \infty$

$\Rightarrow$ velocity $\to \infty$

according to the model.

Exercise 9G

1 A 2.5 m long ladder is leaning against a wall on a building construction site. It starts to slip horizontally along the ground at a rate of 4 cms^{-1}. How fast is the angle between the ladder and the ground changing when the bottom of the ladder is 1 m away from the wall?

2 Two planes A and B are flying to their destinations. At a given moment plane A is 25 km north and 18 km east of plane B. Plane A flies west at a speed of 200 ms^{-1}, whilst plane B flies direction north at a speed of 160 ms^{-1}.

a How fast are the two planes approaching each other after 0.5 minutes?

b How fast is the bearing of plane B from plane A changing after 1 minute?

3 A professional cameraman on a safari is at a spot 30 metres from a tree, following birds that are moving at a speed of 95 kmh^{-1}. The birds are moving perpendicularly to the line joining the tree and the spot. How fast does he need to turn the camera when filming a bird:

a that is directly in front of the camera

b one second later?

4 An isosceles triangle with the sides 6, 5 and 5 cm is going through a transformation where the longest side is decreasing at a rate of 0.1 cms^{-1}.

a Find the rate of change of the angle opposite to the decreasing side at the start.

b Find the rate of change of the angle opposite to the decreasing side when the triangle is equilateral.

5 A balloon has a spherical shape. There is a hole in the balloon and the air is leaking at 2 cm^3 min^{-1}.

a Find the rate at which the radius is decreasing when $r = 12$ cm.

b Find the rate at which the surface area is decreasing when $r = 4$ cm.

6 A scientist is pointing with a laser to a flying object whose trajectory is vertically above her. The object is flying at a constant height of 10 000 m and maintaining a speed of 1025 kmh^{-1}. Find the rate in degrees per second of the rotating laser when

a the horizontal distance of the object is 8 km from the scientist,

b the object is directly above the scientist.

7 A train is moving along a straight track at 75kmh^{-1} due east. A camera positioned 2 km from the track west of the train is focused on the train.

a Find the rate of change of the distance between the camera and the train when the train is 4 km from the camera.

b At what rate is the camera rotating when the train is 4 km from the camera? Give your answer in degrees per second correct to the nearest tenth of a degree.

8 An observer is watching a fireworks rocket from a distance of 10 metres. He uses a laser to measure the distance to the rocket which is changing at a rate of 5 ms⁻¹. At a particular moment the distance measured to the rocket is 20 metres.

   a Find the rate of increase of the angle of elevation at that moment.

   b Find the speed of the rocket at that moment.

EXAM-STYLE QUESTION

9 A Ferris wheel 15 metres in diameter makes two revolutions per minute. Assume that the wheel is tangential to the ground and let P be the point of tangency.

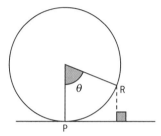

At what rate is the distance between P and a rider R changing, when she is 5 metres above the ground and going up?

9.3 Integration of trigonometric functions

Basic integrals of trigonometric functions

Since integration is a process of finding the antiderivative of the integrand function you can deduce some standard integrals:

$$\rightarrow \int \cos x \, dx = \sin x + c, \, c \in \mathbb{R} \quad \text{since } \frac{d(\sin x)}{dx} = \cos x$$

$$\int \sin x \, dx = -\cos x + c \quad \text{since } \frac{d(-\cos x)}{dx} = \sin x$$

$$\int \sec^2 x \, dx = \tan x + c \quad \text{since } \frac{d(\tan x)}{dx} = \sec^2 x$$

> More integrals of the trigonometric functions will be found later using methods of substitution and integration by parts.

Chapter 7 introduces the compound formula, and here it can be used to obtain other antiderivatives.

$$\rightarrow \int f(ax + b) dx = \frac{1}{a} F(ax + b) + c$$

You can find all the integrals of the form $\int f(ax + b) \, dx$ where f can be any of the three functions mentioned above.

Example 18

Find these integrals.

a $\displaystyle\int \cos 5x\,dx$ **b** $\displaystyle\int 2\sin(5-3x)\,dx$ **c** $\displaystyle\int \frac{1}{2}\sec^2\frac{x}{4}\,dx$

Answers

a $\displaystyle\int \cos 5x\,dx = \frac{1}{5}\sin 5x + c$ *Use compound formula.*

b $\displaystyle\int 2\sin(5-3x)\,dx$

$\displaystyle = -\frac{2}{3}(-\cos(5-3x)) + c$ *Use compound formula.*

$\displaystyle = \frac{2}{3}\cos(5-3x) + c$ *Simplify the expression.*

c $\displaystyle\int \frac{1}{2}\sec^2\frac{x}{4}\,dx = \frac{\frac{1}{2}}{\frac{1}{4}}\tan\frac{x}{4} + c$ *Use compound formula.*

$\displaystyle = 2\tan\frac{x}{4} + c$ *Simplify the expression.*

There are some more complicated integrals that can be determined using the trigonometric identities from Chapter 8.

Example 19

Use trigonometric identities to find these integrals.

a $\displaystyle\int 2\sin x\cos x\,dx$ **b** $\displaystyle\int (2\cos^2 3x - 1)\,dx$ **c** $\displaystyle\int \left(\tan^2\frac{x}{3} + 1\right)dx$

> Recap all the trigonometric identities from Chapter 8.

Answers

a $\displaystyle\int 2\sin x\cos x\,dx = \int \sin 2x\,dx$ *Use: $\sin 2\theta = 2\sin\theta\cos\theta$.*

$\displaystyle = -\frac{1}{2}\cos 2x + c$ *Use compound formula.*

b $\displaystyle\int (2\cos^2 3x - 1)\,dx$ *Use: $\cos 2\theta = 2\cos^2\theta - 1$.*

$\displaystyle = \int \cos 6x\,dx$

$\displaystyle = \frac{1}{6}\sin 6x + c$ *Use compound formula.*

c $\displaystyle\int \left(\tan^2\frac{x}{3} + 1\right)dx$ *Use $\tan^2\theta + 1 = \sec^2\theta$.*

$\displaystyle = \int \sec^2\frac{x}{3}\,dx$ *Use compound formula.*

$\displaystyle = 3\tan\frac{x}{3} + c$

Exercise 9H

1 Find these integrals.

   a $\displaystyle\int \sin 3x\,dx$ **b** $\displaystyle\int \cos(2x+1)\,dx$ **c** $\displaystyle\int \sec^2 3x\,dx$

   d $\displaystyle\int \sec^2(1-x)\,dx$ **e** $\displaystyle\int \sin\left(\frac{5x-1}{3}\right)dx$ **f** $\displaystyle\int \cos\left(\frac{3x+2}{7}\right)dx$

2 Solve these integrals using trigonometric identities.

   a $\displaystyle\int (1-2\cos^2 x)\,dx$ **b** $\displaystyle\int (1+\tan^2 x)\,dx$

   c $\displaystyle\int \sin^2 x\,dx$ **d** $\displaystyle\int \cos^2 x\,dx$

   e $\displaystyle\int (1-2\sin^2(2x))\,dx$ **f** $\displaystyle\int (2+2\tan^2(5x))\,dx$

   g $\displaystyle\int (1+\tan^2 x)(1-\sin^2 x)\,dx$ **h** $\displaystyle\int 4\sin^2 x\cos^2 x\,dx$

When you integrate a linear combination of functions you get a linear combination of the integrals.

> Recap properties of integrals in Chapter 4.

Example 20

Find these integrals.

a $\displaystyle\int (4-x^3+5\cos 2x)\,dx$

b $\displaystyle\int (7e^x-3x^2+1-2\sin 2x)\,dx$

- -

Answers

a $\displaystyle\int (4-x^3+5\cos 2x)\,dx$ *Integrate.*

   $= 4x - \dfrac{x^4}{4} + 5\cdot\dfrac{1}{2}\sin 2x + c$ *Simplify. Don't forget the constant.*

   $= 4x - \dfrac{x^4}{4} + \dfrac{5}{2}\sin 2x + c$

b $\displaystyle\int (7e^x-3x^2+1-2\sin 2x)\,dx$ *Integrate.*

   $= 7e^x - 3\cdot\dfrac{x^3}{3} + x - 2\dfrac{1}{2}(-\cos 2x) + c$ *Simplify.*

   $= 7e^x - x^3 + x + \cos 2x + c$

Exercise 9I

1 Integrate these functions.

a $f(x) = 2\sin x - 3\cos x$

b $f(x) = x^2 - 7\sin x$

c $f(x) = 4e^x - \dfrac{1}{3}\sec^2 x$

d $f(x) = 1 - \sqrt{2x} + 7\sin 3x$

e $f(x) = \dfrac{5}{2x} + \sec^2\left(\dfrac{x}{3}\right)$

f $f(x) = \dfrac{x}{x+1} - \sin\left(\dfrac{3x}{4}\right)$

g $f(x) = 2^x + 5\sin\dfrac{x}{2} - \cos\dfrac{2x}{3}$

h $f(x) = 3^{-2x} - 11\sec^2(11x)$

Finding a particular antiderivative

In Chapter 7, you found that there is no unique antiderivative function, but a family of functions that are distinguished by a constant. In order to find a particular function you need to be given a certain initial condition that the function must satisfy.

Example 21

Given that $f'(x) = 2 - 3\sin x$ find the function f such that $f(0) = -2$.

Answer

$f(x) = \displaystyle\int (2 - 3\sin x)\,dx$

$\quad = 2x - 3 \cdot (-\cos x) + c$

$\quad = 2x + 3\sin x + c$ *Simplify.*

$f(0) = -2x$ *Use the given condition.*

$\quad \Rightarrow 2\cdot 0 + 3\underset{1}{\underline{\cos 0}} + c$

$\quad = -2$ *Solve for the constant and write the*

$c = -5 \Rightarrow f(x) = 2x + 3\cos x - 5$ *function.*

When higher derivatives are involved, you need as many initial conditions as the order of the derivative given.

Example 22

Find the function g that satisfies these conditions
$g''(x) = \cos x - e^x$, $g'(0) = 2$ and $g(0) = 3$.

Answer

$g''(x) = \cos x - e^x \Rightarrow$ $g'(x) = \int (\cos x - e^x)\,dx$ $= \sin x - e^x + c_1$ $\Rightarrow g'(0) = 1 \Rightarrow \sin 0 - e^0 + c_1 = 1$	*Integrate the second derivative function to find the first derivative. Use the condition for the first derivative.*
$c_1 = 2 \Rightarrow g'(x) = \sin x - e^x + 2 \Rightarrow$	*Substitute for c_1 in $g'(x)$.*
$g(x) = \int (\sin x - e^x + 2)\,dx$ $= -\cos x - e^x + 2x + c_2$ $\Rightarrow g(0) = 3$	*Integrate the first derivative function to find $g(x)$.*
$\Rightarrow -\cos 0 - e^0 + 2 \cdot 0 + c_2 = 3$ $c_2 = 5$	*Use the condition for $g(x)$.*
$\Rightarrow g(x) = -\cos x - e^x + 2x + 5$	*Substitute for c_2 in $g(x)$.*

Exercise 9J

1 Find $f(x)$ given these conditions:

a $f'(x) = 5 - 2\cos x$, $f(0) = 0$

b $f'(x) = 4x - 6\sin(2x)$, $f(0) = 1$

c $f'(x) = 3\cos x - 2\sec^2 x$, $f\left(\dfrac{\pi}{6}\right) = -\dfrac{2\sqrt{3}}{3}$

d $f'(x) = 3x^2 - 2e^x + \cos 4x$, $f(0) = -5$

e $f'(x) = \dfrac{3}{x} + \cos(3x) - 4$, $f(1) = \dfrac{\sin(1)}{3}$

f $f'(x) = \dfrac{7}{3 - 4x} - 8x + 4e^{2x-1}$, $f\left(\dfrac{1}{2}\right) = -1$

2 Find $f(x)$ given these conditions:

a $f''(x) = 4\sin x$, $f'\left(\dfrac{\pi}{3}\right) = 0$, $f(0) = 1$

b $f''(x) = 1 + \cos x$, $f'(0) = 3$, $f(1) = -\cos(1)$

c $f''(x) = e^{1-x} + \sin(1 - x)$, $f'(1) = 2$, $f(1) = 2$

d $f''(x) = e^{2x} + \sin(2x) + x^3 - 2x + 1$, $f'(0) = 2$, $f(0) = 2$

Definite integrals

→ To evaluate definite integrals, apply the fundamental theorem of calculus.

$$\int f(x)\,dx = F(x) + c \Rightarrow \int_a^b f(x)\,dx = F(b) - F(a)$$

The fundamental theorem of calculus (FTC) was introduced in Section 7.3.

Example 23

Evaluate these integrals.

a $\displaystyle\int_0^\pi (x + \sin 2x)\,dx$

b $\displaystyle\int_0^{\frac{\pi}{2}} (e^{2x} + \cos(3x))\,dx$

c $\displaystyle\int_{-\frac{5\pi}{2}}^{\frac{5\pi}{2}} \left(1 - \frac{1}{5}\sec^2 \frac{x}{10}\right)dx$

Answers

a $\displaystyle\int_0^\pi (x + \sin 2x)\,dx = \left[\frac{x^2}{2} - \frac{1}{2}\cos 2x\right]_0^\pi$

$$= \left(\frac{\pi^2}{2} - \frac{1}{2}\underbrace{\cos 2\pi}_{1}\right) - \left(\frac{0^2}{2} - \frac{1}{2}\underbrace{\cos 0}_{1}\right)$$

$$= \frac{\pi^2}{2} - \frac{1}{2} + \frac{1}{2} = \frac{\pi^2}{2}$$

Integrate and apply the FTC.

Simplify.
Check on GDC.

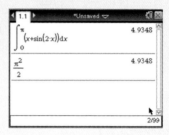

b $\displaystyle\int_0^{\frac{\pi}{2}} (e^{2x} + \cos(3x))\,dx$

$$= \left[\frac{1}{2}e^{2x} + \frac{1}{3}\sin(3x)\right]_0^{\frac{\pi}{2}}$$

$$= \left(\frac{1}{2}e^{2\cdot\frac{\pi}{2}} + \frac{1}{3}\sin\left(3\cdot\frac{\pi}{2}\right)\right)_{-1} - \left(\frac{1}{2}e^0 + \frac{1}{3}\underbrace{\sin 0}_{0}\right)$$

$$= \frac{1}{2}e^\pi - \frac{1}{3} - \frac{1}{2} = \frac{e^\pi}{2} - \frac{5}{6}$$

Apply the FTC.

Simplify.
Check on GDC.

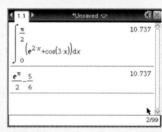

c $\displaystyle\int_{-\frac{5\pi}{2}}^{\frac{5\pi}{2}} \left(1 - \frac{1}{5}\sec^2 \frac{x}{10}\right)dx = \left[x - \frac{1}{5}\cdot\frac{1}{\frac{1}{10}}\tan\frac{x}{10}\right]_{-\frac{5\pi}{2}}^{\frac{5\pi}{2}}$

$$= \left[x - 2\tan\frac{x}{10}\right]_{-\frac{5\pi}{2}}^{\frac{5\pi}{2}}$$

$$= \left(\frac{5\pi}{2} - 2\tan\frac{\frac{5\pi}{2}}{10}\right) - \left(-\frac{5\pi}{2} - 2\tan\left(\frac{-\frac{5\pi}{2}}{10}\right)\right)$$

$$\frac{5\pi}{2} - 2\underbrace{\tan\frac{\pi}{4}}_{1} + \frac{5\pi}{2} - 2\underbrace{\tan\frac{\pi}{4}}_{1} = 5\pi - 4$$

Integrate and apply the FTC.
Simplify the expression before evaluating.
Apply the formula.

Simplify.
Check on GDC.

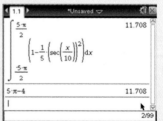

GDC help on CD:
Alternative demonstrations for the TI-84 Plus and Casio fx-9860GII GDCs are on the CD.

Exercise 9K

1 Evaluate these integrals. Check each solution using a GDC

a $\displaystyle\int_{-\frac{\pi}{3}}^{\frac{\pi}{2}} (2x - \sin x)\,dx$

b $\displaystyle\int_{\frac{\pi}{6}}^{\frac{\pi}{2}} (5 + \cos x)\,dx$

c $\displaystyle\int_{0}^{\frac{\pi}{4}} (2\sec^2 x + 1)\,dx$

d $\displaystyle\int_{0}^{\frac{\pi}{3}} (e^x + 2\sin x)\,dx$

e $\displaystyle\int_{-2\pi}^{2\pi} \left(3^{-x} + \frac{\cos\frac{x}{4}}{4}\right)\,dx$

f $\displaystyle\int_{0}^{\frac{\pi}{2}} \left(\frac{e^{3x}}{3} - \frac{2\sin 2x}{5}\right)\,dx$

g $\displaystyle\int_{-\frac{\pi}{4}}^{\frac{\pi}{4}} \left(1 - \frac{x}{2} + 2\sin 2x\right)\,dx$

h $\displaystyle\int_{0}^{\frac{\pi}{12}} (2^x + 3\cos 6x)\,dx$

i $\displaystyle\int_{-\frac{\pi}{8}}^{\frac{\pi}{8}} (x^2 + 2\sec^2 2x)\,dx$

j $\displaystyle\int_{0}^{\pi} (16e^{8x} + 9\sin 3x)\,dx$

9.4 Integration by substitution

This section introduces the method of **substitution**. It comes
from the chain rule (composite function rule)

$$\frac{dy}{dx} = \frac{dy}{dt} \cdot \frac{dt}{dx}$$

To find the integral $\displaystyle\int (2x + 3)^6\,dx$ it would be easier to have a single
variable to the power of 6 rather than the expansion of the binomial
expression $(2x + 3^6)$.

Once you write the substitution equation you need to differentiate
both sides with respect to x.

$$t = 2x + 3$$

Let $\dfrac{dt}{dx} = 2 \Rightarrow dx = \dfrac{1}{2}dt$

> Use the substitution to obtain the new simpler integral for t.

So the **new integral** must be in terms of the **new variable only**.
Take care not to mix the variables.

$$\Rightarrow \int (2x + 3)^6\,dx = \int t^6 \times \frac{1}{2}dt = \frac{1}{2}\int t^6\,dt$$

> Solve the new integral for t.

$$= \frac{1}{2} \times \frac{t^7}{7} + c$$

> Substitute for t to obtain the final answer in terms of x.

$$= \frac{(2x + 3)^7}{14} + c, \, c \in \mathbb{R}$$

Example 24

> Find the integral $\int (2x + 1)\,e^{x^2+x}\ dx$.

Answer

Let $\left.\begin{array}{l} x^2 + x = u \\ 2x + 1 = \dfrac{du}{dx} \Rightarrow (2x+1)dx = du \end{array}\right\} \Rightarrow$	*Notice that $2x + 1$ is the derivative of $x^2 + x$ so if you define it as a new variable u you will have its derivative du too.*
$\displaystyle\int (2x + 1)\,e^{x^2+x}\ dx = \int e^u\,du$	*Use the substitution and solve the new integral for u.*
$= e^u + c = e^{x^2+x} + c$	*Leave your answer in terms of x.*

Example 25

> Find the integral $\int \cot x\,dx$.

Answer

$\displaystyle\int \cot x\,dx = \int \dfrac{\cos x}{\sin x}\,dx$	*Rewrite in terms of sine and cosine. Notice that $\cos x$ is the derivative of $\sin x$ so you have the new variable v and its derivative dv.*
Let $\left.\begin{array}{l} \sin x = v \\ \cos x = \dfrac{dv}{dx} \Rightarrow \cos x\ dx = dv \end{array}\right\}$	
$\Rightarrow \displaystyle\int \dfrac{\cos x}{\sin x}\,dx = \int \dfrac{dv}{v}$	*Use the substitution and integrate with respect to v.*
$= \ln|v| + c = \ln|\sin x| + c$	*Give your answer in terms of x.*

Example 26

> Use an appropriate substitution to find $\int x^2 \sin(x^3 - 2)\,dx$.

Answer

Let $\left.\begin{array}{l} x^3 - 2 = t \\ 3x^2 = \dfrac{dt}{dx} \Rightarrow x^2 dx = \dfrac{1}{3}dt \end{array}\right\} \Rightarrow$	*Notice that x^2 is not exact derivative of x^3, but $\dfrac{1}{3}$ of it.*
$\displaystyle\int x^2 \sin(x^3 - 2)\,dx$	*Use the substitution.*
$= \displaystyle\int \sin t \times \dfrac{1}{3}dt = \dfrac{1}{3}\int \sin t\ dt$	*Simplify and integrate with respect to t.*
$= -\dfrac{1}{3}\cos t + c = -\dfrac{1}{3}\cos\left(x^3 - 2\right) + c$	*Give your answer in terms of x.*

Exercise 9L

1 Find these integrals by the method of substitution.

a $\displaystyle\int 2x\sin x^2\,dx$

b $\displaystyle\int 3x^2\sqrt{x^3+3}\,dx$

c $\displaystyle\int (3-4x)e^{1+3x-2x^2}\,dx$

d $\displaystyle\int \tan x\,dx$

e $\displaystyle\int 2\cos 2x\,e^{\sin 2x}\,dx$

f $\displaystyle\int \frac{e^{\sqrt{x}}}{2\sqrt{x}}\,dx$

g $\displaystyle\int 2^x\ln 2\sin(2^x)\,dx$

h $\displaystyle\int \frac{\arcsin x}{\sqrt{1-x^2}}\,dx$

i $\displaystyle\int \frac{2\arctan 2x}{1+4x^2}\,dx$

> An appropriate substitution for **1a** would be $u=x^2$. For **1b** it would be $u=x^3+3$

2 Use an appropriate substitution to find these integrals.

a $\displaystyle\int x\cos x^2\,dx$

b $\displaystyle\int x^5\sqrt[3]{x^6-1}\,dx$

c $\displaystyle\int (x+2)e^{3x^2+12x-7}\,dx$

d $\displaystyle\int \frac{\tan(5x+4)}{5}\,dx$

e $\displaystyle\int \sin 3x\cdot 3^{\cos 3x}\,dx$

f $\displaystyle\int \frac{\sin\sqrt[4]{x}}{\sqrt[4]{x^3}}\,dx$

g $\displaystyle\int 5^x\cos(5^x)\,dx$

h $\displaystyle\int \frac{e^{2x}+e^{-2x}}{e^{-2x}-e^{2x}}\,dx$

i $\displaystyle\int \frac{\sqrt{\arctan\dfrac{x}{3}}}{9+x^2}\,dx$

j $\displaystyle\int (x^2+x)\cos\left(x^3+\frac{3}{2}x^2\right)dx$

k $\displaystyle\int \frac{\arcsin^2(2x+1)}{\sqrt{-x-x^2}}\,dx$

Definite integrals and integration by substitution

When solving definite integrals there are two methods.

Method I Solve the original integral by using substitution and then just apply the fundamental theorem of calculus on the given boundaries to the solution.

Method II When substituting a new variable in the process to obtain a simpler integral we use the substitution to change the boundaries. Use new boundaries and apply the fundamental theorem to the new integral.

Example 27

Evaluate these integrals.

a $\displaystyle\int_{1}^{7} 2x\sqrt{x^2-1}\,dx$ **b** $\displaystyle\int_{-\frac{\pi}{6}}^{\frac{\pi}{4}} \cos x \sin x\,dx$

Answers

a Method 1

Let $\left.\begin{array}{l} x^2-1=t \\ 2x\,dx=dt \end{array}\right\} \Rightarrow$

Identify the substitution.

$$\int 2x\sqrt{x^2-1}\,dx = \int \sqrt{t}\,dt$$

Use the substitution and as $2x\,dx = 2dt$ you can simplify.

$$= \frac{t^{\frac{3}{2}}}{\frac{3}{2}} + c = \frac{2}{3}\left(x^2-1\right)^{\frac{3}{2}} + c$$

$$\int_{1}^{7} 2x\sqrt{x^2-1}\,dx = \left[\frac{2}{3}\left(x^2-1\right)^{\frac{3}{2}}\right]_{1}^{7}$$

Solve the integral and give the answer in terms of the original variable.

$$= \frac{2}{3}\left((7^2-1)^{\frac{3}{2}} - \underbrace{(1^2-1)^{\frac{3}{2}}}_{0}\right) = \frac{2}{3}\cdot 48^{\frac{3}{2}}$$

Use the FTC for definite integral.

$$= \frac{2}{3}\cdot 64\cdot 3\sqrt{3} = 128\sqrt{3}$$

Calculate and simplify.

Method 2

Let $\left.\begin{array}{ll} x^2-1=t & 1^2-1=0 \\ 2x\,dx=dt & 7^2-1=48 \end{array}\right| \Rightarrow$

Identify the substitution and find new boundaries.

$$\int_{1}^{7} 2x\sqrt{x^2-1}\,dx = \int_{0}^{48} \sqrt{t}\,dt$$

Use the substitution and apply the new boundaries.

$$= \left[\frac{2}{3}t^{\frac{3}{2}}\right]_{0}^{48} = \frac{2}{3}\cdot 48^{\frac{3}{2}} = 128\sqrt{3}$$

Use the FTC and calculate the answer.
Check on GDC.

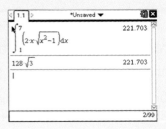

GDC help on CD:
Alternative demonstrations for the TI-84 Plus and Casio fx-9860GII GDCs are on the CD.

b Method 1

Let $\left.\begin{array}{l} \sin x = t \\ \cos x\,dx = dt \end{array}\right\} \Rightarrow$

Identify the substitution and $dx = \dfrac{dt}{\cos x}$

$$\int \cos x \sin x\,dx$$

$$= \int t\,dt = \frac{t^2}{2} + c = \frac{\sin^2 x}{2} + c$$

Solve the integral and give the answer in terms of x.

▶ Continued on next page

$$\int_{-\frac{\pi}{6}}^{\frac{\pi}{4}} \cos x \sin x \, dx = \left[\frac{\sin^2 x}{2} \right]_{-\frac{\pi}{6}}^{\frac{\pi}{4}}$$

Use the FTC for definite integral.

$$= \frac{\left(\sin\left(\frac{\pi}{4}\right) \right)^2}{2} - \frac{\left(\sin\left(-\frac{\pi}{6}\right) \right)^2}{2}$$

Evaluate the definite integral.

$$= \frac{\left(\frac{\sqrt{2}}{2} \right)^2}{2} - \frac{\left(-\frac{1}{2} \right)^2}{2} = \frac{1}{4} - \frac{1}{8} = \frac{1}{8}$$

Calculate and simplify.

Method 2

Let $\begin{matrix} \sin x = t \\ \cos x \, dx = dt \end{matrix} \left. \begin{matrix} \sin\left(-\frac{\pi}{6}\right) = -\frac{1}{2} \\ \sin\left(\frac{\pi}{4}\right) = \frac{\sqrt{2}}{2} \end{matrix} \right\} \Rightarrow$

Identify the substitution and find new boundaries.

$$\int_{-\frac{\pi}{6}}^{\frac{\pi}{4}} \cos x \sin x \, dx = \int_{-\frac{1}{2}}^{\frac{\sqrt{2}}{2}} t \, dt = \left[\frac{t^2}{2} \right]_{-\frac{1}{2}}^{\frac{\sqrt{2}}{2}}$$

Use the substitution and apply the new boundaries.

$$= \frac{\left(\frac{\sqrt{2}}{2} \right)^2}{2} - \frac{\left(-\frac{1}{2} \right)^2}{2}$$

$$= \frac{1}{4} - \frac{1}{8} = \frac{1}{8}$$

Evaluate the definite integral.
Notice that you could have used $\cos x = t$; the boundaries would be changed but the final result will remain the same.
Check on GDC.

> The work in both methods is similar, but it is slightly simpler in Method 2, so you could use this method this in paper 1 to gain some time. In paper 2, unless otherwise stated in the question simply use a GDC.

Exercise 9M

Find the **exact values** of these integrals:

1 $\displaystyle\int_0^1 3x^2 \left(x^3 - 1 \right)^4 \, dx$

2 $\displaystyle\int_0^3 \frac{2x}{x^2 + 1} \, dx$

3 $\displaystyle\int_0^{\frac{\pi}{6}} \cos x \sqrt{\sin x} \, dx$

4 $\displaystyle\int_1^{e^3} \frac{\ln x}{x} \, dx$

5 $\displaystyle\int_0^{\ln 2} \frac{e^x}{e^x + 1} \, dx$

6 $\displaystyle\int_0^{\frac{\pi}{6}} 2 \tan(2x) \, dx$

7 $\displaystyle\int_0^1 \left(x^2 + x \right) \cos\left(x^3 + \frac{3}{2} x^2 \right) dx$

8 $\displaystyle\int_0^3 2^x \sqrt{2^x + 1} \, dx$

9 Repeat questions 1 to 8 using a GDC.

9.5 Integration by parts

Integration by parts is related to the product rule for differentiation.

$$\frac{d(u \cdot v)}{dx} = \frac{du}{dx} \cdot v + u \cdot \frac{dv}{dx}$$

Integrating this identity with respect to x you find:

$$\int \frac{d(u \cdot v)}{dx} dx = \int \frac{du}{dx} \cdot v \, dx + \int u \cdot \frac{dv}{dx} dx \Rightarrow uv = \int v \, du + u \, dv$$

This gives:

$$\rightarrow \int u \frac{dv}{dx} dx = uv - \int v \frac{du}{dx} dx$$
where u and v are functions of x.

These examples give typical integrals that can be calculated using integration by parts.

> Integration by parts allows you to convert an integral into another one that is simpler. The method is a kind of **reduction formula**.

Example 28

Find the integral $\int 2x e^x dx$.

Answer

Let
$$u = 2x \Rightarrow \frac{du}{dx} = 2 \Rightarrow du = 2 \, dx,$$

$$\frac{dv}{dx} = e^x \Rightarrow v = \int e^x dx = e^x$$

$$\int 2x e^x dx = 2x e^x - \int e^x \cdot 2 \, dx$$

$$= 2x e^x - 2 \int e^x \, dx$$

$$= 2x e^x - 2e^x + c$$
$$= (2x - 2)e^x + c, c \in \mathbb{R}$$
$$= 2e^x(x - 1) + c$$

In these cases, always differentiate the polynomial and integrate the exponential. You could try it the other way around.

Choose the variables and apply the formula. Use integral properties to simplify it. Simplify the final answer. Notice that in the process of integrating dv you do not add a constant at the end, but only add the constant to the final answer.

> Normally, let $\frac{dv}{dx}$ be the more complicated function that is still integrable. Considering $y = 2x$ and $y = e^x$, $y = e^x$ is the more complicated function of the two whose integral you can still find.

Exercise 9N

Find these integrals using integration by parts.

1 $\displaystyle\int x\mathrm{e}^x\,\mathrm{d}x$

2 $\displaystyle\int (2x+9)\cos x\,\mathrm{d}x$

3 $\displaystyle\int (2-5x)\sin x\,\mathrm{d}x$

4 $\displaystyle\int (3x-1)\mathrm{e}^{3x}\,\mathrm{d}x$

5 $\displaystyle\int (4x-7)\mathrm{e}^{(4x-1)}\,\mathrm{d}x$

6 $\displaystyle\int \frac{x+3}{2}\sin(2x+3)\,\mathrm{d}x$

7 $\displaystyle\int \frac{3-x}{4}\cos\!\left(\frac{x}{4}\right)\mathrm{d}x$

8 $\displaystyle\int x\cdot 2^x\,\mathrm{d}x$

9 $\displaystyle\int (1-x)\cdot 5^x\,\mathrm{d}x$

10 $\displaystyle\int \frac{(2-x)}{7\cdot 3^x}\,\mathrm{d}x$

11 $\displaystyle\int \frac{4x\cdot 3^x}{5^x}\,\mathrm{d}x$

Example 29

Integrate the expression $y = (4x + 5)\ln x$ with respect to x.

Answer

Let

$u = \ln x \Rightarrow \dfrac{\mathrm{d}u}{\mathrm{d}x} = \dfrac{1}{x} \Rightarrow \mathrm{d}u = \dfrac{1}{x}\,\mathrm{d}x$

Choose the variables.

$\dfrac{\mathrm{d}v}{\mathrm{d}x} = (4x+5) \Rightarrow v = \displaystyle\int (4x+5)\,\mathrm{d}x$

$\quad = 2x^2 + 5x$

$\displaystyle\int (4x+5)\ln x\,\mathrm{d}x$

Apply the formula and simplify the integral.

$= (2x^2+5x)\ln x - \displaystyle\int (2x^2+5x)\,\dfrac{1}{x}\,\mathrm{d}x$

$= (2x^2+5x)\ln x - \displaystyle\int (2x+5)\,\mathrm{d}x$

$= (2x^2+5x)\ln x - (x^2+5x) + c$

Simplify, if possible.

$= x^2(2\ln x - 1) + 5x(\ln x - 1) + c$

Notice that in this case you differentiate ln x and integrate $(4x + 5)$ even though polynomials are considered more simple than logarithms. This is the only such case.

Exercise 9O

Find these integrals using integration by parts.

1 $\displaystyle\int x\ln x\,\mathrm{d}x$

2 $\displaystyle\int (3x+2)\ln x\,\mathrm{d}x$

3 $\displaystyle\int (1-x)\ln x\,\mathrm{d}x$

4 $\displaystyle\int x\ln(4x)\,\mathrm{d}x$

5 $\displaystyle\int (3x-2)\ln\!\left(\frac{x}{5}\right)\mathrm{d}x$

6 $\displaystyle\int (3+4x)\ln(3+4x)\,\mathrm{d}x$

7 $\displaystyle\int (5+7x)\ln(4-11x)\,\mathrm{d}x$

8 $\displaystyle\int x^2\ln x\,\mathrm{d}x$

9 $\displaystyle\int (2-x+x^2)\ln(3x)\,\mathrm{d}x$

In this section you will use integration by parts to find some special integrals.

Example 30

Integrate these functions.

a $f(x) = \ln x$ **b** $f(x) = \arcsin x$

Answers

a Let

$u = \ln x \Rightarrow du = \dfrac{1}{x}dx$

$\dfrac{dv}{dx} = 1 \Rightarrow v = x$

$\displaystyle\int \ln x\,dx = x\ln x - \int x \cdot \dfrac{1}{x}dx$

$= x\ln x - \displaystyle\int 1 \cdot dx = x\ln x - x + c$

$= x(\ln x - 1) + c$

$\ln x = 1 \times \ln x$
Choose the variables.

Apply the formula.

Simplify and integrate.

b Let $u = \arcsin x \Rightarrow du = \dfrac{1}{\sqrt{1-x^2}}dx$

$\dfrac{dv}{dx} = 1 \Rightarrow v = x$

$\displaystyle\int \arcsin x\,dx$

$= x\arcsin x - \displaystyle\int x \cdot \dfrac{1}{\sqrt{1-x^2}}dx$

$= x\arcsin x + \displaystyle\int \dfrac{-x}{\sqrt{1-x^2}}dx$

$t = 1-x^2 \Rightarrow \dfrac{dt}{dx} = -2x$

$\Rightarrow \dfrac{1}{2}dt = -x\,dx$

$\displaystyle\int \dfrac{-x}{\sqrt{1-x^2}}dx$

$= \dfrac{1}{2}\displaystyle\int t^{-\frac{1}{2}}dt = \left(\dfrac{1}{2}\dfrac{t^{\frac{1}{2}}}{\frac{1}{2}}\right) + c = \sqrt{1-x^2} + c$

$\displaystyle\int \arcsin x\,dx$

$= x\arcsin x + \sqrt{1-x^2} + c$

Choose the variables.

Apply the formula.

Simplify the integral.

Use substitution to solve the new integral.

Apply the result and find the final answer.

Exercise 9P

Find these integrals.

1 $\displaystyle\int \log x \, dx$ **2** $\displaystyle\int \log_a x \, dx$ **3** $\displaystyle\int \arctan x \, dx$

4 $\displaystyle\int \arccos x \, dx$ **5** $\displaystyle\int 2x \arctan x \, dx$ **6** $\displaystyle\int x^2 \arcsin x \, dx$

Sometimes you have to apply integration by parts more than once until you reach a simple integral. That often occurs with polynomials of a higher degree.

Example 31

Find $\displaystyle\int (3x^2 - x + 1) \sin x \, dx$.

Answer

$u = 3x^2 - x + 1 \Rightarrow \dfrac{du}{dx} = 6x - 1$ — *Choose the variables.*

$\Rightarrow du = (6x - 1) \, dx$

$\dfrac{dv}{dx} = \sin x \Rightarrow v = \displaystyle\int \sin x \, dx = -\cos x$

$\displaystyle\int (3x^2 - x + 1) \sin x \, dx$ — *Apply the formula.*

$= (3x^2 - x + 1) \cdot (-\cos x) - \displaystyle\int (6x - 1) \cdot (-\cos x) \, dx$ — *Simplify the integral and identify the new integral.*

$= -(3x^2 - x + 1) \cdot \cos x + \displaystyle\int (6x - 1) \cdot \cos x \, dx$

$u = 6x - 1 \Rightarrow \dfrac{du}{dx} = 6 \Rightarrow du = 6 \, dx$ — *Choose the variables for the new integral to be solved by parts.*

$\dfrac{dv}{dx} = \cos x \Rightarrow v = \displaystyle\int \cos x \, dx = \sin x$

$\displaystyle\int (6x - 1) \cdot \cos x \, dx = (6x - 1)\sin x - \displaystyle\int 6 \cdot \sin x \, dx$ — *Apply the formula.*

$= (6x - 1)\sin x - 6(-\cos x)$ — *Simplify the integral.*

$= (6x - 1)\sin x + 6\cos x$

$\displaystyle\int (3x^2 - x + 1)\sin x \, dx =$ — *Continue with integration of the original integral.*

$-(3x^2 - x + 1) \cdot \cos x + \displaystyle\int (6x - 1) \cdot \cos x \, dx$ — *Use the result.*

Simplify.

$= -(3x^2 - x + 1) \cdot \cos x + (6x - 1)\sin x + 6\cos x + c$

$= (-3x^2 + x + 5) \cdot \cos x + (6x - 1)\sin x + c$

Exercise 9Q

Integrate:

1 $\displaystyle\int x^2 e^x \, dx$

2 $\displaystyle\int (x^2 + 1)\sin x \, dx$

3 $\displaystyle\int (2x - x^2)\cos x \, dx$

4 $\displaystyle\int (1 + x - x^2) e^{2x} \, dx$

5 $\displaystyle\int (2x^2 + x + 3)\cos(2x) \, dx$

6 $\displaystyle\int x^2 \sin(1 - 2x) \, dx$

7 $\displaystyle\int x^2 3^x \, dx$

8 $\displaystyle\int (1 + x^3) e^{\frac{x}{2}} \, dx$

9 $\displaystyle\int (x^3 + x^2)\sin(5x) \, dx$

10 $\displaystyle\int x^4 \cos x \, dx$

11 $\displaystyle\int x^5 e^{2x} \, dx$

Multiple applications of the method will occur in a product of an exponential and sine or cosine function.

Example 32

Find $\displaystyle\int e^x \cos x \, dx$.

Answer

$u = e^x \Rightarrow \dfrac{du}{dx} = e^x \cdot 3 \Rightarrow du = e^x dx$	*Choose the variables.*
$\dfrac{dv}{dx} = \cos x \Rightarrow v = \displaystyle\int \cos x \, dx = \sin x$	
$\displaystyle\int e^x \cos x \, dx = e^x \sin x - \int \sin x \cdot e^x \, dx$	*Apply the formula.*
$u = e^x \Rightarrow \dfrac{du}{dx} = e^x \cdot \Rightarrow du = e^x dx$	*Simplify the integral and identify the new integral.*
$\dfrac{dv}{dx} = \sin x \Rightarrow v = \displaystyle\int \sin x \, dx = -\cos x$	*Choose the variables for the new integral to be solved by parts.*

Exercise 9R

Find these integrals.

1 $\displaystyle\int \sin x\, e^x\, dx$

2 $\displaystyle\int e^{2x} \cos x\, dx$

3 $\displaystyle\int \cos 3x e^{4x}\, dx$

4 $\displaystyle\int \frac{\sin(2x)}{e^x}\, dx$

5 $\displaystyle\int \sin x\, e^x\, dx$

This process of multiple application of integration by parts can be shown in a table. The expression for u is successively differentiated in a column, whilst dv is successively integrated in the other column as many times as needed.

Example 33

Find the integral $\displaystyle\int x^3 e^{3x}\, dx$.

Answer

	$dv = e^{3x}\, dx$	sign
$u = x^3$	$v = \dfrac{1}{3}e^{3x}$	+
$3x^2$	$\dfrac{1}{9}e^{3x}$	−
$6x$	$\dfrac{1}{27}e^{3x}$	+
6	$\dfrac{1}{81}e^{3x}$	−

Notice that the second integral in the integration by parts formula has a minus in front of it, so the sequence of the products of derivatives and integrals signs alternate.

So the result is

$\displaystyle\int x^3 e^{3x}\, dx$

$= x^3 \cdot \dfrac{1}{3}e^{3x} - 3x^2 \cdot \dfrac{1}{9}e^{3x} + 6x \cdot \dfrac{1}{27}e^{3x} - 6 \cdot \dfrac{1}{81}e^{3x} + c$

$= \dfrac{x^3}{3}e^{3x} - \dfrac{x^2}{3}e^{3x} + \dfrac{2x}{9}e^{3x} - \dfrac{2}{27}e^{3x} + c$

$= \dfrac{e^{3x}}{27}\left(9x^3 - 9x^2 + 6x - 2\right) + c$

You can use the method from Example 33 to verify your solutions to Exercise 9R.

9.6 Special substitutions

In this section you are going to study some special substitutions that are not immediately obvious.

Substitution in radical expressions

To simplify a radical linear expression, substitute the entire radical expression with a new variable and then integrate with respect to that new variable.

Example 34

Find the integral $\int x\sqrt{2x+1}\,dx$.

Answer

Let $\sqrt{2x+1} = t \Rightarrow 2x+1 = t^2$ $$\Rightarrow x = \frac{1}{2}(t^2 - 1)$$	*Express x in terms of t.*
$dx = \frac{1}{2} \cdot 2t\,dt \Rightarrow dx = t\,dt$	*Find dx in terms of t and simplify it.*
$\int x\sqrt{2x+1}\,dx = \int \frac{t^2 - 1}{2} \cdot t \cdot t\,dt$	*Use the substitutions to obtain the integral in terms of t only.*
$= \frac{1}{2}\int (t^4 - t^2)\,dt = \frac{1}{2}\left(\frac{t^5}{5} - \frac{t^3}{3}\right)$	*Simplify the new integral and integrate it with respect to t.*
$= \frac{(2x+1)^{\frac{5}{2}}}{10} - \frac{(2x+1)^{\frac{3}{2}}}{6} + c$	*Use the substitution to return the old variable x.*
$= \frac{(2x+1)^{\frac{3}{2}}(3(2x+1)-5)}{30} + c$	*Simplify.*
$= \frac{(2x+1)^{\frac{3}{2}}(3x-1)}{15} + c, c \in \mathbb{R}$	

You can also take only the radicand for the new variable and then integrate with respect to it.

Example 35

Find the integral $\int 3x^2 \sqrt[3]{2-3x}\, dx$.

Answer

Let $2-3x = u \Rightarrow 2-u = 3x$
$$\Rightarrow x = \frac{1}{3}(2-u)$$

Express x in terms of u.

$$dx = -\frac{1}{3}du$$

Find dx in terms of the variable u.

$$x^2 = \frac{1}{9}(2-u)^2 \Rightarrow x^2 = \frac{1}{9}\left(4-4u+u^2\right)$$

Express the remaining factor in the integral x^2 in terms of u.

$$\int 3x^2 \sqrt[3]{2-3x}\, dx = \int 3 \cdot \frac{1}{9}\left(4-4u+u^2\right) \cdot u^{\frac{1}{3}}\left(-\frac{1}{3}\right)du$$

Use all the substitutions to obtain the integral in terms of u.

$$= -\frac{1}{9}\int \left(4u^{\frac{1}{3}} - 4u^{\frac{4}{3}} + u^{\frac{7}{3}}\right)dt$$

Simplify the new integral and integrate it with respect to u.

$$= -\frac{1}{9}\left(\frac{4u^{\frac{4}{3}}}{\frac{4}{3}} - \frac{4u^{\frac{7}{3}}}{\frac{7}{3}} + \frac{u^{\frac{10}{3}}}{\frac{10}{3}}\right) + c$$

$$= -\frac{u^{\frac{4}{3}}}{3} + \frac{4u^{\frac{7}{3}}}{21} - \frac{u^{\frac{10}{3}}}{30} + c$$

$$= -\frac{u^{\frac{4}{3}}\left(70-40u+u^2\right)}{210} + c$$

Factorize and simplify.

$$= -\frac{(2-3x)^{\frac{4}{3}}\left(70-40(2-3x)+(2-3x)^2\right)}{210} + c$$

Substitute for x.

$$= -\frac{(2-3x)^{\frac{4}{3}}\left(70-80+120x+28-84x+63x^2\right)}{210} + c$$

Expand and simplify.

$$= -\frac{(2-3x)^{\frac{4}{3}}\left(18+36x+63x^2\right)}{210} + c$$

Factorize and simplify.

$$= -\frac{(2-3x)^{\frac{4}{3}}\, 9 \cdot \left(2+4x+7x^2\right)}{210} + c$$

$$= -\frac{3(2-3x)^{\frac{4}{3}}\left(7x^2+4x+2\right)}{70} + c$$

Exercise 9S

Find these integrals.

1 $\displaystyle\int x\sqrt{x+2}\,dx$

2 $\displaystyle\int 3x\sqrt{1-2x}\,dx$

3 $\displaystyle\int 5x^2\sqrt{3+4x}\,dx$

4 $\displaystyle\int x\sqrt[3]{x+3}\,dx$

5 $\displaystyle\int x^2\sqrt[4]{x+1}\,dx$

6 $\displaystyle\int x^3\sqrt[5]{1-x}\,dx$

You have used trigonometric identities to solve the integrals of squares of the sine and cosine functions. Here you can find out what happens with higher powers.

Note that there are two different methods when solving the integrals in Examples 36 and 37. One method is by substitution if the exponent is odd. The other is by using the double angle formula if the exponent is even.

Example 36

Find the integral $\displaystyle\int \sin^3 x\,dx$.

Answer

$\displaystyle\int \sin^3 x\,dx = \int \sin^2 x \sin x \cdot dx$

$\displaystyle\qquad = \int (1 - \cos^2 x)\cdot \sin x\,dx$

Rewrite sine in terms of cosine so that you have a new variable $\cos x$ and its differential $-\sin x$.

Let

$v = \cos x \Rightarrow \dfrac{dv}{dx} = -\sin x$

$\qquad \Rightarrow -dv = \sin x\,dx$

$\displaystyle\int (1 - v^2)\cdot (-dv)$

Use the substitution.

$\displaystyle = \int (v^2 - 1)\,dv$

$\displaystyle = \frac{v^3}{3} - v + c$

$\displaystyle = \frac{\cos^3 x}{3} - \cos x + c$

Return the original variable x.

Example 37

Find the integral $\int \sin^4 x \, dx$.

Answer

$$\int \sin^4 x \, dx = \int (\sin^2 x)^2 \, dx$$

$$= \int \left(\frac{1 - \cos 2x}{2} \right)^2 dx \qquad \textit{Use double angle formula.}$$

$$= \int \frac{1 - 2\cos 2x + \cos^2 2x}{4} \, dx \qquad \textit{Expand the expression.}$$

$$= \int \left(\frac{1}{4} - \frac{1}{2}\cos 2x + \frac{1}{4} \cdot \frac{1 + \cos 4x}{2} \right) dx \qquad \textit{Again use double angle formula.}$$

$$= \frac{1}{4}x - \frac{1}{2} \cdot \frac{\sin 2x}{2} + \frac{1}{8}\left(x + \frac{\sin 4x}{4} \right) + c \qquad \textit{Simplify.}$$

$$= \frac{3}{8}x - \frac{1}{4}\sin 2x + \frac{1}{32}\sin 4x + c$$

Exercise 9T

Find these integrals.

1 $\int \cos^3 x \, dx$ **2** $\int \cos^4 x \, dx$ **3** $\int \sin^5 \left(\frac{x}{5} \right) dx$ **4** $\int 48\cos^6 (2x) \, dx$

Investigation – recursive formula

Use integration by parts to find a **recursive** formula for the integrals of the forms:

1 $\int \sin^n x \, dx$, where n is a positive integer

2 $\int \cos^n x \, dx$, where n is a positive integer.

Example 38

Find the integral $\int \sin^3 x \cos^2 x \, dx$.

Answer

$$\int \sin^3 x \cos^2 x \, dx = \int \sin^2 x \cdot \sin x \cdot \cos^2 x \, dx \; = \int (1 - \cos^2 x) \cdot \sin x \cdot \cos^2 x \, dx$$

$$= \int (\cos^2 x - \cos^4 x) \cdot \sin x \, dx$$

Let $u = \cos x \Rightarrow \dfrac{du}{dx} = -\sin x \Rightarrow -du = \sin x \, dx$

$$= \int (u^2 - u^4) \cdot (-du) = -\frac{u^3}{3} + \frac{u^5}{5} + c = \frac{\cos^5 x}{5} - \frac{\cos^3 x}{3} + c$$

Example 39

Find the integral $\int \sin^4 x \cos^7 x \, dx$.

Answer

$$\int \sin^4 x \cos^7 x \, dx = \int \sin^4 x \cdot \cos^6 x \cdot \cos x \, dx$$

$$= \int \sin^4 x \cdot (1 - \sin^2 x)^3 \cdot \cos x \, dx \qquad \text{Use } \cos^2 x = 1 - \sin^2 x$$

Let $t = \sin x \Rightarrow \dfrac{dt}{dx} = \cos x \Rightarrow dt = \cos x \, dx$

$$= \int t^4 \cdot (1 - t^2)^3 \, dt$$

$$= \int t^4 \cdot (1 - 3t^2 + 3t^4 - t^6) \, dt$$

$$= \int t^4 - 3t^6 + 3t^8 - t^{10}) \, dt$$

$$= \frac{t^5}{5} - \frac{3t^7}{7} + \frac{3t^9}{9} - \frac{t^{11}}{11} + c$$

$$= \frac{\sin^5 x}{5} - \frac{3\sin^7 x}{7} + \frac{3\sin^9 x}{9} - \frac{\sin^{11} x}{11} + c \qquad \text{Substitute back in terms of } x.$$

Notice that in Examples 38 and 39 you always factorized the trigonometric function raised to an odd power and expressed everything in terms of the trigonometric function raised to an even power.

Investigation – more recursive formula

Find recursive formulae for the integrals of the form $\int \sin^n x \cos^m x \, dx$, where n and m are positive integers.

Trigonometric substitutions

→ When the integrand contains a quadratic radical expression use one of these trigonometric substitutions to transform the integral.

1 If the form is $\sqrt{a^2 - x^2}$ use the substitution $x = a \sin \theta$.

2 If the form is $\sqrt{x^2 - a^2}$ use the substitution $x = a \sec \theta$.

3 If the form is $\sqrt{x^2 + a^2}$ use the substitution $x = a \tan \theta$.

Example 40

Use an appropriate trigonometric substitution to find $\int \sqrt{25-x^2}\,dx$.

Answer

Let $\sqrt{25-x^2} = \sqrt{5^2-x^2} \Rightarrow$

$\sqrt{25-x^2} = \sqrt{25-25\sin^2\theta}$

$\qquad = 5\sqrt{1-\sin^2\theta} = 5\cos\theta$

Let $x = 5\sin\theta \Rightarrow dx = 5\cos\theta\,d\theta$

$\int \sqrt{25-x^2}\,dx = \int 5\cos\theta \cdot 5\cos\theta\,d\theta$

$\qquad = 25\int \cos^2\theta\,d\theta = 25\int \frac{1+\cos 2\theta}{2}\,d\theta$

$\qquad = \frac{25}{2}\theta + \frac{25}{4}\sin 2\theta + c$

$\qquad = \frac{25}{2}\theta + \frac{25}{4}\cdot 2\sin\theta\cos\theta + c$

$x = 5\sin\theta \Rightarrow \theta = \arcsin\left(\frac{x}{5}\right)$

$= \frac{25}{2}\arcsin\left(\frac{x}{5}\right) + \frac{25}{4}\cdot 2\cos\left(\arcsin\left(\frac{x}{5}\right)\right) + c$

$= \frac{25}{2}\arcsin\left(\frac{x}{5}\right) + \frac{5x}{2}\cdot\sqrt{1-\sin^2\left(\arcsin\left(\frac{x}{5}\right)\right)} + c$

$= \frac{25}{2}\arcsin\left(\frac{x}{5}\right) + \frac{5x}{2}\sqrt{1-\left(\frac{x}{5}\right)^2} + c$

$= \frac{25}{2}\arcsin\left(\frac{x}{5}\right) + \frac{x}{2}\sqrt{25-x^2} + c,\ c\in\mathbb{R}$

Identify the substitution $x = 5\sin\theta$.
Express radical expression in terms of θ.

Find dx in terms of the variable θ.

Use the substitutions to obtain the integral in terms of the variable θ.
Use double angle formula to simplify the integral.
Integrate with respect to θ.

Use double angle formula to simplify the primitive function.
In order to return the variable x express θ in terms of x.

Now proceed in substituting θ in term of x.

Return to x.

Express cosine in terms of sine.

Simplify the trigonometric expressions.

Simplify the radical expression.

> In this problem you could also use the substitution $x = a\cos\theta$. The radical expression would be equal to $\sqrt{a^2-x^2} = a\sin\theta$ and the differential $dx = a\cos\theta\,d\theta$.

Example 41

Use an appropriate trigonometric substitution to find $\displaystyle\int \frac{1}{\sqrt{3x^2-48}}\,dx$

Answer

$\displaystyle\int \frac{1}{\sqrt{3x^2-48}}\,dx = \frac{1}{\sqrt{3}}\int \frac{1}{\sqrt{x^2-16}}\,dx$

Identify the substitution $x = 4\sec\theta$.

Let $\sqrt{x^2-16} = \sqrt{16\sec^2\theta-16}$

Express x in terms of θ.

$\qquad\qquad = 4\sqrt{\sec^2\theta-1} = 4\tan\theta$

Use the formula $\sec^2\theta - 1 = \tan^2\theta$.

$dx = 4\cdot\left(-\dfrac{\sin\theta}{\cos^2\theta}\right)d\theta$

Find dx in terms of the variable θ.

$\displaystyle = \frac{1}{\sqrt{3}}\int \frac{1}{\sqrt{x^2-16}}\,dx = \frac{1}{\sqrt{3}}\int \frac{1}{4\tan\theta}\cdot\left(-4\frac{\sin\theta}{\cos^2\theta}\right)d\theta$

Now proceed to solving the original integral.

$\displaystyle = -\frac{1}{\sqrt{3}}\int \sec\theta\,d\theta$

$\displaystyle = -\frac{1}{\sqrt{3}}\left(\ln\left(\frac{\sin x+1}{\cos\theta}\right)\right)+c$

Use the substitutions to obtain the integral in terms of the variable θ.

$x = 4\sec\theta \Rightarrow \cos\theta = \left(\dfrac{4}{x}\right) \Rightarrow \theta = \arccos\left(\dfrac{4}{x}\right)$

Now substitute θ in terms of x.

$\displaystyle = -\frac{1}{\sqrt{3}}\left(\ln\left(\frac{\sin\left(\arccos\dfrac{4}{x}\right)+1}{\dfrac{4}{x}}\right)\right)+c$

Return to the variable x.

$\displaystyle = -\frac{1}{\sqrt{3}}\left(\ln\left(\frac{\sqrt{1-\left(\cos\left(\arccos\dfrac{4}{x}\right)\right)^2}+1}{\dfrac{4}{x}}\right)\right)+c$

Express sine in terms of cosine and apply the inverse function.

$\displaystyle = -\frac{1}{\sqrt{3}}\left(\ln\left(\frac{\sqrt{1-\left(\dfrac{4}{x}\right)^2}+1}{\dfrac{4}{x}}\right)\right)+c = -\frac{1}{\sqrt{3}}\left(\ln\left(\frac{\dfrac{\sqrt{x^2-16}}{x}+1}{\dfrac{4}{x}}\right)\right)+c$

Simplify the radical expressions.

Simplify the fraction.

$\displaystyle = -\frac{1}{\sqrt{3}}\cdot\ln\left(\frac{\sqrt{x^2-16}+x}{4}\right)+c$

$\displaystyle = -\frac{1}{\sqrt{3}}\cdot\left(\ln\left(\sqrt{x^2-16}+x\right)-\ln 4\right)+c$

$\displaystyle = -\frac{1}{\sqrt{3}}\cdot\ln\left(\sqrt{x^2-16}+x\right)\underbrace{-\frac{\ln 4}{\sqrt{3}}+c}_{k\in\mathbb{R}}$

Simplify the constant.

$\displaystyle = -\frac{1}{\sqrt{3}}\cdot\ln\left(\sqrt{x^2-16}+x\right)+k$

Example 42

Use an appropriate trigonometric substitution to find $\displaystyle\int \frac{1}{\sqrt{x^2+2}}\,dx$

Answer

Notice that 2 is not a perfect square, but you can write it as

$$x^2 + 2 = x^2 + \left(\sqrt{2}\right)^2$$

Let $\sqrt{x^2+2} = \sqrt{2\tan^2\theta + 2}$

$$= \sqrt{2}\cdot\sqrt{\tan^2\theta + 1} = \sqrt{2}\sec\theta$$

Identify the substitution $x = \sqrt{2}\tan\theta$.

Use the formula $\tan^2\theta + 1 = \sec^2\theta$.

$x = \sqrt{2}\tan\theta \Rightarrow dx = \sqrt{2}\sec^2\theta\,d\theta$

$$\int \frac{1}{\sqrt{x^2+2}}\,dx = \int \frac{1}{\sqrt{2}\sec\theta}\cdot\sqrt{2}\sec^2\theta\,d\theta$$

$$= \int \sec\theta\,d\theta$$

Find dx in terms of the variable θ. Now solve the original integral.
Use the substitutions to obtain the integral in terms of the variable θ and simplify it.

$x = \sqrt{2}\tan\theta \Rightarrow \theta = \arctan\left(\dfrac{x}{\sqrt{2}}\right)$

$\sec\theta = \sqrt{1+\tan^2\theta}$

In order to return the variable x express θ in terms of x.

Also express sec in terms of tan.

$$\int \frac{1}{\sqrt{x^2+2}}\,dx = \ln\left(\frac{x}{\sqrt{2}} + \sqrt{1+\left(\frac{x}{\sqrt{2}}\right)^2}\right) + c$$

Now substitute θ in terms of x.

$$= \ln\left(\frac{x}{\sqrt{2}} + \sqrt{\frac{2+x^2}{2}}\right) + c$$

$$= \ln\left(\frac{x+\sqrt{2+x^2}}{\sqrt{2}}\right) + c$$

$$= \ln\left(x+\sqrt{2+x^2}\right) \underbrace{-\ln\sqrt{2} + c}_{k}$$

$$= \ln\left(x+\sqrt{2+x^2}\right) + k$$

Exercise 9U

Use an appropriate trigonometric substitution to find these integrals:

1 $\displaystyle\int \sqrt{4-x^2}\,dx$

2 $\displaystyle\int \frac{1}{\sqrt{x^2-1}}\,dx$

3 $\displaystyle\int \sqrt{x^2+9}\,dx$

4 $\displaystyle\int \frac{3}{\sqrt{36-x^2}}\,dx$

5 $\displaystyle\int 3\sqrt{x^2-16}\,dx$

6 $\displaystyle\int \frac{5}{\sqrt{x^2+121}}\,dx$

7 $\displaystyle\int \frac{2}{\sqrt{81-4x^2}}\,dx$

8 $\displaystyle\int \sqrt{3x^2-75}\,dx$

9 $\displaystyle\int \frac{7}{\sqrt{7x^2+28}}\,dx$

9.7 Applications and modeling

You can now revisit some of the applications of calculus studied in the previous sections of this chapter.

Minima and maxima problems

Example 43

A rectangle ABCD is inscribed under a curve $y = \cos 2x$, $-\dfrac{\pi}{4} \le x \le \dfrac{\pi}{4}$.

Two vertices, A and B on the x-axis and two vertices, C and D on the curve, are shown in the diagram.

a Given that the coordinates of vertex C are $(x, \cos 2x)$, $x > 0$ find the area of the rectangle in terms of x.

b Find the value of x for which the area is maximum. What is the maximum area?

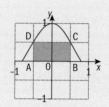

Answers

a $A(x) = \underbrace{2x}_{\text{length}} \cdot \underbrace{\cos 2x}_{\text{width}}$

b **Method 1**

$A'(x) = 2\cos 2x + 2x(-\sin 2x \cdot 2)$
$\quad\;\;\, = 2\cos 2x - 4x \sin 2x$

Find the derivative and simplify the expression.

$A'(x) = 0 \Rightarrow 2\cos 2x - 4x \sin 2x = 0$

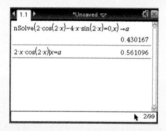

Find the x-coordinate of the maximum point.
Make A′ (x) = 0 for maximum (and minimum) points.
Since you cannot solve this equation algebraically, use a GDC to find the values of x and the corresponding A(x).

So the area will reach a maximum for $x = 0.430$ and its maximum value will be $A_{max} = 0.561$, both given correct to 3 sf.

Method 2

Since there are no demands on showing workings in this case, simply use the graphical method on a GDC, apply it to the area function and find the maximum.

Notice that the coordinates give both the answers.
$x = 0.430$ *and* $A_{max} = 0.561$

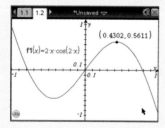

Example 44

Find the minimum distance between the point A(0,2) and the curve $y = \sin 2x$.

Answer

Method 1

Let P(x, $\sin 2x$) be any point on the curve.

$$AP = \sqrt{(x-0)^2 + (\sin 2x - 2)^2}$$

Use the distance formula to find the length of AP.

$$AP^2 = x^2 + \sin^2 2x - 4\sin 2x + 4$$

Square the equation to obtain an expression for a simpler calculation.
Differentiate it with respect to x.

$$\frac{d}{dx}(AP^2) = 2x + 4\sin 2x \cos 2x - 8\cos 2x$$

$$2x + 2\sin 2x \cos 2x - 8\cos 2x = 0$$

Since you cannot solve this equation algebraically use a GDC to find the value of x and the corresponding distance.

Find the x-coordinate of the minimum point.

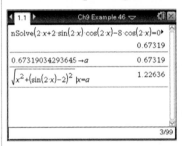

So the minimum distance between the point A and the curve $y = \sin 2x$ is 1.22, correct to 3 sf. The distance is obtained at the point P(0.632, 0.953) where the coordinates are also given correct to 3 sf.

Notice that squaring the distance had no effect on the minimum x-value.

Method 2

Since there are no demands on showing working in this case, use the graphical method on a GDC, apply it to the distance function and find the minimum.

Notice that in this method you can immediately read out the answer $AP_{min} = 1.22$.
To find the actual point, read out the x-coordinate but you still need to find the y-coordinate.

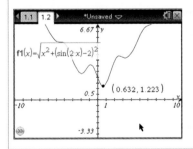

GDC help on CD:
Alternative demonstrations for the TI-84 Plus and Casio fx-9860GII GDCs are on the CD.

Exercise 9V

In this exercise use a GDC where appropriate.

1 A rectangle is inscribed under a curve $y = \sin x$, $0 \le x \le \pi$ in such a way that one side is on the x-axis. Find the dimensions of such a rectangle that has a maximum area.

2 Find the minimum distance between the point A(1, −1) and the curve $y = \cos x$.

3 A circle has a radius of 10 cm. A tangent is drawn through a point A on the circumference. Chord BC is parallel to the tangent.

   a Show that the area of the triangle ABC is given by the formula $A(\alpha) = 100(1 + \cos\alpha)\sin\alpha$.

   b Find the value of α for which the area is maximum.

4 A roller coaster track will have a gap of 15 metres between the first two posts. The part of the track is straight and going downhill. The formula to calculate the time needed for a car to pass that part of the track is $t = 2\sqrt{\dfrac{15}{g \cdot \sin 2\theta}}$, where g is the Earth's acceleration and θ is the angle of inclination to the horizontal position.

   a What is the value of the angle θ such that the speed of the car would be maximum?

   b What is the length of the track is that case? Give your answer correct to the nearest millimetre.

5 An object's displacement, d metres, from a fixed point F is given by the formula $d(t) = \sin\left(\dfrac{\pi t}{6}\right) + \cos\left(\dfrac{\pi t}{6}\right)$, $t > 0$ at time t seconds.

   a Show that the acceleration of the object is proportional to the displacement.

   b Find the maximum speed of the object and the first time when is achieved.

6 A metal chain is hanging between two poles that are 3 metres apart. The height (in metres) of the chain is given by the formula

$$h(x) = e^{-\frac{x}{2}} + e^{\frac{x}{5}}, \, 0 \le x \le 3$$

where x is the distance to the first pole. What is the minimum height of the chain and which pole is closer to the point of minimum height? Justify your answer.

Areas and volumes of revolution

Example 45

Shade the region between the curve $y = \sin x$, $0 \leq x \leq 2\pi$ and the x-axis and find its area.

Answer

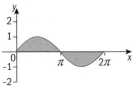

$$A = \int_0^{2\pi} |\sin x| \; dx$$

$$= 2 \cdot \int_0^{\pi} \sin x \, dx = 2 \cdot [-\cos x]_0^{\pi}$$

$$= 2 \cdot (-\cos \pi - (-\cos 0))$$

$$= 2(1 + 1) = 4$$

Shade the region required, and notice that the region above the x-axis is symmetrical to the region below the x-axis.

So the required area is twice the area of the curve between 0 and π.
Evaluate.

Example 46

Find the area of the region bounded by the curve $y = \tan\left(\dfrac{x}{2}\right)$, the line $y = 1$ and the y-axis.

Answer

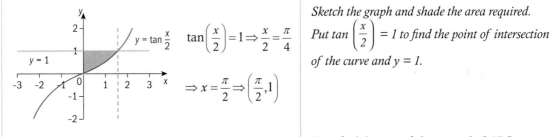

$$\tan\left(\frac{x}{2}\right) = 1 \Rightarrow \frac{x}{2} = \frac{\pi}{4}$$

$$\Rightarrow x = \frac{\pi}{2} \Rightarrow \left(\frac{\pi}{2}, 1\right)$$

$$\text{Area} = \overbrace{\frac{\pi}{2} \cdot 1}^{\text{Rectangle}} - \overbrace{\int_0^{\frac{\pi}{2}} \tan\left(\frac{x}{2}\right) dx}^{\text{Area under curve}}$$

$$= \frac{\pi}{2} - \left[2 \cdot \ln\left|\sec\frac{x}{2}\right|\right]_0^{\frac{\pi}{2}}$$

Sketch the graph and shade the area required. Put $\tan\left(\dfrac{x}{2}\right) = 1$ to find the point of intersection of the curve and y = 1.

First find the area of the rectangle OABC. Subtract the area under the curve from this.

▶ Continued on next page

$$= \frac{\pi}{2} - 2 \cdot \left(\ln \left| \sec \frac{\pi}{4} \right| - \ln \left| \sec 0 \right| \right)$$

Calculate trigonometric values.

$$= \frac{\pi}{2} - 2 \cdot \left(\ln \sqrt{2} - \ln 1 \right)$$

$$= \frac{\pi}{2} - 2 \cdot \left(\frac{1}{2} \ln 2 \right) = \frac{\pi}{2} - \ln 2$$

Simplify logarithmic expressions.

Note that you could alternatively integrate with respect to y. Try it and you should get the same answer.

Example 47

Shade the finite regions enclosed by the curves $y = \cos(2x)$ and $y = \ln(x - 1)$. Find the area of the shaded region.

Answer

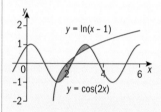

To find points of intersections between these two curves, solve the equation $\cos(2x) = \ln(x - 1)$. It cannot be solved algebraically, so use a GDC this time.

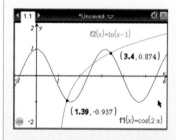

Store the x-coordinates of the points of intersections in the variables a and b.

You don't need to find the middle point of intersection since you are using the absolute value of the difference of two functions.

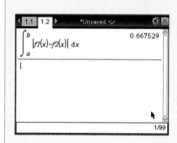

Exercise 9W

1 Shade the region enclosed by each of these curves and find its area.

a $y = \cos x$, $y = 0$, $x = -\pi$, $x = \dfrac{\pi}{3}$

b $y = \sec^2 x$, $y = 0$, $x = -\dfrac{\pi}{4}$, $x = \dfrac{\pi}{4}$

c $y = 2\sin 2x$, $y = 0$, $x = -\dfrac{\pi}{3}$, $x = \dfrac{\pi}{6}$

d $y = \dfrac{1}{3}\cos 3x$, $y = 0$, $x = \dfrac{5\pi}{18}$, $x = \dfrac{2\pi}{3}$

e $y = 4\tan\dfrac{x}{2}$, $y = 0$, $x = -\dfrac{\pi}{2}$, $x = -\dfrac{\pi}{3}$

2 Given that x is positive, shade the region enclosed by each of these curves and find its area.

a $y = \cos x, x = 0, y = \dfrac{1}{2}$

b $y = \tan 2x, x = 0, y = \sqrt{3}$

c $y = 3\sin\dfrac{x}{2}, x = 0, y = -\dfrac{3\sqrt{3}}{2}$

d $y = 2\cos\left(2x - \dfrac{\pi}{6}\right), x = 0, y = \sqrt{2}$

e $y = \tan\dfrac{x}{3}, x = 0, y = \dfrac{1}{\sqrt{3}}$

3 Shade the first region enclosed by the curves $y = \cos\dfrac{x}{2}$ and $y = \cos 2x, x > 0$. Find the area of the shaded region.

4 Shade the finite region enclosed by the curve $y = \tan x$, $0 \le x \le \dfrac{\pi}{2}$, its tangent at a point where $x = \dfrac{\pi}{4}$ and the x-axis. Find the area of the shaded region.

5 Shade the regions enclosed by the curves $y = 2\sin x$, $x > 0$ and $y = e^{\frac{x}{2}-4} + 1$. Find the area of the shaded region.

6 Consider the curves $y = \dfrac{8}{4 + x^2}$ and $y = \dfrac{x^2}{4}$.

a Find the points of intersection.
b Write down the integral that represents the area of the region enclosed by the curves.
c Calculate the area of the region.

The volume V of a solid formed by a curve $y = f(x)$, between $x = a$ and $x = b$ rotated through 2π radians about the x-axis is given by

$$V = \pi \int_a^b y^2 \, dx$$

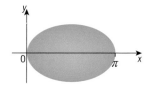

You met this formula in Section 7.3

Example 48

Find the volume of a solid obtained by rotating the curve $y = \sqrt{\sin x}$, $0 \le x \le \pi$ through 2π radians about the x-axis.

Answer

$V = \pi \int_0^\pi \left(\sqrt{\sin x}\right)^2 dx$	Use the formula $V = \pi \int_a^b y^2 \, dx$
$= \pi \int_0^\pi \sin x \, dx = \pi \left[-\cos x\right]_0^\pi$	Simplify.
$= \pi(-\cos \pi - \cos 0) = 2\pi \text{ units}^3$	Evaluate.

The volume of a solid formed by rotating a curve, $y = f(x)$, through 2π radians about the y-axis is given by $V = \pi \int_a^b x^2 \, dy$.

To use this formula directly you must first find $x = f^{-1}(y)$.

Example 49

Find the volume of a solid that is obtained by rotating the curve $y = \arccos x$, $0 \le x \le 1$ through 2π radians about the y-axis.

Answer

$y = \arccos x$, $0 \le x \le 1$	Express x in terms of y and
$\Rightarrow x = \cos y$, $0 \le y \le \dfrac{\pi}{2}$	find the domain of y values.
$V = \pi \int_0^{\frac{\pi}{2}} (\cos y)^2 \, dy$	Use the volume formula.
$= \pi \int_0^\pi \dfrac{1 - \cos 2y}{2} \, dy$	Use the half angle formula.
$= \pi \left[\dfrac{y}{2} - \dfrac{\sin 2y}{4}\right]_0^\pi$	
$= \pi \left(\dfrac{\pi}{2} - \dfrac{\sin 2\pi}{4}\right) - \left(0 - \dfrac{\sin 0}{4}\right) = \dfrac{\pi^2}{2} \text{ units}^3$	Evaluate.

You can subtract volumes of two different curves.

Example 50

Find the volume of a solid that is obtained by rotating the finite region enclosed by the curves $y = \ln x + 1$ and $y = \tan\dfrac{x}{2}$ through 2π radians about the x-axis.

- -

Answer

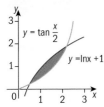

Sketch the graph to identify the finite region and the points of intersection.

The equation $\ln x = \tan\dfrac{x}{2}$ cannot be solved algebraically so use a GDC.

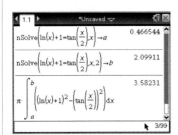

Solve the equation $\ln x = \tan\dfrac{x}{2}$ and store the solutions as the variables a and b. Apply the formula for the volume of rotating region between two curves.

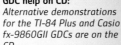 **GDC help on CD:**
Alternative demonstrations
for the TI-84 Plus and Casio
fx-9860GII GDCs are on the
CD.

Vol. = 3.58 units³

Exercise 9X

1 Find the volume of a solid generated by rotating the region bounded by the given curves through 2π radians about the x-axis.

   a $y = \sqrt{\cos x}$, $x = 0$, $x = \dfrac{\pi}{2}$, $y = 0$

   b $y = \sec x$, $x = 0$, $x = \dfrac{\pi}{4}$, $y = 0$

   c $y = \cos x$, $x = \dfrac{\pi}{6}$, $x = \dfrac{5\pi}{6}$, $y = 0$

   d $y = \sin x$, $x = \dfrac{\pi}{3}$, $x = \dfrac{2\pi}{3}$, $y = 0$

2 Find the volume of a solid generated by rotating the region bounded by the given curves through 2π radians about the y-axis.

   a $y = \arcsin x$, $x = 0$, $y = 1$

   b $y = \arcsin x$, $x = 0$, $x = 1$, $y = 0$

   c $y = \tan x$, $x = 0$, $x = \dfrac{\pi}{4}$, $y = 0$

   d $y = \tan x$, $x = 0$, $y = 1$

3 Find the volume of a solid generated by rotating the region enclosed by the curves

a $y = \sin x$, $y = \cos x$, $\dfrac{\pi}{4} \le x \le \dfrac{5\pi}{4}$ through 2π radians about the x-axis.

b $y = \sin 2x$, $y = \sin x$, $0 \le x \le \pi$ through 2π radians about the x-axis.

c $y = e^{\frac{x}{3}} - 1$ and $y = \arctan x$ through 2π radians about the x-axis.

d $y = e^{\frac{x}{3}} - 1$ and $y = \arctan x$ through 2π radians about the y-axis.

Extension material on CD:
Worksheet 9

Investigation – volume of a torus

Find the volume of a torus that is obtained by rotating a circle with the centre at (h, k) and a radius r, $k, r > 0$ and $k > r$, about the x-axis.

> Look back at the chapter introduction on page 435.

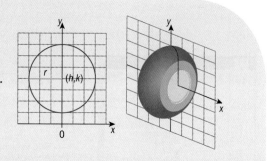

Exercise 9Y

1 Use volume of revolution to find the volume of a torus obtained by rotating the circle $(x - 4)^2 + (y + 3)^2 = 4$ through 2π radians about the x-axis.

2 Use volume of revolution to find the volume of a torus obtained by rotating the circle $(x - 4)^2 + (y + 3)^2 = 4$ through 2π radians about the y-axis.

3 Use volume of revolution to find the volume of a sphere obtained by rotating the circle $x^2 + y^2 = 9$ through 2π radians about the x-axis.

4 Find the volume of a solid obtained by rotating the ellipse $4x^2 + 9y^2 = 36$ through 2π radians about the x-axis.

5 Find the volume of a solid obtained by rotating the ellipse $4x^2 + 9y^2 = 36$ through 2π radians about the y-axis.

6 Find the volume of a solid obtained by rotating the ellipse $(5 - x)^2 + 9y^2 = 36$ through 2π radians about the x-axis.

Review exercise

1 Differentiate with respect to x:

  a $f(x) = (2x + 3) \sin x$

  b $g(x) = e^x \cos 3x$

  c $h(x) = \dfrac{\tan x}{2x^2}$

2 Find the equation of a tangent to the curve $\sin y + e^{2x} = 1$ at the origin.

3 Find the value of m that satisfies this equation

$$\int_{\frac{\pi}{4}}^{m} \sec^2 x \, dx = 2\left(\cos\frac{\pi}{6} - \sin\frac{\pi}{6}\right).$$

4 Use the method of integration by parts to solve:

  a $\displaystyle\int (2x - 5) e^{2x} \, dx;$

  b $\displaystyle\int (x^2 - 5x) \cos x \, dx;$

  c $\displaystyle\int e^x \cos 3x \, dx.$

5 The diagonal of a square is increasing at a rate of 0.2 cm s^{-1}. Find the rate of change of the area of the square when the side has a length of 5 cm.

6 The curve $y = e^{2x-1}$ is given.

  a Find the equation of the tangent to the curve that passes through the origin.

  b Find the area, in terms of e, of the region bounded by the curve, the tangent and the y-axis.

  c Find the volume of the revolution, in terms of π, obtained by rotating the region in part **b** about the x-axis.

7 Use the substitution $x = 3\cos\theta$ to find $\displaystyle\int \sqrt{9 - x^2}\, dx$.

8 The region bounded by the curve $y = \ln(2x)$, the vertical line $x = 1$ and the x-axis is rotated through 2π radians about the y-axis.

  a Sketch the region in the coordinate system.

  b Find the exact value of the volume of revolution obtained by this rotation.

9 The velocity, v, of an object, at a time t, is given by $v = 5e^{-\frac{2t}{3}}$, where t is in seconds and v is in m s^{-1}.

  a Find the distance traveled in the first k seconds, $k > 0$.

  b What is the total distance travelled by the object?

10 Find the equation of the normal to the curve $x^2y^3 - \cos(\pi x)$ at the point $(1, -1)$.

Review exercise

1 Find the points of inflexion of the curve $y = x^2 \sin 2x$, $-1 \le x \le 1$.

2 Given the curve $y^3 = \cos x$, find the equation of the tangent at the point where $x = 1$.

3 Find the value of a, $0 < a < 1$, such that $\displaystyle\int_{a^2}^{0} \frac{1}{\sqrt{1-x^2}}\, dx = 0.2709$

4 An airplane is flying at a constant speed at a constant altitude of 10 km in a straight line directly over an observer. At a given moment the observer notes that the angle of elevation θ to the plane is 54° and is increasing at 1° per second. Find the speed, in kilometres per hour, at which the airplane is moving towards the observer.

5 The region in the first quadrant bounded by the curves $y = \cos x$ and $y = e^x - 1$ is rotated by the x-axis by 2π radians. Find the volume of revolution of the solid generated.

CHAPTER 9 SUMMARY

Derivatives of trigonometric functions

$\displaystyle\lim_{h \to 0} \frac{\sin h}{h} = 1$

$\dfrac{d}{dx}(\sin x) = \cos x$

$\dfrac{d}{dx}(\cos x) = -\sin x$

Derivatives of inverse trigonometric functions

If $y = \arcsin x$ then $\dfrac{dy}{dx} = \dfrac{1}{\sqrt{1-x^2}}$

If $y = \arcsin \dfrac{x}{a}$ then $\dfrac{dy}{dx} = \dfrac{1}{\sqrt{a^2 - x^2}}$

Basic integrals of trigonometric functions

$\displaystyle\int \cos x\, dx = \sin x + c, \, c \in \mathbb{R}$ since $\dfrac{d(\sin x)}{dx} = \cos x$

$\displaystyle\int \sin x\, dx = -\cos x + c$ since $\dfrac{d(-\cos x)}{dx} = \sin x$

$\displaystyle\int \sec^2 x\, dx = \tan x + c$ since $\dfrac{d(\tan x)}{dx} = \sec^2 x$

$\displaystyle\int f(ax + b)\, dx = \frac{1}{a}F(ax + b) + c$

Continued on next page

Definite integrals

$$\int f(x)\, dx = F(x) + c \Rightarrow \int_a^b f(x) dx = F(b) - F(a)$$

Integration by parts

$$\int u\, \frac{dv}{dx}\, dx = uv - \int v\, \frac{du}{dx}\, dx$$

Trigonometric substitutions

If an integral contains a quadratic radical expression use one of the following substitutions.

If the form is $\sqrt{a^2 - x^2}$ use the substitution $x = a\sin\theta$.

If the form is $\sqrt{x^2 - a^2}$ use the substitution $x = a\sec\theta$.

If the form is $\sqrt{x^2 + a^2}$ use the substitution $x = a\tan\theta$.

The nature of mathematics

Mathematics in nature

French–American mathematician Benoît Mandelbrot (1924–2010) discovered the self-symmetry of complex patterns, sometimes described as 'worlds within worlds', and coined the term 'fractal' to describe a shape where when you enlarge a complex pattern, the same complex pattern emerges. Mandelbrot discovered how to generate this complexity using simple mathematical rules based on fractional dimensions.

> You researched fractals in the theory of knowledge section at the end of chapter 7.

> The word 'fractal' comes from the Latin 'frangere' which means to break or to fragment.

A line has one dimension and a square has two dimensions. Fractals are shapes with dimensions between 1 and 2. How can we explain this?

If you take a line and cover it with a sheet of transparent squared paper you can count how many 'squares long' the line is, e.g. n squares long. If we repeat this but use squared paper with squares half the size of the first squared paper, the line is now $2n$ squares long, that is increased by a factor of 2.

Now repeat the process for a square. Consider a square with sides the length of the original line above. If we place the original sheet of squared paper on this square we would observe that the square is covered by n^2 squares. Now if we place the second sheet of squared paper we would find that the number of squares has increased to $2n \times 2n = 4n^2$, that is, by a factor of 2^2.

If we repeat the same process on a complex line – say part of the coastline of Australia, we find that the number of squares increases by a factor of $2^{1.13}$.

For part of the coastline of England the multiplier is $2^{1.26}$. The fractal dimension of the Australian coastline is therefore 1.13 and that of England is 1.26. This shows that the coastline of Australia is 'smoother' than the coastline of England. The coastline of Norway has a fractal dimension of 1.52.

> The fractional dimension of a shape tells us how much space this complex shape occupies.
>
> ■ Fractal geometry is used to model complex natural structures.
>   Can we use mathematics to describe all the complexities of the Universe?
>
> ■ Does this mean that there is a finite amount of mathematics to be discovered/invented?
>
> ■ Mathematics models the real world. Do we create mathematics to explain natural processes or is the world intrinsically mathematical, meaning just have to find the mathematics?

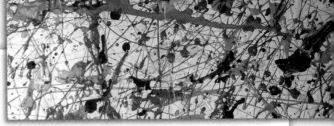

Mathematics in art

The American artist Jackson Pollock produced abstract paintings by splashing paint across canvas. In 1999 a group of mathematicians analysed Pollock's work and found that the technique he used created fractal shapes. When magnified, some sections of Pollock's work look very similar to the original full-size version. The image shown here was inspired by Pollock's *Blue Poles*.

■ Is mathematics art? Can art be created mathematically? What are the ethical issues around this?

> *"The process of creation isn't the same in Art as it is in Nature"*
>
> Wassily Kandinsky,
> Russian artist (1866–1944)

Mathematical paradoxes – unnatural mathematics

Evangelista Toricelli (1606–47) studied under Galileo and is renowned for his work as a physicist. He was amazed to discover that mathematically you can have a cylinder of infinite area but with finite volume.

Consider a rectangle of length L^2 cm and width $\dfrac{2\pi}{L}$ cm rolled into a cylinder.

Curved surface area = area of rectangle = $\dfrac{2\pi}{L} \times L^2 = 2\pi L$.

Circumference of cylinder = $\dfrac{2\pi}{L}$, so radius is $\dfrac{1}{L}$

Volume of cylinder is $\pi r^2 h = \pi \dfrac{1}{L^2} L^2 = \pi$.

As $L \to \infty$ we get a cylinder with infinite surface area but with finite volume (π).

We could fill it up with less than $4\,cm^3$ of paint but we would never have enough paint to paint the curved surface!

Toricelli lived before calculus was discovered, so he came across this paradox intuitively. When calculus was developed this result was confirmed using integration to find the volume and area formed when rotating the curve $f(x) = \dfrac{1}{x}$ by 2π about the x-axis for values of x between 1 and ∞. The shape formed is called Toricelli's Trumpet or Gabriel's Horn.

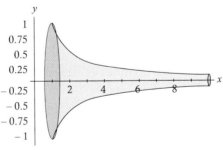

■ Does mathematics always reflect reality?

■ Can mathematics be right and wrong at the same time?

■ Do mathematical explanations hinder our understanding of the real world?

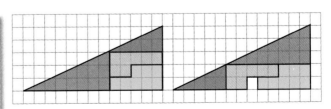

▲ The pieces from the first triangle are rearranged to make the second triangle. Why is there a gap in the second one?

10 Modeling randomness

Before you start

You should know how to:

1 Use the basic probability laws.
$P(A \cup B) = P(A) + P(B) - P(A \cap B)$
$P(A') = 1 - P(A)$
e.g. If $P(A) = 0.7$, $P(B) = 0.2$ and
$P(A \cap B) = 0.3$,
$P(A') = 1 - 0.7 = 0.3$
and $P(A \cup B) = 0.7 + 0.2 - 0.3 = 0.6$

2 Calculate probabilities of dependent and independent events.

$$P(A|B) = \frac{P(A \cap B)}{P(B)}.$$

When A and B are independent,
$P(A \cap B) = P(A) \times P(B)$.
e.g. If A and B are independent
and $P(A) = 0.4$ and $P(B) = 0.1$, find
$P(A \cap B)$, $P(A \cup B)$ and $P(A|B)$.
$P(A \cap B) = 0.4 \times 0.1 = 0.04$
$P(A \cup B) = 0.4 + 0.1 - 0.04 = 0.46$
$P(A|B) = \frac{0.04}{0.1} = 0.4$

Skills check

1 Given two events A and B such that
$P(A) = 0.2$ and $P(B) = 0.3$, find:
a $P(A')$
b $P(A \cup B)$ if $P(A \cap B) = 0.1$

2 Given two events A and B such that
$P(A) = 0.2$ and $P(B) = 0.3$, find:
a $P(A \cap B)$ and $P(A \cup B)$ if
A and B are independent
b $P(A|B)$ if $P(A \cap B) = 0.1$

> Note that when A and B are independent events, $P(A|B) = P(A)$.

Random variables and distributions

In a production factory, the weight of each unit of a product must be within a given range of a target weight (usually the weight stated on the packaging). This is monitored using a control chart. The quality control manager uses a probability model to calculate acceptable limits for the weight of each unit. If these limits are exceeded too often, this indicates inefficiency in the production process.

Control charts

Manufacturing processes use control charts to ensure that standards are kept within acceptable limits. A control chart is a graphical display showing data obtained from the process being monitored. A horizontal center line shows the desired statistic, a line shows the **upper control limit** (UCL) and a line shows the **lower control limit** (LCL).

The LCL and UCL are determined from historical data. By comparing current data to these limits, conclusions can be drawn about whether any variation in the process is consistent (in control) or is unpredictable (out of control, affected by special causes of variation).

Walter Shewhart
(1891–1967), American physicist, engineer and statistician, invented the control chart (a type of graph) while working for Bell telephone laboratories in the 1920s. This helped to impove the reliability of their telephony transmission systems.

→ The outcome of many random phenomena can be measured or represented by numerical values. To describe the behavior of these outcomes, use a random variable with its associated probability distribution.

10.1 Discrete random variables and distributions

In many statistical experiments the data has **discrete** values. For example

- if you roll a dice, the possible scores are 1, 2, 3, 4, 5 and 6
- the possible number of times, n, that you roll the dice until you get a 6 forms, in theory, an infinite discrete set: $\mathbb{Z}^+$.

You can model both these situations using a **discrete random variable**. Its values are the possible outcomes of the experiment and its behavior is described by a probability distribution function.

> A discrete variable can only have certain values. For example, the number of rainy days in November can only be an integer between 0 and 30.

→ The **probability distribution function** of a discrete random variable assigns a probability to *each* value x of the variable X

$$f(x) = P(X = x).$$

> Use upper case letters to represent random variables and lower case to represent their particular values. $P(X = x)$ is sometimes written p_x.

→ A probability distribution function has these properties:

- $0 \le f(x) \le 1$

- $\sum_{x \in A} f(x) = 1$ where A is the set of all possible values of X.

Example 1

X is the random variable 'the number of heads obtained when you toss two coins simultaneously'.

a What are the possible values of X?

b Define the probability distribution of X.

- -

Answers

a X can take the values 0, 1 and 2.

List the possible numbers of heads.

b The possible outcomes are HH, HT, TH and TT.

List the 4 possible outcomes.

$P(X = 0) = P(\text{TT}) = \dfrac{1}{4} = 0.25$

$P(X = 1) = P(\text{HT}) + P(\text{TH})$

$\qquad = \dfrac{2}{4} = 0.5$

$P(X = 2) = P(\text{HH}) = \dfrac{1}{4} = 0.25$

You could also put the information in a table

x	0	1	2
$P(X = x)$	0.25	0.5	0.25

> This means 'give the probability for each possible value of X'.

You can represent a probability distribution function of a discrete random variable with a bar chart where the height of each bar represents the probability.

The probability distribution function is sometimes called the probability mass function.

Example 2

A discrete random variable Y has a probability distribution function given by $f(y) = ky^2$, where $y = 0, 1, 2, 3$ and 4.
Find the value of k and hence draw a bar chart to represent f.

Answer

$\sum_{y=0}^{4} f(y) = 1$ so $k \times 0 + k \times 1 + k \times 4$

$\qquad + k \times 9 + k \times 16 = 1$

$30k = 1$

$k = \dfrac{1}{30}$

So, the values of $f(y)$ are

y	0	1	2	3	4
$f(y)$	0	$\dfrac{1}{30}$	$\dfrac{4}{30}$	$\dfrac{9}{30}$	$\dfrac{16}{30}$

The total of the probabilities is 1

Substitute the values of y in the expression $f(y) = ky^2$

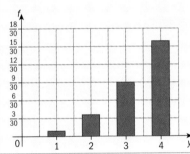

The heights of the bars represent the probability of each value of x.

Exercise 10A

1   State with a reason whether or not each of these functions can be a probability distribution of a random variable.

Check that both properties are true for each probability distribution function:
• Each value of x has a probability between 0 and 1
• The sum of all the probabilities equals 1.

a

x	1	2	3	4
$f(x)$	0.2	0.3	0.4	0.2

b

x	−1	−2	−3	−4
$f(x)$	0.2	0.3	0.4	0.2

c

x	−1	2	3	4
$f(x)$	−0.2	0.5	0.4	0.3

2 A discrete random variable X has this probability distribution

x	1	2	3	4	5
$f(x)$	0.21	0.25	0.41	a	0.01

Find
a the value of a **b** $P(1 \le X \le 3)$ **c** $P(X \le 3)$

3 The discrete random variable T has probability distribution function given by
$P(T = t) = f(t) = k(4 - t)$ for $t = 0, 1, 2$ and 3.
a Find the value of the constant k.
b Hence calculate $P(1 \le T < 3)$.

4 There are eight red socks and six blue socks in a drawer. A sock is taken out, its color noted and it is put back in the drawer. This procedure is performed three times.
Let R be the random variable 'number of red socks taken'.
Determine the probability distribution of R and represent it in a table and on a bar chart.

5 Do question 4 again, this time assuming that each sock is not put back in the drawer after it has been taken out.

Parameters of a discrete random variable

Probability distribution functions are described by characteristics called **parameters**. These parameters have the same names as the statistical measures you studied in Chapter 6: the mean, mode, median, variance and standard deviation.

> → For a random variable X the **mean** μ, also called the expected value $E(X)$, is given by $\mu = \Sigma\, x\, P(X = x)$
>
> The **variance** σ^2 is given by $\sigma^2 = \Sigma (x - \mu)^2\, P(X = x)$ or
> $\sigma^2 = \text{Var}(X) = E(X^2) - (E(X))^2 = \Sigma x^2 P(X = x) - \mu^2$

$\sigma = \sqrt{\text{Var}(X)}$ is called the standard deviation of X.

> → The **mode** of a probability distribution function is the value of x for which the probability distribution function has a maximum.

If there is no maximum value, then the PDF has no mode. A PDF with more then one maximum is multi-modal.

An expected value $\mu = E(X)$ represents a cluster point for the entire distribution.
$\mu = E(X)$ is an **average value** and therefore it may not be one of the values of the variable.
A more formal treatment of expectation algebra is part of the option *Statistics and Probability*.

'PDF' is shorthand for 'probability distribution function'.

Example 3

Let X be the random variable 'the number shown when a dice is rolled'. Find the values of the mean, the variance and the standard deviation of X.

Answer

$P(X = x) = \dfrac{1}{6}$ for $x = 1, 2, 3, 4, 5$ and 6

$E(X) = \displaystyle\sum_{x=1}^{6} \dfrac{x}{6} = \dfrac{21}{6} = \dfrac{7}{2} = 3.5$

Use $E(X) = \sum x\,P(X = x)$

$E(X^2) = \displaystyle\sum_{x=1}^{6} \dfrac{x^2}{6} = \dfrac{1 + 4 + \cdots + 36}{6} = \dfrac{91}{6}$

Use $E(X^2) = \sum x^2\,P(X = x)$

$Var(X) = \dfrac{91}{6} - \left(\dfrac{7}{2}\right)^2 = \dfrac{35}{12}$

Use $Var(X) = E(X^2) - (E(X))^2$

$\sigma = \sqrt{\dfrac{35}{12}} = 1.71\,(3\,\text{sf})$

Use $\sigma = \sqrt{Var(X)}$

Example 3 is typical of a situation where the expected value is not one of the possible values of the variable.

The next example shows a simple application of probability models: how to decide whether or not a game is fair using the expected value of the variable 'amount gained'.

The game is considered fair if the expected value of the winnings is zero.

Example 4

Tom tosses three fair coins. He wins \$9 if he gets three heads, \$5 if he gets two heads, and \$2 if he gets one head. If he doesn't get any heads he pays \$30. Is the game is fair?

What do we mean by 'a fair game'?

Answer

There are 8 equally likely possible outcomes: HHH, HHT, HTH, HTT, THH, THT, TTH, TTT

List all the outcomes.

Let A be the amount gained or lost.

$P(A = 9) = P(\text{HHH}) = \dfrac{1}{8}$

$P(A = 5) = P(\text{HHT, HTH, THH}) = \dfrac{3}{8}$

$P(A = 2) = P(\text{HTT, THT, TTH}) = \dfrac{3}{8}$

$P(A = -30) = P(\text{TTT}) = \dfrac{1}{8}$

The random variable A can take four values: 9, 5, 2 and −30.

This is an example of a PDF with two modes: 2 and 5.

Therefore,

$E(A) = 9 \times \dfrac{1}{8} + 5 \times \dfrac{3}{8} + 2 \times \dfrac{3}{8} - 30 \times \dfrac{1}{8} = 0$

$E(A) = \sum x\,P(A = x)$

So this game is fair.

Exercise 10B

1 A discrete random variable R has the probability distribution given in this table.

r	1	5	10
$P(R = r)$	$\dfrac{1}{5}$	$\dfrac{2}{5}$	$\dfrac{2}{5}$

Find the value of

a $E(R)$ **b** $E(R^2)$ **c** $Var(R)$ **d** standard deviation of R

2 A discrete random variable X can take only the values 0, 1, 2, 3, 4 and 5.

The probability distribution of X is given in this table.

x	0	1	2	3	4	5
$P(X = x)$	a	a	a	b	b	b

Given that $P(X \geq 2) = 3P(X < 2)$

a find the values of a and b

b calculate $E(X)$ and $E(X^2)$

c find the value of $Var(X)$.

3 A pack of 10 cards numbered from 1 to 6 is shuffled.

a Find the probability that the number on the bottom card is larger than the number on the top card. Justify your answer.

Let S be the random variable 'sum of the numbers on the top and bottom cards'.

b Find $P(S = 4)$.

c Construct a table for the probability distribution of S.

d Find $E(S)$ and $Var(S)$.

4 Find the fair price to pay to enter a game in which you can win £20 with probability 0.2, win £10 with probability 0.4 or lose the amount you paid.

EXAM-STYLE QUESTION

5 A random variable T takes only integer values and has probability distribution function defined by:

$$f(t) = P(T = t) = \begin{cases} kt^2 & \text{if } t = 1, 2, 3 \\ k(8-t)^2 & \text{if } t = 4, 5, 6, 7 \\ 0 & \text{otherwise} \end{cases}$$

where k is a constant.

a Find the value of k.

b Calculate $P(T = 4)$, $P(T \leq 4)$ and $P(T = 4 \mid T \leq 4)$.

c Find $E(T)$ and $Var(T)$.

d Determine the mode of T.

> There are three equations here for $f(t)$. Choose the equation that corresponds to each value of t to work out $\Sigma f(t)$, then find k.

6 Use the properties of addition and multiplication to show that

$$\sum (x-\mu)^2 \, P(X=x) = \sum x^2 P(X=x) - \mu^2$$

This shows that
$\mathrm{Var}(X) = E(X)^2 - (E(X))^2$

Cumulative distribution function

> → The **cumulative distribution function (CDF)**, F, of a discrete random variable X is defined by:
> $$F(x) = P(X \le x) = \sum_{t \le x} P(X = t)$$

The **CDF**, sometimes just called the **distribution function**, models the cumulative relative frequencies of the outcomes of an experiment.

$F(x)$ gives the probability that the variable X takes values that do not exceed x. To find its value we need to add together the values of the probabilities $P(X=t)$ for all values of t less than or equal to x.

The **median** of X is defined as $m = \dfrac{x_1 + x_2}{2}$ where

- x_1 is the maximum value for which $F(x_1) \le 0.5$ and

- x_2 is the minimum value for which $F(x_2) \ge 0.5$

Example 5

Construct the cumulative distribution table for the number of heads when four coins are tossed.
Hence find the median of the distribution.

The number of heads is a discrete random variable.

Answer

Let X be the random variable 'number of heads obtained when four coins are tossed'.

X can take the values 0, 1, 2, 3 and 4. There are 16 possible outcomes, HHHH, HHHT, and so on.

$$P(X=0) = \frac{1}{16}$$

P(X = 0) = P(TTTT)

$$P(X=1) = \frac{4}{16}$$

4 of the outcomes have just one head.

$$P(X=2) = \frac{6}{16}$$

6 outcomes have exactly two heads.

$$P(X=3) = \frac{4}{16}$$

4 outcomes have exactly three heads.

$$P(X=4) = \frac{1}{16}$$

P(X = 4) = P(HHHH)

The CDF of X is

x	0	1	2	3	4
$F(x)$	$\dfrac{1}{16}$	$\dfrac{5}{16}$	$\dfrac{11}{16}$	$\dfrac{15}{16}$	1

Median of $X = \dfrac{1+2}{2} = 1.5$

F(1) < 0.5 and F(2) > 0.5

Example 6

Construct the cumulative distribution table for the larger number obtained when two dice are thrown and find the median of the distribution.

Answer

Let X be the random variable 'the larger number obtained when two dice are thrown'.

$P(X = 1) = \dfrac{1}{36}$	$P(X = 4) = \dfrac{7}{36}$
$P(X = 2) = \dfrac{3}{36}$	$P(X = 5) = \dfrac{9}{36}$
$P(X = 3) = \dfrac{5}{36}$	$P(X = 6) = \dfrac{11}{36}$

x	1	2	3	4	5	6
$F(x)$	$\dfrac{1}{36}$	$\dfrac{4}{36}$	$\dfrac{9}{36}$	$\dfrac{16}{36}$	$\dfrac{25}{36}$	1

See p 507 for more details on this.

Median of $X = \dfrac{4+5}{2} = 4.5$

X can take the values 1, 2, 3, 4, 5 and 6.

P(X = 6) is the probability of rolling either a double 6, or 6 on the first dice and a number between 1 and 5 on the second or 6 on the second dice and a number between 1 and 5 on the first.

Add together the values of the probabilities P(X = t) for all values of t less than or equal to x to obtain the CDF table.

F(4) < 0.5 and F(5) > 0.5

Exercise 10C

1 A discrete random variable X has the probability distribution shown in the table.

x	5	10	15	20	25	30
$P(X = x)$	$\dfrac{1}{15}$	$\dfrac{2}{15}$	$\dfrac{3}{15}$	$\dfrac{4}{15}$	$\dfrac{3}{15}$	$\dfrac{2}{15}$

a Find $P(X \le 15)$.

b Construct the CDF table for this distribution.

c Hence find the median of the distribution.

2 An emergency call centre has five service lines that operate 24 hours every day.

A random variable L denotes the number of lines in use during any specific 5-minute period.

Data collected over a long period of time shows that L has this PDF:

n	0	1	2	3	4	5
$P(L = n)$	0.07	0.21	0.25	0.31	0.12	0.04

a What is the probability that at least three lines are in use simultaneously?

b Find $E(L)$ and $Var(L)$.

c Construct the CDF table for this distribution.

d Hence find the value of the median of the distribution.

3 A fair dice has faces numbered 1, 2, 2, 3, 3 and 3. The dice is thrown twice. S is the random variable 'sum of numbers on the uppermost face of the dice'

a Construct tables for the probability distribution f and cumulative distribution F of S.

b Find the mean, median and mode of S.

c Calculate the standard deviation of S.

4 The probability distribution of a discrete random variable X is given by

$f(x) = kx$ where $x = 1, 2, \ldots, n$ and k is a parameter.

a Show that $k = \dfrac{2}{n^2 + n}$ **b** Hence find $E(X)$

5 The probability distribution of a discrete random variable X is given by $f(x) = 3^{a-x}$ where $x = 1, 2, 3, \ldots$ and a is a parameter.

a Show that $a = \log_3 2$

b Hence find an expression for the cumulative distribution of S.

10.2 Binomial distribution

In a TV game show the final round consists of a Wheel of Fortune game. Each contestant that reaches this round has the chance to double the amount won in previous rounds. The Wheel of Fortune is a roulette wheel with 10 slots, one of which is gold. The contestant spins the wheel and wins if the ball lands in the gold slot. The producer of the show wants to know roughly how much will be paid in prize money so he needs a probability model that allows him to estimate the number of times the contestants will win in the last stage during the 50 planned sessions of the contest.

This Wheel of Fortune game is a typical example of a real-world situation that you can model mathematically.

→ The characteristic features of this type of problem are:

- There are a fixed number of trials n.
- The trials are independent and are done under exactly the same conditions.
- Each trial has exactly two possible outcomes: **success** and **failure**.
- For each trial the probability of success p is constant.
- The probability of failure is denoted by $q = 1 - p$ which is also constant.

Jacob Bernoulli (1654–1705), a member of the famous Swiss family of mathematicians, was the first person to study problems like this extensively.

Probability models make interesting topics for the Extented Essay.

Experiments with these characteristics are called **Bernoulli** or **binomial** experiments. The **binomial distribution** is a theoretical probability distribution that models situations that meet the Bernoulli experiment specifications listed above.

Example 7

For each of these situations decide whether or not they are Bernoulli experiments. In the case of a Bernoulli experiment, indicate the values of n, p and q.

a A fair coin is tossed ten times and the number of heads obtained is noted.

b A bag contains ten blue balls and eight red balls. Five balls are removed from the bag and are not replaced. The number of blue balls is noted.

c A bag contains ten blue balls and eight red balls. A ball is removed from the bag and its color is noted. The ball is then replaced. The experiment is repeated five times and the total number of blue balls is recorded.

Answers

a This is a Bernoulli experiment.
$n = 10$ and $p = q = \dfrac{1}{2}$

10 independent trials, two outcomes, probability of success (H) is constant.

b This is not a Bernoulli experiment.

Probability of getting a blue ball changes each time.

c This is a Bernoulli experiment
$n = 5$,
$p = \dfrac{10}{18} = \dfrac{5}{9}$ and
$q = \dfrac{8}{18} = \dfrac{4}{9}$

5 independent trials, two outcomes, probability of success (blue ball) is constant because the ball is replaced.

> In general, when you make selections from a population **without** replacement, the trials are not independent and the experiment is not a Bernoulli experiment. However, when the size of population is much larger than the number of trials the change in p is negligible. This can be approximated to a binomial experiment as the change in probability of success is very small.

Investigation – the Galton Board

This mechanical device was invented by the British mathematician **Sir Francis Galton** (1822–1911), one of the pioneers of statistical theory. The original model was built in 1873 and is kept at University College, London.

You can to use a simplified version to discover the probability function of a binomial random variable X that represents the number of successes x out of n trials, each of them with probability of success p.

The Galton Board has an array of pins mounted on a vertical board. Balls are dropped on to the pin at the top of the array. When a ball hits a pin, it bounces to the left or right with equal probability, and then falls down one level, where it hits one of the two closest pins. At the bottom of the board it falls into one of the bins below the bottom row.

▶ Continued on next page

a The diagram shows a Galton board with four levels. The circles represent the pins. Suppose that a ball is dropped onto the device. Write inside each circle the probability that the ball hits that pin, as a fraction.

Look at the coefficients and the variables in the expressions.

b Suppose that the pins are manipulated so that the probability that a ball bounces to the left is double the probability that it bounces to the right. Recalculate the probability that the ball hits each pin. Copy the diagram and write the new probabilities on it.

c Consider a general case when the probability of bouncing left is p and of bouncing right is $q = 1 - p$. Find an expression for the probability of the ball hitting each pin in the diagram. Observe the expressions written in each level of the diagram and describe any patterns observed.

d Now consider a Galton board with n levels and a random variable X that counts the number of balls in the rth bin counted from the left to the right. Find an expression for $P(X = r)$ in terms of n, r, p and q.

From this investigation, you have discovered the formula for the binomial distribution which is:

$$\rightarrow P(X = r) = \binom{n}{r} p^r q^{n-r} \text{ where } r = 0, 1, \ldots, n$$

You can write $P(X = r)$ or P_r

The binomial probability distribution is built into most GDCs but, for small values of n, you can calculate binomial probabilities using the formula for combinations of r elements selected out of n

$$\binom{n}{r} = \frac{n!}{r!(n-r)!}$$

You can also use Pascal's Triangle, as $\binom{n}{r}$ is simply the $(r + 1)$th entry of its nth row.

For more on Pascal's triangle and combinations, see Chapter 1.

The rows of Pascal's Triangle are numbered such that row $n = 0$ is at the top. The entries are numbered from the left, begining with $r = 0$

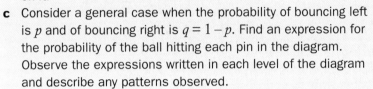

$$\rightarrow P(X = r) = \binom{n}{r} p^r q^{n-r} \text{ is the probability distribution of the}$$

random variable X, where X is the number of successful outcomes in n Bernoulli trials. In this case, X follows a binomial distribution with **parameters** n and p, and we write

$$X \sim B(n, p)$$

The third parameter is q, where $q = 1 - p$

The parameters are the values that stay fixed through the experiment.

Example 8

In an inspection scheme, a sample of ten items is selected at random from a large batch and the number of defective items is noted. If the number of defective items is less than two, the batch is accepted. Data collected over the year shows that the proportion of defective items in a batch is 2%. Write down an expression for the probability that a batch is accepted.

<table>
<tr><td>

Answer

Let X be the random variable 'number of defective items in the sample'. The batch is accepted when $X < 2$ which means $X = 0$ or $X = 1$.

$X \sim B(10, 0.02)$

$P(X < 2) = \binom{10}{0}(0.02)^0 (1-0.02)^{10}$

$\qquad + \binom{10}{1}(0.02)^1 (1-0.02)^9$

</td><td>

10 trials, p = 0.02

$P(X < 2) = P(X = 0) + P(X = 1)$

</td></tr>
</table>

> You have to decide what counts as a success when you define the variable. This might not be the usual concept of success. For example, you might decide that finding a defective piece is a success, as in Example 8.

In Example 8, you assumed that X followed a binomial distribution. This is reasonable as the batch is large and therefore, although the items are not replaced, the change in the probability of selecting a defective item remains very small, giving the accuracy needed for calculation purposes. The trials are independent as the items are selected randomly.

The binomial distribution is an important discrete distribution as you can use it to model many real-life situations. For this reason your GDC has built-in binomial PDF and binomial CDF functions that allow you to calculate binomial probabilities quickly, as shown in this example:

Example 9

X is the random variable 'number of sixes obtained when you roll a dice seven times'.

a Identify, with reasons, the distribution followed by X and state clearly its parameters.

b Calculate $P(X = 3)$ **c** Calculate $P(X \le 3)$

<table>
<tr><td>

Answers

a Each time the dice is rolled the probability of obtaining a 6 is constant and each outcome is independent of previous ones.

X follows a binomial distribution with parameters $n = 7$ and $p = \dfrac{1}{6}$

$X \sim B\left(7, \dfrac{1}{6}\right)$

</td><td>

Verify requirements of a binomial distribution.

Number of times the dice is rolled and probability of getting 6.

</td></tr>
</table>

▶ Continued on next page

b $P(X=3) = \dfrac{21875}{279\,936}$

$\qquad = 0.0781 \,(3\,\text{sf})$

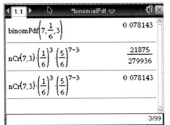

Use either the GDC binomial distribution function or

$$P(X=r) = \frac{n!}{r!(n-r)!}\,p^r q^{n-r}$$

$$= \frac{7!}{3!(7-3)!}\left(\frac{1}{6}\right)^3\left(\frac{5}{6}\right)^4$$

Look carefully at **b** and **c**. In **b** you use the GDC bulit in binomial PDF and in **c** you use binomial CDF. Why?

c $P(X \le 3) = \dfrac{34\,375}{349\,992}$

$\qquad = 0.982 \,(3\,\text{sf})$

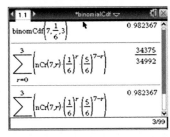

Use GDC cumulative distribution function or

$$P(X \le 3) = \sum_{r=0}^{3} P(X=r)$$

GDC help on CD:
Alternative demonstrations for the TI-84 Plus and Casio fx-9860GII GDCs are on the CD.

Parameters of the binomial distribution

The random variable, X, in a binomial distribution is a discrete variable.

→ The **mode** of X is the x-value or values for which the probability distribution function has a maximum.

See Section 10.1 for definition of mode and median of a discrete random variable.

→ If F denotes the binomial CDF of X

The **median** of X is defined as $m = \dfrac{x_1 + x_2}{2}$ where x_1 is the maximum value for which $F(x_1) \le 0.5$ and x_2 is the minimum value for which $F(x_2) \ge 0.5$

Recall that a PDF may have more than one mode.

The next example shows you how to use the GDC built-in functions to calculate the mode and the median of a binomial random variable.

Example 10

Let $X \sim B(4, 0.25)$
Find **a** The mode of X **b** The median of X.

Answers

x	0	1	2	3	4
$P(X = x)$	0.316	0.421	0.210	0.046...	0.003
$P(X \leq x)$	0.316	0.738	0.949	0.996	1

a mode of $X = 1$ as this is the maximum PDF value.

b $P(X \leq 0) < 0.5$ and $P(X \leq 1) > 0.5$

median of $X = \dfrac{0+1}{2} = 0.5$

PDF is maximum when x = 1

GDC help on CD:
Alternative demonstrations for the TI-84 Plus and Casio fx-9860GII GDCs are on the CD.

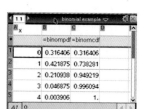

Exercise 10D

1 A fair coin is tossed ten times. Find the probability of getting:
  a exactly four heads   **b** at least six heads
  c not more than five heads.

2 A coin is biased so that the probability of obtaining head is 0.6. Find the probability of getting:
  a exactly 2 heads if the coin is tossed 5 times
  b at least 3 heads if the coin is tossed 7 times
  c more heads than tails if the coin is tossed 9 times.

EXAM-STYLE QUESTIONS

3 In the mass production of light bulbs the probability that one bulb is defective is 1%. Bulbs are selected at random and put in packs of eight.

> What assumption do you need to make to answer question 3?

  a If a pack is selected at random, what is the probability that it will contain:
    i at least one defective bulb
    ii not more than two defective bulbs?
  b Given that a pack selected at random contains at least one defective bulb, what is the probability that it contains exactly two defective bulbs?

4 Let $X \sim B(6, 0.35)$. Find
  a The mode a of X
  b The median b of X
  c $P(X < 2a \,|\, X > b)$

5 Let $X \sim B(n, 0.4)$
  a Construct three tables for the binomial CDF of X when $n = 2$, 5 and 10
  b Given that $P(X \leq 10) > 0.5$, find the largest possible value of n.

6 If $X \sim B(n, 0.3)$ and $P(X > 3) > 0.7$, find the least possible value of n.

7 The probability that it rains in Raincity on any day of the year is 0.45. Calculate the probability that in any given week:

a it rains every day   **b** it rains on at least two days

c it rains on exactly three consecutive days.

8 The probability that Joerg hits a target is 0.6. Find the fewest number of attempts Joerg needs to make to ensure that the probability of hitting the target at least once is more than 95%.

→ • The **mean** μ of a discrete random variable X (including binomial) is given by

$$\mu = \mathrm{E}(X) = \sum x\, \mathrm{P}(X = x)$$

• The **variance** of X is given by

$$\sigma^2 = \mathrm{Var}(X) = \mathrm{E}(X^2) - (\mathrm{E}(X))^2 = \sum x^2 \mathrm{P}(X = x) - \mu^2$$

• σ is the **standard deviation** of X

The next investigation discovers formulae for the parameters μ and σ^2 of a binomial variable $X \sim \mathrm{B}(n, p)$ in terms of n and p.

Investigation – parameters of a binomial variable

Suppose that a coin is tossed n times. The probability of getting a tail each time the coin is tossed is constant and is denoted by p.

For each value of n and p, consider the random variable X which represents the number of tails obtained when the coin is tossed n times. Assuming that $X \sim \mathrm{B}(n, p)$, use your GDC or a spreadsheet to complete the table:

n	p	$q = 1 - p$	$\mathrm{E}(X)$ $= \sum_{x=0}^{n} x p^x q^{n-x}$	$\mathrm{E}(X^2)$ $= \sum_{x=0}^{n} x^2 p^x q^{n-x}$	$\mathrm{Var}(X)$ $= \mathrm{E}(X^2) - (\mathrm{E}(X))^2$
1	0.5				
5	0.5				
10	0.5				
1	0.2				
5	0.2				
10	0.2				
1	0.8				
5	0.8				
10	0.8				

Observe the table and conjecture a formula for $\mathrm{E}(X)$ and $\mathrm{Var}(X)$ in terms of the parameters n, p and q. Test your conjecture for other values of the parameters.

→ If $X \sim B(n, p)$ then $E(X) = \mu = np$ and $Var(X) = \sigma^2 = npq$
where $q = 1 - p$

Example 11

An apple is picked from a large number of apples in a container.
The probability that the apple is bad is 0.05
A sample of 15 apples is selected at random.
Find the expected number of bad apples.

Answer	
Let X be the random variable 'number of bad apples in the sample'. Assume that $X \sim B(15, 0.05)$	*Sample is selected randomly and the probability of selecting a bad apple is constant.*
$E(X) = 15 \times 0.05 = 0.75$	$E(X) = \mu = np$

Example 12

A coin is biased so that it is twice as likely to show tails as heads.
The coin is tossed five times and the number of tails is noted.
The random variable X represents the number of tails shown.
a Draw a table showing the probability distribution of X
b Hence find $E(X)$ and $Var(X)$.

Answers

a $X \sim B\left(5, \dfrac{2}{3}\right)$

The probability distribution function of X is

$$P(X = r) = \binom{5}{r}\left(\frac{2}{3}\right)^r\left(\frac{1}{3}\right)^{5-r}$$

where $r = 0, 1, 2, 3, 4$ and 5

x	$P(X = x)$
0	0.0041
1	0.0412
2	0.1646
3	0.3292
4	0.3292
5	0.1317

The trials are independent and the probability of obtaining a tail is constant. The probability of a tail is $\dfrac{2}{3}$ and of a head is $\dfrac{1}{3}$.

Find the table of values for the probability distribution using your GDC and the built-in binomial probability distribution function.

Some GDCs have built-in spreadsheets that are useful in exercises like this. If your GDC does not have a spreadsheet, you can also use lists to obtain the probability distribution.

► Continued on next page

b

x	$P(X = x)$	$x\,P(X = x)$	$x^2\,P(X = x)$
0	0.0041	0	0
1	0.0412	0.0412	0.0412
2	0.1646	03292	0.6284
3	03292	0.9877	2.9630
4	0.3292	1.3169	5.2675
5	0.1317	0.6584	3.2922
Total	1	3.3333	12.2222

$E(X) = 3.33$ (3 sf)
$E(X^2) = 12.22\ldots$
$Var(X) = 12.22\ldots -(3.33)^2$
$\quad\quad = 1.11$ (3 sf)

Use your GDC to obtain the table.

$E(X)$ is the total of the third column.
$E(X^2)$ is the total of the fourth column.
$Var(X) = E(X^2) - (E(X))^2$

As this question says 'hence' you must use the table to calculate the parameters of the distribution. Here the formulae for E(X) and Var(X) can only be used to check the answers.

You can check your answers using the formulae

$E(X) = np$ and $Var(X) = npq$

$E(X) = 5 \times \dfrac{2}{3} = \dfrac{10}{3} = 3.33$ (3 sf)

$Var(X) = 5 \times \dfrac{2}{3} \times \dfrac{1}{3} = \dfrac{10}{9} = 1.11$ (3 sf)

Exercise 10E

1 Given that $X \sim B(8, 0.4)$, find:
  a $P(X = 5)$ **b** $P(X \le 5)$ **c** $P(X < 5)$
  d the mean of X **e** the variance of X.

Only round at the final values. Early rounding can give you incorrect answers.

2 Given that $Y \sim B(7, 0.3)$, find:
  a $P(Y = 1) + P(Y = 2)$ **b** $P(Y \le 2)$
  c $P(Y \ge 2)$ **d** the median of Y.

3 Given that $T \sim B\left(5, \dfrac{1}{2}\right)$

  a Show that $P(T = 5) = \dfrac{1}{32}$.

  b Construct a table for the probability distribution function of T. Hence state the mode of T.

  c Construct a table for the distribution function F of T.

  d Write down the value of the median of T.

4 The probability that it rains on any given day in June of any given year in Drycity is 0.02.

  a What is the probability that it rains on exactly three days in June in a given year?

  b What is the probability that it does not rain on the first five days in June in a given year?

  c Find the expected number of rainy days in Drycity in June.

⋮ EXAM-STYLE QUESTION
⋮ **5** A random variable R follows a binomial distribution $B(n, p)$ with
⋮   mean 2 and variance 1.5. Find the values of n and p.

6 In a multiple choice test there are 20 questions. For each question there is a choice of four answers and only one of these is correct.
 a If a student guesses each answer find the probability that he gets
  i none correct **ii** more than ten correct
  iii not more than five correct.
 b Calculate the mean and standard deviation of the number of correct answers.
 c Suppose that five students guess the answers to the test. What is the probability that at least two of them get more than ten answers correct?

7 In a large city 18 % of the people are left-handed.
 If a random sample of ten people from this city is selected:
 a find the probability that exactly two of them will be left-handed
 b find the probability that at least one person in the sample is left-handed
 c find the most likely number of left-handed people in the sample.
  If another sample of 25 people from the city is selected,
 d find the expected number of left-handed people in the sample.
 e find the variance of the number of left-handed people in this sample.
  If a sample of size n is to be selected randomly,
 f Find the minimum value of n for which the probability that it contains at least two left-handed people will be greater than 95%.
 g In the same city the percentage of left-handed women is 16% and the percentage of left-handed men is 22%.
  A random sample of five women and five men is selected from the population of the city. What is the probability that the sample contains at least one left-handed woman and one left-handed man?

8 On a TV news channel, the evening news starts at the same time every day. The probability that Mr Li gets home from work in time to watch the news is 0.3
 a Calculate the probability that, in a particular week of five working days, he gets home in time to watch the news:
  i on exactly four days **ii** on at least three days.
 b What is the probability that Mr Li gets home in time to watch the news on three consecutive days?

9 Let $X \sim B(9, p)$
 a Draw bar graphs to represent the probability distribution of X when $p = 0.1, 0.3, 0.5, 0.7$ and 0.9.
 b Compare the graphs obtained in part **a** and comment on their symmetries.
 c For each value of p in part **a**, find the mean, median and mode of the distribution and comment on their values in relation to the symmetries (or asymmetries) of the graph.

10 Consider two independent random variables X and Y such that $X \sim B(2, a)$ and $Y \sim B(2, b)$.

Let W be the random variable that represents the product of each value of X with each value of Y.

a Construct a table showing the probability distribution of W.

b Hence find an expression for $E(W)$ in terms of a and b.

Extension material on CD:
Worksheet 10

10.3 Poisson distribution

The Poisson distribution is named after **Siméon-Denis Poisson** (1781–1840) who first discovered this model as an approximation to the binomial distribution when the number of trials n gets larger and larger while the probability of success p gets smaller and smaller.

→ A discrete random variable X follows a Poisson distribution when it models situations that satisfy these conditions:

- The occurrence of an event at a particular point in space or in time is **independent** of what happens elsewhere.
- The probability of an event occurring within a **small fixed interval** (or in a small region of space) is constant.
- There is no chance that two events will occur at precisely the same moment or at the same place.

Poisson was an extremely hard-working mathematician. His major work on probability was a book with over 400 pages where just one was dedicated to the derivation of the Poisson distribution.

In theory, the Poisson distribution has no upper value, and in this way it differs from the binomial distribution.

→ If an event is randomly scattered in time or space, the discrete random variable X that models the number of its occurrences in a given interval follows a **Poisson distribution** with parameter m, $X \sim \text{Po}(m)$

The probability distribution function (PDF) of X is

$$f(x) = P(X = x) = \frac{e^{-m} m^x}{x!} \text{ where } x = 0, 1, 2, \ldots.$$

The Poisson cumulative distribution function (CDF) is

$$F(x) = P(X \leq x) = \sum_{t=0}^{x} \frac{e^{-m} m^t}{t!}$$

Example 13

The random variable X follows a Poisson distribution such that
$X \sim \text{Po}(2)$
Calculate $P(X = x)$ for $x = 0, 1, 2, 3, 4, 5$ and draw a bar graph to
illustrate the distribution.

Answer

X	$P(X = x)$
0	0.1353
1	0.2707
2	0.2707
3	0.1804
4	0.09022
5	0.03609

Use GDC and

$$P(X = x) = \frac{e^{-2} \times 2^x}{x!}$$

*or use built-in Poisson PDF function
to calculate the probabilities.*

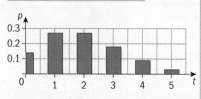

Alternatively, you can
use software like
Autograph to create
the required graph and
probability values.

Example 14

$X \sim \text{Po}(m)$ and $P(X = 0) = 0.2$
a Find the value of m.
b Calculate $P(X \leq 4)$.

Answers

a $P(X = 0) = 0.2 \Rightarrow \dfrac{e^{-m} m^0}{0!} = 0.2$

$e^{-m} = 0.2$

$m = -\ln(0.2)$ or $1.61 (3\,\text{sf})$

b $P(X \leq 4) = 0.976\ (3\,\text{sf})$

$P(X = x) = \dfrac{e^{-m} m^x}{x!}$

$m^0 = 1$ and $0! = 1$

*Apply logarithms to both sides
or use GDC.*

$P(X \leq 4)$
$= P(X = 0) + P(X = 1) + P(X = 2)$
$\quad + P(X = 3) + P(X = 4)$

GDC help on CD:
*Alternative demonstrations
for the TI-84 Plus and Casio
fx-9860GII GDCs are on the
CD.*

1.1 ▶	*Unsaved* ▽
nSolve(poissPdf(*m*,0)=0.2,*m*,0)	1.60944
m:=1.6094379124341	1.60944
poissCdf(*m*,0,4)	0.975794
sum(seq(poissPdf(*m*,x),x,0,4))	0.975794

4/99

Exercise 10F

1 $X \sim \text{Po}(m)$. Calculate $P(X = x)$ for $x = 0, 1, 2, 3, 4, 5$ and draw
bar graphs to illustrate the distributions for $m = 1, 3$ and 5.

2 $Y \sim \text{Po}(3)$. Find the probability that
  a $P(Y = 3)$ **b** $P(Y < 3)$
  c $P(Y > 3)$ **d** $P(Y = 4 \mid Y > 3)$

Examples of typical real-life situations that can be modeled by a Poisson distribution include

- the number of phone calls received during a given period of time
- the number of car accidents at a certain location during a given period of time
- the number of particles emitted by a radioactive source in a given time
- the number of typing errors on a randomly chosen page of a book
- the number of flowers in a randomly chosen area of a field
- the number of flaws in a given length of material.

Example 15

The number of accidents in a randomly chosen day at a road crossing can be modeled by a Poisson distribution with parameter 0.5.
Find the probability that, on a randomly chosen day, there are
a exactly two accidents
b at least two accidents.

> You should work with several significant figures and avoid early rounding as it may lead to incorrect answers.

Answers
If X represents the number of accidents on the given day then
$X \sim \text{Po}(0.5)$ and

$$P(X = x) = \frac{e^{-0.5} \times (0.5)^x}{x!}$$

Substitute 0.5 for m in

$$P(X = x) = \frac{e^{-m} m^x}{x!}$$

Make x = 2 or use your GDC

a $P(X = 2) = \dfrac{e^{-0.5} \times (0.5)^2}{2!}$

$= 0.0758\,(3\text{ sf})$

b $P(X \geq 2) = 1 - P(X \leq 1)$

$P(X \geq 2) = 0.0902$

$$P(X \leq 1) = \frac{e^{-0.5} \times (0.5)^0}{0!} + \frac{e^{-0.5} \times (0.5)^1}{1!}$$

or use the GDC built-in Poisson CDF.

If you have to calculate the parameter of a Poisson variable, the nature of the expression of a Poisson PDF might lead to an equation or inequality that you can solve numerically only with your GDC.

Example 16

> If $X \sim Po(\lambda)$ find the value of λ, correct to 4 decimal places, given that $P(X = 1) = 0.25$

GDC help on CD:
Alternative demonstrations for the TI-84 Plus and Casio fx-9860GII GDCs are on the CD.

Answer	
$P(X = 1) = 0.25$	$P(X = x) = \dfrac{e^{-\lambda}\lambda^x}{x!}$
$\dfrac{e^{-1}\lambda^1}{1!} = 0.25$	
$l = 0.3574$ (4 dp)	*Use GDC numerical solver.*

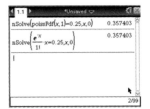

Exercise 10G

1 The number of accidents in a randomly chosen week at Safe School can be modeled by a Poisson distribution with parameter 0.7. Find the probability that, in a randomly chosen week, there are

a exactly two accidents

b at least two accidents.

2 The number of bacteria per millilitre of a certain liquid follows a Poisson distribution with parameter 3. Find the probability that in a millilitre of the liquid there will be

a at least 4 bacteria

b not more than 2 bacteria.

3 $X \sim Po\ (m)$. Find the value of m if $P(X = 1) = 0.1$

4 $X \sim Po\ (m)$. Find the value of m if $P(X \le 1) = 0.5$

Parameters of the Poisson distribution

The variable X in a Poisson distribution is a discrete variable, and therefore you can use the formulae

$$E(X) = \sum_{x=0}^{\infty} x\,P(X = x) \text{ and } Var(X) = E(X^2) - (E(X))^2$$

to calculate the mean and variance. The variable X can take an infinite number of values: 0, 1, 2, 3, 4,… so instead of a finite sum for $E(X)$ and $Var(X)$ you get infinite series. As the study of infinite series is beyond the scope of this course, you cannot directly deduce formulae for these parameters.

This example shows you how to estimate the values of the mean and variance of a Poisson variable using your GDC.

> The exact calculation of the expected value and variance of Poisson variables requires knowledge of techniques that are part of the *Calculus* option.

Example 17

$X \sim \text{Po}(m)$ and $P(X = 0) = 0.2$
Use a GDC to estimate $\text{E}(X)$ and $\text{Var}(X)$.

Answer

$\text{E}(X) = \sum_{x=0}^{\infty} x\, \text{P}(X = x)$

$\qquad = 1.60944 = 1.61 \,(3\,\text{sf})$

$\text{Var}(X) = \text{E}(X^2) - (\text{E}(X))^2$

$\qquad = 4.19973\ldots - (1.60944\ldots)^2$

$\qquad = 1.60944\ldots$

So $\text{E}(X) = \text{Var}(X) = 1.61\,(3\,\text{sf})$

Estimate the value of this series by adding a great number of its terms, e.g. 1000

GDC help on CD:
Alternative demonstrations for the TI-84 Plus and Casio fx-9860GII GDCs are on the CD.

Properties of Poisson distributions

Example 17 suggests an interesting relation between the parameter of the Poisson variable and its mean and variance, which is explored in this investigation.

Investigation – parameters of a Poisson distribution

1 Let $X \sim \text{Po}(m)$

2 Copy and complete the table.

m	$\displaystyle\sum_{x=0}^{100} x\,\text{P}(X = x)$	$\displaystyle\sum_{x=0}^{100} x^2\,\text{P}(X = x)$	$\displaystyle\sum_{x=0}^{100} x^2\,\text{P}\!\left(X = x - \left(\sum_{x=0}^{100} x\,\text{P}(X = x)\right)\right)$
0.1			
0.2			
0.3			
...			

3 Compare the results in the second and fourth columns and write down your conjecture.

4 Increase the number of terms of each series and test your conjecture.

5 Write down your conclusions.

Discussion point: how many terms do you need to consider to be sure of the result?

From the investigation you have discovered an important characteristic of the Poisson distribution.

> → If $X \sim \text{Po}(m)$ then $\text{E}(X) = \text{Var}(X) = m$

Another important property of the Poisson distribution is the relation between the mean and the length of the associated interval.

> → If $X \sim \text{Po}(m)$ and X is the number of successes in an interval of length l, the number of successes in an interval of length kl ($k > 0$) is modeled by another Poisson variable with parameter (or mean) km.

Bortkiewicz was born in Saint Petersburg where he graduated from the Law Faculty in 1890. In 1898 he published a book about the Poisson distribution, *The Law of Small Numbers*. In this book, which made the Prussian horse-kick data famous, he showed that events with low frequency in a large population follow a Poisson distribution even when the probabilities of the events vary. The data collected includes the number of soldiers killed by being kicked by a horse each year in each of 14 cavalry corps over a 20-year period. Bortkiewicz showed that those numbers follow a Poisson distribution

Example 18

In a book of 520 pages there are 135 misprints.
a What is the mean number of misprints per page?
b Find the probability that pages 444 and 445 do not contain any misprints.

Answers

GDC help on CD:
Alternative demonstrations for the TI-84 Plus and Casio fx-9860GII GDCs are on the CD.

a Let $X \sim \text{Po}(m)$

$$m = \frac{135}{520} = \frac{27}{104} = 0.2596...$$

b Let X be the number of misprints found on 2 pages

For two pages the mean is $2m$

$X \sim \text{Po}(2m)$

Use $\text{P}(X = 0)$

$$= \frac{e^{-2m} \times (2m)^0}{0!}$$

$\text{P}(X = 0) = 0.595\,(3\,\text{sf})$

or Poisson PDF on GDC.

Example 19

On Sunday mornings, cars arrive at a petrol station at an average rate of 30 per hour. Assuming that the number of cars arriving at the petrol station follows a Poisson distribution, find the probability that:

a in a half-hour period 12 cars arrive

b no cars arrive during a particular 5-minute interval

c more than 5 cars arrive in a 15-minute interval.

Answers

a In a half-hour period, the number of cars arriving at the petrol station can be modeled by a Poisson random variable $Y \sim Po(15)$
$P(Y = 12) = 0.0829$ (3 sf)

Mean is $\frac{1}{2} \times 30 = 15$

Use $P(Y = 12) = \dfrac{e^{-15} \times (15)^{12}}{12!}$

Use built-in functions on your GDC to calculate these probabilities.

b In a 5-minute interval, the number of cars arriving is modeled by $Y \sim Po(2.5)$
$P(Y = 0) = 0.0821$ (3 sf)

The mean is $5 \times \dfrac{30}{60} = 2.5$

Use $P(Y = 0) = \dfrac{e^{-2.5} \times (2.5)^{0}}{0!}$

GDC help on CD:
Alternative demonstrations for the TI-84 Plus and Casio fx-9860GII GDCs are on the CD.

c In a 15-minute interval, the number of cars arriving is modeled by $Y \sim Po(7.5)$
$P(Y > 5) = 1 - P(Y \le 4)$
$\qquad = 0.868$ (3 sf)

Use $P(Y \le 4)$
$= P(Y \le 0) + P(Y = 1) + P(Y = 2)$
$\quad + P(Y = 3) + P(Y = 4)$

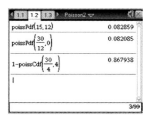

Exercise 10H

1 The mean number of bacteria per millilitre of a given liquid is 3.5 Find the probability that

  a in 2 ml of the liquid there will be fewer than 7 bacteria

  b in 0.5 ml of the liquid there will be at least 2 bacteria.

2 The mean number of flaws per square metre of fabric produced on a machine is 0.01. If flaws occur randomly and their number is modeled by a Poisson variable find the probability that

  a in a randomly chosen 100 square metres of fabric there will be exactly two flaws

  b in 25 square metres of chosen randomly fabric there will be at least one flaw.

3 The number of phone calls received by a school between 8:00 and 9:00 on any weekday is modeled by a Poisson distribution. If the mean number of phone calls per hour is 12, calculate

   a the expected number of phone calls received between 8:00 and 8:15 in a given day

   b the probability that more than five calls are received between 8:00 and 8:10 in a given day.

4 $X \sim \text{Po}(3.5)$

   a Calculate

   i $P(X = 3)$ **ii** $P(X > 3)$ **iii** $P(X < 5 \,|\, X > 3)$

   b Write down the values of $E(X)$ and $Var(X)$

   c Hence find the value of $E(X^2)$.

EXAM-STYLE QUESTION
5 The random variable X is Poisson distributed with mean m and satisfies

$$P(X = 0) + P(X = 1) - P(X = 4) = 0$$

   a Find the value of m correct to four decimal places

   b Hence calculate $P(2 \le X \le 4)$.

6 Let X be a random variable with a Poisson distribution, such that $P(X > 3) = 0.555$

Find $P(X < 3)$

EXAM-STYLE QUESTION
7 The random variable P has a Poisson distribution with mean $\lambda > 0$.

Let p be the probability that P takes the value 0, 1 or 2.

   a Write down an expression for p in terms of λ.

   b Show that $p = p(\lambda)$ is a decreasing function.

   c Sketch the graph of p for $0 < \lambda \le 6$, showing clearly concavities and any points of inflexion.

10.4 Continuous random variables

In the previous section you studied **discrete random variables** whose density functions are suitable models for situations where the outcomes have distinct values, usually integers. In many other situations, when we you need to model the behavior of variables like height, weight, mass and time, you use **continuous random variables**.

> → A variable X is **random** when its value is the result of a random experiment and it is **continuous** when it is not possible to list all of its values but only the range of values it can take.

The values of a continuous random variable form an uncountable set. What is the meaning of uncountable? Is the set of rational numbers uncountable? What is the difference between countable and uncountable?

Example 20

State whether each of these is a discrete or a continuous random variable

- the heights H of the students in an IB school
- the grades G of the students of a Math HL class
- the time T spent by IB students preparing for a particular exam
- the number of exercises N done by each student in an IB class
- the age A of the students in an IB class.

Answer

H and T are continuous random variables.	*H and T can take any value on an interval of real numbers.*
G is a discrete random variable.	*G can only take the values 1, 2, 3, 4, 5, 6 and 7*
N is a discrete random variable.	*N can take the values 0, 1, 2, 3, ..., n.*
A is a special type of continuous random variable.	

> Ages of people can be treated as discrete variables – for convenience their values are rounded down to an integer but in fact they can take any value in an interval of non-negative real numbers.

Similarities

Both discrete and continuous random variables

- are used to model the results of random experiments or phenomena
- have probability distributions that describe the behavior of the variable and allow us to make prediction
- have parameters.

Discrete	Continuous
• Discrete random variables always model situations whose outcomes have distinct values.	• Continuous random variables model situations whose outcomes are measured.
• Each possible value $d_1, d_2, ...$ of a discrete random variable D has an associated probability $p_1, p_2, ...$ where $0 < p_i \leq 1$	• For a continuous random variable C the probability that C has a particular value is always **zero.**
• The characteristics of the probability distributions of discrete and continuous random variables are different.	• The characteristics of the probability distributions of discrete and continuous random variables are different.

> A typical mistake is to apply properties of continuous distributions to discrete distributions. Learn the differences between them!

Probability density function of a continuous random variable

The behavior of a continuous random variable X is described by a function with domain $\mathbb{R}$ called the probability density function of X or simply the PDF of X.

> → If a function f is a **probability density function (PDF)** of a random variable X, it has these properties
>
> - $f(x) \geq 0$ for all values $x \in \mathbb{R}$
> - $\displaystyle\int_{-\infty}^{+\infty} f(x)\, dx = 1$

The definition above is a general definition. In most situations the PDF is only positive for values in an interval $[a, b]$ and takes the value zero elsewhere. For this reason, it is common to restrict the domain of PDF functions to the interval $[a, b]$. This is particularly convenient when you need to test the second condition. By considering the restriction, you only need to verify that

$$\int_a^b f(x)\, dx = 1$$

Example 21

Consider the function defined by $f(x) = \begin{cases} 3x^2 & \text{if } 0 \leq x \leq 1 \\ 0 & \text{elsewhere} \end{cases}$

Show that f is a well-defined probability density function.

You may want to convince yourself of the truth of these results. Although you can use your GDC to graph f and evaluate the integral, in 'show that' questions you will need to give the algebraic argument.

Answer

$f(x) \geq 0$ for all real values of x.

$$\int_0^1 f(x)\, dx = \int_0^1 3x^2\, dx$$

$$= \left[x^3 \right]_0^1 = 1 - 0 = 1$$

Therefore f is a well-defined probability density function.

The range of $y = 3x^2$ is $[0, +\infty[$
The function is positive only for values on the interval $[0, 1]$, so you only need to show that

$$\int_0^1 f(x)\, dx = 1$$

$$\int 3x^2\, dx = x^3 + C$$

Both conditions are satisfied.

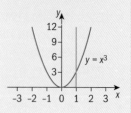

Parameters of a continuous random variable

Consider a continuous random variable X whose PDF function is defined on the interval $[a, b]$. The table shows you the formulae you can use to calculate the values of the parameters of X.

Parameter	Formula
Mean μ (expected value)	$\mu = E(X) = \int_a^b x f(x)\,dx$
Median	Value of m that is the solution of the equation $\int_n^m f(x)\,dx = \dfrac{1}{2}$
Mode	Value(s) of x for which f has a maximum
Variance σ^2	$\sigma^2 = Var(X) = \int_a^b (x - \mu)^2 f(x)\,dx$
Standard deviation σ	$\sigma = \sqrt{Var(X)}$

See p 525 for an alternative formula for variance.

Example 22

A continuous random variable X has a PDF given by

$$f(x) = \begin{cases} \dfrac{3}{4}x(2-x) & \text{if } 0 \le x \le 2 \\ 0 & \text{elsewhere} \end{cases}$$

Find the values of

a $E(X)$ **b** $Var(X)$

c median of X **d** mode of X

Differential and integral calculus is covered in Chapters 4, 7 and 9.

Answers

a $E(X) = \int_0^2 \dfrac{3}{4}x^2(2-x)\,dx$ $E(X) = \int_a^b x\,f(x)\,dx$

$\qquad = \dfrac{3}{4}\int_0^2 (2x^2 - x^3)\,dx$

$\qquad = \dfrac{3}{4}\left[\dfrac{2}{3}x^3 - \dfrac{1}{4}x^4\right]_0^2 = 1$

b $Var(X)$

$\qquad = \int_0^2 \underbrace{(x-1)^2}_{(x-\mu)^2}\underbrace{\left(\dfrac{3}{4}x(2-x)\right)}_{f(x)}dx$ $Var(X) = \int_a^b (x-\mu)^2 f(x)\,dx$

$\qquad = \dfrac{3}{4}\int_0^2 (-x^4 + 4x^3 - 5x^2 + 2x)\,dx$ *Expand and simplify*

$\qquad = \dfrac{3}{4}\left[-\dfrac{1}{5}x^5 + x^4 - \dfrac{5}{3}x^3 + x^2\right]_0^2 = 0.2$

▶ Continued on next page

Chapter 10   523

c $f(x) = \dfrac{3}{4}x(2-x)$ where $0 \le x \le 2$

$\dfrac{3}{4}\displaystyle\int_0^m (2x - x^2)\,dx = \dfrac{1}{2}$

$\dfrac{3}{4}\left[x^2 - \dfrac{1}{3}x^3\right]_0^m = \dfrac{1}{2}$

$\dfrac{3}{4}\left(m^2 - \dfrac{1}{3}m^3\right) = \dfrac{1}{2}$

$m = 1$

median of $X = 1$

$\displaystyle\int_a^m f(x)\,dx = \dfrac{1}{2}$

Use GDC solver

d $f(x) = \dfrac{3}{4}x(2-x)$

$f'(x) = \dfrac{3}{4}(2 - 2x)$

$f'(x) = 0$ when $x = 1$

Find the x-coordinate(s) of points where f has a maximum.

As $f'(x)$ is a decreasing function,
$f''(1) < 0$
So the mode is 1.

Example 22 is typical of a continuous distribution where the mean, mode and median take the same value. If you draw the graph of the PDF of X you can see why this happens.

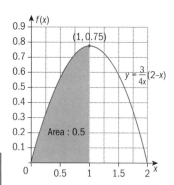

The graph is symmetrical about the vertical line $x = 1$ and has a maximum at $(1, 0.75)$ which is why the parameters, mean, mode and median, all have the same value of 1.

→ Helpful tips for obtaining parameters of continuous variables:

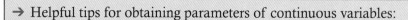

- Before using calculus to calculate the parameters of continuous random variables, it is worth checking for symmetries on the graph.
- In situations when a GDC is not available you should know the shapes of some typical functions like linear, quadratic, simple cubics, exponential and trigonometric functions. If you are familiar with the graph of the PDF of the variable, sketch it and look for symmetries or easy ways of calculating the area under the graph.
- If a GDC is available, you can quickly plot the graph of the PDF of X and use built-in functions that allow you to calculate the parameters numerically.

> • An alternative way of finding Var(X) is given by the formula
>
> $$\text{Var}(X) = E(X^2) - (E(X))^2 \text{ where } E(X^2) = \int_a^b x^2 f(x)\, dx$$

In Example 22, part **b**, you could have used

$$E(X^2) = \int_0^2 x^2 f(x)\, dx$$

$$= \frac{3}{4} \int_0^2 2x^3 - x^4\, dx$$

$$= \frac{3}{4} \int_0^2 \left[\frac{x^4}{2} - \frac{x^5}{5} \right]_0^2 = \frac{6}{5} = 1.2$$

From **a** E(X) = 1

$$\text{Var}(X) = 1.2 - 1 = 0.2$$

Exercise 10I

1 State whether each of these is a discrete or a continuous random variable.

    a The number of books sold per day in a bookshop.

    b The time Mrs Smith spends reading every day.

    c The age of each person who visits the local museum.

    d The amount of milk each customer buys in a supermarket.

2 Consider the function defined by $f(x) = \begin{cases} \dfrac{1}{2} x & \text{if } 0 \le x \le 2 \\ 0 & \text{elsewhere} \end{cases}$

    Show that f is a well-defined probability density function.

EXAM-STYLE QUESTION

3 A continuous random variable X has PDF given by $f(x) = 6x(1 - x)$ where $0 < x < 1$.

    Find the values of:

    a E(X)

    b Var(X)

    c median of X

    d mode of X.

Example 23

A continuous random variable X has PDF given by

$$f(x) = \frac{\sin(x)}{2} \text{ where } 0 \le x \le \pi$$

Without using your GDC, sketch the graph of f. Hence write down the values of the median, the mean and the mode of X.

Answer

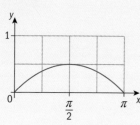

The graph is symmetrical about $x = \dfrac{\pi}{2}$

The median, mean and mode of this distribution are

all equal to $\dfrac{\pi}{2}$

Example 24

A continuous random variable X has PDF

$$f(x) = \frac{a}{x^2 - 4x} \text{ where } 1 \le x \le 3$$

and a is a constant.

a Use your GDC to find the value of a.

b Hence calculate $E(X)$ and $Var(X)$.

Answers

a

$$\int_1^3 \frac{a}{x^2 - 4x} \, dx = 1$$
$$a = -1.82 \text{ (3 sf)}$$

Use GDC numerical solver to find the value of a.

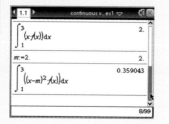

b $E(X) = 2$

$Var(X) = 0.359$ (3 sf)

GDC help on CD:
Alternative demonstrations for the TI-84 Plus and Casio fx-9860GII GDCs are on the CD.

For some GDC models the instructions and syntax may be different.

Exercise 10J

EXAM-STYLE QUESTIONS

1 A continuous random variable X has PDF given by

$f(x) = 2\cos(2x)$ where $0 \le x \le \dfrac{\pi}{4}$

> You can use your GDC to calculate or check your answers to these questions.

a Sketch the graph of f.

b Find the values of the median, mean and mode of X.

2 A continuous random variable X has PDF defined by

$f(x) = ax(1-x)^2$ where $0 \le x \le 1$ and a is a constant.

a Use your GDC to find the value of a.

b Hence, calculate $E(X)$ and $Var(X)$.

3 Consider the function defined by

$$f(x) = \begin{cases} k \text{ for } 0 < x < 2 \\ 2k \text{ for } 2 \le x \le 3 \\ 0 \text{ otherwise} \end{cases}$$

a Determine the value of k given that f is a PDF of a random variable X.

b Sketch the graph of f and use it to find the median of X.

c Find the values of $E(X)$ and $Var(X)$.

4 A continuous random variable X has PDF defined by

$f(x) = ax + b$ where $0 \le x \le 3$ where a and b are constants.

a Find, in terms of a, the value of b.

b Given that the median of X is 1, determine the values of a and b.

c Find the values of $E(X)$ and $Var(X)$.

5 A continuous random variable T has PDF defined by

$f(t) = \dfrac{6}{t^2}$ where $2 \le t \le 3$

Find the values of

a $E(T)$ b $Var(T)$

c median of T d mode of T.

Cumulative distribution function and calculation of probabilities

If X is a continuous random variable with probability density function f defined on the interval $[a, b]$, then the cumulative distribution function of X (CDF of X) is

$$F(X \le x) = \int_a^x f(t)\, dt \text{ where } a \le x \le b$$

One of the main differences between discrete and continuous random variables is the way probabilities are calculated.

For a discrete random variable $P(X = x) = f(x)$, but for a continuous random variable $P(X = x) = 0$ for every value in the domain of the variable X.

For a continuous variable you can only calculate the probability that the value of X lies *in* an interval, which can be quite small but is still a range of values.

To calculate the probabilities of a continuous variable X use the formula

$$P(r \le X \le s) = F(s) - F(r) = \int_r^s f(t)\, dt$$

As the probability of any individual value of X is zero,

$$P(r \le X \le s) = P(r < X \le s) = P(r \le X < s) = P(r < X < s)$$

The next two examples show you how to

- determine expressions for the CDF of a continuous random variable given the expression of the PDF
- use the CDF to calculate probabilities with and without a GDC.

Example 25

A continuous random variable X has probability density function

$$f(x) = \frac{k}{x^2 + x} \text{ for } 1 \le X \le 2$$

a Use a GDC to determine the value of k correct to 4 decimal places.
b Hence find the value of $P(1 \le X \le 1.75)$

Answers

a

$$\int_1^2 \frac{k}{x^2 + x}\, dx = 1$$

Rearrange equation to solve for k

$$k = \frac{1}{\displaystyle\int_1^2 \frac{1}{x^2 + x}\, dx}$$

$k = 3.4761$ (4 dp)

Use GDC numerical integration.

b

$P(1 \le X \le 1.75) = 0.838$ (3 sf)

Use GDC numerical integration

Example 26

A continuous random variable X has PDF defined by

$$f(x) = \frac{1 + \cos(x)}{\pi} \text{ where } 0 \leq x \leq \pi$$

a Find an expression for the distribution function F of X, (CDF of X).

b Determine the exact value of $P\left(\frac{\pi}{6} \leq X \leq \frac{\pi}{2}\right)$

See Chapter 9 for integration of trigonometric functions.

Answers

a $F(x) = \int_0^x \frac{1 + \cos(t)}{\pi} \, dt$

$\qquad F(X \leq x) = \int_0^x f(t) \, dt$

$\quad\quad = \frac{1}{\pi}\left[t + \sin(t)\right]_0^x$

$\quad F(x) = \frac{x + \sin(x)}{\pi}$

b

$P\left(\frac{\pi}{6} \leq X \leq \frac{\pi}{2}\right) = F\left(\frac{\pi}{2}\right) - F\left(\frac{\pi}{6}\right)$

$\qquad P(r \leq T \leq s) = F(s) - F(r)$

$= \dfrac{\dfrac{\pi}{2} + \sin\left(\dfrac{\pi}{2}\right)}{\pi} - \dfrac{\dfrac{\pi}{6} + \sin\left(\dfrac{\pi}{6}\right)}{\pi}$

$\qquad$ *Substitute values of* $\sin\dfrac{\pi}{2}$ *and* $\sin\dfrac{\pi}{6}$

$= \frac{1}{2} + \frac{1}{\pi} - \frac{1}{4} - \frac{1}{2\pi} = \frac{1}{4} - \frac{1}{2\pi}$

Example 27

Data collected over a long period of time suggests that the time T that vehicles have to wait until they can enter the main road of Straightcity has PDF

$$f(t) = \frac{5}{6}\left(1 - \frac{2}{5}t\right) \text{ where } 0 \leq t \leq 2 \text{ minutes}$$

a Find an expression for the distribution function of T (CDF of T).
b Calculate $P(0.5 \leq T \leq 1.5)$.

When a GDC is not allowed, you can make calculations easier by using fractions instead of decimals to represent values.

Answers

a $F(T \leq t) = \int_0^t \frac{5}{6}\left(1 - \frac{2}{5}x\right) dx$

$\qquad P(T < t) = \int_0^t f(x) \, dx$

$\quad = \frac{5}{6}\left[x - \frac{x^2}{5}\right]_0^t = \frac{5}{6}\left(t - \frac{t^2}{5}\right)$

b $P(0.5 \leq T \leq 1.5) = F(1.5) - F(0.5)$

$\qquad P(r \leq T \leq s) = F(s) - F(r)$

$\quad = \frac{5}{6}\left(1.5 - \frac{1.5^2}{5} - 0.5 + \frac{0.5^2}{5}\right) = 0.6$

Exercise 10K

1 Find an expression for the cumulative distribution of the random variable X with density function

   a $f(x) = \dfrac{x^2}{72}$ where $0 \le x \le 2$

   b $f(x) = \dfrac{1}{9}(x+1)^2$ where $0 \le x \le 2$

   c $f(x) = \cos(x)$ where $0 \le x \le \dfrac{\pi}{2}$

2 The probability density function of a continuous random variable X is

$$f(x) = \begin{cases} \dfrac{1}{4}x(4 - x^2), & 0 \le x \le 2 \\ 0, & \text{otherwise.} \end{cases}$$

   a Find an expression for the cumulative distribution of X

   b Calculate $P\left(1 \le X \le \dfrac{3}{2}\right)$ and $P(X > 1)$.

3 The probability density function of a continuous random variable X is
$f(x) = k|x^2 - x|$ for $0 \le x \le 2$ and where k is a constant.

   a Sketch the graph f

   b Find the value of k

   c Calculate $P(1 \le X \le 2)$.

> Usually when exact answers are required, you need to calculate the values by hand.

EXAM-STYLE QUESTIONS

4 The continuous random variable T has probability density function f where
      $f(t) = k(e - e^{kt})$, $0 \le t \le 1$, and k is a constant

   a Show that $k = 1$

   b What is the probability that the random variable T has a value that lies between $\dfrac{1}{3}$ and $\dfrac{2}{3}$?
      Give your answer in terms of e.

   c Find the mean and variance of the distribution.
      Give your answers *exactly*, in terms of e.
      The random variable T above represents the lifetime, in years, of a certain type of cell.

   d Find the probability that a cell lives more than six months.

5 The probability density function $f(x)$ of the continuous random variable X is defined on the interval $[0, a]$ by

$$f(x) = \begin{cases} \dfrac{x}{4} & \text{for } 0 \le x \le 2 \\ \dfrac{5}{x^2} & \text{for } 2 \le x \le a \end{cases}$$

   a Find the value of a.

   b Define the distribution function of X.

6 A continuous random variable Y has PDF

$$f(y)=\begin{cases}\lambda(y+3) & \text{for } -3\le y\le 3 \text{ where } \lambda \text{ is a constant.}\\ 0 & \text{otherwise}\end{cases}$$

a Show that $\lambda=\dfrac{1}{18}$

b Find an expression for the cumulative distribution F of Y.

c Hence or otherwise, calculate $P(0\le Y<1)$ and $P(Y>1)$.

d Find the values of $E(Y)$ and $E(Y^2)$. Hence calculate $\text{Var}(Y)$.

e Solve the equation $F(y)=0.25$. Which statistic does this value model?

EXAM-STYLE QUESTIONS

7 A continuous variable X has PDF defined by $f(x)=ax^2+bx+c$ for $1\le x\le 4$.

Given that the mode of X is 2 and the mean is 1, find the values of the constants a, b and c.

8 The probability density function of X is defined by

$f(x)=\alpha(1+\sin x)$ for $\dfrac{\pi}{2}\le x\le\dfrac{5\pi}{2}$, where α is a constant to be determined.

a Show that $\alpha=\dfrac{1}{2\pi}$

b Hence, calculate $P(X<\pi)$ and $P(X<2\pi)$, giving your answers exactly.

c Sketch the graph of f. What is the value of the median of X?

d Find an expression for the distribution function F of X (CDF of X).

e Find the inter-quartile range of X, i.e., the value of $a-b$ where the values of a and b are such that $F(a)=0.75$ and $F(b)=0.25$

Extension material on CD:
Worksheet 10

Reflect and discuss:

● What is the geometrical meaning of the definition of the distribution function of a continuous random variable?

● What is the relation between the graph of the probability density function of a continuous random variable that models the heights of a group of n students and the histogram that would represent the set of data if we measure their heights?

● How does the relation between the graph of a density function and the histogram of the actual data explain the way probabilities of continuous variables are calculated?

● Sometimes, histograms are used to represent discrete distribution functions, for example, when ages, heights or weights are treated as discrete variables. How different are histograms that represent probability distribution of discrete and continuous variables? Which characteristics are relevant in each case for the calculation of probabilities?

10.5 Normal distribution

The **normal distribution** is the most important continuous theoretical probability distribution.

> → The PDF of the normal distribution is
>
> $$f(x) = \frac{1}{\sigma\sqrt{2\pi}} e^{-\frac{(x-\mu)^2}{2\sigma^2}}, \text{ where } x \in \mathbb{R}$$

The values of $f(x)$ depend on two parameters, μ and σ.

> → The random variable X described by the PDF is a **normal variable** that follows a normal distribution with mean μ and variance σ^2. We write $X \sim N(\mu, \sigma^2)$

The PDF of the normal distribution is built into your GDC. To obtain the graph of a normal PDF you just insert the values of its parameters.

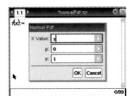 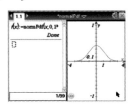

The graph of a normal PDF is called a **normal curve**. Normal curves have interesting properties that you will discover in this investigation.

The normal distribution was first studied by the French mathematician **Abraham De Moivre** (1667–1754) and later by **Carl Friedrich Gauss** (1777–1855), who deduced an expression of its probability density function. For this reason the normal distribution is also known as the **Gaussian distribution**.

GDC help on CD:
Alternative demonstrations for the TI-84 Plus and Casio fx-9860GII GDCs are on the CD.

Investigation – the normal curve

1 Use your GDC to graph the normal curves for each of these normal variables.

   a $X \sim N(1, 1)$ **b** $X \sim N(2, 1)$
   c $X \sim N(3, 1)$ **d** $X \sim N(4, 1)$

   Describe differences and similarities between these normal curves. How does the value of the parameter μ affect the normal curve?

2 Graph the normal curves associated with these normal variables.

   e $X \sim N(0, 1)$ **f** $X \sim N(0, 2)$
   g $X \sim N(0, 3)$ **h** $X \sim N(0, 4)$

   Describe differences and similarities between these normal curves. How does the value of the parameter σ affect the normal curve?

3 Explore some other normal curves and write down your conclusions about the effects of the parameters parameters μ and σ on the graphs of the normal variables.

4 The normal curves are defined by just two parameters, the mean and the variance of the distribution. Based on your knowledge of the shape of a normal curve, explain why it is not necessary to include the values of the median and the mode of the distribution.

Properties of a normal curve

→ A normal curve
  - is a smooth bell-shaped graph
  - is symmetrical about the vertical line $x = \mu$
  - has a maximum point at $x = \mu$
  - has a horizontal asymptote, the x-axis
  - has two inflexion points at $x = \mu \pm \sigma$

> Use your GDC to confirm these properties

Example 28

State, giving reasons, whether or not each of these graphs could be a normal curve.

a   **b**   **c**

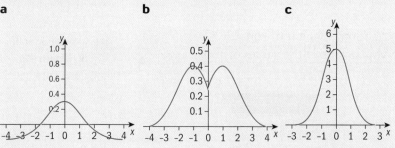

> *Challenge:*
> Use the expression of the normal PDF to prove the properties of the normal curve.

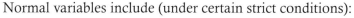

Answers

a No
  The curve goes below the x-axis.

b No
  The curve has two maximum points

c No
  The area under the curve is more than 1. The triangle ABC with A(1, 0), B(0, 4) and C(−1, 0) fits in the region under the graph and has area 2.

Why is the normal distribution so important?

As the characteristics of the normal curve and its shape suggest, the normal distribution is a suitable model for situations where very large and very small values are rather rare, but values in the middle of the range are common. This is typical of many populations in everyday life so the normal distribution makes a good model.

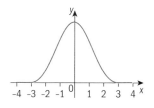

Normal variables include (under certain strict conditions):
  - height, weight, age
  - mass, volume
  - time taken to complete an activity
  - scores in tests and exams
  - random errors in experiments or processes.

> In statistics the population is the universal set from which you take a sample.

Calculating probabilities of normal variables

Suppose that $X \sim N(\mu, \sigma^2)$. As X is a continuous random variable, you can calculate the probability that X takes values in an interval $[a, b]$.

→ As in any continuous distribution

- $P(a \leq X \leq b) = \int_a^b f(x)dx$, where $f(x)$ is the PDF of X

Alternatively, if you consider the cumulative normal distribution function F,

- $P(a \leq X \leq b) = F(b) - F(a)$

You can calculate the probabilities of normal variables using the normal CDF built-in to your GDC as in this example.

Example 29

Given $X \sim N(3, 2^2)$, calculate $P(1 < X < 3)$

Answer

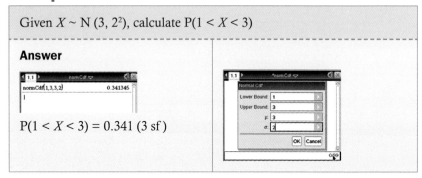

$P(1 < X < 3) = 0.341$ (3 sf)

Use the built-in normal CDF and enter the limits and parameters.

The normal PDF f is defined for all real values. To obtain the values of the corresponding normal CDF, you need an improper integral,

$$F(x) = \int_{-\infty}^x f(t)dt, \text{ where } x \in \mathbb{R}$$

The study of this type of integral is part of the option *Calculus*. However, you can still calculate with great accuracy the probability of normal variables of the form $P(X < a) = F(a)$ and $P(X > b) = 1 - F(b)$ using your GDC as shown in this example.

Example 30

Given $X \sim N(3, 2^2)$, calculate
a $P(X < 2)$ **b** $P(X > 4)$

Answers

a
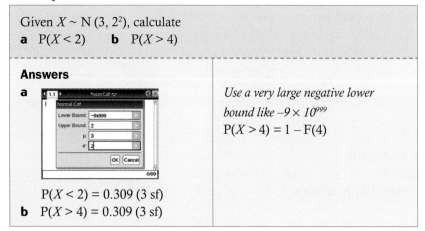

Use a very large negative lower bound like -9×10^{999}

$P(X > 4) = 1 - F(4)$

GDC help on CD:
Alternative demonstrations for the TI-84 Plus and Casio fx-9860GII GDCs are on the CD.

$P(X < 2) = 0.309$ (3 sf)

b $P(X > 4) = 0.309$ (3 sf)

Exercise 10L

1 Given $X \sim N(1, 2^2)$, calculate
  a $P(0 < X < 1.5)$
  b $P(0 < 0.5)$
  c $P(X \geq 3)$

2 If $X \sim N(50, 20^2)$, find the values of
  a $P(X < 45)$
  b $P(37 \leq X < 65)$
  c $P(X \geq 52)$

3 Given that $X \sim N(35, 49)$
  a state the value of the mean and standard deviation of X
  b calculate $P(X < 25)$, $P(29 \leq X \leq 41)$ and $P(X \geq 45)$

The next example shows a typical situation modeled by a normal variable and solved using a GDC.

Example 31

The heights of Grade 6 boys at a large school follow a normal distribution with mean 151.6 cm and standard deviation 7.9 cm. If a boy of this group is selected at random, what is the probability that he has height

a less than 152 cm
b between 150 and 157 cm.

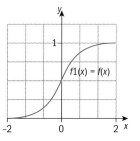

Answers

a Let H be the random variable 'height of grade 6 boys at the school'.

Define the variable and indicate clearly its parameters

GDC help on CD:
Alternative demonstrations for the TI-84 Plus and Casio fx-9860GII GDCs are on the CD.

$H \sim N(151.6, 7.9^2)$
$P(H < 152) = 0.520$ (3 sf)

b $P(150 < H < 157) = 0.333$ (3 sf)

If F is the CDF of a normal variable X, then F is an increasing function and therefore one-to-one.

This means that F has inverse F^{-1} which is useful in determining values of the normal variable X when you are given information about the probability that X is less or greater than certain values.

Example 32

Given $X \sim N(1, 3^2)$ find the values of a and b such that
a $P(X < a) = 0.506$ **b** $P(X > b) = 0.198$

The inverse of the normal CDF is another built-in function on your GDC.

Answers

a

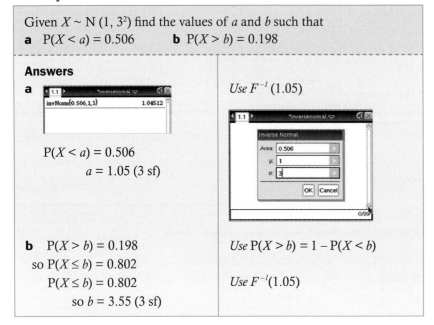

$P(X < a) = 0.506$

$\quad a = 1.05$ (3 sf)

Use $F^{-1}(1.05)$

GDC help on CD:
Alternative demonstrations for the TI-84 Plus and Casio fx-9860GII GDCs are on the CD.

b $\quad P(X > b) = 0.198$
so $P(X \leq b) = 0.802$

$\quad P(X \leq b) = 0.802$

$\quad\quad$ so $b = 3.55$ (3 sf)

Use $P(X > b) = 1 - P(X < b)$

Use $F^{-1}(1.05)$

Exercise 10M

1 A packing machine produces bags of flour whose weights are normally distributed with mean 150 kilograms and standard deviation 0.5 kilograms.
If a bag produced by this machine is selected at random, what is the probability that its weight is
  a less than 149 kilograms
  b more than 151.5 kilograms
  c between 149 and 151 kilograms?

2 A farmer loads 850 cabbages to sell in the local market.
The weight of this type of cabbage follows a normal distribution with mean 1.1 kilograms and standard deviation 150 grams.

Be careful with the units you use.

  a Consider the random variable M. If M denotes the weight of a cabbage from this load, define the distribution of M indicating clearly its parameters.
  b If the farmer picks one cabbage at random, what is the probability that it will have a weight between 1.2 and 1.3 kilograms?
  c Estimate how many cabbages will have a weight greater than 1.4 kilograms.

3 A continuous random variable T follows a normal distribution with mean 20.4 and standard deviation 3.5
   a Find the values of $P(T < 18.1)$ and $P(T > 17.9)$
   b Hence calculate $P(T < 18.1 \mid T > 17.9)$
   c Find the value of t if $P(T < t) = 0.444$.

Although most of the time you will be expected to use your GDC to calculate normal probabilities, the normal distribution has interesting properties that allow you to estimate those probabilities and, in some cases, even find them.

In this investigation you are going to discover some of the most useful properties of the normal curves.

In general, you use your GDC to calculate normal probabilities as you cannot integrate the normal PDF. This is beyond the level of this course.

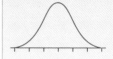

Investigation – further properties of the normal curve

1 Use your GDC to calculate these probabilities.
   a $P(1 < X < 3)$ when $X \sim N(2, 1^2)$
   b $P(0 < X < 4)$ when $X \sim N(2, 2^2)$
   c $P(-1 < X < 5)$ when $X \sim N(2, 3^2)$
   d $P(-1 < X < 3)$ when $X \sim N(2, 2^2)$

   What do these examples have in common?
   How does it affect the values of the probabilities you calculated?

2 If $X \sim N(\mu, \sigma^2)$ what is the value of $P(\mu - \sigma < X < \mu + \sigma)$?

3 Investigate normal curves further to find the values of
   a $P(\mu - 2\sigma < X < \mu + 2\sigma)$
   b $P(\mu - 3\sigma < X < \mu + 3\sigma)$
   c $P(\mu - 4\sigma < X < \mu + 4\sigma)$
      when $X \sim N(\mu, \sigma^2)$

Remember that the normal curve is the graph of a continuous PDF and that the area under a density function is always 1.

4 Sketch the graph of the normal curve with mean μ and variance σ^2. If a is positive, use your sketch to compare these probabilities.
   a $P(X > \mu + a)$ and $P(X < \mu - a)$
   b $P(X > \mu - a)$ and $P(X < \mu + \sigma)$
   c $P(X > \mu - a)$ and $P(\mu - \sigma X < \mu + a)$

You may want to use your GDC to explore a few examples before attempting the general case.

The next example shows you how to use some of the properties of the normal curves you investigated.

Example 33

A variable X follows a normal distribution with mean 2.
Given that $P(X < 3) = 0.8$, find the values of these probabilities.
  a $P(X > 3)$
  b $P(X < 1)$
  c $P(1 < X < 3)$

Answers

a $P(X > 3) = 1 - 0.8 = 0.2$	$P(X > 3) = 1 - P(X < 3)$
b $P(X < 1) = 0.2$	$P(X < 1) = P(X > 3)$
c $P(1 < X < 3) = 0.8 - 0.2 = 0.6$	$P(1 < X < 3) = P(X < 3) - P(X < 1)$

Exercise 10N

1 A continuous random variable X follows a normal distribution with mean 5.
Given that $P(X < 3) = 0.3$, and without using a calculator, find these probabilities.

> Use the properties you discovered in the previous investigation.

  a $P(X \geq 7)$ **b** $P(X < 7)$ **c** $P(3 \leq X < 7)$

2 A continuous random variable Y follows a normal distribution with mean 12.
Given that $P(10 \leq Y < 14) = 0.6$, and without using a calculator, find these probabilities.
  a $P(Y \geq 14)$
  b $P(Y < 10)$
  c $P(12 \leq Y < 14)$
  d $P(Y < 14 \mid Y > 12)$

3 If $X \sim N(-5, \sigma^2)$ and $P(X < -3) = 0.8$, write down the values of these probabilities.
  a $P(X < -7)$
  b $P(-7 \leq X < -5)$
  c $P(X < -7) + P(X > -3)$

4 If $X \sim N(10, 25)$
  a State the mean and standard deviation of X.
  b Calculate $P(X < 5)$ and $P(X \geq 15)$. Comment on the values you obtain.
  c Find the value of a for which $P(X < a) = 0.223$
    Hence state the value of b for which $P(X > b) = 0.223$

Standardized normal distribution and its importance

These normal curves were graphed on the same axes

$$N(\mu, 1) \text{ for } \mu = -2, -1, 0, 1, 2$$

Notice that the only change in the graphs of the normal curve in the diagram is its position.

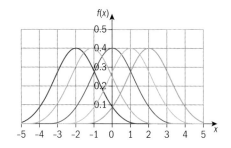

> → The parameter μ determines the position of the normal curve in relation to the y-axis.

Now look at a family of normal curves that have the same mean but different standard deviations.

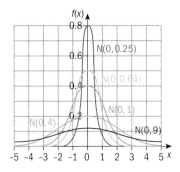

You can see that the normal curve gets wider as the standard deviation *increases* and narrower as the standard deviation *decreases*.

In fact, changes in the parameters μ and σ of a normal curve can be regarded as changes in position and scale as in these diagrams.

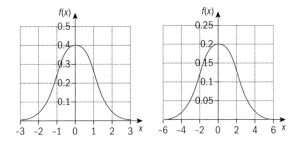

μ is the same in both cases but σ^2 is different, so the positions of both curves are the same but the scales are different.

In the first diagram at the top of the page, the five normal curves have different values of μ but the same value of σ^2 so the positions are different but the scale is the same for all the curves.

Use dynamic graph software to produce more examples of normal curves and to study the effects of the parameters μ and σ on the normal curves.

This means that

→ All normal curves can be related to a single reference distribution called the **standard normal distribution** which has mean 0 and standard deviation 1.

→ The standard normal variable is denoted by $Z \sim N(0, 1)$. The PDF of Z is always denoted by ϕ and its CDF by Φ.

GDC help on CD:
Alternative demonstrations for the TI-84 Plus and Casio fx-9860GII GDCs are on the CD.

Both ϕ and Φ are very important in statistics and for this reason when you use your GDC to calculate normal probabilities, by default, your calculator assumes that $\mu = 0$ and $\sigma = 1$.

ϕ is the lower case Greek letter phi and Φ is the upper case.

Example 34

Consider $Z \sim N(0, 1)$. Find the value of
a $\Phi(1)$ **b** $\Phi^{-1}(0.8)$

Answers

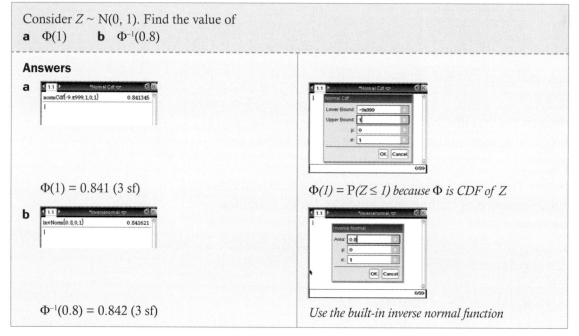

a

$\Phi(1) = 0.841$ (3 sf)

$\Phi(1) = P(Z \le 1)$ *because Φ is CDF of Z*

b

$\Phi^{-1}(0.8) = 0.842$ (3 sf)

Use the built-in inverse normal function

Standardized normal variable

If you have to solve a problem where you do not know the mean or the variance of a normal variable you have to use the **standardized normal variable**.

To standardize a random variable $X \sim N(\mu, \sigma^2)$ into the standardized normal variable $Z \sim N(0, 1)$ you use the transformation

$$Z = \frac{x - \mu}{\sigma}$$

which is the algebraic translation of the relation observed between the normal curves of X and Z.

Example 35

Consider $X \sim N(2, 3^2)$

a Calculate $P(X \le 1)$ and $P\left(Z \le -\dfrac{1}{3}\right)$
Compare your results.

b Calculate $P(-1 \le X \le 2)$
Hence state the value of $P(-1 \le Z \le 0)$

GDC note:
The instructions may vary for different GDC models. For example, if you use any of the TI-83 or TI-84 models, you can obtain your answers without entering the values of the parameters of Z.

```
normalcdf(-10^99
,1)
       .8413447404
invNorm(.8)
       .8416212335
```

Answers

a

$P(X \le 1) = 0.36944...$

$P\left(Z \le -\dfrac{1}{3}\right) = 0.36944...$

The results are the same.

b $P(-1 \le X \le 2) = 0.34134...$

When $X = -1$, $Z = \dfrac{-1-2}{3} = -1$

when $X = 2$, $Z = \dfrac{2-2}{3} = 0$

$P(-1 \le Z \le 0) = 0.34134....$

Use GDC built-in normal CDF.
Upper bound $-\dfrac{1}{3}$, $\mu = 0$, $\sigma = 1$

When $X = 1$, $Z = \dfrac{1-2}{3} = -\dfrac{1}{3}$

Upper bound $= 1$, $\mu = 2$, $\sigma = 3$
Use GDC built-in normal CDF.

GDC help on CD:
Alternative demonstrations for the TI-84 Plus and Casio fx-9860GII GDCs are on the CD.

Exercise 10O

1 Consider $Z \sim N(0, 1)$. Find, correct to 4 decimal places, the values of:

a $\Phi(-1.2)$ and $\Phi(1.2)$

b $\Phi(-2.3)$ and $\Phi(2.3)$

c $\Phi(-2.6)$ and $\Phi(2.6)$

Comment on the values obtained.

2 Consider $Z \sim N(0, 1)$. Find the values of

a $\Phi^{-1}(0.3)$ and $\Phi^{-1}(0.7)$ **b** $\Phi^{-1}(0.4)$ and $\Phi^{-1}(0.6)$

Comment on the values obtained.

3 Consider $Z \sim N(0, 1)$

a Find the value of $\phi(0.45)$ and $\phi^{-1}(0.45)$

b Write down the values of a and b such that
$P(Z \le a) = 0.5$ and $P(Z \le 0.5) = b$

4 Consider $X \sim \text{N}(1, 2^2)$

   a Calculate $\text{P}(X \le 3)$ and $\text{P}(Z \le 1)$. Compare your results.

   b Calculate $\text{P}(0 \le X \le 2)$.

     Hence, using the transformation

     $Z = \dfrac{x - \mu}{\sigma}$, write down the value of $\text{P}\left(-\dfrac{1}{2} \le z < \dfrac{1}{2}\right)$

5 Consider $X \sim \text{N}(-6, 3^2)$ and $Z \sim \text{N}(0, 1)$

   Calculate $\text{P}(0 \le Z \le 1.5)$.

   Hence, using the transformation $Z = \dfrac{x - \mu}{\sigma}$, write down the value of $\text{P}(-6 \le X \le -1.5)$

6 Explain why each of these statements is false.

   a $X \sim \text{N}(7, 2^2)$ and $\text{P}(X \le 5) > \text{P}(X > 9)$

   b $X \sim \text{N}(15, 25)$ and $\text{P}(X \le 10) > \text{P}(10 \le X \le 20)$

   c $X \sim \text{N}(-2, 2^2)$ and $Y \sim \text{N}(2, 2^2)$ and $\text{P}(X > 0) > \text{P}(Y < 0)$

> Explore further the relation between normal curves and standardized normal curve and prove that the transformation
> $$Z = \frac{X - \mu}{\sigma} \text{ is always}$$
> valid.

The next two examples show you how to use the standardized normal curve to determine one or both parameters of the normal variable.

Example 36

Consider $X \sim \text{N}(\mu, 4)$

Find the value of μ given that $\text{P}(X \le 2) = 0.556$

Answer

$$\text{P}(X \le 2) = 0.556$$

so $\text{P}\left(Z \le \dfrac{2 - \mu}{2}\right) = 0.556$

$\dfrac{2 - \mu}{2} = \Phi^{-1}(0.556)$

$\dfrac{2 - \mu}{2} = 0.14083...$

$\mu = 1.72 \,(3\,\text{sf})$

$Z = \dfrac{X - \mu}{\sigma}$

$\text{P}(Z \le a) = b \Rightarrow a = \Phi^{-1}(b)$

Use GDC built-in inverse normal.

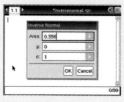

Solve for μ

> Avoid early rounding as it can lead to incorrect answers. Do not use calculator notation.

> **GDC help on CD:**
> *Alternative demonstrations for the TI-84 Plus and Casio fx-9860GII GDCs are on the CD.*
>

Example 37

Consider $X \sim N(\mu, \sigma^2)$
Find the values of μ, and σ given that $P(X \le 2) = 0.546$ and
$P(X \le 3) = 0.743$

- -

Answer

$$P(X \le 2) = 0.546$$

so $P\left(Z \le \dfrac{2-\mu}{\sigma}\right) = 0.546$

$$P(X \le 3) = 0.743$$

so $P\left(Z \le \dfrac{3-\mu}{\sigma}\right) = 0.743$

$\dfrac{2-\mu}{\sigma} = \Phi^{-1}(0.546)$ and

$\dfrac{3-\mu}{\sigma} = \Phi^{-1}(0.743)$

$\mu + \Phi^{-1}(0.546)\,\sigma = 2$ and
$\mu + \Phi^{-1}(0.743)\,\sigma = 3$

$\mu = 1.78$ (3 sf) and
$\sigma = 1.86$ (3 sf)

$Z = \dfrac{X - \mu}{\sigma}$

$P(Z \le a) = b \Rightarrow a = \Phi^{-1}(b)$

Rearrange equations and use GDC to solve simultaneously.

Exercise 10P

1 If $X \sim N(\mu, 9)$, find the value of μ given that $P(X \le 5) = 0.754$

2 Consider $X \sim N(\mu, \sigma^2)$. Find the values of μ and σ given that
$P(X \le 1) = 0.345$ and $P(X \le 3) = 0.943$

EXAM-STYLE QUESTIONS

3 The random variable X is normally distributed with mean μ and
variance σ^2.
Given that $P(X > 58.44) = 0.022$ and $P(X < 48.84) = 0.012$ find
the values of μ and σ.

4 A machine is used to fill 1 kg bags of flour. When the bags are
checked it is found that their average weight is 1.03 kg. Assuming
that the weights of the bags are normally distributed, find the
standard deviation if 1.8% of the bags weigh below 1 kg.
Give your answer correct to the nearest 0.1 gram.

5 A random variable X is normally distributed with mean μ and
standard deviation σ, such that $P(X > 50.1) = 0.119$, and
$P(X < 43.6) = 0.305$
   a Find the values of μ and σ **b** Hence find $P\left(|X - \mu| < \dfrac{\sigma}{2}\right)$

6 The marks in an exam were normally distributed with mean μ and standard deviation σ.

If 10% of the candidates scored at least 80 marks and 20% scored less than 45, what are the values of μ and σ?

7 The masses of the lettuces sold at Freshgreens supermarket are normally distributed with mean 550 grams and standard deviation 20 grams.

   a If a lettuce is chosen at random, what is the probability that its mass lies between 500 g and 600 g?

   b Find the mass that is exceeded by 10% of the lettuces.

   c In one day, 1200 lettuces are sold. Estimate the number of lettuces that weigh more than 540 grams.

   d At the nearby Goodvalue supermarket, 15% of the lettuces sold weigh at least 600 grams and not more than 10% of them weigh less than 540 grams. Assuming that the mass M of these lettuces follows a normal distribution find the expected value and variance of M.

8 Sweetworld company produces one-kilogram bags of sugar. Assuming that the mass M of the bags follow a normal distribution with mean 1.02 kg, what is the maximum value of the variance of M if less than 1% of the bags are underweight?

10.6 Modeling and problem solving

This section looks at some typical exam style questions involving the probability distributions studied in this chapter. When tackling these problems it is important that you keep in mind the characteristics of the distributions you use and the assumptions you make so that you can comment on the validity of your results.

Example 38

The weights W, in grams, of female sparrows is modeled by a normal distribution with mean μ and standard deviation σ.

a Experimental data shows that 84% of the birds weigh at least 20 grams and 44% of the birds weigh more than 22.5 grams. Calculate the values of μ and σ, correct to 5 decimal places.

b A random sample of five sparrows is collected. If B denotes the number of birds in the sample that weigh more than 23 grams calculate $P(2 \leq B \leq 4)$.

c A researcher thinks that the number of eggs E laid by these sparrows is modeled by the Poisson distribution with mean m. Given that $P(E \geq 4) = 0.9071$, correct to 4 decimal places,

   i determine the value of m

   ii give a reason why this distribution cannot be an exact model in this case.

▶ Continued on next page

Answers

a
$$P(W \geq 20) = 0.84$$

$$\text{so } P\left(Z < \frac{20 - \mu}{\sigma}\right) = 0.16$$

$$P(W \geq 22.5) = 0.44$$

$$\text{so } P\left(Z < \frac{22.5 - \mu}{\sigma}\right) = 0.66$$

$$\frac{20 - \mu}{\sigma} = \Phi^{-1}(0.16)$$

$$\frac{20 - \mu}{\sigma} = -0.9944578\ldots$$

$$\frac{22.5 - \mu}{\sigma} = \Phi^{-1}(0.66)$$

$$\frac{22.5 - \mu}{\sigma} = 0.4124631\ldots$$

$\mu = 21.76708$ and
$\sigma = 1.77693$ (5 dp)

b B follows a binomial
distribution with $n = 5$ and
$p = P(W > 23) = 0.243888\ldots$
and $P(2 \leq B \leq 4) = 0.353$ (3 sf)

c i $m = 6.7984$ (4 dp)

ii A Poisson variable is
theoretically defined for
all $n \in \mathbb{N}$ and a bird cannot
lay more than a fairly
small number of eggs.

Standardize the distribution

*Use the built-in GDC inverse
function of the standardized normal
CDF to obtain simultaneous
equations in μ and σ*

*Solve the simultaneous equations
using a GDC*

*Assume independence of the weights
of the 5 birds and use GDC and the
answer to part **a** to calculate p and
the required probability:*

Use GDC numerical solver.

*Compare the assumption you need to
make to apply a distribution with the
real situation.*

Dr Clio Creswell,
an Australian
mathematician,
has been one of
the first to apply
mathematics to
human relationships,
to predict the optimal
number of partners to
have before settling
down, the amount of
compromise needed
to sustain a marriage,
how dating services
determine your perfect
match, and why we
find certain people
attractive. How do you
think mathematics
can be used to predict
human behavior?

GDC help on CD:
*Alternative demonstrations
for the TI-84 Plus and Casio
fx-9860GII GDCs are on the
CD.*

Exercise 10Q

1 *Live Better*, a consumer magazine, commissioned a study about on-time arrivals of airlines at the country's main airports. The study showed that 5% of the flights arrived before the scheduled arrival time and 2.3% arrived more than 30 minutes after the scheduled time.

   a Let T be the random variable 'difference between scheduled time and actual arrival time' of the flights at these airports. Assuming that T follows a normal distribution, determine the value of its parameters.

   b Calculate $P(|T| \leq 5)$

   c On a given day, 500 flights are due to arrive at these airports. What is the probability that at least 50 flights arrive within 5 minutes of the scheduled time?

   d State any assumptions you have made in your answer to part **c**.

   e Comment on the limitations of the models used in this problem.

2 Mr Jones, the manager of the restaurant Dolce Vita, collected data about customers' habits during one year.

   a According to the data, 6% of the customers who make reservations do not show up. On a given day, 200 people made reservations. What is the probability that 2 people will not show up? State any assumption you made.

   b Mr Jones' data also shows that the mean time customers stay in the restaurant is 52 minutes with standard deviation 15 minutes. What is the minimum interval of time between two reservations of the same table that Mr Jones should allow if he wants the probability of double booking to be less than 1%? State any assumptions you made.

   c On a busy day, Dolce Vita can operate well with a minimum of 15 waiters. The data shows that in average 7.5% of the waiters miss work on any day. Calculate the minimum number of waiters Mr Jones needs to hire if he wants to reduce the probability of having less than 15 waiters to less than 5%. State any limitation of the probability model you used.

3 Records over the past years show that 5% of the loans made by the bank Easylife have defaulted. At present Easylife has 5520 customers with loans. What is the probability that not more than 250 loans will default?

State any assumptions and limitations of the model you used.

4 A large exhibition center collected data about attendance at its exhibitions. This weekend a major event is taking place at this centre and historical data shows that the average daily attendance for this type of event is 7850 people with standard deviation 367 people.

a What is the probability that more than 7000 people will try to attend the event on Saturday?

b If the capacity of the exhibition centre is 8500 people, what is the probability that the centre will reach its full capacity on Sunday?

The data also shows that, on each day, the mean arrival time of the visitors after opening time is 155 minutes.

c Assuming that the arrival times of the visitors to the centre follow a Poisson distribution, calculate the percentage of visitors that will arrive in the first 3 hours after the opening of the exhibition on Saturday.

5 A coffee vending machine automatically pours different types of coffee into cups. Statistical historical data shows that the amount of coffee dispensed by this type of machines can be modeled by a normal distribution with mean 120 ml and standard deviation 8.3 ml.

a If cups with a capacity of 130 ml are used, what is the probability that a cup overflows?

b If the machine is loaded with 500 of these cups, how many of these do you expect will overflow when served?

c The data also shows that the machine successfully dispenses a cup 99% of the times it is used. Mr Li uses the machine twice a day and pays 2 yuan each time. How many days is he likely to use the machine before he can expect to lose more than 5 yuan due to error in dispensing of the cups?

6 A sample of two hundred sheets of aluminum alloy were examined for surface flaws. The number of sheets with a given number of flaws per sheet is recorded in this table.

a Calculate the mean m of the number of flaws on the sheets of the sample.

b Assuming that the number of surface flaws per sheet of aluminum alloy produced by the same machine is modeled by a Poisson distribution with mean m, calculate the probability that the number of flaws on the surface of a randomly chosen sheet exceeds 5.

c Give reasons to support the claim that the variable number of flaws follows a Poisson distribution.

Number of flaws	Frequency
0	28
1	42
2	53
3	29
4	31
5	9
6	8

Review exercise

1 The table shows the probability distribution of a discrete random variable X.

X	1	2	3	4
$P(X = x)$	$2a^2$	$3a$	$3a^2 + 2a$	$2a^2 + a$

a Find the value of a.

b Hence, calculate the mean, the mode and the median of X.

2 A continuous random variable X has probability density function defined by

$$f(x) = \begin{cases} \dfrac{1-|x|}{2} & \text{when } -1 \le x \le 1 \\ \dfrac{2-|3-x|}{k} & \text{when } 1 < x \le 5 \\ 0 & \text{otherwise} \end{cases}$$

a Sketch the graph of f and use it to find the value of k.

b State the median and modal values of X.

c Find $P(0 \le X \le 3 \,|\, X \ge 1)$

3 The probability that a student entering Toprank College will graduate is $\dfrac{4}{5}$. If five of these students are selected at random, find the probability that

a none will graduate

b all will graduate

c at least two will graduate.

4 A geography quiz consists of ten true/false questions. A student who knows the correct answers to five of the questions decides to choose at random the answers to the remaining questions.

a What is the probability that the student answers all the questions correctly?

For each correct answer the student scores two marks, for each incorrect answer one mark is subtracted whereas no marks are gained or lost if no answer is given.

b Calculate the student's expected number of marks and compare it with the score he can obtain if he answers only the questions for which he knows the correct answers.

5 If $T \sim \text{Po}(m)$ and $\text{E}(T^2) = 6$, find
   a the value of m
   b $\text{P}(X = 0)$

EXAM-STYLE QUESTION

6 A continuous random variable X has CDF function given by

$$F(x) = \begin{cases} \tan x & \text{for } 0 \leq x \leq \dfrac{\pi}{4} \\[2mm] 0 & \text{for } x < 0 \\[2mm] a & \text{for } x > \dfrac{\pi}{4} \end{cases}$$

   a State the value of a.
   b Find the median value of X.
   c Find expressions for f, the PDF of X.
   d Calculate $\text{P}\left(X \leq \dfrac{\pi}{6}\right)$

Review exercise

EXAM-STYLE QUESTIONS

1 Data collected over a long period of time shows the mean width of the pieces of wood produced by a lathe is 20.05 mm with a standard deviation of 0.02 mm. Assume that widths of pieces of wood follow a normal distribution.
Find the probability that a piece selected at random has width
   a between 20.02 mm and 20.06 mm
   b less than 20.00 mm

2 The average life of a certain type of motor is 15 years, with a standard deviation of two years. If the manufacturer is willing to replace not more than 0.1% of motors that fail, how many complete years of guarantee should he offer?
Assume that the lives of the motors follow a normal distribution.

3 A continuous random variable X has probability density function

$$f(x) = \begin{cases} \dfrac{k}{2 + x^2} & \text{for } 0 \leq x \leq \sqrt{2} \\[2mm] 0 & \text{elsewhere} \end{cases}$$

Find
   a the value of k

   b $\text{P}\left(X \leq \dfrac{1}{2}\right)$

   c $\text{E}(X)$

4 According to empirical data the number of monkeys seen on a river bank is, on average, two per hour. Assume that the number of monkeys can be modeled by Poisson distribution. If Kathy takes a two-hour tour along this river, what is the probability that she will see

   a at least one monkey

   b more than five monkeys?

5 A continuous random variable X has PDF given by

$$f(x) = \frac{a}{4x - x^2} \text{ where } 1 \le x \le 3 \text{ and } a \text{ is a constant.}$$

   a Find the value of a correct to 5 decimal places

   b Calculate the mean and variance of X

   c Find $P(X < 2)$.

6 Mr Kalt, the manager of a new ice cream store conducted a survey to collect information about the preferences of 160 potential customers:

The results are shown in the table.

Favorite ice cream flavor	Votes
Chocolate	71
Vanilla	30
Cherry	1
Strawberry	14
Cookies and Cream	16
Mint	9
Coffee	4
Lime	1
Other	11
Do not like ice cream	3

Assuming that the results represent the preferences of Mr Kalt's customers:

   a Estimate how many customers will buy chocolate ice cream in a day when 250 customers go to Mr Kalt's shop.

   b If five customers arrive at Mr Kalt's shop, what is the probability that

      i three of them buy vanilla ice-cream

      ii two of them do not buy any ice cream?

   c A customer arrives at the shop and asks for an ice cream cone with two flavors. What is the probability that the customer orders strawberry and cherry? Which assumptions have you made to answer this question?

Extension material on CD:
Worksheet 10

CHAPTER 10 SUMMARY

Random variables and distributions

- The **probability distribution function** (PDF) of a discrete random variable X has the properties
$$0 \leq f(x) \leq 1 \text{ and } \sum f(x) = 1$$
- The **mean** or **expected value** of X is given by $\mu = \sum x\, P(X = x)$
- The **variance** of X is given by $\sigma^2 = \sum(x - \mu)^2\, P(X = x)$
- $\sigma^2 = \text{Var}(X) = E(X^2) - E(X)^2$ where $E(X^2) = \sum x^2\, P(X = x)$
- $\sigma = \sqrt{\text{Var}(X)}$ is called the standard deviation of X.
- The **mode** of a probability distribution function is the value of x for which the probability distribution function has a maximum.
- In cases where the PDF of X has a maximum value, this value is called the mode of X. X may have more than one mode.

Binomial distribution

- If $X \sim B(n, p)$, then $P(X = r) = \binom{n}{r} p^r q^{n-r}$ where $r = 0, 1, ..., n$ and $q = 1 - p$
- $E(X) = \mu = np$ and $\text{Var}(X) = \sigma^2 = npq$ where $q = 1 - p$

Poisson distribution

- $X \sim \text{Po}(m)$ has PDF given by $f(x) = P(X = x) = \dfrac{e^{-m} m^x}{x!}$ where $x = 0, 1, 2, ...$
- If $X \sim \text{Po}(m)$ then $E(X) = \text{Var}(X) = m$

Continuous random variables and distributions

- The **probability density function** of a random variable X (PDF of X) has the properties:
$$f(x) \geq 0 \text{ for all values of } x \in \mathbb{R} \text{ and } \int_{-\infty}^{+\infty} f(x)\, dx = 1$$
- $\mu = E(x) = \displaystyle\int_a^b xf(x)\, dx$ and $\sigma^2 = \text{Var}(X) = \displaystyle\int_a^b (x - \mu)^2 f(x)\, dx$, where $f(x)$ is defined on the interval $[a, b]$.
- $\sigma^2 = \text{Var}(X) = E(X^2) - E(X)^2$ where $E(X^2) = \displaystyle\int_a^b x^2 f(x) dx$
- The **median** m is the solution of the equation $\displaystyle\int_a^m f(x) dx = \dfrac{1}{2}$ and the **mode** is the value(s) of x for which f has a maximum value.

Normal distribution

- If $X \sim N(\mu, \sigma^2)$ and $Z = \dfrac{X - \mu}{\sigma}$ then $Z \sim N(0, 1)$

Decisions, decisions

- How should we proceed before making important decisions?
- Should we use historical statistical data or probability models?
- How reliable are statistical methods?
- Should we just trust our intuition?

Statistical evidence on trial

In the UK in 1999, Sally Clark was convicted of the murder of her two children, both of whom died in early infancy – one in 1996 and one in 1998. She was convicted on the basis of medical evidence now recognized to be flawed, and a statistical statement made by a paediatrician, Professor Roy Meadows. His crucial mistake was to assume that Sudden Infant Death Syndrome (SIDS) deaths in the same family were independent events, though as genetic and environment factors seem to be implicated in SIDS, this is unlikely. Thus his estimate, that the probability of two children in the same family dying of SIDs was one in 73 million, was probably very inaccurate.

This is a mistake that would have been picked up very easily if anyone in the courtroom had had a sound understanding of statistics. Unfortunately, Clark's defence counsel's response was along the lines of Mark Twain's "lies, damned lies, and statistics", and the judge actually directed the jury's attention to the statistical 'evidence' in his summing up. The jury reasoned that the probability of two siblings dying by SIDs was so small, Clark had to be guilty. But they should also have considered that statistically it is more rare for a mother to kill both her children.

Clark was finally freed in 2003.

- How can we be sure that juries are properly directed, and statistical evidence is clearly explained in court?
- Should statisticians alone be allowed to handle statistical evidence?

"Nothing is more difficult, and therefore more precious, than to be able to decide."
Napoleon Bonaparte (1769–1821)

In 2011 a UK judge ruled that Bayes' theorem could no longer be used in court cases to analyze statistical evidence. The Royal Statistical Society's working group on statistics and the law challenged this decision, which they believed was based on an inaccurate understanding of probability theory.

"One must either accept some theory or else believe one's own instinct or follow the world's opinion."
Gertrude Stein, American writer and poet (1874–1946)

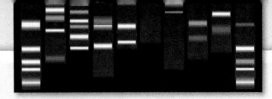

Genetic fingerprinting

In a fiction series like *CSI*, the evidence that often closes the case is a DNA match. As an individual's DNA is almost unique, once the police have matched a suspect's DNA to evidence from the crime scene, then the case looks closed. However, some statistical thinking is required to understand exactly what a match is...

Genetic fingerprinting was developed in 1984 by Professor Alec Jeffreys at the University of Leicester, in the UK. Each one of us has a unique genetic make-up which is contained in the DNA we inherit from our parents. The DNA can be extracted from cells and body fluids and analyzed to produce our 'genetic fingerprint'.

Comparing the characteristic bands in genetic fingerprints for a match has led to criminal convictions, but the field is under constant scrutiny due to its reliance on probability. Usually between 10 and 20 bands are examined and compared. Experimental evidence has suggested that the probability of one band matching by coincidence is $\frac{1}{4}$ (although this figure is subject to debate). The probability of two bands matching will therefore be $\frac{1}{16}$.

- Explore the role of probability in this field.

> You can find out more by searching 'DNA matching' at http://plus.maths.org

Predicting the future

Paul the Octopus, who lived in the Sea Life Center in Oberhausen, Germany, correctly predicted the results of 12 out of 14 football matches between 2008 and 2010. Two boxes, each containing a mussel and marked with the flag of one of the national teams in an upcoming match were placed in his tank. His choice of which mussel to eat first was interpreted as predicting that the country with that flag would win.

Paul's predictions were correct 86% of the time.

- Why do people want to believe that something or someone can predict the future, when rationally predicting the future seems to be illogical?

Lehman Brothers, a US investment bank filed for bankruptcy in September 2008. Despite its large assets – $639 billion – it had a huge $619 billion debt. As the fourth-largest US investment bank, it had 25 000 employees worldwide.

- How did incorrect assumptions in probability models contribute to the fall of Lehman Brothers investment bank?

- How reliable are predictions of future performance based on past performance?

- What are the dangers of extrapolation?

> http://plus.maths.org/content/how-maths-killed-lehman-brothers

"Life is just an endless chain of judgments...
The more imperfect our judgment, the less perfect our success."

B. C. Forbes, Scottish financial journalist and author, founder of *Forbes* magazine (1880–1954)

11

Inspiration and formalism

CHAPTER OBJECTIVES:

4.1 Concept of a vector; representation of vectors using directed line segments; unit vectors; base vectors **i, j, k**; components of a vector
Algebraic and geometric approaches to the following: the sum and difference of two vectors; multiplication by a scalar, $k\mathbf{v}$; position vectors

4.2 The definition of the scalar product of two vectors; the angle between two vectors; perpendicular vectors; parallel vectors

4.3 Vector equation of a line in 2 and 3 dimensions; simple applications to kinematics; the angle between two lines

4.4 Coincident, parallel, intersecting and skew lines; points of intersection.

4.5 The definition and properties of the vector product of two vectors; geometric interpretation of $|\mathbf{v} \times \mathbf{w}|$

4.6 Vector equation of a plane. Cartesian equation of a plane: $ax + by + cz = d$

4.7 Intersections of: a line with a plane; two planes; three planes
Angle between: a line and a plane; two planes

Before you start

You should know how to:

1 Calculate the distance between two points using their coordinates, and find the midpoint of the line segments joining them e.g., given that A(−1, 3) and B(3, −5), AB = $\sqrt{(-1-3)^2 + (3-(-5))^2}$
$= \sqrt{80} = 4\sqrt{5}$.

The midpoint M $\left(\dfrac{-1+3}{2}, \dfrac{3+(-5)}{2} \right) = (1, -1)$.

2 Represent a line in the plane by a Cartesian equation e.g., given A(−1, 3) and B(3, −5), the equation of the line AB is
$y - 3 = -2(x + 1)$ i.e., $y = -2x + 1$.

Skills check

1 **a** Find the distance between P(−1, 5) and Q(2, 1).

b Find the coordinates of the midpoint of PQ when P(−2, −3) and Q(4, −1).

2 A is the point (1, 3) and B is the point (4, 9).

a Find the gradient of the line segment [AB].

b Write down an equation for the line AB.

From vector geometry to vector algebra

The computer animations used in movies and video games are a result of an ingenious application of mathematics – vector geometry. The process may begin with an artist's sculpture which is scanned to produce a highly detailed three-dimensional digital image known as a wireframe model. The resulting model, whose surface can be made up of as many as six million triangles, can be viewed and colored in real three-dimensional life.

When filming a scene for a movie, the position and direction of the light source and the camera lens are known. This makes it possible to calculate the position of each point on the model's surface that is highlighted by the light source and color it accordingly.

The relationship between the orientation of the surface of the model and the light source can be described algebraically using vector algebra – the area of mathematics explored in this chapter.

▲ Most modern computer animations allow movement. Animators can interact with their images, manipulating them with a mouse or controller.

For more detail on how vectors are used in the creation of computer-generated movies, use your internet search engine and the term 'maths and movies'.

11.1 Geometric vectors and basic operations

In previous chapters, you have studied many important concepts and explored several of their applications to solve real-life problems. Further applications of mathematics to areas such as mechanics require a deeper knowledge of mathematics – particularly, a class of mathematical objects called **vectors**.

The idea of a vector is closely related to translations. For example, the diagram shows a triangle, ABC, and its image, A′B′C′, after a translation.

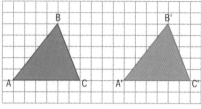

To describe this translation you say: 'Triangle ABC is translated 10 units to the right'. You can also draw a **directed line segment** with an arrowhead to illustrate the effect of the translation.

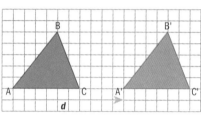

When describing the effects of a translation, you need to mention the magnitude **and** the direction to define it, as illustrated by the directed line segment.

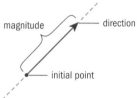

> → A vector is defined by direction and magnitude.

This chapter uses two different approaches to define and study vectors: geometric and analytic.

A **geometric vector** is represented by an arrow or directed line segment, for example, $\overrightarrow{AB}$.

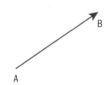

Geometric vectors are frequently used in physics because they are convenient for representing quantities such as force, velocity or acceleration, which possess magnitude and direction. In such cases, it is useful to adopt a simple notation such as **F** or $\overrightarrow{F}$ to represent a force and **v** or $\overrightarrow{v}$ to represent velocity.

$\overrightarrow{AB}$ can be interpreted in different ways according to the context.

→ In mathematics, $\overrightarrow{AB}$ usually represents either the **position vector** of B relative to A or the **displacement** from A to B.

A is called the **initial point** and B the **terminal point**.

This representation shows the properties of the vector, that is,
- the direction of the directed line segment with initial point A and terminal point B (as indicated by the arrow), and
- the magnitude given by the length AB.

Magnitude or size is represented by a number called a **scalar**.

→ When we work with vectors you need to distinguish the vector from its magnitude, which is a non-negative scalar.
The magnitude of $\overrightarrow{AB}$ is simply the length AB and is denoted by $|\overrightarrow{AB}|$. If $\overrightarrow{AB} = \mathbf{a}$ then you can represent its magnitude by $|\mathbf{a}|$ or simply by a.

To distinguish vectors from other quantities, we write them in bold, $\mathbf{d}$, or use a right-pointing arrow above the letter, $\vec{d}$, or a tilde below the letter, $\underset{\sim}{d}$. When you handwrite vectors, you cannot use the bold typeface, so use the arrow or tilde notation.

Equal or equivalent geometric vectors

The vectors $\overrightarrow{AB}$ and $\overrightarrow{CD}$ are equal or equivalent as they have the same direction and magnitude. In this case, $\overrightarrow{AB} = \overrightarrow{CD}$.

$\overrightarrow{BA}$ has the same length but the opposite direction to $\overrightarrow{AB}$ and so they are not equal. Vectors with the same magnitude and opposite direction are called **opposite vectors**.

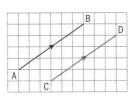

The opposite of a vector is also called its **negative**.

Example 1

ABCD is a rectangle. Using the given points, write down pairs of vectors that
a have the same direction and magnitude
b have the same magnitude and are parallel but have different directions
c are non-parallel and have the same magnitude.

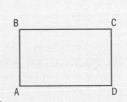

Answers

a $\overrightarrow{AB}$ and $\overrightarrow{DC}$
or
$\overrightarrow{AD}$ and $\overrightarrow{BC}$

The opposite sides of a rectangle have the same length and are parallel.

b $\overrightarrow{AB}$ and $\overrightarrow{CD}$
or
$\overrightarrow{AD}$ and $\overrightarrow{CB}$

Modify the answer from part a by keeping one of the vectors and taking the opposite of the other.

c $\overrightarrow{AC}$ and $\overrightarrow{BD}$

The diagonals of a rectangle have the same length but are not parallel.

A zero vector is also called a trivial vector.
A **non-trivial vector** has a magnitude greater than zero.

Exercise 11A

1 The diagram shows a regular hexagon ABCDEF.

  a Use the properties of the polygon to show that $\overrightarrow{AB} = \overrightarrow{ED}$.

  b Using the given points, write down all the vectors that have

   i the same direction as $\overrightarrow{AB}$

   ii the same magnitude as $\overrightarrow{CF}$.

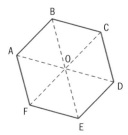

Draw an accurate diagram of a hexagon with sides 3 cm. Which of the geometric properties of this shape have you used?

2 Consider the regular pentagon ABCDE and the vectors shown in the diagram.

  a Use the properties of regular pentagons to show that the vectors in the diagram are all distinct.

  b Is it possible to define a pair of **non-trivial** equivalent vectors using the points given? Give reasons.

Investigation – regular polygons

Explore other regular polygons and the vectors defined by their vertices. Find examples of regular polygons whose vertices do not define a pair of non-trivial equivalent vectors.

Basic operations with geometric vectors

→ The sum of two vectors is determined by the parallelogram law: $\overrightarrow{BA} + \overrightarrow{BC} = \overrightarrow{BD}$ where ABCD is a parallelogram.

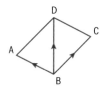

The parallelogram law is a practical rule used in physics that works very well when applied to vectors with different directions but the same initial point.

Is the parallelogram law a good definition? What about $\overrightarrow{AB} + \overrightarrow{AB}$ or $\overrightarrow{AB} + \overrightarrow{BD}$?

In mechanics, looking at the effect of forces, the initial point is very important as it represents the point of application. The effect of the force is determined not only by its intensity but also by the point at which it is applied. Since the initial point is not one of the properties that defines a vector, a more general definition is needed for the addition of geometric vectors which may not have the same initial point.

Here are two vectors $\overrightarrow{AB}$ and $\overrightarrow{CD}$.

These vectors have different initial points. To determine $\overrightarrow{AB} + \overrightarrow{CD}$ consider the vector $\overrightarrow{BE} = \overrightarrow{CD}$ and obtain $\overrightarrow{AB} + \overrightarrow{CD} = \overrightarrow{AE}$, the geometric vector with initial point A and terminal point E.

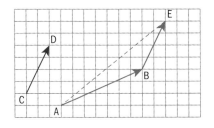

→ This definition is called the **triangle law** and has the advantage that it can easily be applied to the addition of several vectors.

You may think of the triangle law as the 'elephant rule'. Why?

Example 2

Consider the vectors shown in the diagram.
Find $\overrightarrow{AB} + \overrightarrow{CD} + \overrightarrow{EF}$.

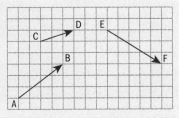

Answer

Draw $\overrightarrow{BP} = \overrightarrow{CD}$ and $\overrightarrow{PQ} = \overrightarrow{EF}$.
$\overrightarrow{AB} + \overrightarrow{CD} + \overrightarrow{EF} = \overrightarrow{AQ}$ (represented in red in the diagram).
$\overrightarrow{AB} + \overrightarrow{CD} + \overrightarrow{EF} = \overrightarrow{AQ}$

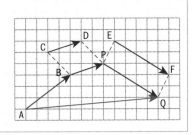

The triangle law allows you to geometrically verify the properties of vector addition.

→ **Properties of vector addition**

Commutative property: $\mathbf{u} + \mathbf{v} = \mathbf{v} + \mathbf{u}$

Associative property: $(\mathbf{u} + \mathbf{v}) + \mathbf{w} = \mathbf{u} + (\mathbf{v} + \mathbf{w})$

Additive identity property (the zero vector): $\mathbf{u} + \mathbf{0} = \mathbf{0} + \mathbf{u} = \mathbf{u}$

Additive inverse property (the opposite vector):
$\mathbf{u} + (-\mathbf{u}) = (-\mathbf{u}) + \mathbf{u} = \mathbf{0}$

Later in the chapter these properties can be verified algebraically.

Example 3

Draw a diagram to show that the addition of vectors is commutative.

Answer

Draw two vectors **u** and **v** with the same initial point. Then use the triangle law to obtain **u** + **v** and **v** + **u** which are the same vector.

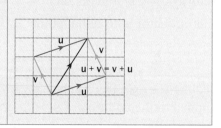

Why are the parallelogram law and the triangle law equivalent?

Example 4

Draw a diagram to show the relationship between the zero vector and opposite vectors.

Answer

Draw two parallel vectors **u** and **v** with the same magnitude and opposite directions. Then add them using the triangle law to obtain a vector represented by a point – the zero vector.

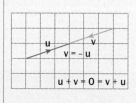

The existence of opposite vectors provides a natural definition for the difference of two vectors.

→ The difference of two vectors **u** and **v** is the vector obtained when you add **u** to the opposite of **v**.

Another operation with vectors that arises naturally from vector addition is the multiplication of a vector by a scalar (in general, a real number) – **scalar multiplication**.

→ The product of a vector **u** and a positive scalar k is another vector **v** with the same direction as **u** and magnitude $|\mathbf{v}| = k|\mathbf{u}|$.

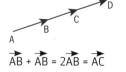

$$\overrightarrow{AB} + \overrightarrow{AB} = 2\overrightarrow{AB} = \overrightarrow{AC}$$
$$\overrightarrow{AB} + \overrightarrow{AB} + \overrightarrow{AB} = 3\overrightarrow{AB} = \overrightarrow{AD}$$

The symbol | | has different meanings dependent on the context. For example, when applied to a scalar it signifies absolute value, but when applied to vectors it signifies magnitude. Are there other mathematical symbols that also have different meanings in different contexts?

→ In general, the product of a vector **u** and a negative scalar k is another vector **v** with the opposite direction to **u** and magnitude
$$|\mathbf{v}| = |k||\mathbf{u}| = -|k\mathbf{u}|.$$

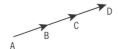

$-\overrightarrow{AB} = \overrightarrow{BA}$,

$-2\overrightarrow{AB}$ can be defined as

$2(-\overrightarrow{AB}) = \overrightarrow{CA}$.

Example 5

The diagram shows the vectors **a** and **b**.
Draw the vectors

a 2**a** **b** −3**b** **c** 2**a** − 3**b**

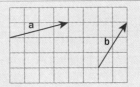

What is the result of the multiplication of the scalar 0 by any vector?

Answers

a

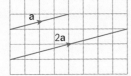

Draw a vector in the same direction as **a** *with magnitude equal to twice the magnitude of* **a**.

b

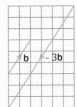

Draw a vector in the opposite direction to **b** *with magnitude equal to three times the magnitude of* **b**.

c

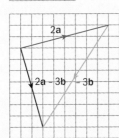

Use the triangle rule to add the vectors found in parts **a** *and* **b**.

→ **Properties of scalar multiplication**

Commutative property: $\alpha\mathbf{v} = \mathbf{v}\alpha$

Associative property: $\alpha(\beta\mathbf{v}) = (\alpha\beta)\mathbf{v}$

Distributive property (1): $\alpha(\mathbf{u} + \mathbf{v}) = \alpha\mathbf{u} + \alpha\mathbf{v}$

Distributive property (2): $(\alpha + \beta)\mathbf{v} = \alpha\mathbf{v} + \beta\mathbf{v}$

Multiplicative identity property: $1\mathbf{v} = \mathbf{v}$

Property of zero: $0\mathbf{v} = \mathbf{0}$ and $\alpha\mathbf{0} = \mathbf{0}$

When writing expressions that involve both scalars and vectors, it is common to use a Greek letter to denote the scalar quantity.

Exercise 11B

1 Draw a diagram to show that the addition of vectors is associative.

2 The diagram shows a regular hexagon ABCDEF.

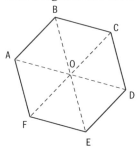

Using only vectors defined by the vertices of the hexagon, copy and complete these statements.

a $\overrightarrow{AF} + \overrightarrow{BC} = \ldots$

b $\frac{1}{2}\overrightarrow{AD} + \overrightarrow{ED} = \ldots$

c $2\overrightarrow{FE} - \overrightarrow{AF} - \overrightarrow{FE} = \ldots$

d $\frac{1}{2}(\overrightarrow{AD} + \overrightarrow{BE}) = \ldots$

e $-\frac{1}{2}\overrightarrow{FC} + \overrightarrow{BC} = \ldots$

f $-2\overrightarrow{ED} - \overrightarrow{AF} + \overrightarrow{AB} = \ldots$

3 The diagram shows a parallelepiped ABCDEFGH.

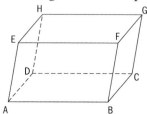

a Let $\mathbf{u} = \overrightarrow{AB}$, $\mathbf{v} = \overrightarrow{AD}$ and $\mathbf{w} = \overrightarrow{AG}$. Express each of these vectors in terms of $\mathbf{u}$, $\mathbf{v}$ and $\mathbf{w}$.

   i $\overrightarrow{AC}$ **ii** $\overrightarrow{HB}$ **iii** $\overrightarrow{CE}$ **iv** $\overrightarrow{AF}$

b Given that $|\overrightarrow{AD}| = 3$, $|\overrightarrow{AB}| = 4$ and $|\overrightarrow{AC}| = 6$, find

   i the angle ABC

   ii the area of the parallelogram ABCD.

4 Use the properties of vector addition and scalar multiplication to solve these equations for $\mathbf{x}$

a $3\mathbf{x} - \mathbf{u} = 6\mathbf{v} + 2\mathbf{u}$

b $2(\mathbf{x} - \mathbf{u}) + 3(\mathbf{u} - \mathbf{v}) = 0$

c $\frac{1}{2}(\mathbf{x} - \mathbf{u}) = \frac{1}{3}(\mathbf{x} + \mathbf{v})$

11.2 Introduction to vector algebra

In this section, vectors will be defined analytically in terms of **components**. Vector operations will then be defined in terms of these components and you can explore the properties of the operations using vector algebra.

In 2-D, consider a pair of perpendicular axes that meet at a point O, the **origin**, and use the same unit for both axes. Each axis has an associated **unit vector**: **i** in the positive direction of the x-axis and **j** in the positive direction of the y-axis.

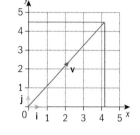

A vector **v** = $\overrightarrow{AB}$ in the plane is represented by two numbers a and b which indicate the horizontal and vertical displacement from A to B respectively, as illustrated in the diagram. The numbers a and b are components of the vector **v** and they define the vector.

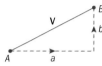

The column vector $\begin{pmatrix} a \\ b \end{pmatrix}$ represents the vector **v** in

component form and means **v** = $a\mathbf{i} + b\mathbf{j}$.
v is a **linear combination** of the vectors **i** and **j**.

→ In 2-D space, $\mathbf{i} = \begin{pmatrix} 1 \\ 0 \end{pmatrix}$, $\mathbf{j} = \begin{pmatrix} 0 \\ 1 \end{pmatrix}$ and O(0, 0).

Example 6

Given A(−1, 2) and B(3, 1),
a write down the vectors $\overrightarrow{OA}$ and $\overrightarrow{OB}$ as column vectors
b find $\overrightarrow{AB}$.

Answers

a $\overrightarrow{OA} = \begin{pmatrix} -1 \\ 2 \end{pmatrix}$ and $\overrightarrow{OB} = \begin{pmatrix} 3 \\ 1 \end{pmatrix}$

b $\overrightarrow{AB} = \overrightarrow{OB} - \overrightarrow{OA} = \begin{pmatrix} 3 \\ 1 \end{pmatrix} - \begin{pmatrix} -1 \\ 2 \end{pmatrix} = \begin{pmatrix} 4 \\ -1 \end{pmatrix}$ *Subtract the components.*

Similarly, in 3-D there are three mutually perpendicular axes: the x-axis, the y-axis and the z-axis. The corresponding unit vectors are **i**, **j** and **k** respectively and a vector is defined by its three components:

$\mathbf{v} = \begin{pmatrix} a \\ b \\ c \end{pmatrix}$ which means $\mathbf{v} = a\mathbf{i} + b\mathbf{j} + c\mathbf{k}$

The analytic treatment of vectors has its roots in the 17th century when **Descartes** (1596–1650) used a pair of numbers (x, y) to locate a point in the plane and a triple of numbers (x, y, z) to represent a point in space. In the 19th century, mathematicians including **Arthur Cayley** (1821–95) realized that there is no mathematical reason to stop with three numbers, but the geometric images that help us to illustrate concepts are not available in greater than three dimensions.

Different texts may use different notations to represent vectors, for example, row vectors <a, b>.

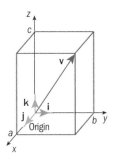

→ In 3-D space, $\mathbf{i} = \begin{pmatrix} 1 \\ 0 \\ 0 \end{pmatrix}$, $\mathbf{j} = \begin{pmatrix} 0 \\ 1 \\ 0 \end{pmatrix}$ and $\mathbf{k} = \begin{pmatrix} 0 \\ 0 \\ 1 \end{pmatrix}$

The origin O has coordinates (0, 0, 0)

Example 7

Given A(1, 3, −1) and B(3, −1, 1),
a write down the vectors $\overrightarrow{OA}$ and $\overrightarrow{OB}$ as column vectors
b express $\overrightarrow{AB}$ as a linear combination of the unit vectors of the Cartesian axes.

Answers

a $\overrightarrow{OA} = \begin{pmatrix} 1 \\ 3 \\ -1 \end{pmatrix}$ and $\overrightarrow{OB} = \begin{pmatrix} 3 \\ -1 \\ 1 \end{pmatrix}$ $\overrightarrow{OA} = \begin{pmatrix} x_A \\ y_A \\ z_A \end{pmatrix}$, $\overrightarrow{OB} = \begin{pmatrix} x_B \\ y_B \\ z_B \end{pmatrix}$

> You will learn more about 3-D vector geometry in Section 11.3.

b $\overrightarrow{AB} = \overrightarrow{OB} - \overrightarrow{OA}$

$= \begin{pmatrix} 3 \\ -1 \\ 1 \end{pmatrix} - \begin{pmatrix} 1 \\ 3 \\ -1 \end{pmatrix} = \begin{pmatrix} 2 \\ -4 \\ 2 \end{pmatrix}$

Therefore, $\overrightarrow{AB} = 2\mathbf{i} - 4\mathbf{j} + 2\mathbf{k}$

Exercise 11C

1 The diagram shows four identical parallelograms ABEF, BCDE, FEHG and EDIH.
Let A(−1, 3), C(5, 4) and I(7, 8).

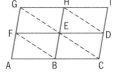

a Find in component form
i $\overrightarrow{AB}$ **ii** $\overrightarrow{AE}$ **iii** $\overrightarrow{CD}$
b Express as a linear combination of the unit vectors of the Cartesian axes
i $\overrightarrow{BF}$ **ii** $\overrightarrow{CH}$ **iii** $\overrightarrow{DG}$
c Find the position vectors of the points B, D, E, F and G.

2 Given P(0, 2, −1) and Q(2, 1, 1)

a write down the vectors $\overrightarrow{OP}$ and $\overrightarrow{OQ}$ in component form
b express $\overrightarrow{PQ}$ as a linear combination of the unit vectors of the Cartesian axes.

3 The diagram shows a parallelepiped ABCDEFGH.

Given $\overrightarrow{OA} = \begin{pmatrix} 1 \\ 1 \\ 1 \end{pmatrix}$, $\overrightarrow{OB} = \begin{pmatrix} 2 \\ 3 \\ 3 \end{pmatrix}$, $\overrightarrow{OD} = \begin{pmatrix} 3 \\ 0 \\ 0 \end{pmatrix}$ and $\overrightarrow{OE} = \begin{pmatrix} 3 \\ 2 \\ -1 \end{pmatrix}$

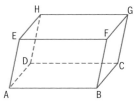

find in component form:

a $\overrightarrow{AB}$ **b** $\overrightarrow{AD}$ **c** $\overrightarrow{AE}$

d $\overrightarrow{AG}$ **e** $\overrightarrow{BD}$ **f** $\overrightarrow{BH}$

4 Let P(−3, 1), Q(5, 7) and R(−1, 5).

> **a** Write down the vectors $\overrightarrow{OP}$, $\overrightarrow{OQ}$ and $\overrightarrow{OR}$ as column vectors.
>
> **b** Find the coordinates of M and N, the midpoints of the line segments PQ and PR respectively.
>
> **c** Show that $\overrightarrow{QR} = 2\overrightarrow{MN}$.

> Use the formula for the midpoint of a line segment.

Vector algebra in two dimensions

→ Given two vectors in the plane, $\mathbf{u} = \begin{pmatrix} u_1 \\ u_2 \end{pmatrix}$ and $\mathbf{v} = \begin{pmatrix} v_1 \\ v_2 \end{pmatrix}$, and a real number λ:

- The sum of the two vectors $\mathbf{u}$ and $\mathbf{v}$ is defined by

$$\mathbf{u} + \mathbf{v} = \begin{pmatrix} u_1 \\ u_2 \end{pmatrix} + \begin{pmatrix} v_1 \\ v_2 \end{pmatrix} = \begin{pmatrix} u_1 + v_1 \\ u_2 + v_2 \end{pmatrix}$$

- The product of a scalar λ and a vector $\mathbf{u}$ is defined by $\lambda\mathbf{u} = \begin{pmatrix} \lambda u_1 \\ \lambda u_2 \end{pmatrix}$

- The zero vector or null vector is $\mathbf{0} = \begin{pmatrix} 0 \\ 0 \end{pmatrix}$

- The opposite vector of $\mathbf{u} = \begin{pmatrix} u_1 \\ u_2 \end{pmatrix}$ is $-\mathbf{u} = \begin{pmatrix} -u_1 \\ -u_2 \end{pmatrix}$

> These definitions are consistent with the previous geometric definitions.

Example 8

Let $\mathbf{a} = \begin{pmatrix} 1 \\ 3 \end{pmatrix}$ and $\mathbf{b} = \begin{pmatrix} 2 \\ 2 \end{pmatrix}$.

a Find $\mathbf{a} + \mathbf{b}$ in component form.

b Draw a diagram to represent $\mathbf{a}$, $\mathbf{b}$ and $\mathbf{a} + \mathbf{b}$.

Answers

a $\mathbf{a} + \mathbf{b} = \begin{pmatrix} 1 \\ 3 \end{pmatrix} + \begin{pmatrix} 2 \\ 2 \end{pmatrix} = \begin{pmatrix} 3 \\ 5 \end{pmatrix}$

b

*Add each component of **a** to the corresponding component of **b**.*
***a** + **b** represents a horizontal displacement of 3 units to the right and a vertical displacement of 5 units up.*

Example 9

Let $\mathbf{a} = \begin{pmatrix} 1 \\ 2 \end{pmatrix}$

a Find $2\mathbf{a}$, $3\mathbf{a}$, $-\mathbf{a}$ and $-4\mathbf{a}$ in component form.

b Draw a diagram to represent all the vectors, showing that the geometric and analytic definition of scalar multiplication are consistent.

Answers

a $2\mathbf{a} = \begin{pmatrix} 2 \\ 4 \end{pmatrix}$, $3\mathbf{a} = \begin{pmatrix} 3 \\ 6 \end{pmatrix}$,

$-\mathbf{a} = \begin{pmatrix} -1 \\ -2 \end{pmatrix}$ and $-4\mathbf{a} = \begin{pmatrix} -4 \\ -8 \end{pmatrix}$

For scalar multiplication, multiply each component by the scalar in front of the vector:

$$2\mathbf{a} = \begin{pmatrix} 2 \times 1 \\ 2 \times 2 \end{pmatrix}$$

Draw the vectors using their components.

b

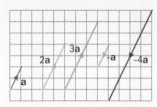

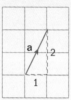

2**a** has the direction of **a** and its magnitude is twice that of **a**.

3**a** has the direction of **a** and its magnitude is three times that of **a**.

−**a** is the opposite of **a**.

−4**a** has the opposite direction of **a** and its length is four times the length of **a**.

Verify that the results obtained algebraically are the same as the results you would obtain using the geometric definition.

The analytic treatment of vectors allows us to verify the properties of operations without the need for diagrams.

Example 10

Use plane vector algebra to show that the addition of vectors is commutative.

Answer

For any two vectors **u** and **v**,

$$\mathbf{u} + \mathbf{v} = \begin{pmatrix} u_1 \\ u_2 \end{pmatrix} + \begin{pmatrix} v_1 \\ v_2 \end{pmatrix}$$

$$= \begin{pmatrix} u_1 + v_1 \\ u_2 + v_2 \end{pmatrix} = \begin{pmatrix} v_1 + u_1 \\ v_2 + u_2 \end{pmatrix}$$

$$= \begin{pmatrix} v_1 \\ v_2 \end{pmatrix} + \begin{pmatrix} u_1 \\ u_2 \end{pmatrix} = \mathbf{v} + \mathbf{u}$$

Use the commutative property of addition of real numbers.

Example 11

> Use plane vector algebra to show that the zero
>
> vector $\mathbf{0} = \begin{pmatrix} 0 \\ 0 \end{pmatrix}$ is the identity element of vector addition.

Use the properties of vector addition on page 565.

Answer

For any vector u,

$$\mathbf{u} + \mathbf{0} = \begin{pmatrix} u_1 \\ u_2 \end{pmatrix} + \begin{pmatrix} 0 \\ 0 \end{pmatrix} = \begin{pmatrix} u_1 + 0 \\ u_2 + 0 \end{pmatrix}$$

Use the fact that 0 is the identity element for the addition of real numbers.

$$= \begin{pmatrix} u_1 \\ u_2 \end{pmatrix} = \mathbf{u} \text{ and}$$

$$\mathbf{0} + \mathbf{u} = \begin{pmatrix} 0 \\ 0 \end{pmatrix} + \begin{pmatrix} u_1 \\ u_2 \end{pmatrix} = \begin{pmatrix} 0 + u_1 \\ 0 + u_2 \end{pmatrix}$$

$$= \begin{pmatrix} u_1 \\ u_2 \end{pmatrix} = \mathbf{u}$$

Exercise 11D

1 Use plane vector algebra to show that
   a $\mathbf{u} + (-\mathbf{u}) = \mathbf{0}$, for any vector $\mathbf{u}$ and its opposite $-\mathbf{u}$.
   b $\mathbf{u} + (\mathbf{v} + \mathbf{w}) = (\mathbf{u} + \mathbf{v}) + \mathbf{w}$, for any vectors $\mathbf{u}$, $\mathbf{v}$ and $\mathbf{w}$.
   c $\alpha\,(\beta\mathbf{u}) = (\alpha\beta)\mathbf{u} = \beta(\alpha\mathbf{u})$, for any scalars α and β.
   d $\alpha\,(\mathbf{u} + \mathbf{v}) = \alpha\mathbf{u} + \alpha\mathbf{v}$, for any vectors $\mathbf{u}$ and $\mathbf{v}$ and any scalar α.
   e $(\alpha + \beta)\mathbf{u} = \alpha\mathbf{u} + \beta\mathbf{u}$, for any vector $\mathbf{u}$ and any scalars α and β.
   f $0\mathbf{u} = \mathbf{0}$, for any vector $\mathbf{u}$.
   g $\alpha\mathbf{0} = \mathbf{0}$, for any scalar α.

2 Solve these equations.

   a $2\begin{pmatrix} x \\ y \end{pmatrix} - 3\begin{pmatrix} y \\ x \end{pmatrix} = 5\begin{pmatrix} 1 \\ -2 \end{pmatrix}$ b $2\left(\begin{pmatrix} 2 \\ y \end{pmatrix} - \begin{pmatrix} x \\ 2 \end{pmatrix}\right) - \begin{pmatrix} 1 \\ 3 \end{pmatrix} = 0$

3 Simplify these expressions, stating all the properties that you use.

   a $\mathbf{u} + (\mathbf{v} + 2\mathbf{u})$ b $(\mathbf{u} - \mathbf{v}) + 2\,(\mathbf{v} - 2\mathbf{u})$ c $3\left(\dfrac{1}{6}(\mathbf{u} - \mathbf{v}) + \dfrac{1}{3}(\mathbf{v} - \mathbf{u})\right)$

EXAM-STYLE QUESTION

4 The vectors $\mathbf{a}$ and $\mathbf{b}$ are given by $\mathbf{a} = 2\mathbf{i} - 3\mathbf{j}$ and $\mathbf{b} = -\mathbf{i} - 2\mathbf{j}$.
   Find the values of the scalars α and β such that $\alpha\mathbf{a} + \beta\mathbf{b} = 3\mathbf{i} - \mathbf{j}$.
   Hence, write $6\mathbf{i} - 2\mathbf{j}$ as a linear combination of $\mathbf{a}$ and $\mathbf{b}$.

Magnitude of a vector

> → If $\mathbf{v} = \begin{pmatrix} v_1 \\ v_2 \end{pmatrix}$ then the magnitude of v is given by
>
> $$v = |\mathbf{v}| = \sqrt{v_1^2 + v_2^2}$$

This formula can be deduced directly by using Pythagoras' theorem.

$$v^2 = |v_1|^2 + |v_2|^2$$

Example 12

Let $\mathbf{u} = \begin{pmatrix} 4 \\ -3 \end{pmatrix}$ and $\mathbf{v} = \begin{pmatrix} 6 \\ k \end{pmatrix}$.

a Calculate the magnitude of $\mathbf{u}$.
b Find the values of k such that the magnitude of $\mathbf{v}$ is 10.

Some GDC models allow you to calculate the magnitude of vectors using the **Norm** command.

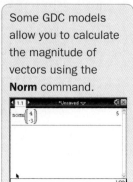

Answers

a $|\mathbf{u}| = \sqrt{4^2 + (-3)^2} = \sqrt{16+9}$ *Use* $|\mathbf{u}| = \sqrt{u_1^2 + u_2^2}$

$\phantom{|\mathbf{u}|} = \sqrt{25} = 5$

b $|\mathbf{v}| = 10 \Rightarrow \sqrt{6^2 + k^2} = 10$ *Use* $|\mathbf{v}| = \sqrt{v_1^2 + v_2^2}$

$36 + k^2 = 100$ *Square both sides of the equation.*
$k^2 = 64 \Rightarrow k = \pm 8$ *Solve for k.*

Given two points A and B, the distance between them, AB, is numerically equal to the magnitude of the vector $\overrightarrow{AB}$.

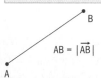

$AB = |\overrightarrow{AB}|$

Example 13

Let A(3, 5) and B(−1, 3).
a Find the distance AB.
b If P(x, y) and AP = BP, express y in terms of x. State the geometric meaning of your result.

Answers

a $\overrightarrow{AB} = \begin{pmatrix} -1-3 \\ 3-5 \end{pmatrix} = \begin{pmatrix} -4 \\ -2 \end{pmatrix}$ *Use* $\overrightarrow{AB} = \overrightarrow{OB} - \overrightarrow{OA}$.

$AB = |\overrightarrow{AB}|$
$ = \sqrt{(-4)^2 + (-2)^2} = \sqrt{16+4}$ *Use* $|\mathbf{u}| = \sqrt{u_1^2 + u_2^2}$
$ = \sqrt{20} = 2\sqrt{5}$

b $\overrightarrow{AP} = \begin{pmatrix} x-3 \\ y-5 \end{pmatrix}$ and $\overrightarrow{BP} = \begin{pmatrix} x+1 \\ y-3 \end{pmatrix}$ *Use* $\overrightarrow{AP} = \overrightarrow{OP} - \overrightarrow{OA}$ *and*
$\overrightarrow{BP} = \overrightarrow{OP} - \overrightarrow{OB}$.

$AP = BP$ *Use* $|\mathbf{u}| = \sqrt{u_1^2 + u_2^2}$ *and the fact that* $AP = BP$.

$\Rightarrow \sqrt{(x-3)^2 + (y-5)^2}$
$= \sqrt{(x+1)^2 + (y-3)^2}$
$\Rightarrow x^2 - 6x + 9 + y^2 - 10y + 25$ *Square both sides, expand, simplify and then solve for y.*
$= x^2 + 2x + 4 + y^2 - 6y + 9$
$\Rightarrow -4y = 8x - 21 \Rightarrow y = -2x + \dfrac{21}{4}$ *Function y is of the form y = ax + b, which represents the equation of a straight line.*

The point P lies on a straight line.

Investigation – locus of a point

Choose two points in the plane, A and B. Let P(x, y) be any point in the plane such that AP = BP. Express y in terms of x, that is, find an equation for the locus of the point P of the form $y = f(x)$. Repeat for other pairs of points A and B.

> For a reminder about triangle facts see page Chapter 14, section 3.

a State your conjecture.

b Prove your conjecture algebraically.

c For each pair of points investigated, plot the points and graph $y = f(x)$. What is the relationship between the graph of y and the line segment AB? Give a geometric argument that justifies this relationship.

Unit and collinear vectors in the plane

A unit vector is any vector with a magnitude of 1 unit.

The **base vectors** $\mathbf{i} = \begin{pmatrix} 1 \\ 0 \end{pmatrix}$ and $\mathbf{j} = \begin{pmatrix} 0 \\ 1 \end{pmatrix}$ are special examples of unit vectors as their directions are the direction of positive coordinate axes x and y. You can obtain a unit vector with the direction of any given vector $\mathbf{v}$.

> Sometimes the unit vector in the direction of $\mathbf{v}$ is represented by $\hat{\mathbf{v}}$.

> → If $\mathbf{v} = \begin{pmatrix} v_1 \\ v_2 \end{pmatrix}$, the **unit vector** in the direction of a non-zero
>
> $$\text{vector } \mathbf{v} \text{ is } \mathbf{u} = \frac{1}{|\mathbf{v}|}\mathbf{v} = \begin{pmatrix} \dfrac{v_1}{\sqrt{v_1^2 + v_2^2}} \\ \dfrac{v_2}{\sqrt{v_1^2 + v_2^2}} \end{pmatrix}.$$

> → Two vectors $\mathbf{u} = \begin{pmatrix} u_1 \\ u_2 \end{pmatrix}$ and $\mathbf{v} = \begin{pmatrix} v_1 \\ v_2 \end{pmatrix}$ are **collinear** if $\mathbf{u} = k\mathbf{v}$ or
>
> $\mathbf{v} = k\mathbf{u}$ for some scalar k
> If $k > 0$, $\mathbf{u}$ and $\mathbf{v}$ have the same direction;
> if $k < 0$, $\mathbf{u}$ and $\mathbf{v}$ have opposite directions.

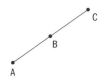

Geometrically, three points are collinear if they lie on a straight line.

Given a non-zero vector $\mathbf{v} = \begin{pmatrix} v_1 \\ v_2 \end{pmatrix}$, you can determine a collinear

vector $\mathbf{u}$ with a given magnitude m using the formula $\mathbf{u} = \pm\dfrac{m}{|\mathbf{v}|}\mathbf{v}$

The plus and minus signs refer to the same or the opposite direction for the collinear vectors.

Example 14

Given $\mathbf{v} = \begin{pmatrix} 3 \\ 4 \end{pmatrix}$, find all possible collinear vectors with magnitude 2.

Answer

$|\mathbf{v}| = 5$

Let $\mathbf{u}$ be a collinear vector.

$|\mathbf{u}| = 2 \Rightarrow \mathbf{u} = \pm\dfrac{2}{5}\mathbf{v}$

So, $\mathbf{u} = \begin{pmatrix} \frac{6}{5} \\ \frac{8}{5} \end{pmatrix}$ or $\mathbf{u} = \begin{pmatrix} -\frac{6}{5} \\ -\frac{8}{5} \end{pmatrix}$

Use $|\mathbf{v}| = \sqrt{v_1^2 + v_2^2}$

Use $\mathbf{u} = \pm\dfrac{m}{|\mathbf{v}|}\mathbf{v}$ with $m = 2$.

Use $\lambda\mathbf{u} = \begin{pmatrix} \lambda u_1 \\ \lambda u_2 \end{pmatrix}$

This example shows you a very important property of the zero vector and justifies the need for the term 'collinear'.

Example 15

Show that the zero vector and **any other** vector are collinear.

Answer

Let $\mathbf{v} = \begin{pmatrix} v_1 \\ v_2 \end{pmatrix}$ be a vector.

$\mathbf{0} = \begin{pmatrix} 0 \\ 0 \end{pmatrix} = \begin{pmatrix} 0v_1 \\ 0v_2 \end{pmatrix} = 0\begin{pmatrix} v_1 \\ v_2 \end{pmatrix} = 0\mathbf{v}$

Therefore, $\mathbf{0}$ and $\mathbf{v}$ are collinear.

Choose and label any vector.

Use $\lambda\mathbf{v} = \begin{pmatrix} \lambda v_1 \\ \lambda v_2 \end{pmatrix}$

> The zero vector has no direction defined. In this case, the term 'parallel' is not appropriate.

Exercise 11E

1 Find unit vectors in the same direction as each of these vectors.

   a $\begin{pmatrix} -1 \\ 5 \end{pmatrix}$ b $\begin{pmatrix} 5 \\ 12 \end{pmatrix}$ c $\begin{pmatrix} -3 \\ 0 \end{pmatrix}$ d $\begin{pmatrix} 1 \\ -1 \end{pmatrix}$

2 Find unit vectors that are collinear with each of these vectors.

   a $\begin{pmatrix} -2 \\ 1 \end{pmatrix}$ b $\begin{pmatrix} -5 \\ -2 \end{pmatrix}$ c $\begin{pmatrix} 0 \\ -1 \end{pmatrix}$ d $\begin{pmatrix} -1 \\ -1 \end{pmatrix}$

3 Find, in component form, the unit vector $\mathbf{v}$ in the same direction as $\mathbf{u} = 2\mathbf{i} - 3\mathbf{j}$.

4 Given that $\mathbf{v}$ and $\mathbf{u}$ are collinear and $|\mathbf{v}| = m$, determine $\mathbf{v}$ when

   a $\mathbf{u} = \begin{pmatrix} 2 \\ 3 \end{pmatrix}$ and $m = 2$

   b $\mathbf{u} = \begin{pmatrix} -2 \\ \sqrt{5} \end{pmatrix}$ and $m = 2$

   c $\mathbf{u} = \begin{pmatrix} 2\sqrt{2} \\ 3\sqrt{2} \end{pmatrix}$ and $m = 13$

5 Find the vector **w** with magnitude $\sqrt{13}$ in the same direction as $\mathbf{u} = -4\mathbf{i} - 6\mathbf{j}$.

6 Given $\mathbf{u} = \mathbf{i} - 3\mathbf{j}$, find the collinear vector **t** with magnitude 5.

7 Show that if $\mathbf{v} = \begin{pmatrix} v_1 \\ v_2 \end{pmatrix}$ the vector $\mathbf{u} = \dfrac{1}{|\mathbf{v}|}\mathbf{v} = \begin{pmatrix} \dfrac{v_1}{\sqrt{v_1^2 + v_2^2}} \\ \dfrac{v_2}{\sqrt{v_1^2 + v_2^2}} \end{pmatrix}$ is the unit vector with the direction of **v**.

> Show that **u** has the direction of **v** and magnitude 1.

8 Given a non-zero vector $\mathbf{v} = \begin{pmatrix} v_1 \\ v_2 \end{pmatrix}$ show that the collinear vectors $\mathbf{u} = \pm\dfrac{m}{|\mathbf{v}|}\mathbf{v}$ have magnitude m ($m \geq 0$).

11.3 Vectors, points and equations of lines

Each point P in the plane can be described by two coordinates: (x, y).

This point can also be described by a position vector $\overrightarrow{OP} = \begin{pmatrix} x \\ y \end{pmatrix}$.

Similarly, a vector $\mathbf{u} = \begin{pmatrix} x \\ y \end{pmatrix}$ can be interpreted as a position vector of a point $P(x, y)$ or as a displacement vector

$\overrightarrow{AB} = \begin{pmatrix} x \\ y \end{pmatrix}$ where $A(x_1, y_1)$, $B(x_2, y_2)$, $x = x_2 - x_1$ and $y = y_2 - y_1$

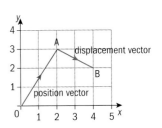

This important fact means you can use coordinate geometry in the study of vectors, and also makes vector algebra a powerful tool in coordinate geometry.

Example 16

Let $A(1, 3)$ and $B(4, 5)$.
 a Draw a diagram to represent $\overrightarrow{AB}$ as a displacement vector.
 b Find the vector $\overrightarrow{AB}$ in component form.
 c Find the length of $\overrightarrow{AB}$.

Answers

a

Plot the points A(1, 3) and B(4, 5) and draw a vector with initial point at A and terminal point at B.

b $\overrightarrow{AB} = \begin{pmatrix} 3 \\ 2 \end{pmatrix}$

The horizontal component is 3 units across. The vertical is 2 units up.

c $|\overrightarrow{AB}| = \sqrt{3^2 + 2^2} = \sqrt{13}$

The displacement from A to B is 3 units to the right and 2 units up. Use $|\mathbf{v}| = \sqrt{v_1^2 + v_2^2}$

Example 17

Given $\overrightarrow{AB} = \begin{pmatrix} -1 \\ 2 \end{pmatrix}$ and A(2, 3), find the coordinates of B when A(2, 3).

Answer

Method 1:

If B(x, y), then $x - 2 = -1$ and
$y - 3 = 2$.

So, $x = 1$ and $y = 5$, that is, B(1, 5).

Method 2:

$\overrightarrow{OB} = \overrightarrow{OA} + \overrightarrow{AB} = \begin{pmatrix} 2 \\ 3 \end{pmatrix} + \begin{pmatrix} -1 \\ 2 \end{pmatrix} = \begin{pmatrix} 1 \\ 5 \end{pmatrix}$ *Rearrange $\overrightarrow{AB} = \overrightarrow{OB} - \overrightarrow{OA}$*

So, B(1, 5).

→ Three points A, B and C in the plane are collinear when $\overrightarrow{AB}$ and $\overrightarrow{AC}$ are collinear vectors, that is when $\overrightarrow{AC} = k\overrightarrow{AB}$ for some scalar k.

Geometrically, points are collinear if they lie on the same line.

Example 18

Show that the points A(−1, 1), B(1, 4) and C(−5, −5) are collinear.

Answer

$\overrightarrow{AB} = \begin{pmatrix} 1-(-1) \\ 4-1 \end{pmatrix} = \begin{pmatrix} 2 \\ 3 \end{pmatrix}$

$\overrightarrow{AC} = \begin{pmatrix} -5-(-1) \\ -5-1 \end{pmatrix} = \begin{pmatrix} -4 \\ -6 \end{pmatrix}$

$\overrightarrow{AC} = \begin{pmatrix} -4 \\ -6 \end{pmatrix} = -2\begin{pmatrix} 2 \\ 3 \end{pmatrix} = -2\overrightarrow{AB}$

Therefore, $\overrightarrow{AC}$ and $\overrightarrow{AB}$ are collinear and hence A, B and C are collinear.

For the points to be collinear, you need to show that $\overrightarrow{AC} = k\,\overrightarrow{AB}$ for some scalar k. First find $\overrightarrow{AB}$ and $\overrightarrow{AC}$ using $\overrightarrow{AB} = \overrightarrow{OB} - \overrightarrow{OA}$ and $\overrightarrow{AC} = \overrightarrow{OC} - \overrightarrow{OA}$

Now show that $\overrightarrow{AC} = k\,\overrightarrow{AB}$ with $k = -2$.

In Example 18, other pairs of vectors could be used to show that the three points are collinear, e.g. $\overrightarrow{AC}$ and $\overrightarrow{BC}$

Exercise 11F

1 The diagram shows a parallelogram ABCD.
   a Write down the coordinates of A, B, C and D.
   b Find $\overrightarrow{AB}$, $\overrightarrow{AC}$ and $\overrightarrow{AD}$ in component form (column vectors).
   c Use your answers to part **b** to determine $\overrightarrow{BD}$ in component form.

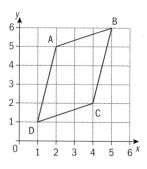

2 Let A(2, 6) and B(−2, 4).

   a Determine $\overrightarrow{AB}$ in component form.

   b Calculate the length of AB. Hence, state the magnitude of $\overrightarrow{AB}$.

   c Find the coordinates of M such that $\overrightarrow{AM} = \overrightarrow{MB}$.
     What is the geometric meaning of $\overrightarrow{AM} = \overrightarrow{MB}$?

   d Find the coordinates of P and Q such that $\overrightarrow{AP} = 2\overrightarrow{PB}$ and $\overrightarrow{AQ} = -2\overrightarrow{QB}$.
     Without calculating them, decide which vector has greater magnitude, $\overrightarrow{AB}$ or $\overrightarrow{PQ}$.

3 Show that the points P(4, −1), Q(6, −3) and R(2, 1) are collinear.

4 Find the value of a such that the points A(a, $a − 1$), B(2, 2a) and C(0, 3a) are collinear.

> Three points define a triangle when they are NOT collinear.

5 Show that the points S(2, −3), U(−1, 2) and N(1, −4) define a triangle.

6 Show that if P(a, b), Q(c, d), R(e, f) and $\dfrac{f - b}{d - b} = \dfrac{e - a}{c - a}$, then P, Q and R are collinear points.

7 Consider the points A($\sin x$, $-1 + \cos x$) and B($\sin 2x$, $\cos 2x$), where $0 < x < \pi$

   a Show that $\overrightarrow{AB}$ is collinear with the vector $\begin{pmatrix} \sin x \\ \cos x \end{pmatrix}$

   b Show that $\overrightarrow{AB} < 1$ for any $0 < x < \pi$

3-D coordinate geometry and vector algebra

In 3-D space, points are represented by three coordinates and vectors by three components. Although all abstract properties of vectors are independent of the dimensions of the space, it is more difficult to visualize points and vectors in 3-D than in 2-D because you cannot draw them on paper. All you can do is draw diagrams where points and vectors are *represented* using perspective rules.

For example, to represent the point A(x_a, y_a, z_a), draw a cuboid with dimensions $|x_a|$ by $|y_a|$ by $|z_a|$ and place the origin O at one of the vertices so that the edges from O lie on the coordinate axes Ox, Oy and Oz. If none of the coordinates of A are zero, OA is a diagonal of the cuboid.

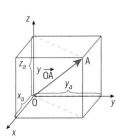

You can draw diagrams with axis in different positions. However, you need to be careful to position the axes so that when you read them counter-clockwise: starting from Ox, you have Oy and then Oz.

A practical way of checking the correct position of the axes is to imagine that you grab the z-axis with your right hand, keeping your thumb pointing up along the positive part of the z-axis. As you close your hand, you should move from the direction of the x-axis to the y-axis.

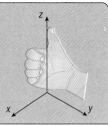

Example 19

Draw a diagram to represent the points A(−1, 1, 1), B(1, −1, 2) and C(1, 1, −1).

Answer

The vector algebra techniques used in 2-D space can be extended to 3-D space.

Points, position vectors and displacement vectors in 3-D space

→ Two points $A(x_1, y_1, z_1)$ and $B(x_2, y_2, z_2)$, have **displacement**

$$\textbf{vector } \overrightarrow{AB} = \begin{pmatrix} x \\ y \\ z \end{pmatrix} \text{ where } x = x_2 - x_1, y = y_2 - y_1 \text{ and } z = z_2 - z_1$$

You can also assign a **position vector** to each point

$$\overrightarrow{OA} = \begin{pmatrix} x_1 \\ y_1 \\ z_1 \end{pmatrix} \text{ and } \overrightarrow{OB} = \begin{pmatrix} x_2 \\ y_2 \\ z_2 \end{pmatrix}$$

Similarly, given a 3-D vector $\mathbf{v} = \begin{pmatrix} x \\ y \\ z \end{pmatrix}$, it can either be

- a position vector of a point with coordinates (x, y, z), or
- a displacement from (x_1, y_1, z_1) to (x_2, y_2, z_2), where $x = x_2 - x_1$, $y = y_2 - y_1$ and $z = z_2 - z_1$

Example 20

Given the points P(3, 1, 2), Q(−1, 3, 1) and R(0, −1, −1),

a find $\overrightarrow{PQ}$ in component form

b express $\overrightarrow{PR}$ as a linear combination of the unit base vectors.

> This example shows that the techniques used in 2-D and 3-D are similar. Compare it to Example 6 on p 563.

Answers

a $\overrightarrow{PQ} = \begin{pmatrix} -1-3 \\ 3-1 \\ 1-2 \end{pmatrix} = \begin{pmatrix} -4 \\ 2 \\ -1 \end{pmatrix}$

$\overrightarrow{PQ}$ *represents the displacement from P to Q, that is,*
$\overrightarrow{PQ} = \overrightarrow{OQ} - \overrightarrow{OP}.$

b $\begin{aligned} \overrightarrow{PR} &= -\mathbf{j} - \mathbf{k} - (3\mathbf{i} + \mathbf{j} + 2\mathbf{k}) \\ &= (0-3)\mathbf{i} + (-1-1)\mathbf{j} + (-1-2)\mathbf{k} \\ &= -3\mathbf{i} - 2\mathbf{j} - 3\mathbf{k} \end{aligned}$

$\overrightarrow{PR} = \overrightarrow{OR} - \overrightarrow{OP}$ *where*
$\overrightarrow{OR} = -\mathbf{j} - \mathbf{k}$ *and*
$\overrightarrow{OP} = 3\mathbf{i} + \mathbf{j} + 2\mathbf{k}.$

Vector addition and scalar multiplication in 3-D space

→ Given two vectors in 3-D space, $\mathbf{u} = \begin{pmatrix} u_1 \\ u_2 \\ u_3 \end{pmatrix}$ and $\mathbf{v} = \begin{pmatrix} v_1 \\ v_2 \\ v_3 \end{pmatrix}$ and a real number λ

- The sum of the two vectors $\mathbf{u}$ and $\mathbf{v}$ is defined by $\mathbf{u} + \mathbf{v} = \begin{pmatrix} u_1 \\ u_2 \\ u_3 \end{pmatrix} + \begin{pmatrix} v_1 \\ v_2 \\ v_3 \end{pmatrix} = \begin{pmatrix} u_1 + v_1 \\ u_2 + v_2 \\ u_3 + v_3 \end{pmatrix}$

- The product of a scalar λ and a vector $\mathbf{u}$ is defined by $\lambda\mathbf{u} = \begin{pmatrix} \lambda u_1 \\ \lambda u_2 \\ \lambda u_3 \end{pmatrix}$

- The zero vector or null vector is $\mathbf{0} = \begin{pmatrix} 0 \\ 0 \\ 0 \end{pmatrix}$

- The opposite vector of $\mathbf{u} = \begin{pmatrix} u_1 \\ u_2 \\ u_3 \end{pmatrix}$ is $-\mathbf{u} = \begin{pmatrix} -u_1 \\ -u_2 \\ -u_3 \end{pmatrix}$

Example 21

Given the vectors $\mathbf{u} = -2\mathbf{i} + 3\mathbf{j} + \mathbf{k}$ and $\mathbf{v} = -\mathbf{i} + 2\mathbf{j} - 3\mathbf{k}$,

a write down $\mathbf{u}$ and $\mathbf{v}$ in component form

b find $\mathbf{u} + \mathbf{v}$, $2\mathbf{u} - 3\mathbf{v}$ and $3(\mathbf{u} - \mathbf{v})$.

Answers

a $\mathbf{u} = \begin{pmatrix} -2 \\ 3 \\ 1 \end{pmatrix}$ and $\mathbf{v} = \begin{pmatrix} -1 \\ 2 \\ -3 \end{pmatrix}$

$\mathbf{u} = u_1\mathbf{i} + u_2\mathbf{j} + u_3\mathbf{k}$ *means* $\mathbf{u} = \begin{pmatrix} u_1 \\ u_2 \\ u_3 \end{pmatrix}$

▶ Continued on next page

b $\mathbf{u} + \mathbf{v} = \begin{pmatrix} -2 \\ 3 \\ 1 \end{pmatrix} + \begin{pmatrix} -1 \\ 2 \\ -3 \end{pmatrix} = \begin{pmatrix} -3 \\ 5 \\ -2 \end{pmatrix}$

or $\mathbf{u} + \mathbf{v} = -3\mathbf{i} + 5\mathbf{j} - 2\mathbf{k}$

$2\mathbf{u} - 3\mathbf{v} = 2\begin{pmatrix} -2 \\ 3 \\ 1 \end{pmatrix} - 3\begin{pmatrix} -1 \\ 2 \\ -3 \end{pmatrix}$

$= \begin{pmatrix} -4 \\ 6 \\ 2 \end{pmatrix} + \begin{pmatrix} 3 \\ -6 \\ 9 \end{pmatrix} = \begin{pmatrix} -1 \\ 0 \\ 11 \end{pmatrix}$

or $2\mathbf{u} - 3\mathbf{v} = -\mathbf{i} + 11\mathbf{k}$

$3(\mathbf{u} - \mathbf{v}) = 3\left(\begin{pmatrix} -2 \\ 3 \\ 1 \end{pmatrix} - \begin{pmatrix} -1 \\ 2 \\ -3 \end{pmatrix} \right) = 3\begin{pmatrix} -1 \\ 1 \\ 4 \end{pmatrix} = \begin{pmatrix} -3 \\ 3 \\ 12 \end{pmatrix}$

or $3(\mathbf{u} - \mathbf{v}) = -3\mathbf{i} + 3\mathbf{j} + 12\mathbf{k}$

For vector addition, use

$\mathbf{u} + \mathbf{v} = \begin{pmatrix} u_1 \\ u_2 \\ u_3 \end{pmatrix} + \begin{pmatrix} v_1 \\ v_2 \\ v_3 \end{pmatrix} = \begin{pmatrix} u_1 + v_1 \\ u_2 + v_2 \\ u_3 + v_3 \end{pmatrix}$ *or*

$\mathbf{u} + \mathbf{v} = (u_1 + v_1)\mathbf{i} + (u_2 + v_2)\mathbf{j} + (u_3 + v_3)\mathbf{k}$

For scalar multiplication, use $\lambda\mathbf{u}\begin{pmatrix} \lambda u_1 \\ \lambda u_2 \\ \lambda u_3 \end{pmatrix}$ *or*

$\lambda\mathbf{u} = (\lambda u_1)\mathbf{i} + (\lambda u_2)\mathbf{j} + (\lambda u_3)\mathbf{k}$
Then add the vectors.

Add $\mathbf{u}$ to the negative of $\mathbf{v}$ and then multiply by the scalar 3.

Magnitude of a vector in 3-D space

→ The magnitude of a vector $\mathbf{v} = \begin{pmatrix} v_1 \\ v_2 \\ v_3 \end{pmatrix}$ is given by $v = |\mathbf{v}| = \sqrt{v_1^2 + v_2^2 + v_3^2}$.

To find the magnitude of a vector in 3-D space, draw a cuboid whose edges are parallel to the coordinate axes and one of whose diagonals represents $\mathbf{v}$. Then apply Pythagoras' theorem to find the length of the diagonal of the cuboid.

$w^2 = v_1{}^2 + v_2{}^2$

$v^2 = w^2 + v_3{}^2$

$\Rightarrow v^2 = v_1{}^2 + v_2{}^2 + v_3{}^2$

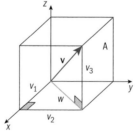

Distance between two points in 3-D space

→ Given two points $A(x_1, y_1, z_1)$ and $B(x_2, y_2, z_2)$, the distance between A and B is given by the magnitude of the vector $\overrightarrow{AB}$

$AB = |\overrightarrow{AB}| = \sqrt{(x_2 - x_1)^2 + (y_2 - y_1)^2 + (z_2 - z_1)^2}$

Unit and collinear vectors in 3-D space

→ The unit vector in the direction of a non-zero vector $\mathbf{v}$ is $\mathbf{u} = \dfrac{1}{|\mathbf{v}|} \mathbf{v} = \begin{pmatrix} \dfrac{v_1}{\sqrt{v_1^2 + v_2^2 + v_3^2}} \\ \dfrac{v_2}{\sqrt{v_1^2 + v_2^2 + v_3^2}} \\ \dfrac{v_3}{\sqrt{v_1^2 + v_2^2 + v_3^2}} \end{pmatrix}$

→ Two vectors $\mathbf{u} = \begin{pmatrix} u_1 \\ u_2 \\ u_3 \end{pmatrix}$ and $\mathbf{v} = \begin{pmatrix} v_1 \\ v_2 \\ v_3 \end{pmatrix}$ are collinear if $\mathbf{u} = k\mathbf{v}$

or $\mathbf{v} = k\mathbf{u}$ for some scalar k. If $k > 0$, $\mathbf{u}$ and $\mathbf{v}$ have the same direction; if $k < 0$, $\mathbf{u}$ and $\mathbf{v}$ have opposite directions.

Given a non-zero vector $\mathbf{v} = \begin{pmatrix} v_1 \\ v_2 \\ v_3 \end{pmatrix}$ you can determine a collinear

vector $\mathbf{u}$ with a given magnitude m using the formula $\mathbf{u} = \pm\dfrac{m}{|\mathbf{v}|} \mathbf{v}$

Example 22

Given the vector $\mathbf{v} = \begin{pmatrix} 1 \\ 2 \\ 3 \end{pmatrix}$ determine in component form

a the unit vector $\mathbf{u}$ in the direction of $\mathbf{v}$
b the unit vectors parallel to $\mathbf{v}$
c the vector with magnitude 7 in the direction of $\mathbf{v}$

Answers

a $|\mathbf{v}| = \sqrt{1 + 4 + 9} = \sqrt{14}$

$\mathbf{u} = \begin{pmatrix} \dfrac{1}{\sqrt{14}} \\ \dfrac{2}{\sqrt{14}} \\ \dfrac{3}{\sqrt{14}} \end{pmatrix}$

Use $|\mathbf{v}| = \sqrt{v_1^2 + v_2^2 + v_3^2}$

Use $\mathbf{u} = \dfrac{1}{|\mathbf{v}|} \mathbf{v} = \begin{pmatrix} \dfrac{v_1}{\sqrt{v_1^2 + v_2^2 + v_3^2}} \\ \dfrac{v_2}{\sqrt{v_1^2 + v_2^2 + v_3^2}} \\ \dfrac{v_3}{\sqrt{v_1^2 + v_2^2 + v_3^2}} \end{pmatrix}$

In this book, we have adopted the convention of column vectors. However, this is not the only the convention in use. Does the country in which you are learning influence the notation used?

▶ Continued on next page

$$\textbf{b} \quad \textbf{u} = \pm \begin{pmatrix} \dfrac{1}{\sqrt{14}} \\ \dfrac{2}{\sqrt{14}} \\ \dfrac{3}{\sqrt{14}} \end{pmatrix}$$

*The unit vectors parallel to **v** are the unit vector in the direction of **v** and its opposite.*

$$\textbf{c} \quad \textbf{u} = \begin{pmatrix} \dfrac{7}{\sqrt{14}} \\ \dfrac{14}{\sqrt{14}} \\ \dfrac{21}{\sqrt{14}} \end{pmatrix}$$

*Multiply each component of the vector found in part **a** by 7.*

Example 23

Let A(−1, 3, 5) and B(3, −5, 1).
a Find the coordinates of M such that $\overrightarrow{AM} = \overrightarrow{MB}$.
b What is M in relation to [AB]?

Answers

a Let M(x, y, z).

$$\overrightarrow{AM} = \overrightarrow{MB} \Rightarrow \begin{pmatrix} x-(-1) \\ y-3 \\ z-5 \end{pmatrix} = \begin{pmatrix} 3-x \\ -5-y \\ 1-z \end{pmatrix}$$

Label the coordinates x, y and z.
Write down the given condition in terms of these coordinates.

$x + 1 = 3 - x \Rightarrow x = 1$
$y - 3 = -5 - y \Rightarrow y = -1$
$z - 5 = 1 - z \Rightarrow z = 3$
Therefore, M(1, −1, 3).

b M is the midpoint of [AB].

Equate the components of the vectors and solve.

M is between A and B because $\overrightarrow{AM}$ and $\overrightarrow{MB}$ have the same direction. M is exactly in the middle because these vectors have the same magnitude.

→ In general, the coordinates of the midpoint M of a line segment [AB], with A(x_1, y_1, z_1) and B(x_2, y_2, z_2), are given by

$$\left(\dfrac{x_1 + x_2}{2}, \dfrac{y_1 + y_2}{2}, \dfrac{z_1 + z_2}{2} \right)$$

Exercise 11G

1 Given the vectors $\textbf{u} = -2\textbf{i} + 3\textbf{j} + \textbf{k}$ and $\textbf{v} = -\textbf{i} + 2\textbf{j} - 3\textbf{k}$, find
  a $\textbf{u} + \textbf{v}$ **b** $-3\textbf{u}$
  c $4\textbf{u} - 2\textbf{v}$ **d** $-2(\textbf{u} - \textbf{v})$

2 Given the vectors $\mathbf{a} = \begin{pmatrix} 2 \\ 3 \\ -1 \end{pmatrix}$, $\mathbf{b} = \begin{pmatrix} -1 \\ 2 \\ -2 \end{pmatrix}$ and $\mathbf{c} = \begin{pmatrix} 0 \\ 1 \\ 3 \end{pmatrix}$ find

   a $\mathbf{a} + \mathbf{b} + \mathbf{c}$ **b** $2\mathbf{a} - \mathbf{b} + \mathbf{c}$ **c** $2(\mathbf{a} - \mathbf{b}) - 3\mathbf{c}$

   d $\frac{1}{2}(\mathbf{a} - 3\mathbf{b})$ **e** $|\mathbf{a}|$ **f** $|\mathbf{b}|$

   g $|\mathbf{a} + \mathbf{b}|$ **h** $|\mathbf{a} - \mathbf{b}|$

3 Let A(0, 2, 1), B(−1, −1, −2) and C(1, −3, 0).

   a Find the vectors $\overrightarrow{AB}$ and $\overrightarrow{AC}$ in component form.

   b Determine $\overrightarrow{AB} - \overrightarrow{AC}$ in component form.

   Hence, write down $\overrightarrow{BC}$ as a linear combination of the base vectors.

4 Given the vector $\mathbf{v} = \begin{pmatrix} 0 \\ 1 \\ -2 \end{pmatrix}$, determine in component form

   a the unit vector $\mathbf{u}$ in the direction of $\mathbf{v}$

   b the unit vectors parallel to $\mathbf{v}$

   c the vector with magnitude 5 in the direction of $\mathbf{v}$

5 Consider this cuboid:

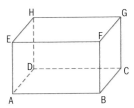

   Let A(4, −1, 3), C(0, −2, 5), D(5, 1, 6) and G(1, −4, 6).

   a Find $\overrightarrow{AC}$, $\overrightarrow{AD}$ and $\overrightarrow{CG}$ in component form.

   b Hence, determine the position vectors of the points B, E, F and H.

Straight lines in 2-D

A straight line is defined by two points. Given two points A and B there is exactly one line that contains both of them, which is the line (AB).

> → A point R is on the line (AB) when $\overrightarrow{AR} = \lambda\overrightarrow{AB}$ for some real value of λ. This is called a **vector equation of the line** AB.
>
> If the point R has position vector $\mathbf{r}$, the point A has position vector $\mathbf{a}$, and $\overrightarrow{AB} = \mathbf{u}$, the direction vector, then the vector equation $\overrightarrow{AR} = \lambda\overrightarrow{AB}$ can be re-written as $\mathbf{r} = \mathbf{a} + \lambda\mathbf{u}$

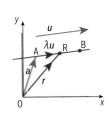

Lines in 2-D

→ If the line (AB) lies in the plane $x \cdot y$, then you can represent

the vector in component form $\mathbf{r} = \begin{pmatrix} x \\ y \end{pmatrix}$, $\mathbf{a} = \begin{pmatrix} x_1 \\ y_1 \end{pmatrix}$ and

$\mathbf{u} = \begin{pmatrix} u_1 \\ u_2 \end{pmatrix}$ This gives a pair of **parametric equations**

$x = x_1 + \lambda u_1$ and $y = y_1 + \lambda u_2$

or $\lambda = \dfrac{x - x_1}{u_1} = \dfrac{y - y_1}{u_2}$

If both components of the vector are non-zero, eliminate the parameter λ to obtain a **Cartesian equation** of the line (AB):

$$y - y_1 = \frac{u_2}{u_1}(x - x_1)$$

This can be reduced to the form $y = mx + c$ where $m = \dfrac{u_2}{u_1}$ is the

gradient of the line and $c = y_1 - \dfrac{u_2}{u_1} x_1$ is the y-intercept.

> $y = mx + c$ is the gradient–intercept form of the equation of a straight line and $y - y_1 = m(x - x_1)$ is the point–slope form.

Example 24

Given the points A(1, 3) and B(2, 5), represent the line (AB) by
a a vector equation **b** parametric equations
c a Cartesian equation.

Answers

a $\overrightarrow{OA} = \begin{pmatrix} 1 \\ 3 \end{pmatrix}$ and $\overrightarrow{OB} = \begin{pmatrix} 2 \\ 5 \end{pmatrix}$

Write down the position vectors of A and B in component form.

$\overrightarrow{AB} = \begin{pmatrix} 2-1 \\ 5-3 \end{pmatrix} = \begin{pmatrix} 1 \\ 2 \end{pmatrix}$

To find the displacement from A to B, use $\overrightarrow{AB} = \overrightarrow{OB} - \overrightarrow{OA}$.

Therefore, $\mathbf{r} = \begin{pmatrix} 1 \\ 3 \end{pmatrix} + \lambda \begin{pmatrix} 1 \\ 2 \end{pmatrix}$.

For the vector equation, use $\mathbf{r} = \mathbf{a} + \lambda \mathbf{u}$ where $\mathbf{a} = \overrightarrow{OA}$ and $\mathbf{u} = \overrightarrow{AB}$.

b $x = 1 + \lambda$ and $y = 3 + 2\lambda$

For the parametric equations, use $x = x_1 + \lambda u_2$ and $y = y_1 + \lambda u_2$ where x_1 and y_1 and u_1 and u_2 are the components of the position vector of point A and vector $\mathbf{u}$ respectively.

c $x - 1 = \dfrac{y-3}{2}$

$\Rightarrow 2(x - 1) = y - 3$

$\Rightarrow y = 2x + 1$

*For the Cartesian equation, eliminate λ from the equations in part **b** and solve for y.*

When one of the components of the vector u is zero, the line is parallel to one of the axes.

- If $\mathbf{u} = \begin{pmatrix} u_1 \\ 0 \end{pmatrix}$, $u_1 \neq 0$, the line is parallel to the x-axis, contains the point (x_1, y_1), and its equation is of the form $y = y_1$.

- If $\mathbf{u} = \begin{pmatrix} 0 \\ u_2 \end{pmatrix}$, $u_2 \neq 0$, the line is parallel to the y-axis, contains the point (x_1, y_1), and its equation is of the form $x = x_1$.

Example 25

Write down vector, parametric and Cartesian equations for the lines through A(−1, 2) parallel to

a $\mathbf{u} = 2\mathbf{i}$ **b** $\mathbf{v} = 3\mathbf{j}$

Answers

a $\mathbf{r} = \begin{pmatrix} -1 \\ 2 \end{pmatrix} + \lambda \begin{pmatrix} 2 \\ 0 \end{pmatrix}$

$x = -1 + 2\lambda$ and $y = 2$

$y = 2$

b $\mathbf{r} = \begin{pmatrix} -1 \\ 2 \end{pmatrix} + \lambda \begin{pmatrix} 0 \\ 3 \end{pmatrix}$

$x = -1$ and $y = 2 + 3\lambda$

$x = -1$

Vector equation: use $\mathbf{r} = \mathbf{a} + \lambda \mathbf{u}$ *where*
$\mathbf{a} = \overrightarrow{OA}$ *and* $\mathbf{u} = 2\mathbf{i}$
Parametric equations: use
$x = x_1 + \lambda u_1$ *and* $y = y_1 + \lambda u_2$.
As $u_2 = 0$, *the Cartesian equation is of the form* $y = y_1$
Vector equation: use $\mathbf{r} = \mathbf{a} + \lambda \mathbf{u}$ *where*
$\mathbf{u} = \mathbf{v} = 3\mathbf{j}$
Parametric equations: use
$x = x_1 + \lambda u_1$ *and* $y = y_1 + \lambda u_2$.
As $u_1 = 0$, *the Cartesian equation is of the form* $x = x_1$.

Straight lines in 3-D space

→ If the line (AB) lies in 3-D space, $\mathbf{r} = \begin{pmatrix} x \\ y \\ z \end{pmatrix}$, $\mathbf{a} = \begin{pmatrix} x_1 \\ y_1 \\ z_1 \end{pmatrix}$ and

$\mathbf{u} = \begin{pmatrix} u_1 \\ u_2 \\ u_3 \end{pmatrix}$, the vector equation $\mathbf{r} = \mathbf{a} + \lambda \mathbf{u}$ can be transformed

into three parametric equations

$x = x_1 + \lambda u_1, y = y_1 + \lambda u_2$ and $z = z_1 + \lambda u_3$

where $\lambda = \dfrac{x - x_1}{u_1} = \dfrac{y - y_1}{u_2} = \dfrac{z - z_1}{u_3}$

If all the components of the vector are non-zero, eliminate the parameter λ to obtain Cartesian equations of the line (AB)

$$\frac{x - x_1}{u_1} = \frac{y - y_1}{u_2} = \frac{z - z_1}{u_3}$$

Different textbooks may use different notation for lines, segments and their lengths. Here we use IB notation: [AB] is the line segment with end points A and B, AB is the length of [AB], and (AB) is the line containing points A and B.

When one of the components of the vector **u** is zero the line is parallel to one of the coordinate planes. If two components are zero, the line is parallel to one of the axes.

A line in 3-D is defined by two Cartesian equations.

Example 26

Given the points A(1, 3, −1) and B(2, 5, 0), represent the line (AB) by
a a vector equation
b parametric equations
c Cartesian equations.

Answers

a $\overrightarrow{AB} = \begin{pmatrix} 2 \\ 5 \\ 0 \end{pmatrix} - \begin{pmatrix} 1 \\ 3 \\ -1 \end{pmatrix} = \begin{pmatrix} 1 \\ 2 \\ 1 \end{pmatrix}$

Write the position vectors of A and B in component form and use
$\overrightarrow{AB} = \overrightarrow{OB} - \overrightarrow{OA}$

$\mathbf{r} = \begin{pmatrix} 1 \\ 3 \\ -1 \end{pmatrix} + \lambda \begin{pmatrix} 1 \\ 2 \\ 1 \end{pmatrix}$

For the vector equation, use
$\mathbf{r} = \mathbf{a} + \lambda\mathbf{u}$ *where* $\mathbf{a} = \overrightarrow{OA}$ *and*
$\mathbf{u} = \overrightarrow{AB}$

b $x = 1 + \lambda$
$y = 3 + 2\lambda$
$z = -1 + \lambda$

For the parametric equations,
use $x = x_1 + \lambda u_1, y = y_1 + \lambda u_2$ *and*
$z = z_1 + \lambda u_3$

where $\begin{pmatrix} x_1 \\ y_1 \\ z_1 \end{pmatrix} = \mathbf{a}$

and $\begin{pmatrix} u_1 \\ u_2 \\ u_3 \end{pmatrix} = \mathbf{u}$

c $x - 1 = \dfrac{y-3}{2} = z + 1$

For the Cartesian equations, eliminate
λ *from the equations in part* **b** *and*
equate.

The equations of the coordinate axes are:
x-axis: $y = z = 0$
y-axis: $x = z = 0$
z-axis: $x = y = 0$

This example shows you how to determine a vector parallel to a line and one of its points given its Cartesian equations.

Example 27

A line *l* has Cartesian equations $\dfrac{x+2}{3} = \dfrac{1-y}{2} = 2z - 1$

Find the coordinates of one of the points on line *l* and a vector parallel to *l*

If a vector is parallel to a line it is called a direction vector of the line.

▶ Continued on next page

Answer

$$\frac{x+2}{3} = \frac{1-y}{2} = 2z - 1$$

$$\Rightarrow \frac{x+2}{3} = \frac{y-1}{-2} = \frac{z-\frac{1}{2}}{\frac{1}{2}}$$

So, $A\left(-2, 1, \frac{1}{2}\right)$ is a point on line l

and $\mathbf{u} = 3\mathbf{i} - 2\mathbf{j} + \frac{1}{2}\mathbf{k}$ is parallel to l.

Re-arrange the equations to the form $\frac{x-x_1}{u_1} = \frac{y-y_1}{u_2} = \frac{z-z_1}{u_3}$ where (x_1, y_1, z_1) are the coordinates of a point on the line and $\mathbf{u} = u_1\mathbf{i} + u_2\mathbf{j} + u_3\mathbf{k}$ has the direction of l

> $\mathbf{u} = 6\mathbf{i} - 4\mathbf{j} + \mathbf{k}$ is another vector parallel to l whose components are integers.

Exercise 11H

1. Given the points $P(1, 3)$ and $Q(2, 5)$, find vector, parametric and Cartesian equations for PQ.

2. Write down vector, parametric and Cartesian equations for the line through the point $A(1, -1, 1)$ in the direction of $\mathbf{u} = 2\mathbf{i} - \mathbf{j} + 3\mathbf{k}$.

3. Given the equation $\frac{x+1}{3} = \frac{2y}{3} = z - 1$ of a line l, write down the coordinates of one of its points and the components of a vector in its direction.

4. A line L has vector equation $\mathbf{r} = \begin{pmatrix} 1 \\ 1 \\ -1 \end{pmatrix} + \lambda \begin{pmatrix} -1 \\ 0 \\ 3 \end{pmatrix}$

   a. Find the coordinates of three distinct points on L.
   b. Show that the point $P(0, 3, 2)$ does not lie on the line.
   c. Write down a vector equation of the line through P parallel to L.

5. A line has vector equation $\mathbf{r} = (1 + k)\mathbf{i} - k\mathbf{j} + 2\mathbf{k}$.
   a. Write down the coordinates of two of its points.
   b. Find, in component form, a vector $\mathbf{u}$ with magnitude 4 parallel to the line.

11.4 Scalar product

Although a vector is an abstract concept, vector algebra has many applications due to a remarkable fact: it is possible to define more than one multiplication between vectors. Moreover, each multiplication has a geometric meaning and provides a useful tool when solving different geometric problems.

Scalar product of two vectors

The **scalar product** of two vectors, also known as the **inner product** or the **dot product**, is defined geometrically as

> → Given two non-zero vectors **u** and **v**, $\mathbf{u} \cdot \mathbf{v} = |\mathbf{u}||\mathbf{v}| \cos \theta$, where θ is the angle between **u** and **v**.

If one of the vectors is the null vector, the scalar (or dot) product is zero.

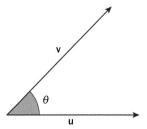

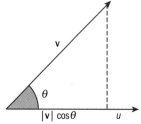

▲ To measure the angle between two vectors, the vectors must have the same initial point.

▲ $|\mathbf{v}| \cos \theta$ is called the projection of **v** in the direction of **u**.

> → Here are some important consequences of the geometric definition of the scalar product.
> - The scalar product of two vectors is always a number.
> - The definition of scalar product does not depend on the dimensions of the space.
> - $\mathbf{u} \cdot \mathbf{v} = 0$ if and only if $\mathbf{u} = \mathbf{0}$, $\mathbf{v} = \mathbf{0}$ or **u** and **v** are **orthogonal**.
> - $\mathbf{u} \cdot \mathbf{v} = \pm|\mathbf{u}||\mathbf{v}|$ if **u** and **v** are parallel.
> - $\mathbf{u} \cdot \mathbf{u} = |\mathbf{u}|^2$
> - $\mathbf{u} \cdot \mathbf{v} > 0$ when θ is acute and $\mathbf{u} \cdot \mathbf{v} < 0$ when θ is obtuse.
> - $\mathbf{u} \cdot \mathbf{v} = \mathbf{v} \cdot \mathbf{u}$
> - $\mathbf{u} \cdot (\mathbf{v} + \mathbf{w}) = \mathbf{u} \cdot \mathbf{v} + \mathbf{u} \cdot \mathbf{w}$
> - $(\lambda \mathbf{u}) \cdot \mathbf{v} = \lambda (\mathbf{u} \cdot \mathbf{v})$

Orthogonal means 'at right angles'.

It is common to use the terms orthogonal, normal and perpendicular when the angle between the directions of the vectors is a right angle.

You can prove all of these properties, for example:

1 $\mathbf{u} \cdot \mathbf{v} = 0$ when **u** and **v** are orthogonal because the angle between them is $\pm 90°$ and, therefore,

$\mathbf{u} \cdot \mathbf{v} = |\mathbf{u}||\mathbf{v}| \cos \theta = |\mathbf{u}||\mathbf{v}| \cos 90° = |\mathbf{u}||\mathbf{v}| \cdot 0 = 0$

2 $\mathbf{u} \cdot \mathbf{v} = \pm |\mathbf{u}||\mathbf{v}|$ when **u** and **v** are parallel because the angle between them is either 0° or 180° and therefore,

$\mathbf{u} \cdot \mathbf{v} = |\mathbf{u}||\mathbf{v}| \cos \theta = |\mathbf{u}||\mathbf{v}| \cos 0° = |\mathbf{u}||\mathbf{v}| \cdot 1 = |\mathbf{u}||\mathbf{v}|$,

or $\mathbf{u} \cdot \mathbf{v} = |\mathbf{u}||\mathbf{v}| \cos 180° = |\mathbf{u}||\mathbf{v}| \cdot (-1) = -|\mathbf{u}||\mathbf{v}|$

Example 28

Find the scalar product of the vectors **u** and **v** given that
$|\mathbf{u}| = 2$, $|\mathbf{v}| = 3$, and the angle between **u** and **v** is 60°.

Answer	
$\mathbf{u} \cdot \mathbf{v} = 2 \times 3 \times \cos 60°$ $= 2 \times 3 \times \dfrac{1}{2} = 3$	*Use* $\mathbf{u} \cdot \mathbf{v} = \|\mathbf{u}\|\|\mathbf{v}\| \cos \theta$

Sometimes vectors are defined by the vertices of a polygon.
Use the properties of the polygon to work out the magnitudes of
vectors and the angles between them.

For a reminder about
the names and
properties of polygons
see Chapter 14,
section 3.

Example 29

Consider the unit square ABCD. Let O be the
point where the diagonals of the square meet.
Find $\overrightarrow{OA} \cdot \overrightarrow{AB}$

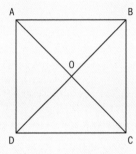

Answer	
$\|\overrightarrow{AB}\| = 1$ $\|\overrightarrow{AC}\| = \sqrt{2}$	[AB] *is a side of the unit square.* *Use Pythagoras' theorem to calculate the length of the diagonal* AC.
$\|\overrightarrow{OA}\| = \dfrac{\sqrt{2}}{2}$	*As O is the midpoint of* AC, $\|\overrightarrow{OA}\| = \dfrac{1}{2}\|\overrightarrow{AC}\|$
$\theta = 45°$	$\triangle ABC$ *is an isosceles right-angled triangle, so the angle between* OA *and* AB *is 45°.*
$\overrightarrow{OA} \cdot \overrightarrow{AB} = \dfrac{\sqrt{2}}{2} \times 1 \times \dfrac{\sqrt{2}}{2} = \dfrac{1}{2}$	*Use* $\mathbf{u} \cdot \mathbf{v} = \|\mathbf{u}\|\|\mathbf{v}\| \cos \theta$ *to find* $\overrightarrow{OA} \cdot \overrightarrow{AB}$

Similarly, you may need to use the properties of polyhedra to
answer questions about the scalar product of vectors in 3-D space.

Example 30

Consider the unit cube ABCDEFGH.
Let O be the point where its four diagonals meet.
Find $\overrightarrow{OA} \cdot \overrightarrow{OB}$

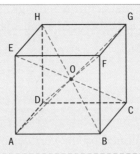

Answer

$|\overrightarrow{AB}| = 1$

$AF = \sqrt{1^2 + 1^2} = \sqrt{2}$

$AG = \sqrt{AF^2 + FG^2}$

$\quad = \sqrt{2 + 1} = \sqrt{3}$

$d = \sqrt{3}$

The length of the side AB of the unit cube is one unit. Use Pythagoras' theorem twice to find the length of the diagonal d of the cube.

$|\overrightarrow{OA}| = |\overrightarrow{OB}| = \dfrac{\sqrt{3}}{2}$

As O is the midpoint of AG, $|\overrightarrow{OA}| = \frac{1}{2}|\overrightarrow{AG}|$

Apply the cosine rule to $\triangle OAB$

$1^2 = \left(\dfrac{\sqrt{3}}{2}\right)^2 + \left(\dfrac{\sqrt{3}}{2}\right)^2 - 2 \times \dfrac{\sqrt{3}}{2} \times \dfrac{\sqrt{3}}{2}\cos\theta$

> To remind yourself of the cosine rule, look back at Chapter 8.

$\Rightarrow \cos\theta = \dfrac{1}{3}$

$|\overrightarrow{OA}| \cdot |\overrightarrow{OB}| = \dfrac{\sqrt{3}}{2} \times \dfrac{\sqrt{3}}{2} \times \dfrac{1}{3} = \dfrac{1}{4}$

$AB^2 = OA^2 + OB^2 - 2(OA)(OB)\cos\theta$
Use $\mathbf{u} \cdot \mathbf{v} = |\mathbf{u}||\mathbf{v}|\cos\theta$ to find $\overrightarrow{OA} \cdot \overrightarrow{OB}$

Exercise 11I

1 Find the scalar product of the vectors **u** and **v** given that $|\mathbf{u}| = 1.5$, $|\mathbf{v}| = 4$ and the angle between **u** and **v** is $30°$.

2 Use the diagram to show that given two non-zero vectors **u** and **v**, $\quad \mathbf{u} \cdot (-\mathbf{v}) = -(\mathbf{u} \cdot \mathbf{v}) = (-\mathbf{u}) \cdot \mathbf{v}$.

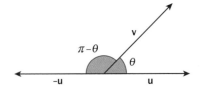

3 Consider an equilateral triangle ABC.
Find $\overrightarrow{AB} \cdot \overrightarrow{BC} + \overrightarrow{BC} \cdot \overrightarrow{AC}$.

4 In right-angled triangle ABC, $\hat{A} = \alpha$, $\hat{B} = 90°$, $AB = x$ and $AC = y$.

Find, in terms of x, y and α

a $\overrightarrow{AB} \cdot \overrightarrow{AC}$ **b** $\overrightarrow{CA} \cdot \overrightarrow{CB}$ **c** $\overrightarrow{AC} \cdot \overrightarrow{CB}$

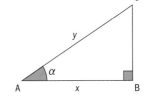

5 A triangle ABC has area 4.
Given that $AB = 2$ and $AC = 5$, find the possible values of $\overrightarrow{AB} \cdot \overrightarrow{AC}$.

6 Show that $\mathbf{u} \cdot \mathbf{u} = |\mathbf{u}|^2$, for any vector $\mathbf{u}$

7 Use the result in question **6** to show that given two vectors $\mathbf{u}$ and $\mathbf{v}$, $\mathbf{u} \cdot \mathbf{v} = 0$ if $|\mathbf{u} + \mathbf{v}| = |\mathbf{u} - \mathbf{v}|$

8 Consider the cuboid ABCDEFGH with centre O shown here.

Using the information given on the diagram, find

a $\overrightarrow{OB} \cdot \overrightarrow{OC}$ **b** $\overrightarrow{AO} \cdot \overrightarrow{OE}$

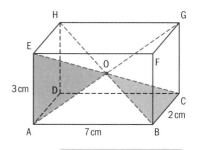

9 The work done by any force $\overrightarrow{F}$ when it produces a displacement $\overrightarrow{d}$ is given by $W = \overrightarrow{F} \cdot \overrightarrow{d}$.

a Kathy is dragging her toy elephant along a corridor. Assuming that she walks in a straight line, and that both the force and angle α between the force and displacement vectors remain constant, find the work done when $|\overrightarrow{F}| = 1.2$ and $|\overrightarrow{d}| = 5$, in terms of α

b If Kathy spins her toy around her, she produces no work. Use the diagram below to explain why this happens.

> To calculate the work done by a force, you only need to consider the component of that force along the direction of the displacement it produces.

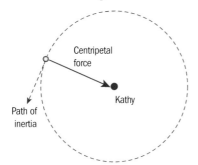

> The path of an object moving due to a centripetal force is a circle. Its velocity vector at each instant has the direction of the tangent to the circle.

Algebraic definitions of scalar product in two and three dimensions

> → Given two vectors in the plane, $\mathbf{u} = u_1\mathbf{i} + u_2\mathbf{j}$ and $\mathbf{v} = v_1\mathbf{i} + v_2\mathbf{j}$,
>
> $$\mathbf{u} \cdot \mathbf{v} = u_1v_1 + u_2v_2$$
>
> In 3-D space, given two vectors, $\mathbf{u} = u_1\mathbf{i} + u_2\mathbf{j} + u_3\mathbf{k}$ and $\mathbf{v} = v_1\mathbf{i} + v_2\mathbf{j} + v_3\mathbf{k}$,
>
> $$\mathbf{u} \cdot \mathbf{v} = u_1v_1 + u_2v_2 + u_3v_3$$

> The scalar product has been defined in two different ways. Do you need to prove their equivalence or can one be an extension of the other?

These algebraic definitions are especially useful for calculating the scalar product of vectors given in component form or when the coordinates of the initial and terminal points are known.

Example 31

Find $\mathbf{u} \cdot \mathbf{v}$ when:

a $\mathbf{u} = \begin{pmatrix} 2 \\ 3 \end{pmatrix}$ and $\mathbf{v} = \begin{pmatrix} -1 \\ -2 \end{pmatrix}$ **b** $\mathbf{u} = \begin{pmatrix} -1 \\ 2 \\ -3 \end{pmatrix}$ and $\mathbf{v} = \begin{pmatrix} -4 \\ -1 \\ 2 \end{pmatrix}$

Answers

a $\mathbf{u} \cdot \mathbf{v} = 2 \times (-1) + 3 \times (-2) = -8$

Use $\mathbf{u} \cdot \mathbf{v} = u_1 v_1 + u_2 v_2$
Some GDC models allow you to calculate the dot product of vectors.

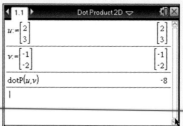

b $\mathbf{u} \cdot \mathbf{v} = -1 \times (-4) + 2 \times (-1) - 3 \times 2 = -4$

Use $\mathbf{u} \cdot \mathbf{v} = u_1 v_1 + u_2 v_2 + u_3 v_3$

GDC help on CD:
Alternative demonstrations for the TI-84 Plus and Casio fx-9860GII GDCs are on the CD.

Example 32

Given the points A(−1, 0, 2), B(0, 1, −2) and C(1, 1, 1), find $\overrightarrow{AB} \cdot \overrightarrow{BC}$.

Answer

$\overrightarrow{AB} = \begin{pmatrix} 0-(-1) \\ 1-0 \\ -2-2 \end{pmatrix} = \begin{pmatrix} 1 \\ 1 \\ -4 \end{pmatrix}$ and

$\overrightarrow{BC} = \begin{pmatrix} 1-0 \\ 1-1 \\ 1-(-2) \end{pmatrix} = \begin{pmatrix} 1 \\ 0 \\ 3 \end{pmatrix}$

$\overrightarrow{AB} \cdot \overrightarrow{BC} = 1 \times 1 + 1 \times 0 - 4 \times 3$
$= -11$

First find $\overrightarrow{AB}$ and $\overrightarrow{BC}$ using
$\overrightarrow{AB} = \overrightarrow{OB} - \overrightarrow{OA}$
and $\overrightarrow{BC} = \overrightarrow{OC} - \overrightarrow{OB}$

Then use $\mathbf{u} \cdot \mathbf{v} = u_1 v_1 + u_2 v_2 + u_3 v_3$

Exercise 11J

1 Find $\mathbf{u} \cdot \mathbf{v}$ when

   a $\mathbf{u} = \begin{pmatrix} -1 \\ -4 \end{pmatrix}$ and $\mathbf{v} = \begin{pmatrix} 12 \\ -6 \end{pmatrix}$ **b** $\mathbf{u} = \begin{pmatrix} 1 \\ 3 \\ 5 \end{pmatrix}$ and $\mathbf{v} = \begin{pmatrix} -1 \\ -1 \\ 2 \end{pmatrix}$

2 Given the points A(−1, 3, −2), B(−1, 1, 2) and C(1, −1, 1), find $\overrightarrow{AB} \cdot \overrightarrow{BC}$ and $\overrightarrow{AC} \cdot \overrightarrow{BC}$.

3 The unit cube OABCDEFG in the diagram has its faces parallel to the coordinate planes.

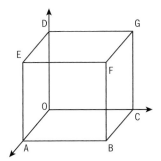

> The coordinate planes are mutually perpendicular and intersect at the origin O.

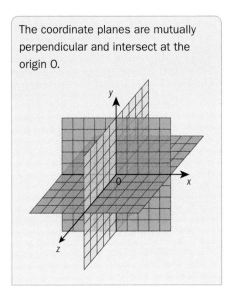

   a Write down the coordinates of its vertices.

   b Hence, find $\overrightarrow{OF} \cdot \overrightarrow{OG}$ and $\overrightarrow{AF} \cdot \overrightarrow{BG}$.

EXAM-STYLE QUESTION

4 Consider a square-based pyramid ABCDE such that the x-axis contains B and D, the y-axis contains A and C and the positive part of the z-axis contains E.

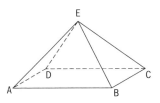

   Given that the area of the base is 4 square units and the volume of the pyramid is $\dfrac{8}{3}$ cubic units, find

   a the coordinates of its vertices

   b $|\overrightarrow{EA}|$ and $\overrightarrow{EA} \cdot \overrightarrow{EB}$

   c the size of angle AÊB.

Equivalence of definitions

In mathematics, different definitions of the same object, operation or property are accepted as long as you can prove that they are consistent.

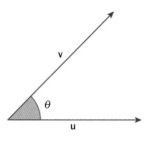

Using trigonometric identities, we can show that the geometric definitions imply the algebraic definition.

1 Consider the plane vectors $\mathbf{u} = u_1\mathbf{i} + u_2\mathbf{j}$ and $\mathbf{v} = v_1\mathbf{i} + v_2\mathbf{j}$

Then $\mathbf{v} - \mathbf{u} = (v_1 - u_1)\mathbf{i} + (v_2 - u_2)\mathbf{j}$.

Let θ be the angle between u and v. Using the cosine rule

$$|v - u|^2 = |u|^2 + |v|^2 - 2|\mathbf{u}||\mathbf{v}|\cos\theta$$

As $|\mathbf{v} - \mathbf{u}|^2 (v_1 - u_1)^2 + (v_2 - u_2)^2$, $|\mathbf{u}|^2 = (u_1)^2 + (u_2)^2$ and $|\mathbf{v}|^2 = (v_1)^2 + (v_2)^2$,

$$-2u_1v_1 - 2u_2v_2 = -2|\mathbf{u}||\mathbf{v}|\cos\theta$$

$$u_1 - v_1 + u_2 - v_2 = |\mathbf{u}||\mathbf{v}|\cos\theta$$

Using the algebraic definition of the scalar product of plane vectors

$$\mathbf{u} \cdot \mathbf{v} = |\mathbf{u}||\mathbf{v}|\cos\theta \quad \text{QED}$$

2 Another important algebraic property of the scalar product is the distributive property

$$(\mathbf{a} + \mathbf{b}) \cdot (\mathbf{c} + \mathbf{d}) = \mathbf{a} \cdot \mathbf{c} + \mathbf{a} \cdot \mathbf{d} + \mathbf{b} \cdot \mathbf{c} + \mathbf{b} \cdot \mathbf{d}$$

Consider the plane vectors $\mathbf{u} = u_1\mathbf{i} + u_2\mathbf{j}$ and $\mathbf{v} = v_1\mathbf{i} + v_2\mathbf{j}$.

$$\mathbf{u} \cdot \mathbf{v} = (u_1\mathbf{i} + u_2\mathbf{j}) \cdot (v_1\mathbf{i} + v_2\mathbf{j})$$

Using the distributive property

$$\mathbf{u} \cdot \mathbf{v} = (u_1\mathbf{i}) \cdot (v_1\mathbf{i}) + (u_1\mathbf{i}) \cdot (v_2\mathbf{j}) + (u_2\mathbf{j}) \cdot (v_1\mathbf{i}) + (u_2\mathbf{j}) \cdot (v_2\mathbf{j})$$

As $\mathbf{u} \cdot \mathbf{v} = |\mathbf{u}||\mathbf{v}|\cos\theta$, where θ is the angle between $\mathbf{u}$ and $\mathbf{v}$

$(u_1\mathbf{i}) \cdot (v_1\mathbf{i}) = u_1v_1$, $(u_2\mathbf{j}) \cdot (v_2\mathbf{j}) = u_2v_2$ and $(u_1\mathbf{i}) \cdot (v_2\mathbf{j}) = (u_2\mathbf{j}) \cdot (v_1\mathbf{i}) = 0$

Therefore, $\mathbf{u} \cdot \mathbf{v} = u_1v_1 + u_2v_2$ as required. QED

> A similar proof can be given for 3-D vectors.

> The distributive property is extremely important as it is used to prove that the geometric definition implies the algebraic definitions.

> **Extension material on CD:**
> *Worksheet 11*

As long as you use an orthogonal Cartesian referential the geometric and algebraic definitions are equivalent, which leads to a very important and useful formula

→ $\cos\theta = \dfrac{\mathbf{u} \cdot \mathbf{v}}{|\mathbf{u}||\mathbf{v}|}$, where θ is the angle between the vectors $\mathbf{u}$ and $\mathbf{v}$.

$$\mathbf{u} = \begin{pmatrix} u_1 \\ u_2 \end{pmatrix} \text{ and } \mathbf{v} = \begin{pmatrix} v_1 \\ v_2 \end{pmatrix} \Rightarrow \cos\theta = \frac{u_1v_1 + u_2v_2}{\sqrt{u_1^2 + u_2^2}\sqrt{v_1^2 + v_2^2}}$$

$$\mathbf{u} = \begin{pmatrix} u_1 \\ u_2 \\ u_3 \end{pmatrix} \text{ and } \mathbf{v} = \begin{pmatrix} v_1 \\ v_2 \\ v_3 \end{pmatrix} \Rightarrow \cos\theta = \frac{u_1v_1 + u_2v_2 + u_3v_3}{\sqrt{u_1^2 + u_2^2 + u_3^2}\sqrt{v_1^2 + v_2^2 + v_3^2}}$$

Example 33

Given $\mathbf{u} = 2\mathbf{i} - 3\mathbf{j} + 4\mathbf{k}$ and $\mathbf{v} = \mathbf{i} - \mathbf{j} - \mathbf{k}$, find the angle θ between the vectors $\mathbf{u}$ and $\mathbf{v}$.

Answer

$\mathbf{u} \cdot \mathbf{v} = 2 + 3 - 4 = 1$

$|\mathbf{u}| = \sqrt{4+9+16} = \sqrt{29}$ and

$|\mathbf{v}| = \sqrt{1+1+1} = \sqrt{3}$

$\cos\theta = \dfrac{1}{\sqrt{29}\sqrt{3}} \Rightarrow$

$\theta = \arccos\left(\dfrac{1}{\sqrt{29}\sqrt{3}}\right) = 83.8°$ to 3 sf

Use $\mathbf{u} \cdot \mathbf{v} = u_1 v_1 + u_2 v_2 + u_3 v_3$

$|\mathbf{u}| = \sqrt{u_1^2 + u_2^2 + u_3^2}$

$\cos\theta = \dfrac{\mathbf{u} \cdot \mathbf{v}}{|\mathbf{u}||\mathbf{v}|}$

Example 34

Find the values of k for which the angle between the vectors $\mathbf{u} = \begin{pmatrix} 1 \\ k \end{pmatrix}$ and $\mathbf{v} = \begin{pmatrix} 3k \\ 2 \end{pmatrix}$ is 60°.

> To learn how to use Solver on a GDC see the GDC Chapter on the CD.

Answer

$\mathbf{u} \cdot \mathbf{v} = 3k + 2k = 5k$

$|\mathbf{u}| = \sqrt{1+k^2}$ and $|\mathbf{v}| = \sqrt{9k^2 + 4}$

$\cos 60° = \dfrac{5k}{\sqrt{1 + k^2}\sqrt{9k^2 + 4}}$

$\dfrac{1}{2} = \dfrac{5k}{\sqrt{1 + k^2}\sqrt{9k^2 + 4}} \Rightarrow k = 0.215$

or $k = 3.10$

Use $\mathbf{u} \cdot \mathbf{v} = u_1 v_1 + u_2 v_2$

$|\mathbf{u}| = \sqrt{u_1^2 + u_2^2}$

$\cos\theta = \dfrac{\mathbf{u} \cdot \mathbf{v}}{|\mathbf{u}||\mathbf{v}|}$

If you square both sides of

$\dfrac{1}{2} = \dfrac{5k}{\sqrt{1 + k^2}\sqrt{9k^2 + 4}}$ *and re-arrange,*

you obtain $9k^4 - 87k^2 + 4 = 0$ *which has four solutions. However, two of them are negative and therefore do not satisfy the original equation.*

Exercise 11K

1 Given $\mathbf{u} = 2\mathbf{i} - 3\mathbf{j}$ and $\mathbf{v} = \mathbf{i} + 2\mathbf{j}$, find the angle θ between the vectors $\mathbf{u}$ and $\mathbf{v}$.

2 Given $\mathbf{u} = \mathbf{i} - 2\mathbf{j} + \mathbf{k}$ and $\mathbf{v} = 2\mathbf{i} - \mathbf{j} + \mathbf{k}$, find the angle θ between the vectors $\mathbf{u}$ and $\mathbf{v}$.

3 Given the points A(−1, 1, 1), B(1, −1, 2) and C(2, 3, −1), find the angles between the vectors
   a $\overrightarrow{AB}$ and $\overrightarrow{AC}$ **b** $\overrightarrow{BC}$ and $\overrightarrow{AC}$
   Hence, write down the sizes of the internal angles of the triangle ABC.

4 Given the vectors $\begin{pmatrix} a \\ a-4 \end{pmatrix}$ and $\begin{pmatrix} a-2 \\ 3 \end{pmatrix}$, find the values of a for which the angle between the vectors is acute.

5 Consider the vectors $\mathbf{u} = \sin(3\alpha)\,\mathbf{i} - \cos(3\alpha)\,\mathbf{j} + 2\mathbf{k}$ and $\mathbf{v} = \cos\alpha\,\mathbf{i} - \sin\alpha\,\mathbf{j} - 2\mathbf{k}$, where $0 < \alpha < 2\pi$. Let θ be the angle between the vectors $\mathbf{u}$ and $\mathbf{v}$.

  a Express $\cos\theta$ in terms of α.

  b Find all possible values of α for which the angle between the two vectors is 150°.

  c Show that the angle between the two vectors is always obtuse.

6 Using the algebraic definition of the scalar product, show that
$$(\mathbf{a} + \mathbf{b}) \cdot (\mathbf{c} + \mathbf{d}) = \mathbf{a} \cdot \mathbf{c} + \mathbf{a} \cdot \mathbf{d} + \mathbf{b} \cdot \mathbf{c} + \mathbf{b} \cdot \mathbf{d}$$
for any vectors a, b, c and d in 3-D Cartesian space.

11.5 Vector (cross) product and properties

In many applications of vector algebra it is necessary to find a vector that is orthogonal to two given vectors. The vector product of 3-D vectors provides a very efficient method for solving this type of problem.

> While the dot product can be defined in spaces of different dimensions, the cross product is just defined in some dimensions. For example, the cross product is not defined in dimension 2.

> → Given $\mathbf{u} = u_1\mathbf{i} + u_2\mathbf{j} + u_3\mathbf{k}$ and $\mathbf{v} = v_1\mathbf{i} + v_2\mathbf{j} + v_3\mathbf{k}$,
> the **vector (cross) product** of $\mathbf{u}$ and $\mathbf{v}$ is the vector given by
> $\mathbf{u} \times \mathbf{v} = (u_2 v_3 - u_3 v_2)\mathbf{i} + (u_3 v_1 - u_1 v_3)\mathbf{j} + (u_1 v_2 - u_2 v_1)\mathbf{k}$.

Example 35

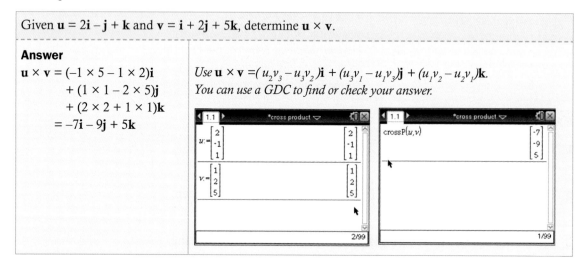

Given $\mathbf{u} = 2\mathbf{i} - \mathbf{j} + \mathbf{k}$ and $\mathbf{v} = \mathbf{i} + 2\mathbf{j} + 5\mathbf{k}$, determine $\mathbf{u} \times \mathbf{v}$.

Answer

$$\begin{aligned}
\mathbf{u} \times \mathbf{v} &= (-1 \times 5 - 1 \times 2)\mathbf{i} \\
&+ (1 \times 1 - 2 \times 5)\mathbf{j} \\
&+ (2 \times 2 + 1 \times 1)\mathbf{k} \\
&= -7\mathbf{i} - 9\mathbf{j} + 5\mathbf{k}
\end{aligned}$$

Use $\mathbf{u} \times \mathbf{v} = (u_2 v_3 - u_3 v_2)\mathbf{i} + (u_3 v_1 - u_1 v_3)\mathbf{j} + (u_1 v_2 - u_2 v_1)\mathbf{k}$.
You can use a GDC to find or check your answer.

The next example shows you how to prove an important property of the vector product.

Example 36

Show that $\mathbf{u} \times \mathbf{v} = \mathbf{0}$ if and only if $\mathbf{u}$ and $\mathbf{v}$ are collinear.

Answer

Let $\mathbf{u} = u_1\mathbf{i} + u_2\mathbf{j} + u_3\mathbf{k}$ *and* $\mathbf{v} = v_1\mathbf{i} + v_2\mathbf{j} + v_3\mathbf{k}$.
If $\mathbf{u} \times \mathbf{v} = \mathbf{0}$ then $u_2v_3 - u_3v_2 = 0$,
$u_3v_1 - u_1v_3 = 0$, and $u_1v_2 - u_2v_1 = 0$.

First prove that if $\mathbf{u} \times \mathbf{v} = \mathbf{0}$, $\mathbf{u}$ and $\mathbf{v}$ are collinear.
Use $\mathbf{u} \times \mathbf{v} = (u_2v_3 - u_3v_2)\mathbf{i} + (u_3v_1 - u_1v_3)\mathbf{j} +$
$(u_1v_2 - u_2v_1)\mathbf{k}$ and make each component of
$\mathbf{u} \times \mathbf{v}$ equal to zero.

If none of the components of $\mathbf{v}$ is zero, this implies that $\dfrac{u_1}{v_1} = \dfrac{u_2}{v_2} = \dfrac{u_3}{v_3}$

Re-arrange the equations to the form $\mathbf{u} = k\mathbf{v}$
where $k = \dfrac{u_1}{v_1} = \dfrac{u_2}{v_2} = \dfrac{u_3}{v_3}$.

Therefore, $\mathbf{u}$ and $\mathbf{v}$ are collinear.

If one of the components of $\mathbf{v}$, say v_1, is zero, then $u_1v_2 = 0$ and $u_1v_3 = 0$ which implies that $u_1 = 0$ or $v_2 = v_3 = 0$. Hence, $\mathbf{u}$ and $\mathbf{v}$ are collinear.

Choose any component of $\mathbf{v}$ to be zero and show that the corresponding component of $\mathbf{u}$ is also zero or $\mathbf{v}$ is the zero vector.

Suppose now that $\mathbf{u}$ and $\mathbf{v}$ are collinear.
Then $u_1 = \lambda v_1$, $u_2 = \lambda v_2$ and $u_3 = \lambda v_3$ and
$\mathbf{u} \times \mathbf{v} = \mathbf{0}$

Now prove that if $\mathbf{u}$ and $\mathbf{v}$ are collinear, $\mathbf{u} \times \mathbf{v} = \mathbf{0}$.
Collinear means that $\mathbf{u} = \lambda\mathbf{v}$, $\lambda \in \mathbb{R}$
Substitute $u_1 = \lambda v_1$, $u_2 = \lambda v_2$ and $u_3 = \lambda v_3$ into
$\mathbf{u} \times \mathbf{v} = (u_2v_3 - u_3v_2)\mathbf{i} + (u_3v_1 - u_1v_3)\mathbf{j}$
$+ (u_1v_2 - u_2v_1)\mathbf{k}$

Investigation – properties of the cross product

Use your GDC to investigate the algebraic properties of the cross product of two vectors.

In each case explore several examples, make a conjecture and then prove it.

1 For any vectors $\mathbf{u}$ and $\mathbf{v}$, is $\mathbf{u} \times \mathbf{v} = \mathbf{v} \times \mathbf{u}$?

2 For any vectors $\mathbf{u}$ and $\mathbf{v}$, is $(\mathbf{u} \times \mathbf{v}) \times \mathbf{w} = \mathbf{u} \times (\mathbf{v} \times \mathbf{w})$?

3 For any vectors $\mathbf{u}$ and $\mathbf{v}$ and for any parameter λ, is $(\lambda\mathbf{u}) \times \mathbf{v} = \lambda (\mathbf{u} \times \mathbf{v})$?

4 For any vectors $\mathbf{u}$, $\mathbf{v}$ and $\mathbf{w}$, is $\mathbf{u} \times (\mathbf{v} + \mathbf{w}) = \mathbf{u} \times \mathbf{v} + \mathbf{u} \times \mathbf{w}$?

Magnitude of the vector u × v

The magnitude of the vector product of two vectors $\mathbf{u}$ and $\mathbf{v}$ is given by $|\mathbf{u} \times \mathbf{v}| = |\mathbf{u}||\mathbf{v}|\sin\theta$, where θ is the angle between the two vectors.

The proof of this result requires some heavy algebraic manipulation.

You can re-arrange this formula to obtain $\sin\theta = \dfrac{|\mathbf{u} \times \mathbf{v}|}{|\mathbf{u}||\mathbf{v}|}$, but as the

angle between two vectors can take any value between $0°$ and $180°$, you cannot use this formula to find the value of θ.

You can now define the cross product of two vectors geometrically.

→ $\mathbf{u} \times \mathbf{v} = (|\mathbf{u}| |\mathbf{v}| \sin \theta) \hat{\mathbf{n}}$ where $\hat{\mathbf{n}}$ is a unit vector orthogonal (normal) to both $\mathbf{u}$ and $\mathbf{v}$ whose direction is given by the right-hand rule illustrated in the diagram.

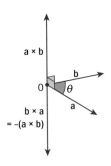

This geometric definition highlights an important property of the cross-product

$$\mathbf{u} \times \mathbf{v} = -(\mathbf{v} \times \mathbf{u})$$

This means that when you calculate the cross product between two vectors you need to be careful – the order you write them is important! With this definition it is also easy to identify the property proved in example 36

$$\mathbf{u} \times \mathbf{v} = \mathbf{0} \Leftrightarrow \mathbf{u} \text{ and } \mathbf{v} \text{ are collinear}$$

However, the most important applications of this definition come from the geometrical interpretation of $|\mathbf{u} \times \mathbf{v}|$.

Geometric interpretation of $|\mathbf{u} \times \mathbf{v}|$

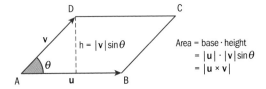

Area = base · height
= $|\mathbf{u}| \cdot |\mathbf{v}| \sin \theta$
= $|\mathbf{u} \times \mathbf{v}|$

→ Given a parallelogram ABCD, if $\mathbf{u} = \overrightarrow{AB} = \overrightarrow{DC}$ and $\mathbf{v} = \overrightarrow{AD} = \overrightarrow{BC}$, the area of ABCD is numerically equal to $|\mathbf{u} \times \mathbf{v}|$

The area of triangle ABD equals $\dfrac{1}{2} |\mathbf{u} \times \mathbf{v}|$

Example 37

Show that the volume of the parallelepiped shown in the diagram is $(\mathbf{u} \times \mathbf{v}) \cdot \mathbf{w}$.

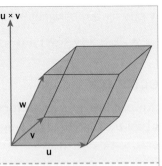

Answer

V = area of base × height Area of base = $|\mathbf{u} \times \mathbf{v}|$	*Write down the formula for the volume of a parallelepiped. Find the area of a parallelogram.*
Height = $|\mathbf{w}| \cos \theta$ = where θ is the angle between $\mathbf{w}$ and $\mathbf{u} \times \mathbf{v}$. $V = |\mathbf{u} \times \mathbf{v}| |\mathbf{w}| \cos \theta = (\mathbf{u} \times \mathbf{v}) \cdot \mathbf{w}$	$\cos \theta = \dfrac{adjacent\ side}{hypotenuse}$ *(θ is acute)* *Use the geometric definition of the scalar product.*

It is common to use the notation $\hat{\mathbf{n}}$ to represent a unit vector normal to a plane or to two given vectors.

Torque, or the moment of a force, is a measure of the turning force on an object.

Pushing or pulling the handle of a spanner connected to a nut or bolt produces a torque that loosens or tightens the nut or bolt. The torque produced by the force can be modeled by a vector product.

$(\mathbf{u} \times \mathbf{v}) \cdot \mathbf{w}$ is called the triple (or mixed) product of these vectors.

If the angle between the vectors $\mathbf{u} \times \mathbf{v}$ and $\mathbf{w}$ is obtuse, the volume is given by the absolute values of the triple product.

→ Summary of the algebraic properties of the vector (cross) product

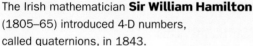

- $\mathbf{u} \times \mathbf{v} = -\mathbf{v} \times \mathbf{u}$
- $\mathbf{u} \times (\mathbf{v} + \mathbf{w}) = \mathbf{u} \times \mathbf{v} + \mathbf{u} \times \mathbf{w}$
- $(\mathbf{u} \times \mathbf{v}) \cdot \mathbf{u} = \mathbf{0}$ ($\mathbf{u} \times \mathbf{v}$ is orthogonal to $\mathbf{u}$)
- $(\mathbf{u} \times \mathbf{v}) \cdot \mathbf{v} = \mathbf{0}$ ($\mathbf{u} \times \mathbf{v}$ is orthogonal to $\mathbf{v}$)
- $\mathbf{u} \times \mathbf{v} = \mathbf{0}$ if and only if $\mathbf{u}$ and $\mathbf{v}$ are collinear.
- $\mathbf{u} \times \mathbf{u} = \mathbf{0}$
- $(\lambda\mathbf{u}) \times \mathbf{v} = \lambda(\mathbf{u} \times \mathbf{v})$
- $|\mathbf{u} \times \mathbf{v}| = |\mathbf{u}||\mathbf{v}| \sin \theta$, where θ is the angle between $\mathbf{u}$ and $\mathbf{v}$.

Are the parentheses in the scalar product necessary – does $(\mathbf{u} \times \mathbf{v}) \cdot \mathbf{u} = \mathbf{u} \times \mathbf{v} \cdot \mathbf{u}$?

The Irish mathematician **Sir William Hamilton** (1805–65) introduced 4-D numbers, called quaternions, in 1843.

Quaternions combine the normal laws of algebra with the exception of multiplication, which is not commutative. You should have found in the Investigation on page 593 that the vector product is not commutative.

Quaternions have many computer graphic applications and allow rotations and movement of objects in 3-D.

Over 150 years later, the development of what seemed a strange theory at the time has been introduced to the general public through movies such as *Jurassic Park* and *Lord of the Rings* – films that would not be possible to create without the aid of mathematics and computer animations.

▲ A quaternion can be used to transform one 3-D vector into another.

Exercise 11L

1 Given $\mathbf{u} = \mathbf{i} - \mathbf{j} - \mathbf{k}$, $\mathbf{v} = \mathbf{i} - 2\mathbf{j} + \mathbf{k}$ and $\mathbf{w} = \mathbf{i} - \mathbf{k}$, determine in component form

   a $\mathbf{u} \times \mathbf{v}$ **b** $\mathbf{v} \times \mathbf{w}$ **c** $\mathbf{u} \times (\mathbf{v} \times \mathbf{w})$

   d $(\mathbf{u} \times \mathbf{v}) \times \mathbf{w}$ **e** $\mathbf{u} \times \mathbf{w}$ **f** $(\mathbf{u} \times \mathbf{v}) \times (\mathbf{u} \times \mathbf{w})$

   g $(\mathbf{u} + \mathbf{v}) \times (\mathbf{u} - \mathbf{w})$

2 Give a **counter-example** to show that the vector product is not associative.

3 Show that if a vector $\mathbf{w}$ is orthogonal to both $\mathbf{u}$ and $\mathbf{v}$, then $\mathbf{w}$ and $\mathbf{u} \times \mathbf{v}$ are collinear.

An interesting topic for an exploration or even an Extended Essay is the applications of vector methods in areas like computer animations which allow the creation of animated 3-D figures that can move like the people or animals they represent.

4 Consider the points A(−1, 3, 4), B(5, 7, 5) and C(3, 9, 6).

  a Given that ABCD is a parallelogram, find the coordinates of D.

  b Find $\overrightarrow{AB} \times \overrightarrow{BC}$.

  c Hence, find the area of parallelogram ABCD.

5 The diagram shows a parallelepiped ABCDEFGH.

  Given that A(1, 0, 2), B(2, 3, 3), C(−3, −1, 2) and E(2, 1, 4), find

  a the coordinates of D

  b the volume of ABCDEFGH.

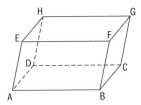

6 Show that the vector product has these algebraic properties.

  a $\mathbf{u} \times \mathbf{v} = -\mathbf{v} \times \mathbf{u}$ **b** $\mathbf{u} \times (\mathbf{v} + \mathbf{w}) = \mathbf{u} \times \mathbf{v} + \mathbf{u} \times \mathbf{w}$

  c $\mathbf{u} \times \mathbf{v} \cdot \mathbf{u} = 0$ **d** $(\lambda\mathbf{u}) \times \mathbf{v} = \lambda(\mathbf{u} \times \mathbf{v})$

Extension material on CD:
Worksheet 11

11.6 Vectors and equations of planes

A plane in 3-D space is defined by three non-collinear points. This means that for any given triangle ABC in 3-D space, there is exactly one plane that contains this triangle – the plane ABC.

A point R is on the plane ABC when $\overrightarrow{RA} = \alpha\overrightarrow{AB} + \beta\overrightarrow{AC}$ for some real values of α and β. This is called the **vector equation** of ABC.

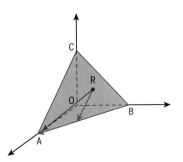

There are several ways of defining a plane ABC. For example, ABC is the only plane that contains the lines [AC] and [AB]. In general, a plane π is defined when one of its points with position vector **a** and two non-collinear vectors **u** and **v** parallel to the plane are known.

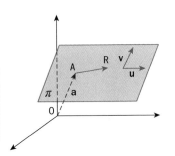

> → In this case, any point on the plane with position vector **r** satisfies the **vector equation of a plane**
>
> $\mathbf{r} = \mathbf{a} + \alpha\mathbf{u} + \beta\mathbf{v}$

This equation can be manipulated to obtain other equations of the plane π.

Let $\mathbf{r} = \begin{pmatrix} x \\ y \\ z \end{pmatrix}$, $\mathbf{a} = \begin{pmatrix} x_1 \\ y_1 \\ z_1 \end{pmatrix}$, $\mathbf{u} = \begin{pmatrix} u_1 \\ u_2 \\ u_3 \end{pmatrix}$ and $\mathbf{v} = \begin{pmatrix} v_1 \\ v_2 \\ v_3 \end{pmatrix}$

The equations $x = x_1 + \alpha u_1 + \beta v_1$, $y = y_1 + \alpha u_2 + \beta v_2$ and $z = z_1 + \alpha u_3 + \beta v_3$ are called **parametric equations of the plane.**

> → If the parameters α and β are eliminated, an equation of the form $ax + by + cz = d$ is obtained.
> This equation is called a **Cartesian equation of the plane**.

Example 38

Given the points A(1, 0, 1), B(−1, 1, 0) and C(0, 1, −1), find the equation of the plane ABC

a in vector form

b in parametric form

c in Cartesian form.

Hence, show that the vector whose components are the coefficients of the Cartesian equation is normal to the plane.

Answers

a $\overrightarrow{AB} = \begin{pmatrix} -1-1 \\ 1-0 \\ 0-1 \end{pmatrix} = \begin{pmatrix} -2 \\ 1 \\ -1 \end{pmatrix}$

Find two non-collinear vectors, $\overrightarrow{AB}$ and $\overrightarrow{AC}$, parallel to the plane ABC.

$\overrightarrow{AC} = \begin{pmatrix} 0-1 \\ 1-0 \\ -1-1 \end{pmatrix} = \begin{pmatrix} -1 \\ 1 \\ -2 \end{pmatrix}$

A vector equation of the plane ABC is

$\mathbf{r} = \begin{pmatrix} 1 \\ 0 \\ 1 \end{pmatrix} + \alpha \begin{pmatrix} -2 \\ 1 \\ -1 \end{pmatrix} + \beta \begin{pmatrix} -1 \\ 1 \\ -2 \end{pmatrix}$

where $\alpha, \beta \in \mathbb{R}$

Use $\mathbf{r} = \mathbf{a} + \alpha\mathbf{u} + \beta\mathbf{v}$ where $\mathbf{a} = \overrightarrow{OA}$, $\mathbf{u} = \overrightarrow{AB}$ and $\mathbf{v} = \overrightarrow{AC}$ are vectors parallel to the plane.

> You could use the point B or C and obtain an equivalent equation.

b $x = 1 - 2\alpha - \beta$
$y = \alpha + \beta$
$z = 1 - \alpha - 2\beta$

Express each component of $\mathbf{r} = \begin{pmatrix} x \\ y \\ z \end{pmatrix}$ in terms of α and β.

c $\left. \begin{array}{l} x = 1 - 2\alpha - \beta \quad (1) \\ y = \alpha + \beta \quad\quad\quad (2) \\ z = 1 - \alpha - 2\beta \quad (3) \end{array} \right\} \Rightarrow$

$\alpha = 1 - x - y \quad (1 + 2)$
$\beta = 1 - y - z \quad (2 + 3)$

Use pairs of equations to find expressions for α and β in terms of x, y and z. Substitute these expressions into one of the equations – so eliminating α and β – to obtain an equation of the form $ax + by + cz = d$

Substitute into (2):

$y = (1 - x - y) + (1 - y - z)$
$\Rightarrow x + 3y + z = 2$

As $\begin{pmatrix} 1 \\ 3 \\ 1 \end{pmatrix} \cdot \begin{pmatrix} -2 \\ 1 \\ -1 \end{pmatrix} = -2 + 3 - 1 = 0$ and

The vector whose components are the coefficients of the Cartesian equation is $\begin{pmatrix} 1 \\ 3 \\ 1 \end{pmatrix}$

$\begin{pmatrix} 1 \\ 3 \\ 1 \end{pmatrix} \cdot \begin{pmatrix} -1 \\ 1 \\ -2 \end{pmatrix} = -1 + 3 - 2 = 0,$

the vector $\begin{pmatrix} 1 \\ 3 \\ 1 \end{pmatrix}$ is normal to the plane.

*Show that the scalar product of the vector whose components are the coefficients of the variables in the Cartesian equation (part **c**) and the two non-collinear vectors parallel to the plane, $\overrightarrow{AB}$ and $\overrightarrow{AC}$, is zero.*

The previous example suggests that the Cartesian equation of the plane can be deduced using the normal direction to the plane.

In fact, as illustrated by the diagram, $\overrightarrow{AR} \perp \mathbf{n}$ if, and only if, R is a point on the plane.

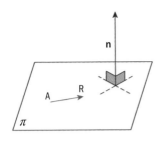

If $\mathbf{n} = \begin{pmatrix} a \\ b \\ c \end{pmatrix}$, the plane is defined by the equation $\begin{pmatrix} x - x_1 \\ y - y_1 \\ z - z_1 \end{pmatrix} \cdot \begin{pmatrix} a \\ b \\ c \end{pmatrix} = 0$

which can be simplified to obtain the Cartesian equation of the plane.

> $\perp$ is shorthand for 'perpendicular'.

→ $a(x - x_1) + b(y - y_1) + c(z - z_1) = 0$
or $ax + by + cz = d$ where $d = ax_1 + by_1 + cz_1$

Equation of the plane (using the normal vector)

$\begin{pmatrix} x - x_1 \\ y - y_1 \\ z - z_1 \end{pmatrix} \cdot \begin{pmatrix} a \\ b \\ c \end{pmatrix} = 0$ this can also be written as $\begin{pmatrix} x \\ y \\ z \end{pmatrix} \cdot \begin{pmatrix} a \\ b \\ c \end{pmatrix} = \begin{pmatrix} x_1 \\ y_1 \\ z_1 \end{pmatrix} \cdot \begin{pmatrix} a \\ b \\ c \end{pmatrix}$

or $\mathbf{r} \cdot \mathbf{n} = \mathbf{a} \cdot \mathbf{n}$

Example 39

a Find the equation of the plane π which contains the point A(−1, 1, 2) and is normal to the vector $\mathbf{n} = 2\mathbf{i} - \mathbf{j} - \mathbf{k}$.
b Hence, show that the point B(0, 4, −1) lies on the plane.

Answers

a $\begin{pmatrix} x \\ y \\ z \end{pmatrix} \begin{pmatrix} 2 \\ -1 \\ -1 \end{pmatrix} = \begin{pmatrix} -1 \\ 1 \\ 2 \end{pmatrix} \begin{pmatrix} 2 \\ -1 \\ -1 \end{pmatrix}$

$\Rightarrow 2x - y - z = -5$

Use $\mathbf{r} \cdot \mathbf{n} = \mathbf{a} \cdot \mathbf{n}$ where
$\mathbf{r} = x\mathbf{i} + y\mathbf{j} + z\mathbf{k}$ and the
position vector of A(−1, 1, 2) is
$\mathbf{a} = -\mathbf{i} + \mathbf{j} + 2\mathbf{k}$

b $x = 0$, $y = 4$ and $z = -1$

$\Rightarrow 2(0) - 4 - (-1) = -5$

Therefore, B(0, 4, −1) lies on the plane π.

*To show that the point lies on the plane, show that the coordinates of B are the solution to the equation from part **a**: $2x - y - z = -5$*

Exercise 11M

1 Given the points A(−3, 1, 1), B(1, 2, 0) and C(1, 1, −2), find the equation of the plane ABC
    a in vector form   **b** in parametric form
    c in Cartesian form.

2 Find the equation of the plane α which contains the point P(1, 0, 1) and is normal to the vector $\mathbf{n} = -\mathbf{i} + 3\mathbf{j} - \mathbf{k}$. Hence, find the coordinates of three distinct points on plane α.

3 Find the Cartesian equation of the plane α if

  a a contains the point (−3, 4, 0) and has the normal vector $\begin{pmatrix} 1 \\ -2 \\ -1 \end{pmatrix}$

  b α contains the point (−1, 1, 1) and the line $x - 1 = \dfrac{y-1}{2} = z$

  c α contains the lines $1 - x = y - 1 = 2z$
  and $x = 2 - t,\ y = 1 + 2t,\ z = t.$

4 Write down the Cartesian equations of the coordinate planes:

  a xz **b** xy **c** yz

5 Consider the pyramid shown where O is the origin and the points A, C and V lie on the x-, y- and z-axes respectively.

Given that ABCO is a square with area 4 square units and the pyramid has a volume of 6 cubic units, find

  a the coordinates of the vertices of the pyramid
  b a vector equation of the plane ABV
  c a Cartesian equation of the plane BCV
  d a vector equation of the line BV
  e Cartesian equations of the line AV.

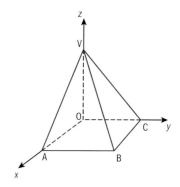

11.7 Angles, distances and intersections

Vector algebra also provides efficient methods for solving practical problems involving angles, distances and intersections of lines and planes. Here are the most usual examples.

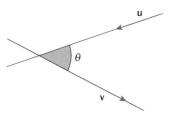

Angle between two lines

The angle θ between a line parallel to $\mathbf{u}$ and a line parallel to $\mathbf{v}$ is given by

$$\theta = \arccos\left(\left|\frac{\mathbf{u} \cdot \mathbf{v}}{\|\mathbf{u}\|\|\mathbf{v}\|}\right|\right)$$

Although the angle between the vectors $\mathbf{u}$ and $\mathbf{v}$ can take any value in the interval [0°, 180°], the angle between two lines can simply take values in the interval [0°, 90°]. When $\theta = 0°$ there are two parallel lines and when $\theta = 90°$ there are two orthogonal lines. In the plane, orthogonal lines always meet at a point and therefore they are perpendicular.

> The difference between the formula for the angle between lines and for the angle between their direction vectors is just the inclusion of the absolute value. Why do we need to include it?

In 3-D space, orthogonal lines may not necessarily meet – when they meet, they are perpendicular. For example, in the diagram, the line [AB] is perpendicular to the line [BC] and orthogonal to the line [FG]. As [AB] and [FG] do not meet, they are not perpendicular.

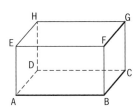

Example 40

Find the angle between the lines $\dfrac{x-2}{3} = y = 1-z$ and $x = 2y = z-3$

Answer

$\dfrac{x-2}{3} = y = 1-z$

$\Rightarrow \dfrac{x-2}{3} = \dfrac{y-0}{1} = \dfrac{z-1}{-1}$

$x = 2y = z-3$

$\Rightarrow \dfrac{x-0}{2} = \dfrac{y-0}{1} = \dfrac{z-3}{2}$

$\mathbf{u} = \begin{pmatrix} 3 \\ 1 \\ -1 \end{pmatrix}$ and $\mathbf{v} = \begin{pmatrix} 2 \\ 1 \\ 2 \end{pmatrix}$ are

direction vectors of the given lines. So,

$\theta = \arccos\left(\left|\dfrac{6+1-2}{\sqrt{9+1+1}\sqrt{4+1+4}}\right|\right)$

$= 59.8°$

(or 1.04 rad) to 3 sf.

Rearrange the equations of the lines and write them in the form

$\dfrac{x-x_1}{u_1} = \dfrac{y-y_1}{u_2} = \dfrac{z-z_1}{u_3}$

$\begin{pmatrix} u_1 \\ u_2 \\ u_3 \end{pmatrix}$ *gives you a direction vector of*

the line, i.e., vector parallel to the line.

Use the formula

$\theta = \arccos\left(\dfrac{\mathbf{u}\cdot\mathbf{v}}{\|\mathbf{u}\|\|\mathbf{v}\|}\right)$ *to determine*

the angle between the two lines.

GDC help on CD:
Alternative demonstrations for the TI-84 Plus and Casio fx-9860GII GDCs are on the CD.

> In coordinate geometry, angles are generally measured in degrees but, unless the question explicitly requires the use of a certain unit, answers can be given in either degrees or radians.

> You can also use a GDC to calculate the angle or check your answer.

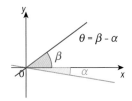

▲ α and β are the angles between the lines and the positive part of the x-axis. Note: α and β may take negative values.

→ In the plane, given two lines with gradients m_1 and m_2, we can calculate the angle θ between them using the formula
$\theta = |\beta - \alpha|$ where $\alpha = \arctan m_1$ and $\beta = \arctan m_2$

Example 41

Find the angle between the lines $\mathbf{r} = \begin{pmatrix} 1 \\ 2 \end{pmatrix} + \lambda \begin{pmatrix} 2 \\ 3 \end{pmatrix}$ and $y = 2x - 1$.

Answers

Method 1

$\begin{pmatrix} 2 \\ 3 \end{pmatrix}$ and $\begin{pmatrix} 1 \\ 2 \end{pmatrix}$ are vectors parallel to the lines given.

So, $\theta = \arccos\left(\left|\dfrac{2+6}{\sqrt{4+9}\sqrt{1+4}}\right|\right) = 7.13°$

(or 0.124 rad) to 3 sf.

$\begin{pmatrix} u_1 \\ u_2 \end{pmatrix}$ *gives you a vector parallel to the line.*

For $y = 2x - 1$,

$m = \dfrac{u_2}{u_1} = \dfrac{2}{1}$; *hence,* $\begin{pmatrix} u_1 \\ u_2 \end{pmatrix} = \begin{pmatrix} 1 \\ 2 \end{pmatrix}$

Use the formula $\theta = \arccos\left|\dfrac{\mathbf{u} \cdot \mathbf{v}}{\|\mathbf{u}\|\|\mathbf{v}\|}\right|$ *to determine the required angle.*

Method 2

The line $\mathbf{r} = \begin{pmatrix} 1 \\ 2 \end{pmatrix} + \lambda \begin{pmatrix} 2 \\ 3 \end{pmatrix}$ has gradient $m_1 = \dfrac{3}{2}$

The line $y = 2x - 1$ has gradient $m_2 = 2$.

So, $\alpha = \arctan\dfrac{3}{2}$ and $\beta = \arctan 2$.

Therefore,

$\theta = \left|\arctan 2 - \arctan\dfrac{3}{2}\right| = 7.13°$

(or 0.124 rad) to 3 sf.

Use $m = \dfrac{u_2}{u_1}$ *to calculate the gradient of the line.*

For equations of the form $y = mx + c$, m gives the gradient of the line.

Then use $\theta = |\beta - \alpha|$ where $\alpha = \arctan m_1$ and $\beta = \arctan m_2$

Exercise 11N

1 Consider the lines with equations $\mathbf{r} = \begin{pmatrix} 0 \\ 3 \end{pmatrix} + \lambda \begin{pmatrix} 1 \\ -1 \end{pmatrix}$ and

$y = \sqrt{3}x - 3$. Find the angle between these lines using two different methods.

EXAM-STYLE QUESTIONS

2 Find the angle between the lines $\dfrac{x-1}{2} = \dfrac{y}{3} = z$ and $2x = y = 3z$.

3 Find the angle between the lines with equations
$\mathbf{r} = (2 - \alpha)\mathbf{i} + (-1 + 2\alpha)\mathbf{j} + (1 - \alpha)\mathbf{k}$ and $\mathbf{r} = (2 + \beta)\mathbf{j} + (3 + \beta)\mathbf{k}$.

4 Consider two lines in the plane with equations
$y = m_1 x + c_1$ and $y = m_2 x + c_2$ where $m_1 > m_2$

a Show that, if θ is the angle between the lines,

$$\cos\theta = \frac{|1 + m_1 m_2|}{\sqrt{1 + m_1^2}\,\sqrt{1 + m_2^2}}$$

b Let $\alpha = \arctan m_1$ and $\beta = \arctan m_2$
Find $\tan(\alpha - \beta)$ in terms of m_1 and m_2

c Hence, calculate $\cos(\alpha - \beta)$

d Compare your answers to parts *a* and *c* and state your conclusion.

Given a plane π there are many non-collinear vectors parallel to π. However all the vectors normal to π are collinear. For this reason, you can define the angle between a line and a plane and the angle between two planes using normal vectors.

Angle between a line and a plane

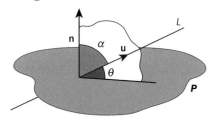

> An equivalent formula is
> $$\theta = 90° - \arccos\left(\left|\frac{\mathbf{u}\cdot\mathbf{n}}{\|\mathbf{u}\|\|\mathbf{n}\|}\right|\right)$$
> $0 \le \theta \le 90°$.

> → The angle θ between a line parallel to **u** and a plane with normal vector **n** is given by
> $$\theta = \arcsin\left(\left|\frac{\mathbf{u}\cdot\mathbf{n}}{\|\mathbf{u}\|\|\mathbf{n}\|}\right|\right),\ 0 \le \theta \le 90°$$

Example 42

Calculate the angle between the line with vector equation $\mathbf{r} = (3 - \lambda)\mathbf{i} + \mathbf{j} + (-1 + 2\lambda)\mathbf{k}$ and the plane with Cartesian equation $x - 3y + z = 1$

Answer

$\mathbf{u} = \begin{pmatrix} -1 \\ 0 \\ 2 \end{pmatrix}$ has the direction of the line and

*The coefficients of λ in the equation of the line are the components of **u**.*

$\mathbf{n} = \begin{pmatrix} 1 \\ -3 \\ 1 \end{pmatrix}$ is normal to the plane.

*The coefficients of the variables in the equation of the plane are the components of **n**.*

$\theta = \arcsin\dfrac{-1 + 2}{\sqrt{1 + 4}\,\sqrt{1 + 9 + 1}} = 7.75°$

(or 0.135 rad) to 3 sf.

Use the formula $\theta = \arcsin\left(\left|\dfrac{\mathbf{u}\cdot\mathbf{n}}{\|\mathbf{u}\|\|\mathbf{n}\|}\right|\right)$

Angle between two planes

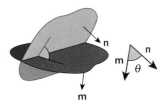

> → The angle θ between a plane with normal vector **m** and a plane with normal vector **n** is equal to the angle between the lines in the direction of the vectors **m** and **n**. It is given by
>
> $$\theta = \arccos\left(\left|\frac{\mathbf{m} \cdot \mathbf{n}}{\|\mathbf{m}\|\|\mathbf{n}\|}\right|\right), \quad 0 \le \theta \le 90°$$

Example 43

Find the angle between these planes.
$\pi_1: 2x - y + z = 4$
$\pi_1: \mathbf{r} = (1 - \alpha + \beta)\mathbf{i} + (2 - 3\alpha)\mathbf{j} + (2\beta)\mathbf{k}$

Answer

$\mathbf{m} = \begin{pmatrix} 2 \\ -1 \\ 1 \end{pmatrix} \perp \pi_1$

$\mathbf{n} = \begin{pmatrix} -1 \\ -3 \\ 0 \end{pmatrix} \times \begin{pmatrix} 1 \\ 0 \\ 2 \end{pmatrix} = \begin{pmatrix} -6 \\ 2 \\ 3 \end{pmatrix}$

$\theta = \arccos\left|\dfrac{-12 - 2 + 3}{\sqrt{4+1+1}\sqrt{36+4+9}}\right| = 50.1°$

(or 0.874 rad) to 3 sf.

The coefficients of the variables in the equation of the plane π_1 are the components of a vector normal to π_2

The coefficients of α and β are the components of vectors parallel to the plane and their cross product is a vector normal to the plane.

Use the formula $\theta = \arccos\left(\left|\dfrac{\mathbf{m} \cdot \mathbf{n}}{\|\mathbf{m}\|\|\mathbf{n}\|}\right|\right)$

Exercise 110

1 Calculate the angle between the line with vector equation
$\mathbf{r} = (1 - 2\lambda)\mathbf{i} + (1 - \lambda)\mathbf{j} + (-2 + \lambda)\mathbf{k}$ and the plane with Cartesian equation $2x - y + z = 5$.

2 Calculate the angle between the line with equation
$\dfrac{x-1}{3} = 2y = 3 - 2z$ and the plane with vector equation
$\mathbf{r} = (2 - 2\alpha - 3\beta)\mathbf{i} + (1 - \alpha + \beta)\mathbf{j} + (-2\alpha + \beta)\mathbf{k}$.

3 Find the angle between the planes defined by these equations.
$x - y + 3z = 1, \mathbf{r} = (4 - 2\alpha + 2\beta)\mathbf{i} + (1 - 3\beta)\mathbf{j} + (2 - \alpha - \beta)\mathbf{k}$

> When calculating angles the magnitude of the vectors is irrelevant. Sometimes you can make a calculation easier by replacing a vector by a collinear one.

4 Consider the points A(1, 0, 1), B(−1, 1, 0), C(2, 3, −1) and D(−1, −1, −1).

a Find a Cartesian equation of the plane ABC.

b Find Cartesian equations of the line AD.

c Find the angle between the line AD and the plane ABC.

Are there any values of *k* for which the line and the plane in question 5 are perpendicular?

⋮ EXAM-STYLE QUESTION

⋮ **5** Determine the values of *k* for which the line $\dfrac{x}{2} = ky = k - z$ and

⋮ the plane $(2k - 1)x - ky + z = 5 + k$ are parallel.

Intersection of a line and a plane

In 3-D space, given a line *l* in the direction of a vector **u** and a plane π normal to a vector **n**,

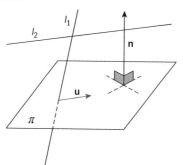

The line l_2 is parallel to the plane π. The line l_1 meets the plane at a point.

- if $\mathbf{u} \perp \mathbf{n}$, either the line *l* lies in the plane π or *l* is parallel to π;
- otherwise, the line and the plane meet at a point.

To determine the intersection between a line and a plane, you need to solve simultaneous equations. This example shows you how to find the intersection between a line defined by a vector equation and a plane defined by a Cartesian equation.

Example 44

The line *l* with vector equation $\mathbf{r} = 2\mathbf{i} - \mathbf{j} + 3\mathbf{k} + \lambda(\mathbf{i} - \mathbf{j} + 2\mathbf{k})$ and the plane π with Cartesian equation $2x - y + z + 2 = 0$ intersect at a point. Find the coordinates of the point of intersection.

Answer

$P \in l \Rightarrow P(2 + \lambda, -1 - \lambda, 3 + 2\lambda)$

The coefficients of the base vectors **i**, **j** *and* **k** *are the coordinates of a general point, P, on l.*

$P \in \pi \Rightarrow 2(2 + \lambda) - (-1 - \lambda) + (3 + 2\lambda) + 2 = 0$
$\lambda = -2 \Rightarrow P(2 - 2, -1 + 2, 3 - 4)$
Therefore, the point of intersection is P(0, 1, −1)

Substitute the coordinates of P into the equation of π and solve for λ.
Finally, substitute your value of l into the expression for the coordinates of P to obtain the coordinates of the point of intersection.

When solving this type of problem, you may be given other types of equations for the line or for the plane. Express the coordinates of a general point on the line in terms of a single parameter and then use the equation of the plane to determine its value.

Distance from a point to a plane

> → The **distance from a point to a plane** is measured along the perpendicular to the plane that contains the point.

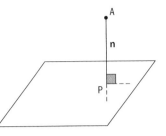

In vector geometry, there are a wide variety of problems that can be solved using the techniques studied so far. This example is a typical multi-step problem where you need to combine and apply several techniques to find the distance from a point to a plane.

P is called the foot of the perpendicular to the plane, dropped from A. The distance from A to the plane is AP.

Example 45

Consider the point A and the plane π as shown in the diagram.
a   Find a vector equation of the line, L, through point A and perpendicular to the plane.
b   Find the point of intersection, P, of the line L and the plane π.
c   Calculate the distance from point A to the plane.

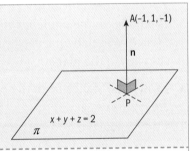

Answers

a $r = (-1 + \lambda)\mathbf{i} + (1 + \lambda)\mathbf{j} + (-1 + \lambda)\mathbf{k}$

Use $\mathbf{r} = \mathbf{a} + \lambda\mathbf{u}$ where a is the position vector of A and u is normal to the plane. The components of $\mathbf{u}$ are the coefficients of the variables in the equation of the plane.

b $P \in L \Rightarrow P(-1 + \lambda, 1 + \lambda, -1 + \lambda)$

$$\underbrace{(-1+\lambda)}_{x}+\underbrace{(1+\lambda)}_{y}+\underbrace{(-1+\lambda)}_{z} = 2$$

$\Rightarrow \lambda = 1$
$\Rightarrow P(-1 + 1, 1 + 1, -1 +1)$
Therefore, L intersects π at P(0, 2, 0).

The coefficients of the base vectors $\mathbf{i}$, $\mathbf{j}$ and $\mathbf{k}$ are the coordinates of a general point, P, on L.
Substitute the coordinates of P into the equation of the plane, π, to find the value of λ.
Finally, substitute your value of λ into the expression for the coordinates of P to obtain the coordinates of the point of intersection.

c   The distance from A to π is given by AP

$$AP = \sqrt{(-1-0)^2 +(1-2)^2 +(-1-0)^2} = \sqrt{3}$$

Use the formula for the distance between two points
$$\sqrt{(x_1 - x_2)^2 +(y_1 - y_2)^2 +(z_1 - z_2)^2}$$

Exercise 11P

EXAM-STYLE QUESTIONS

1 The line **r** with vector equation $\mathbf{r} = \mathbf{i} + \mathbf{k} + \lambda(\mathbf{i} - 2\mathbf{j} + \mathbf{k})$ and the plane π with Cartesian equation $x + y + 2z = 4$ intersect at the point P. Find the coordinates of point P.

2 Find the coordinates of the point of intersection of the line $\dfrac{x-1}{5} = \dfrac{y}{2} = \dfrac{z}{3}$ and the plane $-x - y + 3z = 5$.

3 Consider the line L with parametric equations
   $x = 3k,\ y = 2 - 2k,\ z = 1 - k$
   and the plane π with vector equation
   $\mathbf{r} = (4 - 2\alpha + \beta)\mathbf{i} + (1 - \beta)\mathbf{j} + (2 - \alpha - 2\beta)\mathbf{k}$.
   a Write down parametric equations for the plane π.
   b Hence, determine the coordinates of the point where the line L intersects the plane π.

4 Show that the line L with vector equation
   $\mathbf{r} = (1 - \lambda)\mathbf{i} + (1 + 2\lambda)\mathbf{j} + (1 + \lambda)\mathbf{k}$
   is parallel to the plane π with equation $3x + y + z = 2$.
   Hence, write down a vector equation of a line L' parallel to L that lies in the plane π.

5 Find the distance from the point P(1, 2, 3) to the plane $2x + y - 5z = 1$.

Intersection of two lines

Two lines in a plane are either parallel or intersect at a point. In 3-D space, there may be non-parallel lines that do not have any point in common, as the diagram shows.

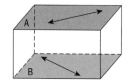

Lines with different directions that do not intersect are called **skew lines**. Skew lines do not define a plane – there is no plane that contains both of them. In fact, they lie in parallel planes as it is always possible to find a pair of parallel planes that each contains one of the lines.

A complicated intersection of roads can be modeled by skew lines – although some roads have a different direction, they are at different levels and do not meet.

Example 46

Find, if possible, the coordinates of the point of intersection of these pairs of lines.

a $x = 2y - 1 = z + 3$ and $x - 1 = \dfrac{y+4}{3} = z + 2$

b $r = (-1 + \lambda)\mathbf{i} + (1 + \lambda)\mathbf{j} + (-1 + \lambda)\mathbf{k}$ and $x = y = z + 1$

Answers

a $\left.\begin{array}{l} x = 2y - 1 \\ x - 1 = \dfrac{y+4}{3} \end{array}\right\} \Rightarrow 2y - 1 - 1 = \dfrac{y+4}{3}$

$\Rightarrow 6y - 6 = y + 4 \Rightarrow y = 2$ and $x = 3$

$3 = 2 \times 2 - 1 = z + 3 \Rightarrow z = 0$ and

$3 - 1 = \dfrac{2+4}{3} = z + 2 \Rightarrow z = 0$

Therefore, $(3, 2, 0)$ is the point of intersection.

Solve simultaneously the equations in x and y for each line.

Then substitute the values of x and y into the original equations to obtain the value of z.
You must obtain the same value for z – otherwise the lines are skew.

b A point P on the line
$r = (-1 + \lambda)\mathbf{i} + (1 + \lambda)\mathbf{j} + (-1 + \lambda)\mathbf{k}$
has coordinates
$(-1 + \lambda, 1 + \lambda, -1 + \lambda)$.
$-1 + \lambda = 1 + \lambda = -1 + \lambda$
$\Rightarrow 1 = -1$ (a contradiction!)
So, the lines do not intersect.

The coefficients of the base vectors in the vector equation of a line give the coordinates of a general point on the line.
Substitute these coordinates into the equation of the second line and solve.
As the first two equations are inconsistent, you have skew lines.

Intersection of two planes

The intersection of two non-parallel planes is a straight line.

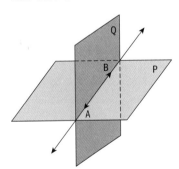

In the diagram, the line AB represents the intersection of the planes P and Q.

The next example shows you how to obtain the equations of the line of intersection of two planes in the form $\dfrac{x - x_1}{u_1} = \dfrac{y - y_1}{u_2} = \dfrac{z - z_1}{u_3}$

Example 47

Determine, in the form $\frac{x - x_1}{u_1} = \frac{y - y_1}{u_2} = \frac{z - z_1}{u_3}$, the equations of the line of intersection of the planes $2x - y + z = 1$ and $x + 2y - z = 0$.

Hence, state the coordinates of a point on this line and the components of a vector parallel to the line of intersection.

Answer

$\left.\begin{array}{r} 2x - y + z = 1 \\ x + 2y - z = 0 \end{array}\right\} \Rightarrow 3x + y = 1$

$\Rightarrow x = \frac{y - 1}{-3}$

$\left.\begin{array}{r} 2x - y + z = 1 \\ x + 2y - z = 0 \end{array}\right\} \Rightarrow 5x + z = 2$

$\Rightarrow x = \frac{z - 2}{-5}$

Therefore, $\frac{x - 0}{1} = \frac{y - 1}{-3} = \frac{z - 2}{-5}$.

So, $(0, 1, 2)$ is a point on the line

$\begin{pmatrix} 1 \\ -3 \\ -5 \end{pmatrix}$ a vector parallel to the line.

Eliminate z from the two equations and solve for x.

Eliminate y from the two equations and solve for x.

Write down the equations obtained in the form $\frac{x - x_1}{u_1} = \frac{y - y_1}{u_2} = \frac{z - z_1}{u_3}$

(x_1, y_1, z_1) is a point on the line and

$\begin{pmatrix} u_1 \\ u_2 \\ u_3 \end{pmatrix}$ *a vector parallel to the line.*

> You could also use a GDC to help solve this problem.

If you select different pairs of equations you may get a solution that looks different but that is equivalent to this one. For example, the GDC solution

shown here gives you the point $\left(\frac{2}{5}, -\frac{1}{5}, 0\right)$ and the vector $\begin{pmatrix} -\frac{1}{5} \\ \frac{3}{5} \\ 1 \end{pmatrix}$.

As $\begin{pmatrix} -\frac{1}{5} \\ \frac{3}{5} \\ 1 \end{pmatrix} = -\frac{1}{5}\begin{pmatrix} 1 \\ -3 \\ -5 \end{pmatrix}$ and $\begin{pmatrix} \frac{2}{5} \\ -\frac{1}{5} \\ 0 \end{pmatrix} = \begin{pmatrix} 0 \\ 1 \\ 2 \end{pmatrix} + \frac{2}{5}\begin{pmatrix} 1 \\ -3 \\ -5 \end{pmatrix}$ both solutions are

equivalent in the sense that each pair point and vector define the same line.

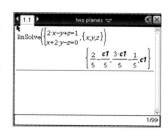

GDC help on CD:
Alternative demonstrations for the TI-84 Plus and Casio fx-9860GII GDCs are on the CD.

Exercise 11Q

1 Find, if possible, the coordinates of the point of intersection of the following pairs of lines.

a $\dfrac{x}{2} = y - 1 = z$ and $x = \dfrac{y+4}{3} = 3 - z$

b $\mathbf{r} = (5 + 2\lambda)\mathbf{i} + (4 + \lambda)\mathbf{j} + (5 - 3\lambda)\mathbf{k}$ and $x = y = z + 1$

2 Determine, in the form $\dfrac{x - x_1}{u_1} = \dfrac{y - y_1}{u_2} = \dfrac{z - z_1}{u_3}$, equations of the line of intersection of the planes $x - 3y + z = 2$ and $-x + y - 2z = 1$.

⋮ EXAM-STYLE QUESTION

3 Consider the plane with equation $3x - y + z = 3$ and the line given by the parametric equations

$$x = 3 - (2 - k)\lambda \quad y = (2k - 1) + \lambda \quad z = -1 + k\lambda$$

Given that the line is perpendicular to the plane, find

a the value of k

b the coordinates of the point of intersection of the line and the plane.

4 Consider the lines

$L_1: y = 2x + 2$ and $z = 3 - x$ and $L_2 : \dfrac{x - 1}{3} = \dfrac{y - 1}{6} = \dfrac{1 - z}{3}$

a Show that the lines L_1 and L_2 are parallel.

b Find the Cartesian equation of the plane defined by the lines L_1 and L_2.

5 Show that the plane with Cartesian equation $x + y + 3z = 1$ contains the line defined by the equations $x = 4 - y$ and $z = -1$.

6 Let π be the plane with equation $2x + 7y + 13z = 19$.

Consider the family of lines L_a defined by the equations

$$x = \dfrac{(y - 3)}{a} = \dfrac{(z + 2)}{2}, \ a \neq 0.$$

a Find, if possible, the coordinates of the point of intersection, in terms of a, of the line L_a with the plane.

b Write down the value of a for which the line is parallel to π.

c For the value of a found in part b, calculate the distance between L_a and π.

Intersection of three planes

Investigation – intersection of three planes

Use a graph plotting program to investigate the intersection of three planes, π_i, with equation $ax_i + by_i + cz_i + d = 0$, $i = 1, 2$ and 3.

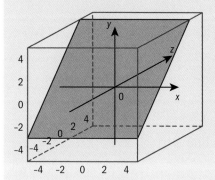

Explore different values for the coefficients and investigate:

1 Under which conditions is the intersection of three planes a point, a line, a plane or the empty set?

2 How many different cases can you identify?

3 How can you use the coefficients of the Cartesian equations of the planes to identify each case?

In the Investigation, you should have discovered that the intersection of three planes can be an empty set, a point, a line or a plane. To determine the intersection, solve a 3 × 3 system of equations and/or look at the directions of the normal vectors to the planes.

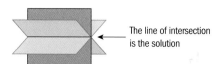

The line of intersection is the solution

▲ One possible arrangement of three planes.

Example 48

Consider the planes defined by the equations
$x + y + 3z = 5$,
$-x + 2y + 2z = 3$ and
$4x + y - 3z = 2$.
Show that the three planes intersect at a point and find its coordinates.

> Solving linear equations in three variables is covered in Chapter 3.

Answer

$$\begin{cases} x + y + 3z = 5 \\ -x + 2y + 2z = 3 \\ 4x + y - 3z = 2 \end{cases} \Rightarrow \begin{cases} x + y + 3z = 5 \\ 3y + 5z = 8 \\ -3y - 15z = -18 \end{cases}$$

Use the 1st equation to eliminate x from the 2nd and 3rd equations.

▶ Continued on next page

$$\begin{cases} x+y+3z=5 \\ 3y+5z=8 \\ -10z=-10 \end{cases} \Rightarrow \begin{cases} x+y+3z=5 \\ 3y+5z=8 \\ z=1 \end{cases}$$

Use the 2nd equation to eliminate y from the 3rd equation and then solve for z.

You can use a GDC to solve this problem or confirm your answer.

$$\begin{cases} x+y+3z=5 \\ 3y+5=8 \\ z=1 \end{cases} \Rightarrow \begin{cases} x+y+3z=5 \\ y=1 \\ z=1 \end{cases}$$

Substitute the value of z into the 2nd equation and solve for y.

$$\begin{cases} x+1+3=5 \\ y=1 \\ z=1 \end{cases} \Rightarrow \begin{cases} x=1 \\ y=1 \\ z=1 \end{cases}$$

Finally, substitute the values of y and z into the 1st equation and solve for x.

Therefore, the system has just one solution.
The intersection point is (1, 1, 1).

The solution of the system gives the coordinates of the point of intersection of the three planes.

GDC help on CD:
Alternative demonstrations for the TI-84 Plus and Casio fx-9860GII GDCs are on the CD.

Example 49

Show that the intersection of these planes is a straight line.
$\pi_1: 3x + 2y + z = 1$
$\pi_2: 7x + 4y + 5z = 3$
$\pi_3: 5x + 3y + 3z = 2$
Find a vector equation of the line of intersection.

Answer

$$\begin{cases} 3x+2y+z=1 \\ 7x+4y+5z=3 \\ 5x+3y+3z=2 \end{cases} \Rightarrow \begin{cases} 3x+2y+z=1 \\ x+3z=1 \\ x+3z=1 \end{cases}$$

$$\Rightarrow \begin{cases} 3x+2y+z=1 \\ x+3z=1 \end{cases}$$

Consider the system formed by the three equations of the planes.

Use the 1st equation to eliminate y from the 2nd and 3rd equations. The 2nd and 3rd equation are the same, so the intersection is defined by the first two equations.

Let $z = \lambda$.
Then $x = 1 - 3\lambda$ and $y = 4\lambda - 1$

Therefore, $\mathbf{r} = \begin{pmatrix} 1 \\ -1 \\ 0 \end{pmatrix} + \lambda \begin{pmatrix} -3 \\ 4 \\ 1 \end{pmatrix}$

Make one of the variables equal to a parameter, e.g. λ. Express the other two variables in terms of λ.
Write the equation of the line in the form $\mathbf{r} = \mathbf{a} + \lambda\mathbf{u}$.

You can use a GDC to solve this problem or confirm your answer.

Example 50

Show that the intersection of these planes is the empty set.
$\pi_1: 3x + 2y + z = 1$
$\pi_2: x + y - z = 3$
$\pi_3: 5x + 3y + 3z = 2$
Does this mean that the three planes are parallel? Explain.

Answer

$\begin{cases} 3x + 2y + z = 1 \\ x + y - z = 3 \\ 5x + 3y + 3z = 2 \end{cases} \Rightarrow \begin{cases} x + y - z = 3 \\ 4x + 3y = 4 \\ -4x - 3y = -1 \end{cases}$

$\Rightarrow \begin{cases} x + y - z = 3 \\ 4x + 3y = 4 \\ 0 = 3 \end{cases}$

Therefore, the planes do not all intersect. The planes are not all parallel as their normal vectors are not parallel.

Consider the system formed by the three equations of the planes.

Use the 1st equation to eliminate z from the 2nd and 3rd equations. If you add the 2nd and 3rd equations, you can see that they are inconsistent.

Here, you can simply use a GDC to confirm your answer. To answer the second part of the question, you could use a GDC and show that each pair of planes intersects in a line.

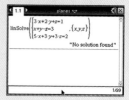

GDC help on CD:
Alternative demonstrations for the TI-84 Plus and Casio fx-9860GII GDCs are on the CD.

Exercise 11R

1 Find the coordinates of the point of intersection of the planes defined by the equations $5x + y + 2z = 3$, $x + y + z = 3$ and $4x + 2y + 2z = 5$.

2 Show that these planes intersect in a straight line.
$2x + y + z = 1$, $3x + y + 2z = 3$, $4x + y + 3z = 5$
Hence, find a vector equation of the line of intersection.

3 Show that the planes defined by these equations have no common point.
$x + y + z = 1$, $x - y + z = 3$, $3x + y + 3z = 1$

4 Consider this system of linear equations in x, y and z
$\begin{cases} x + y + z = 0 \\ ax + y + z = 0, \text{ where } a, b \text{ and } c \text{ are real parameters.} \\ x + by + cz = 0 \end{cases}$

   a Each of the equations in the system defines a plane. Show that the planes always have at least one point in common.

   b Under which condition is the intersection of the planes a straight line?

5 Identify the relative position of the three planes defined by the equations
$\pi_1: x + 2y - 2z = 5$, $\pi_2: 3x - 6y + 3z = 2$ and $\pi_3: x - 2y + z = 7$.

Why are symbolic representations of three-dimensional objects easier to deal with than visual representations? What does this tell us about our knowledge of mathematics in other dimensions?

6 Consider the planes defined by the equations $x + y + z = 2$, $2x - y + z = -1$ and $3x - y + kz = 4$, where k is a real number.

a If $k = -3$, find the coordinates of the point of intersection of the three planes.

b Find the value of k for which the three planes do not have any common point.

7 Prove these properties:

a Two planes are either parallel or they intersect in a line.

b A line is either parallel to a plane, intersects it at a single point, or is contained in the plane.

c Two lines perpendicular to the same plane are parallel to each other.

d Two planes perpendicular to the same line are parallel to each other.

Investigation – coefficient patterns

Consider the system $\begin{cases} a_1 x + a_2 y + a_3 z = a_4 \\ a_5 x + a_6 y + a_7 z = a_8 \\ a_9 x + a_{10} y + a_{11} z = a_{12} \end{cases}$

Use a GDC to study the solutions of the system when

a $a_1, a_2, ..., a_{12}$ are consecutive numbers

b $a_1, a_2, ..., a_{12}$ are consecutive even numbers.

In each case, explore the geometric meaning of the pattern found.

11.8 Modeling and problem solving

Extension material on CD: *Worksheet 11*

Vector geometry can be used to model situations that involve the position and movement of particles. Before exploring a range of applications, you need to distinguish the scalar and vector quantities related to movement.

- **Distance** is a scalar quantity. **Displacement** is a vector quantity.
- **Speed** is a scalar quantity that considers only the magnitude. **Velocity** is a vector quantity that must consider both magnitude and direction.
- **Acceleration** is a 'change in velocity'. This change can be in the magnitude (speed) of the velocity or in the direction of the velocity.

Vector equations of lines provide a method for determining the position of an object when the parameter chosen is time, as shown in the next example.

Example 51

The diagram shows a boat in danger at the point O(0, 0) and the paths of two rescue boats, *Bluespeed* and *Slowmotion*, as they depart from the positions A(10, 5) and B(−5, −2) respectively.

Bluespeed moves at a speed of 15 km h⁻¹ and *Slowmotion* moves at a speed of 8 km h⁻¹.

a Find an equation for the position of each boat *t* hours after departing from A and B respectively.

b Hence, determine how long it takes for each boat to reach the boat in danger.

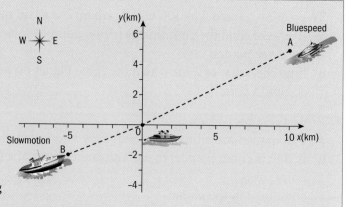

Answers

a *Bluespeed* moves in the direction of $\overrightarrow{AO}$ at a speed of 15 km h⁻¹. So, its velocity vector is

$$\mathbf{v}_1 = \begin{pmatrix} -6\sqrt{5} \\ -3\sqrt{5} \end{pmatrix}$$ and a vector equation for its movement

is $\mathbf{r} = \begin{pmatrix} 10 \\ 5 \end{pmatrix} + t\begin{pmatrix} -6\sqrt{5} \\ -3\sqrt{5} \end{pmatrix}$, $t \geq 0$.

Slowmotion moves in the direction of $\overrightarrow{BO}$ at a speed of 8 km h⁻¹. So, its velocity vector is

$$\mathbf{v}_2 = \begin{pmatrix} \dfrac{40\sqrt{29}}{29} \\ \dfrac{16\sqrt{29}}{29} \end{pmatrix}$$ and a vector equation for its

movement is $\mathbf{r} = \begin{pmatrix} -5 \\ -2 \end{pmatrix} + t\begin{pmatrix} \dfrac{40\sqrt{29}}{29} \\ \dfrac{16\sqrt{29}}{29} \end{pmatrix}$ $t \geq 0$

b $\begin{pmatrix} 0 \\ 0 \end{pmatrix} = \begin{pmatrix} 10 \\ 5 \end{pmatrix} + t\begin{pmatrix} -6\sqrt{5} \\ -3\sqrt{5} \end{pmatrix} \Rightarrow t = \dfrac{\sqrt{5}}{3}$

Bluespeed takes approximately 44 minutes and 43 seconds to reach the rescue site.

$\begin{pmatrix} 0 \\ 0 \end{pmatrix} = \begin{pmatrix} -5 \\ -2 \end{pmatrix} + t\begin{pmatrix} \dfrac{40\sqrt{29}}{29} \\ \dfrac{16\sqrt{29}}{29} \end{pmatrix} \Rightarrow t = \dfrac{\sqrt{29}}{8}$

Slowmotion takes approximately 40 minutes and 23 seconds to reach the rescue site.

Bluespeed:
Find the vector with magnitude 15 in the

direction of $\overrightarrow{AO} = \begin{pmatrix} -10 \\ -5 \end{pmatrix}$ *and*

use $\mathbf{r} = \mathbf{a} + t\mathbf{v}_1$, *where* $a = \overrightarrow{AO}$.

Slowmotion:
Find the vector with magnitude 8 in the

direction of $\overrightarrow{BO} = \begin{pmatrix} 5 \\ 2 \end{pmatrix}$ *and use*

$\mathbf{r} = \mathbf{b} + t\mathbf{v}_2$, *where* $b = \overrightarrow{BO}$.
In both equations, parameter t represents the time, in hours, after departure to the rescue site.

*Using the vector equations obtained in part **a**, substitute the position of the boat in danger for* **r** *and solve for t.*

Convert the time in hours into minutes and seconds to obtain appropriate answers.

> You can use your GDC to do this conversion.

Example 52

The diagram shows the model of a building with dimensions 19 m by 10 m by 12 m. All the floors are 2 m high.
Anne departs from a position A(4, 3, 4) and moves toward the elevator whose path has equation $x = 10$ and $y = 5$. The elevator moves along the intersection of these two planes.

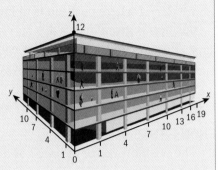

a How many seconds does it take Anne to reach the elevator if she walks at a speed of 1.5 metres per second?

Anne takes the elevator down to the ground floor and then walks in the direction of the vector $\begin{pmatrix} -1 \\ -1 \\ 0 \end{pmatrix}$ towards a door.

b Given that Anne walks at a speed of 1.6 metres per second, determine the coordinates of the location of the door and the time it takes her to get there. State any assumptions you have made.

Answers

a Anne's location: A(4, 3, 4)

Elevator's location: E(10, 5, 4)

$AE = \sqrt{6^2 + 2^2 + 0^2} = 2\sqrt{10}$ metres

$t = \dfrac{2\sqrt{10}}{1.5} = 4.22$ seconds (to 3 sf)

So, Anne takes 4.22 seconds to get to the elevator.

To determine the coordinates of the location of the elevator, use the equation of the path of the elevator and the z-coordinate of point A (i.e. assume that the movement takes place in the plane z = 4).

Use $AE = |\overrightarrow{AE}|$
$$= \sqrt{(x_2 - x_1)^2 + (y_2 - y_1)^2 + (z_2 - z_1)^2}$$

Use $t = \dfrac{distance}{speed}$

b In the plane $x \circ y$, an equation of Anne's movement is:

$$\mathbf{r} = \begin{pmatrix} 10 \\ 5 \end{pmatrix} + t \begin{pmatrix} -0.8\sqrt{2} \\ -0.8\sqrt{2} \end{pmatrix}$$

$0 = 5 - 0.8\sqrt{2}t$

$\Rightarrow t = \dfrac{5}{0.8\sqrt{2}} = 4.42$ (to 3 sf)

$\Rightarrow \mathbf{r} = \begin{pmatrix} 5 \\ 0 \end{pmatrix}$

So, the door is located at (5, 0, 0) and Anne takes 4.42 seconds to get there.
You assumed that the locations of objects and people were represented by points and the floors were represented by planes.

Assume that the ground floor lies in the plane z = 0 and reduce the problem to a 2-D situation. Hence, the location of the elevator is (10, 5) and Anne's velocity has a direction of $\begin{pmatrix} -1 \\ -1 \end{pmatrix}$ and magnitude 1.6.

Anne reaches the door when either x = 0 or y = 0. By inspection, x = 0 gives a negative y-value. So, use y = 0 to obtain the value of t and substitute it in the equation to obtain the coordinates of the door.

Exercise 11S

1 Boat A's position is given by the parametric equations $x = 3 - t$, $y = 2t - 4$ where position is in km and time in hours. Boat B's position is given by $x = 4 - 3t$, $y = 3 - 2t$.

 a Find the initial position of each boat.
 b Find the velocity vector of each boat.
 c What is the angle between the paths of the boats?
 d At what time are the boats closest to each other?

> Why might it be argued that vector equations of lines are superior to Cartesian ones?

EXAM-STYLE QUESTION

2 The position vector at time t of a particle P moving in 3-D space is given by
$$\overrightarrow{OP} = (5+10t)\mathbf{i}+(20-20t)\mathbf{j}+(30t-10)\mathbf{k}, \ t \geq 0.$$

 a Find the coordinates of P when $t = 0$.
 b Show that P moves along the line L with Cartesian equations
 $$\frac{x-5}{1} = \frac{y-20}{-2} = \frac{z+10}{3}.$$

 c i Find the value of t when P lies on the plane with equation $x + y + z = 55$.
   ii State the coordinates of P at this time.
   iii Hence, find the total distance travelled by P before it meets the plane.

   The position vector at time t of another particle, Q, is given by $\overrightarrow{OQ} = \begin{pmatrix} 2t^2 \\ 1-2t \\ 1+t^2 \end{pmatrix} t \geq 0.$

 d i Find the value of t for which the distance between the particles P and Q is a minimum.
   ii Find the coordinates of P and Q at this time.
 e Let $\mathbf{a}$, $\mathbf{b}$ and $\mathbf{c}$ be the position vectors of Q at times $t = 0$, $t = 1$ and $t = 2$ respectively.
   i Show that $\mathbf{a} - \mathbf{b}$ and $\mathbf{b} - \mathbf{c}$ are non-collinear.
   ii What is the geometric meaning of part i?

3 The diagram shows the unit cube OABCDEFG.

Let O be the origin, OA the x-axis, OC the y-axis and OD the z-axis. Let P, Q and R be the midpoints of BC, CG and DG respectively.

a Find the position vectors $\overrightarrow{OP}$, $\overrightarrow{OQ}$ and $\overrightarrow{OR}$ in component form.

b Find the Cartesian equation of the plane PQR.

Let S, T and U be the midpoints of ED, AE and AB respectively.

c Show that PQRSTU is a regular hexagon and find its area.

d Show that the line OF is perpendicular to the plane PQR.

e Determine the coordinates of the point I where the line OF meets the plane PQR.

f Hence, find the distance from F to the hexagon PQRSTU.

> The edges of the unit cube have length 1 unit.

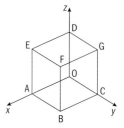

Explore further vectors

In this chapter, we have looked at two approaches to the study of vectors – geometric and analytic. An alternative is the **axiomatic** approach, where no attempt is made to describe the nature of vectors or the algebraic operations on them. Instead, vectors and operations are seen as undefined concepts of which we know nothing except that they satisfy a set of axioms. Such a system, with appropriate axioms, is called a **linear space**. The axiomatic point of view is mathematically the most satisfactory as it is independent of particular geometric representations, systems of coordinates and dimensions. Examples of linear spaces occur in many different branches of mathematics.

1 Examine linear spaces and the axioms that define them.

2 Explore the inner product and its relationship to the distance between points and the way angles are measured.

3 Explore alternative definitions of distance in the plane and the consequences on the geometry of shapes.

> An **axiom** is a statement which is assumed to be true without need for proof, used as a basis for developing an argument.

> What is a distance? can we define distance between two points in different ways?

Review exercise

1 Given $\mathbf{u} = \begin{pmatrix} 3 \\ 4 \end{pmatrix}$ and $\mathbf{v} = \begin{pmatrix} -1 \\ 5 \end{pmatrix}$, find

  a $3\mathbf{u} - 2\mathbf{v}$ **b** $|\mathbf{u}|$, $|\mathbf{v}|$, and $|\mathbf{u} + \mathbf{v}|$

2 Express $\mathbf{u} = 3\mathbf{i} - \mathbf{j} + \mathbf{k}$ as a linear combination of $\mathbf{a} = \mathbf{i} + \mathbf{j}$, $\mathbf{b} = \mathbf{i} + \mathbf{k}$ and $\mathbf{c} = 2\mathbf{i} - \mathbf{j} - \mathbf{k}$.

3 Let $\mathbf{a} = \begin{pmatrix} -5 \\ 4 \\ 3 \end{pmatrix}$

  a Find the unit vector in the direction of $\mathbf{a}$.

  b Find the vectors with magnitude 5 collinear with $\mathbf{a}$.

4 Consider the vector $\mathbf{u} = \cos\alpha\cos\beta\mathbf{i} + \sin\alpha\cos\beta\mathbf{j} + \sin\beta\mathbf{k}$
  Show that $\mathbf{u}$ is a unit vector.

5 Let $\mathbf{u} = \mathbf{i} + \tan\alpha\,\mathbf{j}$ and $\mathbf{v} = \tan\beta\,\mathbf{i} + \mathbf{j}$, where $0 < \alpha, \beta < \dfrac{\pi}{4}$.

Let γ be the angle between the vectors $\mathbf{u}$ and $\mathbf{v}$.

Show that $\alpha + \beta + \gamma = \dfrac{\pi}{2}$.

6 Let θ be the angle between the unit vectors $\mathbf{a}$ and $\mathbf{b}$, where $0 \le \theta \le \pi$.

a Express $|\mathbf{a} - \mathbf{b}|^2$ and $|\mathbf{a} + \mathbf{b}|^2$ in terms of θ.

b Hence, determine the exact value of $\sin\theta$ for which
$|\mathbf{a} + \mathbf{b}| = 2|\mathbf{a} - \mathbf{b}|$

7 Write down, in vector and parametric form, equations of the plane passing through the point A and parallel to the vectors $\mathbf{p}$ and $\mathbf{q}$ when:

a A(2, 3, 4), $\mathbf{p} = 2\mathbf{i} - 3\mathbf{j} + 2\mathbf{k}$ and $\mathbf{q} = \mathbf{j} - 2\mathbf{k}$

b A(1, 0, −2), $\mathbf{p} = -2\mathbf{i} + \mathbf{k}$ and $\mathbf{q} = -\mathbf{j}$

8 The vector equations of the lines L_1 and L_2 are given by:

$$L_1 : \mathbf{r}_1 = \begin{pmatrix} 5 \\ 1 \end{pmatrix} + \lambda \begin{pmatrix} 3 \\ -2 \end{pmatrix}$$

$$L_2 : \mathbf{r}_2 = \begin{pmatrix} -2 \\ 2 \end{pmatrix} + \mu \begin{pmatrix} 4 \\ 1 \end{pmatrix}$$

The lines intersect at the point P. Find the coordinates of P.

Review exercise

1 Find the size of the angle between the vectors $\mathbf{u} = \begin{pmatrix} 1 \\ 1 \\ 3 \end{pmatrix}$ and $\mathbf{v} = \begin{pmatrix} 2 \\ 0 \\ 1 \end{pmatrix}$
Give your answer to the nearest degree.

2 Write down vector, parametric and Cartesian equations of the planes containing the points A, B and C.

a A(1, 1, 0), B(−1, 1, 2) and C(0, 1, −1)

b A(−1, 1, 1), B(−2, −1, 2) and C(0, 1, −1)

3 Write down, in scalar product form, the equation of the plane passing through the point with position vector $\mathbf{a}$ and with normal vector $\mathbf{n}$ when:

a $\mathbf{a} = 2\mathbf{i} - 3\mathbf{j} + 4\mathbf{k}$ and $\mathbf{n} = \mathbf{i} + \mathbf{j} - 2\mathbf{k}$

b $\mathbf{a} = \mathbf{i} + 2\mathbf{k}$ and $\mathbf{n} = \mathbf{i} - \mathbf{j}$

4 Find a Cartesian equation of the plane that:

   a contains the point A(2, −3, 4) and is perpendicular to the position vector of A

   b contains the points A(6, 0, 0), B(0, 0, −3) and C(3, 6, 0)

   c contains the points A(3, 2, −1) and B(4, 4, 0) and is perpendicular to the plane $2x + 4y − 4z = 3$

   d contains the points A(2, −1, −3) and B(4, −3, 2) and is parallel to the x-axis

   e passes through the point (3, 4, 2) and is perpendicular to the x-axis

   f passes through the point (3, 2, 1) and is perpendicular to each of the planes $2x + 3y − z = 5$ and $3x + 3z = 2$

   g passes through the point (3, 1, 1) and contains the line of intersection of the planes $x + y + 5z = 0$ and $2x + 3y + 12z = 0$

EXAM-STYLE QUESTION

5 Calculate the acute angle between the lines with equations

$\mathbf{r} = \mathbf{i} + \mathbf{j} + \alpha(2\mathbf{i} − \mathbf{j} − \mathbf{k})$

and $\mathbf{r} = \mathbf{k} + \beta(\mathbf{i} + \mathbf{j} + 2\mathbf{k})$.

6 The vector equations of the lines L_1 and L_2 are given by:

$L_1: \mathbf{r} = \mathbf{i} + \mathbf{j} + \mathbf{k} + \lambda(\mathbf{i} + 2\mathbf{j} + 3\mathbf{k})$

$L_2: \mathbf{r} = \mathbf{i} + 4\mathbf{j} + 5\mathbf{k} + \mu(2\mathbf{i} + \mathbf{j} + 2\mathbf{k})$

Show that the lines are concurrent.

Hence, determine a vector equation of the plane that contains these lines.

> **Concurrent** lines are lines that all pass through a certain point.

7 Solve the system $\begin{cases} x + y + z = 3 \\ 2x + y − 2z = 0 \\ 3x − 2y + 5z = 23 \end{cases}$

and explain the geometric meaning of your answer.

EXAM-STYLE QUESTION

8 Consider the following system of equations where a and b are constants.

$3x + y + z = 1$,

$x + y − z = 4$

and $2x + y + bz = a$

   a Solve the system in terms of a and b.

   b Hence, write down the values of a and b for which this system of equations has a non-unique solution and state its geometric meaning.

CHAPTER 11 SUMMARY

Geometric vectors and basic operations

- A vector is defined by direction and magnitude.
- In mathematics, $\overrightarrow{AB}$ usually represents either the **position vector** of B relative to A or the **displacement** from A to B. A is called the **initial point** and B the **terminal point**.
- When we work with vectors you need to distinguish the vector from its magnitude, which is a non-negative scalar.

  The magnitude of $\overrightarrow{AB}$ is simply the length AB and is denoted by $|\overrightarrow{AB}|$. If $\overrightarrow{AB} = \mathbf{a}$ then you can represent its magnitude by $|\mathbf{a}|$ or simply by a.
- **Special case**

  The vector $\overrightarrow{AA}$ has no defined direction and is represented by a single point A. This vector is called a **null vector** or **zero vector** and it is the only vector that has a magnitude of zero.
- The sum of two vectors is determined by the parallelogram law: $\overrightarrow{BA} + \overrightarrow{BC} = \overrightarrow{BD}$ where ABCD is a parallelogram.
- This definition is called the **triangle law** and has the advantage that it can easily be applied to the addition of several vectors.
- **Properties of vector addition**

  Commutative property: $\mathbf{u} + \mathbf{v} = \mathbf{v} + \mathbf{u}$

  Associative property: $(\mathbf{u} + \mathbf{v}) + \mathbf{w} = \mathbf{u} + (\mathbf{v} + \mathbf{w})$

  Additive identity property (the zero vector): $\mathbf{u} + \mathbf{0} = \mathbf{0} + \mathbf{u} = \mathbf{u}$

  Additive inverse property (the opposite vector):
  $\mathbf{u} + (-\mathbf{u}) = (-\mathbf{u}) + \mathbf{u} = \mathbf{0}$
- The difference of two vectors $\mathbf{u}$ and $\mathbf{v}$ is the vector obtained when you add $\mathbf{u}$ to the opposite of $\mathbf{v}$.
- The product of a vector $\mathbf{u}$ and a positive scalar k is another vector $\mathbf{v}$ with the same direction as $\mathbf{u}$ and magnitude $|\mathbf{v}| = k|\mathbf{u}|$.
- In general, the product of a vector $\mathbf{u}$ and a negative scalar k is another vector $\mathbf{v}$ with the opposite direction to $\mathbf{u}$ and magnitude
  $$|\mathbf{v}| = |k||\mathbf{u}| = -|k\mathbf{u}|.$$
- **Properties of scalar multiplication**

  Commutative property: $\alpha\mathbf{v} = \mathbf{v}\alpha$

  Associative property: $\alpha(\beta\mathbf{v}) = (\alpha\beta)\mathbf{v}$

  Distributive property (1): $\alpha(\mathbf{u} + \mathbf{v}) = \alpha\mathbf{u} + \alpha\mathbf{v}$

  Distributive property (2): $(\alpha + \beta)\mathbf{v} = \alpha\mathbf{v} + \beta\mathbf{v}$

  Multiplicative identity property: $1\mathbf{v} = \mathbf{v}$

  Property of zero: $0\mathbf{v} = \mathbf{0}$ and $\alpha\mathbf{0} = \mathbf{0}$

Continued on next page

Vector algebra

- In 2-D space, $\mathbf{i} = \begin{pmatrix} 1 \\ 0 \end{pmatrix}$, $\mathbf{j} = \begin{pmatrix} 0 \\ 1 \end{pmatrix}$ and O(0, 0).

- In 3-D space, $\mathbf{i} = \begin{pmatrix} 1 \\ 0 \\ 0 \end{pmatrix}$, $\mathbf{j} = \begin{pmatrix} 0 \\ 1 \\ 0 \end{pmatrix}$ and $\mathbf{k} = \begin{pmatrix} 0 \\ 0 \\ 1 \end{pmatrix}$

  The origin O has coordinates (0, 0, 0)

- Given two vectors in the plane, $\mathbf{u} = \begin{pmatrix} u_1 \\ u_2 \end{pmatrix}$ and $\mathbf{v} = \begin{pmatrix} v_1 \\ v_2 \end{pmatrix}$, and a real number λ:

  - The sum of the two vectors $\mathbf{u}$ and $\mathbf{v}$ is defined by

  $$\mathbf{u} + \mathbf{v} = \begin{pmatrix} u_1 \\ u_2 \end{pmatrix} + \begin{pmatrix} v_1 \\ v_2 \end{pmatrix} = \begin{pmatrix} u_1 + v_1 \\ u_2 + v_2 \end{pmatrix}$$

  - The product of a scalar λ and a vector $\mathbf{u}$ is defined by $\lambda\mathbf{u} = \begin{pmatrix} \lambda u_1 \\ \lambda u_2 \end{pmatrix}$

  - The zero vector or null vector is $\mathbf{0} = \begin{pmatrix} 0 \\ 0 \end{pmatrix}$

  - The opposite vector of $\mathbf{u} = \begin{pmatrix} u_1 \\ u_2 \end{pmatrix}$ is $-\mathbf{u} = \begin{pmatrix} -u_1 \\ -u_2 \end{pmatrix}$

- If $\mathbf{v} = \begin{pmatrix} v_1 \\ v_2 \end{pmatrix}$ then the magnitude of v is given by

  $$v = |\mathbf{v}| = \sqrt{v_1^2 + v_2^2}$$

- If $\mathbf{v} = \begin{pmatrix} v_1 \\ v_2 \end{pmatrix}$, the **unit vector** in the direction of a non-zero

  vector $\mathbf{v}$ is $\mathbf{u} = \dfrac{1}{|\mathbf{v}|}\mathbf{v} = \begin{pmatrix} \dfrac{v_1}{\sqrt{v_1^2 + v_2^2}} \\ \dfrac{v_2}{\sqrt{v_1^2 + v_2^2}} \end{pmatrix}$.

- Two vectors $\mathbf{u} = \begin{pmatrix} u_1 \\ u_2 \end{pmatrix}$ and $\mathbf{v} = \begin{pmatrix} v_1 \\ v_2 \end{pmatrix}$ are **collinear** if $\mathbf{u} = k\mathbf{v}$ or

  $\mathbf{v} = k\mathbf{u}$ for some scalar k
  If $k > 0$, $\mathbf{u}$ and $\mathbf{v}$ have the same direction;
  if $k < 0$, $\mathbf{u}$ and $\mathbf{v}$ have opposite directions.

Continued on next page

Vectors, points and equations of lines

- Three points A, B and C in the plane are collinear when $\overrightarrow{AB}$ and $\overrightarrow{AC}$ are collinear vectors, that is when $\overrightarrow{AC} = k\overrightarrow{AB}$ for some scalar k.

- Two points $A(x_1, y_1, z_1)$ and $B(x_2, y_2, z_2)$, have **displacement vector**

$$\overrightarrow{AB} = \begin{pmatrix} x \\ y \\ z \end{pmatrix} \text{ where } x = x_2 - x_1, y = y_2 - y_1 \text{ and } z = z_2 - z_1$$

  You can also assign a **position vector** to each point

$$\overrightarrow{OA} = \begin{pmatrix} x_1 \\ y_1 \\ z_1 \end{pmatrix} \text{ and } \overrightarrow{OB} = \begin{pmatrix} x_2 \\ y_2 \\ z_2 \end{pmatrix}$$

- The magnitude of a vector $\mathbf{v} = \begin{pmatrix} v_1 \\ v_2 \\ v_3 \end{pmatrix}$ is given by $v = |\mathbf{v}| = \sqrt{v_1^2 + v_2^2 + v_3^2}$.

- Given two points $A(x_1, y_1, z_1)$ and $B(x_2, y_2, z_2)$, the distance between A and B is given by the magnitude of the vector $\overrightarrow{AB}$

$$AB = |\overrightarrow{AB}| = \sqrt{(x_2 - x_1)^2 + (y_2 - y_1)^2 + (z_2 - z_1)^2}$$

- The unit vector in the direction of a non-zero vector $\mathbf{v}$ is

$$\mathbf{u} = \frac{1}{|\mathbf{v}|} \mathbf{v} = \begin{pmatrix} \dfrac{v_1}{\sqrt{v_1^2 + v_2^2 + v_3^2}} \\[2ex] \dfrac{v_2}{\sqrt{v_1^2 + v_2^2 + v_3^2}} \\[2ex] \dfrac{v_3}{\sqrt{v_1^2 + v_2^2 + v_3^2}} \end{pmatrix}$$

- Two vectors $\mathbf{u} = \begin{pmatrix} u_1 \\ u_2 \\ u_3 \end{pmatrix}$ and $\mathbf{v} = \begin{pmatrix} v_1 \\ v_2 \\ v_3 \end{pmatrix}$ are collinear if $\mathbf{u} = k\mathbf{v}$

  or $\mathbf{v} = k\mathbf{u}$ for some scalar k. If $k > 0$, $\mathbf{u}$ and $\mathbf{v}$ have the same direction; if $k < 0$, $\mathbf{u}$ and $\mathbf{v}$ have opposite directions.

- In general, the coordinates of the midpoint M of a line segment [AB], with $A(x_1, y_1, z_1)$ and $B(x_2, y_2, z_2)$, are given by

$$\left(\frac{x_1 + x_2}{2}, \frac{y_1 + y_2}{2}, \frac{z_1 + z_2}{2} \right)$$

Continued on next page

- A point R is on the line (AB) when $\overrightarrow{AR} = \lambda\overrightarrow{AB}$ for some real value of λ. This is called a **vector equation of the line** AB. If the point R has position vector $\mathbf{r}$, the point A has position vector $\mathbf{a}$, and $\overrightarrow{AB} = \mathbf{u}$, the direction vector, then the vector equation $\overrightarrow{AR} = \lambda\overrightarrow{AB}$ can be re-written as $\mathbf{r} = \mathbf{a} + \lambda\mathbf{u}$

- If the line (AB) lies in the plane $x \cdot y$, then you can represent the vector in component form $\mathbf{r} = \begin{pmatrix} x \\ y \end{pmatrix}$, $\mathbf{a} = \begin{pmatrix} x_1 \\ y_1 \end{pmatrix}$ and

  $\mathbf{u} = \begin{pmatrix} u_1 \\ u_2 \end{pmatrix}$ This gives a pair of **parametric equations**

  $x = x_1 + \lambda u_1$ and $y = y_1 + \lambda u_2$

  or $\lambda = \dfrac{x - x_1}{u_1} = \dfrac{y - y_1}{u_2}$

  If both components of the vector are non-zero, eliminate the parameter λ to obtain a **Cartesian equation** of the line (AB):

  $$y - y_1 = \frac{u_2}{u_1}(x - x_1)$$

  This can be reduced to the form $y = mx + c$ where $m = \dfrac{u_2}{u_1}$

  is the gradient of the line and $c = y_1 - \dfrac{u_2}{u_1}x_1$ is the y-intercept.

- If the line (AB) lies in 3-D space, $\mathbf{r} = \begin{pmatrix} x \\ y \\ z \end{pmatrix}$, $\mathbf{a} = \begin{pmatrix} x_1 \\ y_1 \\ z_1 \end{pmatrix}$ and

  $\mathbf{u} = \begin{pmatrix} u_1 \\ u_2 \\ u_3 \end{pmatrix}$, the vector equation $\mathbf{r} = \mathbf{a} + \lambda\mathbf{u}$ can be transformed

  into three parametric equations

  $x = x_1 + \lambda u_1, y = y_1 + \lambda u_2$ and $z = z_1 + \lambda u_3$

  where $\lambda = \dfrac{x - x_1}{u_1} = \dfrac{y - y_1}{u_2} = \dfrac{z - z_1}{u_3}$

  If all the components of the vector are non-zero, eliminate the parameter λ to obtain Cartesian equations of the line (AB)

  $$\frac{x - x_1}{u_1} = \frac{y - y_1}{u_2} = \frac{z - z_1}{u_3}$$

Continued on next page

Scalar product

- Given two non-zero vectors **u** and **v**, $\mathbf{u} \cdot \mathbf{v} = |\mathbf{u}||\mathbf{v}|\cos\theta$, where θ is the angle between **u** and **v**.
- Here are some important consequences of the geometric definition of the scalar product.
    - The scalar product of two vectors is always a number.
    - The definition of scalar product does not depend on the dimensions of the space.
    - $\mathbf{u} \cdot \mathbf{v} = 0$ if and only if $\mathbf{u} = \mathbf{0}$, $\mathbf{v} = \mathbf{0}$ or **u** and **v** are **orthogonal**.
    - $\mathbf{u} \cdot \mathbf{v} = \pm|\mathbf{u}||\mathbf{v}|$ if **u** and **v** are parallel.
    - $\mathbf{u} \cdot \mathbf{u} = |\mathbf{u}|^2$
    - $\mathbf{u} \cdot \mathbf{v} > 0$ when q is acute and $\mathbf{u} \cdot \mathbf{v} < 0$ when θ is obtuse.
    - $\mathbf{u} \cdot \mathbf{v} = \mathbf{v} \cdot \mathbf{u}$
    - $\mathbf{u} \cdot (\mathbf{v} + \mathbf{w}) = \mathbf{u} \cdot \mathbf{v} + \mathbf{u} \cdot \mathbf{w}$
    - $(\lambda\,\mathbf{u}) \cdot \mathbf{v} = \lambda\,(\mathbf{u} \cdot \mathbf{v})$
- Given two vectors in the plane, $\mathbf{u} = u_1\mathbf{i} + u_2\mathbf{j}$ and $\mathbf{v} = v_1\mathbf{i} + v_2\mathbf{j}$,
$$\mathbf{u} \cdot \mathbf{v} = u_1v_1 + u_2v_2$$
In 3-D space, given two vectors, $\mathbf{u} = u_1\mathbf{i} + u_2\mathbf{j} + u_3\mathbf{k}$ and $\mathbf{v} = v_1\mathbf{i} + v_2\mathbf{j} + v_3\mathbf{k}$,
$$\mathbf{u} \cdot \mathbf{v} = u_1v_1 + u_2v_2 + u_3v_3$$
- $\cos\theta = \dfrac{\mathbf{u} \cdot \mathbf{v}}{|\mathbf{u}||\mathbf{v}|}$, where θ is the angle between the vectors **u** and **v**.

Vector product

- Given $\mathbf{u} = u_1\mathbf{i} + u_2\mathbf{j} + u_3\mathbf{k}$ and $\mathbf{v} = v_1\mathbf{i} + v_2\mathbf{j} + v_3\mathbf{k}$,
the **vector (cross) product** of **u** and **v** is the vector given by
$\mathbf{u} \times \mathbf{v} = (u_2v_3 - u_3v_2)\mathbf{i} + (u_3v_1 - u_1v_3)\mathbf{j} + (u_1v_2 - u_2v_1)\mathbf{k}$.
- $\mathbf{u} \times \mathbf{v} = (|\mathbf{u}||\mathbf{v}|\sin\theta)\,\hat{\mathbf{n}}$ where $\hat{\mathbf{n}}$ is a unit vector orthogonal (normal) to both **u** and **v** whose direction is given by the right-hand rule illustrated in the diagram.
- Given a parallelogram ABCD, if $\mathbf{u} = \overrightarrow{AB} = \overrightarrow{DC}$ and $\mathbf{v} = \overrightarrow{AD} = \overrightarrow{BC}$, the area of ABCD is numerically equal to $|\mathbf{u} \times \mathbf{v}|$
The area of triangle ABD equals $\dfrac{1}{2}|\mathbf{u} \times \mathbf{v}|$

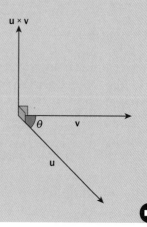

- Summary of the algebraic properties of the vector (cross) product
    - $\mathbf{u} \times \mathbf{v} = -\mathbf{v} \times \mathbf{u}$
    - $\mathbf{u} \times (\mathbf{v} + \mathbf{w}) = \mathbf{u} \times \mathbf{v} + \mathbf{u} \times \mathbf{w}$
    - $(\mathbf{u} \times \mathbf{v}) \cdot \mathbf{u} = \mathbf{0}$ ($\mathbf{u} \times \mathbf{v}$ is orthogonal to **u**)

Continued on next page

- $(\mathbf{u} \times \mathbf{v}) \cdot \mathbf{v} = \mathbf{0}$ ($\mathbf{u} \times \mathbf{v}$ is orthogonal to $\mathbf{v}$)
- $\mathbf{u} \times \mathbf{v} = \mathbf{0}$ if and only if $\mathbf{u}$ and $\mathbf{v}$ are collinear.
- $\mathbf{u} \times \mathbf{u} = \mathbf{0}$
- $(\lambda\mathbf{u}) \times \mathbf{v} = \lambda(\mathbf{u} \times \mathbf{v})$
- $|\mathbf{u} \times \mathbf{v}| = |\mathbf{u}||\mathbf{v}| \sin\theta$, where θ is the angle between $\mathbf{u}$ and $\mathbf{v}$.

Vectors and equations of planes

- In this case, any point on the plane with position vector $\mathbf{r}$ satisfies the **vector equation of a plane**
  $$\mathbf{r} = \mathbf{a} + \alpha\mathbf{u} + \beta\mathbf{v}$$
- If the parameters α and β are eliminated, an equation of the form $ax + by + cz = d$ is obtained.
  This equation is called a **Cartesian equation of the plane**.
- $a(x - x_1) + b(y - y_1) + c(z - z_1) = 0$
  or $ax + by + cz = d$ where $d = ax_1 + by_1 + cz_1$
  Equation of the plane (using the normal vector)

$$\begin{pmatrix} x - x_1 \\ y - y_1 \\ z - z_1 \end{pmatrix} \cdot \begin{pmatrix} a \\ b \\ c \end{pmatrix} = 0 \text{ this can also be written as } \begin{pmatrix} x \\ y \\ z \end{pmatrix} \cdot \begin{pmatrix} a \\ b \\ c \end{pmatrix} = \begin{pmatrix} x_1 \\ y_1 \\ z_1 \end{pmatrix} \cdot \begin{pmatrix} a \\ b \\ c \end{pmatrix}$$

or $\mathbf{r} \cdot \mathbf{n} = \mathbf{a} \cdot \mathbf{n}$

- In the plane, given two lines with gradients m_1 and m_2, we can calculate the angle θ between them using the formula
  $\theta = |\beta - \alpha|$ where $\alpha = \arctan m_1$ and $\beta = \arctan m_2$
- The angle θ between a line parallel to $\mathbf{u}$ and a plane with normal vector $\mathbf{n}$ is given by

$$\theta = \arcsin\left(\left|\frac{\mathbf{u} \cdot \mathbf{n}}{\|\mathbf{u}\|\|\mathbf{n}\|}\right|\right), 0 \le \theta \le 90°$$

- The angle θ between a plane with normal vector $\mathbf{m}$ and a plane with normal vector $\mathbf{n}$ is equal to the angle between the lines in the direction of the vectors $\mathbf{m}$ and $\mathbf{n}$. It is given by

$$\theta = \arccos\left(\left|\frac{\mathbf{m} \cdot \mathbf{n}}{\|\mathbf{m}\|\|\mathbf{n}\|}\right|\right), 0 \le \theta \le 90°$$

- The **distance from a point to a plane** is measured along the perpendicular to the plane that contains the point.

From inspired ideas to formal proof

Abstract nature and applications of mathematics

Any mathematical object (such as a vector) is abstract by nature and the relations between them are built on these abstractions, even if sometimes inspired by physical situations. For example, you have seen in this chapter that the abstract algebraic qualities of vectors are equivalent to physical geometric properties.

Mathematicians are professional 'reasoners' who add new methods and extend results that pass from generation to generation. Each adds a 'story' to the mathematical structure. Results are accepted when a formal proof is given and all theories must be constructed on a solid foundation of proof.

To a non-mathematician, it may seem paradoxical that such an abstract body of knowledge and thought can have so many applications to completely distinct areas of the physical world and that mathematics has such an influence on our cultures.

- Where does this power of mathematics come from? Does it come from its language? From its methods of proof? Or simply from its abstract nature?

- How can an abstract area of knowledge have so many applications?

- Are we discovering the mathematics in the world? Or are we creating mathematics to explain the world?

"Nothing is more impressive than the fact that Mathematics withdrew increasingly into upper regions of ever greater extremes of abstract thought, it returned to earth with a corresponding growth of importance for analysis of concrete fact…"

Alfred Whitehead, English mathematician (1861–1947)

▶ Mathematics in the natural world – the cuboid structure of salt (NaCl) crystals

Communicating mathematics

Until the mid-16th century, mathematics was created and developed by individuals or small groups with a prominent leader. They shared their work in handwritten manuscripts, or orally.

As printing presses became more widespread, in the 17th century mathematicians began to publish their work in printed form. However, printing was expensive and there was no copyright law and little respect for authors' rights. Most mathematicians still preferred to communicate and share their ideas with each other through letters.

Scientific societies and academies began to flourish, promoting the exchange of ideas ●

and making it possible for mathematicians with similar areas of interest to contact each other directly.

Until the 18th century universities only played a small role in mathematical research, as most of them were conservative and dogmatic and very slow to incorporate new knowledge.

- Investigate more about communication in mathematics in the 16th and 17th centuries.
- Which scientific societies played a leading role?
- Which main areas of mathematics were developed in this period?

Nowadays, new research is published in scientific and mathematical journals, either in print or online, where it can reach a global audience of people with similar interests.

- Before publication in a journal, new research has to be 'peer reviewed'. What does this mean?
- How does this ensure published research is of a high quality?

Applying mathematics

The 17th century has been called the century of genius, and the 18th century the century of the ingenious. In the 18th century, mathematics was seen much more as a tool to solve and understand physical problems than it had been before, and mechanics was the major interest. Although mathematicians of this period knew they needed to prove their results, their thinking was often loose and intuitive and most of them did not worry about limits of validity. As a result they made huge mathematical progress, but it was left to 19th and 20th century mathematicians to strengthen the foundations and develop rigorous proofs.

- Investigate some of the major mathematical developments in the 18th century.
- Were any of these proved rigorously by later mathematicians?
- How does the development of new mathematics sometimes make it easier to prove old mathematical theories?

Swiss mathematician Leonhard Euler (1707–83) investigated the Königsberg bridge problem, using counting methods. Research this classic problem, and how Euler solved it.

▼ The French mathematician D'Alambert (1717–83) investigated the behavior of waves in vibrating strings, using differential equations.

▲ The mathematical theories of Isaac Newton (1642–1727) about 'flattening' at the Earth's poles were later proved to be true by experiments with pendulums at different latitudes.

12

Multiple perspectives in mathematics

CHAPTER OBJECTIVES:

1.5 Modulus and argument of a complex number

1.6 Modulus–argument (polar) form; Euler form; the complex plane

1.7 Power of complex numbers; De Moivre's theorem; nth roots of a complex number

Before you start

You should know how to:

1 Represent complex numbers in the Argand diagram.

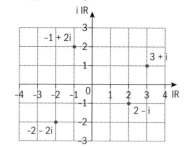

2 Determine conjugate, opposite and reciprocal of a complex number.
e.g., if $z = 1 - 2i$ then the conjugate of z is $z^* = 1 + 2i$, the opposite of z is $-z = -1 + 2i$ and the reciprocal of z is $\dfrac{1}{z} = \dfrac{1}{5} + \dfrac{2}{5}i$.

3 Identify $\text{Re}(z)$, $\text{Im}(z)$ and $|z|$ of a given complex number z. e.g.,
If $z = 3 - 4i$ then $\text{Re}(z) = 3$, $\text{Im}(z) = -4$
and $|z| = \sqrt{3^2 + (-4)^2} = 5$

4 Operations with complex numbers in algebraic form. e.g., if $z_1 = 2 - i$
and $z_2 = 1 + 5i$, $z_1 + 2z_2 = 4 + 9i$ and

$$\frac{z_1}{z_2} = \frac{2-i}{1+5i} = \frac{(2-i)(1-5i)}{(1+5i)(1-5i)}$$

$$= \frac{(2-i)(1-5i)}{1^2 + 5^2} = -\frac{3}{26} - \frac{11}{26}i$$

Skills check

1 Represent the complex numbers in the Argand diagram:

a $1 + 2i$

b $-2i$

c $1 - 3i$

> This is covered in Chapter 3.

2 Given $z = 5 - 4i$,
a write down z^* and $-z$
b find $\dfrac{1}{z}$ in the form $a + bi$, $a, b \in \mathbb{R}$

3 Given $z = -3 + 4i$, write down the values of
a $\text{Re}(z)$ **b** $\text{Im}(z)$ **c** $|z|$

4 Given that $z_1 = 1 - 2i$ and $z_2 = 3 + i$, calculate:
a $z_1 + 2z_2$
b $3z_1z_2 + z_1^2$
c $\dfrac{z_1}{z_2}$

The Mandelbrot set is made up of points plotted on a plane to form a **fractal**: an incredible shape in which each part is actually a miniature copy of the whole. The Mandelbrot set has become popular for both its aesthetic appeal and as an interesting case study of a complex structure arising from the application of simple iteration rules.

Exploring the power of complex numbers

Benoit B. Mandelbrot (1924–2010) was a French–American mathematician. He employed an imaginative, independent approach in his work, applying mathematics to physics, finance and many other fields. He is best known for developing the field of *fractal geometry*, having coined the term in 1975 to refer to a class of shapes whose uneven contours could mimic the irregularities found in nature. Other mathematicians had explored fractals before Mandelbrot, and had dismissed them as curious but unnatural. Mandelbrot sought to counteract this notion by showing that fractal geometry offers a systematic way of approaching phenomena found in nature – such as blood vessels, coastlines and galaxy clusters – that look more elaborate the more they are magnified. Mandelbrot's work was widely publicized through his writing and lectures, which were aimed at the general public as well as the academic community.

Did you know that images can be compressed using fractal codes so that they fit any screen size without the loss of sharpness that occurs with conventional compression methods, such as jpeg and gif? The image of the Mandelbrot set above needs only 7 bytes using fractal data compression. It is a 99.8% compression of a gif image that needs 35 kilobytes! Computer graphics owes its astronomical development in large part to fractal geometry.

Benoit B. Mandelbrot

In this chapter you will investigate the complex plane from a geometrical perspective and discover advanced and powerful techniques that make complex numbers an important part of modern mathematics with multiple applications in areas such as fluid and aerodynamics.

12.1 Complex numbers as vectors

Until the end of the 18th century complex numbers were not fully understood, but the square root of minus one was being used more and more. Numbers of the form $x + yi$ were in fairly frequent use by mathematicians, and it became common to represent them as points in the plane. In 1799, when Gauss published his first proof **The Fundamental Theorem of Algebra,** it became known that complex numbers (as solutions to algebraic equations) were numbers. In one sense all the historical discussion before Gauss was the prehistory of complex numbers.

Since the Gauss proof, it is known that all complex numbers are of the form $x + yi$, where x and y are real numbers. Therefore, you can use the xy-plane to display complex numbers and to explore further the algebra of complex numbers, introduce a new system of coordinates and interpret the meaning of operations.

Look at this representation of the complex plane, an **Argand diagram**:

Each complex number $z = x + yi$ on the Argand diagram is uniquely represented by a point P(x, y) where $x = \text{Re}(z)$ and $y = \text{Im}(z)$ are real numbers. Furthermore, since to each point P on the plane you can associate a position vector $\overrightarrow{OP}$, you can think of complex numbers as 2-D vectors!

Caspar Wessel (1745–1818) worked on surveying, and as part of his work he was led to explore the geometrical significance of complex numbers. In 1799 he published a theory where he gave a geometrical concept of complex number, with length and direction – in another words, complex numbers were described as vectors!

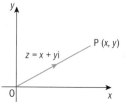

Was the complex plane already there before it was used to represent complex numbers geometrically?

Geometrical meaning of addition of complex numbers

Investigation – the complex plane

Consider the complex numbers $z_1 = 2 + i$ and $z_2 = 3 + 2i$. Represent them on the complex plane as the vectors **u** and **v**, respectively. Find $z_1 + z_2$. Draw the vector **w** that represents the complex number $z_1 + z_2$. What is the relation between **w**, **u** and **v**?

Choose other pairs of complex numbers z_1 and z_2, find their sum and represent the three complex numbers by their position vectors. What do you observe? Prove your conjecture.

From the Investigation on the previous page you should have concluded that the sum of complex numbers corresponds to the sum of two vectors. In fact, it can be interpreted in two ways:

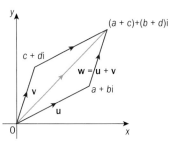

- Consider the vectors that represent each of the complex numbers z_1 and z_2 as position vectors, then use the **parallelogram law** to obtain the position vector of their sum.

- Consider the addition of z_2 as a transformation acting over z_1, then $z_1 + z_2$ is just the image of z_1 under the translation defined by the vector associated to z_2. In this case, the vector **u** associated to z_1 acts as a position vector and the vector **v** associated to z_2 acts as a displacement vector. The triangle law then gives you the position vector **w** associated with $z_1 + z_2$.

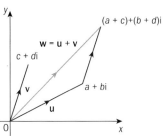

In either case, the addition of complex numbers is an addition of vectors and these properties hold

> - $z_1 + z_2 = z_2 + z_1$ (commutative property)
> - $(z_1 + z_2) + z_3 = z_1 + (z_2 + z_3)$ (associative property)
> - The null vector represents the complex number $0 = 0 + 0i$.
> - If $z = a + bi$ the opposite of z is $-z = -a - bi$; z and $-z$ are represented by vectors with same magnitude and opposite directions.
> - The magnitude of the vector **u** associated with $z = a + bi$ is called modulus of z: $|z| = \sqrt{a^2 + b^2}$.
> - The subtraction of two complex numbers z_1 and z_2 is defined as the sum of z_1 with the opposite of z_2.

Investigation – multiplication by real numbers

Consider the complex number $z = 1 + 2i$. Represent z on the complex number plane as the vector **u**. Calculate in algebraic form $2z$, $2.5z$, $3z$, $-2z$ and $-4z$. Draw the vectors that represent each of these complex numbers.

Choose other complex numbers z and multiply them by different real numbers. What do you observe? Prove your conjecture.

From the Investigation above you should have concluded that the product of a complex number and a real number corresponds to scalar multiplication studied in Chapters 3 and 11. Therefore these properties hold

> - $k(gz) = (kg)z$
> - $k(z_1 + z_2) = kz_1 + kz_2$
> - $(k + g)z = kz + gz$
> where $k, g \in \mathbb{R}$ and $z, z_1, z_2 \in \mathbb{C}$.

Geometrically, the scalar multiplication corresponds to an enlargement with center at the origin and enlargement factor given by the real number.

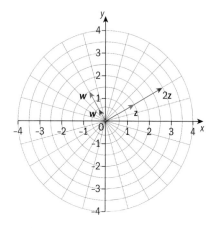

> Some dynamic geometry software uses polar grids to visualise enlargements or rotations with center at the origin. You can also find them in a printable version on the internet.

Other transformations of complex numbers

Given a complex number $z = x + iy$, if you plot it on the complex plane and then reflect this point in the x-axis, you obtain another point that represents the conjugate of z, $z^* = x - iy$. So, the conjugate corresponds to the image of a transformation called a 'reflection in the x-axis' and as a result, the following properties hold

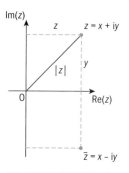

→ • $(z^*)^* = z$
   • $z \in \mathbb{R} \Rightarrow z^* = z$
   • $|z| = |z^*|$

> In Chapter 3, these properties were proven algebraically.

Investigation – transformations of complex numbers

Consider the complex number $z = a + bi$. Choose different pairs of values for a and b and represent z geometrically by a point P.

1 Reflect each point P in the y-axis to obtain a point P′ and write down the corresponding complex number z'.
   What do you observe? Prove your conjecture.

2 If you reflect P′ in the x-axis to obtain another point P″, which complex number do you obtain?

3 What if you reflect P first in the x-axis to obtain Q and then reflect Q in the y-axis to obtain Q′? Which complex number do you obtain? Prove your result.

4 How can you obtain P″ from P using a single geometrical transformation? Describe it.

> Look at properties of transformations in Chapter 14.

> Reflections in circles: Investigate the role of inversion in solving challenging geometrical problems.

From the Investigation on the previous page, you should have concluded that the composition of the reflections on the two axes is simply a 180° rotation around the origin that maps the complex number z onto its opposite –z.

As –z can be seen as the product of −1 and z, this suggests that multiplication of complex numbers may be related to rotations around the origin.

Investigation – multiplication of complex numbers

Consider the complex number $z = a + bi$. Choose different pairs of values for a and b and represent z geometrically by a point P.

1 For each pair of values a and b, multiply the complex number z by i to obtain z' and plot the corresponding point P′. What do you observe?

2 What if you multiply z by $- i$?

3 Investigate the effect on the position of P′ of the multiplication of z by 2i, 3i, −3i... and other purely imaginary numbers. Is there any relation between the geometrical representation of these numbers and their effect on the position of P′? State your conjecture.

From this Investigation, you should have conjectured that multiplication of complex numbers corresponds to a rotation and an enlargement whose scale factor depends on the modulus of the complex number it is multipled by. To prove this conjecture you can look at complex numbers from a different perspective and introduce a new system of coordinates that lets you deal with rotations easily. This is the focus of the next section.

12.2 Complex plane and polar form

Any complex number $z = x + iy$ occupies a position P in the plane specified by its Cartesian coordinates (x, y). Another way of specifying the location of P is to give its distance r from the origin and the angle θ its position vector makes with the positive direction of the x-axis.

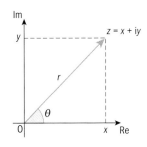

The pair (r, θ) gives the location of any point on the plane. r and θ are called **polar coordinates** of P. In relation to the complex number z, r is the modulus of z, $r = |z|$ and it is always a non-negative real number. θ is called the argument of z, $\theta = \arg(z)$ and, when you use polar coordinates to define z, the complex number is said to be in polar form. θ can be expressed either in radians or in degrees using the usual convention that, if you are not told otherwise, the angle is measured in radians.

When you add multiples of 360° to arg(z) you obtain an equivalent expression for arg(z).

Example 1

Represent these complex numbers in the plane:

a $|z_1| = 2$ and $\arg(z_1) = 60°$ **b** $|z_2| = 3$ and $\arg(z_2) = -\dfrac{\pi}{4}$

c $|z_3| = 1$ and $\arg(z_3) = \dfrac{7\pi}{4}$

Answer

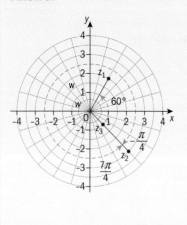

For each complex number draw a circle with centre at the origin with radius equal to its modulus.

Use a protractor to obtain a ray from the origin that makes an angle with the positive x-axis equal to the argument of the complex number.

The intersection of the ray with the circle gives you the location of the complex number.

Alternatively, use a polar grid where the circles and rays are already given!

Equality of complex numbers in polar form

As you may have noticed from the previous example, the argument of a complex number can take different values that always differ by a multiple of 2π (or 360°) but its modulus is uniquely defined. Therefore,

$$z_1 = z_2 \Rightarrow |z_1| = |z_2| \text{ and } \arg(z_1) - \arg(z_2) = 2k\pi,\ k \in \mathbb{Z}$$

Example 2

Find all possible values of r and θ such that $z_1 = z_2$ if $|z_1| = r^2$, $\arg(z_1) = 3\theta$, $|z_2| = 5r - 6$ and $\arg(z_2) = \theta + \pi$ where $r \geq 0$.

Answer

$r^2 = 5r - 6 \Rightarrow r^2 - 5r + 6 = 0$

$(r - 2)(r - 3) = 0 \Rightarrow r = 2$ or $r = 3$

$(3\theta) - (\theta + \pi) = 2k\pi,\ k \in \mathbb{Z}$

$2\theta = \pi + 2k\pi \Rightarrow \theta = \dfrac{\pi}{2} + k\pi$,

$k \in \mathbb{Z}$

$\left(\theta = \pm\dfrac{\pi}{2}, \pm\dfrac{3\pi}{2}, \pm\dfrac{5\pi}{2}, \pm\dfrac{7\pi}{2}, \ldots \right)$

Use $z_1 = z_2 \Rightarrow |z_1| = |z_2|$
Factorize and solve for r.

$z_1 = z_2 \Rightarrow \arg(z_1) - \arg(z_2) = 2k\pi$,
$k \in \mathbb{Z}$
Solve for θ.

Florence Nightingale (1820–1910), although mainly remembered for her pioneering work in the nursing field, developed the polar-area diagram to dramatize the needless number of deaths caused by insanitary conditions for soldiers at war, and hence the need for reform. The polar area diagram is similar to a usual pie chart, except sectors are equal angles and differ in how far each sector extends from the center of the circle. The polar area diagram is used to plot cyclic phenomena (e.g. count of deaths by month).

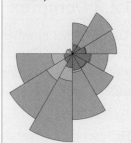

Exercise 12A

1 Represent these complex numbers in the complex plane:

  a $|z_1| = 4$ and $\arg(z_1) = 120°$

  b $|z_2| = 5$ and $\arg(z_2) = \dfrac{5\pi}{4}$

  c $|z_3| = 2$ and $\arg(z_3) = \dfrac{7\pi}{6}$

  d $|z_4| = 3$ and $\arg(z_4) = \dfrac{3\pi}{2}$

  e $|z_5| = 1$ and $\arg(z_5) = \pi$

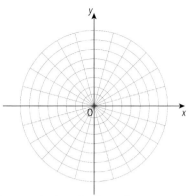

> Although formal treatment of polar coordinates is not a part of the course, you may find polar grids useful to plot complex numbers given in polar form.

2 Given the complex numbers z_1 and z_2 such that $|z_1| = r^3$, $\arg(z_1) = 4\theta$, $|z_2| = r^2 + 2r$ and $\arg(z_2) = \theta + \dfrac{\pi}{2}$ where $r \geq 0$, find all possible values of r and θ for which $z_1 = z_2$. Hence write down the complex numbers that satisfy the conditions given.

3 Find all real values of a and θ such that $|a + ai| = 2$ and $\arg(a + ai) = \theta$. How many distinct complex numbers $a + ai$ verify the conditions given?

4 Give examples to show that, in general,

  a $\arg(z_1 \pm z_2) \neq \arg(z_1) \pm \arg(z_2)$ **b** $|z_1 \pm z_2| \neq |z_1| \pm |z_2|$

Relation between Cartesian and polar coordinates

As complex numbers can be represented in Cartesian or modulus-argument form, it is important to be able to change from one form to the other. Drawing an Argand diagram helps you with this conversion.

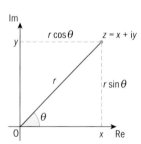

Conversion from Cartesian to polar form

Consider a complex number $z = x + iy$ and its representation on the complex plane. To find the polar coordinates of z, find its modulus using Pythagoras' theorem: $r = |z| = \sqrt{x^2 + y^2}$. From the diagram you can also obtain a relation between the argument θ and the rectangular coordinates x and y: $\tan \theta = \dfrac{y}{x}$. If z is located in the first or fourth quadrants, $\theta = \arctan \dfrac{x}{y}$;

otherwise, $\theta = \pi + \arctan \dfrac{x}{y}$ as you measure the angle θ from the real axis.

> Some texts adopt the convention that the argument of a complex number is always represented by an angle in the interval $-\pi < \theta \leq \pi$ called principal argument; other texts adopt the interval $0 \leq \theta < 2\pi$. On exam papers, unless you are told otherwise, you may give your answer in any interval.

$$\Rightarrow z = x + yi$$
$$\Rightarrow r = |z| = \sqrt{x^2 + y^2} \text{ and } \tan\theta = \dfrac{y}{x}$$

Conversion from polar to Cartesian form

If a complex z has modulus r and argument θ then, using the diagram again and your knowledge of right-angled trigonometry, you can obtain the relations

$$\cos\theta = \frac{x}{r} \Rightarrow x = r\cos\theta \text{ and } \sin\theta = \frac{y}{r} \Rightarrow y = r\sin\theta$$

> In some texts, Cartesian form is also called 'rectangular form' and polar form is called 'trigonometric form'.

→ $z = r\cos\theta + (r\sin\theta)\mathrm{i}$.

This expression can be simplified and rewritten in the form $z = r(\cos\theta + \mathrm{i}\sin\theta)$ called the **modulus-argument form**. Sometimes a short form of the modulus-argument form is used: $z = r\operatorname{cis}\theta$.

Example 3

Express these complex numbers in Cartesian form, giving your answers exactly.

a $|z_1| = 4$ and $\arg(z_1) = 60°$ **b** $z_2 = 2\operatorname{cis}\dfrac{5\pi}{6}$

> Revise trigonometry and learn exact values of trigonometric functions for special angles.

Answers

a $x = 4\cos 60° \Rightarrow x = 2$
$y = 4\sin 60° \Rightarrow y = 2\sqrt{3}$
$\therefore z_1 = 2 + 2\sqrt{3}\,\mathrm{i}$

Use $x = r\cos\theta$, $y = r\sin\theta$

b $x = 2\cos\dfrac{5\pi}{6} \Rightarrow x = -\sqrt{3}$
$y = 2\sin\dfrac{5}{6} \quad y = 1$
$\therefore z_2 = -\sqrt{3} + \mathrm{i}$

	30° or $\dfrac{\pi}{6}$	45° or $\dfrac{\pi}{4}$	60° or $\dfrac{\pi}{3}$
sin	$\dfrac{1}{2}$	$\dfrac{\sqrt{2}}{2}$	$\dfrac{\sqrt{3}}{2}$
cos	$\dfrac{\sqrt{3}}{2}$	$\dfrac{\sqrt{2}}{2}$	$\dfrac{1}{2}$

Example 4

Express these complex numbers in modulus-argument form.

a $z = 1 - \mathrm{i}$ **b** $z = -\sqrt{3} + \mathrm{i}$

Answers

a $r = \sqrt{1^2 + (-1)^2} = \sqrt{2}$

$\tan\theta = -1 \Rightarrow \theta = -\dfrac{\pi}{4}$

$\therefore z = \sqrt{2}\operatorname{cis}\left(-\dfrac{\pi}{4}\right)$

Use $r = |z| = \sqrt{x^2 + y^2}$,
$\arg(z) = \theta = \arctan\dfrac{x}{y}$ for
$\theta \in$ 4th quadrant

b $r = \sqrt{(-\sqrt{3})^2 + 1^2} = \sqrt{4} = 2$

$\tan\theta = -\dfrac{1}{\sqrt{3}} \Rightarrow \theta = -\dfrac{\pi}{6} + \pi = \dfrac{5\pi}{6}$

$\therefore z = 2\operatorname{cis}\left(\dfrac{5\pi}{6}\right)$

and $\theta = \pi + \arctan\dfrac{x}{y}$ for
$\theta \in$ 2nd quadrant.

Write z in the form $z = r\operatorname{cis}\theta$.

Polar form of opposite, conjugate and opposite of conjugate of a complex number

Given a complex number $z = r\operatorname{cis}\theta$, you can obtain expressions for its opposite, conjugate and opposite of the conjugate from their representation in an Argand diagram:

$$-z = r\operatorname{cis}(\pi + \theta),$$
$$z^* = r\operatorname{cis}(-\theta) = r\operatorname{cis}(2\pi - \theta) \text{ and}$$
$$-z^* = r\operatorname{cis}(\pi - \theta)$$

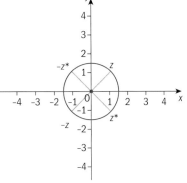

Polar form of the product of a real number by a complex number

In Section 12.1 you discovered that the multiplication of a real number α by a complex number $z = r\operatorname{cis}\theta$ corresponds to an enlargement with centre at the origin and enlargement factor $|\alpha|$.

> What is the geometrical meaning of a negative enlargement factor?

> - $\alpha > 0 \Rightarrow \alpha z = (\alpha r)\operatorname{cis}\theta$
> - $\alpha < 0 \Rightarrow \alpha z = (|\alpha|r)\operatorname{cis}(\pi + \theta)$
> - $\alpha = 0 \Rightarrow \alpha z = 0$

Exercise 12B

1 Express these complex numbers in Cartesian form giving your answers exactly.

a $|z_1| = 6$ and $\arg(z_1) = 45°$ **b** $|z_2| = 10$ and $\arg(z_2) = 135°$

c $z_3 = 4\operatorname{cis}\dfrac{5\pi}{3}$ **d** $z_4 = 5\operatorname{cis}\dfrac{7\pi}{6}$

> Draw a sketch to help you to locate each complex number in an Argand diagram before determining its argument.

2 Express these complex numbers in modulus-argument form.

a $z_1 = -1 - i$ **b** $z_2 = 2\sqrt{3} + 2i$
c $z_3 = 4 - 4i$ **d** $z_4 = -5 + 5i$

3 Represent these complex numbers in an Argand diagram and write them down in modulus-argument form:

a $z_1 = -3i$ **b** $z_2 = 4i$
c $z_3 = 2$ **d** $z_4 = -4$

4 Given $z = 4\operatorname{cis}40°$, write down in modulus-argument form

a z^* **b** $-z$ **c** $-z^*$
d $3z^*$ **e** $-4z^*$

5 Consider the complex numbers $z_1 = -2 - 2\sqrt{3}i$ and $z_2 = 3\sqrt{3} + 3i$.

  a Find in Cartesian form $z_3 = z_1 z_2$.

  b Express z_1, z_2 and z_3 in modulus-argument form.

  c Compare the moduli and arguments of the complex numbers in **b**. What do you notice?

Your GDC allows you to convert complex numbers from rectangular to polar form and vice-versa, but polar form may appear displayed using an alternative notation called **Euler form**.

polar form ▽		
$z = 2 + 2 \cdot i$		$2 + 2 \cdot i$
$(2 + 2 \cdot i) \blacktriangleright$ Polar		$e^{0.785398 \cdot i} \cdot 2.82843$
$\|z\|$		2.82843
angle(z)		0.785398

4/99

cartesian form ▽		
$z = 2 \cdot e^{\frac{\pi}{4} \cdot i}$		$1.41421 + 1.41421 \cdot i$
real(z)		1.41421
imag(z)		1.41421

3/99

> → If $r = |z|$ and $\theta = \arg(z)$, then z representation in Euler's form is $z = re^{\theta i}$

This form is the one preferred by many mathematicians because it allows them to relate five important constants in mathematics using the three main arithmetic operations.

> → $e^{\pi i} + 1 = 0$

Why do mathematicians refer to this equation as beautiful?

Phasor diagrams

Electronics and aircraft engineers use phasor diagrams to visualize complex constants and variables (phasors). Like vectors, arrows drawn on graph paper or computer displays represent phasors. Cartesian and polar representations each have advantages.

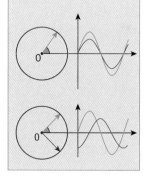

12.3 Operations with complex numbers in modulus-argument form

You are now ready to deduce rules that allow you to multiply complex numbers in polar form.

Consider two complex numbers $z_1 = r_1(\cos\theta_1 + i\sin\theta_1)$ and $z_2 = r_2(\cos\theta_2 + i\sin\theta_2)$. Using the multiplication rule for complex numbers studied in Chapter 3

$$z_1 z_2 = (r_1(\cos\theta_1 + i\sin\theta_1))(r_2(\cos\theta_2 + i\sin\theta_2))$$

$$= (r_1 r_2)(\cos\theta_1 + i\sin\theta_1)(\cos\theta_2 + i\sin\theta_2)$$

$$= (r_1 r_2)\left(\underbrace{\cos\theta_1\cos\theta_2 - \sin\theta_1\sin\theta_2}_{\cos(\theta_1 + \theta_2)} + i\underbrace{(\sin\theta_1\cos\theta_2 + \cos\theta_1\sin\theta_2)}_{\sin(\theta_1 + \theta_2)}\right).$$

Look at Chapter 8 (Trigonometric formulae).

→ Therefore, $|z_1 z_2| = |z_1| \, |z_2|$ and $\arg(z_1 z_2) = \arg(z_1) + \arg(z_2)$

You have just proved this theorem

→ If $z_1 = r_1 \operatorname{cis} \theta_1$ and $z_2 = r_2 \operatorname{cis} \theta_2$ then $z_1 z_2 = r_1 r_2 \operatorname{cis}(\theta_1 + \theta_2)$.

You also proved that the multiplication of complex numbers corresponds to a rotation with an enlargement both with center at the origin. This means when you multiply $z_1 = r_1 \operatorname{cis} \theta_1$ by $z_2 = r_2 \operatorname{cis} \theta_2$ the point P that represents z_1 in the complex plane is rotated by an angle θ_2 around the origin and its distance to the origin changes by a factor r_2. When $r_2 = 1$, the multiplication of z_1 by z_2 corresponds to just a rotation around the origin with angle θ_2.

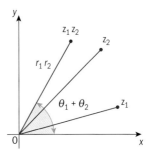

→ Euler form provides a familiar form for this theorem

$$\left(r_1 e^{\theta_1 i}\right) \cdot \left(r_2 e^{\theta_2 i}\right) = r_1 r_2 e^{(\theta_1 + \theta_2)i}$$

This example shows you how to use this theorem to multiply complex numbers in polar form.

Example 5

Multiply these pairs of complex numbers.

a $z_1 = 4 \operatorname{cis}(60°)$ and $z_2 = 2 \operatorname{cis}(50°)$

b $z_1 = 3 \operatorname{cis} \dfrac{5\pi}{6}$ and $z_2 = 4 \operatorname{cis} \dfrac{\pi}{5}$

- -

Answers

a $z_1 z_2 = 8 \operatorname{cis}(110°)$

b $z_1 z_2 = 12 \operatorname{cis}\left(\dfrac{31\pi}{30}\right)$ *Use $z_1 z_2 = r_1 r_2 \operatorname{cis}(\theta_1 + \theta_2)$*

If you are given complex numbers in different forms you need to convert at least one of them before multiplying. You need to decide if you prefer to convert the complex numbers into Cartesian form and operate them this way or convert them to polar form to operate them using the theorem above. In some cases, you are asked to operate them in both forms and use your result to determine exact values of trigonometric functions.

The next example shows you how to deduce exact values of trigonometric ratios of $\dfrac{5\pi}{12}$.

Example 6

Consider the complex numbers $z_1 = 3 \operatorname{cis} \dfrac{\pi}{6}$ and $z_2 = 5 + 5i$.

a Express z_1 in Cartesian form.

b Express z_2 in polar form.

c Calculate $z_1 z_2$ both in Cartesian and polar forms.
Hence find the exact values of $\sin \dfrac{5\pi}{12}$, $\cos \dfrac{5\pi}{12}$ and $\tan \dfrac{5\pi}{12}$.

Answers

a $z_1 = 3 \left(\underbrace{\cos \dfrac{\pi}{6}}_{\frac{\sqrt{3}}{2}} + i \underbrace{\sin \dfrac{\pi}{6}}_{\frac{1}{2}} \right) \Rightarrow z_1 = \dfrac{3\sqrt{3}}{2} + \dfrac{3}{2} i$

Use $r \operatorname{cis} \theta = r(\cos \theta + i \sin \theta)$.

b $z_2 = 5 + 5i \Rightarrow r = \sqrt{5^2 + 5^2} = 5\sqrt{2}$

and $\theta = \dfrac{\pi}{4}$

$\therefore z_2 = 5\sqrt{2} \operatorname{cis} \dfrac{\pi}{4}$

Use $r = \sqrt{x^2 + y^2}$,

$\theta = \arctan \dfrac{x}{y}$ for $\theta \in$ 1st quadrant.

Write it in the form $r \operatorname{cis} \theta$.

c $z_1 z_2 = \left(\dfrac{3\sqrt{3}}{2} + \dfrac{3}{2} i \right)(5 + 5i)$

Use $(a + bi)(c + di) = (ac - bd) + (ad + bc)i$.

$= \dfrac{15\sqrt{3} - 15}{2} + \dfrac{15\sqrt{3} + 15}{2} i$

$z_1 z_2 = \left(3 \times 5\sqrt{2} \right) \operatorname{cis} \left(\dfrac{\pi}{6} + \dfrac{\pi}{4} \right) = 15\sqrt{2} \operatorname{cis} \dfrac{5\pi}{12}$

Use $z_1 z_2 = r_1 r_2 \operatorname{cis} (\theta_1 + \theta_2)$.

$\cos \dfrac{5\pi}{12} = \dfrac{\sqrt{3} - 1}{2\sqrt{2}} = \dfrac{\sqrt{6} - \sqrt{2}}{4}$

Use $\cos \theta = \dfrac{x}{r}$,

$\sin \dfrac{5\pi}{12} = \dfrac{\sqrt{3} + 1}{2\sqrt{2}} = \dfrac{\sqrt{6} + \sqrt{2}}{4}$

$\sin \theta = \dfrac{y}{r}$,

$\tan \dfrac{5\pi}{12} = \dfrac{\sqrt{3} + 1}{\sqrt{3} - 1} = \dfrac{2\sqrt{3} + 4}{2} = \sqrt{3} + 2$

$\tan \theta = \dfrac{y}{x}$ and rationalize the denominators.

Exercise 12C

1 Multiply these pairs of complex numbers.

a $z_1 = \operatorname{cis} 10°$ and $z_2 = 5 \operatorname{cis} 125°$

b $z_1 = \dfrac{1}{2} \operatorname{cis} \dfrac{7\pi}{6}$ and $z_2 = \dfrac{4}{7} \operatorname{cis} \dfrac{\pi}{8}$

2 Consider the complex numbers $z_1 = \operatorname{cis} \dfrac{5\pi}{6}$ and $z_2 = 1 - i$

a Express z_1 in Cartesian form.

b Express z_2 in polar form.

c Calculate $z_1 z_2$ both in Cartesian and polar forms. Hence find
the exact values of $\sin \dfrac{7\pi}{12}$, $\cos \dfrac{7\pi}{12}$ and $\tan \dfrac{7\pi}{12}$.

3 Let $z_1 = 2\text{cis}\dfrac{\pi}{6}$ and $z_2 = r \text{ cis } \theta$, where $r > 0$ and $0 \le \theta < 2\pi$.

Find the range of values of r and θ for which $z_1 z_2$ is

a a real number greater than 5

b a purely imaginary number with modulus less than 1.

4 Represent in modulus-argument form the opposite, the conjugate and the opposite of the conjugate of $z = -3 + \sqrt{3}i$.

5 Let $z = \sin \alpha + \cos \alpha i$ and $w = \sin 2\alpha - \cos 2\alpha i$.

a Write z and w in modulus argument form.

b Find zw in polar form.

Polar form of the reciprocal of a complex number

Given a complex number $z = r \text{ cis } \theta$, if $z \ne 0$ its reciprocal is the complex number $\dfrac{1}{z}$ and it is characterized by the property $z \cdot \dfrac{1}{z} = 1$

Suppose that $\dfrac{1}{z} = \rho \text{ cis } \alpha$ is the polar form of the reciprocal of z.

Then $z \cdot \dfrac{1}{z} = 1 \Rightarrow (r\text{cis}\,\theta)(\rho\,\text{cis}\,\alpha) = 1\,\text{cis}\,0$

After multiplying z and its reciprocal, you obtain $r\rho\,\text{cis}(\theta + \alpha) = 1 \text{ cis } 0$
Using the equality condition of complex numbers

$$r\rho = 1 \Rightarrow \rho = \frac{1}{r} \text{ and } \theta + \alpha = 2k\pi \Rightarrow \alpha = -\theta + 2k\pi, \ k \in \mathbb{Z}.$$

→ So, a polar representation of the reciprocal of $z = r\text{cis}\,\theta$ is

$$\frac{1}{z} = \frac{1}{r}\,\text{cis}(-\theta), r \ne 0$$

Division of complex numbers in polar form

Given $z_1 = r_1 \text{ cis } \theta_1$ and $z_2 = r_2 \text{ cis } \theta_2$ with $r_2 \ne 0$, you can find $\dfrac{z_1}{z_2}$ by simply multiplying z_1 by the reciprocal of z_2

→ $\dfrac{z_1}{z_2} = (r_1 \text{ cis } \theta_1)\left(\dfrac{1}{r_2}\text{cis}(-\theta_2)\right) = \dfrac{r_1}{r_2}\text{cis}(\theta_1 - \theta_2), \ r_2 \ne 0$

→ The Euler form also provides a familiar form for these theorems.

$$\frac{1}{re^{\theta i}} = \frac{1}{r}e^{-\theta i}$$

$$\frac{r_1 e^{\theta_1 i}}{r_2 e^{\theta_2 i}} = \frac{r_1}{r_2}e^{(\theta_1 - \theta_2)i}$$

Example 7

Find in polar form $\dfrac{z_1}{z_2}$ in each of these cases.

a $|z_1| = 6$, $\arg(z_1) = 75°$, $|z_2| = 2$ and $\arg(z_2) = 20°$

b $z_1 = 8\operatorname{cis}\dfrac{5\pi}{7}$ and $z_2 = 4\operatorname{cis}\dfrac{\pi}{3}$

Answers

a $\dfrac{z_1}{z_2} = 3\operatorname{cis}55°$

b $\dfrac{z_1}{z_2} = 2\operatorname{cis}\dfrac{8\pi}{21}$

Use $\dfrac{z_1}{z_2} = \dfrac{r_1}{r_2}\operatorname{cis}\left(\theta_1 - \theta_2\right)$

Exercise 12D

1 Find in polar form $\dfrac{z_1}{z_2}$ in each of these cases.

  a $|z_1| = 10$, $\arg(z_1) = 170°$ and $z_2 = 5\operatorname{cis}125°$

  b $z_1 = \dfrac{1}{2}\operatorname{cis}\dfrac{7\pi}{6}$ and $z_2 = \dfrac{4}{7}\operatorname{cis}\dfrac{\pi}{8}$

2 Let $z_1 = 2\operatorname{cis}\dfrac{11\pi}{6}$ and $z_2 = 2 - 2i$.

  Find, in modulus-argument form, the complex numbers:

  a z_2 **b** $z_2{}^*$ **c** $z_1 z_2$

  d $\dfrac{z_1}{z_2}$ **e** $-\dfrac{1}{z_1 z_2}$

3 Express in modulus-argument form:

  a $\dfrac{4}{\sqrt{3}+i}$ **b** $\dfrac{2-2i}{\sqrt{6}+\sqrt{2}i}$ **c** $\dfrac{1}{\left(\sqrt{21}-\sqrt{7}i\right)^*}$

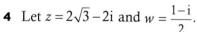

EXAM-STYLE QUESTION

4 Let $z = 2\sqrt{3} - 2i$ and $w = \dfrac{1-i}{2}$.

  a Write z and w in the form $r(\cos\theta + i\sin\theta)$, where $r > 0$ and $-\pi \le \theta \le \pi$.

  b Find $\dfrac{z}{w}$ in modulus-argument form.

  c Find the value of $\dfrac{z}{w}$ in Cartesian form, giving your answer exactly.

  d Hence find the exact values of $\cos\dfrac{\pi}{12}$, $\sin\dfrac{\pi}{12}$ and $\tan\dfrac{\pi}{12}$.

12.4 Powers and roots of complex numbers: De Moivre's theorem and applications

As in the previous section, the product of complex numbers can easily be obtained in polar form and corresponds to an enlargement with a rotation. This section looks at the result of iterating this process, i.e., powers of a complex number $z = r(\cos\theta + i\sin\theta)$. Intuitively, you can predict that the result is an enlargement spiral, as whenever you increase the power of z by one you rotate its image on the complex plane by θ and its distance to the origin changes by a factor r. The diagram illustrates this process for $r > 1$.

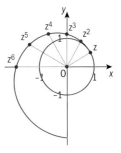

If $r < 1$, the spiral moves inwards and converges to the origin. For $r = 1$, the transformation is just an iterated rotation around the origin.

Formally, you can prove a theorem that allows you to find integral powers of any complex number.

De Moivre's theorem:

> → $[r(\cos\theta + i\sin\theta)]^n = r^n(\cos n\theta + i\sin n\theta)$ where $r \in \mathbb{R}^+$, $\theta \in \mathbb{R}$ and $n \in \mathbb{Z}$.

$\theta \in \mathbb{R}$: This means that θ is expressed in radians but the formula can also be applied with the argument expressed in degrees.

Proof

Case 1: Let $n \in \mathbb{Z}^+$. In this case, use induction to prove the proposition

$P(n)$: $[r(\cos\theta + i\sin\theta)]^n = r^n(\cos n\theta + i\sin n\theta)$ is true for any $r \in \mathbb{R}^+$, $\theta \in \mathbb{R}$ and $n \in \mathbb{Z}^+$:

Start by verifying that P(1) is a true statement:

$[r(\cos\theta + i\sin\theta)]^1 = r^1(\cos(1 \cdot \theta) + i\sin(1 \cdot \theta))$ (verified true)

Next, assume the truth of the proposition for a particular value of n, let's say k:

$P(k)$: $[r(\cos\theta + i\sin\theta)]^k = r^k(\cos k\theta + i\sin k\theta)$ (assumed true)

and consider the proposition for the next value of n, i.e., $n = k + 1$:

$$P(k+1): \underbrace{\left[r(\cos\theta + i\sin\theta)\right]^{k+1}}_{LHS} = \underbrace{r^{k+1}(\cos(k+1)\theta + i\sin(k+1)\theta)}_{RHS}$$

(under consideration)

As $\underbrace{\left[r(\cos\theta+i\sin\theta)\right]^{k+1}}_{LHS} = \left[r(\cos\theta+i\sin\theta)\right]^{k}\left[r(\cos\theta+i\sin\theta)\right]$

$= [r^{k}(\cos k\theta + i\sin k\theta)]\,[r(\cos\theta + i\sin\theta)]$
(using the induction hypothesis)

$= [r^{k}r(\cos(k\theta + \theta) + i\sin(k\theta + \theta))]$
(using the multiplication theorem)

$= \underbrace{r^{k+1}\left(\cos(k+1)\theta + i\sin(k+1)\theta\right)}_{RHS}$ (QED)

You have shown that $P(k)$ true $\Rightarrow P(k + 1)$ true and, as you had established that $P(1)$ is true, by the principle of mathematical induction, you can conclude that $P(n)$:
$[r(\cos\theta + i\sin\theta)]^n = r^n(\cos n\theta + i\sin n\theta)$ is true for any $r \in \mathbb{R}^+$, $\theta \in \mathbb{R}$ and $n \in \mathbb{Z}^+$.

Case 2: If $n = 0$, you have that $P(0)$: $\underbrace{\left[r(\cos\theta+i\sin\theta)\right]^{0}}_{1} = r^{0}\left(\underbrace{\cos 0}_{1}+i\underbrace{\sin 0}_{0}\right)$

is true as long as we use the convention that $z^0 = 1$.

Case 3: If $n \in \mathbb{Z}^-$, than $n = -1 \cdot m$ where $m \in \mathbb{Z}^+$.

Consider $z = (\cos\theta + i\sin\theta)$. Then $\dfrac{1}{z} = \dfrac{1}{r}(\cos(-\theta) + i\sin(-\theta))$.

Now apply the result proved in case 1:

$\left(\dfrac{1}{z}\right)^{m} = \left(\dfrac{1}{r}\right)^{m}(\cos m(-\theta) + i\sin m\,(-\theta))$ as $r \in \mathbb{R}^+$, $\theta \in \mathbb{R}$ and $m \in \mathbb{Z}^+$.

> We have given the proof for $m \in \mathbb{Z}^+$ here, but the result does hold for all $m \in \mathbb{R}$.

> → If you use the convention that $z^{-m} = \dfrac{1}{z^{m}}$, then
>
> $z^{-m} = r^{-m}(\cos(-m\theta) + i\sin(-m\theta))$.
>
> Euler Form: $z^{-m} = (re^{i\theta})^{-m} = r^{-m}e^{-im\theta}$

You may have noticed that, in this proof, we have excluded the case $z = 0$. Obviously $0^n = 0$ for $n \in \mathbb{Z}^+$. What do you think is the meaning of 0^0?

De Moivre's theorem provides a very efficient method to find powers of complex numbers, as shown in this example.

Example 8

Use De Moivre's theorem to calculate $\left(\dfrac{1}{2} - \dfrac{\sqrt{3}}{2}\right)^n$ in polar form for

$n = 1, 2, 3, ..., 12$ and represent the complex numbers obtained in an Argand diagram.

▶ continued on next page

Answer

$$\left(\frac{1}{2}-\frac{\sqrt{3}}{2}\right)^n = \left(cis\left(-\frac{\pi}{6}\right)\right)^n = cis\left(-\frac{n\pi}{6}\right)$$

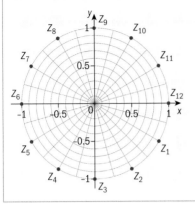

Convert the complex number into modulus-argument form and apply De Moivre's theorem.

Substitute n by 1,2,3... and 12 and represent the complex numbers z_n in the Argand diagram.

As all complex numbers z_n have modulus 1. They lie on the unit circle with center at the origin, equally spaced as

$$arg\,(z_n) - arg(z_{n-1}) = -\frac{\pi}{6}$$

Very often you will need to apply De Moivre's theorem more than once and combine it with other results from this chapter.

Example 9

Let $z_1 = 4cis\,\frac{5\pi}{3}$ and $z_2 = 2\sqrt{3} - 2i$.

Find $(z_1^*)^4 (z_2)^{-3}$

a in polar form

b in Cartesian form, giving your answer exactly.

Answers

a $\quad z_1 = 4cis\frac{5\pi}{3} \Rightarrow z_1^* = 4cis\frac{\pi}{3}$

$\qquad \left(z_1^*\right)^4 = 4^4\,cis\frac{4\pi}{3}$

$\qquad z_2 = 2\sqrt{3} - 2i \Rightarrow z_2 = 4\,cis\left(-\frac{\pi}{6}\right)$

$\qquad \left(z_2\right)^{-3} = 4^{-3}\,cis\frac{\pi}{2}$

$\qquad \left(z_1^*\right)^4 \left(z_2\right)^{-3} = 4\,cis\frac{11\pi}{6}$

b $\quad \left(z_1^*\right)^4 \left(z_2\right)^{-3} = 4\left(\cos\frac{11\pi}{6} + i\sin\frac{11\pi}{6}\right)$

$\qquad\qquad = 4\left(\frac{\sqrt{3}}{2} + \frac{1}{2}i\right) = 2\sqrt{3} + 2i$

Use $z^* = r\,cis\,(2\pi - \theta)$.

Apply De Moivre's theorem.

Convert z_2 into modulus-argument form.

Apply De Moivre's theorem again.

Use $z_1z_2 = r_1r_2\,cis(\theta_1 + \theta_2)$.

Convert the complex number into Cartesian form.

Abraham De Moivre (1667–1754) was a French mathematician who had to leave his country due to religious persecution. After being imprisoned for more than two years, he moved to England where he worked as a private tutor, as his status as a foreigner prevented him from getting a teaching position. He mastered Newton's theory of fluxions and began his own original work on this field that was accepted by the Royal Society of which he was elected a member in 1697.

Exercise 12E

1 Let $z_1 = 2\operatorname{cis}\dfrac{3\pi}{4}$ and $z_2 = \sqrt{3} + \mathrm{i}$.

Find:

a $\left(z_1^*\right)^2\left(z_2\right)^3$ **b** $\left(\dfrac{z_1}{z_2}\right)^4$ **c** $\left(\dfrac{z_1^*}{-z_2}\right)^3$

2 Let $z_1 = 4e^{-\frac{\pi}{4}\mathrm{i}}$ and $z_2 = \dfrac{1}{2}e^{\frac{\pi}{3}\mathrm{i}}$

Find:

a $z_1 z_2$ **b** $\left(z_1\right)^3\left(z_2\right)^{-2}$ **c** $\dfrac{z_1}{z_2}$ **d** $\left(\dfrac{z_1^*}{z_2}\right)^{-3}$

EXAM-STYLE QUESTIONS

3 Determine in Cartesian form $\left(\dfrac{\cos\theta - \mathrm{i}\sin\theta}{\sin\theta + \mathrm{i}\cos\theta}\right)^5$

4 Show that $\left(1 + \mathrm{i}\sqrt{3}\right)^n + \left(1 - \mathrm{i}\sqrt{3}\right)^n = 2^{n+1}\cos\left(\dfrac{n\pi}{3}\right)$, for any $n \in \mathbb{Z}$.

5 Find the smallest positive integer for which $(1 - \mathrm{i})^n$ is
 a a negative real number **b** purely imaginary.

6 Solve the equation $|z|z^3 = 16$, giving your answer both in polar and Cartesian forms.

Roots of complex numbers

An important application of De Moivre's theorem is the method it provides to calculate the nth roots of a complex number $r\operatorname{cis}\theta$. It allows us to find all complex numbers z that are solutions of the equation $z^n = r\operatorname{cis}\theta$. To solve this equation, make $z = \rho\operatorname{cis}\alpha$ and use De Moivre's theorem to deduce an expression for ρ and α in terms of r, θ and n

$z^n = r\operatorname{cis}\theta$ and $z = \rho\operatorname{cis}\alpha \Rightarrow (\rho\operatorname{cis}\alpha)^n = r\operatorname{cis}\theta$.

After applying De Moivre's theorem, $\rho^n\operatorname{cis}n\alpha = r\operatorname{cis}\theta$.

Using the equality of complex numbers in polar form

$\rho^n = r \Rightarrow \rho = \sqrt[n]{r}$ and $n\alpha = \theta + 2k\pi \Rightarrow \alpha = \dfrac{\theta + 2k\pi}{n}$, $k \in \mathbb{Z}$.

> → So, $z^n = r\operatorname{cis}\theta \Rightarrow z = \sqrt[n]{r}\operatorname{cis}\dfrac{\theta + 2k\pi}{n}$, $k \in \mathbb{Z}$.
>
> Euler form: $z^n = r\,e^{\mathrm{i}\theta} \Rightarrow z = \sqrt[n]{r}\,e^{\frac{\theta + 2k\pi}{n}\mathrm{i}}$, $k \in \mathbb{Z}$

Represent the nth roots of the complex number r cis θ in an Argand diagram to see that they all lie in a circle with radius $\sqrt[n]{r}$ centered at the origin and that they divide this circle into n equal arcs corresponding to central angles with size $\dfrac{2\pi}{n}$, as illustrated in the diagram.

This means that, in fact, there are exactly n distinct nth roots of the complex number r cis θ that you can obtain explicitly by giving k the values $0, 1, \ldots (n-1)$.

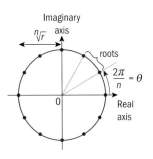

Example 10

Find the fourth roots of $16 \text{ cis} \dfrac{4\pi}{3}$ and represent them in the complex plane.

Answer

Let z be a fourth root of $16 \text{ cis} \dfrac{4\pi}{3}$.

$z = \sqrt[4]{16} \text{ cis} \dfrac{\dfrac{4\pi}{3} + 2k\pi}{4}$,

$k = 0, 1, 2$ and 3.

$z_0 = 2 \text{ cis} \dfrac{\pi}{3},\ z_1 = 2 \text{ cis} \dfrac{5\pi}{6}$,

$z_2 = 2 \text{ cis} \dfrac{4\pi}{3}$ and $z_3 = 2 \text{ cis} \dfrac{11\pi}{6}$

Use $z = \sqrt[n]{r} \text{ cis} \dfrac{\theta + 2k\pi}{n}$,

$k = 0, 1, \ldots, n-1$

with $n = 4$ to find the four fourth roots. Alternatively, k could equal 0, ± 1, 2

Substitute k by $0, 1, 2$ and 3.

Represent the complex numbers z_k in the complex planes.

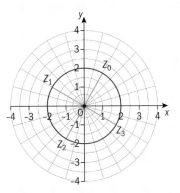

All of these roots have the same modulus (2) and lie on the same circle with center at the origin and radius 2 units.

The next example shows you a very special case: the determination of the **complex nth roots of unity** i.e., the nth roots of the number 1.

It is usual to represent the nth roots of unity by ω.

Example 11

> **a** Solve the equation $z^5 = 1$ in $\mathbb{C}$.
> **b** Let $\omega \neq 1$ be a solution of $z^5 = 1$. Show that
> $1 + \omega + \omega^2 + \omega^3 + \omega^4 = 0$.
>
> **Answers**
>
> **a** $z^5 = \text{cis } 0 \Rightarrow z = \text{cis } \dfrac{2k\pi}{5}$,
>
> $k = 0, \ldots, 4$
>
> *Use* $z = \sqrt[n]{r} \text{ cis } \dfrac{\theta + 2k\pi}{n}$,
>
> $k = 0, 1, \ldots, n-1$
>
> *with* $n = 5$ *to find the five fifth roots.*
>
> **b** $1 + \omega + \omega^2 + \omega^3 + \omega^4 = \dfrac{\omega^5 - 1}{\omega - 1}$
>
> $\omega \neq 1$
>
> *Note that this is the sum of 5 consecutive terms of a geometric progression with common ratio* ω.
>
> solution of $z^5 = 1 \Rightarrow \omega^5 = 1$
>
> $\therefore 1 + \omega + \omega^2 + \omega^3 + \omega^4$
>
> $= \dfrac{1 - 1}{\omega - 1} = 0$
>
> *Note that* $\omega \neq 1$

Recall the sum of consecutive terms of a geometric progression.

Discuss the geometrical meaning of part **b**. If you think of complex numbers as vectors, what is the meaning of this sum?

Exercise 12F

1 Find:

a the cube roots of $8 \text{cis} \dfrac{\pi}{3}$ in modulus-argument form

b the fourth roots of $(-4i)^2$ in Cartesian form

c the fifth roots of $32\, e^{-\pi i}$ in Euler form.

2 Solve these equations in $\mathbb{C}$.

a $z^2 = 1 - i$ **b** $z^4 = -\sqrt{3} + i$

c $z^3 = 27e^{\frac{\pi}{4}i}$ **d** $z^3 - \left(\sqrt{2} - \sqrt{2}i\right)z = 0$

Which regular polygons can be constructed using only straight edges and compasses? How do complex numbers help us to answer this question?

3 **a** Find the sixth roots of unity and represent them on an Argand diagram.

b Label the complex numbers found in a) $z_1, z_2, \ldots, z_6$.
Show that $z_1^3 = z_3^3 = z_5^3$ and $z_2^3 = z_4^3 = z_6^3$.
What is the significance of these results?

EXAM-STYLE QUESTIONS

4 $\dfrac{-1 + i\sqrt{3}}{4}$ is a fifth root of a complex number z. Without

calculating z, find the remaining fifth roots of z.

5 Find the fourth roots of -81 in Cartesian form. Hence, write down all the solutions of the equation $(z - 3)^4 + 81 = 0$.

6 Find the sum of the series $1 + \dfrac{1}{3}e^{2i\theta} + \dfrac{1}{9}e^{4i\theta} + \dfrac{1}{27}e^{6i\theta} + \ldots$

Investigation – properties of roots of unity

1 The equation $z^3 = 1$ has three complex roots called **cube roots of unity**.

a Solve the equation $z^3 = 1$ in $\mathbb{C}$, giving your answers in Cartesian form. One of the roots is a real number. Denote one of the other roots by ω.

b **i** How can you express the other root in terms of ω?

  ii Square each of the complex roots.
  What do you observe?
  Give a geometrical explanation for your observation.

  iii Find the reciprocal of each complex root.
  What do you observe?
  Give a geometrical explanation for your observation.

c Show that $1 + \omega + \omega^2 = 0$ and deduce that $1 + \omega^* + (\omega^*)^2 = 0$
Give a geometrical explanation for your results.

d Hence, evaluate the following expressions, stating clearly the results you use:

  i $\omega^4 + \omega^5 + \omega^6$ **ii** $\dfrac{1}{\omega + \omega^2}$ **iii** $\dfrac{\omega^*}{1 + \omega}$ **iv** $(1 + \omega)^*$

e Use the binomial expansions of $(\omega + 1)^3$ and $(\omega^* + 1)^3$ to show that $(\omega + 1)^3 = (\omega^* + 1)^3$. Hence, deduce that both $\omega + 1$ and $\omega^* + 1$ are cube roots of -1. Represent the cube roots of 1 and -1 on the complex plane and explain the relation between the numbers 1 and -1 and their corresponding cube roots.

> You may find it useful to find the nth roots of unity for some values of n and represent them on an Argand diagram.

2 Extend your results. Consider the nth roots of unity for any $n = 3, 4, \ldots$.

a Show that these roots can be written as $1, \omega, \omega^2, \ldots \omega^{n-1}$ where $\omega = \operatorname{cis}\dfrac{2\pi}{n}$.

b Show that $1 + \omega + \omega^2 + \ldots + \omega^{n-1} = 0$.

c Use de Moivre's theorem to show that:

  i $(\omega^k)^* = \omega^{n-k}$ for $k = 1, 2, \ldots, n-1$.

  ii $1 \cdot \omega \cdot \omega^2 \ldots \cdot \omega^{n-1} = (-1)^{n-1}$

What is the geometrical meaning of these results?

12.5 Mathematical connections

In examinations it is usual that complex numbers questions involve the knowledge of other topics of the course. This example shows you how to use De Moivre's theorem and the binomial theorem to find the expression of a trigonometric integral.

Example 12

Let $z = \text{cis}\,\alpha$.

a Use De Moivre's theorem to show that $z^n + \left(\dfrac{1}{z}\right)^n = 2\cos(n\alpha)$.

b Use the binomial theorem to expand $\left(z + \dfrac{1}{z}\right)^4$.

Hence show that $\cos^4\alpha = \dfrac{1}{8}(\cos 4\alpha + 4\cos 2\alpha + 3)$ and find $\displaystyle\int \cos^4\alpha \, d\alpha$.

Answers

a $z^n = \text{cis}\,(n\alpha) = \cos(n\alpha) + i\sin(n\alpha)$ | *Use De Moivre's Theorem twice.*

$\left(\dfrac{1}{z}\right)^n = (\text{cis}\,(-\alpha))^n = \text{cis}(-n\alpha) = \cos(-n\alpha)$
$\qquad + i\sin(-n\alpha)$

$z^n + \left(\dfrac{1}{z}\right)^n = (\text{cis}\,(n\alpha) + \cos(-n\alpha)) + i\,(\sin(n\alpha) + \sin(-n\alpha))$ | *Cosine is an even function and sine an odd function.*
$\qquad = (\cos(n\alpha) + \cos(n\alpha)) + i\,(\sin(n\alpha) - \sin(n\alpha))$
$\qquad = 2\cos(n\alpha) \quad$ QED

b $\left(z + \dfrac{1}{z}\right)^4 = z^4 + 4z^2 + 6 + \dfrac{4}{z^2} + \dfrac{1}{z^4}$ | *Apply binomial theorem.*

$\left(z + \dfrac{1}{z}\right)^4 = \left(z^4 + \dfrac{1}{z^4}\right) + 4\left(z^2 + \dfrac{1}{z^2}\right) + 6$ | *Rearrange the expression.*

$(2\cos\alpha)^4 = 2\cos 4\alpha + 4(2\cos 2\alpha) + 6\cos^4\alpha$ | *Apply the result deduced in part **a** and simplify the expression.*
$\qquad = \dfrac{1}{8}(\cos 4\alpha + 4\cos 2\alpha + 3)$.

Then $\cos^4\alpha \, d\alpha = \dfrac{1}{8}\displaystyle\int (\cos 4\alpha + 4\cos 2\alpha + 3) \, d\alpha$ | *Integrate each term and simplify.*

$\qquad = \dfrac{1}{8}\left(\dfrac{\sin 4\alpha}{4} + 4\dfrac{\sin 2\alpha}{2} + 3\alpha\right) + c$

$\qquad = \dfrac{\sin 4\alpha}{32} + \dfrac{\sin 2\alpha}{4} + \dfrac{3\alpha}{8} + c$

In this next example, use the tangent addition formula and De Moivre's theorem to deduce the exact value of the trigonometric expression.

Example 13

Consider the complex number $z = \text{cis } 20°$.
Let $z = x + iy$ be the Cartesian form of z.

a Expand and simplify $(x + iy)^3$.

Hence show that $\dfrac{3x^2y - y^3}{x^3 - 3xy^2} = \sqrt{3}$.

b Show that $\tan 40° \tan 80° = \dfrac{3 - \tan^2 20°}{1 - 3\tan^2 20°}$.

c Use parts **a** and **b** to show that $\tan 20° \tan 40° \tan 80° = \sqrt{3}$.

Answers

a $(x + iy)^3 = x^3 + 3x^2y\,i - 3xy^2 - y^3\,i$ *Apply binomial theorem.*
$(x + iy)^3 = x^3 - 3xy^2 + i\,(3x^2y - y^3)$ *Rearrange the expression.*
$(x + iy)^3 = \text{cis } 60°$

Then $\dfrac{3x^2y - y^3}{x^3 - 3xy^2} = \tan 60°$ *Apply De Moivre's theorem and use $\tan 60 = \sqrt{3}$.*

$\Rightarrow \dfrac{3x^2y - y^3}{x^3 - 3xy^2} = \sqrt{3}$

b $\tan 40° \tan 80°$
$= \tan(60° - 20°)\tan(60° + 20°)$

$= \dfrac{\tan 60° - \tan 20°}{1 + \tan 60° \tan 20°} \times \dfrac{\tan 60° + \tan 20°}{1 - \tan 60° \tan 20°}$ *Apply addition formulae.*

$= \dfrac{\sqrt{3} - \tan 20°}{1 + \sqrt{3}\tan 20°} \times \dfrac{\sqrt{3} + \tan 20°}{1 - \sqrt{3}\tan 20°}$ *Simplify using difference of squares.*

$= \dfrac{3 - \tan^2 20°}{1 - 3\tan^2 20°}$ QED

c $\dfrac{3x^2y - y^3}{x^3 - 3xy^2} = \sqrt{3}$ *Rearrange the LHS of the expression.*

$\Rightarrow \dfrac{y}{x} \dfrac{3 - \left(\dfrac{y}{x}\right)^2}{1 - 3\left(\dfrac{x}{y}\right)^2} = \sqrt{3}$ *Substitute $\dfrac{y}{x}$ for $\tan 20°$.*

$\tan 20° \dfrac{3 - \tan^2 20°}{1 - 3\tan^2 20°} = \sqrt{3}$ *Use part **b** to obtain the result.*

$\tan 20° \tan 40° \tan 80° = \sqrt{3}$ QED

Exercise 12G

1 Let $z = \operatorname{cis} \alpha$.

  a Use De Moivre's theorem to show that $z^n - \left(\dfrac{1}{z}\right)^n = 2i\sin(n\alpha)$.

  b Use the binomial theorem to expand $\left(z - \dfrac{1}{z}\right)^5$

    Hence show that $16\sin^5\alpha = \sin 5\alpha - 5\sin$

    $3\alpha + 10\sin\alpha$ and find $\displaystyle\int \sin^5\alpha \, d\alpha$.

2 Use the results $z^n + \left(\dfrac{1}{z}\right)^n = 2\cos(n\alpha)$ and

  $z^n - \left(\dfrac{1}{z}\right)^n = 2i\sin(n\alpha)$ to show that

  $\tan 5\alpha = \dfrac{5\tan\alpha - 10\tan^3\alpha + \tan^5\alpha}{1 - 10\tan^2\alpha + 5\tan^4\alpha}$

  Hence show that the exact value of $\tan\dfrac{\pi}{5}$ is $\sqrt{5 - 2\sqrt{5}}$.

3 Use De Moivre's theorem to show that
$\cos 7\theta = 64\cos^7\theta - 112\cos^5\theta + 56\cos^3\theta - 7\cos\theta$.
Hence solve the equation
$64\cos^7\theta - 112\cos^5\theta + 56\cos^3\theta - 7\cos\theta = 1$
for $0 \le \theta < 2\pi$

4 Let $z = \cos^2\theta + \dfrac{\sin 2\theta}{2}i$ where $-\dfrac{\pi}{2} \le \theta \le \dfrac{\pi}{2}$.

  a Show that $|z| = \cos\theta$ and $\arg z = \theta$.

  b Find z^2 in polar form.

  c Hence find the exact values of θ for which $|2z^2| = |z|$

5 Consider the complex number $w = \dfrac{z+i}{z+2}$, where $z = x + iy$
  and $i = \sqrt{-1}$.

  a Prove that $\operatorname{Re}(w) = \dfrac{x^2 + 2x + y^2 + y}{(x+2)^2 + y^2}$ and $\operatorname{Im}(w) = \dfrac{x + 2y + 2}{(x+2)^2 + y^2}$.

  b Hence show that

    i when $\operatorname{Re}(w) = 1$ the points (x, y) lie on a straight line l_1
      and state its gradient.

    ii when $\operatorname{Im}(w) = 0$ the points (x, y) lie on a straight line l_2
      perpendicular to l_1.

  c Given $\operatorname{Arg}(z) = \operatorname{Arg}(w) = \dfrac{\pi}{4}$, find $|z|$.

Henri Poincaré (1854–1912) was French mathematician, theoretical physicist, engineer, and a philosopher of science. He made many contributions to different fields of pure and applied mathematics, such as celestial mechanics, fluid mechanics, optics, electricity, telegraphy, capillarity, elasticity, thermodynamics, potential theory, quantum theory, theory of relativity and physical cosmology.
In his research on the three-body problem, Poincaré became the first person to discorver a chaotic deterministic system, laying the foundations for modern chaos theory. He is best known for the Poincaré Conjecture, the only one of the seven Millennium Prize Problems to have been solved (so far). It was proved in 2003 by the Russian Grigori Perelman.

Research more about the Poincaré Conjecture, the Millennium Prize Problems and the Fields Medal.

6 Consider a complex number $z = r \operatorname{cis} \theta$, where $r \in \mathbb{R}^+$ and $\theta \in \mathbb{R}$.

a Use mathematical induction to show that $(z^n)^* = (z^*)^n$ for $n \in \mathbb{Z}^+$.

b Use De Moivre's theorem to deduce the result in part **a**. For which values of n is your proof valid? Give reasons.

7 Let 1, ω and ω^* be the cube roots of unity.

a Show that $\dfrac{1}{1+\omega} = -\omega$ and $\dfrac{1}{1+\omega^*} = -\omega^*$.

b Determine the real numbers a, b and c such that 1, $\dfrac{1}{1+\omega}$ and $\dfrac{1}{1+\omega^*}$ are zeros of the polynomial $p(z) = z^3 + az^2 + bz + c$.

c Hence find $p(\omega)$ and $p(\omega^*)$.

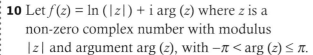

8 Let ω and ω' be consecutive sixth roots of unity.

a Show that $\dfrac{1}{\omega}$ and $\dfrac{1}{\omega'}$ are also consecutive sixth roots of unity.

b Show that ω, ω' and their opposites define a rectangle in the complex plane and find its area.

9 Find the values of n such that $\left(\sqrt{3} - i\right)^n$ is a positive real number.

10 Let $f(z) = \ln(|z|) + i \arg(z)$ where z is a non-zero complex number with modulus $|z|$ and argument $\arg(z)$, with $-\pi < \arg(z) \le \pi$.

a Evaluate $f(i)$, $f(-i)$, $f(1+i)$ and $f(1-i)$.

b Show that $(f(z))^* = f(z^*)$

c Hence show that if $f(z) = f(z^*)$ then z is a real number.

d Find the values of z for which:

   i $f(z)$ is purely imaginary

   ii $f(z)$ is a negative real number

   iii $f(z) = 0$

Georg Riemann (1826–66) was a German mathematician and a student of Gauss. He wrote his PhD dissertation on foundations of a generalized function of one complex variable. The theory of complex functions is considered to be the main achievement of the 19th century, often called the mathematical joy of the 19th century.

Extended Essay suggestions

1 When we study functions of a real variable it usual to graph them to better understand their behavior. How can we visualize a function of a complex variable? Explore Riemann's ideas of dealing with this problem and the concept of the **Riemann surface**.

2 Learn about **Möbius transformations** and explore the **Joukowski aerofoil** and its applications to fluid and aerodynamics.

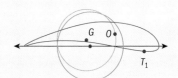

▲ The Joukowski aerofoil and its applications to fluid and aerodynamics

Investigation – paths of chaos

1 Consider the recurrence relation $z_{n+1} = z_n^2$.
   Find the first ten complex numbers obtained when

   a $z_0 = 2\operatorname{cis}\dfrac{\pi}{4}$ **b** $z_0 = \dfrac{1}{2}\operatorname{cis}\dfrac{\pi}{3}$ **c** $z_0 = \operatorname{cis}\dfrac{\pi}{6}$ **d** $z_0 = 0$

A recurrence relation where z_{n+1} is defined in terms of z_n generates a
dynamical system. The sequence generated by each value of z_0 is
called the orbit of z_0.
If the orbit of z_0 consists of a single point, z_0 is called a fixed point.
If an orbit converges towards a fixed point, we say that this point is an attractor.
If the orbit leads away from a fixed point, we say that this point is a repeller.
A periodic point of order n is a point which is returned to after n terms of the sequence.

2 Using the terminology above, describe the orbits of z_0 obtained in part 1.

3 Use technology to explore the orbit of z_0 when $z_0 = \operatorname{cis}(1.5)$. Is z_0 periodic? Why?
   Investigate other values of z_0 for which the behavior of the orbits are similar.
   If you modify the recurrence relation and add a complex constant to it you obtain

   $$z_{n+1} = z_n^2 + c$$

You are now dealing with a famous relation studied by two famous
mathematicians: Gaston Julia and Benoit Mandelbrot. This is part of
a branch of modern mathematics called chaos theory.
Each of these mathematicians is associated with a particular
set of points in the complex plane:

- The Julia set is the set of points whose orbits do not tend to infinity,
  i.e., it is the set of values of z_0 for which the orbit is bounded.
  The importance of the Julia set lies in the amazing changes in
  behavior of the orbits for different choices of c.

- The Mandelbrot set is the set of all the values of c
  for which the orbit of z_0 is bounded.

4 Show that the Julia set of $z_{n+1} = z_n^2$ (i.e. $c = 0$) is
   the unit circle and its interior.

   Challenge:
   Use technology to explore graphically the Mandelbrot
   set for different values of c.
   You may even write your own program!

Chaos theory studies the behavior of dynamical systems that are highly
sensitive to initial conditions. It states that small differences in initial
conditions (sometimes due to rounding errors) yield widely diverging results
making long-term prediction of their outcomes impossible.
Chaotic behavior can be observed in many natural systems, the most well
known being the weather! This is sometimes called the butterfly effect; the
flap of a butterfly's wings (a small initial change) can cause a hurricane to
form weeks later.
Explore applications of chaos theory and their mathematical models in an
area that interests you (e.g. patterns in epilepsy seizures, population growth,
dynamics of fluids).

Gaston Julia
(1893–1978) was
born in the Algerian
town of Bel Abbes,
at the time governed
by the French. During
his youth he studied
mathematics and
music. In the First
World War he suffered
a severe injury, losing
his nose.

Review exercise

1 Write in Cartesian form:

   a $2\operatorname{cis}\dfrac{4\pi}{3}$ **b** $\sqrt{2}\operatorname{cis}135°$ **c** $\dfrac{2}{3\operatorname{cis}\dfrac{5\pi}{6}}$

2 Write in modulus-argument form:

   a $5-5\mathrm{i}$ **b** $\dfrac{2}{1+\sqrt{3}\mathrm{i}}$ **c** $\dfrac{1-\mathrm{i}}{\sqrt{3}-\mathrm{i}}$

3 Let $z=2\operatorname{cis}\dfrac{2\pi}{3}$ and $w=4\operatorname{cis}\dfrac{5\pi}{4}$.

   a Determine in modulus-argument form zw, $\dfrac{z}{w}$ and z^2w^3.

   b Find in Cartesian form $z+w$, $z-w$ and $\dfrac{z^2}{w}$.

4 Let $z=a+\mathrm{i}$ where $a\in\mathbb{R}^+$. Find the exact value of a for which

   a $\arg(z)=\dfrac{\pi}{3}$ **b** z^2 is a real number   **c** $|z-1|=|z-2\mathrm{i}|$

5 Solve the equation $z^5=z$, giving your answers in Cartesian form.

6 Let $z=\dfrac{1}{1+\mathrm{i}\tan\theta}$ where $0<\theta<\dfrac{\pi}{2}$. Find z in modulus-argument form.

EXAM-STYLE QUESTIONS

7 $\dfrac{1-\mathrm{i}}{4}$ and $a+a\mathrm{i}$ are consecutive nth roots of a complex number z.

   a Without calculating z, find all possible values of a and n
   b Hence find the remaining nth roots of z.

8 Let ω be a complex cube root of unity ($\omega\neq1$).
   Show that $(x+y)(x+\omega y)(a+\omega^2 y)=x^3+y^3$, for any $x,y\in\mathbb{R}$.

9 The complex number $z=-\sqrt{3}-\mathrm{i}$.
   a Find the modulus and argument of z, giving the argument in degrees.
   b Find that cube root of z which lies in the first quadrant of the Argand diagram.
   c Find the smallest positive integer n for which z^n is a positive real number.

10 Use De Moivre's theorem and the expansion of $(\cos\alpha+\mathrm{i}\sin\alpha)^4$ to deduce that:
   $$\tan4\alpha\equiv\dfrac{4\tan\alpha-4\tan^3\alpha}{1-6\tan^2\alpha+\tan^4\alpha}$$

11 Given $p(z)=a_nz^n+a_{n-1}z^{n-1}+\ldots+a_0$ where $a_n, a_{n-1},\ldots, a_0\in\mathbb{R}$.
   Show that if $z=r\operatorname{cis}\theta$ is a zero of $p(z)$ then z^* is also a zero of $p(z)$.

EXAM-STYLE QUESTION

12 Given that $z+\dfrac{1}{z}=-1$,

   a Expand $\left(z+\dfrac{1}{z}\right)^2$ and find the value of $z^2+\dfrac{1}{z^2}$.

   b Find the value of $z^3+\dfrac{1}{z^3}$.   **c** Hence find the value of $z^5+\dfrac{1}{z^5}$.

13 Let z and z^* be conjugate complex numbers.

   a Show that $(x-z)(x-z^*) = x^2 - 2\text{Re}(z) + |z|^2$ for any real number x.

   b Find the eighth roots of unity in Cartesian form.

   c Write $x^8 - 1$ as a product of two linear and three quadratic factors.

14 Let $\omega = \text{cis}\,\alpha$ where $0 < \alpha < \dfrac{\pi}{2}$.

   a Show that $|1+\omega| = 2\cos\dfrac{\alpha}{2}$ and $\arg(1+\omega) = \dfrac{\alpha}{2}$.

   b Hence use the binomial expansion of $(1+\omega)^n$ to show that

$$\sum_{k=0}^{n}\binom{n}{k}\cos(k\alpha) \equiv \left(2\cos\dfrac{\alpha}{2}\right)^n \cos\left(\dfrac{n\alpha}{2}\right).$$

Review exercise

1 Let $p(z) = z^4 + az^3 + bz^2 + cz + d$ be a polynomial with real coefficients.

Given that $z = \sqrt{2}\,\text{cis}\left(\dfrac{\pi}{4}\right)$ and $z = 3 - i$ are two complex zeros of $p(z)$, find the values of the real numbers a, b, c and d.

> Most exam questions on complex numbers will be on the non-GDC paper.

CHAPTER 12 SUMMARY

Complex numbers as vectors

- $z_1 + z_2 = z_2 + z_1$
- $(z_1 + z_2) + z_3 = z_1 + (z_2 + z_3)$
- The null vector represents the complex number $0 = 0 + 0i$.
- If $z = a + bi$ the opposite of z is $-z = -a - bi$; z and $-z$ are represented by vectors with the same magnitude and opposite directions.
- The magnitude of the vector **u** associated with $z = a + bi$ is called modulus of z: $|z| = \sqrt{a^2 + b^2}$.
- The subtraction of two complex numbers z_1 and z_2 is defined as the sum of z_1 with the opposite of z_2.
- $k(gz) = (kg)z$
- $k(z_1 + z_2) = kz_1 + kz_2$
- $(k + g)z = kz + gz$

where $k, g \in \mathbb{R}$ and $z, z_1, z_2 \in \mathbb{C}$.

Other transformations of complex numbers

- $(z^*)^* = z$
- $z \in \mathbb{R} \Rightarrow z^* = z$
- $|z| = |z^*|$

Continued on next page

Representations of complex numbers in polar form

Modulus-argument form: $z = r(\cos\theta + i\sin\theta)$ or $z = r\operatorname{cis}\theta$

Euler's form: $z = re^{\theta i}$, where $r = |z|$ is the modulus of z and $\theta = \arg(z)$ is the argument of z.

Product of a real number by complex number

- $\alpha > 0 \Rightarrow \alpha z = (\alpha r)\operatorname{cis}\theta$
- $\alpha < 0 \Rightarrow \alpha z = (|\alpha|r)\operatorname{cis}(\pi + \theta)$
- $\alpha = 0 \Rightarrow \alpha z = 0$

Equality of complex numbers in polar form

$z_1 = z_2 \Rightarrow |z_1| = |z_2|$ and $z_1 = z_2 \Rightarrow \arg(z_1) - \arg(z_2) = 2k\pi, k \in \mathbb{Z}$

Conversion from Cartesian to polar form:

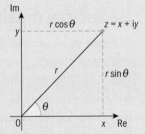

$z = x + yi \Rightarrow r = |z| = \sqrt{x^2 + y^2}$ and $\tan\theta = \dfrac{y}{x}$

Conversion from polar to Cartesian form:

$\cos\theta = \dfrac{x}{r} \Rightarrow x = r\cos\theta$ and $\sin\theta = \dfrac{y}{r} \Rightarrow y = r\sin\theta$

Conjugate, opposite, opposite of conjugate and reciprocal of z in polar form:

$z^* = r\operatorname{cis}(-\theta) = r\operatorname{cis}(2\pi - \theta)$

$-z = r\operatorname{cis}(\pi + \theta)$

$-z^* = r\operatorname{cis}(\pi + \theta)$

$\dfrac{1}{z} = \dfrac{1}{r}\operatorname{cis}(-\theta), r \neq 0$

Multiplication and division of complex numbers in modulus-argument form

If $z_1 = r_1\operatorname{cis}\theta_1$ and $z_2 = r_2\operatorname{cis}\theta_2$ then

$z_1 z_2 = r_1 r_2 \operatorname{cis}(\theta_1 + \theta_2)$ and $\dfrac{z_1}{z_2} = \dfrac{r_1}{r_2}\operatorname{cis}(\theta_1 - \theta_2), r_2 \neq 0.$

De Moivre's theorem:

$[r(\cos\theta + i\sin\theta)]^n = r^n(\cos n\theta + i\sin n\theta)$ where $r \in \mathbb{R}^+$, $\theta \in \mathbb{R}$ and $n \in \mathbb{Z}$.

Roots of complex numbers

$z^n = r\cos\theta \Rightarrow z = \sqrt[n]{r}\operatorname{cis}\dfrac{\theta + 2k\pi}{n}$, $k = 0, 1, 2,\dots, n-1.$

The changing structure of mathematics

Mathematics is one of humankind's greatest intellectual achievements, not just for the mathematical knowledge accumulated through centuries, but also for the power and organization that its language, methods and theories offer. Mathematical methods have dictated technological progress, inspired social and economic thought and fashioned styles in painting, music and architecture.

Beautiful mathematics

In 1988 the scientific magazine *The Mathematical Intelligencer* ran a survey on beauty in mathematics. Its readers selected $e^{i\pi} + 1 = 0$ as the most beautiful theorem in mathematics.

- Why do mathematicians consider this equation beautiful?

- Do the survey results prove that this theorem is the most beautiful result in mathematics? Should the beauty of theorems be established in this way?

- How can we assess the beauty of a mathematical result?

- Investigate mathematical influences in architecture and the arts.

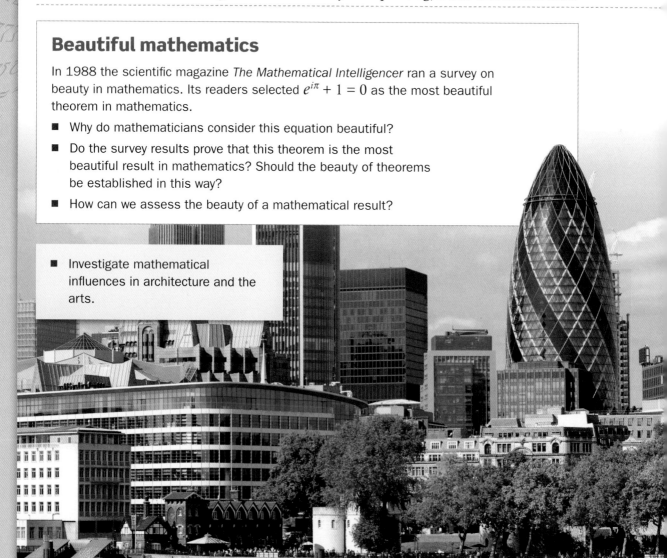

Moving into higher dimensions

In the last 200 years, the nature of mathematical concepts has changed greatly. Generations of mathematicians have realized that old concepts like number and geometrical shapes can be extended beyond the scope of the physical world. Mathematics has become an intellectual activity where the value of a theory transcends its applications. At times mathematical explorations even seem to deny intuition.

- What is hyperbolic space?
- Mathematics allows us to work in higher dimensions.
  Does this mean that these dimensions exist beyond our perception?

> *"There is an amazingly high consensus in mathematics as to what is correct or accepted. But alongside this, equally important, is the issue of what is interesting or important or deep or elegant. These esthetic criteria vary widely from person to person and decade to decade."*
>
> Reuben Hersh, Professor Emeritus at the University of New Mexico

Intuition to formalism

While exploring the possible extension of complex numbers to dimension 3, the Irish mathematician William Rowan Hamilton (1805–65) realized that he could define 'new' numbers in dimension four. Like complex numbers, they could be described by vectors, but their multiplication was not commutative.

This was a remarkable development. Until then all 'new' numbers were manipulated assuming that the 'usual' properties of operations such as multiplication hold. This discovery was of extreme importance for mathematicians, as it left them free to 'invent' new structures and develop new algebras. Also, the need to establish which properties are valid when dealing with new numbers brought more rigor to mathematics.

- Look at the mathematical structures you have studied. Which properties do the operations with numbers, vectors or functions have?
- Which operations are not commutative?
- Do you know of any operation that is not associative?
- Is it possible to 'invent' a non-associative operation?

Here as he walked by
on the 16th of October 1843
Sir William Rowan Hamilton
in a flash of genius discovered
the fundamental formula for
quarternion multiplication
$$i^2 = j^2 = k^2 = ijk = -1$$
and cut it on a stone of this bridge

◄ Inscription on Broome Bridge, Dublin, commemorating Hamilton's discovery of multiplication in four dimensions

You can find out more about Hamilton's discovery at **http://plus.maths.org/content/os/issue33/features/baez/index**

13 Exploration

As part of your Mathematics HL course, you need to write an exploration, which will be assessed internally and counts as 20% of your final grade.
This chapter gives you advice on planning your exploration, hints and tips to help you get a good grade by making sure your exploration satisfies the assessment criteria, as well as suggestions on choosing a topic and getting started on your exploration.

13.1 About the exploration

The exploration is an opportunity for you to show that you can apply mathematics to an area that interests you.

You should aim to spend:

10 hours of class time	10 hours of your own time
<ul><li>Discussing the assessment criteria</li><li>Discussing suitable topics/titles</li><li>Discussing your progress with your teacher</li></ul>	<ul><li>Planning your exploration, doing research to help select an appropriate topic</li><li>Researching, collecting and organising your data and/or information</li><li>Applying mathematical processes:<ul><li>Ensuring that all of your results are derived using logical deductive reasoning</li><li>Ensuring that your proofs (when necessary) are coherent and correct</li></ul></li><li>Demonstrating mathematical communication and presentation:<ul><li>Checking that your notation and terminology are consistently correct</li><li>Adding diagrams, graphs or charts where necessary</li><li>Making sure your exploration is clearly structured and reads well</li></ul></li></ul>

Your school will set you deadlines for submitting a draft and the final piece of work.

If you do not submit an exploration then you receive a grade of "N" for Mathematics HL, which means you will not receive your IB diploma.

13.2 Internal assessment criteria

Your exploration will be assessed by your teacher, against given criteria.

It will then be externally moderated by the IB using the same assessment criteria.

The final mark for each exploration is the sum of the scores for each criterion.

The maximum possible final mark is 20.

This is 20% of your final grade for Mathematics HL.

A good exploration should be clear and easily understood by one of your peers, and self-explanatory all the way through.

The criteria are split into five areas, A to E:

Criterion A	Communication
Criterion B	Mathematical presentation
Criterion C	Personal engagement
Criterion D	Reflection
Criterion E	Use of mathematics

These criteria are explained in more detail, with tips on how to ensure your exploration satisfies them.
Make sure you understand these criteria and consult them frequently when writing your exploration.

Criterion A: Communication

This criterion assesses the organization, coherence, conciseness and completeness of the exploration.

Achievement level	Descriptor
0	The exploration does not reach the standard described by the descriptors below.
1	The exploration has some coherence.
2	The exploration has some coherence and shows some organisation.
3	The exploration is coherent and well organised.
4	The exploration is coherent, well organised, concise and complete.

Your exploration

To get a good mark for Criterion A: Communication

✓ A well organised exploration should have

- An **introduction** in which you should discuss the context of the exploration
- A **rationale** which should include an explanation of why you chose this topic
- A description of the **aim** of the exploration which should be clearly identifiable
- A **conclusion**.

✓ A coherent exploration is logically developed and easy to follow.

✓ Your exploration should "read well".

✓ Any graphs, tables and diagrams that you use should accompany the work in the appropriate place and not be attached as appendices to the document.

✓ A concise exploration is one that focuses on the aim and avoids irrelevancies.

✓ A complete exploration is one in which all steps are clearly explained without detracting from its conciseness.

✓ It is essential that references are cited where appropriate, i.e.,

- Your exploration should contain footnotes as appropriate. For example, if you are using a quote from a publication, a formula from a mathematics book, etc, put the source of the quote in a footnote.
- Your exploration should contain a bibliography as appropriate. This can be in an appendix at the end. List any books you use, any websites you consult, etc.

Criterion B: Mathematical presentation

This criterion assesses to what extent you are able to:

- use appropriate mathematical language (notation, symbols, terminology)
- define key terms, where required
- use multiple forms of mathematical representation such as formulae, diagrams, tables, charts, graphs and models, where appropriate.

Achievement level	Descriptor
0	The exploration does not reach the standard described by the descriptors below.
1	There is some appropriate mathematical presentation.
2	The mathematical presentation is mostly appropriate.
3	The mathematical presentation is appropriate throughout.

Criterion C: Personal engagement

This criterion assesses the extent to which you engage with the exploration and make it your own.

Achievement level	Descriptor
0	The exploration does not reach the standard described by the descriptors below.
1	There is evidence of limited or superficial personal engagement.
2	There is evidence of some personal engagement.
3	There is evidence of significant personal engagement.
4	There is abundant evidence of outstanding personal engagement.

Your exploration

To get a good mark for Criterion C: Personal engagement

✓ You should choose a topic for your exploration that you are interested in as it will be easier to display personal engagement.

✓ You can demonstrate personal engagement by using some of the following different attributes and skills.

- Thinking and working independently
- Thinking creatively
- Addressing your personal interests
- Presenting mathematical ideas in your own way
- Asking questions, making conjectures and investigating mathematical ideas
- Looking for and creating mathematical models for real-world situations
- Considering historical and global perspectives
- Exploring unfamiliar mathematics.

Criterion D: Reflection

This criterion assesses how you review, analyse and evaluate the exploration.

Achievement level	Descriptor
0	The exploration does not reach the standard described by the descriptors below.
1	There is evidence of limited or superficial reflection.
2	There is evidence of meaningful reflection.
3	There is substantial evidence of critical reflection.

Your exploration

To get a good mark for Criterion D: Reflection

✓ Although reflection may be seen in the conclusion to the exploration, it may also be found throughout the exploration.

✓ You can show reflection in your exploration by

- Discussing the implications of your results
- Considering the significance of your findings and results
- Stating possible limitations and/or extensions to your results
- Making links to different fields and/or areas of mathematics.

Criterion E: Use of mathematics

This criterion assesses to what extent and how well you use mathematics in your exploration.

Achievement level	Descriptor
0	The exploration does not reach the standard described by the descriptors below.
1	Some relevant mathematics is used. Limited understanding is demonstrated.
2	Some relevant mathematics is used. The mathematics explored is partially correct. Some knowledge and understanding are demonstrated.
3	Relevant mathematics commensurate with the level of the course is used. The mathematics explored is correct. Good knowledge and understanding are demonstrated.
4	Relevant mathematics commensurate with the level of the course is used. The mathematics explored is correct and reflects the sophistication expected. Good knowledge and understanding are demonstrated.
5	Relevant mathematics commensurate with the level of the course is used. The mathematics explored is correct and reflects the sophistication and rigor expected. Thorough knowledge and understanding are demonstrated.
6	Relevant mathematics commensurate with the level of the course is used. The mathematics explored is precise and reflects the sophistication and rigor expected. Thorough knowledge and understanding are demonstrated.

Your exploration

To get a good mark for Criterion E: Use of mathematics

✓ You are expected to produce work that is commensurate with the level of the course you are studying. The mathematics you explore should either be part of the syllabus, or at a similar level (or beyond).

✓ You should ensure that the mathematics involved is not completely based on mathematics listed in the prior learning.

✓ **If the level of mathematics is not commensurate with the level of the course you can only get a maximum of two marks for this criterion.**

✓ You need to demonstrate within your exploration that you fully understand the mathematics used.

✓ You can demonstrate sophistication of mathematics in your exploration by

   ● Showing that you understand and can use challenging mathematical concepts

   ● Showing that you can extend the applications of mathematics beyond that which you learned in the classroom

   ● Looking at a problem from different mathematical perspectives

   ● Identifying underlying structures to link different areas of mathematics.

✓ Rigor involves clarity of logic and language when making mathematical arguments and calculations.

✓ Precise mathematics is error-free and uses an appropriate level of accuracy at all times.

13.3 How the exploration is marked

Once you have submitted the final version of your exploration, your teacher will mark it. The teacher looks at each criterion in turn, starting from the lowest grade. As soon as your exploration fails to meet one of the grade descriptors, then the mark for that criterion is set.

The teacher submits these marks to the International Baccalaureate, via a special website. A sample of your school's explorations is selected automatically from the marks that are entered and this sample is sent to an external moderator to be checked. This person moderates the explorations according to the assessment criteria and checks that your teacher has marked the explorations accurately.

If your teacher has applied the criteria to the exploration too severely then your school's exploration marks may be increased.

If your teacher has applied the exploration criteria too leniently then your school's exploration marks may be decreased.

13.4 Academic honesty

This is extremely important in all your work. Make sure that you have read and are familiar with the IB academic honesty document.

Academic honesty means:

- that your work is authentic
- that your work is your own intellectual property
- that you conduct yourself properly in written examinations
- that any work taken from another source is properly cited.

Authentic work:

- is work based on your own original ideas
- can draw on the work and ideas of others, but this must be fully acknowledged (e.g. in footnotes and a bibliography)
- must use your own language and expression – for both written or oral assignments.
- must acknowledge all sources fully and appropriately (e.g. in a bibliography).

> Your teacher or IB Diploma Programme coordinator will be able to give you this document.

Malpractice

The IB defines **malpractice** as 'behavior that results in, or may result in, the candidate or any other candidate gaining an unfair advantage in one or more assessment components'.

Malpractice includes:

- plagiarism – copying from others' work, published or otherwise
- collusion – working secretly with at least one other person in order to gain an undue advantage. This includes having someone else write your exploration, and passing it off as your own
- duplication of work
- any other behavior that gains an unfair advantage.

> 'Plagiarism' is a word derived from Latin, meaning 'to kidnap'.

Advice to schools:

- A school-wide policy must be in place to promote academic honesty
- All candidates must clearly understand this policy
- All subject areas must promote the policy
- Candidates must be clearly aware of the penalties for academic dishonesty
- Schools must enforce penalties, if incurred.

Acknowledging sources

Remember to acknowledge all your sources. Both teachers and moderators can usually tell when a project has been plagiarised. Many schools use computer software to check for plagiarism. If you are found guilty of plagiarism then you will not receive your diploma. It is not worth taking the risk.

> You will find a definition of plagiarism in the IB academic honesty document.

13.5 Record keeping

Throughout the course, it would be a good idea to keep an exploration journal, either manually or online. Keeping a journal will help you to focus your search for a topic, and also remind you of deadlines.

> Keeping a journal while you write your exploration will also help you to demonstrate its academic honesty.

If you use a journal for theory of knowledge you will probably appreciate how much help it is when writing your essays. In the same way, keeping a journal for your exploration will be a great assistance in focusing your efforts.

- Make notes of any books or websites you use, as you go along, so you can include them in your bibliography.
- There are different ways of referencing books, websites, etc. Make sure that you use the style advised by your school and **be consistent**.
- Keep a record of your actions so that you can show your teacher how much time you are spending on your exploration. Include any meetings you may have with your teacher about your exploration.
- Remember to follow your teacher's advice and meet the school's deadlines.
- The teacher is there to help you – so do not be afraid to ask for guidance. The more focused your questions are, the better guidance your teacher can give you.

13.6 Choosing a topic

You need to choose a topic that interests you, because then you will enjoy working on your exploration, you will put more effort into the exploration, and you will be able to demonstrate authentic personal engagement more effectively. You should discuss the topic with your teacher before you put too much time and effort into writing your exploration.

These questions may help you to find a topic for your exploration:
- What areas of the syllabus am I enjoying the most?
- What areas of the syllabus am I performing best in?
- Which mathematical skills are my strengths?
- Do I prefer pure mathematics, or applied problems and modeling?
- Have I discovered, either through reading or the media, mathematical areas outside the syllabus that I find interesting?
- What career do I eventually want to enter, and what mathematics is important in this field?
- What are my own special interests or hobbies? Where is the mathematics in these areas?

Your teacher might give your class a set of stimuli – general areas from which you could choose a topic. Alternatively they might encourage you to find your own topic based on your interests and level of mathematical competence.

Each chapter of this book suggests some ideas for explorations, which could be starting points for you to choose a topic.

Mind mapping may help you choose a topic. See pages 670–671.

13.7 Getting started

Once you have chosen your topic, the next step is to do some research. The purpose of the research is to determine the suitability of your topic.

> These questions will help you to decide if your chosen topic is suitable.
> - What areas of mathematics are contained in my topic?
> - Which of these areas are accessible to me or are part of the syllabus?
> - Is there mathematics outside the syllabus that I would have to learn in order to complete the exploration successfully? Am I capable of doing this?
> - Can I show personal engagement in my topic, and how?
> - Can I limit my work to the recommended length of 6 to 12 pages if I choose this topic?

Do not limit your research to the internet. Your school library will have books on mathematics that are interesting and related to a variety of different fields.

Once you think you have a workable topic, write a brief outline covering:

- why you chose this topic
- how your topic relates to mathematics
- the mathematical areas in your topic, e.g. algebra, geometry, trigonometry, calculus, probability and statistics, etc.
- the key mathematical concepts in your topic, e.g. areas of irregular shapes, curve fitting, modeling data, etc.
- the mathematical skills you will need, e.g. writing formal proofs, integration, operations with complex numbers, graphing piecewise functions, etc.
- any mathematics outside the syllabus that you will need
- possible technology and software that can help in the design of your exploration and in doing the mathematics
- key mathematical terminology and notation required in your topic.

If your original choice of topic is not suitable, has your research suggested another, better topic? Otherwise could you either widen out or narrow down your topic to make it more suitable for the exploration?

Now you are ready to start writing the topic in detail.

Remember that your fellow students (your peers) should be able to read and understand your exploration. You could ask one of your classmates to read your work and comment on any parts which are unclear, so you can improve them.

Make sure you keep every internal deadline that your teacher assigns. In this way, you will receive feedback in time for you to be able to complete your exploration successfully.

Mind map

One way of choosing a topic is to start with a general area of interest and create a mind map. This can lead to some interesting ideas on applications of mathematics to explore.

The mind map below shows how the broad topic 'Geography' can lead to suggestions for explorations into such diverse topics as the spread of disease, earthquakes or global warming.

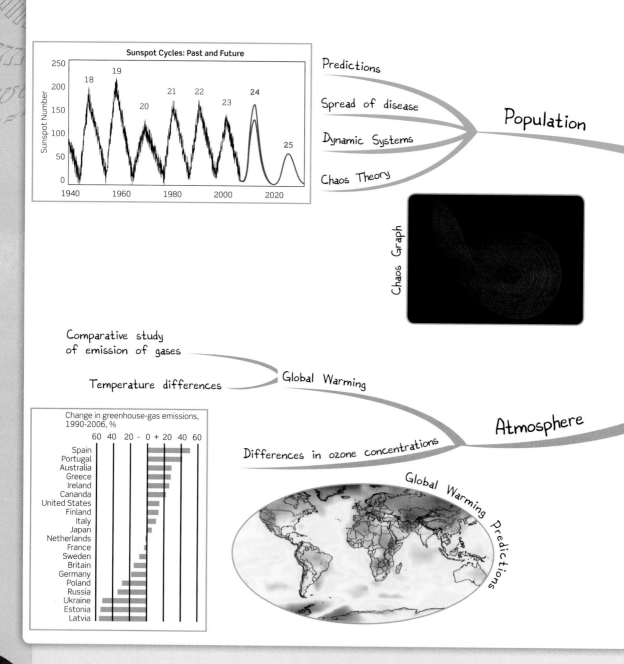

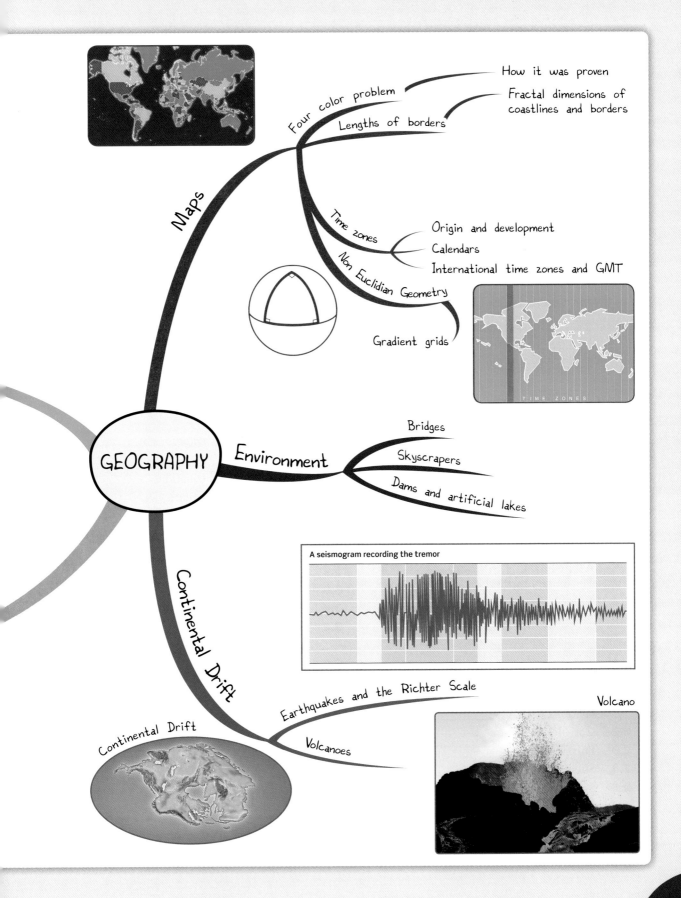

Four color problem
How it was proven

Lengths of borders
Fractal dimensions of coastlines and borders

Maps

Time zones
Origin and development
Calendars
International time zones and GMT

Non Euclidian Geometry

Gradient grids

GEOGRAPHY

Environment
Bridges
Skyscrapers
Dams and artificial lakes

Continental Drift

Continental Drift

A seismogram recording the tremor

Earthquakes and the Richter Scale

Volcanoes

Volcano

14 Prior learning

CHAPTER OBJECTIVES:

This chapter contains a number of short topics that you should know before starting the course. You do not need to work through the whole of this chapter in one go. For example, before you start work on an algebra chapter in the book, make sure you have covered the algebra prior learning in this chapter.

The IB Higher Level examination questions will expect you to know all the topics in this chapter. Make sure you have covered them.

Chapter contents

1 Number

1.1 Calculation

There are several versions of the rules for the order of operations. They all amount to the same thing:

- Brackets or parentheses are calculated first.
- Next come exponents, indices or orders.
- Then multiplication and division, in order from left to right.
- Finally additions and subtractions.

A fraction line or the line above a square root counts as a bracket too.

Your GDC follows the rules, so if you enter a calculation correctly you should get the correct answers.

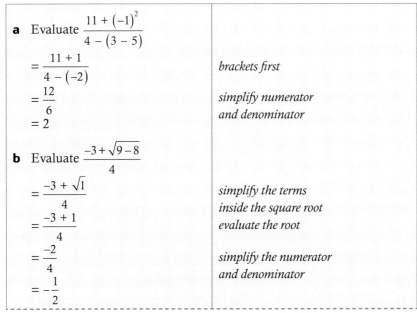

GDC help on CD:
Alternative demonstrations for the TI-84 Plus and Casio fx-9860GII GDCs are on the CD.

The GDC shows divisions as fractions, which makes the order of operations clearer.

BEDMAS:	Brackets, exponents, division, multiplication, addition, subtraction.
BIDMAS:	Brackets, indices, division, multiplication, addition, subtraction.
BEMDAS:	Brackets, exponents, multiplication, division, addition, subtraction.
BODMAS:	Brackets, orders, division, multiplication, addition, subtraction.
BOMDAS:	Brackets, orders, multiplication, division, addition, subtraction.
PEMDAS:	Parentheses, exponents, multiplication, division, addition, subtraction.

Simple calculators, like the ones on phones, do not always follow the calculation rules.

Example 1

a Evaluate $\dfrac{11 + (-1)^2}{4 - (3 - 5)}$

$= \dfrac{11 + 1}{4 - (-2)}$ *brackets first*

$= \dfrac{12}{6}$ *simplify numerator and denominator*

$= 2$

b Evaluate $\dfrac{-3 + \sqrt{9 - 8}}{4}$

$= \dfrac{-3 + \sqrt{1}}{4}$ *simplify the terms inside the square root evaluate the root*

$= \dfrac{-3 + 1}{4}$

$= \dfrac{-2}{4}$ *simplify the numerator and denominator*

$= -\dfrac{1}{2}$

▶ Continued on next page

On your GDC you can either use templates for the fractions and roots or you can use brackets.

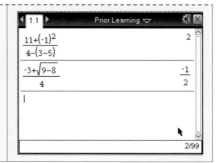

GDC help on CD:
Alternative demonstrations for the TI-84 Plus and Casio fx-9860GII GDCs are on the CD.

Exercise 1A

Do the questions by hand first, then check your answers with your GDC.

1 Calculate

  a $12 - 5 + 4$ **b** $6 \div 3 \times 5$ **c** $4 + 2 \times 3 - 2$

  d $8 - 6 \div 3 \times 2$ **e** $4 + (3 - 2)$ **f** $(7 + 2) \div 3$

  g $(1 + 4) \times (8 - 4)$ **h** $1 - 3 + 5 \times (2 - 1)$

2 Find

  a $\dfrac{6 + 9}{4 - 1}$ **b** $\dfrac{2 \times 9}{3 \times 4}$ **c** $\dfrac{2 - (3 + 4)}{4 \times (2 - 3)}$ **d** $\dfrac{6 \times 5 \times 4}{3 \times 2 - 1}$

3 Determine

  a $3 \times (-2)^2$ **b** $2^2 \times 3^3 \times 5$ **c** $4 \times (5 - 3)^2$ **d** $(-3)^2 - 2^2$

4 Calculate

  a $\sqrt{3^2 + 4^2}$ **b** $\left(\sqrt{4}\right)^3$ **c** $\sqrt{4^3}$ **d** $\sqrt{2 + \sqrt{2 + 2}}$

5 Find

  a $\sqrt{\dfrac{13^2 - \left(3^2 + 4^2\right)}{2 \times 18}}$ **b** $2\sqrt{\dfrac{3 + 5^2}{7}}$ **c** $2(3^2 - 4(-2)) - (2 - \sqrt{7 - 3})$

1.2 Simplifying expressions involving roots

$\sqrt{2}, 2 - \sqrt{3}, 2\sqrt{5}, \dfrac{\sqrt{3}}{3}$, are **irrational numbers** that involve square roots. They are called **surds** or **radicals**.

In calculations, you can use approximate decimals for these types of irrational number, but for more accurate results you can use surds.

Surds are written in their **simplest form** when:

- there is no surd in the denominator
- the smallest possible whole number is under the $\sqrt{\ }$ sign.

> If a question asks for an exact value, it means leave your answer in surd form. In examinations, surds may be left in the denominator.

→ **Rules of surds**

$$\sqrt{a}^{\,2} = a \qquad \sqrt{a \times b} = \sqrt{a} \times \sqrt{b} \qquad \sqrt{\dfrac{a}{b}} = \dfrac{\sqrt{a}}{\sqrt{b}}$$

Example 2

Simplify

a $\dfrac{4}{\sqrt{5}}$ **b** $\dfrac{3}{\sqrt{3}}$

Answers

a $\dfrac{4}{\sqrt{5}} = \dfrac{4}{\sqrt{5}} \times \dfrac{\sqrt{5}}{\sqrt{5}}$

$\quad = \dfrac{4\sqrt{5}}{\left(\sqrt{5}\right)^2}$

$\quad = \dfrac{4\sqrt{5}}{5}$

Multiply numerator and denominator by $\sqrt{5}$

b $\dfrac{3}{\sqrt{3}} = \dfrac{3}{\sqrt{3}} \times \dfrac{\sqrt{5}}{\sqrt{3}}$

$\quad = \dfrac{3\sqrt{3}}{\left(\sqrt{3}\right)^2}$

$\quad = \dfrac{3\sqrt{3}}{3}$

$\quad = \sqrt{3}$

Multiply numerator and denominator by $\sqrt{3}$

Cancel the common factor

Example 3

Simplify

a $\sqrt{20}$ **b** $\sqrt{8} - \sqrt{18}$

Answers

a $\sqrt{20} = \sqrt{4} \times \sqrt{5} = 2\sqrt{5}$

b $\sqrt{8} - \sqrt{18} = \sqrt{4 \times 2} - \sqrt{9 \times 2}$

$\quad\quad\quad\quad\quad = 2\sqrt{2} - 3\sqrt{3}$

$\quad\quad\quad\quad\quad = -\sqrt{2}$

$\sqrt{a \times b} = \sqrt{a} \times \sqrt{b}$

Look for square numbers that divide into 8 and 18. Use these to write 8 and 18 as products
Use $\sqrt{a \times b} = \sqrt{a} \times \sqrt{b}$

Example 4

Expand the brackets and simplify $\left(1 + \sqrt{2}\right)\left(1 - \sqrt{2}\right)$

Answer

$\left(1 + \sqrt{2}\right)\left(1 - \sqrt{2}\right) = 1 - \sqrt{2} + \sqrt{2} - \left(\sqrt{2}\right)^2$

$\quad\quad\quad\quad\quad\quad\quad = 1 - 2$

$\quad\quad\quad\quad\quad\quad\quad = -1$

(a +b)(c + d)
= ac + ad + bc + bd

Example 5

Rewrite the fraction $\dfrac{1}{\left(1+\sqrt{3}\right)}$ without surds in the denominator

Answer

$$\frac{1}{\left(1+\sqrt{3}\right)} = \frac{1}{\left(1+\sqrt{3}\right)} \times \frac{\left(1-\sqrt{3}\right)}{\left(1-\sqrt{3}\right)}$$

$$= \frac{1-\sqrt{3}}{1-3} = \frac{1-\sqrt{3}}{-2}$$

Multiply numerator and denominator by
$1-\sqrt{3}$

Exercise 1B

1 Simplify

a $\dfrac{1}{\sqrt{2}}$ **b** $\dfrac{6}{\sqrt{3}}$ **c** $\dfrac{5}{\sqrt{5}}$ **d** $\dfrac{10\sqrt{2}}{\sqrt{5}}$ **e** $\sqrt{\dfrac{2}{5}}$

2 Simplify

a $\sqrt{12}$ **b** $\sqrt{75}$ **c** $\sqrt{72}$ **d** $3\sqrt{8}$ **e** $5\sqrt{27}$

3 Simplify

a $\sqrt{3}\times\sqrt{12}$ **b** $\sqrt{3}\times\sqrt{27}$ **c** $\sqrt{24}\times\sqrt{32}$

d $2\sqrt{3}\times3\sqrt{2}$ **e** $3\sqrt{5}\times5\sqrt{75}$

4 Simplify

a $3\sqrt{5}+2\sqrt{5}$ **b** $5\sqrt{2}-3\sqrt{2}$ **c** $2\sqrt{3}+\sqrt{12}$

d $\sqrt{2}-\sqrt{8}$ **e** $\sqrt{12}-2\sqrt{3}$

5 Expand and simplify

a $\left(3+\sqrt{2}\right)^2$ **b** $\left(\sqrt{2}+\sqrt{3}\right)^2$ **c** $\left(3+\sqrt{2}\right)\left(1-\sqrt{2}\right)$

d $\left(4+\sqrt{3}\right)\left(1-\sqrt{2}\right)$ **e** $\left(2+\sqrt{2}\right)\left(2-\sqrt{2}\right)$

6 Simplify

a $\dfrac{1+\sqrt{3}}{\sqrt{7}}$ **b** $\dfrac{1}{1-2\sqrt{3}}$ **c** $\dfrac{\sqrt{5}}{1+\sqrt{5}}$ **d** $\dfrac{4+\sqrt{2}}{3-2\sqrt{2}}$

7 Write these without a surd on the denominator.
Simplify as much as possible.

a $\dfrac{2}{\sqrt{3}}+3\sqrt{3}$ **b** $\dfrac{\sqrt{3}}{2}+\dfrac{5}{\sqrt{3}}$ **c** $\sqrt{20}+\dfrac{2}{\sqrt{5}}$

1.3 Primes, factors and multiples

A **prime** number is an integer, greater than 1, that is not a multiple of any other number apart from 1 and itself.

In 2009, the largest known prime was a 12 978 189 digit number.
Prime numbers have become big business because they are used in cryptography.

Example 6

List all the factors of 42.	
Answer $42 = 1 \times 42$, $42 = 2 \times 21$, $42 = 3 \times 14$, $42 = 6 \times 7$ The factors of 42 are 1, 2, 3, 6, 7, 14, 21 and 42.	*Write 42 as a product of two numbers every way you can.*

Example 7

Write the number 24 as a product of prime factors.		
Answer		
$2\overline{)24}$ $2\overline{)12}$ $2\overline{)6}$ $3\overline{)3}$ 1	$24 = 2 \times 2 \times 2 \times 3$ $= 2^3 \times 3$	*Begin dividing by the smallest prime number. Repeat until you reach an answer of 1.*

The lowest common multiple of a pair of numbers is the smallest number that is a multiple of both of them.

Example 8

Find the **lowest common multiple** (LCM) of 12 and 15.	
Answer The multiples of 12 are 12, 24, 36, 48, 60, 72, 84, 96, 108, 120, 132, 144... The multiples of 15 are 15, 30, 45, 60, 75, 90, 105, 120, 135... The common multiples are 60, 120... The LCM is 60.	*List all the multiples until you find some in both lists. The LCM is the smallest number in each of the lists.*

The highest common factor of a pair of numbers is the largest number that is a factor of both of them.

Example 9

Find the **highest common factor** (HCF) of 36 and 54.

Answer

$2\overline{)36}$ $36 = 2 \times 2 \times 3 \times 3$ $2\overline{)54}$ $54 = 2 \times 3 \times 3 \times 3$

$2\overline{)18}$ $3\overline{)27}$

$3\overline{)9}$ $3\overline{)9}$

$3\overline{)3}$ $3\overline{)3}$

   1 1

The HCF of 36 and 54 is $2 \times 3 \times 3 = 18$

> Write each numbers as a product of prime factors. Find the product of all the factors that are common to both numbers.

Exercise 1C

1 List all the factors of
   a 18   **b** 27   **c** 30   **d** 28   **e** 78

2 Write as products of prime factors.
   a 36   **b** 60   **c** 54   **d** 32   **e** 112

3 Find the LCM of
   a 8 and 20   **b** 6, 10 and 16

4 Find the HCF of
   a 56 and 48   **b** 36, 54 and 90

Some GDCs are able to perform these operations, as in these examples.

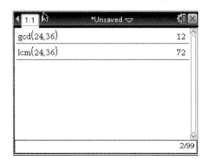

> The HCF is also called the 'greatest common divisor' or 'gcd'. Most GDCs use 'gcd'.

GDC help on CD: *Alternative demonstrations for the TI-84 Plus and Casio fx-9860GII GDCs are on the CD.*

1.4 Fractions and decimals

There are two types of fraction:

- **common** fractions (often just called 'fractions')

  like $\dfrac{4}{5}$ $\dfrac{\text{numerator}}{\text{denominator}}$

- **decimal** fractions (often just called 'decimals')

  like 0.125.

Fractions can be:

proper such as $\frac{2}{3}$, where the numerator is less than the denominator

improper such as $\frac{4}{3}$ where the numerator is greater than the denominator

mixed numbers such as $6\frac{7}{8}$.

Fractions where the numerator and denominator have no common factor are in their **lowest terms**.

$\frac{1}{3}$ and $\frac{4}{12}$ are **equivalent** fractions.

0.675 is a **terminating** decimal.

0.32... or $0.\overline{32}$ or $0.\dot{3}\dot{2}$ are different ways of writing the **recurring** decimal 0.3232 323 232...

Non-terminating, non-recurring decimals are **irrational** numbers, like π or $\sqrt{2}$.

Using a GDC, you can either enter a fraction using the fraction template $\frac{\square}{\square}$ or by using the divide key ÷. Take care – you will sometimes need to use brackets.

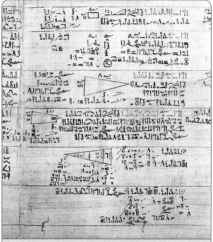

The Rhind Papyrus from ancient Egypt in around 1600 BCE shows calculations using fractions. Egyptians used **unit** fractions so for $\frac{4}{5}$ they would write $\frac{1}{2} + \frac{1}{4} + \frac{1}{20}$. This is not generally regarded as a very helpful way of writing fractions.

$\pi \approx 3.14159265358979323846264$
$3383279502884197169399375...$
$\sqrt{2} \approx 1.4142135623730950488016$
$8872420969807856967187537...$
They do not terminate and there are no repeating patterns in the digits.

Example 10

a Evaluate

$$\frac{1}{2} + \frac{3}{8} \times \frac{4}{9}$$

$$= \frac{1}{2} + \frac{1}{6} = \frac{4}{6}$$

$$= \frac{2}{3}$$

× *before +*

simplify

b Evaluate

$$\frac{\dfrac{1}{2} + \dfrac{1}{3}}{\dfrac{1}{2} \times \dfrac{1}{3}}$$

$$= \frac{\dfrac{5}{6}}{\dfrac{1}{6}} = 5$$

evaluate numerator and denominator first

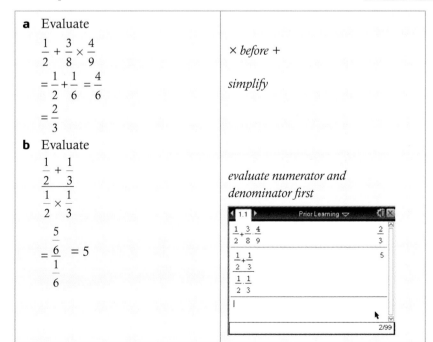

GDC help on CD:
Alternative demonstrations for the TI-84 Plus and Casio fx-9860GII GDCs are on the CD.

Example 11

a Convert $\dfrac{7}{16}$ to a decimal.   **b** Write $3\dfrac{7}{8}$ as an improper fraction.

Answers

a $\dfrac{7}{16} = 0.4375$

b $3\dfrac{7}{8} = \dfrac{24}{8} + \dfrac{7}{8}$

$\quad = \dfrac{31}{8}$

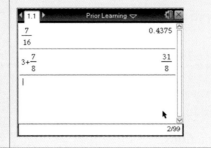

GDC help on CD:
Alternative demonstrations for the TI-84 Plus and Casio fx-9860GII GDCs are on the CD.

Exercise 1D

1 Calculate

a $\dfrac{1}{2} + \dfrac{3}{4} \times \dfrac{5}{9}$ **b** $\dfrac{2}{3} \div \dfrac{5}{6} \times 1\dfrac{1}{3}$

c $\sqrt{\left(\dfrac{3}{5}\right)^2 + \left(\dfrac{4}{5}\right)^2}$ **d** $\dfrac{1 - \left(\dfrac{2}{3}\right)^5}{1 - \dfrac{2}{3}}$

2 Write the following fractions in their lowest terms.

a $\dfrac{16}{36}$ **b** $\dfrac{35}{100}$ **c** $\dfrac{34}{51}$ **d** $\dfrac{125}{200}$

There are some useful tools for working with fractions. Look in [menu] 2:Number.

3 Write these mixed numbers as improper fractions.

a $3\dfrac{3}{5}$ **b** $3\dfrac{1}{7}$ **c** $23\dfrac{1}{4}$ **d** $2\dfrac{23}{72}$

4 Write these improper fractions as mixed numbers.

a $\dfrac{32}{7}$ **b** $\dfrac{100}{3}$ **c** $\dfrac{17}{4}$ **d** $\dfrac{162}{11}$

5 Convert to decimals.

a $\dfrac{8}{25}$ **b** $\dfrac{5}{7}$ **c** $3\dfrac{4}{5}$ **d** $\dfrac{45}{17}$

To convert a fraction to a decimal, divide the numerator by the denominator. Pressing [ctrl] $\approx$ will give the result as a decimal instead of a fraction.

1.5 Percentages

A percentage is a way of expressing a fraction or a ratio as part of a hundred.

For example 25% means 25 parts out of 100.

As a fraction, $25\% = \dfrac{25}{100} = \dfrac{1}{4}$.

As a decimal, $25\% = 0.25$.

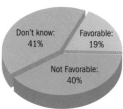

Don't know: 41%   Favorable: 19%   Not Favorable: 40%

Example 12

Lara's mark in her maths test was 25 out of 40. What was her mark as a percentage?

Answer	
$\dfrac{25}{40} \times 100 = 62.5\%$	*Write the mark as a fraction.* *Multiply by 100.* *Use your GDC.*

Example 13

There are 80 students taking the IB in a school. 15% take Maths Standard level. How many students is this?

Answer	
Method 1 $\dfrac{15}{100} \times 80 = 12$	*Write the percentage as a fraction out of a hundred and then multiply by 80.*
Method 2 $15\% = 0.15$ $0.15 \times 80 = 12$	*Write the percentage as a decimal.* *Multiply by 80.*

Exercise 1E

1 Write as percentages

  a 13 students from a class of 25

  b 14 marks out of 20

2 Find the value of

  a 7% of 32 CHF

  b $4\frac{1}{2}\%$ of 12.00 GBP

  c 25% of 750.28 EUR

  d 130% of 8000 JPY

> 7% = 0.07

Percentage increase and decrease

Consider an increase of 35%.

The new value after the increase will be 135% of the original value.

So to increase an amount by 35%, find 135% of the amount.

Multiply by $\dfrac{135}{100}$ or 1.35.

Now consider a decrease of 15%.

After a 15% decrease, the new value will be 85% of the original. So to decrease an amount by 15% find 85%. Multiply by $\dfrac{85}{100}$ or by 0.85.

Example 14

a The manager of a shop increases the prices of CDs by 12%.
A CD originally cost 11.60 CHF.
What will it cost after the increase?

b The cost of a plane ticket is decreased by 8%.
The original price was 880 GBP. What is the new price?

c The rent for an apartment has increased from 2700 EUR to
3645 EUR per month.
What is the percentage increase?

Answers

a $11.60 \times 1.12 = 12.99$ CHF (to the nearest 0.01 CHF)	*After a 12% increase, the amount will be 112% of its original value.*
b $880 \times 0.92 = 809.60$ GBP	*After an 8% decrease, the amount will be 92% of its original value.*

c **Method 1**

The increase is $3645 - 2700$
$= 945$ EUR

The percentage increase
is $\dfrac{945}{2700} \times 100 = 35\%$

Method 2

$\dfrac{3645}{2700} = 1.35 = 135\%$

Percentage increase is 35%.

Find the increase.
Work out the increase as a percentage of the original amount.
Percentage increase =

$\dfrac{actual\ increase}{original\ amount} \times 100\%$

Calculate the new price as a percentage of the old price.

Example 15

In a shop, an item's price is given as 44 AUD, *including* tax.
The tax rate is 10%.
What was the price without the tax?

Answer

Call the original price x.
After tax has been added, the price
will be $1.10x$.
Hence $1.10x = 44$
$\qquad x = 44 \div 1.10$
$\qquad\quad = 40$
The price without tax is 40 AUD.

110% = 1.10
Solve for x.
Divide both sides by 1.10.

Exercise 1F

1 In the UK, prices of some goods include a government tax
called VAT, which is at 20%.
A TV costs 480 GBP before VAT. How much will it cost
including VAT?

2 In a sale in a shop in Tokyo, a dress that was priced at 17 000 JPY is reduced by 12.5%. What is the sale price?

3 The cost of a weekly train ticket goes up from 120 GBP to 128.40 GBP. What is the percentage increase?

4 Between 2004 and 2005, oil production in Australia fell from 731 000 to 537 500 barrels per day. What was the percentage decrease in the production?

5 Between 2005 and 2009 the population of Venezuela increased by 7%. The population was 28 400 000 in 2009. What was it in 2005 (to the nearest 100 000)?

6 An item appears in a sale marked as 15% off with a price tag of 27.20 USD. What was the original price before discount?

7 The rate of GST (goods and service tax) that is charged on items sold in shops was increased from 17% to 20%. What would the price increase be on an item that costs 20 GBP before tax?

8 A waiter mistakenly adds a 10% service tax onto the cost of a meal which was 50.00 AUD. He then reduces the price by 10%. Is the price now the same as it started? If not, what was the percentage change from the original price?

1.6 Ratio and proportion

The **ratio** of two numbers r and s is $r : s$. It is equivalent to the fraction $\frac{r}{s}$. Like the fraction, it can be written in its lowest terms.

> When you write a ratio in its lowest terms, both numbers in the ratio should be positive whole numbers.

For example, 6 : 12 is equivalent to 1 : 2 (dividing both numbers in the ratio by 6).

In a **unitary ratio**, one of the terms is 1.

For example 1 : 4.5 or 25 : 1.

> When you write a unitary ratio, you can use decimals.

If two quantities a and b are in **proportion**, then the ratio $a : b$ is fixed.

We also write $a \propto b$ (a is proportional to b).

Example 16

200 tickets were sold for a school dance. 75 were bought by boys and the rest by girls. Write down the ratio of boys to girls at the dance, in its lowest terms.

Answer
The number of girls is $200 - 75 = 125$
The ratio of boys to girls is $75 : 125 = 3 : 5$

> Always give the ratio in its lowest terms.

Map scales are often written as a ratio. A scale of $1:50\,000$ means that 1 cm on the map represents $50\,000$ cm = 0.5 km on the earth.

Example 17

An old English map was made to the scale of 1 inch to a mile.
Write this scale as a ratio.

Answer

1 mile = 1760 × 3 × 12 = 63 360 inches The ratio of the map is 1 : 63 360	*Always make sure that the units in ratios match each other.*

12 inches = 1 foot
3 feet = 1 yard
1760 yards = 1 mile

Example 18

Three children, aged 8, 12 and 15 win a prize of 140 USD.
They decide to share the prize money in the ratio of their ages.
How much does each receive?

Answer

140 USD is divided in the ratio 8 : 12 : 15. This is a total of 8 + 12 + 15 = 35 parts. 140 ÷ 35 = 4 USD 8 × 4 = 32, 12 × 4 = 48 and 15 × 4 = 60 The children receive 32 USD, 48 USD and 60 USD.	*Divide the money into 35 parts.* *One part is 4 USD.*

Exercise 1G

1 Aspect ratio is the ratio of an image's width to its height.
A photograph is 17.5 cm wide by 14 cm high.
What is its aspect ratio, in its lowest terms?

2 Gender ratio is expressed as the ratio of men to women in the form $n : 100$. Based on the figures for 2008, the gender ratio of the world was 102 : 100. In Japan, there were 62 million men and 65.2 million women in 2008. What was the gender ratio in Japan?

3 Ryoka was absent for a total of 21 days during a school year of 32 weeks. What is the ratio of the number of days that she was absent to the number of possible days she could have spent at the school during the year in its simplest terms? (A school week is 5 days.)

Leonardo da Vinci drew this famous drawing of Vitruvian Man around 1487. The drawing is based on ideal human proportions described by the ancient Roman architect Vitruvius.

4 A model airplane has a wingspan of 15.6 cm. The model is built to a scale of 1 : 72. What was the wingspan of a full-sized airplane (in metres)?

5 On a map, a road measures 1.5 cm. The actual road is 3 km long. What is the scale of the map and how long would a footpath that is 800 m long be on the map?

6 A joint collection is made for two charities and it is agreed that the proceeds should be split in the ratio 5 : 3 between an animal charity and one for sick children. 72 USD is collected. How much is donated to the two charities?

7 For a bake sale, a group of students decide to make brownies, chocolate chip cookies and flapjacks in the ratio 5 : 3 : 2. They plan to make 150 items all together. How many of each will they need to make?

1.7 The unitary method

In the unitary method, you begin by finding the value of **one** part or item.

Example 19

A wheelbarrow full of concrete is made by mixing together 6 spades of gravel, 4 spades of sand, 2 spades of cement and water as required. When there are only 3 spades of sand left, what quantities of the other ingredients will be required to make concrete?

Since the value you want to change is the sand, make sand equal to 1 by dividing through by 4. Then multiply through by 3 to make the quantity of sand equal to 3.

Answer

The ratio   gravel : sand : cement

is $6 : 4 : 2$

or $\dfrac{6}{4} : \dfrac{4}{4} : \dfrac{2}{4}$

$= \dfrac{3}{2} : 1 : \dfrac{1}{2} = \dfrac{9}{2} : 3 : \dfrac{3}{2}$

Hence the mixture requires $4\dfrac{1}{2}$ spades of gravel to 3 spades of sand to $1\dfrac{1}{2}$ spades of cement.

Exercise 1H

1 Josh, Jarrod and Se Jung invested 5000 USD, 7000 USD and 4000 USD to start up a company. In the first year, they make a profit of 24 000 USD which they share in the ratio of the money they invested. How much do they each receive?

2 Amy is taking a maths test. She notices that there are three questions worth 12, 18 and 20 marks. The test lasts one hour and fifteen minutes. She decides to allocate the time she spends on each question in the ratio of the marks. How long does she spend on each question?

1.8 Number systems

Throughout this course, you will be working with **real numbers**. There are two types of real numbers – rational numbers and irrational numbers.

→ **Rational numbers** are numbers that can be written in the form $\frac{a}{b}$, where a and b are both integers, and $b \neq 0$.

$\frac{2}{5}$, $-\frac{17}{8}$, 0.41, $1.\dot{3}$, and 9 are rational numbers.

$\frac{2}{5}$ and $-\frac{17}{8}$ are written in the form $\frac{a}{b}$.

0.41 can be written in the form $\frac{a}{b}$, because $0.41 = \frac{41}{100}$.

$1.\dot{3}$ can be written in the form $\frac{a}{b}$, because $1.\dot{3} = \frac{4}{3}$.

9 can be written in the form $\frac{a}{b}$, because $9 = \frac{9}{1}$.

> Repeating or terminating decimals can be written as fractions, so they are rational numbers.

Within the set of rational numbers are sets of numbers called **natural numbers** {0, 1, 2, 3, ...} and **integers** {–4, –3, –1, 0, 1, 2, 3, ...}.

$\mathbb{R}$ represents the set of real numbers, $\mathbb{Q}$ rational numbers, $\mathbb{N}$ natural numbers, and $\mathbb{Z}$ integers.

→ **Irrational numbers** are real numbers that can be written as decimals that never terminate or repeat.

$\sqrt{3}$, π, e, and $\sqrt{117}$ are irrational numbers.

$\sqrt{3} = 1.7320508...$ $\pi = 3.14159265...$

$e = 2.7182818...$ $\sqrt{117} = 10.8166538...$

Example 20

Classify each of these real numbers as rational or irrational.

$0.75, -2, \sqrt{37}, \sqrt{25}, 0, \dfrac{2\pi}{3}$

Answer

0.75 is a rational number	0.75 can be written in the form $\dfrac{3}{4}$, and
–2 is a rational number	-2 can be written as $-\dfrac{2}{1}$.
$\sqrt{37}$ is an irrational number	$\sqrt{37} = 6.08276...$ This decimal does not repeat or terminate.
$\sqrt{25}$ is a rational number	$\sqrt{25}$ is a rational number, since it is equal to 5.
0 is a rational number	
$\dfrac{2\pi}{3}$ is an irrational number	Even though it is written in fractional form, $\dfrac{2\pi}{3}$ is not a rational number. Multiples of π are irrational.

Example 21

Write the rational number $0.8\dot{3}$ in the form $\dfrac{a}{b}$.

Answer

Let $x = 0.8\dot{3}$.

$100x = 83.\dot{3}$, and $10x = 8.\dot{3}$.

$100x - 10x = 83.\dot{3} - 8.\dot{3}$

$90x = 75$

$x = \dfrac{75}{90} = \dfrac{5}{6}$

Multiply by powers of 10 to change the position of the decimal point.

Subtracting these values cancels out the repeating 3s.

Exercise 1l

1 Classify each of these real numbers as rational or irrational.

  a 83   **b** $\dfrac{4}{9}$ **c** $\dfrac{2\pi}{3}$ **d** -0.96

  e $-0.4\dot{5}$ **f** e^5 **g** $-4\sqrt{81}$ **h** $\dfrac{\sqrt{5}}{7}$

  i $1.24\dot{7}$ **j** $\sqrt{18}$

2 Which of the numbers from question **1** are:
  a integers
  b natural numbers?

3 Write each rational number from question **1** in the form $\dfrac{a}{b}$, where a and b are integers, and $b \neq 0$.

Properties of real numbers

Real number arithmetic uses three important properties.

Commutative property

> → When adding or multiplying two or more numbers, the order does not matter.

For example:

- $a + b = b + a$
- $15 + 7 = 7 + 15$
- $xy = yx$
- $3(8) = 8(3)$

These properties may seem like common sense, but you should think about when you can or can't use them.

Addition and multiplication are commutative. Subtraction and division are not.

Associative property

> → When adding or multiplying three or more numbers, you can group the numbers in different ways for the calculation without changing their order.

For example:

- $a+b+c=(a+b)+c=a+(b+c)$
- $5+9+16=(5+9)+16=5+(9+16)$
- $xyz=(xy)z=x(yz)$
- $6 \times 4 \times 10=(6 \times 4) \times 10=6 \times (4 \times 10)$

The commutative and associative properties do not work for subtraction.
- $20 - 7 \neq 7 - 20$
- $(18 - 9) - 3 \neq 18 - (9 - 3)$

Use BIDMAS – calculate the value in the brackets first.

Distributive property

> → $a(b+c)=ab+ac$ and $a(b-c)=ab-ac$.

1.9 Rounding and estimation

To round to a given number of **decimal places**:

- Look at the figure in the next decimal place.
- If this figure is less than 5, round down.
- If this figure is 5 or more, round up.

To round to a given number of **significant figures**:

- For any number, read from left to right and ignore the decimal point.
- The first significant figure is the first non-zero digit, the second significant figure is the next digit (which can be zero or otherwise), and so on.

We use this when expanding brackets in algebra or simplifying multiplication.
For example $5 \times 32 = (5 \times 30) + (5 \times 2)$

An exam question might tell you to give your answer to two decimal places, for example.

3	5	.	2	7	1		0	.	5	3	9
1st	2nd		3rd	4th	5th				1st	2nd	3rd
sf	sf		sf	sf	sf				sf	sf	sf

Example 22

Write the number 8.0426579 to

a 2 decimal places   **b** 1 significant figure   **c** 1 decimal place
d 4 decimal places   **e** 6 significant figures

Answers

a	8.04	*8.042 next digit less than 5 so round down*
b	8	*8.0 next digit less than 5 so round down*
c	8.0	*8.04 next digit less than 5 so round down*
d	8.0427	*8.04265 next digit 5 so round up*
e	8.04266	*8.042657 next digit greater than 5 so round up*

> When a question asks for a number of decimal places, write them down even if some of the values are zero.

Example 23

Round 42536 to 3 significant figures.

Answer

42500	*42536 next digit (3) less than 5 so round down.* *Replace any other digits before the decimal place with zeros.*

Estimation

To **estimate** the value of a calculation, write all the numbers to one significant figure.

For example, to estimate the value of $197.2 \div 3.97$, calculate $200 \div 4 = 50$

> Estimating the answer to a calculation first gives you an idea of the answer to expect. If your GDC gives you a very different answer, you can then check if you keyed in the values correctly.

Exercise 1J

1 Write each number to the nearest number given in the bracket.

   a 2177 (ten)   **b** 439 (hundred)   **c** 3532 (thousand)
   d 20.73 (unit)   **e** 12.58 (unit)

2 Write down each number correct to the number of decimal places given in the bracket.

   a 0.6942 (2)   **b** 28.75 (1)   **c** 0.9999(2)
   d 77.984561 (3)   **e** 0.05876 (2)

3 Write down each number in question 1 correct to 2 significant figures.

4 Write down each number in question 2 correct to 3 significant figures.

5 Write each fraction as a decimal to 3 sf.

a $\dfrac{2}{3}$ **b** $\dfrac{3}{46}$ **c** $\dfrac{5}{13}$

> Use your GDC to convert each fraction to a decimal.

6 Write down an estimate for the value of the following calculations

a $54.04 \div 9.89$ **b** $\dfrac{2.8 \times 3.79}{1.84}$ **c** $\dfrac{7.08 - 0.7556}{(8.67)^2}$

7 Use your GDC to evaluate each part of question 6 to 3 significant figures.

1.10 Standard form

Very large and very small numbers can be written in standard form as

$A \times 10^n$ where n is an integer and $1 \le A < 10$

- First write the number with the place values adjusted so that it is between 1 and 10.
- Then work out the value of the index, n, the number of columns the digits have moved.

> For example, 37300 is 3.73×10^4 in standard form.

Example 24

Write **a** 89 445   **b** 0.000 000 065 in standard form

Answers

a $89\,445 = 8.9445 \times 10^4$

> *Write $89\,445$ as 8.9445×10^n*
> *The digits have moved 4 places to the right so $n = 4$.*

b $0.000\,000\,065 = 6.5 \times 10^{-8}$

> *Write 6.5×10^n*
> *The digits have moved 8 places to the left, so $n = -8$*

Exercise 1K

1 Write these in standard form

     a 1475   **b** 231000

     c 2.8 billion   **d** 0.35×10^6

     e 73.5×10^5

> Here 1 billion = 1 thousand million. The UK and some other countries used to use 1 billion = 1 million million, but the 1 thousand million definition is becoming standard.

2 Write these as ordinary numbers
  a 6.25×10^4 **b** 4.2×10^8
  c 3.554×10^2

3 Write these in standard form
  a 0.0001232   **b** 0.00004515
  c 0.617   **d** 0.75×10^{-5}
  e 34.9×10^{-5}

4 Write these as ordinary numbers
  a 3.5×10^{-7} **b** 8.9×10^{-8}
  c 1.253×10^{-2}

5 Light travels about 3×10^5 metres per second. Find the time it takes to travel 1 metre. Give your answer in standard form.

1.11 Sets

A **set** is a group of items. We generally use a capital letter to name a set, and the brackets { } to enclose the items of the set.
For example, if P is the set of all the prime numbers less than 20, then P = {2, 3, 5, 7, 11, 13, 17, 19}.

> The curly brackets used for sets are sometimes called 'braces'.

Each item in the set is called an **element** of the set.

- The symbol $\in$ means 'is an element of'.
  For example, $3 \in P$ means '3 is an element of the set P'.
- The symbol $\notin$ means 'is not an element of'.
  For example, $8 \notin P$ means '8 is not an element of the set P'.

We use a lower-case n for the number of items in a set.
The set P has 8 elements, so $n(P) = 8$.

If the number of items in a set is zero, then that set is an **empty set**, or *null set*. We represent the empty set with empty brackets, { }, or with the symbol $\varnothing$.

A set which contains all relevant items is called the **universal set**, and is represented by the letter U. In some cases, the universal set can be assumed. For example, a common universal set is 'all real numbers'.

> The universal set can also be thought of as the *reference set*.

Set builder notation

To fully specify a set, you can use this notation:
$$A = \{x \mid x \in \mathbb{Z},\ 10 < x < 15\}$$

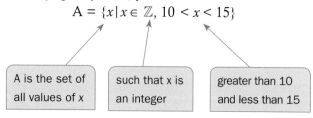

| A is the set of all values of x | such that x is an integer | greater than 10 and less than 15 |

The elements of this set are $A = \{11, 12, 13, 14\}$.

Example 25

Write the elements of each set, and give the number of items in each set.
a B, the set of all multiples of 5 that are less than 30.
b $T = \{x \mid x \in \mathbb{N},\ x \geq 7\}$

Answers

a $B = \{5, 10, 15, 20, 25\}$
$n(B) = 5$

b $T = \{7, 8, 9, 10, 11, ...\}$

*This set is **infinite**, which means it goes on forever.*
We cannot count the number of items in the set.

You can use ellipses (3 dots ...) to show that a series continues.

Exercise 1L

1 List the elements in each set.
  a A, the set of all the factors of 72.
  b B, the set of all the prime factors of 72.
  c C, the set of all even prime numbers.
  d D, the set of all the even multiples of 7.
  e $E = \{x \mid x \in \mathbb{Z},\ |x| < 4\}$
  f $F = \{x \mid x \in \mathbb{N},\ x \geq 20\}$
  g G, the set of all prime numbers that are multiples of 4.

2 State the number of items in each of the sets from question 1.

Subsets, intersections, and unions

→ We say that a set B is a **subset** of set A if all the elements of set B are also elements of set A.

Let $A = \{1, 2, 3, 4, 5, 6\}$, and $B = \{2, 3, 4\}$.
Since B is a subset of A, we write $B \subseteq A$.

The symbol $\subseteq$ means 'is a subset of'.

There are many other subsets of A, such as $\{1, 3, 5, 6\}$, $\{2, 5\}$, $\{4\}$, and even the empty set $\{\}$, as well as the set $\{1, 2, 3, 4, 5, 6\}$ itself.

→ A set C is called a **proper subset** of set A if C is a subset of A, but has fewer elements than A.

$\mathbb{Z} \subset \mathbb{R}$, since all integers belong to the set of real numbers.

For example, C = {2, 5} is a proper subset of the set A = {1, 2, 3, 4, 5, 6}. We write this as $C \subset A$.

The symbol $\subset$ means 'is a subset of'.

→ Two sets that share elements in common have an **intersection**. We use the symbol $\cap$ to represent the intersection of two sets.

For example, let D = {2, 4, 6, 8, 10} and E = {1, 2, 3, 4, 5}. Both sets contain the elements 2 and 4, so $D \cap E = \{2, 4\}$

They only have 2 and 4 in common.

→ The **union** of two sets is the set of all the elements of both sets.
We use the symbol $\cup$ to represent the union of two sets.

For example, if D = {2, 4, 6, 8, 10} and E = {1, 2, 3, 4, 5}, the union of these sets is $D \cup E = \{1, 2, 3, 4, 5, 6, 8, 10\}$.

Numbers which appear in both sets should only be listed once.

Example 26

Let A = {the odd natural numbers less than 16} and
B = {$x \mid x$ is a factor of 15}.
a List the elements of each set.
b Is B a subset of A? Explain.
c Give the intersection and the union of sets A and B.

Answers

a A = {1, 3, 5, 7, 9, 11, 13, 15} B = {1, 3, 5, 15}	*You could write* $B \subseteq A$
b Yes, B is a subset of A. All the elements of B are elements of A. You could write $B \subseteq A$.	*B is also a proper subset of A.* *You could write* $B \subset A$.
c $A \cap B = \{1, 3, 5, 15\}$	*These numbers are elements of both sets.*
$A \cup B = \{1, 3, 5, 7, 9, 11, 13, 15\}$	*This set includes all the elements of A and all the elements of B, once only.*

There are two types of sets which have no intersections.

→ **Disjoint sets** contain no elements in common.

For example, if A = {2, 4, 6, 8}, and B = {1, 3, 5, 7}, A and B are disjoint sets. We write $A \cap B = \{\}$, or $A \cap B = \varnothing$.

→ Sets are **complements** if they have no elements in common, and they contain *all* the elements of U between them.

For example, let U = {all positive integers}, and
A = { 2, 4, 6, 8, 10, ...}.
The complement of A is the set {1, 3, 5, 7, 9, ...}.
We write A′ = {1, 3, 5, 7, 9, ...}. Together, sets A and A′ contain all the positive integers, but they have no elements in common.

> The complement of a set A is written A′, called 'A prime'.

Example 27

Let U = {multiples of 5} and M = {10, 20, 30, ...}.
What is the complement of M?

Answer

M′ = { 5, 15, 25, ...}	Since M contains all the even multiples of 5, M′ must contain all the odd multiples of 5. Together, $M \cup M' = U$.

Exercise 1M

1 Let A = {1, 2, 3, 4, 5, 6}, and let B = {4, 5}.
  a Is B a subset of A? Explain.
  b Are the sets A and B disjoint? Explain.
  c List the intersection of sets A and B.
  d List the union of sets A and B.

2 Let A = {$x | x$ is a factor of 36} and B = {$x | x$ is a factor of 15}.
  a List the elements of each set.
  b Is B a subset of A? Explain.
  c Are the sets A and B disjoint? Explain.
  d List the intersection of sets A and B.
  e List the union of sets A and B.

3 Let A = {$x | x \in \mathbb{Z}, x > 16$} and B = {$x | x$ is a multiple of 20}.
  a List the elements of each set.
  b Is B a subset of A? Explain.
  c Are the sets A and B disjoint? Explain.
  d List the intersection of sets A and B.
  e List the union of sets A and B.

4 Let U = {positive integers} and D = { $x | x$ is a multiple of 3}.
List the elements of the complement of D.

5 Let U = {multiples of 10}, and let B = {10, 20, 30}. List the elements of B′

6 Give two sets A and B such that

  a $A \cap B = \{\}$

  b $A \cap B = \{4, 7, 10\}$

  c $A \cup B = \{1, 2, 3, 4, 5\}$

  d $n(A \cap B) = 2$

  e $n(A \cup B) = 8$

  f $n(A \cup B) = 7$ and $n(A \cap B) = 3$

  g $B \subseteq A$, and $n(A \cap B) = 3$

Sets related to number lines and inequalities

Subsets of the set of real numbers can be represented as intervals on a **real number line**. These intervals can also be expressed using set notation and inequalities.

Example 28

Write each interval using set notation and inequalities.

a

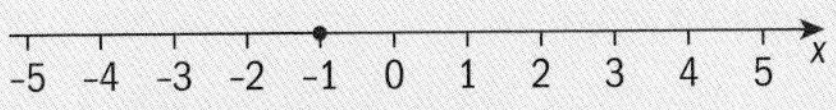

b

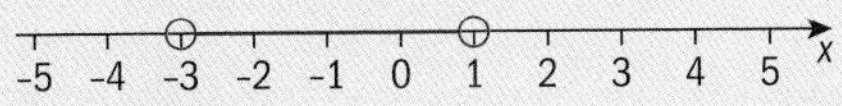

Answers

a $\{x \mid x \in \mathbb{R}, x \geq -1\}$

The numbers greater than −1 are shaded on the number line. The solid circle at −1 tells us that −1 is included.

b $\{x \mid x \in \mathbb{R}, -3 < x < 1\}$

The numbers between −3 and 1 are shaded on the number line. The open circles at −3 and 1 tell us that −3 and 1 are not included.

Example 29

Shade the number line to indicate the interval of real numbers given by the set.

a $\{x \mid x \in \mathbb{R}, x < 2\}$ **b** $\{x \mid x \in \mathbb{R}, 0 < x \leq 4\}$

Answers

a $\{x \mid x \in \mathbb{R}, x < 2\}$

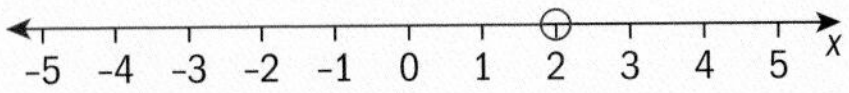

2 is not included, so use an open circle at 2.

b $\{x \mid x \in \mathbb{R}, 0 < x \leq 4\}$

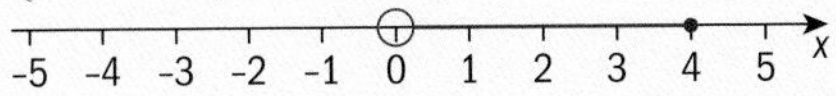

Draw the line between 0 and 4.
0 is not included, so use an open circle.
4 is included, so use a filled circle.

Exercises 1N

1 Write each interval using set notation and inequalities.

a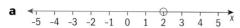

b
number line from -5 to 5, closed point at -2, open circle at 5

c
number line from -5 to 5, closed point at 2

d
number line from -5 to 5, closed points at -4 and 2

2 Shade the number line to indicate the interval of real numbers given by the set.

a $\{x \mid x \in \mathbb{R}, x \leq 0\}$ **b** $\{x \mid x \in \mathbb{R}, -3 \leq x < 2\}$

c $\{x \mid x \in \mathbb{R}, x > -1\}$ **d** $\{x \mid x \in \mathbb{R}, -5 < x < 1\}$

Mappings

You can show the mathematical **relations** between two sets in several different ways.

Example 30

Each member of $\{x \mid x \in \mathbb{R}, -5 < x < 1\}$ is mapped to its square. Express this relation as:

a a mapping diagram   **b** a table

c a set of ordered pairs   **d** a graph.

Answers

a   Input   Output

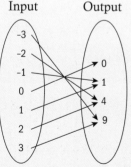

Write the integers −3, −2, −1, 0, 1, 2 and 3 in the input set. Write the squares of the input values, 0, 1, 4 and 9, in the output set.

Draw arrows to map each input value to an output value.

b

x	−3	−2	−1	0	1	2	3
y	9	4	1	0	1	4	9

Use the variable x for the input values and the variable y for the output values.

c   {(−3, 9), (−2, 4), (−1, 1), (0, 0), (1, 1), (2, 4), (3, 9)}

Write each input value as the first member of an ordered pair and its corresponding output value as the second member.

▶ Continued on next page

d

Represent each input value on the horizontal axis and each output value is represented on the vertical axis.

Exercise 10

Express each relation as:
  a a mapping diagram
  b a table
  c a set of ordered pairs
  d a graph.

1 Each member of $\{x \mid x \in \mathbb{N}, x \leq 5\}$ is mapped to 2 more than the number.

2 Each member of $\{x \mid x \in \mathbb{Z}, -4 < x < 4\}$ is mapped to the absolute value of the number.

2 Algebra

The word **algebra** comes from the title of a book *Hisab al-jabr w'al-muqabala* written by Abu Ja'far Muhammad ibn Musa Al-Khwarizmi in Baghdad around 800 CE. It is regarded as the first book to be written about algebra.

2.1 Expanding brackets and factorization

The **distributive law** is used to expand brackets and factorize expressions.

$$a(b + c) = ab + ac$$

Two other laws used in algebra are the **commutative law** $ab = ba$ and the **associative law** $(ab)c = a(bc)$.

Example 31

Expand $2y(3x + 5y - z)$

Answer

$$2y(3x + 5y - z) = 2y \cdot 3x + 2y \cdot 5y + 2y(-z)$$
$$= 6xy + 10y^2 - 2yz$$

Example 32

Factorise $6x^2y - 9xy + 12xz^2$

Answer

$6x^2y - 9xy + 12xz^2 = 3x(2xy - 3y + 4z^2)$

> Look for a common factor. Write this outside the bracket. Find the terms inside the bracket by dividing each term by the common factor.

Exercise 2A

1 Expand

   a $3x(x-2)$ **b** $\dfrac{x}{y}(x^2y - y^2 + x)$ **c** $a(b-2c) + b(2a+b)$

2 Factorize

   a $3pq - 6p^2q^3r$ **b** $12ac^2 + 15bc - 3c^2$ **c** $2a^2bc + 3ab^2c - 5abc^2$

Products resulting in quadratic expressions

The product of two **binomials**, such as $x + a$ and $x + b$, results in a **quadratic expression**.

$(x + a)(x + b) \equiv (x + a)x + (x + a)b \equiv x^2 + ax + bx + ab \equiv x^2 + (a + b)x + ab$

Here is a shorter method to find the product of two binomials.

= **F**irst terms + **O**uter terms + **I**nner terms + **L**ast terms

= x^2 + $\qquad$ bx + $\qquad$ ax + $\qquad$ ab

= $x^2 + (a + b)x + ab$

Example 33

Find each product.
a $(x + 2)(x + 5)$ **b** $(x + 6)(x - 4)$
c $(2x - 3)(3x + 1)$

Answers

a $(x + 2)(x + 5) = x^2 + 5x + 2x + 10$
$\qquad\qquad\qquad = x^2 + 7x + 10$

b $(x + 6)(x - 4) = x^2 + 4x + 6x - 24$
$\qquad\qquad\qquad = x^2 + 2x - 24$

▶ Continued on next page

c $(2x-3)(3x+1) = 6x^2 + 2x - 9x - 3$
$$= 10x^2 - 7x - 3$$

Exercise 2B

Find each product and simplify your answer.

1 $(x+7)(x-4)$ **2** $(x-3)(x-2)$ **3** $(3x-4)(x+2)$

4 $(2x-5)(3x+2)$ **5** $(3x+2)(3x+1)$

→ Consider the following special products.

$$(x+a)^2 = (x+a)(x+a) = x^2 + ax + ax + a^2 = x^2 + 2ax + a^2$$
$$(x-a)^2 = (x-a)(x-a) = x^2 - ax - ax + a^2 = x^2 - 2ax + a^2$$
$$(x+a)(x-a) = x^2 - ax + ax + a^2 = x^2 - a^2$$

> The first two are called squares of binomials. The last one is called the **difference of two squares**.

Example 34

Find each product.
a $(x+4)^2$ **b** $(3x-2)^2$
c $(2x+3)(2x-3)$

Answers

a $(x+4)^2 = x^2 + 8x + 16$	*Square the first term: $(x)^2 = x^2$* *Double the product of the two terms:* $2(4x) = 8x$ *Square the last term: $(4x)^2 = 16$*
b $(3x-2)^2 = 9x^2 - 12x + 4$	*Square the first term: $(3x)^2 = 9x^2$* *Double the product of the two terms:* $2(-6x) = -12x$ *Square the last term: $(-2)^2 = 4$*
c $(2x+3)(2x-3) = 4x^2 - 9$	*Square the first term: $(2x)^2 = 4x^2$* *Square the last term: $(-3x) = 9$* *Write the difference of the squares:* $4x^2 - 9$

Exercise 2C

Find each product and simplify your answer.

1 $(x+5)^2$ **2** $(x-4)^2$ **3** $(x+2)(x-2)$

4 $(3x-4)^2$ **5** $(2x+5)^2$ **6** $(2x+7)(2x-7)$

Factorizing quadratic expressions

The reverse is also possible – to express a quadratic expression as the product of two linear expressions.

$(x + 2)(x + 5) = x^2 + 7x + 10$

$(x + 6)(x - 4) = x^2 + 2x - 24$

To factorize quadratics of the form $x^2 + bx + c$, where the coefficient of x^2 is 1, look for pairs of factors of c whose sum is b.

> 10 is the product of 2 and 5 **and** 7 is the sum of 2 and 5

> –24 is the product of 6 and –4, 2 is the sum of 6 and –4

Example 35

Factorize

a $x^2 - 15x + 14$

b $x^2 + 5x + 6$

c $x^2 - 5x - 24$

Answers

a $x^2 - 15x + 14 = (x - 1)(x - 14)$

Factors of 14	Sum of factors
1 and 14	15
–1 and –14	–15 ←
2 and 7	14
–2 and –7	–14

b $x^2 + 5x + 6 = (x + 2)(x + 3)$

Factors of 6	Sum of factors
1 and 6	7
–1 and –6	–7
2 and 3	5 ←
–2 and –3	–5

c $x^2 + 5x - 24 = (x + 3)(x - 8)$

Factors of –24	Sum of factors
1 and –24	–23
–1 and 24	23
2 and –12	–10
–2 and 12	10
3 and –8	–5 ←
–3 and 8	5
4 and –6	–2
–4 and 6	2

Factorizing quadratics of the form $ax^2 + bx + c$, where $a \neq 0$

Use trial and error to find the correct pair of factors. Try factors that give the correct product for the first and last terms, until you find the one that gives the correct product for the middle term.

Example 36

Factorize
a $2x^2 + 5x + 3$
b $6x^2 + x - 15$

Answers

a $2x^2 + 5x + 3 = (2x + 3)(x + 1)$

Factors of $2x^2$: $2x$, x
Factors of 3:1, 3; –1, –3

Possible factors	*Linear term*
(2x + 1)(x + 3)	*6x + 1x = 7x*
(2x – 1)(x – 3)	*–6x – 1x = –7x*
(2x + 3)(x + 1)	*2x + 3x = 5x* ←
(2x – 3)(x – 1)	*–2x² – 3x = –5x*

b $6x^2 + x - 15 = (2x - 3)(3x + 5)$

Factors of $6x^2$: $6x$, x; $2x$, $3x$
Factors of –15 : 1, –15; –1, 15; 3, –5; –3, 5

Possible factors	*Linear term*
(6x + 1)(x – 15)	*–90x + 1x = –89x*
(6x – 1)(x + 15)	*90x – 1x = 89x*
(6x + 3)(x – 5)	*–30x + 3x = –27x*
(6x – 3)(x + 5)	*30x – 3x = 27x*
(2x + 1)(3x – 15)	*–30x + 3x = –27x*
(2x – 1)(3x + 15)	*30x – 3x = 27x*
(2x + 3)(3x – 5)	*–10x + 9x = –x*
(2x – 3)(3x + 5)	*10x – 9x = x* ←

Factorizing the difference of two squares

Remember that $a^2 - b^2 = (a + b)(a - b)$.

Example 37

Factorize
a $x^2 - 16$
b $9x^2 - 25y^2$

Answers

a $x^2 - 16 = (x + 4)(x - 4)$

$a^2 = x^2$, so $a = x$
$b^2 = 16$, so $b = 4$
Substitute values into $(a + b)(a - b)$.

b $9x^2 - 25y^2 = (3x + 5y)(3x - 5y)$

$a^2 = 9x^2$, so $a = 3x$
$b^2 = 25y^2$, so $b = 5y$
Substitute values into $(a + b)(a - b)$.

Exercise 2D

1 Factorize these quadratic expressions.

a $x^2 + 11x + 28$ **b** $x^2 - 14x + 13$ **c** $x^2 - x - 20$
d $x^2 + 2x - 8$ **e** $x^2 + 13x + 36$ **f** $x^2 - 7x - 18$

2 Factorize these quadratic expressions.

a $2x^2 - 9x + 9$ **b** $3x^2 + 7x + 2$ **c** $5x^2 - 17x + 6$
d $4x^2 - x - 3$ **e** $3x^2 - 7x - 6$ **f** $14x^2 - 17x + 5$

3 Factorize these quadratic expressions.

a $x^2 - 9$ **b** $x^2 - 100$ **c** $4x^2 - 81$
d $25x^2 - 1$ **e** $m^2 - n^2$ **f** $16x^2 - 49y^2$

2.2 Completing the square

A quadratic expression has the form $ax^2 + bx + c$, where a, b, and c are real numbers, and $a \neq 0$.

You can express a quadratic equation in an equivalent form to make it more convenient, for example when graphing quadratic functions and solving quadratic equations.

'Completing the square' means writing a quadratic expression in the form $y = a(x - h)^2 + k$, where h and k are real numbers. The point (h, k) is the vertex of the graph of the quadratic function, that is, the maximum or minimum point on the quadratic curve.

Some quadratics factorize as the square of a binomial.
For example:

$x^2 + 2x + 1$ is equivalent to $(x + 1)^2$

$x^2 - 4x + 4$ is equivalent to $(x - 2)^2$.

In both examples, the value of h is half the value of b in the original quadratic. We will use this observation to change the following quadratics into the completed square form.

> $2x^2 - 12x + 19 =$
> $2(x - 3)^2 + 1$ so
> $h = 3$ and $k = 1$ and
> the vertex of the graph
> is at $(3, 1)$

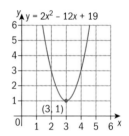

> A binomial is an
> expression with two
> terms.

Example 38

Write the quadratic expression $x^2 + 2x - 5$ in the form $a(x - h)^2 + k$.
Hence find the vertex of the quadratic function.

Answer

$x^2 + 2x - 5 = (x^2 + 2x + 1) - 5 - 1$	*In the original quadratic, $b = 2$,* *so $\dfrac{b}{2} = 1$* *Add 1 to the original expression, and* *then subtract it.*
$x^2 + 2x + 5 = (x + 1)^2 - 6$ The vertex is at $(-1, -6)$	*Simplify the expression, then write it* *as the square of a binomial.*

Example 39

Change the quadratic $2x^2 - 4x + 3$ into the form $a(x - h)^2 + k$.
Hence find the vertex of the quadratic.

Answer

$2x^2 - 4x + 3 = 2(x^2 - 2x) + 3$	*In the original expression, a = 2,*
$2(x^2 - 2x) + 3 = 2(x^2 - 2x + 1)$ $+ 3 - 2$	*Take out a factor of 2 from the first two terms.*
	In the bracket, the coefficient of x is −2. $\dfrac{-2}{2} = -1$
$2x^2 - 4x + 3 = 2(x - 1)^2 + 1$	*Multiplying the 1 in the bracket by 2 gives 2, so subtract 2 to the end of the expression.*
The vertex is at $(1, 1)$	*Simplify and write the expression in the required form.*

In these examples, the leading coefficient of the quadratic was a positive real number. In the following example, the leading coefficient is a negative real number.

Example 40

Change $y = -x^2 + 3x - 2$ into the form $y = a(x - h)^2 + k$, and determine its vertex.

Answer

$y = -x^2 + 3x - 2 = -(x^2 - 3x) - 2$	*In the original expression, a = −1.* *Take out a factor of −1 from the first two terms.*
$y = -\left(x - \dfrac{3}{2}\right)^2 + \dfrac{1}{4}$	
	In the bracket, the coefficient of x is −3. $\left(\dfrac{-3}{2}\right)^2 = \dfrac{9}{4}$
	Add $\dfrac{9}{4}$ to the end of the expression.
The vertex is at $(1.5, 0.25)$	*Simplify and write the expression in the required form.*

Exercise 2E

Change the following quadratics into the form $y = a(x - h)^2 + k$, and find its vertex.

1 $y = x^2 + 2x - 3$

2 $y = x^2 - 6x + 1$

3 $y = x^2 + x - 1$

4 $y = x^2 - 3x + 2$

5 $y = x^2 + 5x - 2$

6 $y = 2x^2 + 2x - 3$

7 $y = 3x^2 - 6x + 1$

8 $y = 2x^2 + 3x - 1$

9 $y = -x^2 + 4x - 3$

10 $y = -2x^2 + 4x - 3$

2.3 Formulae

Rearranging formulae

Example 41

The formula for the area of a circle is $A = \pi r^2$, where A is the area and r is the radius.

The **subject** of the formula is A.

Rearrange the formula to make r the subject.

The subject of a formula is the letter on its own on one side of the = sign.

Answer

$A = \pi r^2$

$r^2 = \dfrac{A}{\pi}$

$r = \sqrt{\dfrac{A}{\pi}}$

Use the same techniques as for solving equations. Whatever you do to one side of the formula, you must do to the other.
Divide both sides by π.
Take the square root of both sides.

You can use this formula to work out the radius of a circle when you know its area.

Example 42

a Einstein's theory of relativity gives the formula $E = mc^2$, where m is the mass, c is the speed of light, and E is the energy equivalent of the mass. Rearrange the formula to make m the subject.

b The formula for gross profit margin is:

$$\text{Gross profit margin} = \frac{\text{Gross profit}}{\text{Sales revenue}} \times 100.$$

Rearrange the formula so that *Sales revenue* is the subject.

Answers

a $E = mc^2$

$m = \dfrac{E}{c^2}$

▶ Continued on next page

b $\text{Gross profit margin} = \dfrac{\text{Gross profit}}{\text{Sales revenue}} \times 100$

$\dfrac{\text{Gross profit margin}}{100} = \dfrac{\text{Gross profit}}{\text{Sales revenue}}$

$\text{Sales revenue} \times \text{Gross profit margin} = \text{Gross profit} \times 100$

$\text{Sales revenue} = \dfrac{\text{Gross profit}}{\text{Gross profit margin}} \times 100$

Exercise 2F

Rearrange the following formulae to make the quantity shown in brackets the subject.

1 $v = u - gt$ (t) **2** $a = \sqrt{b^2 + c^2}$ (c) **3** $c = 2\pi r$ (r)

4 $\dfrac{\sin A}{a} = \dfrac{\sin B}{b}$ (b) **5** $a^2 = b^2 + c^2 - 2bc\cos A$ $(\cos A)$

6 To change temperature from degrees Fahrenheit, F, to degrees Celsius, C, you can use the formula $C = \dfrac{5}{9}(F - 32)$.

> Rearrange the formula to make F the subject.

7 The acid test ratio measures the ability of a company to use its current assets to retire its current liabilities immediately. The formula is given by:

$\text{Acid ratio test} = \dfrac{\text{Current assets} - \text{Stock}}{\text{Current liabilities}}.$

> Rearrange the formula to make *Stock* the subject.

Substituting into formulae

You can always use your GDC in Mathematics Higher Level.

When using formulae, let the calculator do the calculation for you. You should still show your working.

> **1** Find the formula you are going to use (from the formula booklet, from the question or from memory) and write it down.
> **2** Identify the values that you are going to substitute into the formula.
> **3** Write out the formula with the values substituted for the letters.
> **4** Enter the formula into your calculator. Use templates to make the formula look the same on your GDC as it is on paper.
> **5** If you think it is necessary, use brackets. It is better to have too many brackets than too few!
> **6** Write down, with units if necessary, the result from your calculator (to the required accuracy).

Example 43

x and y are linked by the formula $y = \dfrac{x^2 + 1}{2\sqrt{x+1}}$.
Find y when x is 3.1.

Answer

$y = \dfrac{3.1^2 + 1}{2\sqrt{3.1+1}}$ Write the formula
with 3.1 instead of x.

$y = 2.62$

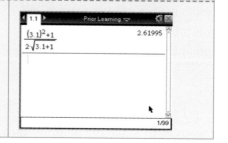

GDC help on CD:
Alternative demonstrations for the TI-84 Plus and Casio fx-9860GII GDCs are on the CD.

Exercise 2G

1 If $a = 2.3$, $b = 4.1$ and $c = 1.7$, find d where

$$d = \frac{3a^2 + 2\sqrt{b}}{ac + b}$$

2 If $b = 8.2$, $c = 7.5$ and $A = 27°$, find a where

$$a = \sqrt{b^2 + c^2 - 2bc \cos A}$$

3 If $u_1 = 10.2$, $r = 0.75$ and $n = 14$, find the value of S, where

$$S = u_1 \frac{1 - r^n}{1 - r}$$

2.4 Solving linear equations

'Solve an equation' means 'find the value of the unknown variable' (the letter).

Rearrange the equation so that the unknown variable x becomes the subject of the equation. To keep the equation 'balanced' always do the same to both sides.

Example 44

Solve the equation $3x + 5 = 17$

Answer

$3x + 5 = 17$	
$3x + 5 - 5 = 17 - 5$	*subtract 5*
$3x = 12$	
$\dfrac{3x}{3} = \dfrac{12}{3}$	*divide by 3*
$x = 4$	

Add, subtract, multiply or divide both sides of the equation until the x is by itself on one side. (This can be either the left or the right hand side.)

Example 45

Solve the equation $4(x - 5) = 8$		
Answer $4(x - 5) = 8$ $\dfrac{4(x-5)}{4} = \dfrac{8}{4}$ $x - 5 = 2$ $x - 5 + 5 = 2 + 5$ $x = 7$	*divide by 4* *add 5*	

Always take care with minus signs.

Example 46

Solve the equation $7 - 3x = 1$		
Answer $7 - 3x = 1$ $7 - 3x - 7 = 1 - 7$ $-3x = -6$ $\dfrac{-3x}{-3} = \dfrac{-6}{-3}$ $x = 2$	*subtract 7* *divide by −3*	

An alternative method for this equation would be to start by *adding* $3x$. Then the x would be positive, but on the right-hand side.

Example 47

Solve the equation $3(2 + 3x) = 5(4 - x)$		
Answer $3(2 + 3x) = 5(4 - x)$ $6 + 9x = 20 - 5x$ $6 + 9x + 5x = 20 - 5x + 5x$ $6 + 14x = 20$ $6 + 14x - 6 = 20 - 6$ $14x = 14$ $\dfrac{14x}{14} = \dfrac{14}{14}$ $x = 1$	*add 5x* *subtract 6* *divide by 14*	

Sometimes it can be quicker to **divide** first rather than expanding the brackets.

Exercise 2H

Solve these equations.

1 $3x - 10 = 2$

2 $\dfrac{x}{2} + 5 = 7$

3 $5x + 4 = -11$

4 $3(x + 3) = 18$

5 $4(2x - 5) = 20$

6 $\dfrac{2}{5}(3x - 7) = 8$

7 $21 - 6x = 9$

8 $12 = 2 - 5x$

9 $2(11 - 3x) = 4$

10 $4(3 + x) = 3(9 - 2x)$

11 $2(10 - 2x) = 4(3x + 1)$

12 $\dfrac{5x + 2}{3} = \dfrac{3x + 10}{4}$

2.5 Simultaneous linear equations

Simultaneous equations involve two (or more) variables.
There are two methods which you can use, called substitution and elimination.

Example 48

Solve the equations $3x + 4y = 17$ and $2x + 5y = 16$.

Answer

Geometrically you could consider these two linear equations as the equations of two straight lines. Finding the solution to the equation is equivalent to finding the point of intersection of the lines. The coordinates of the point will give you the values for x and y.

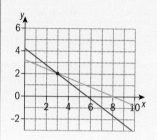

Substitution method

$3x + 4y = 17$

$2x + 5y = 16$

$5y = 16 - 2x$

$y = \dfrac{16}{5} - \dfrac{2}{5}x$

Rearrange one of the equations to make y the subject.

$3x + 4\left(\dfrac{16}{5} - \dfrac{2}{5}x\right) = 17$

Substitute for y in the other equation.

$3x + \dfrac{64}{5} - \dfrac{8}{5}x = 17$

$15x + 64 - 8x = 85$

$15x - 8x = 85 - 64$

$7x = 21$

$x = 3$

Solve the equation for x.

$3(3) + 4y = 17$

$9 + 4y = 17$

$4y = 8$

$y = 2$

Substitute for x in one of the original equations and solve for y.

The solution is $x = 3$, $y = 2$.

Elimination method

$3x + 4y = 17 \longrightarrow (1)$

$2x + 5y = 16 \longrightarrow (2)$

Multiply equation (1) by 2 and equation (2) by 3.

$6x + 8y = 34 \longrightarrow (3)$

$6x + 15y = 48 \longrightarrow (4)$

This is to make the coefficients of x equal.

Subtract the equations. [(4)–(3)]

$7y = 14$

$y = 2$

Subtracting now eliminates x from the equations.

▶ Continued on next page

$$3x + 4(2) = 17$$
$$3x + 8 = 17$$
$$3x = 17 - 8$$
$$3x = 9$$
$$x = 3$$
The solution is $x = 3$, $y = 2$.

| | *Substitute for y in one of the original equations and solve for x.* |

Exercise 2I

1 Solve these simultaneous equations using substitution.
   a $y = 3x - 2$ and $2x + 3y = 5$ **b** $4x - 3y = 10$ and $2y + 5 = x$
   c $2x + 5y = 14$ and $3x + 4y = 7$

2 Solve these simultaneous equations using elimination.
   a $2x - 3y = 15$ and $2x + 5y = 7$ **b** $3x + y = 5$ and $4x - y = 9$
   c $x + 4y = 6$ and $3x + 2y = -2$ **d** $3x + 2y = 8$ and $2x + 3y = 7$
   e $4x - 5y = 17$ and $3x + 2y = 7$

2.6 Exponential expressions

Repeated multiplication can be written as an **exponential** expression. For example, squaring a number $3 \times 3 = 3^2$ or $5.42 \times 5.42 = 5.42^2$.

If we multiply a number by itself three times then the exponential expression is a cube. For example
$$4.6 \times 4.6 \times 4.6 = 4.6^3.$$

You can also use exponential expressions for larger integer values. So, for example, $3^7 = 3 \times 3 \times 3 \times 3 \times 3 \times 3 \times 3$.

Where the exponent is not a positive integer, these rules apply:
$$a^0 = 1, \ a \neq 0 \text{ and } a^{-n} = \frac{1}{a^n}$$

> **Index** and **power** are other names for **exponent**.

> You use squares in Pythagoras' theorem $a^2 = b^2 + c^2$ or in the formula for the area of a circle $A = \pi r^2$. You use a cube in the volume of a sphere $V = \frac{4}{3}\pi r^3$.

Example 49

Write down the values of 10^2, 10^3, 10^1, 10^0, 10^{-2}, 10^{-3}.

Answer
$10^2 = 10 \times 10 = 100$
$10^3 = 10 \times 10 \times 10 = 1000$
$10^1 = 10$
$10^0 = 1$
$10^{-2} = \dfrac{1}{10^2} = \dfrac{1}{100} = 0.01$
$10^{-3} = \dfrac{1}{10^3} = \dfrac{1}{1000} = 0.001$

To evaluate an exponential function with the GDC use either the ⌃ key or the template key ⊞ and the exponent template.

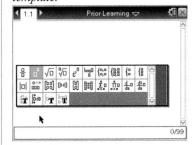

> **GDC help on CD:**
> *Alternative demonstrations for the TI-84 Plus and Casio fx-9860GII GDCs are on the CD.*

Exercise 2J

Evaluate these expressions.

1 a $2^3 + 3^2$ **b** $4^2 \times 3^2$ **c** 2^6

2 a 5^0 **b** 3^{-2} **c** 2^{-4}

3 a 3.5^5 **b** 0.495^{-2} **c** $2\dfrac{(1-0.02)^{10}}{1-0.02}$

2.7 Solving inequalities

Inequalities behave much like equations and can be solved in the same way.

Properties of inequalities

> → When you add or subtract a real number from both sides of an inequality the direction of the inequality is unchanged.

For example:

- $4 > 6 \Rightarrow 4 + 2 > 6 + 2$

- $15 \leq 20 \Rightarrow 15 - 6 \leq 20 - 6$

- $x - 7 \geq 8 \Rightarrow x - 7 + 7 \geq 8 + 7$

- $x + 5 < 12 \Rightarrow x + 5 - 5 < 12 - 5$

> → When you multiply or divide both sides of an inequality by a positive real number the direction of the inequality is unchanged.
> When you multiply or divide both sides of an inequality by a negative real number the direction of the inequality is reversed.

For example:

- $4 > 5 \Rightarrow 2(4) > 2(5)$

- $6 \leq 10 \Rightarrow -2(6) \geq -2(10)$

- $10 \leq 30 \Rightarrow \dfrac{10}{5} \leq \dfrac{30}{5}$

- $18 > 24 \Rightarrow \dfrac{18}{-3} < \dfrac{24}{-3}$

- $-12 > -20 \Rightarrow \dfrac{-12}{4} > \dfrac{-20}{4}$

Example 50

Solve the inequalities **a** $2x + 5 < 7$ **b** $3(x - 2) \geq 4$

Answers

a $2x + 5 < 7$

$\quad 2x < 2$

$\quad\quad x < 1$

b $3(x - 2) \geq 4$

$\quad x - 2 \geq 1\frac{1}{3}$

$\quad\quad x \geq 3\frac{1}{3}$

Take great care with + and – signs.

Example 51

Solve the inequalities $7 - 2x \leq 5$

Answer	
$7 - 2x \leq 5$	
$\quad -2x \leq -2$	*Divide by –2*
$\quad\quad x \geq 1$	*Change $\leq$ to $\geq$*

> If you either multiply or divide an inequality by a negative value, the signs on both sides of the inequality will change. The inequality will also be reversed.

Example 52

Solve the inequalities $19 - 2x > 3 + 6x$

Answer	
$19 - 2x > 3 + 6x$	
$\quad 19 > 3 + 8x$	
$\quad 16 > 8x$	
$\quad\quad 2 > x$	
$\quad\quad x < 2$	*Reverse the inequalities*

> Sometimes the x ends up on the right hand side of the inequality. In this case reverse the inequality as in the example.

Exercise 2K

1 Solve the inequality for x and represent it on the number line.

    a $3x + 4 \leq 13$ **b** $5(x - 5) > 15$ **c** $2x + 3 < x + 5$

2 Solve for x.

    a $2(x - 2) \geq 3(x - 3)$ **b** $4 < 2x + 7$ **c** $7 - 4x \leq 11$

2.8 Absolute value

The absolute value (or modulus) of a number, $|x|$, is the numerical part of the number without its sign. It can be written as

$$|x| = \begin{cases} -x, \text{ if } x \leq 0 \\ x, \text{ if } x \geq 0 \end{cases}$$

Example 53

Write down $|a|$ where $a = -4.5$ and $a = 2.6$

Answer

If $a = -4.5$ then $|a| = 4.5$

If $a = 2.6$ then $|a| = 2.6$

Example 54

Write the value of $|p - q|$ where $p = 3$ and $q = 6$.

Answer

$|p - q| = |3 - 6|$

$\quad\ = |-3| = 3$

Exercise 2L

1 Write the value of $|a|$ when a is

  a 3.25   **b** -6.18 **c** 0

2 Write the value of $|5 - x|$ when $x = 3$ and when $x = 8$.

3 If $x = 6$ and $y = 4$, write the values of

  a $|x - y|$ **b** $|x - 2y|$ **c** $|y - x|$

2.9 Adding and subtracting algebraic fractions

To add or subtract fractions, first write them over a common denominator.

Example 55

Combine these fractions, simplifying your answer.

a $\dfrac{x}{2x+1} + \dfrac{5x+3}{2x+1}$ **b** $\dfrac{2x-3}{4x-5} - \dfrac{6x-2}{4x-5}$

c $\dfrac{3x}{3x-1} + \dfrac{3x+1}{2x+5}$ **d** $\dfrac{5x}{x+3} - \dfrac{2x+1}{2x-1}$

Answers

a $\dfrac{x}{2x+1} + \dfrac{5x+3}{2x+1} = \dfrac{x + (5x+3)}{2x+1}$ *Keep the common denominator and add the numerators.*

$\qquad\qquad\qquad\quad = \dfrac{6x+3}{2x+1}$ *Combine like terms.*

$\qquad\qquad\qquad\quad = \dfrac{3(2x+1)}{2x+1}$ *Factorize and simplify whenever possible.*

$\qquad\qquad\qquad\quad = 3$

▶ Continued on next page

b $\dfrac{2x-3}{4x-5}-\dfrac{6x-2}{4x-5}=\dfrac{(2x-3)-(6x-2)}{4x-5}$

Keep the common denominator and subtract the numerators.
Be sure to distribute the negative.

$\qquad\qquad\qquad =\dfrac{2x-3-6x+2}{4x-5}$

$\qquad\qquad\qquad =\dfrac{-4x-1}{4x-5}$

Combine like terms.

c $\dfrac{3x}{3x-1}+\dfrac{3x+1}{2x+5}=\dfrac{3x}{3x-1}\cdot\dfrac{2x+5}{2x+5}+\dfrac{3x+1}{2x+5}\cdot\dfrac{3x-1}{3x-1}$

Multiply each fraction by 'one' to get a common denominator.

$\qquad\qquad\qquad =\dfrac{3x(2x+5)}{(3x-1)(2x+5)}+\dfrac{(3x+1)(3x-1)}{(2x+5)(3x-1)}$

$\qquad\qquad\qquad =\dfrac{\left(6x^2+15x\right)}{(3x-1)(2x+5)}+\dfrac{\left(9x^2-1\right)}{(3x-1)(2x+5)}$

Expand the brackets

$\qquad\qquad\qquad =\dfrac{15x^2+15x-1}{(3x-1)(2x+5)}$

Combine like terms.

d $\dfrac{5x}{x+3}-\dfrac{2x+1}{2x-1}=\dfrac{5x}{x+3}\cdot\dfrac{2x-1}{2x-1}-\dfrac{2x+1}{2x-1}\cdot\dfrac{x+3}{x+3}$

Multiply each fraction by 'one' to get a common denominator.

$\qquad\qquad\qquad =\dfrac{5x(2x-1)}{(x+3)(2x-1)}-\dfrac{(2x+1)(x+3)}{(2x-1)(x+3)}$

$\qquad\qquad\qquad =\dfrac{\left(10x^2-5x\right)-\left(2x^2+7x+3\right)}{(x+3)(2x-1)}$

Expand the brackets

$\qquad\qquad\qquad =\dfrac{10x^2-5x-2x^2-7x-3}{(x+3)(2x-1)}$

Watch out for negative signs.

$\qquad\qquad\qquad =\dfrac{8x^2-12x-3}{(x+3)(2x-1)}$

Combine like terms.

Exercise 2M

Combine these fractions, simplifying your answer.

1 $\dfrac{2}{x+7}+\dfrac{3x-1}{x+7}$ **2** $\dfrac{4x}{2x+2}-\dfrac{3x-1}{2x+2}$

3 $\dfrac{3x+9}{3x+4}+\dfrac{3x-1}{3x+4}$ **4** $\dfrac{2x}{x+5}+\dfrac{x+1}{2x-1}$

5 $\dfrac{4}{x}+\dfrac{2x+1}{x+2}$ **6** $\dfrac{2x-1}{x-2}-\dfrac{3x}{4x+3}$

7 $\dfrac{x+1}{5x+1}+\dfrac{2x}{2x-5}$ **8** $\dfrac{x+5}{x-4}-\dfrac{x-2}{x+2}$

Solving equations with rational coefficients

To solve equations with rational coefficients, multiply both sides of the equation by the least common multiple of all the denominators.

Example 56

Solve these equations.

a $\dfrac{x}{6} = \dfrac{5}{4} - \dfrac{x}{2}$ **b** $\dfrac{1}{15} + \dfrac{1}{x} = \dfrac{1}{6}$

Answers

a $\dfrac{x}{6} = \dfrac{5}{4} - \dfrac{x}{2}$

$12\left(\dfrac{x}{6}\right) = 12\left(\dfrac{5}{4} - \dfrac{x}{2}\right)$ *LCM of 6, 4 and 2 is 12*

$2x = 14 - 6x$

$8x = 14$

$x = \dfrac{14}{8}$ or $\dfrac{7}{4}$

b $\dfrac{1}{15} + \dfrac{1}{x} = \dfrac{1}{6}$

$30x\left(\dfrac{1}{15} + \dfrac{1}{x}\right) = 30x\left(\dfrac{1}{6}\right)$ *LCM of 15, 6 and x is 30 x*

$2x + 30 = 5x$

$-3x = -30$

$x = 10$

Exercises 2N

Solve the equations.

1 $\dfrac{x}{3} + \dfrac{1}{6} = \dfrac{x}{4} + \dfrac{1}{4}$

2 $\dfrac{1}{k} + \dfrac{1}{4} = \dfrac{9}{4k}$

3 $\dfrac{1}{6} = \dfrac{5}{6} - \dfrac{1}{x}$

4 $\dfrac{3}{5} - \dfrac{2x}{4} = \dfrac{x-1}{2}$

5 $\dfrac{3x}{4} + \dfrac{x+2}{3} = \dfrac{x-1}{8}$

2.10 Solving quadratic equations by factorizing

Consider the quadratic equation $x^2 + 2x - 8 = 0$.

Factor the left side into $(x + 4)(x - 2) = 0$.

Since we have a product that is equal to zero, one of the factors must equal 0, hence either $x + 4 = 0$ and $x = -4$, or $x - 2 = 0$ and $x = 2$.

There are two unique solutions, $x = -4$ or $x = 2$.

$A \cdot B = 0 \Rightarrow A = 0$ or $B = 0$

Example 57

By factoring, solve the quadratic equation $2x^2 + 5x - 3 = 0$.	
Answer	
$2x^2 + 5x - 3 = (2x - 1)(x + 3) = 0$	*Factorize*
$2x - 1 = 0$ or $x + 3 = 0$	*Set each factor equal to 0*
$x = \dfrac{1}{2}$ or $x = -3$	*Solve each linear equation*

Example 58

By factoring, solve the quadratic equation $4x - x^2 = 4$.	
Answer	
$4x - x^2 = 4$ so $-x^2 + 4x - 4 = 0$	*Set the equation equal to 0*
$x^2 - 4x + 4 = 0$	*Multiply both sides by -1*
$x^2 - 4x + 4 = (x - 2)^2 = 0$	*Factor the trinomial*
$x - 2 = 0$	*Set the linear factor equal to 0*
$x = 2$	*Solve*
In this case there is only one unique solution, or a repeated solution.	

Exercise 20

Solve the following quadratics by first factoring.

1 $x^2 - 8x + 15 = 0$ **2** $x^2 + 6x - 16 = 0$

3 $x^2 - 8x + 16 = 0$ **4** $28 + 3x = x^2$

5 $6x^2 + 7x - 3 = 0$ **6** $-2x^2 = 3x - 2$

2.11 Solving quadratic equations by completing the square

Example 59

Solve the quadratic $x^2 + 4x - 5 = 0$ using the method completing the square.

Answer

$x^2 + 4x - 5 = 0$ so $x^2 + 4x = 5$	Bring the constant term to the RHS
$x^2 + 4x + 4 = 5 + 4$	Complete the square on the LHS, and add the constant term to the LHS.
$(x + 2)^2 = 9$	Write the trinomial as the square of a binomial
$x + 2 = \pm 3$	Take the square root of both sides
$x = -2 \pm 3$	Solve for x
$x = -2 + 3 = 1$, or $x = -2 - 3 = -5$	Separate both solutions

Example 60

Solve the quadratic $2x^2 - 4x = 3$ using the method completing the square.

Answer

$2x^2 - 4x = 2(x^2 - 2x) = 3$	Take out a factor of 2 from first two terms
$x^2 - 2x = \dfrac{3}{2}$	Divide both sides by 2
$x^2 - 2x + 1 = \dfrac{3}{2} + 1 = \dfrac{5}{2}$	Complete the square on LHS, and add the constant term to LHS
$(x - 1)^2 = \dfrac{5}{2}$	Change the trinomial to the square of the binomial
$x - 1 = \pm\sqrt{\dfrac{5}{2}} = \pm\dfrac{\sqrt{10}}{2}$	Take square root of both sides
$x = 1 \pm \dfrac{\sqrt{10}}{2}$	Add one to both sides
$x = 1 + \dfrac{\sqrt{10}}{2}$, or $x = 1 - \dfrac{\sqrt{10}}{2}$	Separate solutions

Some quadratic equations have no real solutions.

Example 61

Solve the quadratic $x^2 + 2x = -4$ by completing the square.	

Answer	
$x^2 + 2x + 1 = -4 + 1 = -3$	*Complete the square on the LHS, and add the c constant term to the LHS*
$(x + 1)^2 = -3$ No solution.	*You cannot take the square root of the RHS.*

Exercise 2P

Solve these equations by completing the square.

1 $x^2 + 4x = 3$ **2** $x^2 + 3x = 2$
3 $2x^2 - 2x = 1$ **4** $3x^2 + 6x = -2$

2.12 Quadratic inequalities

When you solve quadratic inequalities using the method of factoring, you obtain at most two linear factors.

Consider the inequality $x^2 - 4x < 0$.

Factorise the LHS: $x(x - 4) < 0$.

If the product $AB < 0$, then the factors must have different signs, i.e., either the factor $A < 0$ and factor $B > 0$, or factor $A > 0$ and factor $B < 0$.

There are two cases to consider:

Case 1: $x > 0$ and $x - 4 < 0$,

i.e. $x < 4$.

The set of values that satisfy both linear inequalities $x > 0$ and $x < 4$ is $0 < x < 4$.

Case 2: $x < 0$ and $x - 4 > 0$, or $x > 4$.

There are no real numbers that are less than 0 and greater than 4.

Hence, the solution to the inequality $x^2 - 4x < 0$ is $0 < x < 4$.

> You can use a GDC to solve quadratic inequalities, or solve them by drawing a graph.

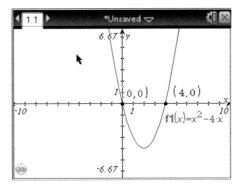

> **GDC help on CD:**
> *Alternative demonstrations for the TI-84 Plus and Casio fx-9860GII GDCs are on the CD.*

Example 62

Solve the quadratic inequality $x^2 - 3x \geq 10$.

Answer

$x^2 - 3x \geq 10$	
So $x^2 - 3x - 10 \geq 0$	*Bring all terms to one side*
$(x - 5)(x + 2) \geq 0$	*Factor the trinomial*
Case 1: $(x - 5) \geq 0$ and $(x + 2) \geq 0$	*A product is positive if both factors are positive*
$x - 5 \geq 0$, $x \geq 5$ and $x + 2 \geq 0$, $x \geq -2$	*Find the solution set satisfying both linear inequalities*
$x \geq 5$ and $x \geq -2$	
Case 2: $x - 5 \leq 0$ and $(x + 2) \leq 0$	*A product is also positive if both factors are negative.*
$x \leq 5$ and $x \leq -2$	*Find the solution set satisfying both inequalities.*
Either $x \geq -2$ or $x \leq 5$	*Solution*

Exercise 2Q

Solve the following quadratic inequalities.

1 $x^2 - x > 6$

2 $x^2 + 5x < -6$

3 $x^2 + 2x - 8 \leq 0$

4 $3x^2 - 5x - 2 \geq 0$

3 Geometry

3.1 Pythagoras' theorem

→ In a right-angled triangle' ABC with sides a, b and c, where a is the **hypotenuse**:

$$a^2 = b^2 + c^2$$

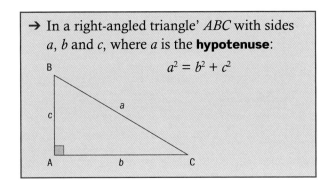

Although the theorem is named after the Greek mathematician Pythagoras, it was known several hundred years earlier to the Indians in their Sulba Sutras and thousands of years before to the Chinese as the Gougu Theorem.

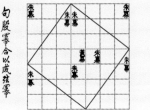

Example 63

Find the length marked a.

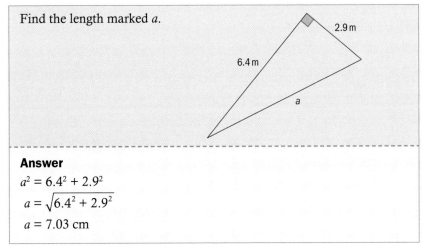

2.9 m

6.4 m

a

You can use Pythagoras' theorem to calculate the length of one side of a right-angled triangle when you know the other two.

Answer

$a^2 = 6.4^2 + 2.9^2$

$a = \sqrt{6.4^2 + 2.9^2}$

$a = 7.03$ cm

Sometimes you have to find a shorter side.

Example 64

Find the length marked b.

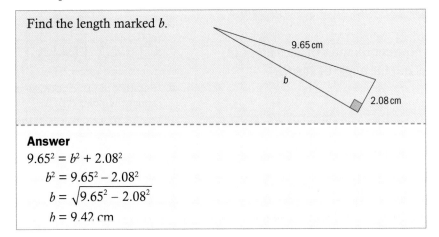

9.65 cm

b

2.08 cm

Answer

$9.65^2 = b^2 + 2.08^2$

$b^2 = 9.65^2 - 2.08^2$

$b = \sqrt{9.65^2 - 2.08^2}$

$b = 9.42$ cm

Check your answer by making sure that the hypotenuse is the longest side of the triangle.

Exercise 3A

In each diagram, find the length of the side marked x. Give your answer to 3 significant figures.

1
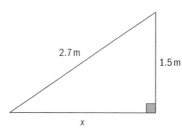

2
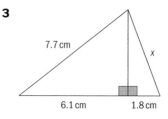

3
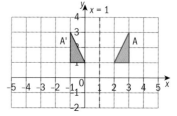

3.2 Geometric transformations

A transformation can change the position as well as the size of an object. A transformation maps an object to its image.

There are four main types of transformation:

- Reflection
- Rotation
- Translation
- Enlargement

Reflection

When an object is **reflected** in a mirror line, the object and its image are symmetrical about the mirror line. Every point on the image is the same distance from the mirror line as the corresponding point on the object.

To describe a reflection, state the equation of the mirror line.

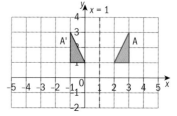

▲ Reflection in line $x = 1$

Rotation

A **rotation** moves an object around a fixed point called the center of rotation, in a given direction through a particular angle.

To describe a rotation give the coordinates of the center of rotation, and the direction and the angle of turn.

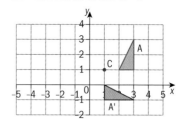

▲ Rotation of 90° clockwise about the point (1,1)

Translation

A **translation** moves every point a fixed distance in the same direction.

To describe a translation write the column vector $\begin{pmatrix} x \\ y \end{pmatrix}$,

where x is the movement in the x direction and y is the movement in the y direction.

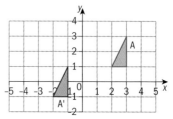

▲ Translation of $\begin{pmatrix} -4 \\ -2 \end{pmatrix}$

Enlargement

Enlargement increases or decreases the size of an object by a given scale factor.

To describe an enlargement give the coordinates of the centre of enlargement and the scale factor.

The image after an enlargement is mathematically **similar** to the original object.

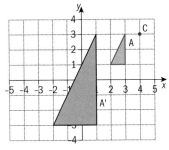

▲ Enlargement scale factor 3 centre (4, 3)

Example 65

The grid contains five shapes A to E. Describe the single transformation that takes:

a A to B
b A to C
c A to D
d A to E
e C to D

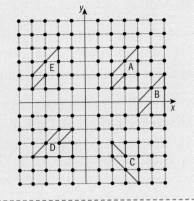

For more on similarity, see page 724

Answers

A ⟶ B: Translation; vector $\begin{pmatrix} 2 \\ -2 \end{pmatrix}$

A ⟶ C: Reflection; line $y = -1$

A ⟶ D: Reflection; line $y = -x$

A ⟶ E: Translation; vector $\begin{pmatrix} -7 \\ 0 \end{pmatrix}$

D ⟶ C: Rotation; center (1, −1), 90° clockwise

Exercise 3B

1 The grid contains four shapes A to D.

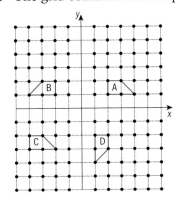

Describe the single transformation that takes:

a A to B   **b** A to C   **c** A to D   **e** B to D

2 Copy this diagram on to graph paper.

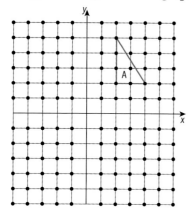

  a Reflect shape A in the line $y = -x$. Label the image B.
  b Reflect shape B in the line x-axis. Label the image C.
  c Describe fully a single transformation that would take A to C.

3 Draw a set of axes from -10 to 10 on both x and y axes.
  a Draw the triangle with vertices at $(2, 1)$ $(4, 1)$ $(4, 4)$.
    Label it A.
  b Reflect A in the x axis. Label the image B.
  c Enlarge B by scale factor 2 centre $(0, 0)$. Label the image C.
  d Rotate C by $180°$ center $(0, 0)$. Label the image D.
  e Reflect D in the x axis. Label the image E.
  f Rotate E by $180°$ center $(0, 0)$. Label the image F.
  Describe the single transformation that maps

  g $C \longrightarrow F$
  h $A \longrightarrow F$
  i $E \longrightarrow A$
  j $C \longrightarrow E$

3.3 Congruence

> → Two figures that are exactly the same shape and size are
> **congruent**.
> In congruent shapes
>
> • Corresponding lengths are equal
> • Corresponding angles are equal

Objects and images after rotations, reflections or translations are
congruent to each other.

To prove that two triangles are congruent, you need to show that
they satisfy one of four sets of conditions.

▼ Three sides are the same (SSS)

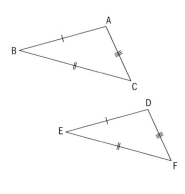

▼ Two sides and the included angle are the same (ASS)

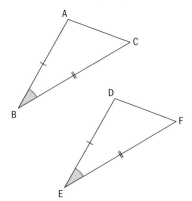

▼ Two angles and the included angle are the same (SAA)

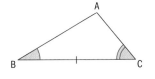

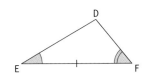

▼ Right-angled triangles with hypotenuse and one other side the same (RHS)

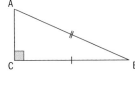

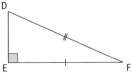

> This is a special case of SSS, as it follows from Pythagoras' theorem that the remaining sides are equal.

Example 66

State whether the shapes in each pair are congruent.
List the vertices in corresponding order and give reasons for congruence.

a

b

c

▶ continued on next page

Exercise 3C

1 Show that $\triangle DEF$ is congruent to $\triangle ABC$. Find the length of each of the sides.

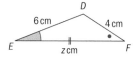 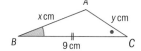

2 Give a brief reason why $\triangle DEF$ and $\triangle ABC$ are congruent. Find the value of each of the angles.

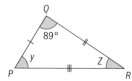 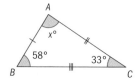

3 Prove that $\triangle DEF$ is congruent to $\triangle ABC$. Find the values of x and y.

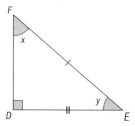 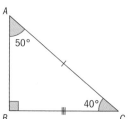

3.4 Similarity

> → Two figures are **similar** if they are the same shape.
>
> They are not necessarily the same size, so generally one is an enlargement of the other.

After an enlargement the image is always similar to the object.
Enlarging a shape leaves the angles the same but changes all the lengths by the same scale factor.

> → The scale factor of an enlargement is the ratio of
>
> $$\frac{\text{length of a side on one shape}}{\text{length of corresponding side on other shape}}$$

Similar triangles

In similar triangles, corresponding angles are equal and corresponding sides are in the same ratio.

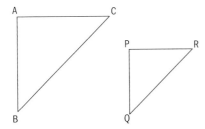

Triangles ABC and PQR are similar because

$$\hat{A} = \hat{P}, \hat{B} = \hat{Q}, \hat{C} = \hat{P}$$

$$\frac{AB}{PQ} = \frac{BC}{QR} = \frac{AC}{PR} = \text{scale factor}$$

To prove that two triangles are similar, show that **one** of these three statements is true:

1 ▼ The three angles of one triangle are equal to the three angles of the other triangle

2 ▼ The corresponding sides of each triangle are in the same ratio

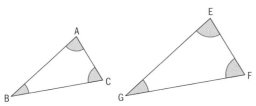

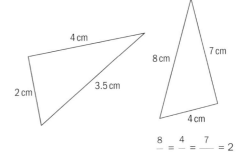

$$\frac{8}{4} = \frac{4}{2} = \frac{7}{3.5} = 2$$

3 ▼ There is one pair of equal angles and the sides containing these angles are in the same ratio.

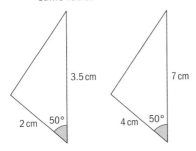

Example 67

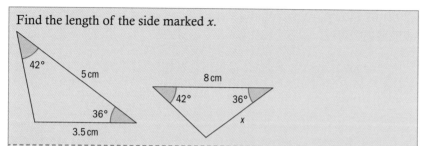

Find the length of the side marked x.

42°
5 cm
8 cm
42° 36°
36° x
3.5 cm

Answer

Two pairs of angles are equal, so the third pair must be equal.
Hence the triangles are similar.

The scale factor of the enlargement is $\frac{8}{5} = 1.6$

So $x = 3.5 \times 1.6 = 5.6$ cm

Prove similarity

Exercise 3D

1 Which pairs of rectangles are similar?

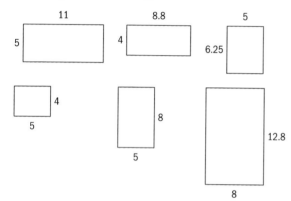

11
5

8.8
4

5
6.25

4
5

8
5

12.8
8

> Note the shapes in this exercise are not drawn to scale.

2 These shapes are similar. Calculate the lengths marked by letters.

a

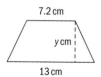

7.2 cm
y cm
13 cm

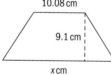

10.08 cm
9.1 cm
x cm

b

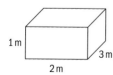

1 m
2 m
3 m

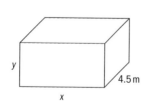

y
x
4.5 m

3 Which triangles are similar?

a

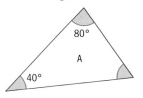 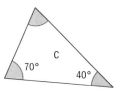

b

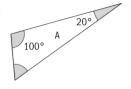 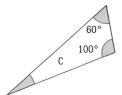

c

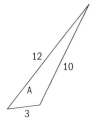 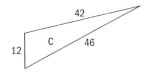

d

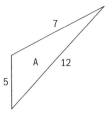 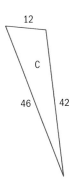

e

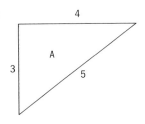 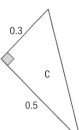

4 Show that triangles ABC and APQ are similar.
Calculate the length of AC and BP.

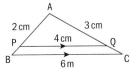

5 In the diagram AB and CD are parallel. AD and BC meet at X.

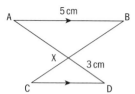

   a Prove that triangles ABX and DCX are similar.
   b Which side in triangle DCX corresponds to AX in triangle ABX?
   c Calculate the length of AX.

3.5 Points, lines, planes and angles

The most basic ideas of geometry are points, lines and planes. A **straight line** is the shortest distance between two points. Planes can be **finite** like the surface of a desk or a wall or can be **infinite**, continuing in every direction.

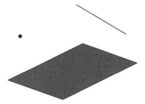

We say that a point has zero dimensions, a line has one dimension and a plane has two dimensions.

Angles are often measured in degrees.

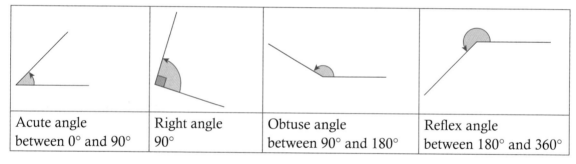

Acute angle between 0° and 90°	Right angle 90°	Obtuse angle between 90° and 180°	Reflex angle between 180° and 360°

Exercise 3E

1 Draw a sketch of:
   a a reflex angle   **b** an acute angle
   c a right angle   **d** an obtuse angle.

2 State whether the following angles are acute, obtuse or reflex.

   a 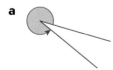   **b**    **c**

3 State whether the following angles are acute, obtuse or reflex.
   a 173°   **b** 44°   **c** 272°
   d 82°   **e** 308°   **f** 196°

3.6 Two-dimensional shapes

Triangles

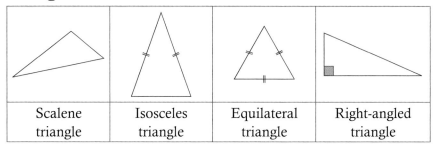

| Scalene triangle | Isosceles triangle | Equilateral triangle | Right-angled triangle |

The small lines on these diagrams show equal lines and the arrows show parallel lines.

Quadrilaterals

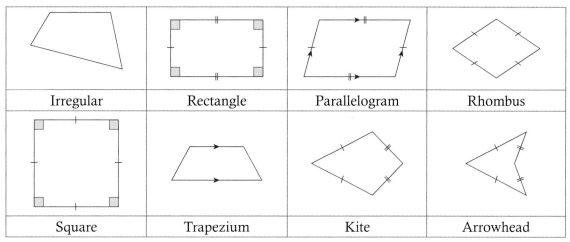

| Irregular | Rectangle | Parallelogram | Rhombus |

| Square | Trapezium | Kite | Arrowhead |

Polygons

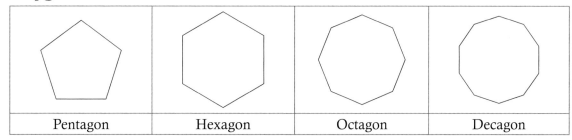

| Pentagon | Hexagon | Octagon | Decagon |

Exercise 3F

1 Sketch the quadrilaterals in the table above with their diagonals.
Copy and complete the following table.

Diagonals	Irregular	Rectangle	Parallelogram	Rhombus	Square	Trapezium	Kite
Perpendicular					✓		
Equal					✓		
Bisect					✓		
Bisect angles					✓		

For example, the diagonals of a square are perpendicular to
each other, equal in length, bisect each other and bisect the angles of the square.

2 List the names of all the shapes that are contained in the following figures.

a

b

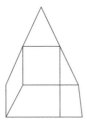

3.7 Circle definitions and properties

You should be familiar with these definitions related to circles.

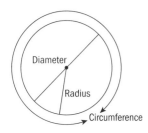

The distance from the center of the circle to any point of the circle is called the **radius**, usually denoted by r.

A **diameter** goes through the centre and is twice as long as the radius. The diameter is usually denoted by D.

$$D = 2r$$

The distance around the circle is called the **circumference**. The circumference of a circle, C, is found using the formulae $C = 2\pi r$ or $C = \pi d$

Here are some other properties and definitions that you should know.

- The **area**, A, of a circle can be calculated using the formula $A = \pi r^2$

- A **chord** of a circle is a straight line drawn between two points on the circumference of the circle.
  A chord divides a circle into two segments – a **minor segment** and a **major segment**.

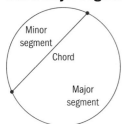

- Any continuous part of the circumference of a circle is called an **arc**.

- A **semicircle** has an arc that is half the length of the circumference.

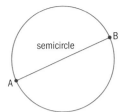

- The area lying between two radii is a **sector**.

- A **tangent** to a circle is a straight line that touches the circumference of the circle at a single point called the **point of tangency**. The angle between a tangent and the radius at that point is 90°.

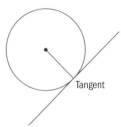

3.8 Perimeter

The **perimeter** of a figure is defined as the length of its boundary. The perimeter of a polygon is found by adding together the sum of the lengths of its sides.

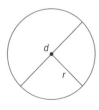

The perimeter of a circle is called its **circumference**.

In the circle on the left, r is the radius and d is the diameter. If C is the circumference.

$C = 2\pi r$

or

$C = \pi d$

$\pi = 3.141592653589793238462...$
Many maths enthusiasts around the world celebrate Pi day on March 14 (3/14). The use of the symbol π was popularised by the Swiss mathematician **Leonhard Euler** (1707–83).

Example 68

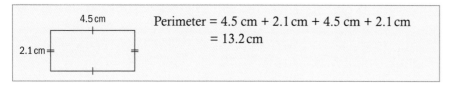

Perimeter = 4.5 cm + 2.1 cm + 4.5 cm + 2.1 cm
$$= 13.2 \text{ cm}$$

Example 69

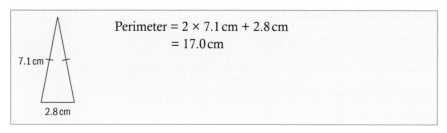

Perimeter = 2 × 7.1 cm + 2.8 cm
$$= 17.0 \text{ cm}$$

Exercise 3G

Find the perimeters of these shapes.

a

3.2 cm
4.3 cm

b

5.5 cm
2.7 cm

c
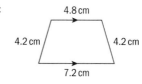
4.8 cm
4.2 cm   4.2 cm
7.2 cm

d
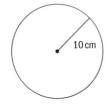
10 cm

e
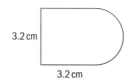
3.2 cm
3.2 cm

f

2.6 cm

3.9 Area

These are the formulae for the areas of a number of plane shapes.

Square	Rectangle	Parallelogram	Triangle
a a	b a	h b	h b
$A = a^2$	$A = ab$	$A = bh$	$A = \dfrac{1}{2}bh$

Trapezium	Kite	Circle
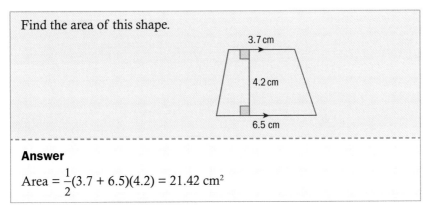		
$A = \dfrac{1}{2}(a + b)\,h$	$A = \dfrac{1}{2}ab$	$A = \pi r^2$

Example 70

Find the area of this shape.

3.7 cm

4.2 cm

6.5 cm

Answer

Area $= \dfrac{1}{2}(3.7 + 6.5)(4.2) = 21.42$ cm^2

Example 71

Find the area of this shape giving your answer to 3 signifigant figures.

Use the π button on your calculator to enter π.

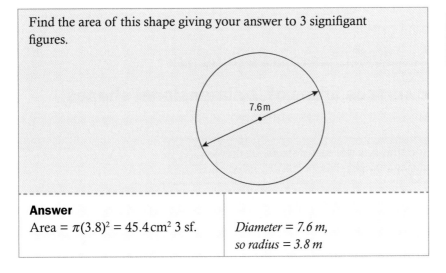

7.6 m

Answer

Area $= \pi(3.8)^2 = 45.4$ cm^2 3 sf.

Diameter = 7.6 m, so radius = 3.8 m

Exercise 3H

Find the areas of these shapes. Give your answer to 3 signifigant figures.

1

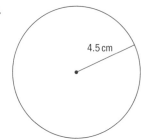

4.5 cm

2

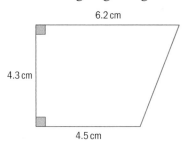

6.2 cm

4.3 cm

4.5 cm

3

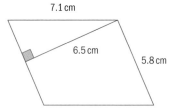

7.1 cm

6.5 cm

5.8 cm

4

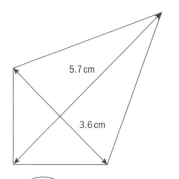

5.7 cm

3.6 cm

5

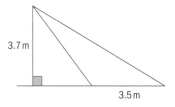

3.7 m

3.5 m

6

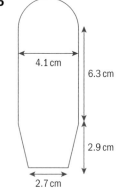

4.1 cm

6.3 cm

2.9 cm

2.7 cm

3.10 Volumes and surface areas of 3-dimensional shapes

Prism

> → A **prism** is a solid shape that has the same shape or cross-section all the way along its length.

A prism takes its name from the shape of its cross-section

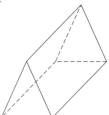

Triangular prism

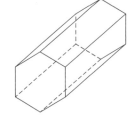

Hexagonal prism

> → To find the volume of a prism, use the formula
>
> $V = Area\ of\ cross\text{-}section \times height$

To find the surface area of a prism, calculate the area of each face and add them together.

Cylinder

A **cylinder** is like a prism, with cross-section a circle.

> → The volume of a cylinder where the radius of
> the circular cross section is r and the height is h is
>
> $V = \pi r^2 \times h$

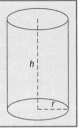

To calculate the surface area of a cylinder, open out the curved surface into a rectangle:

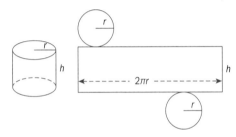

To find the curved surface area use the formula $CSA = 2\pi rh$

> → To find the surface area of a whole cylinder, find the curved
> surface area and add on the areas of the two circular ends:
>
> Total surface area $= 2\pi rh + 2\pi r^2$

Sphere

> → The formula for the volume of a **sphere** with
> radius r is
>
> $V = \dfrac{4}{3}\pi r^3$
>
> The formula for the surface area of a sphere is
>
> $SA = 4\pi r^2$

Pyramid

Any solid that has a flat base and which comes up to a point
(the **vertex**) is a **pyramid**.

A pyramid takes its name from the shape of its base.

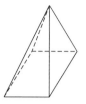

Square based prymid

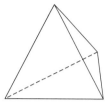

Triangle-based pyramid

A triangle-based pyramid with all edges of equal length is called a tetrahedron.

→ To find the surface area of a pyramid, add together the areas of all the faces.

The volume of a pyramid with height h is
$$V = \frac{1}{3} \times base\ area \times h$$

Cone

A **cone** is a special type of pyramid with a circular base.

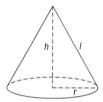

→ The volume of a cone with a circular base of radius r and **perpendicular height** h is given by the formula
$$V = \frac{1}{3} \times \pi r^2 \times h$$

The curved surface area of a cone uses the length of the **slanted height** l
$$CSA = \pi r \times l$$

To find the whole surface area of the cone, add the area of the circular base:
$$SA = \pi r \times l + \pi r^2$$

Example 72

ABCDEF is a wedge
Angle ABC = 90°
AB = 5 cm, BC = 8 cm and CD = 12 cm
Calculate the volume of ABCDEF.

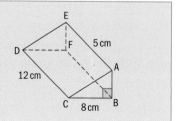

A wedge is a prism with triangular cross-section

▶ Continued on next page

Answer	
Area of triangular cross section $= \frac{1}{2} \times 5 \times 8 = 20$	*Calculate the area of the cross-section*
Volume of wedge $= 20 \times 12 = 240$ cm².	*Volume of prism = area of cross section x length*

Exercise 3I

1 Find the surface area of each shape.

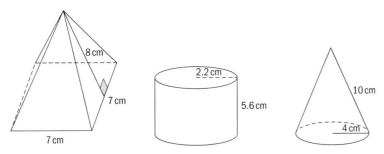

2 Calculate the volume of each shape.

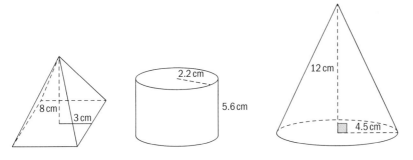

3 Find the height of a cone that has a radius of 2 cm and a volume of 23 cm³.

4 A cylinder has a volume of 2120.6 cm³ and a base radius of 5 cm. What is the volume of a cone with the same height but a base radius of 2.5 cm?

5 Determine the surface area and volume of each sphere.

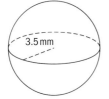

a 3.5 mm

b 15 cm

6 A hemisphere sits on top of a cylinder. Find the surface area and volume.

—6 cm—

5 cm

7 Eight basketballs are put into a holding container. The radius of each basketball is 10 cm. The container is shaped like a square based pyramid with each side of the base measuring 40 cm and with a height of 70 cm. How much space is left in the container?

8 A cylindrical can has a diameter of 9 cm and is 14 cm high. Calculate the volume and surface area of the can to the nearest tenth of a centimetre.

9 Calculate the height of a cylinder that has a volume of 250 cm³ and a radius of 5.5 cm.

10 A cylindrical cardboard tube is 60 cm long and open at both ends. Its surface area is 950 cm². Calculate its radius to the nearest tenth of a centimetre.

3.11 Coordinate geometry

Coordinates

Coordinates describe the position of points in the plane. Horizontal positions are shown on the x-axis and vertical positions on the y-axis.

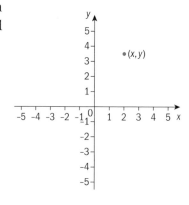

René Descartes introduced the use of coordinates in a treatise in 1637. You may see axes and coordinates described as Cartesian axes and Cartesian coordinates.

Example 73

Draw axes for and $-10 \leq x \leq 10$ and $-10 \leq y \leq 10$.
Plot the points with coordinates: (4, 7), (3, –6), (–5, –2) and (–8, 4).

Answer

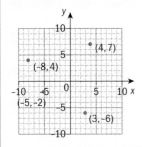

Exercise 3J

1 Draw axes for $-8 \leq x \leq 8$ and $-5 \leq y \leq 10$.
Plot the points with coordinates:
(5, 0), (2, –2), (–7, –4) and (–1, 9).

2 Write down the coordinates of the points shown in this
diagram.

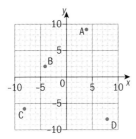

Midpoints

The midpoint of the line joining the points with

coordinates (x_1, y_1) and (x_2, y_2) is given

by $\left(\dfrac{x_1 + x_2}{2}, \dfrac{y_1 + y_2}{2} \right)$.

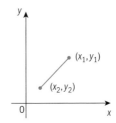

Example 74

Find the midpoint of the line joining the points with coordinates
(1, 7) and (–3, 3).

Answer

The midpoint is $= \left(\dfrac{1+(-3)}{2}, \dfrac{7+3}{2} \right) = (-1,\ 5)$

Exercise 3K

Calculate the midpoints of the lines joining the following pairs of
points.

1 (2, 7) and (8, 3)   **2** (–6, 5) and (4, –7)   **3** (–2, –1) and (5, 6).

Distance between two points

The distance between points with coordinates

$(x_1,\ y_1)$ and $(x_2,\ y_2)$ is given by $\sqrt{(x_2 - x_1)^2 + (y_2 - y_1)^2}$.

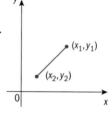

Example 75

Find the distance between the points with coordinates
(2, –3) and (–5, 4).

Answer

Distance $= \sqrt{(-5-2)^2 + (4-(-3))^2} = \sqrt{(-7)^2 + 7^2} = 9.90$

Exercise 3L

Calculate the distance between the following pairs of points. Give
your answer to 3 signifigant figures where appropriate.

1 (1, 2) and (4, 6)

2 (–2, 5) and (3, –3)

3 (–6, –6) and (1, 7)

The gradient of a straight line

The **gradient** of a straight line is a measure of how steep it is.
It is also called the slope.

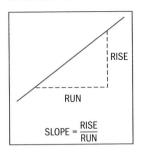

Another way of saying this is

$$\text{gradient} = \frac{\text{Change in } y \text{ values}}{\text{Change in } x \text{ values}}$$

> To find the gradient, measure the vertical increase (rise) between two points and divide by the horizontal increase (run).

Positive gradient

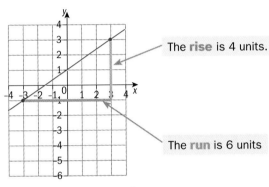

The **rise** is 4 units.

The **run** is 6 units

▲ Gradient = $\dfrac{\text{Rise}}{\text{Run}} = \dfrac{4}{6} = \dfrac{2}{3}$

Negative gradient

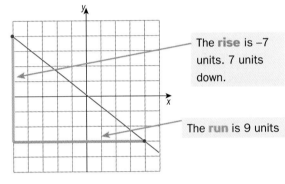

The **rise** is −7 units. 7 units down.

The **run** is 9 units

▲ Gradient = $\dfrac{\text{Rise}}{\text{Run}} = -\dfrac{7}{9}$

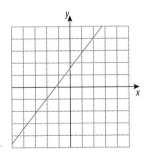

Positive Slope

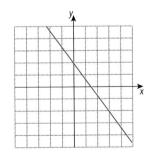

Negative slope

▼ Horizontal lines have a gradient of zero because the rise is zero.

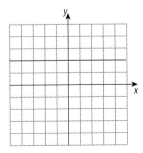

Zero Slope

▼ Vertical lines have an undefined gradient, as the run is zero

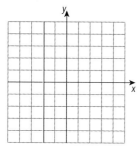

Undefined Slope

Exercise 3M

Find the gradient of each line.

1
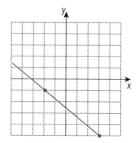

2
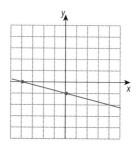

3
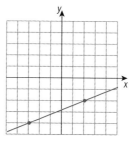

4
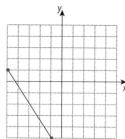

5
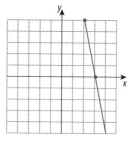

6
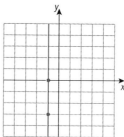

7
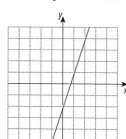

8
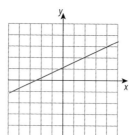

9
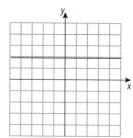

Finding the gradient of a line given two points

→ The gradient of a line is $\dfrac{Rise}{Run}$ which is $\dfrac{\text{The change in } y}{\text{The change in } x}$.

Given two points (x_1, y_1) and (x_2, y_2), $\dfrac{\text{The change in } y}{\text{The change in } x} = \dfrac{y_2 - y_1}{x_2 - x_1}$

Example 76

Find the gradient of the line joining (–3, –2) and (4, –1)

Answer

Gradient $= \dfrac{y_2 - y_1}{x_2 - x_1} = \dfrac{1 - (-2)}{4 - (-3)} = \dfrac{3}{7}$

Exercise 3N

Find the gradient of the line through each pair of points.

1 (19, –16) and (–7, –15) **2** (1, –19) and (–2, –7)

3 (–4, 7) and (–6, –4) **4** (20, 8) and (9, 16)

5 (17, –13) and (17, 7) **6** (14, 3) and (1, 3)

7 (3, 0) and (–11, –15) **8** (19, –2) and (–11, 10)

9 (6, –10) and (–15, 15) **10** (12, –18) and (18, –18)

Parallel and perpendicular lines

Parallel lines have the same gradient

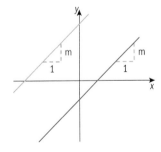

Both of these lines have slope *m*

Perpendicular lines have slope *m* and $-\dfrac{1}{m}$

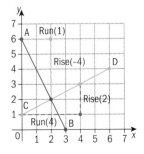

Line CD has slope $\dfrac{1}{2}$

Line AB has slope –2

Notice that the product of perpendicular gradients is –1.

$-2 \times \dfrac{1}{2} = -1$

Exercise 30

1 **a** Which of these gradients would give parallel lines?
  b Which are perpendicular?

$$3, -3, \frac{1}{3}, 4.5, \frac{2}{3}, \frac{2}{9}, \frac{9}{2}, -\frac{2}{9}, -1.5, \frac{6}{2}$$

2 State if the lines in each pair are parallel, perpendicular or neither.

  a Line A through (2, 5) and (0, 1) and line B though (4, 10) and (5, 12).
  b Line C through (3, 14) and (–2, –6) and line D though (12, –3) and (20, –5)
  c Line E through (1, 10) and (5, 15) and line F through (2, 2) and (4, 2).
  d Line G through (5, 7) and (2, 4) and line H through (8, –5) and (4, –1).
  e Line I through (4, 11) and (10, 20) and line J through (2, 1) and (6, 7)

Equations of lines

A straight line is defined by a linear equation of the form

$$y = \mathbf{m}x + \mathbf{c}$$

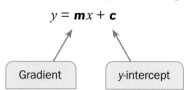

> This is called the gradient-intercept form. Some people use $y = ax + b$

Example 77

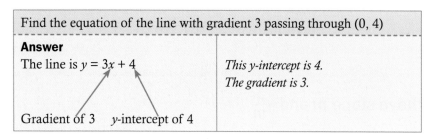

Find the equation of the line with gradient 3 passing through (0, 4)

Answer
The line is $y = 3x + 4$

This y-intercept is 4.
The gradient is 3.

Gradient of 3   y-intercept of 4

Using the gradient formula to find the equation of a line

Consider a line with a fixed point (x_1, y_1) and a general point (x, y).

Then $m = \dfrac{y - y_1}{x - x_1}$

or $y - y_1 = m(x - x_1)$.

Example 78

Find the equation of the line with gradient $m = 3$ passing through $(x_1, y_1) = (6, 12)$

$y - y_1 = m(x - x_1)$.
$y - 12 = 3(x - 6)$
$y - 12 = 3x - 18$
$\quad\quad y = 3x - 6$

Exercise 3P

Find the equation of each line in gradient-intercept form

1 Gradient 3, passing through (1, 5)

2 Gradient 4, passing through (5, 11)

3 Gradient 2.5, passing through (4, 12)

4 Gradient $\dfrac{1}{2}$, passing through (12, 20)

5 Gradient 5, passing through (–2, –13)

6 Gradient –3, passing through (1, 1)

7 Gradient –2, passing through (–3, –1)

8 Gradient $-\dfrac{1}{2}$, passing though (–4, –3)

9 Find the equation of the line passing through (2, 7) and (5, 19).

10 Find the equation of the line passing through (–1, –3) and (–5, –11)

4 Statistics

4.1 Statistical graphs

In a statistical investigation we collect information, known as **data**. To represent this data in a clear way we can use graphs. Three types of statistical graph are bar charts, pie charts and pictograms.

Bar charts

A **bar chart** is a graph made from rectangles, or bars, of equal width whose length is proportional to the quantity they represent, or frequency. Sometimes we leave a small gap between the bars.

Example 79

Juliene collected some data about the ways in which her class travel to school.

Type of transport	Bus	Car	Taxi	Bike	Walk
Frequency	7	6	4	1	2

Represent this information in a bar chart.

Answer

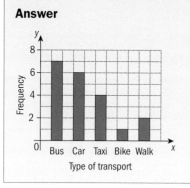

Example 80

Lakshmi collected data from the same class about the number of children in each of their families.

No. of children	1	2	3	4	6
Frequency	3	9	5	2	1

Represent this information in a bar chart.

Answer

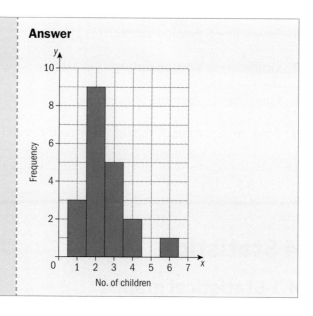

Pie charts

A **pie chart** is a circle divided into sectors, like slices from a pie.
The sector angles are proportional to the quantities they represent.

Example 81

Use Juliene's data from Example 78 to construct a pie chart.

Answer

Type of transport	Frequency		Sector angle
Bus	7	$\frac{7}{20} \times 360°$	126°
Car	6	$\frac{6}{20} \times 360°$	108°
Taxi	4	$\frac{4}{20} \times 360°$	72°
Bike	1	$\frac{1}{20} \times 360°$	18°
Walk	2	$\frac{2}{20} \times 360°$	36°

The total of the frequencies is 20. The total angle for the whole circle is 360°.

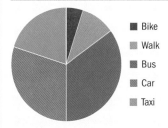

- Bike
- Walk
- Bus
- Car
- Taxi

Start by drawing a radius and then measure, with your protractor, each angle in turn. The total of the sector angles should be 360°.

Pictograms

Pictograms are similar to bar charts, except that pictures are used. The number of pictures is proportional to the quantity they represent. The pictures can be relevant to the items they show or just a simple character such as an asterisk.

Example 82

Use Juliene's data from Example 79 to construct a pictogram.

Answer

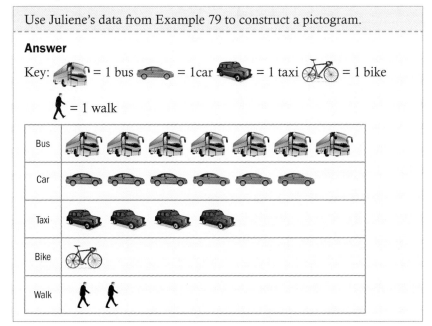

In this pictogram, different symbols are used for each category but the symbols describe the category as well.

Example 83

Use this data on the number of children in a sample of families to construct a pictogram.

Number of children	1	2	3	4	6
Frequency	4	9	6	2	1

Answer

No. of children

1	△△△△
2	△△△△△△△△△
3	△△△△△△
4	△△
6	△

Key: △ = 1 child

Exercise 4A

1 Adam carried out a survey of the cars passing by his window on the road outside. He noted the colors of the cars that passed by for 10 minutes and collected the following data.

Color	Black	Red	Blue	Green	Silver	White
Frequency	12	6	10	7	14	11

Draw a bar chart, a pie chart and a pictogram to represent the data.

2 Ida asked the members of her class how many times they had visited the cinema in the past month. She collected the following data.

Number of times visited	1	2	3	4	8	12
Number of students	4	7	4	3	1	1

Draw a bar chart, a pie chart and a pictogram to represent the data.

Stem and leaf diagrams

Stem and leaf diagrams provide a simple means of organizing raw data without losing any of the detail.

They are also called stem plots.

Here is some data on the weights of 20 people (in kg).

50, 47, 53, 88, 75, 62, 49, 83, 57, 69, 71, 73, 73, 66, 51, 44, 78, 66, 54 and 80

You can draw a stem and leaf diagram for this data.

The 'stem' is the tens, and the 'leaves' are the units.

You must give a key for a stem and leaf diagram.

4	4 7 9
5	0 1 3 4 7
6	2 6 6 9
7	1 3 3 5 8

Key

6|2 means 62 kg

The key explains what the stem and leaf data means

The leaves are the units digits written in order

The stem is the 10s digit

Exercise 4B

1   The test scores out of 50 for a math class are:
21, 23, 25, 26, 28, 30, 30, 30, 33, 36, 37, 39, 39, 40, 41, 42, 42,
42, 42, 46, 49, 50, 54.
Show this on a stem and leaf diagram.

2   The number of advertisements in different issues of a magazine
are:
164, 176, 121, 185, 148, 149, 177, 151, 157, 152, 163, 145,
123, 176
Show this on a stem and leaf diagram.

> Use this key:
> 16|4 means 164 advertisements

3   The waiting time, in minutes, at the dentist's surgery was
recorder for 24 patients as:
55, 26, 27, 53, 19, 28, 30, 29, 22, 44, 48, 48, 37, 46, 62, 57, 49,
42, 25, 34, 58, 43, 52, 36.
Show this on a stem and leaf diagram.

4   The number of tomatoes produced on different plants in a
garden is given below:
11, 34, 14, 23, 56, 36, 28, 19, 26, 35, 24, 30, 51, 18, 14, 16, 27,
29, 38, 26.
Show this on a stem and leaf diagram.

5   The times, in seconds, for scouts to tie a knot were:
4.6, 2.2, 3.1, 4.2, 5.2, 4.3, 6.0, 7.3, 7.4, 3.2,
3.3, 6.3, 3.2, 2.3, 2.5, 6.4, 5.2, 2.5, 2.9, 5.2, 5.4, 4.3, 4.8, 4.7
Show this on a stem and leaf diagram.

> Use the whole number part as the stem, and the tenths as the leaves.

4.2 Data analysis

> → **Discrete data** can only take specific values. Discrete data is often counted.

For example:
- the number of children in your family – the values can only be whole numbers.
- UK shoe sizes – $2, 2\frac{1}{2}, 3, 3\frac{1}{2}, 4, 4\frac{1}{2}, 5, 5\frac{1}{2}, 6, 6\frac{1}{2}, \ldots$

> → **Continuous data** can take any value within a certain range. Continuous data is measured, and its accuracy depends on the measuring instrument used.

For example:

- the time taken to run 100 m may be 14.4 sec or 14.43 secs or 14.428 sec etc. depending on the measuring instrument.

Exercise 4C

State whether each set of data is discrete or continuous.

1 The number of cars in a school car park.

2 The number of books in a library.

3 The length of your pencil.

4 The time that it takes you to rum 400 m.

5 The speed of a car.

6 The number of friends that you have.

7 The number of shoes that you own.

8 The mass of a table.

9 The distance from the Earth to the Sun.

Measures of central tendency

A measure of central tendency, or **average**, describes a typical value for a set of data.

There are three common types of average:

- The **mode** – this is the data value that occurs most often.
- The **median** – this is the middle item when the data is arranged in order of size.
- The **mean** – this is what most people mean when they use the word 'average'. It is found by adding up all of the data and dividing by the number of pieces of data.

Example 84

Find **a** the mode **b** the median and **c** the mean of this data set:

2, 5, 4, 9, 1, 3, 2, 6, 9, 2, 5, 13, 4

Answers

a The mode is 2

2 occurs the most often

b 1, 2, 2, 2, 3, 4, 4, 5, 5, 6, 9, 9, 13

write them in order and find the middle one

The median is 4

c Mean $= \dfrac{1+2+2+2+3+4+4+5+5+6+9+9+13}{13}$

Add them all together. There are 13 pieces of data, so divide by 13.

$= \dfrac{65}{13} = 5$

Exercise 4D

1 Find **a** the mode **b** the median and **c** the mean of
 a 1, 4, 1, 5, 6, 7, 3, 1, 8
 b 4, 7, 5, 12, 5, −3, −2
 c 2, 3, 8, 2, 1, 7, 9, 8, 5
 d 25, 28, 29, 21, 25, 20, 27
 e 7.4, 10.2, 12.5, 6.8, 10.2

2 Fifteen students were asked how many brothers and sisters they had. The results were:
 2, 2, 1, 0, 3, 5, 2, 1,1, 0, 1, 4, 1, 0, 2.

 Find **a** the mode, **b** the median and **c** the mean number of brothers and sisters.

3 My last nine homework scores, marked out of 10, were:
 8, 7, 9, 10, 8, 9, 6, 8, 7

 Find **a** the mode **b** the median and **c** the mean homework score.

4 A sprinter's times in seconds for the 40 m dash were:
 5.13, 4.82, 5.25, 4.94, 5.06, 4.82, 5.12

 Find **a** the mode, **b** the median and **c** the mean of the times.

5 Seven farmers own different numbers of chickens.
 These numbers are:
 253, 78, 497, 166, 710, 497 and 599

 Find **a** the mode, **b** the median and **c** the mean number of chickens.

Measures of dispersion

A measure of dispersion is a value that describes the spread of a set of data.

The **range** and **interquartile range** are two measures of dispersion.

The range shows how spread out the data is.

> → Range = highest value – lowest value

The **quartiles** divide a set of data into four equal amounts.

> → The **lower quartile** Q_1 is 25% of the way through the data and
> its position is found using the formula:
>
> $Q_1 = \left(\dfrac{n+1}{4}\right)^{th}$ where n is the number of items in the data set.
>
> The **upper quartile** Q_3 is 75% of the way through the data and
> its position is found using the formula
>
> $Q_3 = 3\left(\dfrac{n+1}{4}\right)^{th}$
>
> The **interquartile range** shows how spread out the middle
> 50% of the data is.
>
> Inter quartile range = $Q_3 - Q_1$

Example 85

Here are the shoe sizes of fifteen boys:

42, 42, 38, 40, 42, 40, 34, 46, 44, 36, 38, 40, 42, 36, 42

Find   **a**   the range and   **b**   the interquartile range.

Answers

a 34, 36, 36, 38, 38, 40, 40, 40,
42, 42, 42, 42, 42, 44, 46

range = 46 – 34 = 12

lower quartile = $\dfrac{16}{4}$th

value = 4th value

= 38

Upper quartile = 3 × 4th
value = 12th value

= 42

So interquartile range
= 42 – 38 = 4

To find the interquartile range, first arrange the data in order of size

n = 15

Exercise 4E

1 Here are the shoe sizes of fifteen girls:

26, 28, 28, 36, 34, 32, 30, 34, 32, 28, 36, 38, 34, 32, 30

Find **a** the range and **b** the interquartile range of the shoe sizes.

2 23 students were asked how many pets they had at home. Here are the replies:

1, 4, 3, 5, 3, 2, 8, 0, 2, 1, 3, 2, 4, 2, 1, 0, 1, 2, 6, 7, 2, 8, 2

Find **a** the range and **b** the interquartile range for the number of pets.

3 The average daily temperatures in °C in Chillton during January were

−6, −4, −4, −2, −1, 0, 4, 5, 7, 4, 2, 1, 0, −3, −4, −6,
−7, −5, −3, −1, 1, 3, 4, 7, 7, 8, 3, −2, 0, −2, −5

Find **a** the range and **b** the interquartile range for the daily temperatures.

4 The grocer sells potatoes by the kilogram.
I bought 1kg of potatoes every day of the week and counted the number of potatoes each time. Here are the results:

Day	Monday	Tuesday	Wednesday	Thursday	Friday	Saturday	Sunday
Potatoes	18	15	20	17	14	12	15

Find **a** the range and **b** the interquartile range for the number of potatoes in 1 kilogram.

5 The time (in seconds) taken for eleven players in a soccer team to prepare for a free kick is given.

12.4, 2.45, 3.75, 10, 3.5, 8.4, 9.6, 23.5, 2.48, 15.6, 5.2

Find **a** the range and **b** the interquartile range for the time taken.

15 Practice paper 1

Time allowed: 1 hour 30 minutes
- Answer all the questions
- Unless otherwise stated in the question, all numerical answers must be given exactly or correct to three significant figures.

Full marks are not necessarily awarded for a correct answer with no working. Answers must be supported by working and/or explanations. Where an answer is incorrect, some marks may be given for a correct method, provided this is shown by written working. You are therefore advised to show all working.

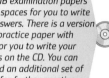

SECTION A

1 Given that $4\ln 2 - 3\ln 4 + \ln k = 0$, find the value of k. *[4 marks]*

2 a Show that $p(x) = 2x^3 - 3x^2 + 8x + 5$ is divisible by $2x + 1$. *[4 marks]*
 b Hence find all the zeros of $p(x)$. *[4 marks]*

3 The sum of the first two terms of a geometric sequence is $\dfrac{8}{9}$ and the sum of the first three terms is $\dfrac{26}{27}$.
Find possible values of the first term and the sum of all its terms. *[8 marks]*

4 Consider the events A and B such that $P(A) = 0.3$ and $P(B) = 0.2$.

Given that $P(A \cup B) = 3P(A \cap B)$, find $P(A|B)$ and $P(A \cap B')$. *[6 marks]*

5 Show that for any complex number z,
 a $z + z^* = 2\mathrm{Re}(z)$ **b** $z - z^* = 2\mathrm{i}\,\mathrm{Im}(z)$ **c** $\mathrm{Re}(z) \le |z|$ *[6 marks]*

6 Find a vector equation of the line of intersection of the planes with equations $x + 2y - z = 5$ and $-3x - y + z = 1$. *[6 marks]*

7 A curve is defined by the equation $x^2 + 4y^2 - 2x + 16y + 13 = 0$.
Find the coordinates of the points on the curve where the tangent to the curve is parallel to the x-axis. *[6 marks]*

8 Use integration by parts to find the rational values of a and b such that

$$\int_1^9 \sqrt{x} \ln x \, dx = a \ln 3 + b.$$

[5 marks]

9 Consider the function f defined by $f(x) = \begin{cases} 0.1x & \text{if } 0 \le x \le 1 \\ 0.1(5x-4) & \text{if } 1 < x \le 2 \\ ax+b & \text{if } 2 < x \le c \\ 0 & \text{otherwise} \end{cases}$

  a Given that f is a continuous pdf of a variable X, find the values of a, b and c.

  b Hence state the value of the mode of X. *[7 marks]*

10 Consider the function defined by $f(x) = 12\sin x - 5\cos x$.
  Find the range of f. *[4 marks]*

SECTION B

11 Consider the lines

$$L_1 : \frac{x-2}{1} = \frac{y-1}{2} = \frac{z}{3} \text{ and } L_2 : \frac{x-1}{4} = \frac{y-2}{1} = \frac{z-3}{-2}$$

  a Show that the lines intersect and find their point of intersection. *[5 marks]*
  b Hence find the equation of the plane that contains both lines. *[4 marks]*
  c Show that the point A(1, −1, 0) does not lie on the plane π. *[2 marks]*
  d Write down the equation of the line L_3 perpendicular to
    the plane π that contains the point A. *[1 mark]*
  e Hence find the distance from A to the plane π. *[7 marks]*

12 a Prove by mathematical induction that

$$0^2 + 1^2 + 2^2 + \ldots + n^2 = \frac{n(n+1)(2n+1)}{6}, \text{ for all } n \in \mathbb{N}.$$

[7 marks]

  b Hence find an expression for $3^2 + 6^2 + \ldots + (3n)^2$. *[4 marks]*

  c Given that $A_n = 1^2 + 4^2 + \ldots + (3n-2)^2$ and $B_n = 2^2 + 5^2 + 8^2 + \ldots + (3n-1)^2$,
    prove that $A_n + B_n = 6n^3 - n$ and $A_n - B_n = -3n^2$
    Hence find A_n and B_n in terms of n. *[9 marks]*

13 Let $f : x \rightarrow e^{\cos x}$, where $-\dfrac{\pi}{2} < x < \dfrac{\pi}{2}$

    a State with a reason whether or not the function f is even. *[2 marks]*

    b Find $f'(x)$. *[2 marks]*

    c Given that the graph of f has a maximum point, find its coordinates. *[5 marks]*

    d Show there is a point of inflexion on the graph of f, for $0 < x < \dfrac{\pi}{2}$ and find its coordinates. *[6 marks]*

    e Sketch the graph of f. *[1 mark]*

    f A rectangle is drawn so that its lower vertices are on the x-axis and its upper vertices are on the curve $y = e^{\cos x}$ where $-\dfrac{\pi}{2} < x < \dfrac{\pi}{2}$

        i Write down an expression for the area of the rectangle. *[1 mark]*

        ii Show that there is a positive value $x = a$ for which the area of the rectangle reaches a maximum.

           Hence show that its value is given by $2ae^{\frac{\sqrt{a^2-1}}{a}}$ *[4 marks]*

> Use the mark scheme in the Answer section at the back of this book to mark your answers to this practice paper.

Practice paper 2

Time allowed: 1 hour 30 minutes
- Answer all the questions
- Unless otherwise stated in the question, all numerical answers must be given exactly or correct to three significant figures.

Full marks are not necessarily awarded for a correct answer with no working. Answers must be supported by working and/or explanations. Where an answer is incorrect, some marks may be given for a correct method, provided this is shown by written working. You are therefore advised to show all working.

Practice exam papers on CD: *IB examination papers include spaces for you to write your answers. There is a version of this practice paper with space for you to write your answers on the CD. You can also find an additional set of papers for further practice.*

Worked solutions on CD: *Detailed worked solutions for this practice paper are given as a PowerPoint presentation on the CD.*

SECTION A

1 A rental agreement says that the yearly rent on an office shall increase by €600 each year. In the fifth year of the agreement the rent was €12200.
Find:
  a the rent paid in the first year *[2 marks]*
  b the total amount paid in the first 5 year. *[2 marks]*
  c the first year that the annual rent will exceed €15000 *[3 marks]*

2 The complex numbers w and z are such that $w = \dfrac{az+b}{z+c}$, $a, b, c \in \mathbb{R}$.

  a Given that $w = 3i$ when $z = -3i$ and $w = 1 - 4i$ when $z = 1 + 4i$, show that $b = 9$ and find the values of a and c. *[5 marks]*
  b Hence, show that if Re $z = 4$ then Re $w = 4$. *[2 marks]*

3 A random variable X is normally distributed with mean and variance both equal to a. Given that $P(X < 2) = 0.3$, find the value of a. *[4 marks]*

4 Consider the function defined by $f(x) = \dfrac{1-x^2}{x}$
  a Find the first and second derivatives and hence show that the graph of f has no maxima, no minima or points of inflexion. *[5 marks]*
  b Hence sketch the graph of f, showing clearly the intercepts and any asymptotes. *[3 marks]*

5 Given the points A(1, −3, −1) and B(−5, 2, −4), find the coordinates of the point P that lies on the segment [AB] and is such that AP : PB = 1 : 2. *[6 marks]*

6 Given that $(1 + x)^5 (1 + ax)^6 \equiv 1 + bx + 10x^2 + \ldots + a^6 x^{11}$, find the values of integers a and b. *[6 marks]*

7 Use substitution to find $\displaystyle\int \frac{e^{2z}}{4 + e^{4x}}\,dx$ *[6 marks]*

8 The hands of a clock are 20 cm and 15 cm long. Let θ be the angle between the hands at any time t between 14:45 and 15:15.

  a Express the distance d between the tips of the hands at the time t in terms of θ. *[2 marks]*

  b Assuming that the movement of the hands of the clock is continuous, find the rate of change of θ in radians per minute. Hence find the rate of change of d at three o'clock, in cm per minute. *[6 marks]*

9 Consider the sequence (u_n) defined by $\begin{cases} u_1 = 2 \\ u_{n+1} = \dfrac{u_n + 1}{3}, \ n \in \mathbb{Z}^+ \end{cases}$

Investigate the numerical behaviour of the terms of the sequence and deduce that $u_n = \dfrac{3^{2-n} + 1}{2}$ *[8 marks]*

SECTION B

10 A box contains a very large number of ribbons of which 25% are red, 30% are white and the rest are blue. Twelve ribbons are selected at random from the box.

  a Find the expected number of red ribbons selected. *[1 mark]*

  b Find the probability that exactly six of these ribbons are blue. *[3 marks]*

  c Find the probability that at least two of these ribbons are blue. *[2 marks]*

  d Find the most likely number of white ribbons selected. State any assumptions you have made about the probability of selecting a white ribbon. *[4 marks]*

There are two other boxes with large number of coloured ribbons: one where 25% of the ribbons are blue, 25% white and the others red, and another box where 50% of the ribbons are white, 20% blue and the rest red.

  e Kathy picks a box at random and, without looking, takes a ribbon out. If she takes a white ribbon out, what is the probability that this ribbon was taken from the first box? State any assumptions made. *[4 marks]*

11 The points A(1, 2, −3), B(2, −1, 0) and C(−1, 0, 3) are given.

    a Find the vector equation of a line (AB) that passes through the points A and B. *[3 marks]*

    b Find the midpoint M of the line segment [AB]. Hence show that the equation of the plane α perpendicular to [AB] that bisects the line segment [AB] is

$$2x - 6y + 6z = -9$$ *[3 marks]*

    c Show that the equation of the plane β perpendicular to [AC] that bisects the line segment [AC] is

$$x + y - 3z = 1.$$ *[3 marks]*

    d Find the angle between the planes α and β.

      The plane γ perpendicular to [BC] that bisects the line segment [BC] has an equation

$$6x - 2y - 6z = -5.$$ *[4 marks]*

    e Show that the planes α, β and γ intersect and find the vector equation of the line of intersection. *[4 marks]*

    f Consider the plane π defined by the equation $x + y + z = 0$
Find the coordinates of the point P on the plane π that is at the same distance from the points A, B and C. *[6 marks]*

12 Let $f(x) = \cos(2x) + 1$ and $g(x) = \dfrac{e^x + e^{-x}}{2}$

    a Show that both functions are even. *[3 marks]*

    b Find the derivatives $f'(x)$ and $g'(x)$. *[4 marks]*

    c Show that both derivative functions are odd. *[2 marks]*

    d Sketch the curves $y = f(x)$ and $y = g(x)$ and find their points of intersection. *[4 marks]*

    e Show that the tangents to the curves $y = f(x)$ and $y = g(x)$ at the point of intersection in the first quadrant have equations $y = -1.95x + 2.53$ and $y = 0.719x + 0.751$ respectively. *[3 marks]*

    f Find the area of the region enclosed by all four tangents to the curves $y = f(x)$ and $y = g(x)$ at the points of intersection. *[3 marks]*

    g The region enclosed by the curves $y = f(x)$ and $y = g(x)$ is rotated by 2π about the x-axis. Find the volume of revolution generated. *[4 marks]*

Answers

Chapter 1

Skills check

1. a $\{1, 2, 3, 4, 5\}$
   b $\{-4, -3, -2, -1, 0, 1\}$
   c $\{1, 2, 3, 4, 5, 6\}$
2. a 26   b –12   c 2
3. a $3\sqrt{3}-7$ b $\dfrac{13\sqrt{2}}{2}$
   c $-2-\sqrt{3}$
4. a $x = 5$ b $x = -\dfrac{1}{2}$
5. a 35   b –10

Investigation - curious numbers

Tiles along diagonal	Tiles along each edge	Tiles total
9	5	25
13	7	49
133	67	4489
1333	667	444889
13333	6667	44448889

Exercise 1A

1. a $0, 1.5, 3$
   b $\dfrac{9}{10}, \dfrac{11}{12}, \dfrac{13}{14}$
   c $\dfrac{1}{99}, \dfrac{1}{143}, \dfrac{1}{195}$
2. a $r(r+1)$
   b $\dfrac{1}{r^2+1}$
   c $2r-3$
3. a $1, 5, 9, 13$
   b $\dfrac{1}{3}, \dfrac{2}{5}, \dfrac{3}{7}, \dfrac{4}{9}$
   c $1, \dfrac{1}{4}, \dfrac{1}{9}, \dfrac{1}{16}$
4. a $2 + 6 + 12 + 20$
   b $\dfrac{1}{3} + \dfrac{2}{5} + \dfrac{3}{7} + \dfrac{4}{9} + \dfrac{5}{11}$
   c $-1 + 4 - 9 + 16 - 25$
5. a $\displaystyle\sum_{r=1}^{\infty}(4r-5)$ b $\displaystyle\sum_{r=1}^{10}(-1)^r$
   c $\displaystyle\sum_{r=1}^{6}6\times(-2)^{r-1}$

Investigation - quadratic sequences

$n^2 - 2n + 3$

$n = p - 1 \Rightarrow n^2 - 2n + 3$
$\qquad = (p-1)^2 - 2(p-1) + 3$

$n = p \Rightarrow n^2 - 2n + 3 = p^2 - 2p + 3$

$n = p + 1 \Rightarrow n^2 - 2n + 3$
$\qquad = (p+1)^2 - 2(p+1) + 3$

Differences $2p - 3$ and $2p - 1$
Second difference
$(2p - 1) - (2p - 3) = 2$

$\underline{2n^2 + 2n + 1}$

Entering data for $2n^2 + 2n + 1$ into the GDC gives the second difference is always equal to 4

$\underline{-n^2 + 3n - 4}$

The second difference is always –2

Conjecture: The second difference appears to be constant and equal to twice the coefficient of n^2.

Proof:
Three consecutive terms of $N = an^2 + bn + c$ are N_{p-1}, N_p, N_{p+1}

$N_{p-1} = a(p-1)^2 + b(p-1) + c$

$N_p = ap^2 + bp + c$

$N_{p+1} = a(p+1)^2 + b(p+1) + c$

First differences:
$N_p - N_{p-1} = 2ap - a + b$

$N_{p+1} - N_p = 2ap + a + b$

Second difference equals $2a$ which proves the conjecture.

Investigation - triangular numbers

Second difference is 1 which indicates that the numbers are generated by a quadratic formula $an^2 + bn + c$ where $a = \dfrac{1}{2}$

$N_1 = \dfrac{1}{2} + b + c = 1 \Rightarrow b + c = \dfrac{1}{2}$

$N_2 = \dfrac{1}{2} \times 4 + 2b + c = 3 \Rightarrow 2b + c = 1$

$\therefore b = \dfrac{1}{2}, c = 0$

For triangular numbers:

$N = \dfrac{1}{2}n(n+1)$

Investigation - more number patterns

Repeating the procedure we obtain:

Polygon numbers	N	Number of sides of polygon	N
Triangular	$\dfrac{n(n+1)}{2}$	3	$\dfrac{n}{2}(n+1)$
Square	n^2	4	$\dfrac{n}{2}(2n+0)$
Pentagonal	$\dfrac{n(3n-1)}{2}$	5	$\dfrac{n}{2}(3n-1)$
Hexagonal	$n(2n-1)$	6	$\dfrac{n}{2}(4n-2)$
Heptagonal	$\dfrac{n(5n-3)}{2}$	7	$\dfrac{n}{2}(5n-3)$
Polygon with k sides		k	$\dfrac{n}{2}[(k-2)n$ $- (k-4)]$

Term	1st	2nd	3rd	4th	5th	6th	7th	8th
Triangular numbers	1	3	6	10	15	21	28	36
Square numbers	1	4	9	16	25	36	49	64
Pentagonal numbers	1	5	12	22	35	51	70	92
Hexagonal numbers	1	6	15	28	45	66	91	120
Heptagonal numbers	1	7	18	34	55	81	112	148
Octagonal numbers	1	8	21	40	65	96	133	176
Nonagonal numbers	1	9	24	46	75	111	154	204

Each column in the table is made up of numbers with a constant difference.

Exercise 1B

1. a $u_n = 6n - 1$
   b $u_n = -7n + 17$
   c $u_n = a - 2 + 2n$
2. a 128
   b $\dfrac{51}{4}$
   c $4n - 1$
3. $33, 38 - 5n$
4. $4, -12$
5. €55000, 48 years

Exercise 1C

1. a. 522
   b. 108
   c. −1870
2. a. 345
   b. −285
3. 120
4. −1, 2, 5, 8, 11
5. $d = 4$ 5, 9, 13, 17

Exercise 1D

1. a. $u_6 = 32$
      $un = 1 \times 2n^{-1}$
   b. $u_6 = \dfrac{1}{27}$
      $u_n = \left(\dfrac{1}{3}\right)^{n-3}$
   c. $u_6 = \left(\dfrac{1}{x}\right)^2$
      $u_n = \left(\dfrac{1}{x}\right)^{n-4}$
2. a. $\dfrac{1}{2}, \dfrac{3}{32}$
   b. $-\dfrac{1}{6}, \dfrac{1}{243}$
3. a. 7
   b. 9
4. $\pm 3, \pm \dfrac{2}{3}$
5. $\dfrac{27}{4}$
6. $-\dfrac{1}{3}, \dfrac{5}{2}$

Investigation – convergent series

n	3^n	2^n	$\left(\dfrac{3}{2}\right)^n$	$\left(\dfrac{1}{2}\right)^n$	$\left(-\dfrac{1}{5}\right)^n$	$\left(\dfrac{3}{4}\right)^n$
1	3	2	1.5	0.5	−0.2	0.75
2	9	4	2.25	0.25	0.04	0.5625
3	27	8	3.375	0.125	−0.008	0.421875
4	81	16	5.0625	0.0625	0.0016	0.31640625
5	243	32	7.59375	0.03125	−0.00032	0.237304688
6	729	64	11.390625	0.015625	0.000064	0.177978516
7	2187	128	17.0859375	0.0078125	−0.0000128	0.133483887
8	6561	256	25.62890625	0.00390625	0.00000256	0.100112915
9	19683	512	38.44335938	0.001953125	−0.000000512	0.075084686
10	59049	1024	57.66503906	0.000976563	1.024×10^{-7}	0.056313515

Exercise 1E

1. a. $\dfrac{63}{16}$ b. $-\dfrac{1261}{64}$
   c. $\dfrac{683}{512}$ or 1.33 (3 sf)
   d. 0.125 (3 sf)
2. a. $\dfrac{3906}{25}$ b. $10^n - 1$
3. $\dfrac{1365}{32}$ or $\dfrac{819}{32}$
4. a. $\dfrac{1}{2}, \dfrac{3}{4}, \dfrac{9}{8}$
5.
7. $\sqrt{2}, \dfrac{2}{7}\left(3 - \sqrt{2}\right)$ metres
8. $-4 < x < 2, \dfrac{185}{216}$
9. $r = \dfrac{k}{k+1}$

Exercise 1F

1. a. $\dfrac{1}{2}$
   b. 3.23%
2. $4000
3. 62, 19
4. a. $47.07   b. $1130

Exercise 1G

4.

3	4	$3 \times 4 + 4$	16
7	8	$7 \times 8 + 4$	64
−6	−5	$-6 \times -5 + (-5)$	25
11	12	$11 \times 12 + 12$	144
8	9	$8 \times 9 + 9$	81

$n(n + 1) + (n + 1) = (n + 1)^2$

Exercise 1H

all proofs

Exercise 1I

all proofs

Exercise 1J

1.

$8! - 7!$	$7 \times 7!$
$10! - 9!$	$9 \times 9!$
$5! - 4!$	$4 \times 4!$
$95! - 94!$	$94 \times 94!$
$(n+1)! - n!$	$n \times n!$

2. a. $\dfrac{1}{30}$ b. 1
   c. 241 920
3. a. $\dfrac{1}{n}$ b. $(n-1)^2$
   c. $(n! - 1)$

Exercise 1K

1. 15 600
2. a. 479 001 600   b. 41 472
3. 70 weeks
4. a. 4845   b. 4280
5. a. 1176   b. 294   c. 420
6. 17 576 000

Exercise 1L

1. proofs
2. a. $1 + 22x + 220x^2 + 1320x^3 + \dots$
   b. $1 - 21x + 189x^2 - 945x^3 + \dots$
   c. $32 + 400x + 2000x^2 + 5000x^3 + \dots$
   d. $512 - 768x + 512x^2 - \dfrac{1792}{9}x^3$
3. a. $-2240x^3$ b. $\dfrac{95}{2}x^2$
   c. $-1792a^5b^3$
4. 126 720
5. $32 + 16x + \dfrac{16x^2}{5} + \dfrac{8x^3}{25} + \dfrac{2x^4}{125} + \dfrac{x^5}{3125}$, 32.80804
6. a. $49 - 20\sqrt{6}$
   b. $\dfrac{13}{5}\sqrt{2} + \dfrac{31}{5}\sqrt{5}$
   c. $248\sqrt{7}$
7. a, b Proof
   c. $(a - b)(a + b)(a^2 + b^2)$
   d, e Proof

Review exercise (non-GDC)

1 4, 16, 64 and 64, 16, 4

2 736

3 $3, -\dfrac{3}{2}, -6$

4 1, 3, 7, 15, 31, 63

5 proof

6 a $\dfrac{(n+1)!}{4!(n-3)!}$

   b $\dfrac{(n-1)!}{2!(n-3)!}$

   c 8

7 a 2^n b 0

Review exercise (GDC)

1 a $\dfrac{1}{\sqrt{2}}, \dfrac{1}{2}, \dfrac{1}{2\sqrt{2}}$

   b 4.86   c 5.12

   d 0.249   e 0.25

2 a 908 107 200

   b 14688

   c 384

3 66

4 20412

5 1001, 2002, 3003

Chapter 2

Skills check

1 $y = \left(x - \dfrac{3}{2}\right)^2 - \dfrac{13}{4}$;

   vertex $\left(\dfrac{3}{2}, -\dfrac{13}{4}\right)$;

   axis of symmetry $x = \dfrac{3}{2}$

2 a $x = -\dfrac{4}{3}$ b $x = -\dfrac{1}{3}$ or 1

3 a $x = (y - 3)^2 + 2$

   b $x = \dfrac{2y+1}{2-3y}$

Exercise 2A

1 a Yes. D = {0, 1, 2, 3};
      R = {−1, 1, 2, 3}

   b Yes. D = {−3, −2, −1, 0};
      R = {0}

   c Not a function

   d Not a function

2 a Yes.
      Domain = $\{x \mid -3 \le x \le 3\}$;
      Range = $\{y \mid 0 \le y \le 3\}$

   b not a function

   c not a function

   d Yes.
      Domain = $\{x \mid x \ge -1\}$;
      Range = $\{y \mid y \ge 0\}$

Investigation – quadratic graphs

(h, k) is the vertex. When $h < 0$, the graph moves h units to the left, and when $h > 0$ the graph moves h units to the right. When $k > 0$, the graph moves k units upward, and when $k < 0$, it moves k units downward.

Exercise 2B

1 $y^2 = x$ is not a function since for two different y values there is the same x value, e.g., when $x = 1$ $y = \pm 1$. $y = \sqrt{x}$ is the positive square root of x so for each value of x, there is only one value of y.

2 a $y = (x - 2)^2 - 2$; Domain is the set of Reals, and Range = $\{y \mid y \ge -2\}$

   b Domain is the Reals,
      Range = $\{y \mid y \le -3\}$

   c Domain = $\{x \mid x \ge -2\}$,
      Range is the set of non-negative Reals.

   d Domain = $\{x \mid x \le 3\}$;
      Range is the set of non-negative Reals.

   e Domain is the set of Reals; Range = $\{y \mid y \le 2\}$.

   f Domain = $\{x \mid x \le 2\}$,
      Range = non-negative Reals.

Investigation – absolute value functions

a When $h < 0$, the graph moves h units to the left, and when $h > 0$ the graph moves h units to the right. When $k > 0$, the graph moves k units upward, and when $k < 0$, it moves k units downward.

b For $a > 1$, as a increases in value, the graph gets

closer to the y-axis. When $0 < a < 1$, the width of the graph increases and moves further away from the y-axis. Similarly for $a < 0$, but the graph will be inverted.

c The coordinates of the vertex are (h, k). When $a < 0$, the vertex is a maximum point, and when $a > 0$ it is a minimum point.

Exercise 2C

1 Domain = Reals; Range is non-positive Reals.

2 Domain = Reals; Range is non-negative Reals.

3 Domain = Reals; Range is non-positive Reals.

4 Domain = Reals; Range is non-negative Reals.

5 Domain = Reals; Range is non-positive Reals.

6 Domain = Reals; Range = $\{y \mid y \ge -2\}$

7 Domain = Reals; Range = $\{y \mid y \le 1\}$

8 Domain = Reals; Range = $\{y \mid y \ge -2\}$

Exercise 2D

1 $D = \{x \mid x \ne -\dfrac{2}{3}\}$ $R = \{y \mid y \ne 0\}$

2 $D = \{x \mid x \ne 2\}$ $R = \{y \mid y \ne 0\}$

3 $D = \{x \mid x \ne 3\}$ $R = \{y \mid y \ne 0\}$

4 $D = \{x \mid x \ne -\dfrac{1}{2}\}$ $R = \{y \mid y \ne 0\}$

5 $D = \{x \mid x \ne \dfrac{1}{2}\}$ $R = \{y \mid y \ne -1\}$

6 $D = \{x \mid x \ne -1\}$ $R = \{y \mid y \ne 3\}$

Exercise 2E

1 $D = \{x \mid x \ne -1\}$ $R = \{y \mid y > 0\}$

2 $D = \{x \mid x \ne 1\}$ $R = \{y \mid y < 0\}$

3 $D = \{x \mid x \ne 0\}$ $R = \{-1, 1\}$

4 $D = \{x \mid x < 1\}$ $R = \{y \mid y < 0\}$

5 a
   $D = \left\{x \mid -\dfrac{1}{\sqrt{2}} < x < \dfrac{1}{\sqrt{2}}, x \ne 0\right\}$
   $R = \{y \mid y > 0\}$

Exercise 2F

1 a $-1; 1; 1; 1$

b $f(x) = \begin{cases} 1, x \geq 0 \\ -1, x < 0 \end{cases}$

c Reals; $y = \pm 1$

2 a $4; 2.5; 2; 5$

b
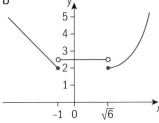

c Reals: $y \geq 2$

3 a $-4; -1; \sqrt{2}; 3$

b
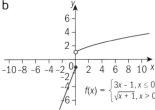

$f(x) = \begin{cases} 3x - 1, x \leq 0 \\ \sqrt{x + 1}, x > 0 \end{cases}$

c Reals; $\{y \mid y \leq -1 \text{ or } y > 1\}$

4 a $5, \dfrac{1}{2}, 0, -1$

b
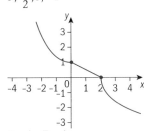

c Reals; Reals

Investigation – odd and even functions

a Even

b Odd

c Neither

d Even

e Even

f Odd

Exercise 2G

1 many-to-one; even

2 one-to-one; odd

3 one-to-one; odd

4 one-to-one; neither

5 many-to-one; neither

6 many-to-one; odd

7 many-to-one; odd

8 many-to-one; neither

9 $f(x) = 0$

Investigation – composite functions

a Not commutative

b Is associative

c i Even;   ii Odd;

iii Even

d May be neither even nor odd if one of the constituent functions is neither even nor odd.

Exercise 2H

1 a $x \geq 0$

b $x \geq 0$

c $x > 0$

2 $x < -2$ or $x > 2$

3 a 2 b -7 c -2

d $-4 + 16x + 12x^2 - 36x^3 - 27x^4$

4 a i $3 - 2x^2$

ii $2x + 3$

iii $1 - 2\sqrt{2x + 4}$

iv $\sqrt{2x^2 + 2}$

b i Reals; $y \leq 3$

ii $x \geq -2, y \geq -1$

iii $x \geq -2; y \leq 1$

iv Reals; $\{y \mid y \geq \sqrt{2}\}$

d $1, \sqrt{6}$

5 e.g., $f(x) = x - 2$; $g(x) = x^2$

6 e.g., $h(x) = \sqrt{x}$; $g(x) = 2x - 3$

Exercise 2I

1 $\dfrac{(x + 1)}{3}$

2 $3x + 1$

3 $\sqrt{x + 2}, x \geq -2$

4 $-\sqrt{x - 1}; x \geq 1$

5 $\sqrt{x + 5} - 2; x \geq -4$

6 No

8 $\dfrac{1 + x}{x - 2}; x \neq 2$

Investigation – self-inverse functions

$f(x) = a - x$, a is a real number;

$f(x) = \dfrac{b}{x}$, b is a real number.

The graph of the function will be symmetric about the line $y = x$.

Exercise 2J

1 a
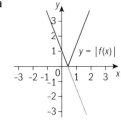

b
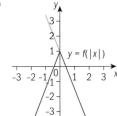

2 a
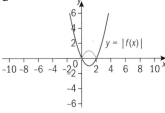

b
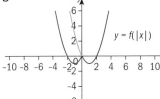

3 a
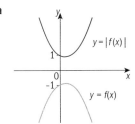

b

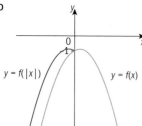

4 a

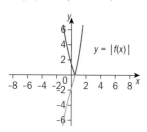

b

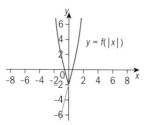

5 a

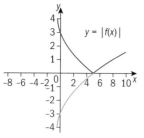

b

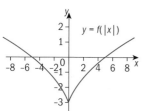

Exercise 2K

1

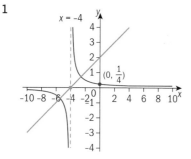

2

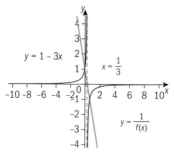

3

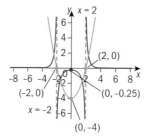

4

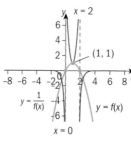

5

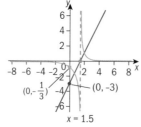

6 a

b

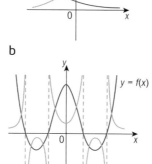

Exercise 2L

1 a Translation $\begin{pmatrix} 3 \\ 2 \end{pmatrix}$ i.e.

$g(x) = f(x - 3) + 2$

b Translation $\begin{pmatrix} -2 \\ -1 \end{pmatrix}$ i.e.

$g(x) = f(x + 2) - 1$

c $g(x) = -f(x) - 1$, reflected in the x-axis and translated by $\begin{pmatrix} 0 \\ -1 \end{pmatrix}$

d horizontal compression by a factor of $\frac{1}{2}$. i.e.

$g(x) = f(2x)$

e Reflection in the x-axis and vertical stretch of 2 i.e. $g(x) = -2f(x)$

f Reflection in the y-axis and horizontal stretch of 2. i.e. $g(x) = f(\frac{-1}{2} x)$

2 $g(x) = h(-x + 3)$

3 a

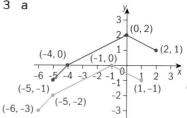

b

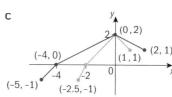

c

d

e

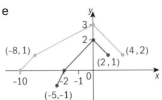

4 $(-5, -2) \rightarrow (-6, 1)$
  $(-4, 0) \rightarrow (-5, -3)$
  $(-3, 2) \rightarrow (-4, -7)$
  $(-1, -1) \rightarrow (-2, -1)$
  $(3, -3) \rightarrow (2, 3)$
  $(8, 2) \rightarrow (7, -7)$

6 a $g(x) = -f(x)$
  b $g(x) = f(-x)$
  c $g(x) = f(x + 3) - 1$
  d $g(x) = -f(x) + 1$
  e $g(x) = \dfrac{1}{f(x)}$
  f $g(x) = -f(2x)$
  g $g(x) = f(-x) + 2$

7 Proof

Review exercise (GDC)

1 a $3x^2 - 1$ b $\dfrac{1}{x^2 + 2}$

  c $\sqrt{3x - 1}$ d $\dfrac{3 - 7x}{x}; x \neq 0$

2 $f(x) = \dfrac{x - 1}{x + 3}; g(x) = x^2$

4 a

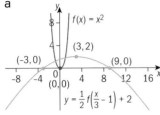

b

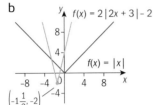

c

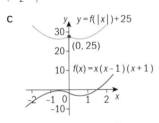

5 a

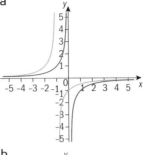

3

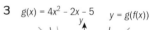

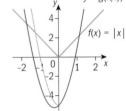

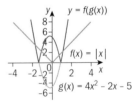

5 $y = \dfrac{9x + 20}{3x + 2}$ or $y = \dfrac{2}{3(x + 2)} + 3$

  Domain $= \{x \mid x \neq -2\}$
  Range $= \{y \mid y \neq 3\}$

6 a $x = -\dfrac{1}{2}$

  b $(0, 5), (-\dfrac{5}{4}, 0)$

b

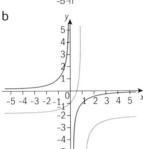

c

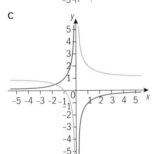

c

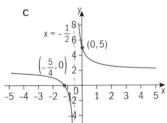

  d $f(x) = 3g(2x + 1) + 2$

Review exercise - (non-GDC)

1 a Reals, $y \leq 2$
  b No c No
  d $x \neq \pm 1, y \leq -0.25$ or
    $y > 0$

2 3; 3

3 $\dfrac{x - 1}{2 - x}; \dfrac{2x + 1}{x + 1}$

d

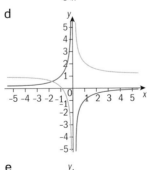

e

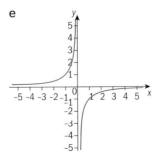

4 Vertical stretch of $\dfrac{9}{5}$ followed

  by a translation of $\begin{pmatrix} 0 \\ 32 \end{pmatrix}$. Same

  when $x = -40$ i.e.
  $-40°F = -40°C$.

5 $y = \dfrac{-1 - 4x}{2(x + 1)}$

Chapter 3

Skills check

1 a $-3, 1$ b $1, 10$

  c $\dfrac{-3}{2}, 1$

2 a $2x^3 - 3$
  b $6x^4 - 8x^3 + 5x^2 - 27x + 11$
  c $\dfrac{3}{2}x^4 - \dfrac{4}{5}x^3 - \dfrac{3}{5}x^2 - \dfrac{6}{5}x - \dfrac{9}{10}$

Exercise 3A

1 a $0, \dfrac{3}{2}$ b ± 5

   c $0, \dfrac{4}{5}$ d No real roots

   e $0, -\dfrac{1}{121}$ f $\pm\sqrt{2}$

   g $0, \dfrac{11}{\pi}$ h $\pm\sqrt{\dfrac{\sqrt{3}}{e}}$

2 a $-2, -\dfrac{1}{2}$ b $3, \dfrac{1}{3}$

   c $-1, \dfrac{2}{5}$ d $-\dfrac{2}{3}, \dfrac{3}{7}$

   e No real roots f $\dfrac{4}{13}, \dfrac{6}{11}$

3 a $-2 \pm \sqrt{2}$

   b $\dfrac{3 \pm \sqrt{14}}{5}$

   c $\dfrac{1 \pm \sqrt{37}}{6}$

   d $\dfrac{-11 \pm \sqrt{17}}{4}$

   e $\dfrac{23 \pm \sqrt{221}}{22}$

   f $\dfrac{29 \pm \sqrt{1661}}{10}$

4 a $-2p, p$

   b $-\dfrac{2}{k}, -1$

   c $\dfrac{6}{a}, \dfrac{1}{2}$

   d $-2a + b, a - b$

Exercise 3B

1 a 2 real roots
   b one real root
   c no real roots
   d 2 real roots
   e no real roots
   f one real root

2 a $k = -1$

   b $k > -\dfrac{9}{8}$

   c $k > \dfrac{37}{24}$

   d $k = -\dfrac{2}{5}$ or -2

   e $k > 0$

   f $\dfrac{-12 - 3\sqrt{14}}{4} < k < \dfrac{-12 + 3\sqrt{14}}{4}$

Exercise 3C

1 a 3 b $\dfrac{19}{3}$

   c $-\dfrac{29}{9}$ d -12

e -1 f 47

g $\dfrac{7}{64}$ h $\dfrac{24336}{2401}$

Exercise 3D

1 a $\text{Re}\,(z) = 0,\ \text{Im}\,(z) = 3$

   b $\text{Re}\,(z) = -7,\ \text{Im}\,(z) = 0$

   c $\text{Re}\,(z) = \dfrac{9}{4},\ \text{Im}\,(z) = -\dfrac{3}{2}$

   d $\text{Re}\,(z) = \dfrac{11}{4},\ \text{Im}\,(z) = \dfrac{\sqrt{7}}{5}$

   e $\text{Re}\,(z) = \dfrac{-2}{3\pi^2},\ \text{Im}\,(z) = \dfrac{4}{3\pi^2}$

2 a 13 b 25

   c $\sqrt{13}$ d 1

   e $\dfrac{5}{\pi}$

Exercise 3E

1 a $3 - 2i$

   b $-1 + 11i$

   c $\dfrac{21}{10} - \dfrac{16i}{5}$

   d $12i$

   e $8 - 6i$

   f $\dfrac{52}{5} - \dfrac{13}{10}i$

   g $-\dfrac{401}{15} - \dfrac{46i}{5}$

Exercise 3F

1 a $-\dfrac{2}{5} + \dfrac{9}{5}i$

   b $-\dfrac{15}{17} - \dfrac{8}{17}i$

   c $-\dfrac{25}{39} + \dfrac{5}{39}i$

   d $\dfrac{19}{2} + \dfrac{15}{2}i$

   e $-\dfrac{13}{25} + \dfrac{84}{25}i$

2 a $a = 4, b = -3$

   b $a = 4, b = 19$

   c $a = 6, b = -5$

   d $a = 0, b = 4$

3 a $\dfrac{3}{4}, \dfrac{-1}{2}$

   b $\dfrac{5}{3}, \dfrac{2}{3}$

   c $1, -\dfrac{4}{3}$

   d $0, -\dfrac{24}{13}$

4 a $6 + 18i$

   b $6 + 18i$

   c 12

5 a $\dfrac{7}{10} - \dfrac{19}{10}i$

   b $2 - i$

   c $\dfrac{31}{13} - \dfrac{12}{13}i$

   d $\dfrac{111}{505} + \dfrac{302}{505}i$

6 $7\,\text{Re}\,(z) + 2\,\text{Im}\,(z) = 0$

7 $3\,\text{Re}\,(z) + 5\,\text{Im}\,(z) = 0$

8 a $-\dfrac{7}{8} - 3i$

   b $-1 - \dfrac{3}{4}i$

   c $0, 1, -\dfrac{1}{2} + \dfrac{\sqrt{3}}{2}i, -\dfrac{1}{2} - \dfrac{\sqrt{3}}{2}i$

Exercise 3G

1 a 0 b $2 - 2i$

   c $8 + i$ d $\dfrac{7}{29} + \dfrac{26}{29}i$

   e 1 f 1

2 a $-20 + 4i$

   b 10

   c -18

   d -8

3 a $\pm (2 + i)$

   b $\pm (2 + 3i)$

   c $\pm\left(\dfrac{3}{2} + i\right)$

   d $\pm\left(\dfrac{2}{3} - \dfrac{1}{4}i\right)$

   e $\pm\left(\dfrac{1}{\sqrt{2}} + \dfrac{1}{\sqrt{2}}i\right)$

   f $\pm\left(\dfrac{1}{\sqrt{2}} - \dfrac{1}{\sqrt{2}}i\right)$

5 a $n = 4k$

   b $n = 2(2k + 1)$, where $k \in \mathbb{N}$

Exercise 3H

1 a $\lambda = 3, \mu = -2$

   b $\lambda = 10, \mu = 2$

2 a $x^5 - 4x$

   b $81x^5 + 81x^4 - 256x - 256$

3 $a = 2, b = 2$

4 $a = -1$, $b = -3$, $c = 2$

5 $a = 13$, $b = 12$,
$f(x) = (x^2 + 3x + 2)^2$ or
$a = 5$, $b = -12$,
$f(x) = (x^2 + 3x - 2)^2$

6 $g(x) = x^3 + 6x^2 - 30x + 31$

7 $f(x) = x^4 - 3x^2 - 2x$

8 $f(x) = 3x^4 + 2x^3 + 5x^2 + 8x + 4$

Exercise 3I

1 a $x^3 + 3x^2 + 2x - 1$
   b $x^3 + 3x^2 + 2x - 1$
   c $2x^3 - 5x^2 + 4x - 1$

2 a $q(x) = 2x^3 + 3x^2 + x + 3$,
      $r(x) = 0$
   b $q(x) = 3x^2 - 2x + 1$,
      $r(x) = 2x + 3$
   c $q(x) = x^4 - x^3 + x - 1$,
      $r(x) = x$

Exercise 3J

1 a $q(x) = x^2 + 2x + 2$, $r(x) = 1$
   b $q(x) = 2x^2 + 3x + 1$, $r(x) = 2$
   c $q(x) = x^4 - 2x^3 + x^2 - 2x + 2$
      $r(x) = -3$
   d $q(x) = 3x^5 + 3x^4 + x^3 + x^2$
      $+ 6x + 6$
      $r(x) = 4$

2 Proof

Exercise 3K

1 a $q(x) = x^4 - x^3 + x^2 + 2x + 1$,
      $r(x) = -2$
   b $q(x) = x^3 + x^2 + x - 1$,
      $r(x) = 7$

2 $f(x) = 3x^3 + 2x^2 - 10x + 6$

3 $a = -9$

4 $b = 0$

5 $a = -5$, $b = 2$

6 Remainder $= \dfrac{3}{2}x - \dfrac{5}{2}$

7 Remainder $= 0$

Exercise 3L

1 a $(x + 2)(x + 2)(2x - 1)$
      $(x - 2)$
   b $(2x - 1)^2 (3x - 5)$

2 a $x^3 - 9x^2 + 23x - 15$
   b $x^4 + 2x^3 - x^2 - 2x$
   c $3x^4 - 16x^3 + 21x^2 +$
      $4x - 12$

3 a $x^4 - 5x^2 + 6$
   b $8x^4 - 2x^3 - 43x + 10x + 15$

c $5x^6 - 7x^5 - 11x^4 - 18x^3$
   $+ 21x^2 + 33x + 9$

4 a $(x - 1)(x - 3)(x + 2)$
   b $(x - 1)(2x - 3)(x + 2)$
   c $(x - 1)(x + 1)(5x + 3)$
      $(x - 3)$

Exercise 3M

1 a $-2i$, -3
   b $1 + 2i$, 4
   c $-\dfrac{3}{2} - \dfrac{\sqrt{3}}{2}i$, $\dfrac{-2}{5}$
   d i, 2, 2
   e $-1 + 3i$, $-\dfrac{1}{2}$, 1
   f $-2 - i$, $-\dfrac{3}{2}$, 1
   g $-\dfrac{1}{2} - \dfrac{\sqrt{5}}{2}i$, $-\dfrac{1}{3}$, -3
   h $\dfrac{1}{3} - \dfrac{\sqrt{2}}{3}i$, i, $-i$

2 a $a = -12$; -3 and 4
   b $a = 17$; $2 \pm i$
   c $a = -8$; $-1 + i$, 2, -2
   d $a = -16$, $b = 20$; $2i$, $2 \pm i$

Exercise 3N

1 a $\dfrac{2}{3}$ b $\dfrac{4}{3}$
   c $-\dfrac{5}{3}$ d $-\dfrac{15}{2}$
   e 34

2 a 3 b -6
   c 2 d 4
   e -2 f 1

3 a -2, 5
   b $-\dfrac{1}{4}$, 0
   c 0, 2
   d $\dfrac{4}{5}$, $-\dfrac{8}{5}$

Exercise 3O

1 a 1, 2, 3
   b 1, 1, -4
   c 2, -2, -3
   d 3, $\dfrac{5}{2}$, -3

Exercise 3P

1 a -1, $-\dfrac{3}{4}$, $\dfrac{1}{3}$
   b -2, $3 \pm \sqrt{2}$

c $-\dfrac{2}{3}$
   f -1, 2, 3, -3

2 a $a = 3$
   b 1, -1

3 a $a = 2$, $b = 4$
   b $-\dfrac{1}{2}$

5 b $-a$

Exercise 3Q

1 a $[1, 2] \cup [3, \infty[$
   b $]-\infty, -4]$ or $x = 1$
   c $]-\infty, -3[\cup]-2, 2[$
   d $]-3, 2.5[\cup]3, \infty[$
   e $]-\infty, -1] \cup \left[-\dfrac{3}{4}, \dfrac{1}{3}\right]$
   f $]-2, 3 - \sqrt{2}[\cup]3 + \sqrt{2}, \infty[$
   g $]-\infty, -\dfrac{2}{3}[$
   h $]-\infty, -3] \cup [-1, 2] \cup$
      $[3, \infty[$

2 $]-\dfrac{1}{3}, 2[\cup]\dfrac{5}{2}, \infty[$

3 a $[-1, -0.921] \cup [1.26, \infty[$
   b $]-\infty, 0]$

4 a $]1, \infty[$
   b $]-\infty, -1] \cup [-0.366,$
      $1.37]$
   c $[-1, -0.544]$

Exercise 3R

2 a $(1, -2, 3)$
   b $(1, 1, 1)$
   c $\left(-\dfrac{1}{19}, \dfrac{18}{19}, -\dfrac{8}{19}\right)$
   d $\left(2y - 2, y, \dfrac{4 - 2y}{3}\right)$
   e No solutions
   f No solutions

3 a $k = 1$
   b $k = 1$ or 4

4 a $k = 2$, $(1 - 2y, y, 0)$
   b $k = 4$, $(y + 5, y, -4 - 2y)$

5 $m \neq 1$ $\left(\dfrac{m - m^2 - 1}{1 - m}, -\dfrac{m}{1 - m}, \dfrac{1 + m}{1 - m}\right)$

Review exercise (non-GDC)

1 $a = -\dfrac{55}{4}$

2 $x = i, y = 1 + i$

3 $m < -\dfrac{1}{12}$

4 $1 + 2i, -3i, 3i$

5 $m = -4$ or 2

6 $a = 3$ or $\pm \sqrt{3}$

9 $\dfrac{2}{3}$

10 b $2, -1 \pm i\sqrt{6}$

   c 2

Review exercise (GDC)

1 $x \in [\,1.67, \infty[$

2 $m \in \,]-\infty, -1.05\,[$

3 $x = \dfrac{1}{2}, y = \dfrac{2}{7}, z = -\dfrac{4}{3}$

4 $\dfrac{143}{27}$

5 0.833

Chapter 4

Skills check

1

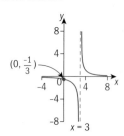

2 10

Exercise 4A

1 -2

2 3

3 Limit does not exist

4 Limit does not exist

5 0

6 Limit does not exist

Exercise 4B

1 Not continuous

2 Continuous

3 Not continuous

4 $k = \dfrac{4}{3}$

5 $a = \dfrac{1}{2}$

6 a Not continuous at $x = \pm 1$

   b Not continuous at $x = \pm 2$

   c Continuous for all x

   d Not continuous at $x = -4$, $x = 1$

   e Not continuous at $x = 1$

   f Continuous for all x

Exercise 4C

1 a 7 b -3

   c 0 d does not exist

   e 2 f 0

   g $\dfrac{2}{3}$ h $2a$

2 a 2 b 3

   c $\dfrac{2}{3}$ d 0

   e 0 f 4

3 a $y = 3$ b $y = \dfrac{1}{2}$

   c $y = 0$ d $y = -1$

   e No horizontal asymptote

Exercise 4D

1 a converges

   b converges

   c converges

   d does not converge

   e converges

2 a converges; sum $= \dfrac{2}{3}$

   b does not converge

   c converges; sum $= \dfrac{5}{2}$

   d converges; sum $= \dfrac{1}{3}$

   e converges; sum $= -\dfrac{7}{20}$

   f converges; sum $= \dfrac{5}{2}$

3 a $x < 0$

   b $x = -3$

4 $-\dfrac{1}{4} < x < \dfrac{1}{2}$

Exercise 4E

1 a 4

   b $-\dfrac{1}{2}$

   c 3

   d -2

   e 1

   f $-\dfrac{1}{4}$

2 $(-1, 1)$

3 $f'(x) = 4x - \dfrac{1}{x^2}; (1, 3)$

Exercise 4F

1 a $2x + 2, 2$

   b $3x^2, 3$

   c $-\dfrac{2}{x^2}, -\dfrac{2}{9}$

   d $\dfrac{1}{2\sqrt{x-1}}, \dfrac{1}{2}$

   e $\dfrac{1}{4}$

   f $-\dfrac{1}{16}$

2 a $-10a - 5h$

   b $-10a$

Exercise 4G

1 a 2

   b $y = 2x + 10$

   c $y = -\dfrac{1}{2}x + \dfrac{15}{2}$

2 $(2, 1)$ and $(0, -1); y = -x + 3;$
   $y = -x - 1$

3 a $\left(\dfrac{1}{2}, \dfrac{19}{4}\right)$

   b $(0, 1)$

   c none

   d $\left(\dfrac{3}{2}, -\dfrac{9}{4}\right)$

   e none

4 $y = 2$. Normal is $x = 1$

Exercise 4H

1 a $-1 - 6x$

   b $8x^3 - 3$

   c $12x^2 + \dfrac{3}{x^4} + 4x - \dfrac{4}{3x^3}$

   d $-\dfrac{2}{x^2} - 3 + 15x^2$

2 $y = 8x - 6$

3 $y = -\dfrac{1}{3}x + \dfrac{11}{3}$

Exercise 4I

1 a $10(2x + 3)^4$

   b $-\dfrac{3}{2\sqrt{2-3x}}$

   c $-\dfrac{2}{x^2} - 3 + 15x^2$

   d $\dfrac{15x}{\sqrt{(5x^2+1)^3}}$

e $\dfrac{1-2\sqrt{x}}{4\sqrt{x-x\sqrt{x}}}$

2 $y = 2x - 1$

3 $y = -\dfrac{1}{3}x - \dfrac{5}{3}$

4 $\left(1, -\dfrac{1}{2}\right)$

5 $-\dfrac{1}{4\sqrt{\sqrt{x}-x}}$

Exercise 4J

1 $4x(x+3)^2$

2 $8(2x-3)(5x-4)(4x+1)^2$

3 $-\dfrac{2}{(x-1)^2}$

4 $\dfrac{(1-3x)}{(1-2x)^{\frac{1}{2}}}$

5 $\dfrac{(3-4x^3)}{(x^4-3x+1)^2}$

6 $\dfrac{2(x-1)^3(7x-5)}{(3x-2)^{\frac{1}{3}}}$

Exercise 4K

1 **a** $\dfrac{21-x^2}{x^4}$

b $\dfrac{1}{(x^2+1)^{\frac{3}{2}}}$

c $\dfrac{3-4x^3}{(x^4-3x+1)^2}$

d $\dfrac{1}{\sqrt{x}(1-\sqrt{x})^2}$

e $\dfrac{3x^2(2-\sqrt{x})}{2(1-\sqrt{x})^4}$

f $\dfrac{7x-12x^2-3}{x^4(1-2x+3x^2)^{\frac{3}{2}}}$

2 0

3 $y = \dfrac{25}{16}x + \dfrac{3}{80}$

4 $\dfrac{2}{3(2+x)^2\sqrt[3]{1-\dfrac{1}{2+x}}}$

Exercise 4L

1 $\dfrac{2}{x^3}$

2 $-2, 12$

3 $x = 2$

4 $r = 2, s = -8, t = 5$

5 **a** 0 ms^{-1}

b 0 ms^{-2}

c 870 ms^{-3}

6 $f^{(n)}(x) = \dfrac{n!(-1)^n}{x^{n+1}}$

Investigation - Leibniz's formula

$f'''(x) = u'''v + u''v' + 2u''v' + 2u'v'' + u'v'' + uv'''$

$f^{(4)}(x) = u^{(4)}v + 4u^{(3)}v' + 6u''v'' + 4u'v^{(3)} + uv^{(4)}$

$f^{(n)}(x) = \displaystyle\sum_{k=0}^{n} \binom{n}{k} u^{(k)}v^{(n-k)}$

$f^{(5)}(x) = u^{(5)}v + 5u^{(4)}v' + 10u^{(3)}v'' + 10u''v^{(3)} + 5u'v^{(4)} + uv^{(5)}$

Exercise 4M

1 **a** Minimum value $= -\dfrac{5}{4}$ (at $x = \dfrac{3}{2}$)

b Minimum value $= -3$ (at $x = 0$)

Maximum value $= 5$ (at $x = 2$)

c Minimum values are $\dfrac{59}{16}$ (at $x = -\dfrac{1}{2}$) and 2 (at $x = 1$) Maximum value is 4 (at $x = 0$)

d Horizontal point of inflexion at $(0, 0)$. Minimum value of -27 (at $x = 3$)

Exercise 4N

1 **a** **i** $x = \pm 1;$

ii $]-\infty, -1[$ and $]1, \infty[$

iii $]-1, 1[$

b **i** $x = -1, 0.5, 1.5$

ii $]-1, 0.5[$ and $]1.5, \infty[$

iii $]-\infty, -1[$ and $]0.5, 1.5[$

2 **a** **i** $x = 0.5$

ii $]-0.5, \infty[$

iii $]-\infty, -0.5[$

b **i** $x = \pm\dfrac{1}{2}, x = \dfrac{3}{2}$

ii $-\dfrac{1}{2} < x < \dfrac{1}{2}, x > \dfrac{3}{2}$

iii $x < -\dfrac{1}{2}, \dfrac{1}{2} < x < \dfrac{3}{2}$

3 **a** $(1, 2)$ maximum; $x < 1; x > 1$

b $(-1, 1)$ minimum, $(1, 1)$ maximum; $-1 < x < 1; -\sqrt{2} < x < -1$ and $1 < x < \sqrt{2}$

c $(-1, -\dfrac{1}{2})$ minimum, $(1, \dfrac{1}{2})$ maximum; $-1 < x < 1;$ $x < -1$ and $x > 1$

d $\left(\dfrac{1}{2}, -\dfrac{3}{2^{\frac{4}{3}}}\right)$ minimum; $x > \dfrac{1}{2};$ $x < \dfrac{1}{2};$

e Maximum at $\left(\pm\dfrac{2}{3}, \dfrac{4}{3}\sqrt{\dfrac{2}{3}}\right)$

Minimum at $(0, 0);$

$-\sqrt{2} < x < \dfrac{-2}{3}$ and $0 < x < \dfrac{2}{3};$

$-\dfrac{2}{3} < x < 0$ and $\dfrac{2}{3} < x < \sqrt{2}$

Exercise 4O

1 **a** max $(-2, 17)$; min $(1, -10)$

b max $(0.794, 0.191)$

c max $(-1, 4)$; min $(1, -4)$

d max $(-1, -3)$

e max $(-3, \dfrac{1}{3})$; min $(1, 3)$

2 **i** max $(4, -256)$

ii $(-\infty, 0]$ and $(4, \infty)$; $(0, 4)$

iii

b **i** max $(-2, 0.25)$; min $(4, -0.125)$

ii $]-\infty, -2[$ and $]4, \infty[$ $]-2, 4[$

iii

Exercise 4P

1 a i $(0, 0)$;
   ii concave up $]0, \infty[$,
      concave down $]-\infty, 0[$
  b i none;
   ii concave up: $]-\infty, 0[$
      and $]0, \infty[$
  c i none
   ii concave down for
      $0 \le x \le 4$
  d i none
   ii concave down for all x
      $(x \ne 1)$
  e i none
   ii concave up for $]1, \infty[$
      concave down for
      $]-\infty, 1[$

2 a max $x = 0, 2$;
     min $x = 1, 1$
  b i $-1 < x < 0; 1 < x < 2$
   ii $x < -1; 0 < x < 1; x > 2$
  c i $x < \dfrac{1}{2}, \dfrac{1}{2} < x < \dfrac{3}{2}$
   ii $-\dfrac{1}{2} < x < \dfrac{1}{2}; x > \dfrac{3}{2}$
  d

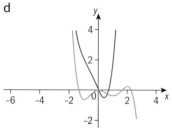

Investigation - cubic
polynomials

Exercise 4Q

1 a 10 m   b 2 s
  c -15 ms^{-1}; -10 ms^{-2}
     The diver is moving
     downwards and speeding
     up as he hits the water.

2 a 41.7 m
  b ± 36.1 ms^{-1}; 36.1 ms^{-1}
  c -30 ms^{-2} d $\dfrac{31}{3}$ s

---

3 a 7 ms^{-1}; 10 ms^{-2}
  b 3 ms^{-1}; -14 ms^{-2}
4 a 21 ms^{-1}
  b 33 ms^{-1}; 2 ms^{-2}
  c speeding up
  d $\dfrac{20}{3}$ secs
5 a $v(t) = t^2 - 6t + 8$;
     $a(t) = 2t - 6$
  b i $t = 2, 4$
   ii $2 < t < 3; t > 4$
   iii $0 < t < 2; 3 < t < 4$
  c at $t = 2$, $a = -2$ ms^{-2};
     at $t = 4$, $a = 2$ ms^{-2}
  d $t = 2, 4$
  e 9.33 m

Exercise 4R

1 a $c'(x) = 180 - 0.2x$
  b 160 euros/tank
  c 159.90 euros
     The cost of producing
     1 extra tank is nearly the
     same as the marginal cost
     function.

2 a i $0 < x < 3500$
   ii $c'(x) = 3$ euros,
      so it will cost an
      extra 3 euros at any
      production level to
      make an extra memory
      stick
   iii $r(x) = x(7 - 0.002x)$
  b $134 < x < 1870$

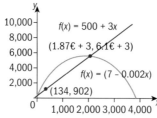

3 141
4 a 272   b 367

Exercise 4S

1 3.65; 6.67
2 80 000 m^2
3 rectangle is 0.778 by 1 metres;
   $r = 0.903$
4 14.7 cm, 34.7, 2630 cm^3

---

5 9.43 cm
6 a $l = 14.5$ cm
     $w = 3.46$ cm
  b $l = 18 - \sqrt{12}$
     $w = \sqrt{12}$

Exercise 4T

1 a $\dfrac{dy}{dx} = -\dfrac{x}{3y}$
  b $\dfrac{dy}{dx} = \dfrac{3x^2}{4y^3}$
  c $\dfrac{dy}{dx} = \dfrac{3 - 2x}{2y + 4}$
  d $\dfrac{dy}{dx} = \dfrac{3xy^2 - 2x}{y - 3x^2y}$
  e $\dfrac{dy}{dx} = -\dfrac{(1 + x + y)}{x + y}$
  f $\dfrac{dy}{dx} = \dfrac{1 - 3x^2 - 2xy}{x^2 + 1}$

2 $y = \dfrac{5}{4}x - \dfrac{9}{4}$

3 $y = \dfrac{4}{3}x - 10$

4 $y = 2, x = \sqrt{3}$

5 $(3, -1); (3, 9)$

6 1,6

7 $(\sqrt{3}, 0)$ and $(-\sqrt{3}, 0)$

8 $x = 6.63; y = 2.84$

9 $\dfrac{dy}{dx} = \dfrac{2x - 2y - 1}{2x - 2y + 1}$

Exercise 4U

1 $\dfrac{dA}{dt} = 2\pi r \dfrac{dr}{dt}$

2 $\dfrac{dA}{dt} = 4\pi r \dfrac{dr}{dt} + 2\pi (r\dfrac{dh}{dt} + h\dfrac{dr}{dt})$

3 $\dfrac{l\dfrac{dl}{dt} + w\dfrac{dw}{dt} + h\dfrac{dh}{dt}}{\sqrt{l^2 + w^2 + h^2}}$

4 i -14 cm^{2}s^{-1}
  ii 0
  iii $\dfrac{14}{13}$ cm s^{-1}

5 1.39 m^{2}s^{-1}

6 a 0.375 ms^{-1}
  b 0.4345 m^{2}s^{-1}

7 $\dfrac{1}{5\pi}$ cm s^{-1}

8 160 mph

9 0.866 ms^{-1}

10 18.6π m^{2}s$^{-1} \approx 58.4$ m^{2}s^{-1}

11 $f(x) = x^2$ units/s

12 $\dfrac{-90}{20\pi}$ m/min

Review exercise (non-GDC)

1 a -1 **b** none

   c $\ln(3) = (1.10)$ **d** 4

   e 0   **f** 0

2 No

3 Yes

4 Yes, $\dfrac{5}{2}$

5 $a \neq 0,\ 1 + a^2$

6 a $y = -1$

   b $\left(\pm\sqrt{5}, -1\right)$

7 $y = 2x + 1,\ y = -\dfrac{1}{2}x + 1$

8 If f is an even function, then f is symmetrical about the y-axis, hence there is a stationary point at $x = 0$, so its gradient is parallel to the x-axis.

9 $\left(-\dfrac{8}{9}, -\dfrac{8}{27}\right)$

10 $(3.08, -2.20)$

11 $y = 2x - \dfrac{1}{8}$

12 a $-6\,(3x - 1)^6\,(3x + 5)^2$
     $(15x + 16)$

   b $\dfrac{5}{2}\left(4x^2 - 3x + 1\right)^{\frac{3}{2}}(8x - 3)$

   c $\dfrac{3x^2 + 4x + 3}{2(x+1)^{\frac{3}{2}}}$

   d $\dfrac{\sqrt{\sqrt{x^2+1}+x}}{2\sqrt{x^2+1}}$

   e $3\,(x + 2 + (x - 3)^8)^2$
     $(1 + 8\,(x - 3)^7)$

13 $a = -2;\ b = -18$

14 a $(0, 0),\ (3\sqrt{3}, 0),\ (-3\sqrt{3}, 0)$

   b max $(-1, 2)$; min $(1, -2)$

   c none

   d **i** $]-\infty, -1\,[\cup]\,1, \infty[$

     ii $]-1, 1\,[$

15 a $x = \pm 1;\ y = 0$

   b $f(-x) = -f(x)$

   c $\dfrac{dy}{dx} = -\dfrac{2(x^2+1)}{(x^2-1)^2}$, negative for all x.

d

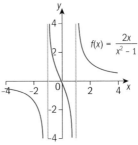

16 a $(3, 0),\ (0, -3);\ y = 1;$
     $x = \pm\sqrt{3}$

   b min $(3, 0)$; max $(1, -2)$

   c $(4.20, 0.0979)$

   d **i** $]-\infty, -1.73\ldots\,[,$
      $]-1.73\ldots, 1\,[,]\,3, \infty[$

     ii $]\,1, 1.73\ldots\,[,$
      $]1.73\ldots, 3\,[$

   e

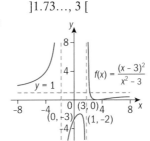

17 $-1, \dfrac{1}{4}$

Review exercise GDC

1 $\sqrt{1.25}$

2 $\dfrac{20}{\pi + 2}$

3 324π cm^{3}min^{-1}, increasing

4 16.4 cm; 11.0 cm

5 $-\dfrac{2}{125}$

Chapter 5

Skills check

1 The line $y = x$ is a line of symmetry.

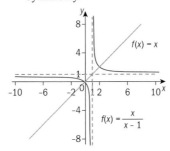

2 $(f \circ g)(x) = (g \circ f)(x) = x$

Exercise 5A

1 $1, \dfrac{1}{2}, \dfrac{1}{3}, \dfrac{1}{4}, \dfrac{1}{5}$

   $u_n = \dfrac{1}{n}$

2 $2, -2, -\dfrac{2}{3}, -\dfrac{2}{5}, -\dfrac{2}{7}, -\dfrac{2}{9}$

   $u_n = -\dfrac{2}{2n-3}$

3 a $2, \dfrac{3}{2}, \dfrac{5}{4}, \dfrac{9}{8}, \dfrac{17}{16}, \dfrac{33}{32}$

   b $u_n = \dfrac{2^n + 1}{2^n}$

4 a $1, 2, 5, 10, 17$

Investigation - sequences

1 $1, 1, 2, 3, 5, 8, 13, 21$

3 $\varphi^2 = \varphi + 1$

   $\varphi^3 = F_3\varphi + F_2$

   $\varphi^4 = F_4\varphi + F_3$

   Conjecture: $\varphi^n = F_n\varphi + F_{n-1}$

4 $\left(-\dfrac{1}{\phi}\right)^2 = F_2\left(-\dfrac{1}{\phi}\right) + F_1$

   $\left(-\dfrac{1}{\phi}\right)^3 = F_3\left(-\dfrac{1}{\phi}\right) + F_2$

   $\left(-\dfrac{1}{\phi}\right)^4 = F_4\left(-\dfrac{1}{\phi}\right) + F_3$

   Conjecture:

   $\left(-\dfrac{1}{\phi}\right)^n = F_n\left(-\dfrac{1}{\phi}\right) + F_{n-1}$

5 Binet's formula:

   $F_n = \sqrt{5}\left(\dfrac{1+\sqrt{5}}{2}\right)^n - \sqrt{5}\left(\dfrac{1-\sqrt{5}}{2}\right)^n$

Exercise 5B

1 a 16   **b** $\dfrac{2}{3}$ **c** $\dfrac{8}{27}$

3 $y^{\frac{5}{6}}$ **4** $\dfrac{x^2 y^3}{z^9}$ **5** 0

6 $x = 1,\ x = 0$

Investigation – music and indices

G# below middle C:

$220 \times 2^{\frac{15}{12}}$ Hz

G# above middle C:

$220 \times 2^{\frac{11}{12}}$ Hz

C in next octave above middle C:

$220 \times 2^{\frac{15}{12}}$ Hz

Lowest note on grand piano:
27.5 Hz

Highest note on grand piano:
4186 Hz

Exercise 5C

1   3%
2   a   19.6%   b   6.12%
    c   2.47%
3   Samira $3280
    Hemanth $3340
4   €16700

Investigation – compound interest

$$A = \left(1 + \frac{1}{n}\right)^n$$

n	A
1	2
500	2,715569
1000	2,716924
1500	2,717376
2000	2,717603
2500	2,717738
3000	2,717829
3500	2,717894
4000	2,717942
4500	2,71798
5000	2,71801
5500	2,718035
6000	2,718055
6500	2,718073
7000	2,718088
7500	2,718101
8000	2,718112
8500	2,718122
9000	2,718131
10000	2,718146

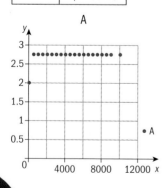

Exercise 5D

1   2.5; 0.25
2   $2^a f(x)$
3

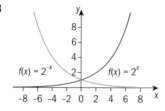

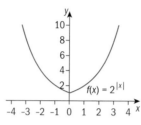

4   $x = \dfrac{\log\left(\dfrac{1}{e-1}\right)}{\log e}$ or

$x = \ln\left(\dfrac{1}{e-1}\right) \approx 0.541$

5   a   Reflection in the y-axis
    b   Reflection in the x-axis
    c   Reflection in the y-axis
        followed by a reflection in
        the x-axis

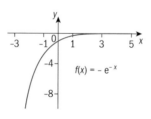

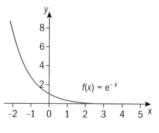

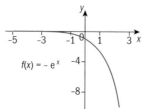

Investigation – the graph of $y = a^x$

a	Graph of a^x	$\dfrac{dy}{dx}\Big|_{x=0}$
1	$f(x) = 1^x$	$\dfrac{dy}{dx}\Big|_{x=0} = 0$
2	$f(x) = 2^x$; 0.693	$\dfrac{dy}{dx}\Big|_{x=0} = 0.693$
2.5	$f(x) = (2.5)^x$	$\dfrac{dy}{dx}\Big|_{x=0} = 0.916$
3	$f(x) = 3^x$	$\dfrac{dy}{dx}\Big|_{x=0} = 1.098$
3.5	$f(x) = (3.5)^x$	$\dfrac{dy}{dx}\Big|_{x=0} = 1.252$
4	$f(x) = 4^x$	$\dfrac{dy}{dx}\Big|_{x=0} = 1.386$

Exercise 5E

1   a $\log_5 125 = 3$
    b $\log_{10} 1000 = 3$
    c $\log_{27} 3 = \dfrac{1}{3}$
    d $\log_{10} 0.001 = -3$
    e $\log_n m = 2$
    f $\log_a 2 = b$
2   a $3^2 = 9$
    b $10^6 = 1000000$

c $49^{\frac{1}{2}} = 7$

d $a^0 = 1$

e $4^b = a$

f $p^r = q$

3 a 2   b $\dfrac{1}{2}$

   c -2 d $\dfrac{1}{2}$

   e 0   f $\dfrac{1}{3}$

4 a 9   b 81

   c 2   d 125

   e 64   f $\dfrac{1}{2}$

Exercise 5F

1 a $2\log_a p - \log_a q$

   b $\dfrac{1}{3}\log_a p - \dfrac{2}{3}\log_a q$

2 a $\log 6$ b $\log_a \sqrt{pq}$

   c $\log 20$

3 a 1   b 4   c 3

4 a $y = x^{\frac{2}{3}}$

   b $y = 2x$

   c $y = 2x^3$

   d $y = 10^{2+3x}$

Exercise 5G

1 a 4   b 1   c $\dfrac{1}{3}$

   d 1   e 2   f 1

3 $\log_x a = \dfrac{1}{p}$

   $\log_y a = \dfrac{1}{q}$

Exercise 5H

1 a 1.21   b 0.896

2 -2

Exercise 5I

1 a 0.774   b 1.55

2 a $\sqrt{3}$ or $\dfrac{1}{\sqrt{3}}$ b $\dfrac{2}{3}$

3 1.89

4 64; 256

5 -1; 0.861

6 3; 9

7 1.71

8 $x = \sqrt{2}$, $y = 4$

9 $x = 25$, $y = 5$

10 $x = 5$, $y = \dfrac{5}{3}$

11 a $x = 1$, $y = 0$

    b $x = 2$, $y = -1$

Exercise 5J

1 $f(x) = e^x$, $x \in \mathbb{R}$, $y \in \mathbb{R}$, $y > 0$

   $g(x) = \ln x$, $x \in \mathbb{R}$, $x > 0$, $y \in \mathbb{R}$

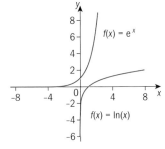

2 $f^{-1}(x) = \log_a x$, $f_0 f^{-1}(x) = x$

3

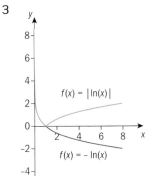

4

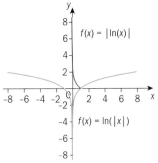

5

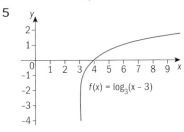

$f(x) = \log_3(x - 3)$

Translation of $\begin{pmatrix} 3 \\ 0 \end{pmatrix}$

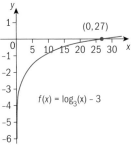

$f(x) = \log_3(x) - 3$

Translation of $\begin{pmatrix} 0 \\ -3 \end{pmatrix}$

6

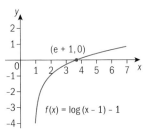

a Vertical asymptote at $x = 1$

   $x \in \mathbb{R}$, $x > 1$

   $(e+1, 0)$

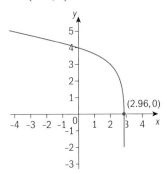

b Vertical asymptote at $x = 3$

   $x \in \mathbb{R}$, $x < 3$

   $\left(\dfrac{80}{27}, 0\right)$, $(0, 4)$

Exercise 5K

1 a $3xe^{x^2}$

   b $15e^{-3x+1}$

   c $4e^{4x-1}$

   d $e^x - \dfrac{1}{e^x}$

   e $3e^{3x-1}$

   f $\dfrac{1}{\sqrt{x}}e^{\sqrt{x}}$

2 a $e^x(x+1)$

　 b $xe^{-x}(2-x)$

　 c $\dfrac{e^{2x}}{\sqrt{x}}\left(2-\dfrac{1}{2x}\right)$

　 d $\dfrac{1}{2}e^{\sqrt{x}}\left(1+\dfrac{1}{\sqrt{x}}\right)$

3 a $\dfrac{e^{2x}}{2x\sqrt{x}}(4x-1)$

　 b $\dfrac{x^2-2x-1}{e^x}$

　 c $\dfrac{e^{3x}(2+3x)}{(1+x)^2}$

　 d $\dfrac{2e^x}{(1-e^x)^2}$

4 a $\dfrac{e^x(x+1+e^x)}{(1+e^x)^2}$

　 b $2e^x(1+e^x)$

　 c $\dfrac{-1}{2e^x\sqrt{(1+e^{-x})}}$

　 d $e^x(x+1+2e^x)$

　 e $\dfrac{-4}{(e^x-e^{-x})^2}$

5 a Minimum at $\left(-1,-\dfrac{1}{e}\right)$

　 b POI at $\left(-2,\dfrac{-2}{e^2}\right)$

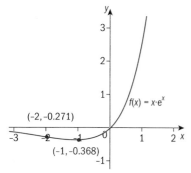

(-2,-0.271)

(-1,-0.368)

$f(x) = x\cdot e^x$

　 c Equation of tangent
$$y=-\dfrac{x}{e^2}-\dfrac{4}{e^2}$$

　 d $(-4,0)$

　 e Area of triangle $=\dfrac{8}{e^2}$

Exercise 5L

1 a $(3\ln 5)5^{3x}$

　 b $\dfrac{4}{(4x+1)}$

2 a $\dfrac{2}{x}$

　 b $-\dfrac{1}{x(\ln x)^2}$

Exercise 5M

1 a $x(1+2\ln x)$

　 b $a^x(x\ln a+1)$

2 a $-\dfrac{1}{x}$　　 b $\dfrac{2}{x}$

　 c $\dfrac{1-\ln x}{x^2}$

3 a $\dfrac{dy}{dx}=\dfrac{\ln x-1}{(\ln x)^2}$

　 b $2x^{2x}(1+\ln x)$

4 b Minimum at $(0,-1)$

　 c x-intercept $(1,0)$

　 d

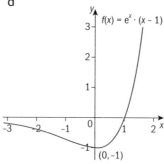

$f(x) = e^x \cdot (x - 1)$

(0,-1)

5 a $x\in\mathbb{R}$

　 b $f(-x)=f(x)$

　 c $f'(x)=\dfrac{2x}{1+x^2}$

$$f''(x)=\dfrac{2(1-x^2)}{(1+x^2)^2}$$

　 d

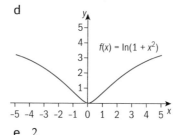

$f(x) = \ln(1 + x^2)$

　 e 2

$2\pi;\ \pi;\ -\pi;$ i $\dfrac{\pi}{2}$,　 ii $\dfrac{3\pi}{2}$

Angle measure conversion table	
Degrees	Radians
30	$\dfrac{\pi}{6}$
45	$\dfrac{\pi}{4}$
60	$\dfrac{\pi}{3}$
75	$\dfrac{5\pi}{12}$
90	$\dfrac{\pi}{2}$
120	$\dfrac{2\pi}{3}$
135	$\dfrac{3\pi}{4}$
150	$\dfrac{5\pi}{6}$
180	π
210	$\dfrac{7\pi}{6}$
225	$\dfrac{5\pi}{4}$
240	$\dfrac{4\pi}{3}$
270	$\dfrac{3\pi}{2}$
285	$\dfrac{19\pi}{12}$
300	$\dfrac{5\pi}{3}$
315	$\dfrac{7\pi}{4}$
330	$\dfrac{11\pi}{6}$
360	2π

Investigation - areas and perimeters of sectors

1	Number of diameters	Number of congruent sectors	Angle subtended by minor arc (radians)	Area of one sector	Length of minor arc
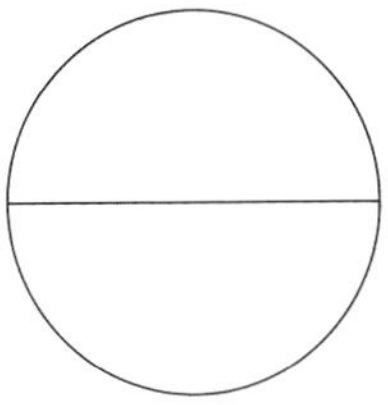	1	2	π	$\dfrac{1}{2}\left(\pi r^2\right) = \left(\dfrac{\pi}{2}\right)r^2$	$\dfrac{1}{2}\left(2\pi r\right) = \pi r$
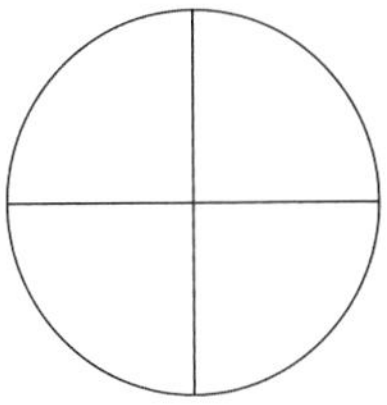	2	4	$\dfrac{\pi}{2}$	$\dfrac{1}{4}\left(\pi r^2\right) = \left(\dfrac{\pi}{4}\right)r^2$	$\dfrac{1}{4}\left(2\pi r\right) = \dfrac{\pi}{2}r$
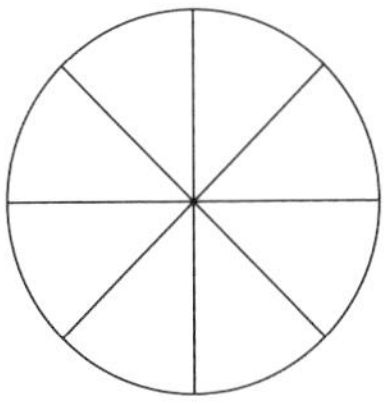	4	8	$\dfrac{\pi}{4}$	$\dfrac{1}{8}\left(\pi r^2\right) = \left(\dfrac{\pi}{8}\right)r^2$	$\dfrac{1}{8}\left(2\pi r\right) = \dfrac{\pi}{4}r$
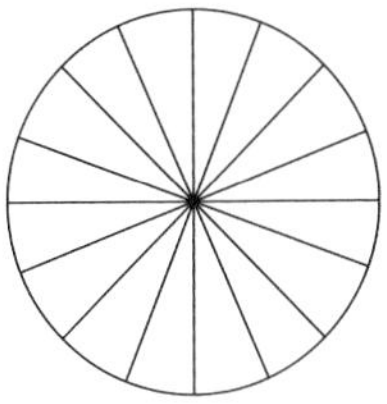	8	16	$\dfrac{\pi}{8}$	$\dfrac{1}{16}\left(\pi r^2\right) = \left(\dfrac{\pi}{16}\right)r^2$	$\dfrac{1}{16}\left(2\pi r\right) = \dfrac{\pi}{8}r$
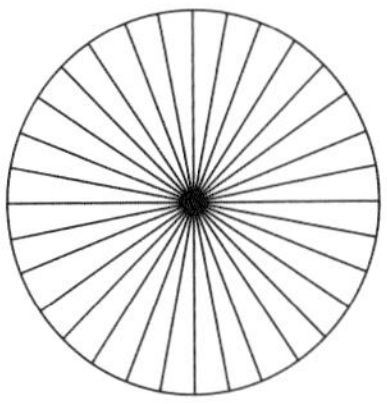	16	32	$\dfrac{\pi}{16}$	$\dfrac{1}{32}\left(\pi r^2\right) = \left(\dfrac{\pi}{32}\right)r^2$	$\dfrac{1}{32}\left(2\pi r\right) = \dfrac{\pi}{16}r$

Exercise 5N

1 a $\dfrac{5\pi}{4}; \dfrac{25\pi}{8}$

 b $\dfrac{5\pi}{3}; \dfrac{10\pi}{3}$

 c $26.9\,\text{cm}; 72.7\,\text{cm}^2$

2 a $22.4\,\text{m}^2; 19.2\,\text{m}$

3 $69.6\,\text{cm}$

4 a $46250\,\text{km}$

 b 13.5%

5 Sum = area of triangle ABC

Review exercise (non-GDC)

1 3^n

2 $\dfrac{3}{2}$

3 a 2; 1

 b 2

4 a 28

 b 160

 c 27; $\dfrac{1}{9}$

 d 49; 7

5 a $x = 27, y = 3$

 b $x = 8, y = 4$

 c $x = 14, y = 8$

6 a 0

 b $\dfrac{6}{7}$

 c $\ln 2$

7 64

10 $\ln(2+\sqrt{5})$

11 $\ln(2\pm\sqrt{3})$

12 a $x \in \mathbb{R}, x > 0$

b $f'(x) = \dfrac{2\ln x}{x}$,

$f''(x) = \dfrac{2(1 - \ln x)}{x^2}$

c Minimum at $(1, 0)$

POI at $(e, 1)$

d $y = \dfrac{2x}{e} - 1$

e Area of triangle = e

13 $S_1 = \dfrac{a^2}{4}\left(2\pi - 3\sqrt{3}\right)$

$S_2 = \dfrac{3a^2}{16}\left(2\pi - 3\sqrt{3}\right)$

$S_3 = \dfrac{9a^2}{64}\left(2\pi - 3\sqrt{3}\right)$

Total area = $a^2\left(2\pi - 3\sqrt{3}\right)$

14 $96\sqrt{3} - 48\pi \cong 15.5\,\text{cm}^2$

Chapter 6

Skills check

1 a 96.5 kg

b 103 kg

c 98.6 kg

d 29 kg

e $LQ = 91.75$ kg

$UQ = 106$ kg

f $IQR = 14.25$ kg

2 a 56

b 920

Exercise 6A

1 a discrete

b

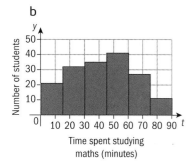

2 a continuous

b 17

c

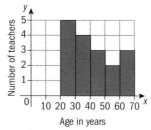

3 a continuous

b

Mass of chickens (kg)	Number of chickens
$1 \le w < 2$	8
$2 \le w < 3$	24
$3 \le w < 4$	50
$4 \le w < 5$	14

c 96

4 a continuous

b

Time to get home (mins)	Number of students
$5 \le t < 10$	1
$10 \le t < 15$	2
$15 \le t < 20$	4
$20 \le t < 25$	4
$25 \le t < 30$	2
$30 \le t < 35$	2
$35 \le t < 40$	1
$40 \le t < 45$	1

c 5 mins.

5 D, A, C

Exercise 6B

1 a 1 goal

b $170 \le h < 180$

2 a

t (minutes)	Frequency	CF
$0 \le t < 1$	8	8
$1 \le t < 2$	11	19
$2 \le t < 3$	10	29
$3 \le t < 4$	7	36
$4 \le t < 5$	8	44
$5 \le t < 6$	4	48
$6 \le t < 7$	1	49
$7 \le t < 8$	1	50
	50	

b

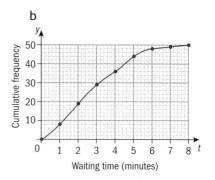

12 %

c Mean ≈ 2.8 mins, Median ≈ 2.6 mins

Modal class is $1 \le t < 2$ mins

3 a

Height (m)	Frequency
$0 \le h < 5$	15
$5 \le h < 10$	20
$10 \le h < 20$	15
$20 \le h < 30$	10
	60

b

Height (m)	CF
5	15
10	35
20	50
30	60

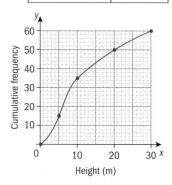

c 22%

d Mean ≈ 11.0 m, Median ≈ 9 m, Modal class is $5 \le h < 10$ m.

4 a mode is 3

median is 3

b $a = 6$

The new data set is bimodal with modes 3 and 6

5 **a** $r = \dfrac{1}{2}$ sum is $2\ln a$ $\sigma(\ln a)^2$

b $\mu = \ln a\dfrac{2^n - 1}{n \times 2^{n-1}}$

c $n = 200$

Investigation – what happens to the measures of central tendency when the data values are adjusted?

	Data	Mean	Mode	Median
Data set	6, 7, 8, 10, 12, 14, 14, 15, 16, 20	12.2	14	13
Add 4 to each data set	10, 11, 12, 14, 16, 18, 18, 19, 20, 24	16.2	18	17
Multiply the original data set by 2	12, 14, 16, 20, 24, 28, 28, 30, 32, 40	24.4	28	26

Exercise 6C

1 **a** 95 cm

 b 67.5 cm

 c 56.25 cm

 d 101.25 cm

 e 45 cm

2 **a** 75 cm

 b 5.5 cm

3

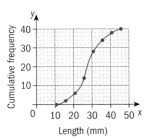

4 **a**

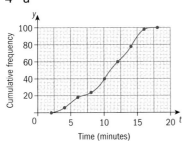

 i 11 mins

 ii 5.6 mins

 b $p = 32$, $q - 8$

5 **a** 1100

 b $a = 39$, $b = 64$

 c 7.1%

6 **a** 23 mins

 b 15 mins

 c 37 mins

Exercise 6D

1 $a = 1$, $b = 2$

2 **a** $a + 1$, 2.5

 b $a + 4$, 2.5

3 **a** 9.4, 1.41

 b 1

4 5 and 11, 9, 6.75

5 **a** $\dfrac{11k + 3}{5}$

 b $\dfrac{34k^2}{25} - \dfrac{56k}{25} + \dfrac{96}{25}$

 c $\dfrac{11k - 7}{5}$

 d unchanged

6 **c** $\dfrac{1}{3}n^2 - \dfrac{1}{3}$

Exercise 6E

1 **a** $\dfrac{1}{2}$ **b** $\dfrac{1}{4}$

 c $\dfrac{1}{4}$ **d** $\dfrac{3}{4}$

 e $\dfrac{3}{8}$

2 $\dfrac{1}{5}$

3 **a** $\dfrac{1}{2}$ **b** $\dfrac{2}{3}$

 c $\dfrac{5}{6}$ **d** $\dfrac{1}{3}$

 e $\dfrac{5}{6}$

4 $\dfrac{11}{20}$

5 **a** $\dfrac{3}{4}$ **b** $\dfrac{1}{2}$

 c $\dfrac{3}{4}$ **d** $\dfrac{1}{2}$

 e $\dfrac{1}{2}$

Exercise 6F

1 $\dfrac{8}{25}$

2 **a** $\dfrac{11}{32}$

 b $\dfrac{9}{32}$

3 **a** 0.33

 b 0.24

 c 0.3

4 **a** $\dfrac{1}{4}$

 b $\dfrac{3}{4}$

5 **a** 0.6

 b 0.4

 c 0.6

Exercise 6G

1 $\dfrac{2}{15}$

2 **a** $\dfrac{3}{36}$

 b $\dfrac{15}{28}$

3 **a** $\dfrac{7}{44}$

 b $\dfrac{35}{44}$

4 **a** $\dfrac{2}{91}$

 b $\dfrac{24}{91}$

 c $\dfrac{11}{91}$

5 **a** **i** 0.175

 ii 0.0827

 iii $(0.7)^n (0.75)^{n-1} 0.25$

 c 0.368

Exercise 6H

1 **a** 0.3

 b 0.1

 c 0.7

2 0.55

3 a 0.95

b 0.35

c 0.75

4 a 0.7

b 0.35

c 0.15

5 Proof

6 Proof

7 Proof

Exercise 6I

1 a i 0.21

ii 0.33

b 252

2 a 0.27

b No. Biased towards 1.

c 450

3 a $\dfrac{51}{250}$ b $\dfrac{53}{100}$

c $\dfrac{299}{500}$

4 a $\dfrac{2}{5}$ b $\dfrac{3}{5}$

c $\dfrac{1}{2}$

Exercise 6J

1 12

a $\dfrac{8}{27}$ b $\dfrac{23}{27}$

c $\dfrac{4}{5}$

2 a 0.2 b $\dfrac{1}{3}$

c $\dfrac{7}{15}$

3 $\dfrac{13}{16}$

4 a $\dfrac{1}{3}$ b $\dfrac{2}{5}$

c $\dfrac{3}{5}$ d $\dfrac{1}{2}$

5 $\dfrac{61}{95}$

6 $\dfrac{1}{6}$

7 a 0 b 0

c 0.63

8 67.3 %

9 $\dfrac{34}{47}$

10 a $\dfrac{1}{10}$ b $\dfrac{43}{50}$

c $\dfrac{11}{13}$

11 $\dfrac{1}{3}$

Exercise 6K

1 A and B, A and C

2 A and B, B and C

4 Yes

5 a 0.4 b 0.27

c 0.33

6 $P(A) = \dfrac{1}{4}$, $P(B) = \dfrac{1}{2}$

7 a $\dfrac{1}{12}$ b $\dfrac{1}{4}$

c $\dfrac{1}{6}$

8 a $\dfrac{1}{2}$ b $\dfrac{1}{5}$

c $\dfrac{1}{4}$

9 4

10 9

Exercise 6L

1 0.12

2 0.851

3 0.3125

4 a $\dfrac{7}{15}$ b $\dfrac{4}{5}$

5 a $\dfrac{204}{1015}$ b $\dfrac{811}{1015}$

c $\dfrac{663}{1015}$

6 a $\dfrac{2}{17}$ b $\dfrac{11}{850}$

c $\dfrac{22}{425}$ d $\dfrac{1}{5525}$

Exercise 6M

1 a $\dfrac{47}{90}$ b $\dfrac{20}{90}$

2 a 0.038 b 0.789

3 a 0.51 b 0.25

4 a 0.381 b $\dfrac{7}{13}$

5 b i $\dfrac{29}{36}$ ii $\dfrac{20}{29}$

iii $\dfrac{3}{7}$

6 a $\dfrac{5}{12}$ b $\dfrac{5}{7}$

7 a 0.674

b 0.754

8

9 a 0.8875

b 0.482

c 0.178

10 a $\dfrac{19}{33}$ b $\dfrac{6}{49}$

c $\dfrac{18}{49}$

11 a 0.59

b 0.102

c 0.427

12 0.33

13 $\dfrac{28}{171}$

14 0.6

15 $\dfrac{3}{7}$

Review exercise (non-GDC)

1 6, 6, 8, 12

2 $\dfrac{7}{8}$

3 a 68 kg b 61 – 77 kg

c 36

4 a 220 b $\dfrac{11}{22}$ c $\dfrac{4}{11}$

5 b $\dfrac{10}{21}$ c $\dfrac{7}{10}$

6 $\dfrac{4}{11}$

7 a $\dfrac{1}{6}$ b $\dfrac{3}{4}$ c $\dfrac{5}{9}$

d $\dfrac{5}{9}$

8 a i 18 ii 3

b No

9 a $\dfrac{3n+1}{2}$ b $\dfrac{n+1}{3n+1}$

Review exercise (GDC)

1 170.8 cm

2 a 50400 b $\dfrac{3}{10}$ c $\dfrac{7}{10}$

3 a $\dfrac{1}{2}$ b $\dfrac{71}{500}$ c $\dfrac{31}{1000}$

4 0.1351

5 a 337.5 km b 132.64 km,
602.8 km

6

7 17 times

8 a 4.69; 0.552 b 4.64

c 10

9 a $\dfrac{1}{38760}$ b $\dfrac{7}{38760}$

10 a 0.568 b 0.505

Chapter 7

Skills check

1. a $1 + \ln x$

   b $\dfrac{e^{2x-3}(9-4x)}{2(2-x)^{\frac{3}{2}}}$

   c $4x^3 + \dfrac{4}{x^5}$

2. a $(2, 4), (3, 7)$

   b $(0, 1)$

   c $(-2.95, -10.9)$
   or $(2.95, 10.9)$

3. $v = 12t^3 - 3t^2 + 1$

   $a = 36t^2 - 6t$

Exercise 7A

1. $-x^2 + c$
2. $\dfrac{x^9}{3} + c$
3. $-x^5 + c$
4. $-\dfrac{1}{4x^4} + c$
5. $\dfrac{2x^{\frac{5}{2}}}{5} + c$
6. $-\dfrac{2}{\sqrt{x}} + c$
7. $\dfrac{4x^{\frac{3}{2}}}{3} + c$
8. $\dfrac{4}{21x^{\frac{3}{4}}} + c$

Exercise 7B

1. a $\dfrac{5x^3}{3} + \dfrac{1}{5x} + c$

   b $\dfrac{2x^3}{3} + \dfrac{5x^2}{2} - 3x + c$

   c $\dfrac{-1}{x} + \dfrac{1}{3x^3} + c$

   d $\dfrac{x^3}{3} + 2x - \dfrac{1}{x} + c$

   e $-\dfrac{1}{2x^2} + \dfrac{1}{3x^3} + \dfrac{3}{x^4} + c$

   f $\dfrac{2x^{\frac{3}{2}}}{3} - \dfrac{15x^{\frac{2}{3}}}{2} + c$

2. $y = x^3 - 4x - 1$

3. $f(t) = \dfrac{t^2}{2} + 3t + \dfrac{1}{t} - 5$

4. $y = 2x^4 + 12x^3 + 27x^2 + 27x + 12$

   or $\dfrac{(2x+3)^4 + 15}{8}$

5. $A = \dfrac{x^4}{2} + \dfrac{x^3}{3} - x^2 - x + \dfrac{7}{6}$

6. $s = \dfrac{3t^2}{2} + \dfrac{8}{t} - 8$

7. $y = x^3 - \dfrac{x^2}{2} - 6x + 6$

8. $v(t) = 3t^2 + t + 2$;

   $s(t) = t^3 + \dfrac{t^2}{2} + 2t + 1$

Investigation - integrating $(ax + b)^n$

$\displaystyle\int (ax + b)^n = \dfrac{1}{a(n+1)}(ax + b)^{n+1} + c$

Exercise 7C

1. $\dfrac{(3x-1)^8}{24} + c$
2. $-\dfrac{2(2x+1)^{\frac{3}{2}}}{3} + c$
3. $-\dfrac{1}{16(4x-1)^4} + c$
4. $-\dfrac{8}{3}(3-x)^{\frac{3}{4}} + c$
5. $-\dfrac{3(2-5x)^{\frac{2}{3}}}{5} - \dfrac{3(1-x)^{\frac{4}{3}}}{4} + c$
6. $-\dfrac{8(2-3x)^{\frac{3}{2}}}{9} - \dfrac{6(3x+2)^{\frac{5}{3}}}{5} + c$

Exercise 7D

1. $\dfrac{5e^{-2x}}{2} + c$
2. $\dfrac{-e^{-3x-2}}{3} + c$
3. $2e^{-x-1} + 3e^{\frac{x}{3}} + c$
4. $\dfrac{3^x}{\ln 3} + c$
5. $-\dfrac{\left(\dfrac{1}{3}\right)^{2x}}{2\ln 3} + c$
6. $-\dfrac{4^{1-x}}{\ln 4} + c$

Exercise 7E

1. $\dfrac{\ln|x|}{3} + c$
2. $-6\ln|x| + c$
3. $\dfrac{-\ln|3x-2|}{3} + c$
4. $-\ln|5x - 3| + c$

5. $\dfrac{-2\ln|3x+4|}{3} + c$

Exercise 7F

1. $\dfrac{38}{3}$
2. 13
3. $-\dfrac{2(e^9 - 1)}{3e^5}$
4. $\dfrac{36}{\ln 2}$
5. $13\,072$
6. -1

Exercise 7G

1. $\dfrac{121}{5}$
2. Not possible: $s \neq 0$
3. Not possible: $x \neq 1$; 1 is an element of $[0, 2]$
4. $\dfrac{2}{9}$
5. Not possible: $x \neq -1$
6. $\ln\left(\dfrac{7}{16}\right)$
7. $2 + 4e - \dfrac{4}{e}$
8. $\dfrac{99}{\ln(10)}$

Exercise 7H

1. 1 sq. units
2. $10\dfrac{2}{3}$ sq. units
3. 16 sq. units
4. $e^3 + 6\ln 3 - 14$ sq. units
5. 16.9 sq. units
6. $\dfrac{16}{3}$ sq. units
7. 6.3 sq. units
8. $\dfrac{2}{\ln 2}$ sq. units
9. 3.32 sq. units
10. $2\ln(2)$ sq. units
11. $\ln(3)$ sq. units
12. 101.75 sq. units
13. $\dfrac{5}{6}$ sq. units
14. 3 sq. units

Exercise 7I

1. 18 sq. units

2. $\dfrac{64}{3}$ sq. units

3. $\dfrac{16}{3}$ sq. units

4. $\dfrac{4}{3}$ sq. units

5. $\dfrac{9}{2}$ sq. units

Exercise 7J

1. $\dfrac{9}{2}$ sq. units

2. $\dfrac{1}{12}$ sq. units

3. $\dfrac{9}{2}$ sq. units

4. $\dfrac{1}{5}$ sq. units

5. 72 sq. units

6. $\dfrac{128}{15}$ sq. units

7. 13.2 sq. units

8. 5.41 sq. units

9. 7.00 sq. units

10. $4^{\frac{2}{3}}$

11. $\dfrac{5}{6}$ sq. units

12. $2\left(1 - \dfrac{1}{e}\right)$ sq. units

13. $\dfrac{3}{5} + \ln(3)$ sq. units

Exercise 7K

1. $\dfrac{32}{3}$ m

2. a $\dfrac{20}{3}$ m   b 30 m

3. 3.25 m

4. 26.3 m

Exercise 7L

1. $\dfrac{8\pi}{15}$ cu.units or 1.68 cu.units

2. $\dfrac{4\pi}{3} \cdot \left(2\sqrt{2} + 3\right)$ cu.units or 24.4 cu.units

3. 4π cu.units

4. $\dfrac{2\pi}{3}$ cu.units

5. 16.1 cu.units

6. 7.2π cu.units

Exercise 7M

1. $\dfrac{117\pi}{4}$ cu.units

2. $\dfrac{108\pi}{5}$ cu.units

3. $\dfrac{16\pi}{15}$ cu.units

4. $\dfrac{7\pi}{15}$ cu.units

Review exercise (non-GDC)

1. $\dfrac{41}{4}$ m

2. 2.5 m

3. $8\sqrt{3} - \dfrac{40}{3}$ sq. units

4. $y = \dfrac{2}{5}x^2 - \dfrac{32}{5x} - \dfrac{24}{5}$

Review exercise (GDC)

1. a $\dfrac{x^4 - 6}{x} + c$

   b $\dfrac{x^3}{3} + \dfrac{1}{x} + c$

   c $-\dfrac{\ln|3x - 2|}{3} + c$

   d $-\sqrt{1 - 4x} + c$

   e $\dfrac{-2}{3}e^{-3x} + 3\sqrt[3]{e^x} + c$

2. a $\dfrac{41}{24}$ b 54 c $-\ln(2)$

   d $-\dfrac{\ln\left|\dfrac{4e - 1}{3}\right|}{4}$

3. $\dfrac{2}{(2x - 1)} + x + 2;\ \ln 3 + \dfrac{7}{2}$

4. $4 - \ln(5)$ sq.units

5. $\dfrac{2}{3}$ sq. units

6. 2

7. $\dfrac{1}{3}$ sq. units

8. $\dfrac{64\pi}{9\ln 3}$ cu.units

9. 41.3 cu.units

Chapter 8

Skills check

1. 2.48 m

Exercise 8A

1. a $62°$, 3.76 cm, 7.06 cm

   b 5.6 cm, $36.9°$, $53.1°$

2. 38.8 m

3. 2.10 radians

4. 13.3 m³

Exercise 8B

1. a $\sin 36°$

   b $-\cos 30°$

   c $\tan 50°$

   d $\sin\left(\dfrac{\pi}{8}\right)$

   e $\tan\left(\dfrac{\pi}{3}\right)$

   f $-\cos\left(\dfrac{\pi}{6}\right)$

2. $\dfrac{-12}{13}, \dfrac{-5}{12}$

3. $\dfrac{-4}{5}, \dfrac{3}{4}, \dfrac{-3}{5}$

4. a Max $= 6$ at $\theta = 2\pi$,
      Min $= -2$ at $\theta = \pi$

   b Max $= 8$ at $\theta = \dfrac{3\pi}{2}$,
      Min $= 2$ at $\theta = \dfrac{\pi}{2}$

   c Max $= 1$ at $\theta = \dfrac{\pi}{2}$,
      Min $= -3$ at $\theta = \dfrac{3\pi}{2}$

   d Max $= -1$ at $\theta = \pi$,
      Min $= -5$ at $\theta = 2\pi$

Investigation – trigonometric identites

a $\sin\theta = \cos(90° - \theta)$
   $\cos\theta = \sin(90° - \theta)$
   $\tan\theta = \cot(90° - \theta)$

b $\tan\theta = \dfrac{\sin\theta}{\cos\theta}$

c $\sin^2\theta + \cos^2\theta = 1$

e $\tan^2 + 1 = \sec^2\theta$

f $\cot^2\theta + 1 = \csc^2\theta$

Investigation – exact values of sin, cos and tan

a $\sin\dfrac{\pi}{4} = \dfrac{\sqrt{2}}{2} = \cos\dfrac{\pi}{4}$

$\tan\dfrac{\pi}{4} = 1$

b $\sin\dfrac{\pi}{3} = \dfrac{\sqrt{3}}{2}$

$\cos\dfrac{\pi}{3} = \dfrac{1}{2}$

$\tan\dfrac{\pi}{3} = \sqrt{3}$

$\sin\dfrac{\pi}{6} = \dfrac{1}{2}$

$\cos\dfrac{\pi}{6} = \dfrac{\sqrt{3}}{2}$

$\tan\dfrac{\pi}{6} = \dfrac{1}{\sqrt{3}}$

Exercise 8C

1 $\cos\theta = -\dfrac{\sqrt{15}}{4}$

$\tan\theta = -\dfrac{1}{\sqrt{15}}$

2 $\sin\theta = \dfrac{5}{13}$

$\tan\theta = -\dfrac{5}{12}$

Exercise 8D

1 a $\dfrac{1+\sqrt{3}}{2}$

b $\dfrac{\sqrt{3}-1}{1+\sqrt{3}}$

c $-\dfrac{2\sqrt{2}}{\sqrt{3}-1}$

2 a $\dfrac{1}{2}$

b $\dfrac{\left(1+\sqrt{3}\right)^4}{4}$

3 $\dfrac{3}{4}$

6 b $\dfrac{3\pi}{4}$

Exercise 8E

2 $\dfrac{4}{5}$ or $-\dfrac{44}{125}$

3 $-\dfrac{7}{9}, \dfrac{17}{81}$

5 a $\dfrac{1}{2}(\cos 2A + 3)$

b $\dfrac{1}{4}(\cos 2A + 1)^2$

c $\dfrac{1}{4}(1 - \cos 2A)^2$

Exercise 8F

1
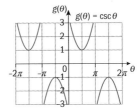

2 a
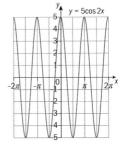

b
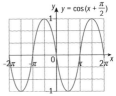

c
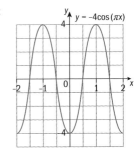

d
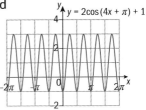

3 a
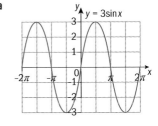

b
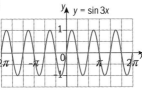

c
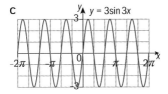

d
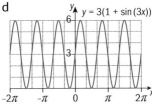

4
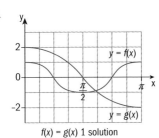

$f(x) = g(x)$ 1 solution

5 a $f(x) = 4\sin x \cos x = 2\sin 2x$

$f(x)$ is odd, period $= \pi$

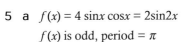

b $g(x) = 1 - 2\sin^2 x = \cos 2x$

$g(x)$ is even, period $= \pi$

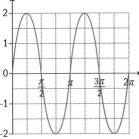

c $h(x) = x\sin x$

$h(x)$ is even, not periodic

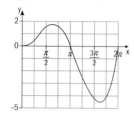

Exercise 8G

1 i a $7, \dfrac{\pi}{3}, \dfrac{\pi}{12}$

b $-4, 10$

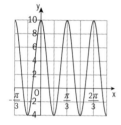

ii a $3, \pi, \dfrac{\pi}{4}$

b $-8, -2$

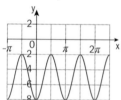

2 a 220 b -220

c 220

d $\dfrac{1}{60}$

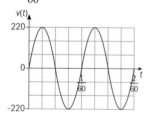

3 a $a = 6.6$

$b = \dfrac{\pi}{6}$

$c = -5.25$

$d = 7.8$

b 02.15

c 00.00 to 00.05

04.25 to 12.05

16.25 to 24.00

4 $a = 1.05$

$b = \dfrac{2\pi}{365}$

$c = -80.75$

$d = 11.7$

12.7 hours

Exercise 8H

1 a $\dfrac{\sqrt{2}}{2}$ b $\dfrac{\sqrt{5}}{2}$

c $\dfrac{1}{2}$ d $\dfrac{5\pi}{6}$

e $\dfrac{\pi}{4}$ f $\dfrac{\pi}{6}$

2 a 1 b $-\dfrac{7}{25}$

c $\dfrac{24}{7}$

Exercise 8I

1 $0, \pm 0.841, \pm\pi$

2 0.170

3 $\pm\dfrac{\pi}{6}, \pm\dfrac{5\pi}{6}$

4 $\dfrac{5}{3}$ or 3

5 $\dfrac{\pi}{2}$ or π

6 $\dfrac{\pi}{2}$

7 $\dfrac{2}{3}$

a $\dfrac{12}{5}$

b $\dfrac{-3\pm\sqrt{13}}{2}$

8 $0, \pi, 5.17, 2\pi$

9 $0, 0.294, 0.536, 1.02, 1.32$

Exercise 8J

1 a 4.44, 116°, 34.3°

b 6.67, 48.3°, 36.7°

c 30.8°, 24.1°, 125°

2 89.2°

3 29.0°

4 $x = \dfrac{8}{3}$, 32.2°, 87.8°

Exercise 8K

1 a 25°, 16.4 cm, 8.45 cm

b 95°, 9.86 cm, 6.36 cm

c 30.0°, 110°, 13.2 cm

2 37 secs

3 299 m

4 MC = 20.5 m, MB = 8.93 m,
MA = 8.03 m

5 55°, 55°, 70° or 55°, 15°, 110°

Exercise 8L

1 197 sq. units

2 1350 cm²

3 19.7 cm²

4 156°, 11.8 m²

5 a $\dfrac{\sqrt{3}r^2}{4}, \dfrac{r^2}{4}$

b $\dfrac{r^2}{12}(4\pi - 3\sqrt{3})$

c $\dfrac{r^2}{12}(\pi - 3)$

Review exercise (non-GDC)

1 $0, \dfrac{2\pi}{3}, 2\pi$

2 a $\dfrac{\sqrt{6}-\sqrt{2}}{4}$

b $\dfrac{\sqrt{3}+1}{1-\sqrt{3}}$

c $\dfrac{\sqrt{6}-\sqrt{2}}{4}$

d $\sqrt{2}-1$

4 a $\dfrac{4+3\sqrt{3}}{10}$

b $-\dfrac{24}{25}$

c $\dfrac{7\sqrt{2}}{10}$

5

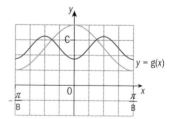

Review exercise (GDC)

1 $\dfrac{\pi}{2}$

2 a $y = \arcsin(1.1 - \sin x)$

$y = \arccos(1.8 - \sin 2x)$

b $x = 0.619, y = 0.546$

or $x = 1.09, y = 0.216$

3 a 110°

b 1.90 m

c 0.428 m

4 b $\left(0, \dfrac{2}{7}\right)$

c $p = 3.87$, $q = 5.55$

d

e $0.510 < x < 3.53$
$3.99 < x < 5.49$

f Max value is 2.39
(when $x = 1.88$)

5 a 8.60 cm

b 11.3 cm

c 7.30 cm

d 49.4°

e 31.4 cm²

6 b 17.3 m, 10 m, 26.5 m,
3.27 m

7 83.6 cm

Chapter 9

Skills check

1 Proof

2 a $f'(x) = 6e^{2x} - 4x$

b $g'(x) = 2\ln(x+1) + 2$

c $h'(x) = \dfrac{(2x^2 + 2x - 1)e^{x^2}}{(x+1)^2}$

Exercise 9A

Proofs

Exercise 9B

1 a $-\csc^2 x$ **b** $-\cot x \cdot \csc x$

c $3\cos 3x$ **d** $5\sec^2(5x - 3)$

e $3\sin(8 - 3x)$

f $-\dfrac{1}{4}\cot\left(\dfrac{x-3}{4}\right)\csc\left(\dfrac{x-3}{4}\right)$

g $\dfrac{2}{13}\csc^2\left(\dfrac{7 - 2x}{13}\right)$

2 a $5x^4 \cos(x^5 - 3)$

b $-e^x \sin(e^x)$

c $-2x\cot(x^2 + 11)$
$\csc(x^2 + 11)$

d $-(12x^2 - 4x + 7)$
$\csc^2(4x^3 - 2x^2 + 7x + 17)$

e $\dfrac{2\sec^2(\ln(2x+1))}{2x+1}$

f $\dfrac{e^x \sin(\sqrt{e^x + 1})\sec^2(\sqrt{e^x + 1})}{2\sqrt{e^x + 1}}$

g $-\cos(\cos(\tan x)) \cdot$
$\sin(\tan x) \cdot \sec^2 x$

Exercise 9C

1 a $2\cos x - (2x - 1)\sin x$

b $(3 - 2x)\sin 2x +$
$2(3x - x^2)\cos 2x$

c $e^{1-x}(\sec^2 x - \tan x)$

d $\dfrac{x\cos x - \sin x}{x^2}$

e $\dfrac{2\sin 2x - (4x + 6)\cos 2x}{\sin^2 2x}$

f $\dfrac{2(2 - x)\sec^2 x + \tan x}{2(2 - x)\sqrt{2 - x}}$

2 a 1 **b** $\dfrac{3\sqrt{2}}{2}$

c -2 **d** -2

e $\dfrac{3\pi}{2}$ **f** $\dfrac{9\pi^2 - 12\pi}{8}$

g 1

3 a 0

b $\tan\beta \cdot \sec\beta$

c $4\sec^2 4\theta$

d $\dfrac{3}{2}\sec^2\dfrac{3\rho}{2}$

e $\cos\varphi - \sin\varphi$

Exercise 9D

1 a $f'(x) = -\dfrac{1}{\sqrt{1 - x^2}}$

b $f'(x) = \dfrac{3}{\sqrt{1 - 9x^2}}$

c $f'(x) = \dfrac{1}{2x^2 + 2x + 1}$

2 a $\dfrac{dy}{dx} = 2\arcsin x + \dfrac{2x}{\sqrt{1 - x^2}}$

b $\dfrac{dy}{dx} = -\dfrac{x + \arccos x \cdot \sqrt{1 - x^2}}{x^2\sqrt{1 - x^2}}$

c $\dfrac{dy}{dx} = 2\arctan x + \dfrac{2x + 1}{1 + x^2}$

d $\dfrac{dy}{dx} = 1 - \dfrac{x\arcsin x}{\sqrt{1 - x^2}}$

e $\dfrac{dy}{dx} = 8x\arctan 2x + 2$

3 a $\arcsin x + \arccos x = \dfrac{\pi}{2}$

b $\arctan x + \arctan(-x) = 0$

c $2\arctan x - \arcsin\left(\dfrac{2x}{x^2 + 1}\right) = 0$

4 a $y' = \sec y$

b $y' = \cot^2 y$

c $y' = \dfrac{1 + \cos x}{1 - \sin y}$

d $y' = 2x \cdot e^{-\sin y}\sec y$

e $\dfrac{1}{\cos y - y\sin y}$

f $\dfrac{y}{2xy\sec^2 2y - x}$

Exercise 9E

1 a $y = 3x$

b $y = -x + \dfrac{\pi}{3} + \dfrac{\sqrt{3}}{2} - 1$

c $y = -2x + 1$

d $y = \dfrac{2}{3}x - \dfrac{\pi}{2} + 2$

2 a $x = 0$

b $y = -\dfrac{1}{8}x + \dfrac{\pi}{128} + 1$

c $y = -x$

d $y = x - \pi - 3$

3 $y = x - 1$

4 $y = \dfrac{1}{2}x - 1$

5 a $(0.974, 0.583)$

b $T_1 : y = -1.58x + 2.12$,
$T_2 : y = 5.03x - 4.32$

c $\theta = 0.760$ or $\theta = 43.5°$

6 π^2

Exercise 9F

1 a $8\sqrt{3}$

b 2

c 1

d $-\dfrac{1}{2}\left(\cos\dfrac{1}{2} + \sin\dfrac{1}{2}\right)$

e $-3e^{\frac{\pi}{4}}$

f -2π

2 a $f^{(n)}(x) = \sin\left(x + \dfrac{(n+1)\pi}{2}\right)$,
$n \in \mathbb{Z}^+$

b $f^{(n)}(x) = 3^n\sin\left(3x + \dfrac{n\pi}{2}\right)$,
$n \in \mathbb{Z}^+$

c $f^{(n)}(x) =$
$a^n\sin\left(ax + b + \dfrac{(n+1)\pi}{2}\right)$,
$n \in \mathbb{Z}^+$

3 a $\dfrac{1}{\sqrt{2}}, \sqrt{2}, -2\sqrt{2}, -4\sqrt{2}$

b $\dfrac{615\sqrt{2}}{2}$

Exercise 9G

1 $\dfrac{d\theta}{dt} = -0.0175\,\text{c sec}^{-1}$

2 a $-240\,\text{ms}^{-1}$

   b $\dfrac{d\theta}{dt} = 0.466\,\text{c min}^{-1}$

3 a $\dfrac{d\theta}{dt} = 0.880\,\text{c sec}^{-1}$

   b $\dfrac{d\theta}{dt} = 0.496\,\text{c sec}^{-1}$

4 a $-0.0125\,\text{c sec}^{-1}$

   b $-0.0231\,\text{c sec}^{-1}$

5 a $\dfrac{dr}{dt} = -0.00111\,\text{cm min}^{-1}$

   b $\dfrac{dA}{dt} = -1\,\text{cm}^2\,\text{min}^{-1}$

6 a $0.995°\,\text{s}^{-1}$

   b $1.63°\,\text{s}^{-1}$

7 a $65.0\,\text{km h}^{-1}$

   b $0.1°\,\text{sec}^{-1}$

8 a $0.144\,\text{c s}^{-1}$ or $8.27°\,\text{s}^{-1}$

   b $5.77\,\text{m s}^{-1}$

9 $1.28\,\text{m s}^{-1}$

Exercise 9H

1 a $-\dfrac{1}{3}\cos 3x + c$

   b $\dfrac{1}{2}\sin(2x+1) + c$

   c $\dfrac{1}{3}\tan 3x + c$

   d $-\tan(1-x) + c$
     $= \tan(x-1) + c$

   e $-\dfrac{3}{5}\cos\dfrac{5x-1}{3} + c$

   f $\dfrac{7}{3}\sin\left(\dfrac{3x+2}{7}\right) + c$

2 a $-\dfrac{1}{2}\sin 2x + c$

   b $\tan x + c$

   c $\dfrac{1}{2}x - \dfrac{1}{4}\sin 2x + c$

   d $\dfrac{1}{2}x + \dfrac{1}{4}\sin 2x + c$

   e $\dfrac{1}{4}\sin 4x + c$

   f $\dfrac{2}{5}\tan 5x + c$

   g $x + c$

   h $\dfrac{1}{2}x - \dfrac{1}{8}\sin 4x + c$

Exercise 9I

1 a $-2\cos x - 3\sin x + c$

   b $\dfrac{1}{3}x^3 + 7\cos x + c$

   c $4e^x - \dfrac{1}{3}\tan x + c$

   d $x - \dfrac{2x\sqrt{2x}}{3} - \dfrac{7}{3}\cos 3x + c$

   e $\dfrac{5}{2}\ln |x| + 3\tan\dfrac{x}{3} + c$

   f $x - \ln |x+1| +$
     $\dfrac{4}{3}\cos\left(\dfrac{3x}{4}\right) + c$

   g $\dfrac{2^x}{\ln 2} - 10\cos\dfrac{x}{2} -$
     $\dfrac{3}{2}\sin\left(\dfrac{2x}{3}\right) + c$

   h $-\dfrac{3^{-2x}}{2\ln 3} - \tan(11x) + c$

Exercise 9J

1 a $f(x) = 5x - 2\sin x$

   b $f(x) = 2x^2 + 3\cos 2x - 2$

   c $f(x) = 3\sin x - 2\tan x - \dfrac{3}{2}$

   d $f(x) = x^3 - 2e^x + \dfrac{1}{4}\sin 4x - 3$

   e $f(x) = 3\ln |x| + \dfrac{1}{3}\sin 3x$
     $- 4x + 4$

   f $f(x) = -\dfrac{7}{4}\ln |3 - 4x|$
     $- 4x^2 + 2e^{2x-1} - 2$

2 a $f(x) = -4\sin x + 2x + 1$

   b $f(x) = \dfrac{1}{2}x^2 + 3x - \cos x - \dfrac{7}{2}$

   c $f(x) = e^{1-x} - \sin(1-x)$
     $+ 2x - 1$

   d $f(x) = \dfrac{1}{4}e^{2x} - \dfrac{1}{4}\sin 2x +$
     $\dfrac{1}{20}x^5 - \dfrac{1}{3}x^3 + \dfrac{1}{2}x^2 + 2x + \dfrac{7}{4}$

Exercise 9K

1 a $\dfrac{5\pi^2 - 18}{36}$

   b $\dfrac{10\pi + 3}{6}$

   c $\dfrac{\pi + 8}{4}$

   d $e^{\frac{\pi}{3}}$

   e $2 + \dfrac{3^{2\pi} - 3^{-2\pi}}{\ln 3}$

   f $\dfrac{1}{9}\left(e^{\frac{3\pi}{2}} - 1\right) - \dfrac{2}{5}$

   g $\dfrac{\pi}{2}$

   h $\dfrac{2^{\frac{\pi}{12}} - 1}{\ln 2} + \dfrac{1}{2}$

   i $\dfrac{\pi^3 + 1536}{768}$ j $2e^{8\pi} + 4$

Exercise 9L

1 a $-\cos x^2 + c$

   b $\dfrac{2}{3}(x^3 + 3)\sqrt{x^3 + 3} + c$

   c $e^{1+3x-2x^2} + c$

   d $\ln |\sec x| + c$

   e $e^{\sin 2x} + c$

   f $e^{\sqrt{x}} + c$

   g $-\cos 2^x + c$

   h $\dfrac{1}{2}(\arcsin x)^2 + c$

   i $\dfrac{1}{2}(\arctan(2x))^2 + c$

2 a $\dfrac{1}{2}\sin x^2 + c$

   b $\dfrac{1}{8}(x^6 - 1)\sqrt[3]{x^6 - 1} + c$

   c $\dfrac{1}{6}e^{3x^2 + 12x - 7} + c$

   d $\dfrac{1}{25}\ln |\sec(5x + 4)| + c$

   e $-\dfrac{3^{\cos 3x}}{3\ln 3} + c$

   f $-4\cos\sqrt[4]{x} + c$

   g $\dfrac{1}{\ln 5}\sin 5^x + c$

   h $-\dfrac{1}{2}\ln |e^{-2x} - e^{2x}| + c$

   i $\dfrac{2}{9}\left(\arctan\dfrac{x}{3}\right)^{\frac{3}{2}} + c$

   j $\dfrac{1}{3}\sin\left(x^3 + \dfrac{3x^2}{2}\right) + c$

   k $\dfrac{1}{3}\arcsin^3(2x + 1) + c$

Exercise 9M

1 a $\dfrac{1}{5}$ b $\ln 10$

   c $\dfrac{\sqrt{2}}{6}$ d $\dfrac{9}{2}$

   e $\ln 3 - \ln 2$ f $\ln 2$

   g $\dfrac{1}{3}\sin\dfrac{5}{2}$ h $\dfrac{54 - 4\sqrt{2}}{3\ln 2}$

Exercise 9N

1 $(x - 1)e^x + c$

2 $(2x + 9)\sin x + 2\cos x + c$

3 $(5x - 2)\cos x - 5\sin x + c$

4 $\dfrac{1}{3}(3x - 2)e^{3x} + c,$

5 $(x - 2)e^{4x-1} + c,$

6 $-\dfrac{x+3}{4}\cos(2x + 3) +$
   $\dfrac{1}{8}\sin(2x + 3) + c$

7 $(3-x)\sin\dfrac{x}{4} - 4\cos\dfrac{x}{4} + c$

8 $\left(\dfrac{x}{\ln 2} - \dfrac{1}{\ln^2 2}\right)2^x + c$

9 $\left(\dfrac{1-x}{\ln 5} + \dfrac{1}{\ln^2 5}\right)5^x + c$

10 $\left(\dfrac{x-2}{7\ln 3} + \dfrac{1}{7\ln^2 3}\right)3^{-x} + c$

11 $\left(\dfrac{4x}{\ln 3 - \ln 5} - \dfrac{4}{(\ln 3 - \ln 5)^2}\right)\left(\dfrac{3}{5}\right)^x + c$

Exercise 9O

1 $\dfrac{x^2}{4}(2\ln x - 1) + c$

2 $\left(\dfrac{3}{2}x^2 + 2x\right)\ln x - \dfrac{3}{4}x^2 - 2x + c$

3 $\left(x - \dfrac{1}{2}x^2\right)\ln x - x + \dfrac{1}{4}x^2 + c$

4 $\dfrac{1}{2}x^2\ln 4x - \dfrac{1}{4}x^2 + c$

5 $\left(\dfrac{3}{2}x^2 - 2x\right)\ln\dfrac{x}{5} - \dfrac{3}{4}x^2 + 2x + c$

6 $\left(2x^2 + 3x + \dfrac{9}{8}\right)\ln(3+4x) - x^2$
$- \dfrac{3}{2}x + c$ or $\dfrac{(3+4x)^2}{16}$
$(2\ln(3+4x)-1) + k$

7 $\dfrac{1}{121}\left[\left\{\dfrac{7}{2}(4-11x)^2 - 83(4-11x)\right\}\right.$
$\ln(4-11x) - \dfrac{7}{4}(4-11x)^2$
$\left. + 83(4-11x)\right] + c$

8 $\dfrac{x^3}{9}(3\ln x - 1) + c$

9 $\left(2x - \dfrac{x^2}{2} + \dfrac{x^3}{3}\right)$
$\ln 3x - 2x + \dfrac{x^2}{4} - \dfrac{x^3}{9} + c$

Exercise 9P

1 $x\log x - \dfrac{x}{\ln 10} + c$

2 $x\log_a x - \dfrac{x}{\ln a} + c$

3 $x\arctan x - \dfrac{1}{2}\ln(1+x^2) + c$

4 $x\arccos x - \sqrt{1-x^2} + c$

5 $(x^2 + 1)\arctan x - x + c$

6 $\dfrac{x^3}{3}\arcsin x + \dfrac{(x^2+2)\sqrt{1-x^2}}{9} + c$

Exercise 9Q

1 $(x^2 - 2x + 2)e^x + c$

2 $(1 - x^2)\cos x + 2x\sin x + c$

3 $(2 + 2x - x^2)\sin x + (2 - 2x)$
$\cos x + c$

4 $\dfrac{1}{2}e^{2x}(2x - x^2) + c$

5 $\dfrac{1}{2}(2 + x + 2x^2)\sin 2x +$
$\dfrac{1}{4}(4x + 1)\cos 2x + c$

6 $\dfrac{1}{4}(2x^2 - 1)\cos(1 - 2x) +$
$\dfrac{1}{2}\sin(1 - 2x) + c$

7 $\dfrac{1}{\ln^3 3}3^x(x^2\ln^2 3 - 2x\ln 3 + 2) + c$

8 $2e^{\frac{x}{2}}(x^3 - 6x^2 + 24x - 47) + c$

9 $-\dfrac{\cos 5x}{125}(25x^3 + 25x^2 - 6x - 2)$
$+ \dfrac{\sin 5x}{625}(75x^2 + 50x - 6) + c$

10 $(x^4 - 12x^2 + 24)\sin x +$
$(4x^3 - 24x)\cos x + c$

11 $e^{2x}\left(\dfrac{x^5}{2} - \dfrac{5x^4}{4} + \dfrac{5x^3}{2}\right.$
$\left. - \dfrac{15x^2}{4} + \dfrac{15x}{4} - \dfrac{15}{8}\right) + c$

Exercise 9R

1 $\dfrac{1}{2}e^x(\sin x - \cos x) + c$

2 $\dfrac{1}{5}e^{2x}(2\cos x + \sin x) + c$

3 $\dfrac{1}{25}e^{4x}(4\cos 3x + 3\sin 3x) + c$

4 $-\dfrac{1}{5}e^{-x}(\sin 2x + 2\cos 2x) + c$

5 $\dfrac{e^x}{2}(\sin x + \cos x + c)$

Exercise 9S

1 $\dfrac{2}{15}(3x - 4)(x + 2)^{\frac{3}{2}} + c$

2 $-\dfrac{1}{5}(3x + 1)(1 - 2x)^{\frac{3}{2}} + c$

3 $\dfrac{1}{28}(10x^2 - 6x + 3)(4x+3)^{\frac{3}{2}} + c$

4 $\dfrac{3}{28}(4x - 9)(x + 3)^{\frac{4}{3}} + c$

5 $\dfrac{4}{585}(45x^2 - 40x + 32)(x+1)^{\frac{5}{4}} + c$

6 $-\dfrac{5}{3696}(176x^3 + 165x^2 + 150x + 125)$
$(1-x)^{\frac{6}{5}} + c$

Exercise 9T

1 $\sin x - \dfrac{1}{3}\sin^3 x + c$

2 $\dfrac{3}{8}x + \dfrac{1}{4}\sin 2x + \dfrac{1}{32}\sin 4x + c$

3 $-5\cos\left(\dfrac{x}{5}\right) + \left(\dfrac{10}{3}\right)\cos^3\left(\dfrac{x}{5}\right) -$
$\cos^5\left(\dfrac{x}{5}\right) + c$

4 $15x + 6\sin 4x + \dfrac{9}{8}\sin 8x -$
$\dfrac{1}{2}\sin^3 4x + c$

Investigation - recursive formula

1 $\displaystyle\int \sin^n x\,dx = \dfrac{-1}{n}\cos x\sin^{n-1} x$
$+ \dfrac{n-1}{n}I_{n-2}$

2 $\displaystyle\int \cos^n x\,dx = \dfrac{-1}{n}\sin x\cos^{n-1} x$
$+ \dfrac{n-1}{n}I_{n-2}$

Exercise 9U

1 $2\arcsin\left(\dfrac{x}{2}\right) + \dfrac{1}{2}x\sqrt{4-x^2} + c$

2 $\ln|\sqrt{x^2-1} + x| + c$

3 $\dfrac{9}{2}\ln\left(\sqrt{x^2+9} + x\right) + \dfrac{1}{2}x\sqrt{x^2+9} + c$

4 $3\arcsin\left(\dfrac{x}{6}\right) + c$

5 $-24\ln|\sqrt{x^2-16} + x| +$
$\dfrac{3}{2}x\sqrt{x^2-16} + c$

6 $5\ln|\sqrt{x^2+121} + x| + c$

7 $\arcsin\left(\dfrac{2x}{9}\right) + c$

8 $-\dfrac{25\sqrt{3}}{2}\ln|\sqrt{x^2-25} + x|$
$+ \dfrac{\sqrt{3}}{2}x\sqrt{x^2-25} + c$

9 $\sqrt{7}\ln|\sqrt{x^2+4} + x| + c$

Exercise 9V

1 1.72×0.652
2 1.11
3 b 1.05 rad
4 a $\theta = \dfrac{\pi}{4}$ b 21.213 m

5 a $-\dfrac{\pi^2}{36}\,d(t)$

b $0.740\ \text{ms}^{-1}$, $4.5\ \text{s}$

6 $h_{\min} = 1.82\ \text{m}$, closer to the first pole since $x = 1.31$, $3 - x = 1.69$.

Exercise 9W

1 a 2.87
b 2
c 2
d 0.278
e 1.62

2 a 0.342
b 0.560
c 2.44
d 0.282
e 0.475

3 2.38
4 0.0966
5 1.55
6 a $(2, 1)$, $(-2, 1)$ **c** 4.95

Exercise 9X

1 a π **b** π

c $\dfrac{4\pi^2 - 3\pi\sqrt{3}}{12}$

d $\dfrac{2\pi^2 + 3\pi\sqrt{3}}{12}$

2 a $\dfrac{2\pi - \pi\sin 2}{4}$

b $\dfrac{\pi^2}{4}$

c 1.17

d 0.771

3 a $\dfrac{3\pi}{2} + \dfrac{\pi^2}{4}$ **b** $\dfrac{3\pi\sqrt{3}}{16}$

c 2.35 **d** 4.18

Exercise 9Y

1 $24\pi^2$
2 $32\pi^2$
3 36π
4 16π
5 24π
6 $60\pi^2$

Review exercise (non-GDC)

1 a $2\sin x + (2x + 3)\cos x$

b $e^x + (\cos 3x - 3\sin 3x)$

c $\dfrac{x - \sin 2x}{2x^3\cos^2 x}$

2 $y = -2x$

3 $m = \dfrac{\pi}{3}$

4 a $(x - 3)e^{2x} + c$

b $(x^2 - 5x - 2)\sin x + 2x\cos x + c$

c $\dfrac{e^x}{10}(\cos 3x + 3\sin 3x)$

5 $\dfrac{\sqrt{5}}{5}\ \text{cm}^2\text{s}^{-1}$

6 a $y = 2x$

b $\dfrac{e - 2}{4e}$

c $\dfrac{(e^2 - 3)\pi}{12e^2}$

7 $\dfrac{x}{2}\sqrt{9 - x^2} - \dfrac{9}{2}\arccos\left(\dfrac{x}{3}\right) + c$

8 b $\dfrac{(8\ln 2 - 3)\pi}{8}$

9 a $\dfrac{15}{2}\left(1 - e^{\frac{-2k}{3}}\right)$

b 7.5 m

10 $y = -\dfrac{3}{2}x + \dfrac{1}{2}$

Review exercise (GDC)

1 $(-0.760, -0.577)$, $(0, 0)$, $(0.760, 0.577)$

2 $y = -0.423x + 1.24$

3 $a = 0.500$

4 $960\ \text{km}\,\text{h}^{-1}$

5 $v = 1.31$

Chapter 10

Skills check

1 a $\dfrac{1}{8}$ **b** $\dfrac{5}{36}$

2 a 0.8 **b** 0.4

Exercise 10A

1 a $\displaystyle\sum_{x=1}^{4} f(x) = 0.2 + 0.3 + 0.4 + 0.2 = 1.1 > 1.$ Therefore, f cannot be a probability distribution function.

b $\displaystyle\sum_{x=1}^{4} f(x) = 0.2 + 0.3 + 0.4 + 0.2 = 1.1 > 1$ Therefore, f cannot be a probability distribution function.

c $f(-1) \le 0$ therefore, f cannot be a probability distribution function.

2 a $a = 0.12$ **b** 0.87
c 0.87

3 a $k = \dfrac{1}{10}$ **b** $\dfrac{1}{2}$

4

x	0	1	2	3
$P(X = x)$	$\dfrac{27}{343}$	$\dfrac{108}{343}$	$\dfrac{144}{343}$	$\dfrac{64}{343}$

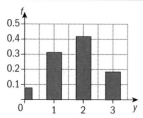

5

x	0	1	2	3
$P(X = x)$	$\dfrac{120}{2184}$	$\dfrac{702}{2184}$	$\dfrac{1008}{2184}$	$\dfrac{336}{2184}$

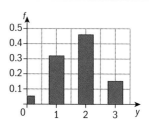

Exercise 10B

1 a 6.2 **b** 50.2

c 11.76 **d** $\sqrt{\dfrac{294}{25}}$ or 3.43 (3sf)

2 a $a = \dfrac{1}{8}$ and $b = \dfrac{5}{24}$

b $E(X) = \dfrac{23}{8}$ and $E(X^2) = \dfrac{265}{24}$

c $\text{Var}(X) = \dfrac{533}{192}$

3 a 0.5. As the cards are distinct $P(\text{bottom} > \text{top}) = P(\text{top} < \text{bottom})$, and as the two events are mutually exclusive and exhaustive $P(\text{bottom} > \text{top}) + P(\text{top} > \text{bottom}) = 1$.

b $P(S = 4) = \dfrac{1}{45}$, $P(S = 8) = \dfrac{3}{45}$ and $P(S = 11) = \dfrac{1}{9}$

c

x	$P(S = x)$
3	$\dfrac{1}{45}$
4	$\dfrac{1}{45}$
5	$\dfrac{2}{45}$
6	$\dfrac{2}{45}$
7	$\dfrac{3}{45}$
8	$\dfrac{3}{45}$
9	$\dfrac{4}{45}$
10	$\dfrac{4}{45}$
11	$\dfrac{1}{9}$
12	$\dfrac{4}{45}$
13	$\dfrac{4}{45}$
14	$\dfrac{3}{45}$
15	$\dfrac{3}{45}$
16	$\dfrac{2}{45}$
17	$\dfrac{2}{45}$
18	$\dfrac{1}{45}$
19	$\dfrac{1}{45}$

d $E(S) = 11$ and

$Var(S) = 14\dfrac{2}{3}$

4 £20

5 a $k = \dfrac{1}{44}$

b $P(T = 4) = \dfrac{4}{11}$,

$P(T \leq 4) = \dfrac{15}{22}$ and

$P(T = 4 | T \leq 4) = \dfrac{8}{15}$

c $E(T) = 4$ and $Var(T) = \dfrac{17}{11}$

d 4

Exercise 10C

1 a 0.4

b

x	5	10	15	20	25	30
$f(x) = P(X \leq x)$	$\dfrac{1}{15}$	$\dfrac{3}{15}$	$\dfrac{6}{15}$	$\dfrac{10}{15}$	$\dfrac{13}{15}$	1

c 20

2 a 0.47

b $E(L) = 2.32$ and

$Var(L) = 1.5376$

c

x	0	1	2	3	4	5
$f(x) = P(X \leq x)$	0.07	0.28	0.53	0.84	0.96	1

d 2

3 a

x	2	3	4	5	6
$f(x) = P(X = x)$	$\dfrac{1}{36}$	$\dfrac{4}{36}$	$\dfrac{10}{36}$	$\dfrac{12}{36}$	$\dfrac{9}{36}$
$f(x) = P(X \leq x)$	$\dfrac{1}{36}$	$\dfrac{5}{36}$	$\dfrac{15}{36}$	$\dfrac{27}{36}$	1

b mean $4\dfrac{2}{3}$, median 5 and

mode 5

c $\dfrac{\sqrt{10}}{3}$, 1.05 (3 sf)

4 b $\dfrac{2n+1}{3}$

5 b $F(x) = 1 - 3^{-x}$

where $x = 1, 2, 3, \ldots$

Investigation – the Galton Board

a

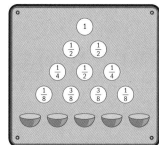

b

c The coefficients are equal to the Pascal triangle entries as each of them is obtained adding the two coefficients above it.

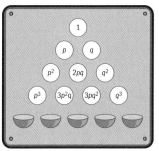

d The rth entry of the nth row

is given by $\dbinom{n}{r} p^r q^{n-r}$ where

$\dbinom{n}{r} = \dfrac{n!}{r!(n-r)!}$

Exercise 10D

1 a 0.205 (3 sf)

b 0.377 (3 sf)

c 0.63 (3 sf)

2 a 0.2304

b 0.903744

c 0.733 (3 sf)

3 a i 0.0773 (3 sf)

ii 0.999946 (3 sf)

b 0.0341 (3 sf)

4 a 2

b 2

c 0.667 (3 sf)

5 a

$P(X \leq x)$	$x = 0$	$x = 1$	$x = 2$	$x = 3$	$x = 4$	$x = 5$	$x = 6$	$x = 7$	$x = 8$	$x = 9$	$x = 10$
$n = 2$	0.36	0.84	1								
$n = 5$	0.07776	0.33696	0.68256	0.91296	0.98976	1					
$n = 10$	0.00605 (3 sf)	0.0464 (3 sf)	0.167 (3 sf)	0.382 (3 sf)	0.633 (3 sf)	0.834 (3 sf)	0.945 (3 sf)	0.98771 (3 sf)	0.99832 (3 sf)	0.99990 (3 sf)	1 (3 sf)

b 26

6 15

7 a 0.00374 (3 sf)

b 0.898 (3 sf)

c 0.0417 (3 sf)

8 4

Investigation – parameters of a binomial variable

n	p	$q = 1-p$	$E(X) = \sum\limits_{x=0}^{n} x \cdot p^x \cdot q^{n-x}$	$E(X^2) = \sum\limits_{x=0}^{n} x^2 \cdot p^x \cdot q^{n-x}$	$VAR(X) = E(X^2) - (E(X))^2$
1	0.5	0.5	0.5	0.5	0.25
5	0.5	0.5	2.5	7.5	1.25
10	0.5	0.5	5	27.5	2.5
1	0.2	0.8	0.2	0.2	0.16
5	0.2	0.8	1	1.8	0.8
10	0.2	0.8	2	5.6	1.6
1	0.8	0.2	0.8	0.8	0.16
5	0.8	0.2	4	16.8	0.8
10	0.8	0.2	8	65.6	1.6

$E(X) = np$ and $Var(X) = npq$

Exercise 10E

1 a 0.124 (3 sf) **b** 0.950 (3 sf)

c 0.826 (3 sf) **d** 3.2

e 1.92

2 a 0.565 (3 sf) **b** 0.647 (3 sf)

c 0.671 (3 sf) **d** 2

3

x	0	1	2	3	4	5
b $f(x) = P(X=x)$	$\frac{1}{32}$	$\frac{5}{32}$	$\frac{5}{16}$	$\frac{5}{16}$	$\frac{5}{32}$	$\frac{1}{32}$
c $F(x) = P(X \leq x)$	$\frac{1}{32}$	$\frac{3}{16}$	$\frac{1}{2}$	$\frac{13}{16}$	$\frac{31}{32}$	1

b mode is 2 and 3

d 2.5

4 a 0.0188 (3 sf)

b 0.904 (3 sf) **c** 0.6 days

5 $n = 8$, $p = 0.25$

6 a i 0.00317 (3 sf)

ii 0.00394 **iii** 0.617

b 5 and 1.94 (3 sf)

c 0.000154 (3 sf)

7 a 0.298 (3 sf)

b 0.863 (3 sf)

c 1 **d** 4.5

e 3.69 **f** 25

g 0.414 (3 sf)

8 a i 0.02835 **ii** 0.16308

b 0.05913

9 a

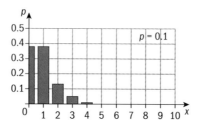

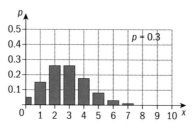

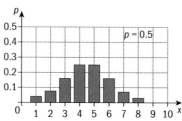

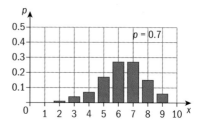

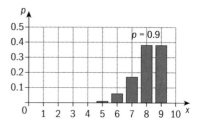

b The graph is symmetrical when $p = 0.5$ and asymmetrical with respect to the line $x = 4.5$ otherwise. For $p < 0.5$ the graph is positively skewed and for $p > 0.5$ it is negatively skewed. The graphs for values of p that add up to 1 are reflections of each other in the line $x = 4.5$

c

	mean	mode	median
$p = 0.1$	0.9	0 and 1	1
$p = 0.3$	2.7	2 and 3	3
$p = 0.5$	4.5	4 and 5	4.5
$p = 0.7$	6.3	6 and 7	6
$p = 0.9$	8.1	8 and 9	8

The values of the parameters of the distributions reflect the symmetries observed (eg. the sum of the means for $p = 0.1$ and $p = 0.9$ is 9).

10

a

w	0	1	2	4
$P(W = w)$	$(1-a)^2(1-2b)$	$4ab$ $(1-a)(1-b)$	$2ab$ $(a+b-2ab)$	a^2b^2

b $E(W) = 4ab$

Exercise 10F

1

$P(X=x)$	0	1	2	3	4	5
$m=1$	0.368	0.368	0.184	0.0613	0.0153	0.00307
$m=3$	0.0498	0.149	0.224	0.224	0.168	0.101
$m=5$	0.0067	0.0337	0.0842	0.140	0.176	0.176

(to 3 sf)

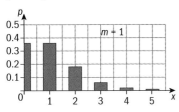

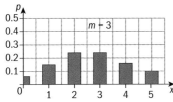

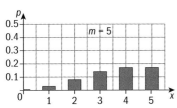

2   **a**   0.224 (3 sf)
    b   0.423 (3 sf)
    c   0.353 (3 sf)
    d   0.476 (3 sf)

Exercise 10G

1   **a**   0.122 (3 sf)
    b   0.156 (3 sf)
2   **a**   0.353 (3 sf)
    b   0.423 (3 sf)
3   0.12, 3.58
4   1.68

Investigation – parameters of a Poisson distribution

1, 2

m	$\sum_{x=0}^{100} x \cdot P(X=x)$	$\sum_{x=0}^{100} x^2 \cdot P(X=x)$	$\sum_{x=0}^{100} x^2 \cdot P(X=x) - \left(\sum_{x=0}^{100} x \cdot P(X=x)\right)^2$
0.1	0.1	0.11	0.1
0.2	0.2	0.24	0.2
0.3	0.3	0.39	0.3

3   The values are the same.
4 $E(X) = \text{Var}(X)$.

Exercise 10H

1   **a**   0.450 (3 sf)   **b**   0.552 (3 sf)
2   **a**   0.184 (3 sf)   **b**   0.221 (3 sf)
3   **a**   3   **b**   0.0166 (3 sf)
4   **a**   **i**   0.216 (3 sf)
        ii   0.463 (3 sf)
        iii   0.407 (3 sf)
    b $E(X) = \text{Var}(X) = 3.5$
    c   15.75
5   **a**   3.1612 (4 dp)
    b   0.611 (3 sf)
6   0.247 (3 sf)
7   **a** $p = e^{-\lambda}\left(1 + \lambda + \dfrac{\lambda^2}{2}\right)$

      (or equivalent)

    b $\dfrac{dp}{d\lambda} = -\dfrac{\lambda^2}{2}e^{-\lambda}$ which is

      negative for all values
      of $\lambda > 0$.

    c

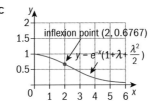

Exercise 10I

1   **a**   discrete   **b**   continuous
    c   continuous (sometimes treated as discrete)
    d   discrete as milk is bought in prepacked containers of fixed sizes
2 $f(x) \geq 0$ for all values of x and

$$\int_0^2 f(x)\,dx = \int_0^2 \frac{1}{2}x\,dx = \left[\frac{x^2}{4}\right]_0^2$$
$$= 1 - 0 = 1.$$

3   **a**   0.5   **b**   0.05
    c   0.5   **d**   0.5

Exercise 10J

1   **a**

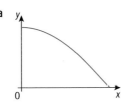

b   0.262 (3 sf), 0.285 (3 sf)
    and 0
2   **a**   12   **b**   0.4 and 0.04
3   **a**   0.25

    b

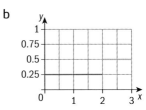

    median = 2

    c $E(X) = \dfrac{7}{4}$ and

       $\text{Var}(X) = \dfrac{37}{48}$

4   **a** $b = \dfrac{1 - 9a}{3}$

    b $a = -\dfrac{1}{16}$ and $b = \dfrac{25}{48}$

    c $E(X) = \dfrac{69}{64} = 1.08$ (3sf)

      and $\text{Var}(X) = \dfrac{9987}{20480}$

             $= 0.488$ (3 sf)

5   **a** $E(T) = 6\ln\dfrac{3}{2} = 2.43$ (3 sf)

    b $\text{Var}(T) = 6 - 36\ln^2\dfrac{3}{2}$

             $= 0.0815$ (3 sf)

    c   2.4   **d**   2

Exercise 10K

1   **a** $f(x) = \dfrac{x^3}{27}$

    b $f(x) = \dfrac{x(x^2 + 3x + 3)}{26}$

    c $f(x) = \sin x$

2   **a** $f(x) = \dfrac{8x^2 - x^4}{16}$

    b $\dfrac{95}{256}$ and $\dfrac{9}{16}$

3   **a**

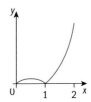

b $k = 1$ **c** $\dfrac{5}{6}$

4 a $\int_0^1 k\left(e - e^{kt}\right) dt = 1$

$ke[t]_0^1 - \left[e^{kt}\right]_0^1 = 1$

$ke - e + 1 = 1$

$k = 1$

b $\dfrac{e}{3} - e^{\frac{2}{3}} + e^{\frac{1}{3}}$

c $E(X) = \dfrac{e-2}{2}$ and

$Var(X) = \dfrac{12 + 4e - 3e^2}{12}$

d 0.290 (3 sf)

5 a $a = \dfrac{5}{2}$

b

$f(x) = \begin{cases} \dfrac{x^2}{8} & \text{for } 0 \le x \le 2 \\ \dfrac{3x-5}{x} & \text{for } 2 \le x \le \dfrac{5}{2} \end{cases}$

6 b

$f(y) = \begin{cases} 0 & \text{for } y < -3 \\ \dfrac{(y+3)^2}{36} & \text{for } -3 \le y \le 3 \\ 1 & \text{for } y > 3 \end{cases}$

c $\dfrac{7}{36}$ and $\dfrac{5}{9}$

d 1, 3 and 2

e 0, lower quartile

7 $a = \dfrac{-2}{9}$, $b = -\dfrac{8}{9}$ and $c = \dfrac{-1}{3}$

8 b $\dfrac{1}{4} + \dfrac{1}{(2\pi)}$, $\dfrac{3}{4} \dfrac{-1}{(2\pi)}$

c

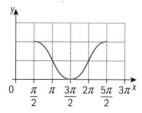

$\text{median} = \dfrac{3\pi}{2}$

d

$F(x) = \begin{cases} 0 & \text{for } x < \dfrac{\pi}{2} \\ \dfrac{(x - \cos x)}{2\pi} - \dfrac{1}{4} & \text{for } \dfrac{\pi}{2} \le x \le \dfrac{5\pi}{2} \\ 1 & \text{for } x > \dfrac{5\pi}{2} \end{cases}$

e 4.62 (3 sf)

Investigation – the normal curve

1

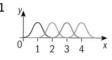

A $X \sim N(1, 1)$

B $X \sim N(2, 1)$

C $X \sim N(3, 1)$

D $X \sim N(4, 1)$

2 E $X \sim N(0, 1)$

F $X \sim N(0, 2)$

G $X \sim N(0, 3)$

H $X \sim N(0, 1)$

Exercise 10L

1 a 0.290 (3 sf) **b** 0.401 (3 sf)

c 0.159 (3 sf)

2 a 0.401 (3 sf) **b** 0.516 (3 sf)

c 0.460 (3 sf)

3 a 35 and 7

b 0.766, 0.609 and 0.0766 (3 sf)

Exercise 10M

1 a 0.0228 (3 sf)

b 0.00135 (3 sf)

c 0.954 (3 sf)

2 a $M \sim N(1.1, 0.15^2)$

b 0.161

c approx. 19

3 a 0.256 and 0.762 (3 sf)

b 0.0236 (3 sf)

c $t = 19.9$ (3 sf)

Investigation – further properties of the normal curve

1 (a–d) 0.682689…. the probability stays the same

2 0.682689…

3 0.954499…, 0.997300…. and 0.999936…

4 a equal

c $2P(X < \mu - a) + P(\mu - a < X < \mu + a) = 1$

Exercise 10N

1 a 0.3   **b** 0.7   **c** 0.4

2 a 0.2   **b** 0.2   **c** 0.3

d 0.6

3 a 0.2   **b** 0.3   **c** 0.4

4 a $\mu = 10$ and $\sigma = 5$

b $P(X < 5) = 0.158655…$ and $P(X \ge 15) = 0.158655…$ values are equal

c $a = 6.189…$ and $b = 13.81…$

Exercise 10O

1 a 0.1151 and 0.8849

b 0.01072 and 0.9893

c 0.0047 and 0.9953

They all add up to 1.

2 a −0.524 and 0.524

b −0.253 and 0.253

They are symmetric.

3 a 0.674 and −0.126

b $a = -0.126$ and $b = 0.674$

4 a both equal to 0.841

b both equal to 0.3829

5 both equal to 0.433

6 a $P(Y < 0)$

b

$P(X \le 10) = \dfrac{1 - P(10 \le X \le 20)}{2}$

$< 0.5 < P(10 \le X \le 20).$

c they are equal

Exercise 10P

1 $\mu = 2.94$ (3 sf)

2 $\mu = 1.40$ (3 sf) and $\sigma = 1.01$ (3 sf)

3 $\mu = 53.9$ (3 sf) and $\sigma = 2.25$ (3 sf)

4 $\mu = 0.0143$ kg

5 a $\mu = 45.6$ (3 sf) and $\sigma = 3.85$ (3 sf)

   b 0.383 (3 sf)

6 $\mu = 58.9$ (3 sf) and $\mu = 16.5$ (3 sf)

7 a 0.988 (3 sf) b 576 g (3 sf)

   c 830 (3 sf)

   d $\mu = 573$ g (3 sf) and $\sigma^2 = 670$ g^2 (3 sf)

8 $\sigma^2 > 7.39 \times 10^{-5}$ kg

Exercise 10Q

1 a $\mu = 13.6$ (3 sf) and $\sigma = 8.24$ (3 sf)

   b 0.137 (3 sf)

   c 0.995 (3 sf)

   d Independence of events

   e Independence may not hold (eg. bad weather conditions, strikes)

2 a 0.000342 (3 sf)

   b allow at least 87 minutes; assumptions: normally distributed, customers arrive on time

   c 18 waiters. The assumption of independence may not hold (eg. contagious diseases)

3 0.0560 (3 sf) Limitations: assumption of independence and validity of historical data

4 a 0.990 (3 sf)

   b 0.0383 (3 sf)

   c 97.8% (3 sf)

5 a 0.114 (3 sf)

   b 57.1 (3 sf)

   c 125 days

6 a 2.26 (3 sf) b 0.0280 (3 sf)

   c flaws randomly scattered on the sheets; independence of number of flaws per sheet

Review exercise (non-GDC)

1 a $a = \dfrac{1}{7}$ b $\mu = \dfrac{131}{49}$, mode 2 and median $\dfrac{5}{2}$

2

   a $k = 8$

   b median = 1, mode = 0

   c 0.5

3 a $\left(\dfrac{1}{5}\right)^5$ b $\left(\dfrac{4}{5}\right)^5$

   c $\dfrac{5^5 - 21}{5^5} = \dfrac{3104}{3125}$

4 a $\left(\dfrac{1}{2}\right)^5 = \dfrac{1}{32}$

   b 12.5; 2.5 marks more than if he answers just the questions he knows

5 a 2 b $\dfrac{1}{e^2}$

6 a 1 b $\arctan\dfrac{1}{2}$

   c $f(x) = \begin{cases} \sec^2 x & \text{for } 0 \le x \le \dfrac{\pi}{4} \\ 0 & \text{otherwise} \end{cases}$

   d $\dfrac{\sqrt{3}}{3}$

Review exercise (GDC)

1 a 0.625 (3 sf)

   b 0.00621 (3 sf)

2 8 years

3 a $\dfrac{4\sqrt{2}}{\pi}$

   b $\dfrac{4}{\pi}\arctan\dfrac{1}{2\sqrt{2}}$

   c $\dfrac{\sqrt{2}}{2\pi}\ln 2$

4 a 0.982 (3 sf)

   b 0.215 (3 sf)

5 a $a = 1.82205$ (5 dp)

   b $\mu = 2$ and $\sigma^2 = 0.359$ (3 sf)

   c 0.5

6 a 111

   b i 0.0435 (3 sf)

     ii 0.00332 (3 sf)

   c 0.00109 (3 sf)

Chapter 11

Skills check

1 a 5 b $(1, -2)$

2 a 2 b $y = 2x + 1$

Exercise 11A

1 b i $\overrightarrow{ED}$ and $\overrightarrow{FC}$

     ii $\overrightarrow{AD}$, $\overrightarrow{DA}$, $\overrightarrow{BE}$, $\overrightarrow{EB}$ and $\overrightarrow{FC}$

2 b No. If we draw all the diagonals we notice that all segments have either different directions or different lengths.

Exercise 11B

1

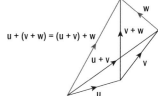

2 a $\overrightarrow{BD}$ b $\overrightarrow{AC}$ c $\overrightarrow{AB}$

   d $\overrightarrow{AE}$ e $\overrightarrow{CD}$ f $\overrightarrow{CB}$

   (other answers are possible)

3 a i $\mathbf{u} + \mathbf{v}$ ii $2\mathbf{u} - \mathbf{w}$

     iii $-2\mathbf{u} - 2\mathbf{v} + \mathbf{w}$

     iv $-\mathbf{v} + \mathbf{w}$

   b i 117° ii 10.7 square units

4 a $\mathbf{x} = \mathbf{u} + 2\mathbf{v}$

   b $\mathbf{x} = \dfrac{1}{2}(-\mathbf{u} + 3\mathbf{v})$

   c $\mathbf{x} = 3\mathbf{u} + 2\mathbf{v}$

Exercise 11C

1 a i $\begin{pmatrix} 3 \\ 1 \\ \frac{1}{2} \end{pmatrix}$ ii $\begin{pmatrix} 4 \\ 5 \\ \frac{5}{2} \end{pmatrix}$ iii $\begin{pmatrix} 1 \\ 2 \end{pmatrix}$

b i $-2\mathbf{i} + \dfrac{3}{2}\mathbf{j}$ ii $-\mathbf{i} + \dfrac{7}{2}\mathbf{j}$

  iii $-5\mathbf{i} + \mathbf{j}$

c $\overrightarrow{OB} = \begin{pmatrix} 2 \\ 7 \\ 2 \end{pmatrix}$, $\overrightarrow{OD} = \begin{pmatrix} 6 \\ 6 \end{pmatrix}$,

  $\overrightarrow{OE} = \begin{pmatrix} 3 \\ 11 \\ 2 \end{pmatrix}$, $\overrightarrow{OF} = \begin{pmatrix} 0 \\ 5 \end{pmatrix}$ and

  $\overrightarrow{OG} = \begin{pmatrix} 1 \\ 7 \end{pmatrix}$

2 **a** $\overrightarrow{OP} = \begin{pmatrix} 0 \\ 2 \\ -1 \end{pmatrix}$ and $\overrightarrow{OQ} = \begin{pmatrix} 2 \\ 1 \\ 1 \end{pmatrix}$

  b $2\mathbf{i} - \mathbf{j} + 2\mathbf{k}$

3 **a** $\begin{pmatrix} 1 \\ 2 \\ 2 \end{pmatrix}$ **b** $\begin{pmatrix} 2 \\ -1 \\ -1 \end{pmatrix}$

  c $\begin{pmatrix} 2 \\ 1 \\ -2 \end{pmatrix}$ **d** $\begin{pmatrix} 5 \\ 2 \\ -1 \end{pmatrix}$

  e $\begin{pmatrix} 1 \\ -3 \\ -3 \end{pmatrix}$ **f** $\begin{pmatrix} 3 \\ -2 \\ -5 \end{pmatrix}$

4 **a** $\overrightarrow{OP} = \begin{pmatrix} -3 \\ 1 \end{pmatrix}$, $\overrightarrow{QC} = \begin{pmatrix} 5 \\ 7 \end{pmatrix}$ and

  $\overrightarrow{OR} = \begin{pmatrix} -1 \\ 5 \end{pmatrix}$

  b M(1, 4) and N(–2, 3)

Exercise 11D

2 **a** $x = 4, y = 1$

  b $x = \dfrac{3}{2}, y = \dfrac{7}{2}$

3 **a** $3\mathbf{u} + \mathbf{v}$ commutative and associative properties

  b $-3\mathbf{u} + \mathbf{v}$ distributive, commutative and associative properties

  c $-\dfrac{1}{2}\mathbf{u} + \dfrac{1}{2}\mathbf{v}$ distributive, commutative and associative properties

4 $\alpha = 1, \beta = -1; 2\mathbf{a} - 2\mathbf{b}$

Exercise 11E

1 **a** $\begin{pmatrix} -\dfrac{\sqrt{26}}{26} \\ \dfrac{5\sqrt{26}}{26} \end{pmatrix}$ **b** $\begin{pmatrix} \dfrac{5}{13} \\ \dfrac{12}{13} \end{pmatrix}$

  c $\begin{pmatrix} -1 \\ 0 \end{pmatrix}$ **d** $\begin{pmatrix} \dfrac{\sqrt{2}}{2} \\ -\dfrac{\sqrt{2}}{2} \end{pmatrix}$

2 **a** $\pm\begin{pmatrix} -\dfrac{2\sqrt{5}}{5} \\ \dfrac{\sqrt{5}}{5} \end{pmatrix}$ **b** $\pm\begin{pmatrix} -\dfrac{5\sqrt{29}}{29} \\ -\dfrac{2\sqrt{29}}{29} \end{pmatrix}$

  c $\pm\begin{pmatrix} 0 \\ 1 \end{pmatrix}$ **d** $\pm\begin{pmatrix} \dfrac{\sqrt{2}}{2} \\ \dfrac{\sqrt{2}}{2} \end{pmatrix}$

3 $\begin{pmatrix} \dfrac{2\sqrt{13}}{13} \\ -\dfrac{3\sqrt{13}}{13} \end{pmatrix}$

4 **a** $\pm\begin{pmatrix} \dfrac{4\sqrt{13}}{13} \\ \dfrac{6\sqrt{13}}{13} \end{pmatrix}$

  b $\pm\begin{pmatrix} -\dfrac{4}{3} \\ \dfrac{2\sqrt{5}}{3} \end{pmatrix}$

  c $\pm\begin{pmatrix} \dfrac{2\sqrt{13}}{3\sqrt{13}} \end{pmatrix}$

5 $\begin{pmatrix} -2 \\ -3 \end{pmatrix}$ or $-2\mathbf{i} - 3\mathbf{j}$

6 $\pm\begin{pmatrix} \dfrac{\sqrt{10}}{2} \\ -\dfrac{3\sqrt{10}}{2} \end{pmatrix}$ or $\pm\left(\dfrac{\sqrt{10}}{2}\mathbf{i} - \dfrac{3\sqrt{10}}{2}\mathbf{j} \right)$

Exercise 11F

1 **a** A(2, 5); B(5, 6); C(4, 2); D(1, 1)

  b $\overrightarrow{AB} = \begin{pmatrix} 3 \\ 1 \end{pmatrix}$, $\overrightarrow{AC} = \begin{pmatrix} 2 \\ -3 \end{pmatrix}$ and

  $\overrightarrow{AD} = \begin{pmatrix} -1 \\ -4 \end{pmatrix}$

  c $\overrightarrow{BD} = \begin{pmatrix} -4 \\ -5 \end{pmatrix}$

2 **a** $\overrightarrow{AB} = \begin{pmatrix} -4 \\ -2 \end{pmatrix}$ **b** $2\sqrt{5}$

  c M(0, 5); M is the midpoint of AB

  d $P\left(-\dfrac{2}{3}, \dfrac{14}{3} \right)$ and Q(–6, 2)

  $\overrightarrow{PQ}$ has greater magnitude

4 $a = 2 \pm \sqrt{6}$

Exercise 11G

1 **a** $\mathbf{u} + \mathbf{v} = \mathbf{u} = -3\mathbf{i} + 5\mathbf{j} - 2\mathbf{k}$

  b $-3\mathbf{u} = 6\mathbf{i} - 9\mathbf{j} - 3\mathbf{k}$

  c $4\mathbf{u} - 2\mathbf{v} = -6\mathbf{i} + 8\mathbf{j} + 10\mathbf{k}$

  d $-2(\mathbf{u} - \mathbf{v}) = 2\mathbf{i} - 2\mathbf{j} - 8\mathbf{k}$.

2 **a** $\begin{pmatrix} 1 \\ 6 \\ 0 \end{pmatrix}$ **b** $\begin{pmatrix} 5 \\ 5 \\ 3 \end{pmatrix}$

  c $\begin{pmatrix} 6 \\ -1 \\ -7 \end{pmatrix}$ **d** $\begin{pmatrix} \dfrac{5}{2} \\ -\dfrac{3}{2} \\ \dfrac{5}{2} \end{pmatrix}$

  e $\sqrt{14}$ **f** 3

  g $\sqrt{35}$ **h** $\sqrt{11}$

3 **a** $\overrightarrow{AB} = \begin{pmatrix} -1 \\ -3 \\ -3 \end{pmatrix}$ and $\overrightarrow{AC} = \begin{pmatrix} 1 \\ -5 \\ -1 \end{pmatrix}$

  b $\overrightarrow{AB} - \overrightarrow{AC} = \begin{pmatrix} -2 \\ 2 \\ -2 \end{pmatrix}$ and

  $\overrightarrow{BC} = 2\mathbf{i} - 2\mathbf{j} + 2\mathbf{k}$

4 **a** $\mathbf{u} = \begin{pmatrix} 0 \\ \dfrac{\sqrt{5}}{5} \\ -\dfrac{2\sqrt{5}}{5} \end{pmatrix}$

  b $\pm\begin{pmatrix} 0 \\ \dfrac{\sqrt{5}}{5} \\ -\dfrac{2\sqrt{5}}{5} \end{pmatrix}$ **c** $\begin{pmatrix} 0 \\ \sqrt{5} \\ -2\sqrt{5} \end{pmatrix}$

5 a $\overrightarrow{AC} = \begin{pmatrix} -4 \\ -1 \\ 2 \end{pmatrix}$, $\overrightarrow{AD} = \begin{pmatrix} 1 \\ 2 \\ 3 \end{pmatrix}$ and

$\overrightarrow{CG} = \begin{pmatrix} 1 \\ -2 \\ 1 \end{pmatrix}$

b $\overrightarrow{OB} = \begin{pmatrix} -1 \\ -4 \\ 2 \end{pmatrix}$, $\overrightarrow{OE} = \begin{pmatrix} 5 \\ -3 \\ 4 \end{pmatrix}$,

$\overrightarrow{OF} = \begin{pmatrix} 0 \\ -6 \\ 3 \end{pmatrix}$ and

$\overrightarrow{OH} = \begin{pmatrix} 6 \\ -1 \\ 7 \end{pmatrix}$

Exercise 11H

1 $\mathbf{r} = \begin{pmatrix} 1 \\ 3 \end{pmatrix} + \lambda \begin{pmatrix} 1 \\ 2 \end{pmatrix}$, $x = 1 + \lambda$,

$y = 3 + 2\lambda$ and $\dfrac{x-1}{1} = \dfrac{y-3}{2}$

(or equivalent)

2 $\mathbf{r} = \begin{pmatrix} 1 \\ -1 \\ 1 \end{pmatrix} + \lambda \begin{pmatrix} 2 \\ -1 \\ 3 \end{pmatrix}$, $x = 1 + 2\lambda$,

$y = -1 - \lambda$, $z = 1 + 3\lambda$ and

$\dfrac{x-1}{2} = \dfrac{y+1}{-1} = \dfrac{z-1}{3}$

(or equivalent)

3 $(-1, 0, 1)$ and any vector

collinear with $\begin{pmatrix} 6 \\ 3 \\ 2 \end{pmatrix}$

4 a $(1, 1, -1)$, $(0, 1, 2)$ and
$(-1, 1, 5)$ [others possible]

c $\mathbf{r} = \begin{pmatrix} 0 \\ 3 \\ 2 \end{pmatrix} + \lambda \begin{pmatrix} -1 \\ 0 \\ 3 \end{pmatrix}$.

5 a $(1, 0, 2)$ and $(2, -1, 2)$
[others possible]

b $\pm \begin{pmatrix} 2\sqrt{2} \\ -2\sqrt{2} \\ 0 \end{pmatrix}$

Exercise 11I

1 $3\sqrt{3}$

3 0

4 a $xy\cos\alpha = x^2$

b $y^2\sin^2\alpha$

c $-y^2\sin^2\alpha$

5 ± 6

8 a 13.5   **b** 11

9 a $6\cos\alpha$

b the centripetal force is
perpendicular to the
displacement

Exercise 11J

1 a 12   **b** 6

2 0 and 9

3 a C(1, 0, 0), G(1, 1, 0), D(0, 1, 0), O(0, 0, 0), A(0, 0, 1), B(1, 0, 1), F(1, 1, 1) and E(0, 1, 1)

b 2 and 1

4 a $B\left(\sqrt{2}, 0, 0\right)$,

$C\left(0, \sqrt{2}, 0\right)$,

$D\left(-\sqrt{2}, 0, 0\right)$,

$A\left(0, -\sqrt{2}, 0\right)$ and

$E(0, 0, 2)$

b $\sqrt{6}$ and 4

c 48.2° (3 sf)

Exercise 11K

1 120° (3 sf)

2 33.6° (3 sf)

3 a 90°   **b** 36.0°
90°, 36.0° and 54.0°

4 $a < -4$ or $a > 3$

5 a $\dfrac{\sin 4\alpha - 4}{5}$

b 0.870, 1.49, 2.44, 3.06, 4.01, 4.63, 5.58, 6.20

Exercise 11L

1 a $\begin{pmatrix} -3 \\ -2 \\ -1 \end{pmatrix}$ **b** $\begin{pmatrix} 2 \\ 2 \\ 2 \end{pmatrix}$

c $\begin{pmatrix} 0 \\ -4 \\ 4 \end{pmatrix}$ **d** $\begin{pmatrix} 2 \\ -4 \\ 2 \end{pmatrix}$

e $\begin{pmatrix} 1 \\ 0 \\ 1 \end{pmatrix}$ **f** $\begin{pmatrix} -2 \\ 2 \\ 2 \end{pmatrix}$

g $\begin{pmatrix} 0 \\ 0 \\ -2 \end{pmatrix}$

2 e.g. the vectors from
question 1

4 a D(−3, 5, 5)

b $\begin{pmatrix} 2 \\ -8 \\ 20 \end{pmatrix}$ **c** $6\sqrt{13}$

5 a D(−4, −4, 1)

b 19

Exercise 11M

1 a $\mathbf{r} = (-3 + 4\alpha)\mathbf{i} + (1 + \alpha - \beta)\mathbf{j} + (1 - \alpha - 2\beta)\mathbf{k}$

b $x = -3 + 4\alpha$, $y = 1 + \alpha - \beta$, $z = 1 - \alpha - 2\beta$

c $-3x + 8y - 4z = 13$

Other forms are possible in a
and b

2 $x - 3y + z = 2$; e.g. (0, 0, 2), (0, 1, 5), (2, 0, 0)

3 a $x - 2y - z = -11$

b $2x - 3y + 4z = -1$

c $y - 2z = 1$

4 a $y = 0$ **b** $z = 0$

c $x = 0$

5 a A(2, 0, 0), B(2, 2, 0), C(0, 2, 0), O(0, 0, 0) and V(0, 0, 4.5)

b $\mathbf{r} = (2 - 2\beta)\mathbf{i} + 2\alpha\mathbf{j} + 4.5\beta\mathbf{k}$

c $9y + 4z = 18$

d $\mathbf{r} = (2 + 4\alpha)\mathbf{i} + (2 + 4\alpha)\mathbf{j} - 9\alpha\mathbf{k}$

e $4z = 36 - 9x$, $y = 0$

Exercise 11N

1 75°

2 6.93° (3 sf)

3 73.2° (3 sf)

4 a student verification

b $\tan(\alpha - \beta) = \dfrac{m_1 - m_2}{1 + m_1 m_2}$

c $\cos(\alpha - \beta) = \dfrac{|1 + m_1 m_2|}{\sqrt{1 + m_1^2}\sqrt{1 + m_2^2}}$

d $\cos(\alpha - \beta) = \cos\theta$
since $\theta = \alpha - \beta$

Exercise 11O

1 $\arcsin\dfrac{1}{3} = 19.5°$ (3 sf)

2 $\arcsin\dfrac{19}{\sqrt{38}\sqrt{90}} = 19.0°$ (3 sf)

3 $\arccos\dfrac{19}{\sqrt{11}\sqrt{61}} = 42.8°$ (3 sf)

4 a $x - 5y - 7z = -6$

b $\dfrac{x-1}{2} = y = \dfrac{z-1}{2}$

c $\arcsin\dfrac{17}{15\sqrt{3}} = 40.9°$ (3 sf)

5 $k = 0$ or $k = 1$

Exercise 11P

1 $(2, -2, 2)$

2 $(16, 6, 9)$

3 a $x = 4 - 2\alpha + \beta$, $y = 1 - \beta$
and $z = 2 - \alpha - 2\beta$

b $\left(\dfrac{9}{5}, \dfrac{4}{5}, \dfrac{2}{5}\right)$

4 $\mathbf{r} = \lambda\mathbf{i} + 2\lambda\mathbf{j} + (\lambda - 2)\mathbf{k}$
(or equivalent)

5 $\dfrac{2}{5}\sqrt{30}$ or 2.19 (3 sf)

Exercise 11Q

1 a $(2, 2, 1)$

b do not intersect

2 $\dfrac{x+5}{5} = \dfrac{y+2}{1} = \dfrac{z-1}{-2}$ $\left(\begin{array}{c}\text{or}\\\text{equivalent}\end{array}\right)$

3 a $k = -1$ **b** $\left(\dfrac{9}{11}, -\dfrac{25}{11}, -\dfrac{19}{11}\right)$

4 b $5x - y + 3z = 7$

6 a $\left(\dfrac{24}{7a+28}, \dfrac{45a+84}{7a+28}, \dfrac{-14a-8}{7a+28}\right)$

b $a = -4$ **c** $\dfrac{11\sqrt{222}}{111}$

Exercise 11R

1 $\left(-\dfrac{1}{2}, \dfrac{3}{2}, 2\right)$

2 $\mathbf{r} = \begin{pmatrix}\frac{3}{4}\\-3\\0\end{pmatrix} + \lambda\begin{pmatrix}0\\1\\1\end{pmatrix}$

4 b $a = 1$ or $c = b$
but a, b, c not all $= 1$

5 π_2 is parallel to π_3. So the
planes do not intersect.

6 a $(1, 2, -1)$ **b** $k = \dfrac{5}{3}$

Exercise 11S

1 a A(3, −4) and B(4, 3)

b $\mathbf{v}_A = \begin{pmatrix}-1\\2\end{pmatrix}$ and $\mathbf{v}_A = \begin{pmatrix}-3\\-2\end{pmatrix}$

c 97.1°

d 1.5 hours

2 a $(5, 20, -10)$

c i $t = 2$

ii $(25, -20, 50)$

iii $20\sqrt{14}$

d i 0.496 (3 sf)

ii P(9.96, 10.1, 4.88)
and Q(0.492,
0.008 02, 1.25)

e ii the particle path is not
a straight line

3 a $\overrightarrow{OP} = \begin{pmatrix}\frac{1}{2}\\1\\0\end{pmatrix}$, $\overrightarrow{OQ} = \begin{pmatrix}0\\1\\\frac{1}{2}\end{pmatrix}$

and $\overrightarrow{OR} = \begin{pmatrix}0\\1\\2\\1\end{pmatrix}$

b $x + y + z = \dfrac{3}{2}$

c $\dfrac{3\sqrt{3}}{4}$

e $\left(\dfrac{1}{2}, \dfrac{1}{2}, \dfrac{1}{2}\right)$

f $\dfrac{\sqrt{3}}{2}$

Review exercise (non-GDC)

1 a $\begin{pmatrix}11\\2\end{pmatrix}$

b 5, $\sqrt{26}$ and $\sqrt{85}$

2 $-\dfrac{1}{4}\mathbf{a} + \dfrac{7}{4}\mathbf{b} + \dfrac{3}{4}\mathbf{c}$

3 a $\begin{pmatrix}-\frac{\sqrt{2}}{2}\\\frac{2\sqrt{2}}{5}\\\frac{3\sqrt{2}}{10}\end{pmatrix}$ **b** $\pm\begin{pmatrix}-\frac{5\sqrt{2}}{2}\\2\sqrt{2}\\\frac{3\sqrt{2}}{2}\end{pmatrix}$

6 a $2 - 2\cos\theta$ and $2 + 2\cos\theta$

b $\dfrac{4}{5}$

7 a $\mathbf{r} = \begin{pmatrix}2\\3\\4\end{pmatrix} + \alpha\begin{pmatrix}2\\-3\\2\end{pmatrix} + \beta\begin{pmatrix}0\\1\\-2\end{pmatrix}$;

$x = 2 + 2\alpha$, $y = 3 - 3\alpha + \beta$
and $z = 4 + 2\alpha - 2\beta$

b $\mathbf{r} = \begin{pmatrix}1\\0\\-2\end{pmatrix} + \alpha\begin{pmatrix}-2\\0\\1\end{pmatrix} + \beta\begin{pmatrix}0\\-1\\0\end{pmatrix}$;

$x = 1 - 2\alpha$, $y = -\beta$, and
$z = -2 + \alpha$

8 $(2, 3)$

Review exercise (GDC)

1 48°

2 a $\mathbf{r} = \begin{pmatrix}1\\1\\0\end{pmatrix} + \alpha\begin{pmatrix}-2\\0\\2\end{pmatrix} + \beta\begin{pmatrix}1\\0\\-3\end{pmatrix}$;

or equivalent
$x = 1 - 2\alpha + \beta$, $y = 1$, and
$z = 2\alpha - 3\beta$
$y = 1$

b $\mathbf{r} = \begin{pmatrix}-1\\1\\1\end{pmatrix} + \alpha\begin{pmatrix}-1\\-2\\1\end{pmatrix} + \beta\begin{pmatrix}2\\2\\-3\end{pmatrix}$;

or equivalent
$x = -1 - \alpha + 2\beta$, $y = 1 - 2\alpha$
$+ 2\beta$ and $z = 1 + \alpha - 3\beta$;
$4x - y + 2z = -3$

3 a $\mathbf{r}\cdot\begin{pmatrix}1\\1\\-2\end{pmatrix} = \begin{pmatrix}2\\-3\\4\end{pmatrix}\cdot\begin{pmatrix}1\\1\\-2\end{pmatrix} = -9$

b $\mathbf{r} \cdot \begin{pmatrix} 1 \\ -1 \\ 0 \end{pmatrix} = \begin{pmatrix} 1 \\ 0 \\ 2 \end{pmatrix} \cdot \begin{pmatrix} 1 \\ -1 \\ 0 \end{pmatrix} = 1$

4 a $2x - 3y + 4z = 29$

   b $-2x - y + 4z = -12$

   c $-2x + y = -4$

   d $5y + 2z = -11$

   e $x = 3$

   f $x - y - z = 0$

   g $x - 2y - z = 0$

5 80.4° (3 sf)

6 $\mathbf{r} = \begin{pmatrix} 3 \\ 5 \\ 7 \end{pmatrix} + \alpha \begin{pmatrix} 1 \\ 2 \\ 3 \end{pmatrix} + \beta \begin{pmatrix} 2 \\ 1 \\ 2 \end{pmatrix}$

7 The planes intersect at a point with coordinates $(3, -2, 2)$.

8 a $x = \dfrac{5 - 2a - 3b}{2b}$,

     $y = \dfrac{4a + 11b - 10}{2b}$

     $z = \dfrac{2a - 5}{2b}$

   b When $a = \dfrac{5}{2}$ and $b = 0$, they meet in a line.

Chapter 12

Skills check

1

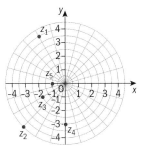

2 a $z^* = 5 + 4i$ and
    $-z = -5 + 4i$

   b $\dfrac{1}{z} = \dfrac{5}{41} + \dfrac{4}{41}i$

3 $\mathrm{Re}\,z = -3$, $\mathrm{Im}\,z = 4$ and
   $|z| = 5$

---

4 a 7   **b** $12 - 19i$

   c $\dfrac{1}{10} - \dfrac{7}{10}i$

Exercise 12A

2 $r = 0$ or $r = 2$;
   $\theta = \dfrac{\pi}{6} + \dfrac{2k\pi}{3}, k \in \mathbb{Z}$.

     0

     $-4 + 4i\sqrt{3}$

     $-4 - 4i\sqrt{3}$

     8

3 $a = \pm\sqrt{2}$ and $\theta = \dfrac{\pi}{4} + k\pi$,
   $k \in \mathbb{Z}$. Two.

4 a and **b** e.g. $z_1 = 1 + i$ and
   $z_2 = -1 + i$

Exercise 12B

1 a $z_1 = 3\sqrt{2} + 3\sqrt{2}i$

   b $z_2 = -5\sqrt{2} + 5\sqrt{2}i$

   c $z_3 = 2 - 2\sqrt{3}i$

   d $z_4 = -\dfrac{5\sqrt{3}}{2} - \dfrac{5}{2}i$

2 a $z_1 = \sqrt{2}\mathrm{cis}\left(-\dfrac{3\pi}{4}\right)$

   b $z_2 = 4\mathrm{cis}\left(\dfrac{\pi}{6}\right)$

   c $z_3 = 4\sqrt{2}\mathrm{cis}\left(-\dfrac{\pi}{4}\right)$

   d $z_4 = 5\sqrt{2}\mathrm{cis}\left(\dfrac{3\pi}{4}\right)$

3 a $z_1 = 3\mathrm{cis}\left(-\dfrac{\pi}{2}\right)$

   b $z_2 = 4\mathrm{cis}\left(\dfrac{\pi}{2}\right)$

   c $z_3 = 2\,\mathrm{cis}\,0$

---

d $z_4 = 4\,\mathrm{cis}(\pi)$

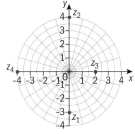

4 a $z^* = 4\,\mathrm{cis}(-40°)$

   b $-z = 4\,\mathrm{cis}(220°)$

   c $-z^* = 4\,\mathrm{cis}(140°)$

   d $3z^* = 12\,\mathrm{cis}(-40°)$

   e $-z^* = 16\,\mathrm{cis}(140°)$

5 a $z_3 = -24i$

   b $z_1 = 4\mathrm{cis}\left(\dfrac{4\pi}{3}\right)$,

     $z_2 = 6\mathrm{cis}\left(\dfrac{\pi}{6}\right)$ and

     $z_3 = 24\mathrm{cis}\left(-\dfrac{\pi}{2}\right)$.

   c $|z_3| = |z_1||z_2|$ and
     $\arg z_3 = \arg z_1 + \arg z_2$.

Exercise 12C

1 a $5\,\mathrm{cis}(135°)$

   b $\dfrac{2}{7}\,\mathrm{cis}\left(\dfrac{31\pi}{24}\right)$

2 a $-\dfrac{\sqrt{3}}{2} + \dfrac{1}{2}i$ **b** $\sqrt{2}\,\mathrm{cis}\left(-\dfrac{\pi}{4}\right)$

   c $z_1z_2 = \dfrac{1-\sqrt{3}}{2} + \dfrac{1+\sqrt{3}}{2}i$

     $z_1z_2 = \sqrt{2}\,\mathrm{cis}\dfrac{7\pi}{12}$;

     $\sin\dfrac{7\pi}{12} = \dfrac{\sqrt{2}+\sqrt{6}}{4}$,

     $\cos\dfrac{7\pi}{12} = \dfrac{\sqrt{2}-\sqrt{6}}{4}$ and

     $\tan\dfrac{7\pi}{12} = -2 - \sqrt{3}$

3 a $r > \dfrac{5}{2}$ and $\theta = \dfrac{11\pi}{6}$

   b $0 < r < \dfrac{1}{2}$ and

     $\theta = \dfrac{\pi}{3}$ or $\theta = \dfrac{4\pi}{3}$

4 $-z = 2\sqrt{3}\,\text{cis}\left(-\dfrac{\pi}{6}\right),$

$z* = 2\sqrt{3}\,\text{cis}\left(-\dfrac{5\pi}{6}\right)$ and

$-z* = 2\sqrt{3}\,\text{cis}\,\dfrac{\pi}{6}$

5 a $z = \text{cis}\left(\dfrac{\pi}{2} - \alpha\right)$

and $w = \text{cis}\left(-\dfrac{\pi}{2} + 2\alpha\right)$

b $zw = \text{cis}\,\alpha$

Exercise 12D

1 a $2\,\text{cis}\,45°$

b $\dfrac{7}{8}\text{cis}\,\dfrac{25\pi}{24}$

2 a $2\sqrt{2}\text{cis}\left(-\dfrac{\pi}{4}\right)$

b $2\sqrt{2}\text{cis}\left(\dfrac{\pi}{4}\right)$

c $4\sqrt{2}\text{cis}\left(\dfrac{19\pi}{12}\right)$

d $\dfrac{\sqrt{2}}{2}\text{cis}\left(\dfrac{\pi}{12}\right)$

e $\dfrac{\sqrt{2}}{8}\text{cis}\left(-\dfrac{7\pi}{12}\right)$

3 a $2\text{cis}\left(-\dfrac{\pi}{6}\right)$

b $\text{cis}\left(-\dfrac{5\pi}{12}\right)$

c $\dfrac{\sqrt{7}}{14}\text{cis}\left(-\dfrac{\pi}{6}\right)$

4 a
$z = 4\left(\cos\left(-\dfrac{\pi}{6}\right) + i\sin\left(-\dfrac{\pi}{6}\right)\right)$ and

$w = \dfrac{\sqrt{2}}{2}\left(\cos\left(-\dfrac{\pi}{4}\right) + i\sin\left(-\dfrac{\pi}{4}\right)\right)$

b $4\sqrt{2}\text{cis}\left(\dfrac{\pi}{12}\right)$

c $\left(2\sqrt{3} + 2\right) + i\left(2\sqrt{3} - 2\right)$

d $\cos\dfrac{\pi}{12} = \dfrac{\sqrt{6} + \sqrt{2}}{4},$

$\sin\dfrac{\pi}{12} = \dfrac{\sqrt{6} - \sqrt{2}}{4}$ and

$\tan\dfrac{\pi}{12} = 2 - \sqrt{3}$

Exercise 12E

1 a $32\,\text{cis}\,\pi = -32$

b $\text{cis}\,\dfrac{\pi}{3}$

c $\text{cis}\,\dfrac{\pi}{4}$

2 a $2e^{\frac{\pi}{12}i}$ **b** $256e^{\frac{7\pi}{12}i}$

c $8e^{-\frac{7\pi}{12}i}$ **d** $\dfrac{1}{512}e^{\frac{\pi}{4}i}$

3 $-i$

5 a 4 **b** 2

6 $2\text{cis}\,0 = 2$ or

$2\text{cis}\left(\pm\dfrac{2\pi}{3}\right) = -1 \pm \sqrt{3}i$

Exercise 12F

1 a $2\text{cis}\dfrac{\pi}{9},\ 2\text{cis}\dfrac{7\pi}{9}$

and $2\text{cis}\dfrac{13\pi}{9}$

b $\sqrt{2} + i\sqrt{2}$
$-\sqrt{2} + i\sqrt{2}$
$-\sqrt{2} - i\sqrt{2}$
$\sqrt{2} - i\sqrt{2}$

c $2e^{-\frac{\pi}{5}i}, 2e^{\frac{\pi}{5}i}, 2e^{\frac{3\pi}{5}i},\ 2e^{\pi i}$

and $2e^{\frac{7\pi}{5}i}$

2 a $\sqrt[4]{2}\text{cis}\left(-\dfrac{\pi}{8}\right)$ or $\sqrt[4]{2}\text{cis}\left(\dfrac{7\pi}{8}\right)$

b $\sqrt[4]{2}\text{cis}\left(\dfrac{5\pi}{24}\right), \sqrt[4]{2}\text{cis}\left(\dfrac{17\pi}{24}\right),$

$\sqrt[4]{2}\text{cis}\left(\dfrac{29\pi}{24}\right)$ or $\sqrt[4]{2}\text{cis}\left(\dfrac{41\pi}{24}\right)$

c $z = 3e^{\frac{\pi}{12}i},\ z = 3e^{\frac{3\pi}{4}i}$ and

$z = 3e^{\frac{17\pi}{12}i}$

d $0, \sqrt{2}\text{cis}\left(-\dfrac{\pi}{8}\right)$ or $\sqrt{2}\text{cis}\left(\dfrac{7\pi}{8}\right)$

3 a $\text{cis}\,0,\ \text{cis}\dfrac{\pi}{3},\ \text{cis}\dfrac{2\pi}{3},\ \text{cis}\,\pi,$

$\text{cis}\dfrac{4\pi}{3}$ and $\text{cis}\dfrac{5\pi}{3}.$

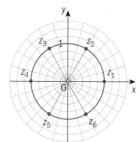

b z_1, z_3, z_5 are cube roots of 1 and z_2, z_4, z_6 are the cube roots of -1

4 $\dfrac{1}{2}\text{cis}\dfrac{4\pi}{15}, \dfrac{1}{2}\text{cis}\dfrac{16\pi}{15}, \dfrac{1}{2}\text{cis}\dfrac{22\pi}{15}$

and $\dfrac{1}{2}\text{cis}\dfrac{28\pi}{15}.$

5 $3\text{cis}\left(\pm\dfrac{\pi}{4}\right) = \dfrac{3\sqrt{2}}{2} \pm \dfrac{3\sqrt{2}}{2}i$

and $3\text{cis}\left(\pm\dfrac{3\pi}{4}\right) = -\dfrac{3\sqrt{2}}{2} \pm \dfrac{3\sqrt{2}}{2}i.$

Hence the solutions of the equation are

$\left(3 + \dfrac{3\sqrt{2}}{2}\right) \pm \dfrac{3\sqrt{2}}{2}i$

and $\left(3 - \dfrac{3\sqrt{2}}{2}\right) \pm \dfrac{3\sqrt{2}}{2}i.$

6 $\dfrac{3}{3 - e^{i\theta}}$

Investigation - properties of roots of unity

1 a 1 and $-\dfrac{1}{2} \pm \dfrac{\sqrt{3}}{2}i$

b i $\omega*$

   ii $\omega^2 = \omega*$ and $(\omega*)^2 = \omega$; twice the argument of one root gives the argument of the other, as the cube roots of unity divide the unit circle into three equal parts.

   iii $\dfrac{1}{\omega} = \omega*$ and $\dfrac{1}{\omega*} = \omega.$ The roots are symmetric with respect to the x-axis.

c The three roots define an equilateral triangle with centre at the origin which means that the origin divides the height with respect to the vertex $(1, 0)$ in the ratio 1:2. Therefore the real parts of the complex roots add up to $-1.$

d i 0 **ii** -1
   iii -1 **iv** $-\omega$

e Both 1 and −1, and their corresponding cube roots are symmetric with respect to the x-axis.

2 The nth roots of the unity appear in pairs of conjugate complex numbers which are represented by points symmetric with respect to the x-axis; when n is even these points can also form pairs of points symmetric with respect to the y-axis.

The list of properties is as follows:

The complex cube roots of unity are conjugate of each other, the reciprocal of each other and if we square one of them we obtain the other one.

Geometrically, the three cube roots of unity are represented by points that divide the unit circle in three equal arcs with length $\dfrac{2\pi}{3}$; the real root corresponds to the point $(1,0)$ and the complex roots are symmetric with respect to the x-axis and correspond to the points $\left(-\dfrac{1}{2} \pm \dfrac{2\pi}{3}\right)$. This symmetry explains the

properties above.

1, ω and ω^2 are consecutive terms of a geometric sequence with common ratio ω whose sum is 0; similarly, 1, $\omega*$ and $(\omega*)^2$ are consecutive terms of a geometric sequence with common ratio $\omega*$ whose sum is 0.

If we reflect the points that represent the cube roots of unity in the y-axis we obtain the points that represent the cube roots of −1.

The properties above make the cube roots of unity a very important case. However, some of these properties can be extended to other roots of unity. In general

the sum of nth roots of unity is 0 and the points that represent these roots divide the unit circle into n equal arcs. Also, if ω is a nth root of unity, $\omega*$ is also a nth root of unity.

Exercise 12G

1 b $z^5 - 5z^3 + 10z - \dfrac{10}{z} + \dfrac{5}{z^3} - \dfrac{1}{z^5}$;

$-\dfrac{1}{80}\cos 5\alpha + \dfrac{5}{48}\cos 3\alpha$

$-\dfrac{5}{8}\cos\alpha + C$

$1.0, \dfrac{2\pi}{7}, \dfrac{4\pi}{7}, \dfrac{6\pi}{7},$

$\dfrac{8\pi}{7}, \dfrac{10\pi}{7}$ and $\dfrac{12\pi}{7}$.

4 b $\cos^2\theta \operatorname{cis} 2\theta$

c $\theta = \pm\dfrac{\pi}{3}$ or $\theta = \pm\dfrac{\pi}{2}$

5 c $\sqrt{2}$

6 b $n \in \mathbb{Z}$ because De Moivre's theorem has been proved for these values.

7 b $a = -2, b = 2$ and $c = -1$

c $2\sqrt{3}i, -2\sqrt{3}i$

8 Proofs

9 $n = 12k, k \in \mathbb{Z}$

10 a $f(i) = \dfrac{\pi}{2}i, \ f(-i) = -\dfrac{\pi}{2}i,$

$f(1+i) = \ln\sqrt{2} + \dfrac{\pi}{4}i$ and

$f(1-i) = \ln\sqrt{2} - \dfrac{\pi}{4}i.$

d i All complex numbers with $|z| = 1; z \neq 1$

ii $0 < |z| < 1$ and $\arg z = 0$ (positive real numbers between 0 and 1)

iii 1

Review exercise (non-GDC)

1 a $-1 - \sqrt{3}i$

b $-1 + i$

c $-\dfrac{\sqrt{3}}{3} - \dfrac{1}{3}i$

2 a $5\sqrt{2} \operatorname{cis}\left(-\dfrac{\pi}{4}\right)$

b $\operatorname{cis}\left(-\dfrac{\pi}{3}\right)$

c $\dfrac{\sqrt{2}}{2} \operatorname{cis}\left(-\dfrac{\pi}{12}\right)$

3 a $zw = 8 \operatorname{cis}\left(\dfrac{23\pi}{12}\right),$

$\dfrac{z}{w} = \dfrac{1}{2} \operatorname{cis}\left(-\dfrac{7\pi}{12}\right)$ and

$z^2 w^3 = 256 \operatorname{cis}\left(\dfrac{13\pi}{12}\right)$

b $z + w = \left(-1 - 2\sqrt{2}\right)$

$+ i\left(\sqrt{3} - 2\sqrt{2}\right)$

$z - w = \left(2\sqrt{2} - 1\right) + i\left(\sqrt{3} + 2\sqrt{2}\right)$

$\dfrac{z^2}{w} = \dfrac{\sqrt{6} + \sqrt{2}}{4} + \dfrac{\sqrt{6} - \sqrt{2}}{4}i$

4 a $\dfrac{\sqrt{3}}{3}$

b 0

c 0.5

5 $0, \pm 1$ and $\pm i$

6 $\cos\theta \operatorname{cis}(-\theta)$

7 a $a = \pm\dfrac{1}{4}, n = 4$

b $\left[\dfrac{-1+1}{4}; \dfrac{-1-i}{4}\right], \left[\dfrac{-1+i}{4}; \dfrac{1+i}{4}\right]$

9 a 2 and 210°

b $\sqrt[3]{2} \operatorname{cis}(70°)$

c 12

12 a −1

b 2

c −1

13 b $\pm 1, \pm i, \ \pm\dfrac{\sqrt{2}}{2} \pm \dfrac{\sqrt{2}}{2}i$

c $(x-1)(x+1)(x^2+1)$

$\left(x^2 - \sqrt{2}x + 1\right)$

$\left(x^2 + \sqrt{2}x + 1\right)$

Review exercise – with GDC

1 $a = -8, b = 24, c = -32, d = 20$

Chapter 15

Mark scheme

When marking scripts, examiners use mark schemes similar to these. There are different types of marks that can be awarded under specific conditions. It is important that you are familiar with these rules as this may help you to maximise your score in your final IB examination.

Types of marks

M   These are method marks awarded for attempting to use correct method. They can just be awarded if the working is seen.

(M) These are implied method marks that can be awarded even if the working is not seen but it is obvious from the subsequent work that the correct method has been used (or attempted).

A   These are answer or accuracy marks. In case they depend on preceding M marks, they can just be awarded if the answer is obtained using a correct method (correct answers from incorrect methods receive no credit).

(A) These are implied answer or accuracy marks. They also may depend on preceding M marks and they are awarded even if you skip an intermediate step but subsequent correct work shows that you have considered this/these values. In general it is better to write down all the intermediate steps as you do not know beforehand which steps may not be considered necessary.

R   These are reasoning marks. It is important that you state your reasons clearly, namely when you write proofs or answer show that questions. It is not enough to quote answers that are given or rephrase those using equivalent expressions.

N   These are marks for correct answers with no working. They appear mostly in paper 2 in questions where the answer can be obtained with GDC.

AG This is an indication for examiners that the answer is given in the question and no credit should be awarded for copying it. Therefore, do not waste time copying answers that are given if you do not know how to obtain them.

Examiners also follow some rules that you should be familiar with:

Follow through marks (FT)

If you give an incorrect answer to a part of a question but it is possible to answer subsequent parts using this answer, you get credit for correct work in these subsequent parts using incorrect values. In general, if the difficulty of subsequent parts remains the same you are not penalized twice for the same mistake. However, if the question becomes much easier due to use of a different value, examiners may decide to give no credit or just partial credit to your work. In case you notice that a question where you need to use a previous answer is either too easy or extremely difficult for the marks available, it is wise to double check previous work for calculation errors. This may save you from wasting time trying to answer something impossible!

Alternative forms

Alternative forms of notation are accepted as long they are correct and used consistently. Simplification of expressions is not always required. The mark schemes in general show the simplified form in brackets as an indication that it is not necessary to simplify the answer to obtain credit. However, you may need to simplify answers even if you are not required to in a situation where it is needed to answer subsequent parts.

Crossed out work

This is not marked at all. Do not cross any answer unless you have written a new answer to the same question and you do not want the first one to be marked. Be aware though that if you give two answers to the same question, the one the examiner marks first will be the only one considered.

Accuracy of answers

Since 2011, the accuracy penalty per paper no longer applies. Read the instruction on the cover page of the exam carefully: you may be told to use a minimum number of significant figures in questions where the rounding criterion is not given. In any case, you should also look carefully and see if your answer makes sense in the context of the question.

Practice paper 1

SECTION A

1 $4\ln 2 - 3\ln 4 + \ln k = 0 \Rightarrow \ln k = -4\ln 2 + 3\ln 4$ M1

   $\ln k = \ln 2^{-4} + \ln 4^{3}$ A1

   $\ln k = \ln 2^{-4} \times 4^{3}$ (A1)

   $k = 2^{-4} \times 4^{3} = 4$ A1   [4 marks]

2 a $2x+1 \overline{\smash{\big)}\, 2x^{3} - 3x^{2} + 8x + 5}$ with quotient $x^{2} - 2x + 5$ M1A2

$$
\begin{array}{r}
x^{2} - 2x + 5 \\
2x+1 \overline{\smash{\big)}\, 2x^{3} - 3x^{2} + 8x + 5} \\
\underline{-2x^{3} - x^{2}} \\
-4x^{2} + 8x + 5 \\
\underline{4x^{2} + 2x} \\
10x + 5 \\
\underline{-10x - 5} \\
0
\end{array}
$$

as the remainder is 0, $p(x)$ is divisible by $2x + 1$.

Alternative method:

$$
\begin{array}{c|cccc}
 & 2 & -3 & 8 & 5 \\
\left(-\dfrac{1}{2}\right) & & + & + & + \\
 & & -1 & 2 & -5 \\
\hline
 & 2 & -4 & 10 & 0
\end{array}
$$

> Use synthetic division for the value of $-\dfrac{1}{2}$.

R1

b $p(x) = (2x+1)(x^{2} - 2x + 5)$ (A1)

   $p(x) = 0 \Rightarrow 2x + 1 = 0$ or $x^{2} - 2x + 5 = 0$ (M1)

   $x = -\dfrac{1}{2}$ or $x = 1 \pm 2i$ A1A1   [8 marks]

3   $a_{1} + a_{2} = \dfrac{8}{9} \Rightarrow a + ar = \dfrac{8}{9} \Rightarrow a(1+r) = \dfrac{8}{9}$ M1

   $\Rightarrow a = \dfrac{8}{9(r+1)}$ A1

   $\underbrace{a_{1} + a_{2}}_{\frac{8}{9}} + a_{3} = \dfrac{26}{27} \Rightarrow \dfrac{8}{9} + ar^{2} = \dfrac{26}{27}$ M1

   $\Rightarrow \dfrac{8}{9} + \dfrac{8}{9(r+1)}r^{2} = \dfrac{26}{27} \Rightarrow 12r^{2} = r + 1$ M1

   $\Rightarrow r = \dfrac{1}{3}$ or $r = -\dfrac{1}{4}$ A1

   $a = \dfrac{2}{3}$ or $a = \dfrac{32}{27}$ A1

   $S = \dfrac{\frac{2}{3}}{1 - \frac{1}{3}} = 1$ or $S = \dfrac{\frac{32}{27}}{1 + \frac{1}{4}} = \dfrac{128}{135}$ A1A1   [8 marks]

4 $\text{P}(A\cup B)=\text{P}(A)+\text{P}(B)-\text{P}(A\cap B)$ and

$3\text{P}(A\cap B)=0.3+0.2-\text{P}(A\cap B)$ M1

$\text{P}(A\cap B)=\dfrac{1}{8}$ (0.125) $\left(\text{accept}\dfrac{5}{40}\right)$ A1

$\text{P}(A|B)=\dfrac{\text{P}(A\cap B)}{\text{P}(B)}=\dfrac{\frac{1}{8}}{0.2}=\dfrac{5}{8}$ M1A1

$\text{P}(A\cap B')=\text{P}(A)-\text{P}(A\cap B)=0.3-0.125=0.175$ M1A1   [6 marks]

5 Let $z=a+ib$ where $a=\text{Re}(z)$ and $b=\text{Im}(z)$,

a $z+z*=(a+ib)+(a-ib)=2a$ M1A1

     $=2\text{Re}(z)$ AG

b $z-z*=(a+ib)-(a-ib)=2b\text{i}$ M1A1

     $=2\text{i}\,\text{Im}(z)$ AG

c $\text{Re}(z)=a\le|a|=\sqrt{a^2}$ R1

     $\le\sqrt{a^2+b^2}=|z|$ A1

     Therefore $\text{Re}(z)\le|z|$ AG   [6 marks]

6 $\begin{cases}x+2y-z=5\\-3x-y+z=1\end{cases}$ eliminate z first,   M1A1

$\Rightarrow-2x+y=6\Rightarrow y=2x+6$, express all in terms of x.   A1

$z=1+3x+y\Rightarrow z=5x-5$ A1

$r=\begin{pmatrix}0\\-6\\-5\end{pmatrix}+t\begin{pmatrix}1\\2\\5\end{pmatrix},t\in\mathbb{Z}$ M1A1

7 $x^2+4y^2-2x+16y+13=0\Rightarrow 2x+8y\dfrac{\text{d}y}{\text{d}x}-2+16\dfrac{\text{d}y}{\text{d}x}=0$ M1A1

$\Rightarrow\dfrac{\text{d}y}{\text{d}x}=\dfrac{2-2x}{8y+16}\left(=\dfrac{1-x}{4y+8}\right)$ A1

$\dfrac{\text{d}y}{\text{d}x}=0\Rightarrow x=1$

$1^2+4y^2-2+16y+13=0\Rightarrow y^2+4y+3=0$ M1

$\Rightarrow y=-3$ or $y=-1$ M1

So the points are $(1,-3)$ and $(1,-1)$ A1   [6 marks]

8 $\displaystyle\int_1^9\sqrt{x}\ln x\,\text{d}x=\left[\dfrac{2}{3}x\sqrt{x}\ln x\right]_1^9-\dfrac{2}{3}\int_1^9\sqrt{x}\text{d}x$ M1A1

$=\left[\dfrac{2}{3}x\sqrt{x}\ln x\right]_1^9-\dfrac{2}{3}\left[\dfrac{2}{3}x\sqrt{x}\right]_1^9=36\ln3-\dfrac{104}{9}$ A1

$\displaystyle\int_1^9\sqrt{x}\ln x\,\text{d}x=a\ln3+b\Rightarrow a=36$ and $b=-11\dfrac{5}{9}$ A1A1   [5 marks]

9 **a** f is a continuous PDF of X $\Rightarrow \lim\limits_{x\to2^-} f(x)=\lim\limits_{x\to2^+} f(x)$ M1

$$\Rightarrow 0.6=2a+b$$ A1

$\lim\limits_{x\to c^-} f(x)=\lim\limits_{x\to c^+} f(x) \Rightarrow ac+b=0$ A1

The area under the graph is 1   M1

$$0.05+\frac{0.1+0.6}{2}\times1+\frac{(2a+b)+(ac+b)}{2}\times(c-2)=1$$

$$\Rightarrow c=4$$ A1

Solve simultaneously to obtain

$a=-0.3$ and $b=1.2$ A1

b 2   A1   [7 marks]

10 $f(x)=12\sin x-5\cos x \Rightarrow f(x)=13\left(\underset{\cos\alpha}{\frac{12}{13}}\sin x-\underset{\sin\alpha}{\frac{5}{13}}\cos x\right)$ M1A1

$\Rightarrow f(x)=13\sin(x-\alpha)$ where $\alpha=\arctan\dfrac{5}{12}$ (A1)

The range of f is $[-13, 13]$ A1   [4 marks]

SECTION B

11 **a** $x-2=\dfrac{y-1}{2}\Rightarrow 2x-y=3$ and $\dfrac{x-1}{4}=y-2\Rightarrow x-4y=-7$ M1

$\Rightarrow x=\dfrac{19}{7}$ and $y=\dfrac{17}{7}$ A1A1

$z=\dfrac{15}{7}$ A1

Therefore, as the system of equations has unique solution   R1

The lines meet at $\left(\dfrac{19}{7},\dfrac{17}{7},\dfrac{15}{7}\right)$

b $\begin{pmatrix}1\\2\\3\end{pmatrix}\times\begin{pmatrix}4\\1\\-2\end{pmatrix}=\begin{pmatrix}-7\\14\\-7\end{pmatrix}$ or $7\begin{pmatrix}-1\\2\\-1\end{pmatrix}$ M1A1

$\begin{pmatrix}-1\\2\\-1\end{pmatrix}\cdot\begin{pmatrix}x\\y\\z\end{pmatrix}=\begin{pmatrix}-1\\2\\-1\end{pmatrix}\cdot\begin{pmatrix}\frac{19}{7}\\\frac{17}{7}\\\frac{15}{7}\end{pmatrix}\Rightarrow -x+2y-z=0$ (or $x-2y+z=0$)   M1A1

c As $-1\times1+2\times(-1)-1\times0\neq0$ M1R1

A(1, −1, 0) does not lie on the plane π AG

d $\dfrac{x-1}{-1}=\dfrac{y+1}{2}=\dfrac{z}{-1}$ (or equivalent)   A1

e $z = t \Rightarrow x = t + 1$ and $y = -2t - 1$ M1A1

$-x + 2y - z = 0 \Rightarrow -(t+1) + 2(-2t-1) - t = 0$ M1A1

$\Rightarrow t = -\dfrac{1}{2}$ A1

The intersection point of L_3 with the plane is $\left(\dfrac{1}{2},\ 0,\ -\dfrac{1}{2} \right)$

So the distance from A to the plane is

$\sqrt{\left(1 - \dfrac{1}{2}\right)^2 + (-1-0)^2 + \left(0 + \dfrac{1}{2}\right)^2} = \dfrac{\sqrt{6}}{2}$ M1A1 [19 marks]

12 a Let $P(n): 0^2 + 1^2 + 2^2 + \ldots + n^2 = \dfrac{n(n+1)(2n+1)}{6}$ where $n \in \mathbb{N}$.

Verify that $P(0)$ is a true statement:

$P(0): 0^2 = \dfrac{0(0+1)(2 \times 0 + 1)}{6}$ (verified true)   A1

Next, assume the truth of the proposition for a particular value of n, say k:

$P(k): 0^2 + 1^2 + 2^2 + \ldots + k^2 = \dfrac{k(k+1)(2k+1)}{6}$ (assumed true)   M1

Consider the proposition for the next value of n, ie, $n = k + 1$:

$P(k+1): 0^2 + 1^2 + 2^2 + \ldots + (k+1)^2 = \dfrac{(k+1)(k+2)(2k+3)}{6}$

(under consideration)   M1

As $0^2 + 1^2 + 2^2 + \ldots + k^2 + (k+1)^2$

$= \dfrac{k(k+1)(2k+1)}{6} + (k+1)^2$ (using the induction hypothesis)   A1

$= (k+1)\left(\dfrac{k(2k+1)}{6} + (k+1) \right)$ M1

$= (k+1)\left(\dfrac{2k^2 + 7k + 6}{6} \right)$ A1

$= \dfrac{(k+1)(k+2)(2k+3)}{6}$ (QED)

We have shown that $P(k)$ true $\Rightarrow P(k + 1)$ true and, as we had established that $P(0)$ is true, by the principle of mathematical induction, we can conclude that $P(n)$ is true for any $n \in \mathbb{N}$.   R1

b $3^2 + 6^2 + \ldots + (3n)^2 = 9 \times 1^2 + 9 \times 2^2 + \ldots + 9 \times n^2$ M1

$= 9 \times \left(1^2 + 2^2 + \ldots + n^2 \right)$ (A1)

$= 9\left(\dfrac{n(n+1)(2n+1)}{6} \right)$ M1

$= \dfrac{3n(n+1)(2n+1)}{2}$ A1

c $\quad A_n + B_n = \left(1^2 + 4^2 + \ldots + (3n-2)^2\right) + \left(2^2 + 5^2 + 8^2 + \ldots + (3n-1)^2\right)$

$\qquad = \left(1^2 + 2^2 + 4^2 + 5^2 + \ldots + (3n-2)^2 + (3n-1)^2\right)$ M1

$\qquad = \left(0^2 + 1^2 + 2^2 + 3^2 \ldots + (3n)^2\right) - \left(3^2 + 6^2 + \ldots + (3n)^2\right)$ A1

$\qquad = \dfrac{3n(3n+1)(6n+1)}{6} - \dfrac{3n(n+1)(2n+1)}{2}$ M1

$\qquad = \dfrac{3n\left(18n^2 + 9n + 1 - 3n^2 - 9n - 3\right)}{6} = 6n^3 - n$ A1AG

$\qquad A_n - B_n = \left(1^2 + 4^2 + \ldots + (3n-2)^2\right) - \left(2^2 + 5^2 + 8^2 + \ldots + (3n-1)^2\right)$

$\qquad = \left(1^2 - 2^2\right) + \left(4^2 - 5^2\right) + \ldots + \left((3n-2)^2 - (3n-1)^2\right)$ M1

$\qquad = -3 - 9 - \ldots - (6n-3)$

$\qquad = \dfrac{-3 + (-6n+3)}{2} \times n$ M1A1

$\qquad = -3n^2$ AG

Solve simultaneously $A_n + B_n = 6n^3 - n$ and $A_n - B_n = -3n^2$ M1

$A_n = \dfrac{6n^3 - 3n^2 - n}{2}$ and $B_n = \dfrac{6n^3 + 3n^2 - n}{2}$ A1   [20 marks]

13 Let $f : x \to e^{\cos x}$, where $-\dfrac{\pi}{2} < x < \dfrac{\pi}{2}$

  a As $\cos(-x) = \cos x$ R1

     $f(-x) = e^{\cos(-x)} = e^{\cos x} = f(x)$ A1

     Therefore f is even.   AG

  b $f'(x) = -\sin x \, e^{\cos x}$ M1A1

  c $f'(x) = 0 \Rightarrow \sin x = 0$ M1A1

     $\Rightarrow x = 0$ (as $-\dfrac{\pi}{2} < x < \dfrac{\pi}{2}$)   A1

     $f'(0^-) > 0$ and $f'(0^+) < 0$ R1

     f has a maximum at $(0, e)$ A1

  d $f''(x) = -\cos x \, e^{\cos x} + \sin^2 x \, e^{\cos x}$ M1A1

        $= \left(-\cos x + \sin^2 x\right) e^{\cos x}$

     $f''(x) = 0 \Rightarrow -\cos x + \sin^2 x = 0$ M1

     $\Rightarrow -\cos x + 1 - \cos^2 x = 0$ (or $\cos^2 x + \cos x - 1 = 0$)   M1

     $\Rightarrow \cos x = -\dfrac{1 + \sqrt{5}}{2}$ (as $\cos x > 0$)   A1

     The inflexion point is $\arccos\left(\dfrac{-1 + \sqrt{5}}{2}, e^{\frac{-1+\sqrt{5}}{2}}\right)$ A1

                                                    A1   (1)

  e

f **i** $A(x) = 2x\,e^{\cos x}$ 　　　　　　　　　　　　A1

　　ii $A'(x) = 2\,e^{\cos x} - 2x\sin x\,e^{\cos x}$ 　　　　M1A1 　(1)

　　　　$= 2(1 - x\sin x)\,e^{\cos x}$

$A'(x) = 0 \Leftrightarrow 1 - x\sin x = 0$ and this equation must have a

zero in $\left[0, \dfrac{\pi}{2}\right]$ because $g(x) = 1 - x\sin x$ is continuous and

changes sign in this interval 　　　　　　　　　　R1

$A'(a) = 0 \Rightarrow \sin a = \dfrac{1}{a}$ and $\cos a = \sqrt{1 - \dfrac{1}{a^2}} = \dfrac{\sqrt{a^2-1}}{a}$ 　　M1

Therefore $A(a) = 2a\,e^{\frac{\sqrt{a^2-1}}{a}}$ 　　　　　　　AG 　[21 marks]

Mark scheme

Practice paper 2

SECTION A

1 **a** $12200 = a + 4\times600 \Rightarrow a = 9800$ 　　　　　M1A1

　b $S_5 = \dfrac{9800 + 12200}{2} \times 5 = 55000$ 　　　　　M1A1

　c $9800 + 600(n+1) > 15000 \Rightarrow n > 7\dfrac{2}{3}$ 　　　M1A1

　　So on the 8th year. 　　　　　　　　　　A1 　[7 marks]

2 **a** $3i = \dfrac{a(-3i) + b}{-3i + c} \Rightarrow 9 + 3ci = b - 3ai$ 　　　M1

　　$\Rightarrow b = 9$ and $c = -a$ 　　　　　　　　　A1

　　$1 - 4i = \dfrac{a(1+4i) + b}{1 + 4i + c} \Rightarrow (1+4i)(1+c-4i) = a + b + 4ai$ 　M1

　　$\Rightarrow c + 17 - 4ci = a + b + 4ai$ 　　　　　A1

　　As $b = 9$ and $c = -a$,

　　$-a + 17 + 4ai = a + 9 + 4ai \Rightarrow a = 4$ and $c = -4$ 　A1

　b $w = \dfrac{4z+9}{z-4} \Rightarrow w = \dfrac{4(4+yi)+9}{4+yi-4}$ 　　　M1

　　　　　　　　　　　　　　　　　　　A1

　　$w = \dfrac{25 + 4yi}{yi} \Rightarrow w = 4 - \dfrac{25}{y}i$

　　So, $\operatorname{Re} w = 4$ 　　　　　　　　　　AG 　[7 marks]

3 $P(X < 2) = 0.3 \Rightarrow \underbrace{\dfrac{2-a}{\sqrt{a}} = \phi^{-1}(0.3)}_{-0.5244\ldots}$ 　　　M1A1

　$a = 2.89\,(3\,\text{sf})$

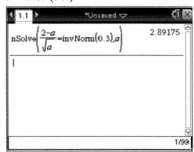

　　　　　　　　　　　(Can be solved on a GDC)

　　　　　　　　　　　　　　　　　(M1)A1 　[4 marks]

4 a   Method 1:

$$f'(x) = \frac{-2x^2 - 1 + x^2}{x^2}$$
M1A1

$$\Rightarrow f'(x) = -\frac{x^2 + 1}{x^2} \neq 0 \text{ in the domain of } f$$
R1

So no maxima/minima points
AG

$$f''(x) = -\frac{2x^3 - 2x(x^2 + 1)}{x^4}$$
A1

$$\Rightarrow f''(x) = \frac{2}{x^3} \neq 0 \text{ in the domain of } f$$
R1

Method 2:

$$f(x) = \frac{1 - x^2}{x} = \frac{1}{x} - x$$
M1

$$f'(x) = -\frac{1}{x^2} - 1 \neq 0$$
A1R1

So no maxima/minima points
AG

$$f''(x) = \frac{2}{x^3} \neq 0$$
A1R1

b

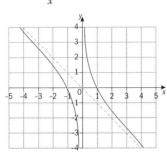

Note: A1 for shape, A1 for zeros, A1 for asymptotes
(no equations required)   A3   [8 marks]

5 $\overrightarrow{AB} = \begin{pmatrix} -6 \\ 5 \\ -3 \end{pmatrix}$ A1

either

$$AP : PB = 1 : 2 \Rightarrow \overrightarrow{AP} = \frac{1}{3}\overrightarrow{AB}$$
M1

$$\overrightarrow{AP} = \begin{pmatrix} -2 \\ \frac{5}{3} \\ -1 \end{pmatrix}$$
A1

$$\overrightarrow{OP} = \overrightarrow{OA} + \overrightarrow{AP} = \begin{pmatrix} 1 \\ -3 \\ -1 \end{pmatrix} + \begin{pmatrix} -2 \\ \frac{5}{3} \\ -1 \end{pmatrix} = \begin{pmatrix} -1 \\ -\frac{4}{3} \\ -2 \end{pmatrix}$$
M1A1

or

$$AP : PB = 1 : 2 \Rightarrow PB = 2PA$$
M1

$$\sqrt{(6 - 6t)^2 + (5t - 4)^2 + (3 - 3t)^2} = 2\sqrt{(-6t)^2 + (5t)^2 + (-3t)^2}$$
M1A1

$$(6 - 6t)^2 + (5t - 4)^2 + (3 - 3t)^2 = 4\left((-6t)^2 + (5t)^2 + (-3t)^2\right)$$

$$\Rightarrow t = \frac{1}{3} \quad (t > 0)$$
A1

Therefore

$$P\left(-1, -\frac{4}{3}, -2\right)$$
A1   [6 marks]

6 $(1+x)^5(1+ax)^6 = (1+5x+10x^2+\ldots+x^5)(1+6ax+10x^2+\ldots+a^6x^{11})$ M1

$$= 1+(6a+5)x+(15a^2+30a+10)x^2+\ldots$$

A1A1

$(1+x)^5(1+ax)^6 \equiv 1+bx+10x^2+\ldots+a^6x^{11}$

Note: award A1 for first two terms, A1 for third term

$\Rightarrow 6a+5=b$ and $15a^2+30a+10=10$ M1

$\Rightarrow a=0$ and $b=5$ A1

$a=-2$ and $b=-7$ A1   [6 marks]

7 $\displaystyle\int \frac{e^{2x}}{4+e^{4x}}\,dx = \int \frac{e^{2x}}{4+\left(e^{2x}\right)^2}\,dx$ M1

$u=e^{2x} \Rightarrow du=2e^{2x}\,dx$ (A1)

$\displaystyle\int \frac{e^{2x}}{4+e^{4x}}\,dx = \frac{1}{2}\int \frac{1}{4+u^2}\,du$ A1

$\displaystyle = \frac{1}{2}\int \frac{\frac{1}{2}}{1+\left(\frac{u}{2}\right)^2}\,du$ (A1)

$\displaystyle = \frac{1}{4}\arctan\frac{u}{2}+C$ (A1)

$\displaystyle = \frac{1}{4}\arctan\frac{e^{2x}}{2}+C$ A1   [6 marks]

8 a Attempt to apply cosine rule   M1

$d=\left(20^2+15^2-2\times20\times15\cos\theta\right)^{\frac{1}{2}}$ (or equivalent)   A1

b Minute hand moves $\dfrac{2\pi}{60}\left(=\dfrac{\pi}{30}\right)$ A1

Hour hand $\dfrac{2\pi}{12\times60}\left(=\dfrac{\pi}{360}\right)$ radians per minute   A1

$\dfrac{d\theta}{dt}=\dfrac{\pi}{360}-\dfrac{\pi}{30}=-\dfrac{11\pi}{360}$ A1

$d'=\dfrac{1}{2}\left(20^2+15^2-2\times20\times15\cos\theta\right)^{-\frac{1}{2}}600\sin\theta\dfrac{d\theta}{dt}$ M1A1

$=-\dfrac{55\pi}{6}\left(20^2+15^2-2\times20\times15\cos\theta\right)^{-\frac{1}{2}}\sin\theta$ (or equivalent)

At 3 o'clock, $\theta=\dfrac{\pi}{2}\Rightarrow d'=-\dfrac{55\pi}{6}\left(20^2+15^2\right)^{-\frac{1}{2}}=1.15$ (3 sf)   A1   [6 marks]

9 For investigating the pattern or deducing that $\left(u_n-\dfrac{1}{2}\right)$ is a GP with   M3

$r=\dfrac{1}{3}$ and $a=\dfrac{3}{2}$ A1A1

$u_n-\dfrac{1}{2}=\dfrac{3}{2}\left(\dfrac{1}{3}\right)^{n-1}$ M1A1

$u_n=\dfrac{1}{2}+\dfrac{3\times3^{1-n}}{2}=\dfrac{3^{2-n}+1}{2}$ A1AG   [8 marks]

SECTION B

10 a $12 \times 0.25 = 3$ A1

   b Let B be the number of blue ribbons taken from the box.

      $B \sim B(12, 0.45)$ A1

      $P(B=6) = \binom{12}{6}(0.45)^6(0.55)^6 = 0.212$ (3 sf)   (M1)A1

   c $P(B \geq 2) = 1 - P(B \leq 1) = 0.992$ (3 sf)   M1A1

   d Let W be the number of white ribbons taken from the box.

      $W \sim B(12, 0.3)$ (A1)

      The value of W for which the pdf takes the maximum value is 3   (M1)A1

      Assumptions: independence of events and   A1

      probability of success constant due to big number of ribbons
      in the box   A1

   e Attempt to use Bayes' theorem or correct tree diagram   M1

      $P(box\ 1|W) = \dfrac{0.3}{0.3 + 0.25 + 0.5}$ A1

      $= 0.286$ (3sf)   A1   [14 marks]

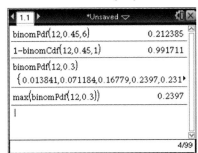

                                      (GDC may be used)

11 a $\mathbf{d} = \overrightarrow{AB} = \begin{pmatrix} 2-1 \\ -1-2 \\ 0+3 \end{pmatrix} = \begin{pmatrix} 1 \\ -3 \\ 3 \end{pmatrix}$ A1

      $\mathbf{r} = \mathbf{a} + \lambda\mathbf{d} \Rightarrow \mathbf{r} = \begin{pmatrix} 1 \\ 2 \\ -3 \end{pmatrix} + \lambda\begin{pmatrix} 1 \\ -3 \\ 3 \end{pmatrix}, \lambda \in \mathbb{R}$ M1A1

   b $M\left(\dfrac{1+2}{2}, \dfrac{2-1}{2}, \dfrac{-3+0}{2}\right) = \left(\dfrac{3}{2}, \dfrac{1}{2}, -\dfrac{3}{2}\right)$ A1

      $\mathbf{n} = \mathbf{d} \Rightarrow \mathbf{r} \cdot \mathbf{n} = \mathbf{m} \cdot \mathbf{n} \Rightarrow \begin{pmatrix} x \\ y \\ z \end{pmatrix} \cdot \begin{pmatrix} 1 \\ -3 \\ 3 \end{pmatrix} = \begin{pmatrix} \frac{3}{2} \\ \frac{1}{2} \\ -\frac{3}{2} \end{pmatrix} \cdot \begin{pmatrix} 1 \\ -3 \\ 3 \end{pmatrix}$ M1A1

      $x - 3y + 3z = \dfrac{3}{2} - \dfrac{3}{2} - \dfrac{9}{2} \Rightarrow 2x - 6y + 6z = -9$ AG

c Let Q be the midpoint of [AC].

$$Q\left(\frac{1-1}{2}, \frac{2-0}{2}, \frac{-3+3}{2}\right) = (0, 1, 0)$$

         A1

$$\overrightarrow{AC} = \begin{pmatrix} -1-1 \\ 0-2 \\ 3+3 \end{pmatrix} = \begin{pmatrix} -2 \\ -2 \\ 6 \end{pmatrix} = -2 \times \begin{pmatrix} 1 \\ 1 \\ -3 \end{pmatrix} \Rightarrow \mathbf{n}_2 = \begin{pmatrix} 1 \\ 1 \\ -3 \end{pmatrix}$$

         A1

$$\mathbf{r} \cdot \mathbf{n}_2 = \mathbf{q} \cdot \mathbf{n}_2 \Rightarrow \begin{pmatrix} x \\ y \\ z \end{pmatrix} \cdot \begin{pmatrix} 1 \\ 1 \\ -3 \end{pmatrix} = \begin{pmatrix} 0 \\ 1 \\ 0 \end{pmatrix} \cdot \begin{pmatrix} 1 \\ 1 \\ -3 \end{pmatrix}$$

         A1

$$x + y - 3z = 1$$

         AG

d $$\cos\theta = \frac{\begin{pmatrix} 1 \\ -3 \\ 3 \end{pmatrix} \cdot \begin{pmatrix} 1 \\ 1 \\ 3 \end{pmatrix}}{\sqrt{1^2 + (-3)^2 + 3^2}\sqrt{1^2 + 1^2 + 3^2}} = \frac{7}{\sqrt{19}\sqrt{11}}$$

         M1A1A1

    $\theta = 1.07(61.0°)$

         A1

e Attempt to solve the system of three simultaneous equations:

         (M2)

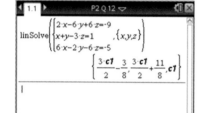

    (Can be solved with GDC)

$$\begin{cases} x = \frac{3}{2}\lambda - \frac{3}{8} \\ y = \frac{3}{2}\lambda + \frac{11}{8}, \lambda \in \mathbb{R} \\ z = \lambda \end{cases}$$

         A2

f Since the point P is equally distant to the points A, B and C therefore it lies in the bisecting planes of the line segments [AB], [AC] and [BC].

         R1

Attempt to find the point of intersection between the line in part e) and the plane $x + y + z = 0$.

$$\frac{3}{2}\lambda - \frac{3}{8} + \frac{3}{2}\lambda + \frac{11}{8} + \lambda = 0$$

         M1 A1

$$4\lambda = -1 \Rightarrow \lambda = -\frac{1}{4}$$

         A1

$$\begin{cases} x = \frac{3}{2}\left(-\frac{1}{4}\right) - \frac{3}{8} \\ y = \frac{3}{2}\left(-\frac{1}{4}\right) + \frac{11}{8} \Rightarrow P\left(-\frac{3}{4}, 1, -\frac{1}{4}\right) \\ z = -\frac{1}{4} \end{cases}$$

         M1A1   [23 marks]

12 a $f(-x)=\cos(-2x)+1=\cos(2x)+1=f(x)$ M1A1

$g(-x)=\dfrac{e^{-x}+e^{x}}{2}=g(x)$ A1

b $f'(x)=-\sin(2x)\cdot 2=-2\sin(2x)$ M1A1

$g'(x)=\dfrac{1}{2}\left(e^{x}+(-1)\times e^{-x}\right)=\dfrac{e^{x}-e^{-x}}{2}$ M1A1

c $f'(-x)=-2\sin(-2x)=2\sin(2x)=-f'(x)$ A1

$g'(-x)=\dfrac{e^{-x}-e^{x}}{2}=-\dfrac{e^{x}-e^{-x}}{2}=-g'(x)$ A1

d Correct shape of the graphs   A1A1

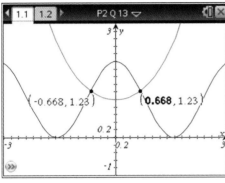

$(-0.668,1.23)$ A1

$(0.668,1.23)$ A1

> Notice that we stored the x-value of the point in the first quadrant to a variable called a, that will be seen in the following parts

e Attempt to use the formula for the tangent $y=f'(x_1)(x-x_1)+y_1$ M1

$y=f'(0.668)(x-0.668)+1.23$ A1

$y=-1.95(x-0.668)+1.23$

$y=-1.95x+2.53$ AG

$y=g'(0.668)(x-0.668)+1.23$

$y=0.719(x-0.668)+1.23$ A1

$y=0.719x+0.751$ AG

> Store all the values found. It saves you time and reduces the risk of errors due to randing and mistakes.

f Since the functions are even, their graphs and the respective tangents are symmetrical with respect to the *y*-axis. They form a kite so the area is half the product of the diagonals.   R1

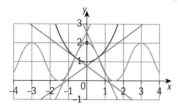

$$A = \frac{e \cdot f}{2} \Rightarrow A = \frac{(2.53 - 0.751) \cdot (2 \cdot 0.668)}{2} = 1.19$$ M1A1

g Either $V = \pi \displaystyle\int_{-0.688}^{0.668} (f^2(x) - g^2(x)) \, dx = 7.88$ M1A1A2

or

$V = 2\pi \displaystyle\int_{0}^{0.668} (f^2(x) - g^2(x)) \, dx = 7.88$ M1A1A2   [23 marks]

Note: M1 for correct formula, A1 for correct limits, A2 for correct final answer

Subject index

data fitting, sine functions, 406–8
De Moivre, Abraham (1667–1754), 532, 645
De Moivre's theorem, 643–9, 657
decagons, 729
decimal fractions, 678
decimal places, 688
decimals
  and fractions, 678–80
  recurring, 679
  terminating, 679
decisions, 552–3
decomposition, unique, 126–7, 162
definite integrals, 459–61, 491
  geometrical significance of, 355–76
  and integration by substitution, 463–5
  properties, 353–5
definite integration, 352–5, 378
definitions, equivalence of, 590–1
degrees, and radians, 268
dependent variables, 50
derivatives, 184
  constant functions, 189
  exponential functions, 261–6
  of functions, 180–9, 227
  and functions, relationships, 205–8
  graphical meaning of, 199–203
  inverse trigonometric functions, 442–3, 490
  logarithmic functions, 261–6
  trigonometric functions, 436–49, 490
  see also higher derivatives
Descartes, René (1596–1650), 106, 140, 344, 563, 738
Descartes' rule of signs, 140
diagrams
  Argand, 107, 630
  cumulative frequency, 285
  phasor, 638
  polar area, 634
  probability tree, 321–6
  stem and leaf, 748–9
  Venn, 302–4, 313
  see also charts; graphs
diameters, 730
dice problems, 299
difference
  common, 10
  of functions, 190–1
difference of two squares, 699
  factorization, 701–2
differential calculus
  applications

economics, 211–15, 229
  kinematics, 208–11, 229
differentiation, 180
  implicit, 444
  of implicit functions, 218–20, 229
  rules, 189–205, 227
dilations, 84–9
  horizontal, 85
  vertical, 85
directed line segments, 556
direction, vectors, 556–7
discontinuous functions, 171, 226
discrete data, 280–1, 749
discrete random variables, 496–503, 520–1
  parameters, 498–501
  probability distribution functions, 496–8
discrete values, 496
discriminants, 160
  of quadratic equations, 102–4
disjoint sets, 307, 693
dispersion, measures of, 291–8, 338, 752–3
displacement, 557, 613, 620
displacement vectors, 622
  in three-dimensional space, 574–5
distances, 599–613
  between points, 740
distribution functions, 501
distributions, 495
  Gaussian, 532
  random, 496–503
  see also binomial distributions; normal distributions Poisson distributions
distributive law, 697
distributive property, 688
divergent sequences, 178
divergent series, 18–21, 179–80
division
  complex numbers, 112–15
    polar form, 641–2
  polynomials, 125–6
domains
  of functions, 50
  restriction, 52
dot product, 584
double angle identities, 401–2, 430
doughnuts, 435
  volume, 488
dynamical systems, 654

e (constant), 245, 276
  see also Euler's number

economics, differential calculus, 211–15, 229
Eddington, Sir Arthur Stanley (1882–1944), 165
Egypt, fractions, 679
Einstein, Albert (1879–1955), 276, 433
elements, 691
elimination method, simultaneous linear equations, 708–9
empty sets, 691
enlargements, 721
equal geometric vectors, 557–8
equations
  cubic, 162
  exponential, 255–8
  logarithmic, 255–8
  of normals, 188, 444–6
  with rational coefficients, 714
  recursive, 10
  systems of, solving, 153–8
  of tangents, 188, 444–6
  see also Cartesian equations; linear equations; parametric equations; polynomial equations; quadratic equations; trigonometric equations; vector equations
equations of lines, 622–3, 744–5
  gradient formula, 744–5
  and vectors, 571–83
equilateral triangles, 729
equivalence, of definitions, 590–1
equivalent fractions, 679
equivalent geometric vectors, 557–8
Eratosthenes of Cyrene (*c.*276 BC–*c.*195 BC), 432
Erdős, Paul (1913–96), 307
Escher, Maurits Cornelis (1898–1972), 277
estimation, 688–90
Euler, Leonard (1707–83), 5, 51, 106, 276, 627, 731
Euler's number, 243–7
  alternative approaches, 248–9
even functions, 67–8, 71, 363, 429
events, 299, 339
  certain, 306
  impossible, 306
  independent, 318–21, 339, 513
  mutually exclusive, 303–4
exhaustion, method of, 344
expected values, 551
experimental probability, 308–12

interquartile range (IQR), 292, 752
intersections, 599–613, 692–5
  between lines and planes, 604–5
  of three planes, 610–13
  of two lines, 606–7
  of two planes, 607–9
intuition, 659
  and luck, 432–3
  and probability, 341
invariance, and exponential
      functions, 248–9
inverse cosine functions, 410
inverse functions, 74–5
  exponential, 259–60
  graphical properties, 75–8
inverse sine functions, 409–10
inverse tangent functions, 410–11
inverse trigonometric functions,
      409–12
  derivatives, 442–3, 490
IQR (interquartile range), 292, 752
irrational numbers, 674, 679, 686
irregular quadrilaterals, 729
irregular shapes, area, 357–65
isosceles triangles, 729

Jeffreys, Sir Alec John (b.1950), 553
Joukowski aerofoil, 653
Julia, Gaston (1893–1978), 654
Julia set, 654

Kandinsky, Wassily (1866–1944),
      493
Kepler, Johannes (1571–1630), 375
kinematics, 379
  area and, 369–71
  differential calculus, 208–11, 229
kites, 729
  area, 733
Kolmogorov, Andrey Nikolaevich
      (1903–87), 306
Kramp, Christian (1760–1826), 31

Lang, Andrew (1844–1912), 340
language
  evolution, 94
  mathematics as, 48–95
  symbolic, 49–50
Laplace, Pierre-Simon, marquis de
      (1749–1827), 300
LCL (lower control limit), 495
LCM (lowest common multiple),
      677–8
legal system, statistics and, 552
Leibniz, Gottfried Wilhelm
      (1646–1716), 51, 180, 208,
      231, 380

Leibniz's formulae, 198
Leonardo da Vinci (1452–1519),
      684
Leonardo de Pisa (*c*.1170–*c*.1250),
      97
limits, 168–80
  control, 495
  finding algebraically, 176–7
  properties, 226–7
  to infinity, 173–5
  trigonometric, 436–7
Lindemann, Carl Louis Ferdinand
      von (1852–1932), 245
line segments, directed, 556
linear combinations, 119, 563
linear equations
  with complex coefficients,
      solving, 153–8
  solving, 706–7
  see also simultaneous linear
      equations
linear functions, 120
linearity, 187–8
lines, 728
  and planes
    angles between, 602
    intersections, 604–5
  skew, 606
  two
    angles between, 599–602
    intersections, 606–7
  in two dimensions, 580–1
  vector equations of, 579, 623
  see also equations of lines; parallel
      lines; perpendicular lines;
      straight lines
Liouville, Joseph (1809–82), 245
loci, 569
logarithm tables
  Briggs, 249–50
  Napier, 249–50
logarithmic equations, 255–8
logarithmic functions
  behavior, 258–61
  derivatives, 261–6
  integration, 351
  properties, 275
logarithms
  and bases, 249–58
  changing base of, 253–4
  definition, 250
  properties, 238–43, 251–3, 275
lower control limit (LCL), 495
lower quartiles, 292, 752
lowest common multiple (LCM),
      677–8
lowest terms, 679

luck, and intuition, 432–3

magnitudes
  vectors, 556–7, 567–9, 593–5
    in three-dimensional space,
      576
major segments, 730
malpractice, in explorations, 667
Mandelbrot, Benoît (1924–2010),
      492, 629, 654
Mandelbrot set, 629, 654
many-to-one functions, 65–6
mappings, 696–7
mathematical induction, 25–31,
      45
Mathematical Intelligencer, The, 658
mathematical paradoxes, 493
mathematical presentation, in
      explorations, 662–3
mathematics
  abstract nature of, 626
  aesthetics in, 232–77
  ancient, and modern methods,
      382–433
  applications, 626, 627
  in art, 493
  beauty in, 276–7, 658
  changing structure of, 658–9
  communicating, 626–7
  connections, 650–4
  invention vs. discovery, 164–5
  as a language, 48–95
  long journey of, 96–165
  multiple perspectives in, 628–59
  in nature, 492–3
  pure, 231
  as science of patterns, 2–47
  use of, in explorations, 665
maxima, 228
  local, 119, 199–200
  problems, 480–2
  tests for, 204–5
mean, 289, 498, 509, 551, 750–1
measures
  of central tendency, 288–91, 338,
      750–1
  of dispersion, 291–8, 338, 752–3
median, 289–90, 338, 501, 507,
      551, 750–1
midpoints, 739–40
mind maps, 670–1
minima, 228
  local, 119, 199–200
  problems, 480–2
  tests for, 204–5
minor segments, 730
mixed numbers, 679

planes (*continued*)

three, intersections, 610–13

two, intersections, 607–9

vector equations of, 596–9, 625

Poincaré, Jules Henri (1854–1912), 652

Poincaré conjecture, 652

points, 622–3, 728

of concavity, 205–7

distance between, 740

in three-dimensional space, 576

of inflexion, 201, 205–7, 228

initial, 557, 620

and planes, distance between, 605

of tangency, 731

terminal, 557, 620

in three-dimensional space, 574–5

and vectors, 571–83

see also stationary points

Poisson distributions, 513–20, 551

parameters, 516–17

properties, 517–20

Poisson, Siméon-Denis (1781–1840), 513

polar area diagrams, 634

polar coordinates, 633

and Cartesian coordinates, 635–6

polar form, 633–8, 641–2, 657

and Cartesian form, 635–6, 657

Pollock, Paul Jackson (1912–56), 493

Pólya, George (1887–1985), 321

polygons, 729

frequency, 284

inscribed, 179

regular, 558

polynomial equations, 140–53

solving, 144–6

polynomial functions

graphs, 118–31

operations, 118–31

product, 131–40

sum, 131–40

zeros, 131–40

polynomial inequalities, 140–53

solving, 146–53

polynomial remainder theorem, 126–30

polynomial roots

products, 136–40

sum, 136–40

polynomials, 119–31

cubic, 208

degree of, 162

division, 125–6

graphs of, 177

higher-degree, 123

integration by parts, 469–70

operations, 123–31

quartic, coefficients, 136

theorems, 140–4

of third degree, 136

use of term, 119

Viète's formulae, 162

zero, 120

populations, and samples compared, 281–6, 338

position vectors, 557, 620, 622

in three-dimensional space, 574–5

positive gradients, 199–200, 201, 741–2

positive integer powers, 189

powers

complex numbers, 116–18, 643–9

positive integer, 189

see also exponents; indices

practice papers, 754–9

predictions, 553

primes, 46–7, 677–8

prior learning, 672–753

prisms

hexagonal, 734

surface area, 735

triangular, 734

volume, 735

probability, 278–341

axioms, 306

conditional, 312–17, 339

and counting methods, 304–6

and cumulative distribution functions, 527–31

experimental, 308–12

and intuition, 341

of normal variables, 534–8

properties, 306–8, 339

and statistics, 341

theoretical, 299–306, 339

probability density functions (PDFs), 551

continuous random variables, 522

probability distribution functions (PDFs), 498–9, 513, 551

discrete random variables, 496–8

normal distributions, 532

probability tree diagrams, 321–6

problem-solving, and modeling, 544–7, 613–17

product rule, 193–5, 228

products

dot, 584

inner, 584

polynomial functions, 131–40

polynomial roots, 136–40

quadratic equation roots, 104–5

resulting in quadratic expressions, 698–9

scalar, 583–92

see also cross product; multiplication; scalar product; vector products

progressions

arithmetic, 10

see also geometric progressions

proofs, 4, 24–5

formal, 626–7

proper fractions, 679

proper subsets, 693

proportion, 683–5

Ptolemy (*c*.90–168 CE), 386

pure mathematics, applications, 231

pyramids

surface area, 735–6

volume, 735–6

Pythagoras (569–500 BC), 719

Pythagoras' theorem, 719–20

Pythagorean identities, 393, 430

quadratic equations

discriminants of, 102–4

roots, 100, 102–3

sum and product, 104–5

solutions, 100

solving

by completing the square, 98–101, 716–17

by factorization, 99–101, 715

using quadratic formula, 97–101

Viète's formulae, 162

quadratic expressions

factorization, 700–1

products resulting in, 698–9

quadratic formula, 97–101, 160

quadratic functions, 120

general, 102

quadratic graphs, 55

quadratic inequalities, 717–18

quadratic sequences, 8

quadratic trigonometric equations, 413–14

quadrilaterals, 729

irregular, 729

qualitative data, 280

quantitative data, 280

quartic functions, 122

quartic polynomials, coefficients, 136

quartiles, 291–3
lower, 292, 752
upper, 292, 752
quaternions, 595, 659
quintic functions, 122
quotient rule, 195–7, 228

radians, 267
and angles, 394–5
and degrees, 268
radical expressions, substitution in, 472–6
radical functions, 56
radicals, 674
radii, 730
random distributions, 496–503
random experiments, 299, 339
random numerical patterns, 234
random samples, 281
random variables, 495, 551
see also continuous random variables; discrete random variables
randomness
exploring, 278–341
modeling, 494–553
range, 291, 752
of functions, 50
interquartile, 292, 752
rates, related, 221–4
rates of change
average, 181–3
trigonometric expressions, 450–5
rational coefficients, equations with, 714
rational functions, 59–63
rational numbers, 686
rational zero theorem, 140, 142–3, 163
ratios, 683–5
common, 15
golden, 233–4
unitary, 683
see also trigonometric ratios
real number lines, 695
real numbers
field of, 115
multiplication of complex numbers by, 631, 637–8, 657
properties, 688
real parts, 106
reasoning, 47
reciprocal trigonometric functions, 439
reciprocal trigonometric ratios, 391–2
reciprocals, complex numbers,

polar form, 641
record keeping, explorations, 667–8
rectangles, 729
area, 732
recurring decimals, 679
recursive equations, 10
recursive formulae, 475–6
recursive functions, 234–8, 274
reduction formulae, 466
reflections, 720
regular polygons, 558
related rates, 221–4
relations
and functions, 50–3
and mappings, 696
relationships, functions and derivatives, 205–8
relative frequency, 309
remainder theorem, 162
polynomial, 126–30
revolution
areas of, 483–8
solids of, volume, 371–5, 379
volumes of, 483–8
Rhind Mathematical Papyrus, 56, 679
rhombuses, 729
Riemann, Georg Friedrich Bernhard (1826–66), 47, 358, 653
Riemann hypothesis, 47
Riemann sums, 358
Riemann surface, 653
right-angled triangles, 729
trigonometric ratios, 384–9
roots
complex numbers, 116–18, 643–9, 657
expressions with, 674–6
quadratic equations, 100, 102–3
sum and product, 104–5
see also polynomial roots
roots of unity
complex nth, 647
properties, 649
rotations, 720
rounding, 688–90
rules
chain, 191–3, 227
differentiation, 189–205, 227
product, 193–5, 228
quotient, 195–7, 228
of surds, 674
Russell, Bertrand (1872–1970), 381

sample space, 299
samples

and populations compared, 281–6, 338
random, 281
scalar multiplication, 560
properties, 561, 620
in three-dimensional space, 575–6
scalar product, 583–92, 624
algebraic definitions, 587–9
of two vectors, 584–6
scalars, 557
scalene triangles, 729
second derivative test, 204–5, 228
sectors, 731
area, 269–72
perimeters, 269–72
segments
directed line, 556
major, 730
minor, 730
self-inverse functions, 78
semicircles, 731
sequences, 5–9, 45, 237–8
arithmetic, 10–14
convergence of, 178–9
convergent, 178
divergent, 178
geometric, 15–24
quadratic, 8
see also series
series, 5–9, 45
convergence of, 179–80, 227
convergent, 18–21, 179–80
divergent, 18–21, 179–80
finite, 5
Fourier, 433
harmonic, 5
infinite, 5
see also arithmetic series; geometric series; sequences
set builder notation, 692
sets, 691–7
disjoint, 307, 693
elements, 691
empty, 691
and inequalities, 695–6
and number lines, 695–6
universal, 691
see also subsets
shapes
irregular, area, 357–65
two-dimensional, 729–30
see also three-dimensional shapes
Shewhart, Walter Andrew (1891–1967), 495
SIDS (sudden infant death syndrome), **552**